Cancer Biomarker Research and Personalized Medicine

Cancer Biomarker Research and Personalized Medicine

Editor

James Meehan

MDPI • Basel • Beijing • Wuhan • Barcelona • Belgrade • Manchester • Tokyo • Cluj • Tianjin

Editor
James Meehan
Translational Oncology Research
Group, Institute of Genetics and
Cancer
The University of Edinburgh
Edinburgh
United Kingdom

Editorial Office
MDPI
St. Alban-Anlage 66
4052 Basel, Switzerland

This is a reprint of articles from the Special Issue published online in the open access journal *Journal of Personalized Medicine* (ISSN 2075-4426) (available at: www.mdpi.com/journal/jpm/special_issues/cancer_biomarker_personalized).

For citation purposes, cite each article independently as indicated on the article page online and as indicated below:

LastName, A.A.; LastName, B.B.; LastName, C.C. Article Title. *Journal Name* **Year**, *Volume Number*, Page Range.

ISBN 978-3-0365-3918-8 (Hbk)
ISBN 978-3-0365-3917-1 (PDF)

© 2022 by the authors. Articles in this book are Open Access and distributed under the Creative Commons Attribution (CC BY) license, which allows users to download, copy and build upon published articles, as long as the author and publisher are properly credited, which ensures maximum dissemination and a wider impact of our publications.

The book as a whole is distributed by MDPI under the terms and conditions of the Creative Commons license CC BY-NC-ND.

Contents

About the Editor vii

James Meehan
Special Issue "Cancer Biomarker Research and Personalized Medicine"
Reprinted from: *J. Pers. Med.* **2022**, *12*, 585, doi:10.3390/jpm12040585 1

Óscar Rapado-González, Cristina Martínez-Reglero, Ángel Salgado-Barreira, Laura Muinelo-Romay, Juan Muinelo-Lorenzo and Rafael López-López et al.
Salivary DNA Methylation as an Epigenetic Biomarker for Head and Neck Cancer. Part I: A Diagnostic Accuracy Meta-Analysis
Reprinted from: *J. Pers. Med.* **2021**, *11*, 568, doi:10.3390/jpm11060568 5

Óscar Rapado-González, Cristina Martínez-Reglero, Ángel Salgado-Barreira, María Arminda Santos, Rafael López-López and Ángel Díaz-Lagares et al.
Salivary DNA Methylation as an Epigenetic Biomarker for Head and Neck Cancer. Part II: A Cancer Risk Meta-Analysis
Reprinted from: *J. Pers. Med.* **2021**, *11*, 606, doi:10.3390/jpm11070606 23

Pei-Hung Chang, Hung-Ming Wang, Yung-Chia Kuo, Li-Yu Lee, Chia-Jung Liao and Hsuan-Chih Kuo et al.
Circulating p16-Positive and p16-Negative Tumor Cells Serve as Independent Prognostic Indicators of Survival in Patients with Head and Neck Squamous Cell Carcinomas
Reprinted from: *J. Pers. Med.* **2021**, *11*, 1156, doi:10.3390/jpm11111156 35

Arnaud Gauthier, Pierre Philouze, Alexandra Lauret, Gersende Alphonse, Céline Malesys and Dominique Ardail et al.
Circulating Tumor Cell Detection during Neoadjuvant Chemotherapy to Predict Early Response in Locally Advanced Oropharyngeal Cancers: A Prospective Pilot Study
Reprinted from: *J. Pers. Med.* **2022**, *12*, 445, doi:10.3390/jpm12030445 53

Carlos Martínez-Pérez, Charlene Kay, James Meehan, Mark Gray, J. Michael Dixon and Arran K. Turnbull
The IL6-like Cytokine Family: Role and Biomarker Potential in Breast Cancer
Reprinted from: *J. Pers. Med.* **2021**, *11*, 1073, doi:10.3390/jpm11111073 65

Carlos Martínez-Pérez, Jess Leung, Charlene Kay, James Meehan, Mark Gray and J Michael Dixon et al.
The Signal Transducer IL6ST (gp130) as a Predictive and Prognostic Biomarker in Breast Cancer
Reprinted from: *J. Pers. Med.* **2021**, *11*, 618, doi:10.3390/jpm11070618 97

James Meehan, Mark Gray, Carlos Martínez-Pérez, Charlene Kay, Jimi C. Wills and Ian H. Kunkler et al.
A Novel Approach for the Discovery of Biomarkers of Radiotherapy Response in Breast Cancer
Reprinted from: *J. Pers. Med.* **2021**, *11*, 796, doi:10.3390/jpm11080796 115

Chantell Payton, Lisa Y. Pang, Mark Gray and David J. Argyle
Exosomes Derived from Radioresistant Breast Cancer Cells Promote Therapeutic Resistance in Naïve Recipient Cells
Reprinted from: *J. Pers. Med.* **2021**, *11*, 1310, doi:10.3390/jpm11121310 141

Tatiana Kalinina, Vladislav Kononchuk, Efim Alekseenok, Grigory Abdullin, Sergey Sidorov and Vladimir Ovchinnikov et al.
Associations between the Levels of Estradiol-, Progesterone-, and Testosterone-Sensitive MiRNAs and Main Clinicopathologic Features of Breast Cancer
Reprinted from: *J. Pers. Med.* **2021**, *12*, 4, doi:10.3390/jpm12010004 165

Shun-Wen Cheng, Po-Chih Chen, Tzong-Rong Ger, Hui-Wen Chiu and Yuan-Feng Lin
GBP5 Serves as a Potential Marker to Predict a Favorable Response in Triple-Negative Breast Cancer Patients Receiving a Taxane-Based Chemotherapy
Reprinted from: *J. Pers. Med.* **2021**, *11*, 197, doi:10.3390/jpm11030197 189

Chia-Jung Li, Li-Te Lin, Pei-Yi Chu, An-Jen Chiang, Hsiao-Wen Tsai and Yi-Han Chiu et al.
Identification of Novel Biomarkers and Candidate Drug in Ovarian Cancer
Reprinted from: *J. Pers. Med.* **2021**, *11*, 316, doi:10.3390/jpm11040316 203

Kyungmi Kim, Jihion Yu, Jun-Young Park, Sungwoon Baek, Jai-Hyun Hwang and Woo-Jong Choi et al.
Low Preoperative Lymphocyte-to-Monocyte Ratio Is Predictive of the 5-Year Recurrence of Bladder Tumor after Transurethral Resection
Reprinted from: *J. Pers. Med.* **2021**, *11*, 947, doi:10.3390/jpm11100947 217

James Meehan, Mark Gray, Carlos Martínez-Pérez, Charlene Kay, Duncan McLaren and Arran K. Turnbull
Tissue- and Liquid-Based Biomarkers in Prostate Cancer Precision Medicine
Reprinted from: *J. Pers. Med.* **2021**, *11*, 664, doi:10.3390/jpm11070664 225

Gianluca Ingrosso, Emanuele Alì, Simona Marani, Simonetta Saldi, Rita Bellavita and Cynthia Aristei
Prognostic Genomic Tissue-Based Biomarkers in the Treatment of Localized Prostate Cancer
Reprinted from: *J. Pers. Med.* **2022**, *12*, 65, doi:10.3390/jpm12010065 263

Ingrid Jenny Guldvik, Lina Ekseth, Amar U. Kishan, Andreas Stensvold, Else Marit Inderberg and Wolfgang Lilleby
Circulating Tumor Cell Persistence Associates with Long-Term Clinical Outcome to a Therapeutic Cancer Vaccine in Prostate Cancer
Reprinted from: *J. Pers. Med.* **2021**, *11*, 605, doi:10.3390/jpm11070605 275

Claudia-Gabriela Moldovanu, Bianca Boca, Andrei Lebovici, Attila Tamas-Szora, Diana Sorina Feier and Nicolae Crisan et al.
Preoperative Predicting the WHO/ISUP Nuclear Grade of Clear Cell Renal Cell Carcinoma by Computed Tomography-Based Radiomics Features
Reprinted from: *J. Pers. Med.* **2020**, *11*, 8, doi:10.3390/jpm11010008 287

Mi-Ae Kang, Jongsung Lee, Chang Min Lee, Ho Sung Park, Kyu Yun Jang and See-Hyoung Park
IL13Rα2 Is Involved in the Progress of Renal Cell Carcinoma through the JAK2/FOXO3 Pathway
Reprinted from: *J. Pers. Med.* **2021**, *11*, 284, doi:10.3390/jpm11040284 303

Mark Gray, Jamie R. K. Marland, Alan F. Murray, David J. Argyle and Mark A. Potter
Predictive and Diagnostic Biomarkers of Anastomotic Leakage: A Precision Medicine Approach for Colorectal Cancer Patients
Reprinted from: *J. Pers. Med.* **2021**, *11*, 471, doi:10.3390/jpm11060471 323

Bader Almuzzaini, Jahad Alghamdi, Alhanouf Alomani, Saleh AlGhamdi, Abdullah A. Alsharm and Saeed Alshieban et al.
Identification of Novel Mutations in Colorectal Cancer Patients Using AmpliSeq Comprehensive Cancer Panel
Reprinted from: *J. Pers. Med.* **2021**, *11*, 535, doi:10.3390/jpm11060535 353

Jose Carlos Benitez, Marc Campayo, Tania Díaz, Carme Ferrer, Melissa Acosta-Plasencia and Mariano Monzo et al.
Lincp21-RNA as Predictive Response Marker for Preoperative Chemoradiotherapy in Rectal Cancer
Reprinted from: *J. Pers. Med.* **2021**, *11*, 420, doi:10.3390/jpm11050420 371

Seunghyup Jeong, Unyong Kim, Myung Jin Oh, Jihyeon Nam, Se Hoon Park and Yoon Jin Choi et al.
Detection of Aberrant Glycosylation of Serum Haptoglobin for Gastric Cancer Diagnosis Using a Middle-Up-Down Glycoproteome Platform
Reprinted from: *J. Pers. Med.* **2021**, *11*, 575, doi:10.3390/jpm11060575 381

Duo Zuo, Haohua An, Jianhua Li, Jiawei Xiao and Li Ren
The Application Value of Lipoprotein Particle Numbers in the Diagnosis of HBV-Related Hepatocellular Carcinoma with BCLC Stage 0-A
Reprinted from: *J. Pers. Med.* **2021**, *11*, 1143, doi:10.3390/jpm11111143 395

About the Editor

James Meehan

James Meehan graduated from Trinity College Dublin in 2011 with a degree in Neuroscience. He moved to the UK after receiving a scholarship to study at The University of Edinburgh, where he completed an M.Sc. in Biomedical Sciences before moving into a Ph.D. in Molecular and Clinical Medicine. His Ph.D. investigated the potential of different pH regulating proteins to act as therapeutic targets in breast cancer, and was carried out as part of the METAOXIA (Metastatic Tumors Facilitated by Hypoxic Tumor Micro-environments) project. Following a short stint within The University of Edinburgh investigating the ability of nanoparticles to act as radiosensitizers, he was employed as a Research Associate at Heriot-Watt University. Here, he worked on the multi-disciplinary IMPACT (Implantable Microsystems for Personalized Anti-Cancer Therapy) project, aimed at the development of miniaturized sensors to monitor the status of individual tumors. He then returned to The University of Edinburgh to work as a Postdoctoral Research Fellow within the Translational Oncology Research Group. His work concentrates on translational studies in breast and prostate cancer, with his projects focusing on the discovery, development and validation of novel biomarkers of response to radiotherapy and chemotherapy. At present, he has authored/co-authored 26 research articles in peer-reviewed journals.

Editorial

Special Issue "Cancer Biomarker Research and Personalized Medicine"

James Meehan

Translational Oncology Research Group, Institute of Genetics and Cancer, Western General Hospital, University of Edinburgh, Edinburgh EH4 2XU, UK; jmeehan@ed.ac.uk

While the term biomarker is thought to have first been used in the 1970s, the concept itself is considered to be much older. The turn of the 21st century saw a dramatic increase in the number of papers published concerning biomarkers. Biomarkers can be described as characteristics that can be assessed and quantified as indicators of standard biological processes, pathogenesis or response to therapy. The treatment of individual patients based on particular factors, such as biomarkers, distinguishes standard, generalized treatment plans from personalized medicine. Even though personalized medicine is applicable to most branches of medicine, the field of oncology is perhaps where it is most easily employed. Cancer is a heterogeneous disease; although patients may be diagnosed histologically with the same cancer type, their tumors can comprise varying tumor microenvironments and molecular characteristics that can impact treatment response and prognosis.

There has been a major drive over the past decade to try and realize personalized cancer medicine through the discovery and use of disease-specific biomarkers. This Special Issue, entitled "Cancer Biomarker Research and Personalized Medicine", encompasses 22 publications from colleagues working on a diverse range of cancers, including prostate, breast, ovarian, head and neck, liver, gastric, bladder, colorectal and kidney. The biomarkers assessed in these studies include genes, intracellular or secreted proteins, exosomes, DNA, RNA, miRNA, circulating tumor cells and circulating immune cells, in addition to radiomic features.

A number of different biomarker subtypes have been delineated according to their recognized applications. Biomarkers can be defined by the mechanisms that lead to disease development, perhaps linked with susceptibility/risk factors that can initiate a pathophysiological process. Susceptibility/risk biomarkers reveal the possibility for developing a disease in those that do not currently have a clinically apparent disease. Work submitted to this Special Issue highlights the potential of DNA methylation as a risk biomarker for head and neck cancer [1].

Diagnostic biomarkers differ in that they detect/confirm the presence of a disease. The early diagnosis of cancer is vital for improving the survival of patients. Several publications within this Special Issue explore diagnostic biomarkers, with research providing new insights into the development of diagnostic biomarkers for prostate cancer [2], hepatocellular carcinoma [3], gastric cancer [4] and head and neck cancer [5]. As we progress further into the precision medicine era, diagnostic biomarkers will continue to evolve; such biomarkers may not only be utilized to identify those with cancer, but also to re-define the classification of cancer [6].

In patients that have already been diagnosed with cancer, it can be challenging to stratify those with tumors that are less likely to progress from patients with tumors that are more aggressive and therefore require treatment intensification; tumor heterogeneity contributes greatly to this problem. While the early diagnosis of cancer is crucial for enhancing the survival of patients, the identification of biomarkers at the time of diagnosis that can give an indication of cancer aggressiveness is possibly the greatest unmet clinical need for many cancer types [2,7]. Prognostic biomarkers identify the likelihood of disease recurrence

Citation: Meehan, J. Special Issue "Cancer Biomarker Research and Personalized Medicine". *J. Pers. Med.* 2022, 12, 585. https://doi.org/10.3390/jpm12040585

Received: 31 March 2022
Accepted: 2 April 2022
Published: 5 April 2022

Publisher's Note: MDPI stays neutral with regard to jurisdictional claims in published maps and institutional affiliations.

Copyright: © 2022 by the author. Licensee MDPI, Basel, Switzerland. This article is an open access article distributed under the terms and conditions of the Creative Commons Attribution (CC BY) license (https://creativecommons.org/licenses/by/4.0/).

or progression; these factors are crucial to decision-making processes in the clinic, helping clinicians determine the most appropriate treatment for each patient. Several publications within this Special Issue explored prognostic biomarkers, focusing on the development of prognostic tissue-based biomarkers in ovarian cancer [8], prostate cancer [2,7] and renal cell carcinoma [9], in addition to prognostic liquid-based biomarkers in prostate cancer [2], bladder cancer [10] and head and neck cancer [11]. Studies published within the Special Issue also show how cancer prognosis is moving towards the use of imaging, rather than relying on tissue/liquid biomarkers alone [12].

While prognostic biomarkers can help identify patients at a higher risk who might benefit from more aggressive treatment, they do not give any information on which patients are likely to gain a clinical benefit from a specific therapy. Conversely, predictive biomarkers are those that can indicate the probability of a patient gaining a therapeutic benefit from a specific treatment. While many of the standard cancer treatments such as radiotherapy and chemotherapy are effective, the use of these treatments in non-responding patients is associated with increased levels of toxicity and can delay the of instigation of alternative treatments that may have a greater effect. As such, predictive biomarkers represent a major research area, with work submitted to this Special Issue showing their potential to predict response to chemotherapy [13,14], radiotherapy [15], chemo-radiotherapy [16] and therapeutic cancer vaccines [17] in various tumor types.

Monitoring biomarkers are those that can be measured serially to evaluate the status of a disease or to assess treatment response. These types of biomarkers are useful to detect evidence of early therapeutic response, or to reveal complications resulting from a therapy [2]. Although biomarkers have been defined according to specific applications, biomarkers may also meet multiple criteria for different uses. Numerous papers contributed to this Special Issue deal with biomarkers that fall into this category, including studies involving colorectal [18] and breast [19–21] cancers.

Cancer biomarker research, translated from the lab to the clinic, has led to a significant improvement in patient management, leading to increased survival rates and improved quality of life, while also lowering healthcare costs. Additional research into the detection of novel mutational variants to identify genes that are driving cancer development is critical for biomarker discovery and the further development of personalized medicine [22]. Detailed genotypic/phenotypic evaluation of individual patients is becoming increasingly available and is occurring in tandem with the development of new and improved treatments. Further advances will allow personalized medicine to become a reality for all cancer types in the decades to come.

Funding: This research received no external funding.

Institutional Review Board Statement: Not applicable.

Informed Consent Statement: Not applicable.

Data Availability Statement: Not applicable.

Acknowledgments: I wish to thank all of the authors who contributed their work, and in so doing made the Special Issue a success. I also wish to thank the staff of JPM for their excellent support throughout the editorial process.

Conflicts of Interest: The author declares no conflict of interest.

References

1. Rapado-González, Ó.; Martínez-Reglero, C.; Salgado-Barreira, Á.; Santos, M.A.; López-López, R.; Díaz-Lagares, Á.; Suárez-Cunqueiro, M.M. Salivary DNA Methylation as an Epigenetic Biomarker for Head and Neck Cancer. Part II: A Cancer Risk Meta-Analysis. *J. Pers. Med.* **2021**, *11*, 606. [CrossRef] [PubMed]
2. Meehan, J.; Gray, M.; Martínez-Pérez, C.; Kay, C.; McLaren, D.; Turnbull, A.K. Tissue- and Liquid-Based Biomarkers in Prostate Cancer Precision Medicine. *J. Pers. Med.* **2021**, *11*, 664. [CrossRef] [PubMed]
3. Zuo, D.; An, H.; Li, J.; Xiao, J.; Ren, L. The Application Value of Lipoprotein Particle Numbers in the Diagnosis of HBV-Related Hepatocellular Carcinoma with BCLC Stage 0-A. *J. Pers. Med.* **2021**, *11*, 1143. [CrossRef] [PubMed]

4. Jeong, S.; Kim, U.; Oh, M.; Nam, J.; Park, S.; Choi, Y.; Lee, D.; Kim, J.; An, H. Detection of Aberrant Glycosylation of Serum Haptoglobin for Gastric Cancer Diagnosis Using a Middle-Up-Down Glycoproteome Platform. *J. Pers. Med.* **2021**, *11*, 575. [CrossRef]
5. Rapado-González, Ó.; Martínez-Reglero, C.; Salgado-Barreira, Á.; Muinelo-Romay, L.; Muinelo-Lorenzo, J.; López-López, R.; Díaz-Lagares, Á.; Suárez-Cunqueiro, M.M. Salivary DNA Methylation as an Epigenetic Biomarker for Head and Neck Cancer. Part I: A Diagnostic Accuracy Meta-Analysis. *J. Pers. Med.* **2021**, *11*, 568. [CrossRef]
6. Kalinina, T.; Kononchuk, V.; Alekseenok, E.; Abdullin, G.; Sidorov, S.; Ovchinnikov, V.; Gulyaeva, L. Associations between the Levels of Estradiol-, Progesterone-, and Testosterone-Sensitive MiRNAs and Main Clinicopathologic Features of Breast Cancer. *J. Pers. Med.* **2021**, *12*, 4. [CrossRef]
7. Ingrosso, G.; Alì, E.; Marani, S.; Saldi, S.; Bellavita, R.; Aristei, C. Prognostic Genomic Tissue-Based Biomarkers in the Treatment of Localized Prostate Cancer. *J. Pers. Med.* **2022**, *12*, 65. [CrossRef]
8. Li, C.-J.; Lin, L.-T.; Chu, P.-Y.; Chiang, A.-J.; Tsai, H.-W.; Chiu, Y.-H.; Huang, M.-S.; Wen, Z.-H.; Tsui, K.-H. Identification of Novel Biomarkers and Candidate Drug in Ovarian Cancer. *J. Pers. Med.* **2021**, *11*, 316. [CrossRef]
9. Kang, M.-A.; Lee, J.; Lee, C.M.; Park, H.S.; Jang, K.Y.; Park, S.-H. IL13Rα2 Is Involved in the Progress of Renal Cell Carcinoma through the JAK2/FOXO3 Pathway. *J. Pers. Med.* **2021**, *11*, 284. [CrossRef]
10. Kim, K.; Yu, J.; Park, J.-Y.; Baek, S.; Hwang, J.-H.; Choi, W.-J.; Kim, Y.-K. Low Preoperative Lymphocyte-to-Monocyte Ratio Is Predictive of the 5-Year Recurrence of Bladder Tumor after Transurethral Resection. *J. Pers. Med.* **2021**, *11*, 947. [CrossRef]
11. Chang, P.-H.; Wang, H.-M.; Kuo, Y.-C.; Lee, L.-Y.; Liao, C.-J.; Kuo, H.-C.; Hsu, C.-L.; Liao, C.-T.; Lin, S.H.-C.; Huang, P.-W.; et al. Circulating p16-Positive and p16-Negative Tumor Cells Serve as Independent Prognostic Indicators of Survival in Patients with Head and Neck Squamous Cell Carcinomas. *J. Pers. Med.* **2021**, *11*, 1156. [CrossRef]
12. Moldovanu, C.-G.; Boca, B.; Lebovici, A.; Tamas-Szora, A.; Feier, D.S.; Crisan, N.; Andras, I.; Buruian, M.M. Preoperative Predicting the WHO/ISUP Nuclear Grade of Clear Cell Renal Cell Carcinoma by Computed Tomography-Based Radiomics Features. *J. Pers. Med.* **2021**, *11*, 8. [CrossRef]
13. Cheng, S.W.; Chen, P.C.; Ger, T.R.; Chiu, H.W.; Lin, Y.F. GBP5 Serves as a Potential Marker to Predict a Favorable Response in Triple-Negative Breast Cancer Patients Receiving a Taxane-Based Chemotherapy. *J. Pers. Med.* **2021**, *11*, 197. [CrossRef]
14. Gauthier, A.; Philouze, P.; Lauret, A.; Alphonse, G.; Malesys, C.; Ardail, D.; Payen, L.; Céruse, P.; Wozny, A.-S.; Rodriguez-Lafrasse, C. Circulating Tumor Cell Detection during Neoadjuvant Chemotherapy to Predict Early Response in Locally Advanced Oropharyngeal Cancers: A Prospective Pilot Study. *J. Pers. Med.* **2022**, *12*, 445. [CrossRef]
15. Meehan, J.; Gray, M.; Martínez-Pérez, C.; Kay, C.; Wills, J.C.; Kunkler, I.H.; Dixon, J.M.; Turnbull, A.K. A Novel Approach for the Discovery of Biomarkers of Radiotherapy Response in Breast Cancer. *J. Pers. Med.* **2021**, *11*, 796. [CrossRef]
16. Benitez, J.; Campayo, M.; Díaz, T.; Ferrer, C.; Acosta-Plasencia, M.; Monzo, M.; Cirera, L.; Besse, B.; Navarro, A. Lincp21-RNA as Predictive Response Marker for Preoperative Chemoradiotherapy in Rectal Cancer. *J. Pers. Med.* **2021**, *11*, 420. [CrossRef]
17. Guldvik, I.J.; Ekseth, L.; Kishan, A.U.; Stensvold, A.; Inderberg, E.M.; Lilleby, W. Circulating Tumor Cell Persistence Associates with Long-Term Clinical Outcome to a Therapeutic Cancer Vaccine in Prostate Cancer. *J. Pers. Med.* **2021**, *11*, 605. [CrossRef]
18. Gray, M.; Marland, J.R.K.; Murray, A.F.; Argyle, D.J.; Potter, M.A. Predictive and Diagnostic Biomarkers of Anastomotic Leakage: A Precision Medicine Approach for Colorectal Cancer Patients. *J. Pers. Med.* **2021**, *11*, 471. [CrossRef]
19. Martínez-Pérez, C.; Kay, C.; Meehan, J.; Gray, M.; Dixon, J.M.; Turnbull, A.K. The IL6-like Cytokine Family: Role and Biomarker Potential in Breast Cancer. *J. Pers. Med.* **2021**, *11*, 1073. [CrossRef]
20. Martínez-Pérez, C.; Leung, J.; Kay, C.; Meehan, J.; Gray, M.; Dixon, J.; Turnbull, A. The Signal Transducer IL6ST (gp130) as a Predictive and Prognostic Biomarker in Breast Cancer. *J. Pers. Med.* **2021**, *11*, 618. [CrossRef]
21. Payton, C.; Pang, L.Y.; Gray, M.; Argyle, D.J. Exosomes Derived from Radioresistant Breast Cancer Cells Promote Therapeutic Resistance in Naïve Recipient Cells. *J. Pers. Med.* **2021**, *11*, 1310. [CrossRef]
22. Almuzzaini, B.; Alghamdi, J.; Alomani, A.; AlGhamdi, S.; Alsharm, A.; Alshieban, S.; Sayed, A.; Alhejaily, A.; Aljaser, F.; Abudawood, M.; et al. Identification of Novel Mutations in Colorectal Cancer Patients Using AmpliSeq Comprehensive Cancer Panel. *J. Pers. Med.* **2021**, *11*, 535. [CrossRef]

Review

Salivary DNA Methylation as an Epigenetic Biomarker for Head and Neck Cancer. Part I: A Diagnostic Accuracy Meta-Analysis

Óscar Rapado-González [1,2,3], Cristina Martínez-Reglero [4], Ángel Salgado-Barreira [4], Laura Muinelo-Romay [2,3], Juan Muinelo-Lorenzo [1], Rafael López-López [3,5], Ángel Díaz-Lagares [3,6] and María Mercedes Suárez-Cunqueiro [1,3,5,*]

1. Department of Surgery and Medical-Surgical Specialties, Medicine and Dentistry School, Universidade de Santiago de Compostela, 15782 Santiago de Compostela, Spain; oscar.rapado@rai.usc.es (Ó.R.-G.); juanmuinelo@hotmail.com (J.M.-L.)
2. Translational Medical Oncology Group (Oncomet), Liquid Biopsy Analysis Unit, Health Research Institute of Santiago (IDIS), 15706 Santiago de Compostela, Spain; lmuirom@gmail.com
3. Centro de Investigación Biomédica en Red de Cáncer (CIBERONC), Instituto de Salud Carlos III, 28029 Madrid, Spain; rafael.lopez.lopez@sergas.es (R.L.-L.); angel.diaz.lagares@sergas.es (Á.D.-L.)
4. Methodology and Statistics Unit, Galicia Sur Health Research Institute (IISGS), 36312 Vigo, Spain; cristina.martinez@iisgaliciasur.es (C.M.-R.); angel.salgado.barreira@sergas.es (Á.S.-B.)
5. Translational Medical Oncology Group (Oncomet), Health Research Institute of Santiago (IDIS), Complexo Hospitalario Universitario de Santiago (SERGAS), 15706 Santiago de Compostela, Spain
6. Cancer Epigenomics, Translational Medical Oncology Group (Oncomet), Health Research Institute of Santiago (IDIS), University Clinical Hospital of Santiago (CHUS/SERGAS), 15706 Santiago de Compostela, Spain
* Correspondence: mariamercedes.suarez@usc.es; Tel.: +34-881-812-437

Abstract: DNA hypermethylation is an important epigenetic mechanism for gene expression inactivation in head and neck cancer (HNC). Saliva has emerged as a novel liquid biopsy representing a potential source of biomarkers. We performed a comprehensive meta-analysis to evaluate the overall diagnostic accuracy of salivary DNA methylation for detecting HNC. PubMed EMBASE, Web of Science, LILACS, and the Cochrane Library were searched. Study quality was assessed by the Quality Assessment for Studies of Diagnostic Accuracy-2, and sensitivity, specificity, positive likelihood ratio (PLR), negative likelihood ratio (NLR), diagnostic odds ratio (dOR), and their corresponding 95% confidence intervals (CIs) were calculated using a bivariate random-effect meta-analysis model. Meta-regression and subgroup analyses were performed to assess heterogeneity. Eighty-four study units from 18 articles with 8368 subjects were included. The pooled sensitivity and specificity of salivary DNA methylation were 0.39 and 0.87, respectively, while PLR and NLR were 3.68 and 0.63, respectively. The overall area under the curve (AUC) was 0.81 and the dOR was 8.34. The combination of methylated genes showed higher diagnostic accuracy (AUC, 0.92 and dOR, 36.97) than individual gene analysis (AUC, 0.77 and dOR, 6.02). These findings provide evidence regarding the potential clinical application of salivary DNA methylation for HNC diagnosis.

Keywords: DNA methylation; epigenetics; head and neck cancer; saliva; biomarkers; liquid biopsy; meta-analysis

1. Introduction

Head and neck cancer (HNC) comprises a heterogenous group of epithelial malignancies arising from mucosal linings of the oral cavity, oropharynx, larynx, and hypopharynx. According to data from the World Health Organization's GLOBOCAN network, HNC is highly prevalent worldwide, accounting for an estimated 890,000 new cases and 450,000 deaths in 2018 [1]. Despite improvements in diagnosis and therapeutic strategies, the 5-year survival rate for HNC has remained around 50% for the last decade. Unfortunately, HNC patients are frequently diagnosed in advanced stages involving a high risk of locoregional recurrence and distant metastasis [2,3]. Therefore, it is necessary to identify new biomarkers for diagnosis and prognosis to allow early detection and improved overall survival. HNC arises through multistep carcinogenic pathways as a result of cumulative genetic and epigenetic aberrations resulting from risk factors including alcohol and tobacco consumption, human papillomavirus infection, chronic inflammation, and genetic predisposition, and leading to reduced tumor suppressor gene function as well as oncogene activation [2,4]. In recent years, accumulating scientific evidence has highlighted the important role in tumorigenesis of epigenetic mechanisms, which represent a cancer hallmark [5,6]. Epigenetic alterations such as DNA methylation, histone covalent modifications, chromatin remodeling, and non-coding RNAs have been implicated in the landscape of phenotypical changes occurring in a wide variety of malignancies, including HNC [7]. DNA hypermethylation has been shown to be an important epigenetic mechanism for gene expression inactivation. The hypermethylation of cytosine–phosphodiester bond–guanine (CpG) islands within promoter regions plays an important role in carcinogenesis through the transcriptional silencing of different tumor suppressor genes or dysfunction in DNA repair genes [7,8]. Several studies have focused on the identification of aberrant promoter methylation patterns in HNC tissue and liquid biopsies [9,10]. Moreover, a number of investigations have detected promoter hypermethylation in various genes using saliva from HNC patients [10–12]. Therefore, salivary DNA hypermethylation represents a promising biomarker for non-invasively diagnosing HNC.

DNA methylation is a heritable and stable epigenetic mechanism implicated in the regulation of gene expression that plays an important role during normal development, regulating X chromosome inactivation, genomic imprinting, and preventing the transcription of DNA repetitive sequences, inserted viral sequences, and transposons [8,13]. DNA methylation is characterized by the covalent addition of a methyl group to the 5'-position of the pyrimidine ring of cytosines by DNA methyltransferases, giving rise to 5-methylcytosine. This enzymatic process occurs predominantly within CpG dinucleotides which are concentrated at CpG-rich DNA stretches named CpG islands (CGIs), which overlap the promoter region of 60–70% of protein-coding genes [14]. In the human genome, approximately 80% of CpG dinucleotides are heavily methylated whereas CGIs in gene promoters are mostly unmethylated, allowing active gene transcription [13]. Dysregulation of DNA methylation has been found to be involved in several diseases, representing an early epigenetic event in carcinogenesis. These alterations of normal DNA methylation patterns in cancer have been characterized as global hypomethylation and gene-specific hypermethylation. The global hypomethylation of repetitive sequences and transposable elements within the genome induces genomic instability and mutagenesis. In this line, the loss of DNA methylation may also activate latent viral sequences, promoting carcinogenesis. By contrast, in addition to global hypomethylation, aberrant promoter hypermethylation can drive the inactivation of key tumor suppressor genes, which are unmethylated in non-malignant tissues [15]. In this sense, although silencing by DNA hypermethylation of some genes is common in many types of tumors, the methylation profile of gene promoters is different for each human cancer, allowing the identification of cancer-specific hypermethylation patterns [16]. Although tissue biopsy of the primary tumor or metastatic lesions remains the gold standard method for diagnosis, DNA methylation biomarkers can be assessed in different liquid biopsies, representing a non-invasive alternative for early cancer detection [10,17,18].

Recently, saliva has emerged as an attractive liquid biopsy for genomic and epigenomic analysis. Saliva-based liquid biopsy is a fast, reliable, cost-effective, and non-invasive approach to analyze epigenetic alterations involved in the onset and course of the disease. Some researchers have found comparable methylation profiles between saliva and tissue, representing a non-invasive alternative for epigenomic profiling [19,20]. Additionally, although similar methylation DNA patterns have been reported between saliva and blood [21,22], methylation differences between both biofluids can be identified due to tumor shedding and tissue-specific methylation [23,24]. Focusing on promoter hypermethylation, a number of investigations have detected various methylated genes in saliva, representing a promising biomarker for non-invasive HNC detection [10–12].

The purpose of this systematic review and meta-analysis was to summarize the results of published clinical studies to assess the overall diagnostic accuracy of salivary DNA hypermethylation for discriminating HNC.

2. Materials and Methods

2.1. Protocol and Registration

This study was conducted according to Preferred Reporting Items for Systematic Reviews and Meta-analysis (PRISMA) guidelines [25], and the protocol was registered with the International Prospective Register of Systematic Reviews (reference No. CRD42020199114).

2.2. Search Strategy and Study Selection

The systematic literature search of eligible articles published up to 27 August 2020 was carried out without language restrictions using PubMed, EMBASE, Web of Science, LILACS, and the Cochrane Library. The search strategy was based on the following combinations of keywords and medical subject headings: (methylation OR hypermethylation OR epigenomics) AND (saliva OR oral rinse OR mouthwash) AND (head and neck cancer OR head and neck neoplasm OR head and neck carcinoma OR head and neck squamous cell carcinoma OR HNSCC OR oral cancer OR pharyngeal cancer OR laryngeal cancer). Studies were screened based on title and abstract, and eligible manuscripts were retrieved for full-text review. In addition, reference lists from each original and review article were searched manually in order to find further relevant studies. The literature search was performed independently by two investigators (ORG and MMSC), and disagreements during the selection process were resolved by consensus. The studies selected by means of the search strategy and other references were managed using RefWorks software (https://www.refworks.com/content/path_learn/faqs.asp, accessed 28 October 2020), and duplicate items were removed using the associated tools.

2.3. Selection Criteria

The inclusion criteria were as follows: (1) studies that evaluated the diagnostic accuracy of gene promoter hypermethylation in saliva samples from HNC patients; (2) inclusion of a control group consisting of healthy controls; (3) sufficient data for generating a two-by-two (2 × 2) contingency table containing true positive (TP), false positive (FP), true negative (TN), and false negative (FN) values. The exclusion criteria were as follows: (1) reviews, letters, personal opinions, book chapters, case reports, conference abstracts, and meetings; (2) duplicate publications; (3) in vitro and in vivo animal experiments.

2.4. Data Extraction

All eligible studies were assessed independently by two investigators (ORG and MMSC) and data were extracted using a pre-established form designed on a Microsoft Excel spreadsheet (Microsoft Corp. Redmond, WA, USA). Any disagreement among reviewers was resolved by consensus. The following information was extracted from each study: name of first author, year of publication, country, anatomic tumor location, number of cases and controls, positive methylated cases, positive methylated controls, method for DNA methylation detection, type of saliva sample (saliva or oral rinse), methylated gene names, and statistical analysis outcomes, including diagnostic accuracy and cut-off values. If the required data were incomplete, attempts were made to contact the authors to obtain the missing information. We defined "study unit" as the analysis of a relationship between gene promoter hypermethylation and HNC. Therefore, a single publication could potentially include more than one study unit as a result of reporting promoter hypermethylation for multiple genes.

2.5. Quality Assessment of Individual Studies

Following Healthcare Research and Quality Agency recommendations, two independent researchers (ADL and LMR) applied the Quality Assessment of Diagnostic Accuracy Studies-2 checklist (QUADAS-2) [26]. Any discrepancies were resolved by a third reviewer (MMSC). The QUADAS-2 checklist assesses study quality by analyzing four key domains: (1) patient selection, (2) index tests, (3) reference tests, and (4) flow and times. Risk of bias and applicability concerns for each domain were assessed as "low", "high", or "unclear". One point was assigned to each item assessed as "low". Thus, articles were grouped into the following quality categories based on their cumulative score: "high" quality (6–7 points), "moderate" quality (4–5 points), and "low" quality (0–3 points).

2.6. Statistical Analysis

MetaDiSc software (v.1.4) [27], free R software (v.3.4.4; https://www.r-project.org, accessed 30 November 2020), and STATA (v.14.0; https://www.stata.com, accessed 30 November 2020) were used to carry out statistical analysis. The numbers of TP, FP, FN, and TN in each study unit in the diagnostic meta-analysis were extracted to calculate pooled sensitivity [TP/(TP + FN)], specificity [TN/(TN + FP)], positive likelihood ratio (PLR) [(sensitivity/(1 − sensitivity)], negative likelihood ratio (NLR), [(1 − specificity)/specificity)], diagnostic odds ratio (dOR), and their corresponding 95% confidence intervals (CIs) using a bivariate random or fixed effect meta-analysis model. The pooled diagnostic performance of salivary DNA promoter hypermethylation for HNC detection was determined by plotting the summary receiver operator characteristic (SROC) curve and calculating the area under the SROC curve (AUC). Heterogeneity analysis was used to identify factors influencing accuracy indicators and the statistical model applied [27]. Spearman's correlation analysis and ROC plane plots were used to assess heterogeneity due to the threshold effect. Cochran's Q statistic test-based chi-squared test and I^2 statistics were used to assess non-threshold heterogeneity. When $I^2 > 50\%$ and/or $p < 0.05$ for the Cochran's Q test, heterogeneity was considered to be significant. The DerSimonian and Laird random effects model was applied when heterogeneity was significant; otherwise, we applied the Mantel–Haenszel fixed effects model. Potential sources of non-threshold heterogeneity were explored by meta-regression and subgroup analyses. In addition, the predictive value of post-salivary DNA promoter hypermethylation for HNC diagnosis was evaluated by Fagan's nomogram. Post-test probability was calculated using Bayes theorem under the assumption of prior probabilities of 25%, 50%, and 75%, respectively [28]. Deeks' funnel plot asymmetry test [29] was used to ascertain publication bias (statistical significance: $p < 0.05$).

3. Results

3.1. Study Selection

A PRISMA flowchart for the literature identification and selection process is shown in Figure 1. A total of 576 studies were identified based on the search strategy across the five electronic databases, which was reduced to 470 after removing duplicates. After title and abstract review, 27 articles were submitted for full-text reading, of which nine were excluded for the following reasons: non-independent cancer group (two articles); reviews, letters, personal opinions, book chapters, case reports, conference abstracts, and meetings (three articles); absence of a healthy control group (one article); saliva enriched with brush oral cytology (two articles); and insufficient information for meta-analysis (one article). In the end, 18 articles met the inclusion criteria for final analysis [10–12,30–44].

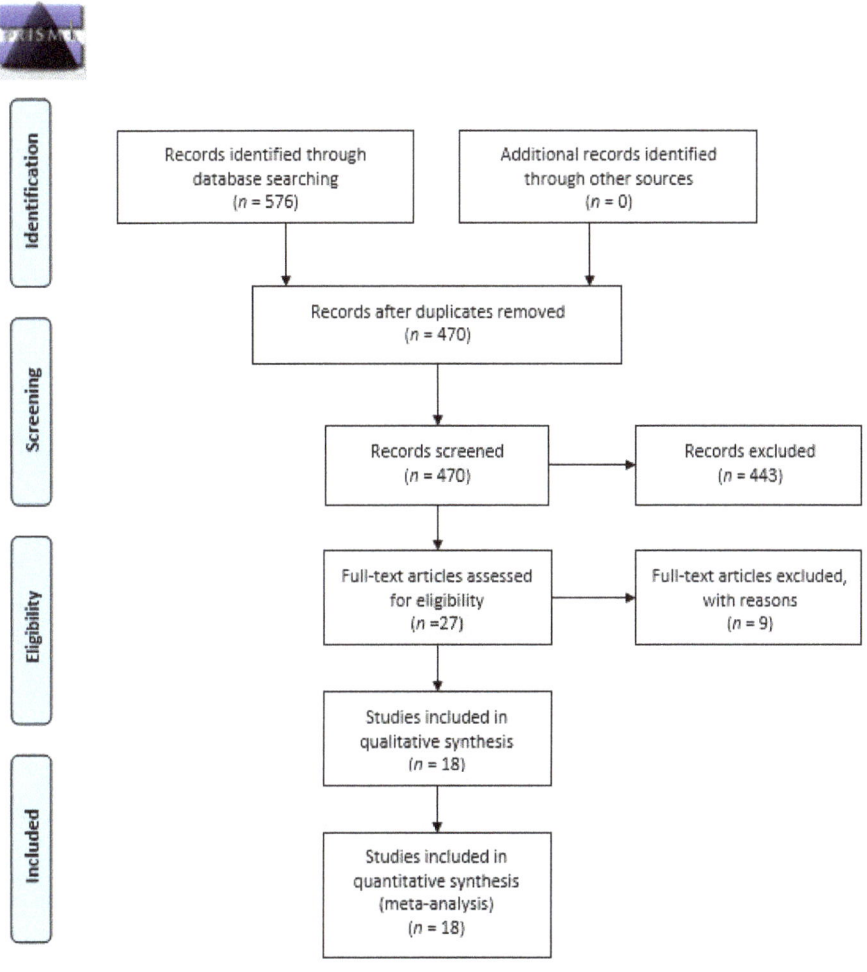

Figure 1. PRISMA flow diagram.

3.2. Characteristics of Included Studies

The characteristics of the included studies are shown in Table 1. The 18 articles comprised a total of 84 study units, including 4758 HNC patients and 3605 healthy individuals (the sample size ranged from 13 [31] to 210 [41]). All articles were published between 2001 and 2020, and studies were conducted in the following geographical regions: the United States (5), Australia (3), Thailand (2), Japan (2), France (1), Brazil (1), India (1), Taiwan (1), Colombia (1), and Italy (1). Saliva samples included oral rinses (12) and whole saliva (6). Salivary DNA methylation was detected by different methods, including methylation-specific polymerase chain reaction (MSP) (11) and quantitative MSP (7). A total of 34 different genes were identified in the studies. Five studies evaluated the methylation status of a single gene [31,34,36,39,41] and 13 studies evaluated two or more genes [10–12,30,32,33,35,37,38,40,42–44]. Ten studies focused on gene promoter methylation panels combining two to four genes, whereas eight studies evaluated only single genes.

Table 1. Summary of descriptive characteristics of included studies.

First Author	Anatomic Tumor Location	Type of Sample	Method	Biomarker	Cancer Group N (M+)	Control Group N (M+)
Rosas 2001	HNC	Oral rinse (NaCl)	MSP	p16	30 (11+)	30 (1+)
				DAPK	30 (6+)	30 (0+)
				MGMT	30 (4+)	30 (1+)
Righini 2007	HNC	Oral rinse (NaCl)	MSP	TIMP3	60 (17+)	30 (0+)
				ECAD	60 (12+)	
				p16	60 (16+)	
				MGMT	60 (13+)	
				DAPK	60 (9+)	
				RASSF1A	60 (10+)	
				p15	60 (7+)	
				p14	60 (2+)	
				APC	60 (4+)	
				FHIT	60 (2+)	
				hMLH1	60 (0+)	
Franzmann 2007	HNC	Oral rinse (NaCl)	MSP	CD44	11 (9+)	10 (0+)
Guerrero-Preston 2011	HNC	Oral rinse (NaCl)	qMSP	HOXA9	32 (20+)	19 (9+)
				NID2	32 (23+)	19 (12+)
				HOXA9+NID2	32 (25+)	19 (6+)
	OC			HOXA9	16 (11+)	19 (9+)
				NID2	16 (14+)	19 (12+)
				HOXA9+NID2	16 (14+)	19 (6+)
	OPC			HOXA9	16 (9+)	19 (9+)
				NID2	16 (9+)	19 (12+)
				HOXA9+NID2	16 (11+)	19 (6+)
Nagata 2011	OC	Oral rinse (NaCl)	MSP	ECAD	34 (32+)	24 (5+)
				TMEFF2	34 (29+)	24 (3+)
				RARβ	34 (28+)	24 (2+)
				MGMT	34 (26+)	24 (5+)
				FHIT	34 (27+)	24 (8+)
				WIF1	34 (24+)	24 (5+)
				DAPK	34 (19+)	24 (6+)
				p16	34 (13+)	24 (2+)
				HIN	34 (10+)	24 (2+)
				TIMP3	34 (8+)	24 (1+)
				p15	34 (22+)	24 (9+)
				APC	34 (18+)	24 (9+)
				SPARC	34 (14+)	24 (8+)
				ECAD+TMEFF2+RARβ+MGMT	34 (34+)	24 (3+)
				ECAD+TMEFF2+MGMT	34 (33+)	24 (2+)
				ECAD+TMEFF2+RARβ	34 (32+)	24 (1+)
				ECAD+RARβ+MGMT	34 (31+)	24 (2+)
Ovchinnikov 2012	OC	Saliva	Nested MSP	p16+RASSF1A+DAPK1	143 (117+)	46 (6+)
Rettori 2012	HNC	Oral rinse (NaCl)	qMSP	DCC	143 (75+)	50 (5+)
				CCNA1	146 (17+)	60 (2+)
				DAPK	146 (12+)	39 (1+)
				MGMT	146 (11+)	57 (2+)
				TIMP3	146 (7+)	60 (2+)
				MINT31	68 (3+)	20 (0+)
				AIM1	71 (2+)	41 (0+)
				SFRP1	71 (2+)	20 (0+)
				APC	62 (2+)	20 (0+)
				CDKN2A	69 (1+)	20 (0+)
				HIN1	134 (16+)	57 (11+)
				CCNA1+DAPK+DCC+MGMT+TIMP3	NA	NA
				CCNA1+DAPK+MGMT+TIMP3	NA	NA
				CCNA1+DAPK+MGMT	NA	NA
				CCNA1+MGMT+TIMP3	NA	NA
				CCNA1+DAPK+TIMP3	NA	NA

Table 1. Cont.

First Author	Anatomic Tumor Location	Type of Sample	Method	Biomarker	Cancer Group N (M+)	Control Group N (M+)
				DAPK+MGMT+TIMP3	NA	NA
				CCNA1+MGMT	NA	NA
				CCNA1+DAPK	NA	NA
				CCNA1+TIMP3	NA	NA
Ksumoto 2012	OC	Oral rinse	MSP	p16	10 (4+)	3 (0+)
Ovchinnikov 2014	HNC	Saliva	MSP	PCQAP5′	62 (42+)	49 (17+)
				PCQAP3′	60 (41+)	45 (19+)
Gaykalova 2015	HNC	Oral rinse	qMSP	ZNF14	59 (5+)	35 (0+)
				ZNF160	59 (10+)	35 (0+)
				ZNF420	59 (8+)	35 (0+)
				ZNF14+ZNF160+ZNF420	59 (13+)	35 (0+)
Lim 2016	HNC	Saliva	MSP	RASSF1α	88 (36+)	122 (10+)
				p16	88 (41+)	122 (38+)
				TIMP3	88 (33+)	122 (22+)
				PCQAP5′	88 (72+)	122 (66+)
				PCQAP3′	88 (30+)	122 (18+)
				RASSF1α+p16+TIMP3+PCQAP5′+PCQAP3′	88 (62+)	122 (24+)
Ferlazzo 2017	OC	Saliva (Oragene DNA kit)	MSP	P16	58 (10+)	90 (5+)
				MGMT	58 (16+)	90 (7+)
				P16 + MGMT	58 (12+)	90 (0+)
Cheng 2017	OC	Oral rinse (0.12% clorhexidine)	qMSP	ZNF582	94 (62+)	65 (10+)
				PAX1	94 (64+)	65 (7+)
				ZNF582+PAX1	94 (75+)	65 (14+)
Puttipanyalears 2018	HNC OC OPC	Oral rinse (NaCl)	qMSP	TRH	66 (57+) 42 (37+) 24 (20+)	54 (4+)
Liyanage 2020	HNC	Saliva	MSP	p16	88 (62+)	NA
				RASSF1 α	88 (59+)	NA
				TIMP3	88 (68+)	NA
				PCQAP/MED15	88 (66+)	NA
				p16+RASSF1α+TIMP3+PCQAP	84 (80+)	60 (5+)
	OC			p16	54 (39+)	NA
				RASSF1α	54 (37+)	NA
				TIMP3	54 (43+)	NA
				PCQAP/MED15	54 (43+)	NA
				p16+RASSF1α+TIMP3+PCQAP	54 (46+)	60 (5+)
	OPC			p16	34 (23+)	NA
				RASSF1α	34 (22+)	NA
				TIMP3	34 (25+)	NA
				PCQAP/MED15	34 (23+)	NA
				p16+RASSF1α+TIMP3+PCQAP	34 (34+)	60 (5+)
Srisuttee 2020	OC	Oral rinse (NaCl)	qMSP	NID2	43 (34+)	90 (0+)
Shen 2020	OPC	Oral rinse (NaCl)	qMSP	EDNRB	21 (15+)	40 (2+)
				PAX5	21 (15+)	40 (4+)
				p16	21 (3+)	40 (0+)
González-Pérez 2020	OC	Saliva	MSP	p16	43 (19+)	40 (4+)
				RASSF1A	43 (10+)	40 (2+)
				p16+RASSF1A	43 (23+)	40 (5+)

Abbreviations: HNC = head and neck cancer; OC = oral cancer; OPC = oropharyngeal cancer; MSP = methylation-specific polymerase chain reaction; qMSP = quantitative MSP; NA = not available.

3.3. Quality Assessment of the Included Studies

All included articles were evaluated for risk of bias using the QUADAS-2 checklist (Figure S1). The major risk of bias in this study was patient selection domain, as 13 out of 18 publications were unclear or lacked detail on whether the patient sample was consecutive or random. Additionally, 14 out of 18 studies did not provide a detailed description of the inclusion/exclusion criteria. Moreover, there was high risk of bias in the index test, since some studies lacked a setting threshold. All domains were considered to have a low risk of bias in terms of applicability concern. All studies were of moderate-to-high quality, with an average QUADAS-2 score of 5.6.

3.4. Diagnostic Accuracy of Salivary DNA Promoter Hypermethylation

The diagnostic accuracy (sensitivity, specificity, and 95% confidence interval) of each study unit ($n = 74$) included in this meta-analysis is shown in Figures 2 and 3. The pooled sensitivity and specificity of salivary DNA hypermethylation genes in the diagnosis of HNC were 0.39 (95% CI: 0.38–0.41) and 0.87 (95% CI: 0.86–0.88), respectively. The PLR and NLR were 3.68 (95% CI: 2.97–4.57) and 0.63 (95% CI: 0.57–0.69) (Figures 4 and 5), respectively; the summary dOR was 8.34 (95% CI: 6.10–11.39) (Figure S2); and the area under the SROC was 0.81 (95% CI 0.77–0.84) (Figure 6). As shown in Fagan's nomogram (Figure S3), given a pre-test probability of 27.8%, a positive measurement leads to a post-test cancer probability of 59%, whereas a negative measurement leads to a post-test probability of 20%.

3.5. Heterogeneity and Subgroup Analysis

As shown in Figures 2–5 and Figure S2, significant heterogeneity was observed regarding the pooled sensitivity ($I^2 = 96.33\%$; $p < 0.001$), specificity ($I^2 = 87.07\%$; $p < 0.001$), PLR ($I^2 = 73.99\%$; $p < 0.001$), NLR ($I^2 = 96.35\%$; $p < 0.001$), and dOR ($I^2 = 71.83\%$; $p < 0.001$). The representation of accuracy estimates from each study in the SROC space revealed a typical pattern of a "shoulder arm", suggesting the presence of a threshold effect (Figure S4). Moreover, Spearman's correlation coefficient between the logit of the true positive rate and the logit of the false positive rate was 0.633 ($p = 0.000$), which showed further indication of a threshold effect. In addition to the variations due to the threshold effect, meta-regression was performed to determine the possible sources of heterogeneity using the following covariates as predictor variables: sample type, sample size, anatomic tumor location, DNA methylation methods, and methylation gene profiling. The results indicated that anatomic tumor location ($p = 0.002$) and gene profiling ($p < 0.001$) were potential sources of heterogeneity in this study (Table S1). Consequently, subgroup analysis based on anatomic tumor location (HNC vs. oral cancer vs. oropharyngeal cancer) and gene profiling (single vs. combination of genes) was performed. As shown in Table S2, the results indicated similar accuracy for salivary methylated genes in oral cancer (sensitivity, 0.63; specificity, 0.87; PLR, 4.02; NLR, 0.40; dOR, 13.07; AUC, 0.88) and oropharyngeal cancer (sensitivity, 0.70; specificity, 0.86; PLR, 3.67; NLR, 0.41; dOR, 13.26; AUC, 0.87). However, differences in diagnostic accuracy were observed in the HNC group (sensitivity, 0.31; specificity, 0.86; PLR, 3.03; NLR, 0.75; dOR, 5.78; AUC, 0.81). When basing the meta-analysis on gene profile, the combination of methylated genes showed higher diagnostic accuracy (sensitivity, 0.73; specificity, 0.88; PLR, 5.76; NLR, 0.22; dOR, 36.97; AUC, 0.92) compared to individual genes (sensitivity, 0.32; specificity, 0.87; PLR, 3.17; NLR, 0.71; dOR, 6.02; AUC, 0.77). Although the meta-regression results were negative for other covariates, we conducted subgroup analyses based on these factors to further explore the diagnostic potential of salivary DNA methylated genes (Table S2).

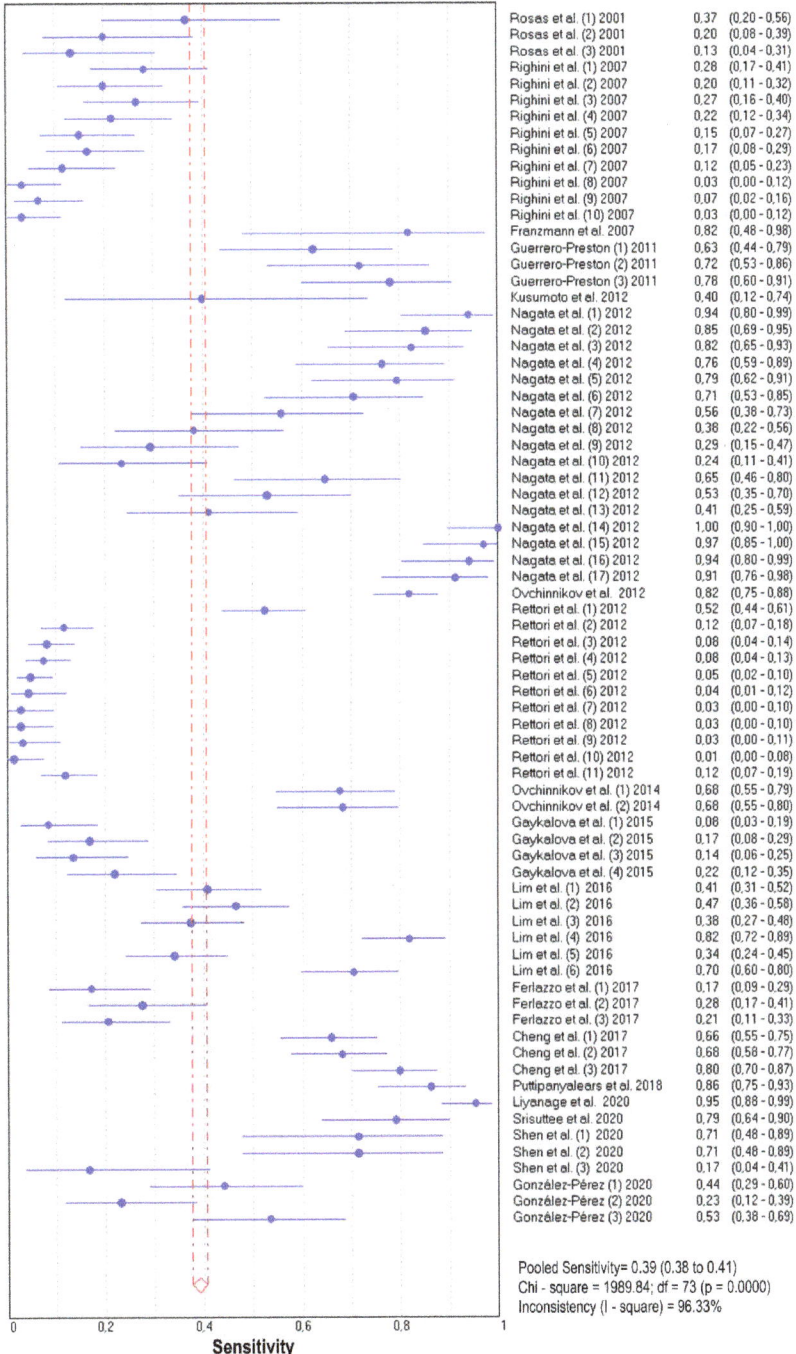

Figure 2. Forest plot of sensitivities from test accuracy studies of salivary DNA methylation for predicting HNC diagnosis.

Study	Specificity	(95% CI)
Rosas et al. (1) 2001	0.97	(0.83 - 1.00)
Rosas et al. (2) 2001	1.00	(0.88 - 1.00)
Rosas et al. (3) 2001	0.97	(0.83 - 1.00)
Righini et al. (1) 2007	1.00	(0.88 - 1.00)
Righini et al. (2) 2007	1.00	(0.88 - 1.00)
Righini et al. (3) 2007	1.00	(0.88 - 1.00)
Righini et al. (4) 2007	1.00	(0.88 - 1.00)
Righini et al. (5) 2007	1.00	(0.88 - 1.00)
Righini et al. (6) 2007	1.00	(0.88 - 1.00)
Righini et al. (7) 2007	1.00	(0.88 - 1.00)
Righini et al. (8) 2007	1.00	(0.88 - 1.00)
Righini et al. (9) 2007	1.00	(0.88 - 1.00)
Righini et al. (10) 2007	1.00	(0.88 - 1.00)
Franzmann et al. 2007	1.00	(0.69 - 1.00)
Guerrero-Preston (1) 2011	0.53	(0.29 - 0.76)
Guerrero-Preston (2) 2011	0.37	(0.16 - 0.62)
Guerrero-Preston (3) 2011	0.68	(0.43 - 0.87)
Kusumoto et al. 2012	1.00	(0.29 - 1.00)
Nagata et al. (1) 2012	0.79	(0.58 - 0.93)
Nagata et al. (2) 2012	0.88	(0.68 - 0.97)
Nagata et al. (3) 2012	0.92	(0.73 - 0.99)
Nagata et al. (4) 2012	0.79	(0.58 - 0.93)
Nagata et al. (5) 2012	0.67	(0.45 - 0.84)
Nagata et al. (6) 2012	0.79	(0.58 - 0.93)
Nagata et al. (7) 2012	0.75	(0.53 - 0.90)
Nagata et al. (8) 2012	0.92	(0.73 - 0.99)
Nagata et al. (9) 2012	0.92	(0.73 - 0.99)
Nagata et al. (10) 2012	0.96	(0.79 - 1.00)
Nagata et al. (11) 2012	0.63	(0.41 - 0.81)
Nagata et al. (12) 2012	0.63	(0.41 - 0.81)
Nagata et al. (13) 2012	0.67	(0.45 - 0.84)
Nagata et al. (14) 2012	0.88	(0.68 - 0.97)
Nagata et al. (15) 2012	0.92	(0.73 - 0.99)
Nagata et al. (16) 2012	0.96	(0.79 - 1.00)
Nagata et al. (17) 2012	0.92	(0.73 - 0.99)
Ovchinnikov et al. 2012	0.87	(0.74 - 0.95)
Rettori et al. (1) 2012	0.90	(0.78 - 0.97)
Rettori et al. (2) 2012	0.97	(0.88 - 1.00)
Rettori et al. (3) 2012	0.97	(0.87 - 1.00)
Rettori et al. (4) 2012	0.96	(0.88 - 1.00)
Rettori et al. (5) 2012	0.97	(0.88 - 1.00)
Rettori et al. (6) 2012	1.00	(0.83 - 1.00)
Rettori et al. (7) 2012	1.00	(0.91 - 1.00)
Rettori et al. (8) 2012	1.00	(0.83 - 1.00)
Rettori et al. (9) 2012	1.00	(0.83 - 1.00)
Rettori et al. (10) 2012	1.00	(0.83 - 1.00)
Rettori et al. (11) 2012	0.81	(0.68 - 0.90)
Ovchinnikov et al. (1) 2014	0.65	(0.50 - 0.78)
Ovchinnikov et al. (2) 2014	0.58	(0.42 - 0.72)
Gaykalova et al. (1) 2015	1.00	(0.90 - 1.00)
Gaykalova et al. (2) 2015	1.00	(0.90 - 1.00)
Gaykalova et al. (3) 2015	1.00	(0.90 - 1.00)
Gaykalova et al. (4) 2015	1.00	(0.90 - 1.00)
Lim et al. (1) 2016	0.92	(0.85 - 0.96)
Lim et al. (2) 2016	0.69	(0.60 - 0.77)
Lim et al. (3) 2016	0.82	(0.74 - 0.88)
Lim et al. (4) 2016	0.46	(0.37 - 0.55)
Lim et al. (5) 2016	0.85	(0.78 - 0.91)
Lim et al. (6) 2016	0.80	(0.72 - 0.87)
Ferlazzo et al. (1) 2017	0.94	(0.88 - 0.98)
Ferlazzo et al. (2) 2017	0.92	(0.85 - 0.97)
Ferlazzo et al. (3) 2017	1.00	(0.96 - 1.00)
Cheng et al. (1) 2017	0.85	(0.74 - 0.92)
Cheng et al. (2) 2017	0.89	(0.79 - 0.96)
Cheng et al. (3) 2017	0.78	(0.67 - 0.88)
Puttipanyalears et al. 2018	0.93	(0.82 - 0.98)
Liyanage et al. 2020	0.92	(0.82 - 0.97)
Srisuttee et al. 2020	1.00	(0.96 - 1.00)
Shen et al. (1) 2020	0.95	(0.83 - 0.99)
Shen et al. (2) 2020	0.90	(0.76 - 0.97)
Shen et al. (3) 2020	1.00	(0.91 - 1.00)
González-Pérez (1) 2020	0.90	(0.76 - 0.97)
González-Pérez (2) 2020	0.95	(0.83 - 0.99)
González-Pérez (3) 2020	0.88	(0.73 - 0.96)

Pooled Specificity = 0.87 (0.86 to 0.88)
Chi - square = 564.89; df = 73 (p = 0.0000)
Inconsistency (I - square) = 87.07%

Figure 3. Forest plot of specificities from test accuracy studies of salivary DNA methylation for predicting HNC diagnosis.

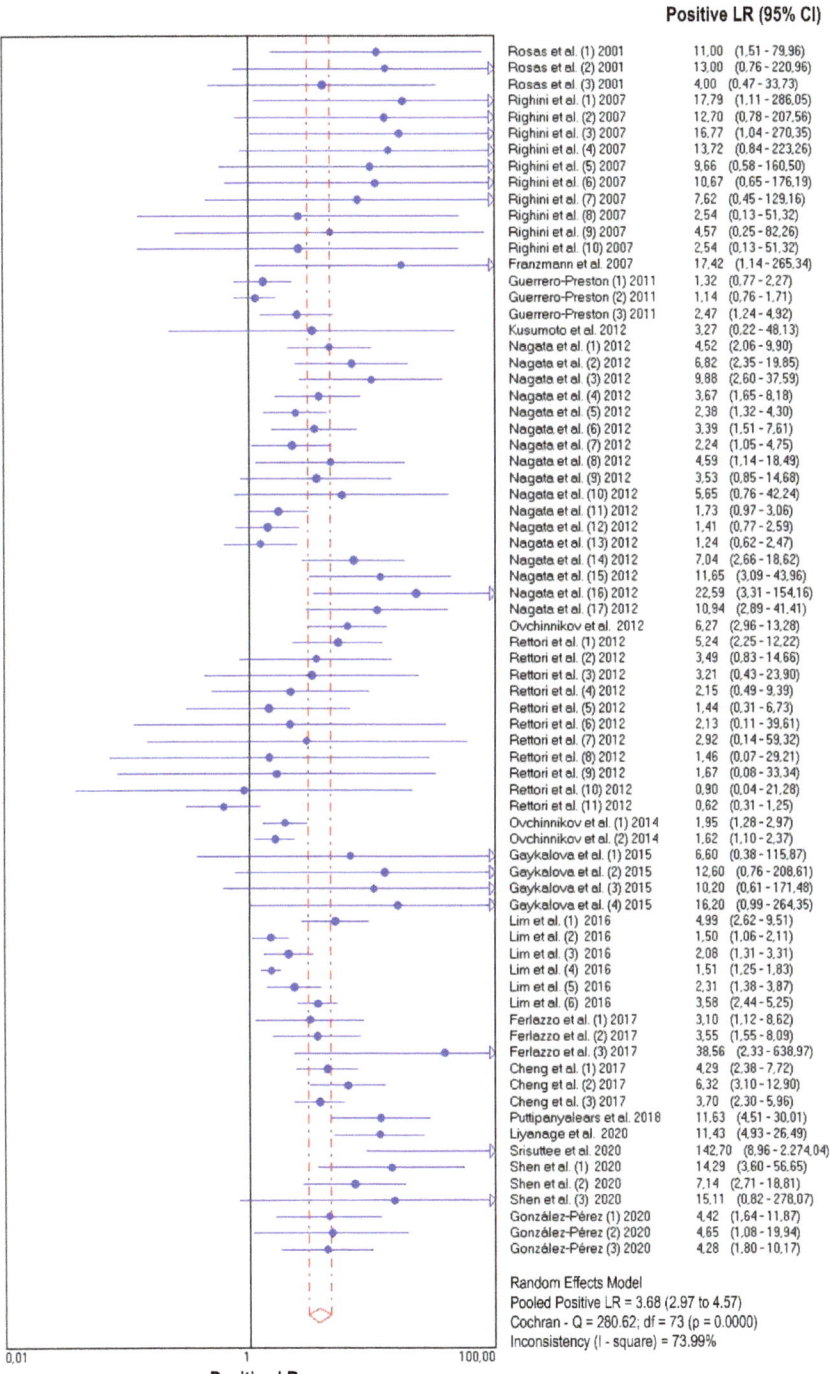

Figure 4. Forest plot of likelihood ratios for positive test results from salivary DNA methylation studies for predicting HNC diagnosis.

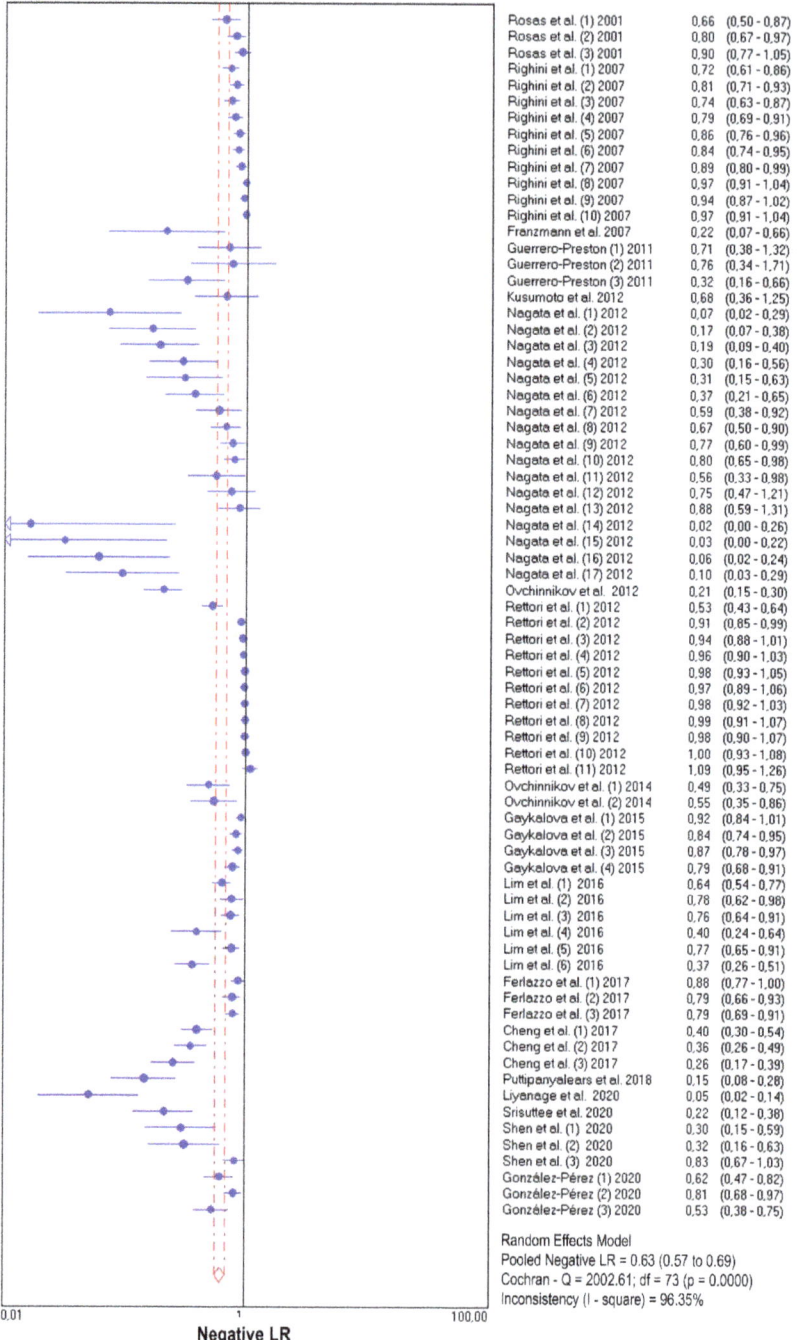

Figure 5. Forest plot of likelihood ratios for negative test results from salivary DNA methylation studies for predicting HNC diagnosis.

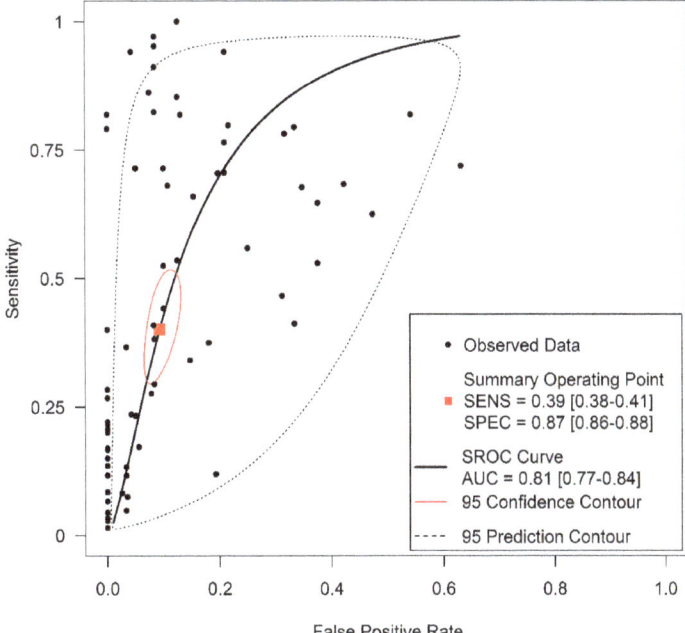

Figure 6. SROC curve with pooled estimates of sensitivity, specificity, and the AUC for all included studies of the salivary DNA methylation studies for detecting HNC.

3.6. Publication Bias

The potential publication bias in each salivary DNA methylation study was explored by Deeks' funnel plot asymmetry test, which yielded a slope coefficient p-value of 0.711 overall. This indication of a symmetric data pattern suggests the absence of publication bias (Figure S5).

4. Discussion

Over the last few years, saliva has aroused great interest in the scientific community due to its potential as a non-invasive liquid biopsy in cancer. Several studies have evidenced the diagnostic capability of salivary biomarkers for diagnosing both HNC [45] and tumors distant from the oral cavity [46]. In this sense, a wide variety of biomolecules have been assessed as tumor biomarkers using saliva-omics approaches, including genomic, epigenomic, transcriptomic, metabolomic, proteomic, and microbiomic technologies [47]. In the field of epigenomics, DNA promoter hypermethylation represents one of the most intensively studied epigenetic alterations in human cancer. Promoter hypermethylation of critical pathway genes has been recognized as an important epigenetic mechanism of carcinogenesis [13]. Its potential role as an early diagnostic biomarker stems from the fact that gene promoter hypermethylation is an early event in cancer development [13]. DNA hypermethylation as a common event in cancer plays an important role in HNC development and progression [7]. The detection of DNA methylation in body fluids has emerged as an opportunity to assess the methylation status non-invasively and cost-effectively. In this line, several studies have investigated the promoter methylation of different tumor-suppressor genes in saliva from HNC patients [10–12,38]. Therefore, salivary DNA methylation biomarkers could be potentially used in the screening and early detection of HNC.

To the best of our knowledge, this study represents the first meta-analysis evaluating the diagnostic accuracy of promoter hypermethylation genes in saliva for differentiating

HNC patients from healthy individuals. The present analysis included a total of 18 articles (84 study units) involving 4758 HNC patients and 3605 healthy individuals. According to QUADAS-2 quality evaluation, most of the included studies were of moderate quality. As for the overall accuracy of salivary hypermethylated genes for discriminating HNC from healthy individuals, the pooled diagnostic parameters of sensitivity, specificity, and AUC values were 0.39, 0.87, and 0.81, respectively. The summary dOR of 8.34 reflects the diagnostic capacity of salivary hypermethylated genes for HNC. The pooled PLR value of 3.68 indicates that a person testing positive had approximately 3.68 times higher probability of having cancer than a healthy individual. On the other hand, the pooled NLR indicated that a person testing negative had a 63% probability of not having cancer. Additionally, given a pre-test probability of 27.8% in Fagan's nomogram, correct HNC diagnosis increased to 59% after a positive test, and reduced to 20% after a negative test. Overall, these results show a low sensitivity for HNC detection, indicating a high false negative rate. Therefore, the salivary gene methylation evaluated in this meta-analysis presented limitations as a screening biomarker for HNC. However, the diagnostic specificity of gene methylation for HNC was very high, suggesting that detection in saliva may aid assessment in HNC diagnosis. New molecular biology techniques such as next-generation sequencing platforms and digital PCR represent an opportunity for improving the sensitivity of methylation assays and for discovering new methylation patterns.

Due to the fact that heterogeneity is inherent to any diagnostic accuracy meta-analysis, an evaluation of the reasons contributing to inconsistencies across studies should be carried out. In the present research, overall heterogeneity among studies was high, so a bivariate random effects model was applied. This significant heterogeneity was reflected numerically in Cochran's Q test and the I^2 statistic. Further exploration of the heterogeneity revealed the presence of a threshold effect. The threshold effect is a major source of heterogeneity in meta-analyses of diagnostic tests. It arises from differences in sensitivities and specificities or likelihood ratios due to different cut-offs among studies for defining positive (or negative) test results [27]. In the present meta-analysis, the threshold effect was suggested by the visual inspection of accuracy estimates in forest plots where increasing specificities with decreasing sensitivities were observed. Later, the ROC plane and the Spearman correlation test also indicated the presence of the threshold effect. The variations in accuracy estimates among different studies could be due to a number of reasons other than the threshold, such as study population (anatomic tumor location, TNM staging), index test (differences in technology, assays), reference standard, and study design. Therefore, heterogeneity should be explored by relating study level co-variates to an accuracy measure by meta-regression techniques [27]. In the present study, meta-regression was performed to test the effect of sample type, sample size, anatomic tumor location, DNA methylation method, and methylated gene profiling. The results point to anatomic tumor location and gene profiling strategy as possible causes of heterogeneity. Stratified analysis by anatomic tumor location showed that salivary gene methylation had a higher diagnostic accuracy for discriminating oral (AUC = 0.88) and oropharyngeal (AUC = 0.87) tumors than overall HNC (AUC = 0.81). The explanation for these findings may be that tumors located in the oral cavity and oropharynx release more tumor cells directly into saliva. With respect to gene profiling strategy, subgroup analysis showed that single genes presented low sensitivity, but were highly specific to cancer tissue. The combination of salivary methylated genes had better diagnostic accuracy than single gene-based tests, with a dOR of 36.97 vs. 6.02 and AUC of 0.92 vs. 0.77, respectively, demonstrating that the use of salivary methylated gene panels as biomarkers may increase HNC detection accuracy without decreasing specificity. We also conducted subgroup analyses based on sample type, sample size, and DNA methylation method but no significant differences in diagnostic accuracy were observed. Future studies should be conducted to clarify the impact of these factors on the diagnostic potential of salivary DNA methylation.

The current meta-analysis is not free of limitations. Firstly, this meta-analysis included case–control studies, but none was multicenter. Moreover, no randomized controlled trials exist on this topic. Secondly, considerable heterogeneity was observed among the included studies. Although we examined the sources of heterogeneity in five variables, we were not able to explore other demographic and clinicopathological factors due to lack of information. Thirdly, studies involving saliva samples only evaluated some of the genes methylated in HNC tissue. The remaining genes should also be tested in saliva to determine their diagnostic potential. Lastly, confounding variables such as gender, age, lifestyle (tobacco and alcohol), and diet were not considered in most of the included studies. Due to these limitations, future research based on large-scale prospective diagnostic studies involving multiple health centers would contribute to further evaluating the clinical utility of salivary gene promoter methylation for HNC diagnosis. Furthermore, better comparison among future studies would benefit from standardization of analytic strategies and cut-off selection.

5. Conclusions

Our meta-analysis suggests that the detection of DNA promoter hypermethylation in saliva is a promising biomarker for HNC diagnosis, mainly in oral and oropharyngeal tumors. The use of salivary hypermethylated gene panels improves diagnostic accuracy with respect to single-gene analysis. This meta-analysis could provide valuable insights into methodology design for further research studies.

Supplementary Materials: The following are available online at https://www.mdpi.com/article/10.3390/jpm11060568/s1, Figure S1: Quality assessment of the included studies according to Quality Assessment of Diagnostic Accuracy Studies-2 (QUADAS-2) criteria, Figure S2: Forest plot of pooled dOR salivary DNA methylation for the diagnosis of HNC, Figure S3: Fagan's monogram evaluating the clinical utility of salivary DNA methylation for differentiating HNC patients, Figure S4: Representation of sensitivity against (1-specificity) in ROC space for each study of salivary methylation in the diagnosis of HNC, Figure S5: Deeks' funnel plot asymmetry test for the assessment of potential bias of included studies, Table S1: Results of meta-regression analysis, Table S2: Subgroup analysis of salivary DNA methylation for HNC detection based on different covariates.

Author Contributions: Conceptualization, Ó.R.-G., L.M.-R., and M.M.S.-C.; methodology, Ó.R.-G. and M.M.S.-C.; software, C.M.-R. and Á.S.-B.; formal analysis, C.M.-R. and Á.S.-B.; investigation, Ó.R.-G., Á.D.-L., and M.M.S.-C.; data curation, Ó.R.-G., Á.D.-L., and M.M.S.-C.; writing—original draft preparation, Ó.R.-G. and M.M.S.-C.; writing—review and editing, L.M.-R., Á.D.-L., and R.L.-L.; visualization, Ó.R.-G. and J.M.-L.; supervision, M.M.S.-C.; project administration, M.M.S.-C.; funding acquisition, M.M.S.-C. All authors have read and agreed to the published version of the manuscript.

Funding: This work was co-funded by the Instituto de Salud Carlos III (ISCIII) (PI20/01449) and the European Regional Development Fund (FEDER). A.D.-L. is funded by a "Juan Rodés" contract from ISCIII (JR17/00016). L.M.-R. is funded by a "Miguel Servet" contract from ISCIII (CP20/00129).

Conflicts of Interest: R.L.-L. reports other from Nasasbiotech, during the study; grants and personal fees from Roche, grants and personal fees from Merck, personal fees from AstraZeneca, personal fees from Bayer, personal fees and non-financial support from BMS, personal fees from Pharmamar, personal fees from Leo, outside the submitted work. The rest of the authors have nothing to disclose. The funders had no role in the design of the study; in the collection, analyses, or interpretation of data; in the writing of the manuscript, or in the decision to publish the results.

References

1. Bray, F.; Ferlay, J.; Soerjomataram, I.; Siegel, R.L.; Torre, L.A.; Jemal, A. Global cancer statistics 2018: GLOBOCAN estimates of incidence and mortality worldwide for 36 cancers in 185 countries. *CA Cancer J. Clin.* **2018**, *68*, 394–424. [CrossRef]
2. Leemans, C.R.; Snijders, P.J.F.; Brakenhoff, R.H. The molecular landscape of head and neck cancer. *Nat. Rev. Cancer* **2018**, *18*, 269–282. [CrossRef]
3. Sacco, A.G.; Cohen, E.E. Current treatment options for recurrent or metastatic head and neck squamous cell carcinoma. *J. Clin. Oncol.* **2015**, *33*, 3305–3313. [CrossRef] [PubMed]
4. Feller, L.; Altini, M.; Lemmer, J. Inflammation in the context of oral cancer. *Oral Oncol.* **2013**, *49*, 887–892. [CrossRef] [PubMed]
5. Jones, P.A.; Baylin, S.B. The fundamental role of epigenetic events in cancer. *Nat. Rev. Genet.* **2002**, *3*, 415–428. [CrossRef] [PubMed]

6. Darwiche, N. Epigenetic mechanisms and the hallmarks of cancer: An intimate affair. *Am. J. Cancer Res.* **2020**, *10*, 1954–1978. [PubMed]
7. Castilho, R.M.; Squarize, C.H.; Almeida, L.O. Epigenetic modifications and head and neck cancer: Implications for tumor progression and resistance to therapy. *Int. J. Mol. Sci.* **2017**, *18*, 1506. [CrossRef]
8. Garinis, G.A.; Patrinos, G.P.; Spanakis, N.E.; Menounos, P.G. DNA hypermethylation: When tumour suppressor genes go silent. *Hum. Genet.* **2002**, *111*, 115–127. [CrossRef]
9. Misawa, K.; Imai, A.; Matsui, H.; Kanai, A.; Misawa, Y.; Mochizuki, D.; Mima, M.; Yamada, S.; Kurokawa, T.; Nakagawa, T.; et al. Identification of novel methylation markers in HPV-associated oropharyngeal cancer: Genome-wide discovery, tissue verification and validation testing in ctDNA. *Oncogene* **2020**, *39*, 4741–4755. [CrossRef]
10. Liyanage, C.; Wathupola, A.; Muraleetharan, S.; Perera, K.; Punyadeera, C.; Udagama, P. Promoter hypermethylation of tumor-suppressor genes p16INK4a, RASSF1A, TIMP3, and PCQAP/MED15 in salivary DNA as a quadruple biomarker panel for early detection of oral and oropharyngeal cancers. *Biomolecules* **2019**, *9*, 148. [CrossRef]
11. Lim, Y.; Wan, Y.; Vagenas, D.; Ovchinnikov, D.A.; Perry, C.F.L.; Davis, M.J.; Punyadeera, C. Salivary DNA methylation panel to diagnose HPV-positive and HPV-negative head and neck cancers. *BMC Cancer* **2016**, *16*, 749. [CrossRef] [PubMed]
12. Guerrero-Preston, R.; Soudry, E.; Acero, J.; Orera, M.; Moreno-López, L.; Macía-Colón, G.; Jaffe, A.; Berdasco, M.; Ili-Gangas, C.; Brebi-Mieville, P.; et al. NID2 and HOXA9 promoter hypermethylation as biomarkers for prevention and early detection in oral cavity squamous cell carcinoma tissues and saliva. *Cancer Prev. Res.* **2011**, *4*, 1061–1072. [CrossRef]
13. Herman, J.G.; Baylin, S.B. Gene silencing in cancer in association with promoter hypermethylation. *N. Engl. J. Med.* **2003**, *349*, 2042–2054. [CrossRef]
14. Borchiellini, M.; Ummarino, S.; Di Ruscio, A. The bright and dark side of DNA methylation: A matter of balance. *Cells* **2019**, *8*, 1243. [CrossRef]
15. Kulis, M.; Esteller, M. DNA methylation and cancer. *Adv. Genet.* **2010**, *70*, 27–56. [CrossRef]
16. Esteller, M.; Corn, P.G.; Baylin, S.B.; Herman, J.G. A gene hypermethylation profile of human cancer. *Cancer Res.* **2001**, *61*, 3225–3229. [PubMed]
17. O'Reilly, E.; Tuzova, A.V.; Walsh, A.L.; Russell, N.M.; O'Brien, O.; Kelly, S.; Dhomhnallain, O.N.; DeBarra, L.; Dale, C.M.; Brugman, R.; et al. epiCaPture: A urine DNA methylation test for early detection of aggressive prostate cancer. *JCO Precis. Oncol.* **2019**, 1–18. [CrossRef] [PubMed]
18. Liang, W.; Zhao, Y.; Huang, W.; Gao, Y.; Xu, W.; Tao, J.; Yang, M.; Li, L.; Ping, W.; Shen, H.; et al. Non-invasive diagnosis of early-stage lung cancer using high-throughput targeted DNA methylation sequencing of circulating tumor DNA (ctDNA). *Theranostics* **2019**, *9*, 2056–2070. [CrossRef]
19. Smith, A.K.; Kilaru, V.; Klengel, T.; Mercer, K.B.; Bradley, B.; Conneely, K.N.; Ressler, K.J.; Binder, E.B. DNA extracted from saliva for methylation studies of psychiatric traits: Evidence tissue specificity and relatedness to brain. *Am. J. Med. Genet. Part B Neuropsychiatr. Genet.* **2015**, *168*, 36–44. [CrossRef]
20. Hearn, N.L.; Coleman, A.S.; Ho, V.; Chiu, C.L.; Lind, J.M. Comparing DNA methylation profiles in saliva and intestinal mucosa. *BMC Genomics* **2019**, *20*, 163. [CrossRef]
21. Wu, H.C.; Wang, Q.; Chung, W.K.; Andrulis, I.L.; Daly, M.B.; John, E.M.; Keegan, T.H.; Knight, J.; Bradbury, A.R.; Kappil, M.A.; et al. Correlation of DNA methylation levels in blood and saliva DNA in young girls of the LEGACY girls study. *Epigenetics* **2014**, *9*, 929–933. [CrossRef]
22. Thompson, T.M.; Sharfi, D.; Lee, M.; Yrigollen, C.M.; Naumova, O.Y.; Grigorenko, E.L. Comparison of whole-genome DNA methylation patterns in whole blood, saliva, and lymphoblastoid cell lines. *Behav. Genet.* **2013**, *43*, 168–176. [CrossRef] [PubMed]
23. Langie, S.A.S.; Szarc Vel Szic, K.; Declerck, K.; Traen, S.; Koppen, G.; Van Camp, G.; Schoeters, G.; Vanden Berghe, W.; De Boever, P. Whole-genome saliva and blood DNA methylation profiling in individuals with a respiratory allergy. *PLoS ONE* **2016**, *11*, e0151109. [CrossRef]
24. Carvalho, A.L.; Jeronimo, C.; Kim, M.M.; Henrique, R.; Zhang, Z.; Hoque, M.O.; Chang, S.; Brait, M.; Nayak, C.S.; Jiang, W.W.; et al. Evaluation of promoter hypermethylation detection in body fluids as a screening/diagnosis tool for head and neck squamous cell carcinoma. *Clin. Cancer Res.* **2008**, *14*, 97–107. [CrossRef]
25. McInnes, M.D.F.; Moher, D.; Thombs, B.D.; McGrath, T.A.; Bossuyt, P.M.; the PRISMA-DTA Group; Clifford, T.; Cohen, J.F.; Deeks, J.J.; Gatsonis, C.; et al. Preferred Reporting Items for a Systematic Review and Meta-analysis of Diagnostic Test Accuracy Studies: The PRISMA-DTA Statement. *JAMA* **2018**, *319*, 388–396. [CrossRef]
26. Whiting, P.F.; Rutjes, A.W.S.; Westwood, M.E.; Mallett, S.; Deeks, J.J.; Reitsma, J.B.; Leeflang, M.M.G.; Sterne, J.A.C.; Bossuyt, P.M.M. Quadas-2: A revised tool for the quality assessment of diagnostic accuracy studies. *Ann. Intern. Med.* **2011**, *155*, 529–536. [CrossRef] [PubMed]
27. Zamora, J.; Abraira, V.; Muriel, A.; Khan, K.; Coomarasamy, A. Meta-DiSc: A software for meta-analysis of test accuracy data. *BMC Med. Res. Methodol.* **2006**, *6*, 31. [CrossRef]
28. Hellmich, M.; Lehmacher, W. A ruler for interpreting diagnostic test results. *Methods Inf. Med.* **2005**, *44*, 124–126.
29. Deeks, J.J.; Macaskill, P.; Irwig, L. The performance of tests of publication bias and other sample size effects in systematic reviews of diagnostic test accuracy was assessed. *J. Clin. Epidemiol.* **2005**, *58*, 882–893. [CrossRef]

30. Rosas, S.L.; Koch, W.; da Costa Carvalho, M.G.; Wu, L.; Califano, J.; Westra, W.; Jen, J.; Sidransky, D. Promoter hypermethylation patterns of p16, O^6-*methylguanine-DNA-methyltransferase*, and *Death-associated protein kinase* in tumors and saliva of head and neck cancer patients. *Cancer Res.* **2001**, *61*, 939–942. [PubMed]
31. Kusumoto, T.; Hamada, T.; Yamada, N.; Nagata, S.; Kanmura, Y.; Houjou, I.; Kamikawa, Y.; Yonezawa, S.; Sugihara, K. Comprehensive Epigenetic Analysis Using Oral Rinse Samples: A Pilot Study. *J. Oral Maxillofac. Surg.* **2012**, *70*, 1486–1494. [CrossRef] [PubMed]
32. Gaykalova, D.A.; Vatapalli, R.; Wei, Y.; Tsai, H.L.; Wang, H.; Zhang, C.; Hennessey, P.T.; Guo, T.; Tan, M.; Li, R.; et al. Outlier Analysis Defines Zinc Finger Gene Family DNA Methylation in Tumors and Saliva of Head and Neck Cancer Patients. *PLoS ONE* **2015**, *10*, e0142148. [CrossRef] [PubMed]
33. Cheng, S.J.; Chang, C.F.; Ko, H.H.; Lee, J.J.; Chen, H.M.; Wang, H.J.; Lin, H.S.; Chiang, C.P. Hypermethylated ZNF582 and PAX1 genes in mouth rinse samples as biomarkers for oral dysplasia and oral cancer detection. *Head Neck* **2018**, *40*, 355–368. [CrossRef]
34. Srisuttee, R.; Arayataweegool, A.; Mahattanasakul, P.; Tangjaturonrasme, N.; Kerekhanjanarong, V.; Keelawat, S.; Mutirangura, A.; Kitkumthorn, N. Evaluation of NID2 promoter methylation for screening of Oral squamous cell carcinoma. *BMC Cancer* **2020**, *20*, 218. [CrossRef]
35. Nagata, S.; Hamada, T.; Yamada, N.; Yokoyama, S.; Kitamoto, S.; Kanmura, Y.; Nomura, M.; Kamikawa, Y.; Yonezawa, S.; Sugihara, K. Aberrant DNA methylation of tumor-related genes in oral rinse. *Cancer* **2012**, *118*, 4298–4308. [CrossRef]
36. Franzmann, E.J.; Reategui, E.P.; Pedroso, F.; Pernas, F.G.; Karakullukcu, B.M.; Carraway, K.L.; Hamilton, K.; Singal, R.; Goodwin, W.J. Soluble CD44 is a potential marker for the early detection of head and neck cancer. *Cancer Epidemiol. Biomarkers Prev.* **2007**, *16*, 1348–1355. [CrossRef]
37. Righini, C.A.; de Fraipont, F.; Timsit, J.F.; Faure, C.; Brambilla, E.; Reyt, E.; Favrot, M.C. Tumor-specific methylation in saliva: A promising biomarker for early detection of head and neck cancer recurrence. *Clin. Cancer Res.* **2007**, *13*, 1179–1185. [CrossRef]
38. Ovchinnikov, D.A.; Cooper, M.A.; Pandit, P.; Coman, W.B.; Cooper-White, J.J.; Keith, P.; Wolvetang, E.J.; Slowey, P.D.; Punyadeera, C. Tumor-suppressor gene promoter hypermethylation in saliva of head and neck cancer patients. *Transl. Oncol.* **2012**, *5*, 321–326. [CrossRef] [PubMed]
39. Ovchinnikov, D.A.; Wan, Y.; Coman, W.B.; Pandit, P.; Cooper-White, J.J.; Herman, J.G.; Punyadeera, C. DNA methylation at the novel CpG sites in the promoter of MED15/PCQAP gene as a biomarker for head and neck cancers. *Biomark. Insights* **2014**, *9*, 53–60. [CrossRef] [PubMed]
40. González-Pérez, L.; Isaza-Guzmán, D.; Arango-Pérez, E.; Tobón-Arroyave, S. Analysis of salivary detection of P16INK4A and RASSF1A promoter gene methylation and its association with oral squamous cell carcinoma in a Colombian population. *J. Clin. Exp. Dent.* **2020**, e452–e460. [CrossRef] [PubMed]
41. Puttipanyalears, C.; Arayataweegool, A.; Chalertpet, K.; Rattanachayoto, P.; Mahattanasakul, P.; Tangjaturonsasme, N.; Kerekhanjanarong, V.; Mutirangura, A.; Kitkumthorn, N. TRH site-specific methylation in oral and oropharyngeal squamous cell carcinoma. *BMC Cancer* **2018**, *18*, 786. [CrossRef] [PubMed]
42. Ferlazzo, N.; Currò, M.; Zinellu, A.; Caccamo, D.; Isola, G.; Ventura, V.; Carru, C.; Matarese, G.; Ientile, R. Influence of MTHFR genetic background on p16 and MGMT methylation in oral squamous cell cancer. *Int. J. Mol. Sci.* **2017**, *18*, 724. [CrossRef] [PubMed]
43. Shen, S.; Saito, Y.; Ren, S.; Liu, C.; Guo, T.; Qualliotine, J.; Khan, Z.; Sadat, S.; Califano, J.A. Targeting viral DNA and promoter hypermethylation in salivary rinses for recurrent HPV-positive oropharyngeal cancer. *Otolaryngol. Neck Surg.* **2020**, *162*, 512–519. [CrossRef] [PubMed]
44. Rettori, M.M.; de Carvalho, A.C.; Bomfim Longo, A.L.; de Oliveira, C.Z.; Kowalski, L.P.; Carvalho, A.L.; Vettore, A.L. Prognostic significance of TIMP3 hypermethylation in post-treatment salivary rinse from head and neck squamous cell carcinoma patients. *Carcinogenesis* **2013**, *34*, 20–27. [CrossRef] [PubMed]
45. Guerra, E.N.S.; Acevedo, A.C.; Leite, A.F.; Gozal, D.; Chardin, H.; De Luca Canto, G. Diagnostic capability of salivary biomarkers in the assessment of head and neck cancer: A systematic review and meta-analysis. *Oral Oncol.* **2015**, *51*, 805–818. [CrossRef]
46. Rapado-González, Ó.; Martínez-Reglero, C.; Salgado-Barreira, Á.; Takkouche, B.; López-López, R.; Suárez-Cunqueiro, M.M.; Muinelo-Romay, L. Salivary biomarkers for cancer diagnosis: A meta-analysis. *Ann. Med.* **2020**, *52*, 131–144. [CrossRef]
47. Yoshizawa, J.M.; Schafer, C.A.; Schafer, J.J.; Farrell, J.J.; Paster, B.J.; Wong, D.T.W. Salivary biomarkers: Toward future clinical and diagnostic utilities. *Clin. Microbiol. Rev.* **2013**, *26*, 781–791. [CrossRef]

Review

Salivary DNA Methylation as an Epigenetic Biomarker for Head and Neck Cancer. Part II: A Cancer Risk Meta-Analysis

Óscar Rapado-González [1,2,3], Cristina Martínez-Reglero [4], Ángel Salgado-Barreira [4], María Arminda Santos [1,5], Rafael López-López [3,6], Ángel Díaz-Lagares [3,7] and María Mercedes Suárez-Cunqueiro [1,3,6,*]

1. Department of Surgery and Medical-Surgical Specialties, Medicine and Dentistry School, Universidade de Santiago de Compostela, 15782 Santiago de Compostela, Spain; oscar.rapado@rai.usc.es (Ó.R.-G.); maría.santos@iucs.cespu.pt (M.A.S.)
2. Translational Medical Oncology Group (Oncomet), Liquid Biopsy Analysis Unit, Health Research Institute of Santiago (IDIS), 15706 Santiago de Compostela, Spain
3. Centro de Investigación Biomédica en Red de Cáncer (CIBERONC), Instituto de Salud Carlos III, 28029 Madrid, Spain; rafa.lopez.lopez@gmail.com (R.L.-L.); angel.diaz.lagares@sergas.es (Á.D.-L.)
4. Methodology and Statistics Unit, Galicia Sur Health Research Institute (IISGS), 36312 Vigo, Spain; cristina.martinez@iisgaliciasur.es (C.M.-R.); angel.salgado.barreira@sergas.es (Á.S.-B.)
5. Department of Oral Rehabilitation, Instituto Universitario de Ciências da Saúde (IUCS), 1317 | 4585-116 Gandra, Portugal
6. Translational Medical Oncology Group (Oncomet), Health Research Institute of Santiago (IDIS), Complexo Hospitalario Universitario de Santiago de Compostela (SERGAS), 15706 Santiago de Compostela, Spain
7. Cancer Epigenomics, Translational Medical Oncology Group (Oncomet), Health Research Institute of Santiago (IDIS), University Clinical Hospital of Santiago (CHUS/SERGAS), 15706 Santiago de Compostela, Spain
* Correspondence: mariamercedes.suarez@usc.es; Tel.: +34-881-812-437

Citation: Rapado-González, Ó.; Martínez-Reglero, C.; Salgado-Barreira, Á.; Santos, M.A.; López-López, R.; Díaz-Lagares, Á.; Suárez-Cunqueiro, M.M. Salivary DNA Methylation as an Epigenetic Biomarker for Head and Neck Cancer. Part II: A Cancer Risk Meta-Analysis. *J. Pers. Med.* **2021**, *11*, 606. https://doi.org/10.3390/jpm11070606

Academic Editor: James Meehan

Received: 3 May 2021
Accepted: 21 June 2021
Published: 26 June 2021

Publisher's Note: MDPI stays neutral with regard to jurisdictional claims in published maps and institutional affiliations.

Copyright: © 2021 by the authors. Licensee MDPI, Basel, Switzerland. This article is an open access article distributed under the terms and conditions of the Creative Commons Attribution (CC BY) license (https://creativecommons.org/licenses/by/4.0/).

Abstract: Aberrant methylation of tumor suppressor genes has been reported as an important epigenetic silencer in head and neck cancer (HNC) pathogenesis. Here, we performed a comprehensive meta-analysis to evaluate the overall and specific impact of salivary gene promoter methylation on HNC risk. The methodological quality was assessed using the Newcastle–Ottawa scale (NOS). Odds ratios (ORs) and 95% confidence intervals (CIs) were calculated to evaluate the strength of the association and Egger's and Begg's tests were applied to detect publication bias. The frequency of salivary DNA promoter methylation was significantly higher in HNC patients than in healthy controls (OR: 8.34 (95% CI = 6.10–11.39; $p < 0.01$). The pooled ORs showed a significant association between specific tumor-related genes and HNC risk: *p16* (3.75; 95% CI = 2.51–5.60), *MGMT* (5.72; 95% CI = 3.00–10.91), *DAPK* (5.34; 95% CI = 2.18–13.10), *TIMP3* (3.42; 95% CI = 1.99–5.88), and *RASSF1A* (7.69; 95% CI = 3.88–15.23). Overall, our meta-analysis provides precise evidence on the association between salivary DNA promoter hypermethylation and HNC risk. Thus, detection of promoter DNA methylation in saliva is a potential biomarker for predicting HNC risk.

Keywords: DNA methylation; epigenetics; head and neck cancer; saliva; biomarkers; liquid biopsy; meta-analysis

1. Introduction

The important role of epigenetic mechanisms in carcinogenesis has been widely reported. Identification of specific genes that are altered by aberrant epigenetic processes contributes to better understanding molecular pathogenesis in HNC [1]. As one of the most important epigenetic alterations, DNA hypermethylation may lead to transcriptional silencing of tumor suppressor genes and, thus, interfere in signaling pathways that control vital cell processes, such as DNA repair, apoptosis, cell proliferation, and cell-to-cell adhesion [2]. Gene promoter methylation is a common epigenetic event in early carcinogenesis, and therefore represents a promising biomarker for high-risk group stratification, early cancer detection, and prognosis prediction [3]. Numerous studies have evaluated DNA methylation as a biomarker in a wide variety of tumors [4–7]. Hypermethylation of tumor-related genes, such as cyclin-dependent kinase inhibitor 2A (*CDKN2A*), E-cadherin (*CDH1*), death-associated protein kinase (*DAPK*), phosphatase and tensin homolog (*PTEN*), and O6-methylguanine-DNA methyltransferase (*MGMT*), have been reported in HNC [8]. Likewise, various studies have focused on the detection of DNA methylation in liquid biopsies in HNC [9–11]. Although evidence suggests a potential association between aberrant salivary DNA methylation patterns and HNC risk, no prior research assessing overall impact is available. Therefore, we conducted a systematic review and meta-analysis to gain better insight into the magnitude of the association between salivary DNA hypermethylation and HNC risk.

2. Materials and Methods

2.1. Protocol and Registration

This study was conducted according to Preferred Reporting Items for Systematic Reviews and Meta-analysis (PRISMA) guidelines [12], and the protocol was registered with the International Prospective Register of Systematic Reviews (reference No. CRD42020199123).

2.2. Search Strategy, Study Selection, and Data Extraction

The search strategy and data extraction were previously described in Part I [13].

2.3. Selection Criteria

The inclusion criteria were as follows: (1) case-control studies; (2) studies based on salivary DNA hypermethylation biomarkers for HNC; and (3) sufficient data to calculate odds ratios (ORs) and corresponding 95% confidential intervals (CIs). The exclusion criteria were as follows: (1) reviews, letters, personal opinions, book chapters, case reports, conference abstracts, and meetings; (2) duplicate publications; (3) incomplete data; and (4) in vitro or in vivo animal experiments.

2.4. Assessment of Study Quality

Independent investigators evaluated methodological quality by applying the Newcastle–Ottawa scale (NOS) [14] to each study selected. Discrepancies were resolved by consensus. For the interpretation of meta-analytic data, the NOS scale was used to score the quality of non-randomized studies based on their design, content, and ease of use. Items were scored according to a "star system" and fell under three broad categories: study group selection, group comparability, and ascertainment of exposure/outcome for case-control or cohort studies. The maximum quality score for each item was one star, except for the comparability item, which had a maximum of two stars. The NOS score ranged from 0 to 9 stars, with 8–9 stars being high quality; 6–7 stars being medium quality; and <5 stars being low quality.

2.5. Statistical Analysis

Statistical analysis was conducted using the meta package of free R software (v.3.4.4; https://www.r-project.org, accessed 30 November 2020). The pooled odds ratios (ORs) and their 95% confidence intervals (CIs) were calculated to assess the strength of the association between salivary promoter methylation and HNC. To evaluate the statistical model applied to the meta analytic database, heterogeneity was assessed on the basis of I-square (I^2) value and Cochran's Q statistic test-based Chi-squared test. Heterogeneity was considered significant when $I^2 > 50\%$ and/or presence of a $p < 0.10$ for the Cochran's Q test. If significant heterogeneity was detected, the DerSimonian and Laird random-effects model was applied to calculate the pooled OR with 95% CIs; otherwise, the Mantel–Haenszel fixed-effects model was used. Meta-regression and subgroup analyses were performed to explore the potential sources of heterogeneity among studies insofar as anatomic tumor location, sample type, sample size, DNA methylation method, and methylation gene profiling. Publication bias was assessed by Begg's and Egger's tests, and funnel plot inspection [15,16]. Begg's rank test examines the correlation between the effect sizes and their corresponding sampling variances. Egger's test regresses the standardized effect sizes on their precisions. In the presence of publication bias, both tests will be statistically significant. Moreover, publication bias was based on visual funnel-plot inspection, which shows the relationship between individual log ORs and their standard errors. The asymmetry of the funnel plot could indicate publication bias.

$p < 0.05$ was considered to be statistically significant.

3. Results

3.1. Study Selection and Characteristics of Included Studies

The main characteristics of the included studies have already been described in Part I [13].

3.2. Study Quality

Bias risk and quality were assessed according to NOS (Table S1). With respect to the selection category, each of the included studies was considered adequate. Regarding comparability, 5 out of the remaining 18 studies matched for age or gender, and 2 studies matched for at least one additional risk factor. Therefore, the median NOS score in our meta-analysis was 7.33 stars.

3.3. Association between Salivary DNA Promoter Hypermethylation and HNC Risk

A total of 7686 subjects, consisting of 4453 patients and 3233 controls, were included in this meta-analysis. As shown in Figure 1, the pooled analysis revealed a significant association between salivary DNA promoter hypermethylation and HNC with an OR of 8.34 (95% CI = 6.10–11.39; $p < 0.01$). A random-effects model was used because heterogeneity among the 18 studies (I^2 = 72%) was identified. The shape of the Begg's funnel plot did not reveal potential asymmetry ($p = 0.271$), although publication bias was detected by Egger's test ($p = 0.002$) (Figure S1).

3.4. Meta-Regression and Subgroup Analysis

Due to the presence of significant heterogeneity in the overall analysis, meta-regression and subgroup analysis were performed in order to reveal potential sources. The outcomes of meta-regression analysis showed that sample type ($p = 0.128$), sample size ($p = 0.349$), and DNA methylation method ($p = 0.275$) were not significant sources of heterogeneity. However, anatomic tumor location ($p = 0.002$) and gene profiling ($p < 0.001$) were, in fact, potential sources of heterogeneity in this study (Table S1—see Part I) [13]. As shown in Table S2, significant heterogeneity was found in all subgroups. With respect to sample type-based subgroup analysis, a significant association between promoter hypermethylation and HNC was found in oral rinse samples (OR: 9.42; 95% CI = 6.30–14.08) and saliva samples (OR: 6.33; 95% CI = 3.90–10.27). In tumor-based subgroup analysis, methylation rates were higher in specific head and neck locations compared to studies that made no differentiation. The pooled OR for oropharyngeal cancer was 13.26 (95% CI = 3.17–5.42) and for oral cancer was 13.07 (95% CI = 8.19–20.88), while for HNC it was 5.78 (95% CI = 3.86–8.67). A significant association between salivary promoter methylation and HNC was found by both MSP (OR: 9.06; 95% CI = 6.30–13.03) and qMSP (OR: 6.81; 95% CI = 3.70–12.54) techniques. With respect to the subgroups categorized by sample size, a significant association was found between salivary promoter methylation and HNC in studies with N < 100 (OR: 9.58; 95% CI = 6.44–14.27) and N > 100 (OR: 8.34; 95% CI = 6.10–11.39). In subgroup analysis based on the gene-profiling approach, salivary promoter hypermethylated gene panels had a significantly higher association to HNC risk (OR: 36.79; 95% CI = 16.81–81.32) than hypermethylated single genes (OR: 6.02; 95% CI = 4.46–8.13).

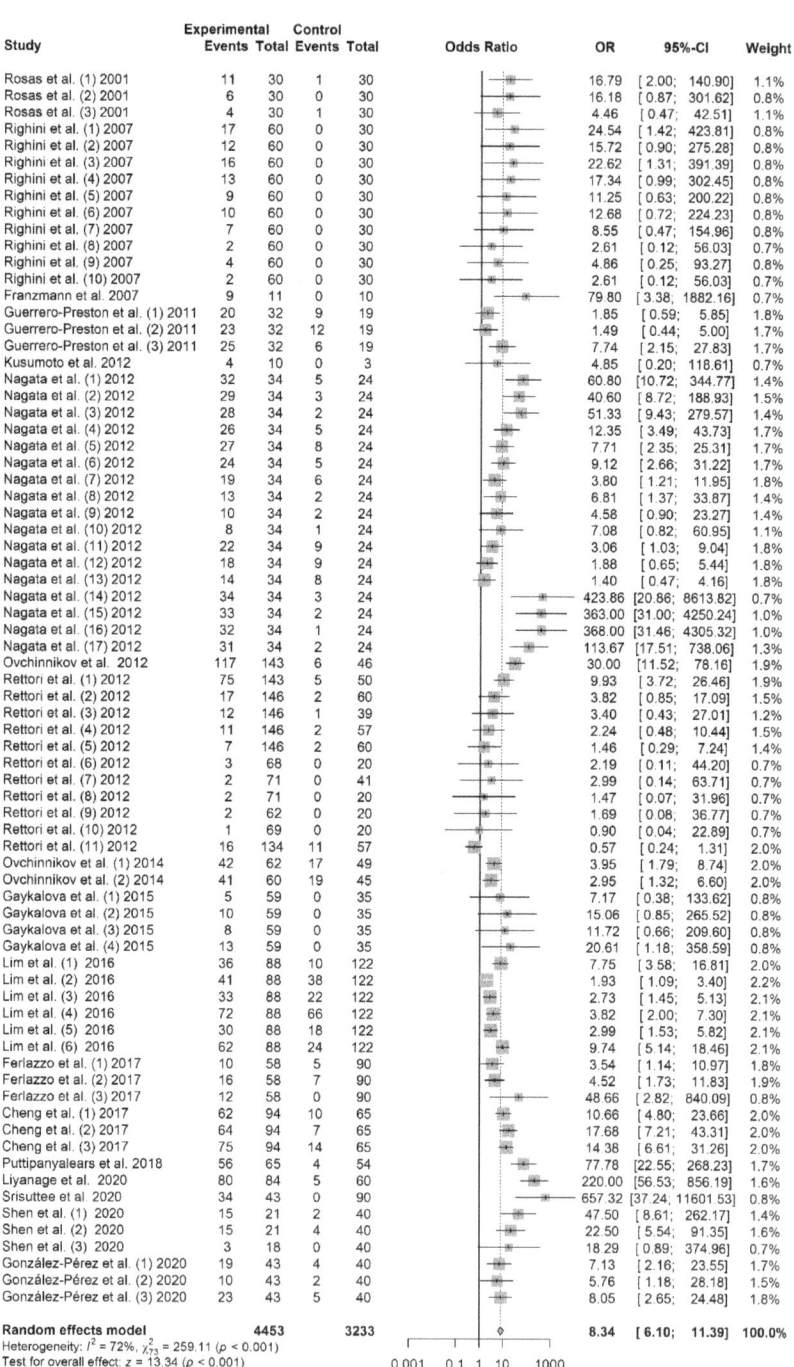

Figure 1. Forest plot for the association between salivary DNA promoter hypermethylation and the HNC risk. The squares represent the ORs for individual studies. Bars represent the 95% CIs. The center of the diamond represents the summary effect size.

3.5. Association between p16 Promoter Hypermethylation and HNC Risk

A total of 410 cases and 399 controls from 9 studies were included to estimate the effect of *p16* promoter hypermethylation on HNC risk. As shown in Figure 2, a significant association was found between salivary *p16* promoter hypermethylation and HNC risk (OR: 3.75; 95% CI = 2.51–5.60). The shape of the Begg's funnel plot did not reveal potential asymmetry ($p = 1$), although publication bias was detected by Egger's test ($p = 0.040$) (Figure S2).

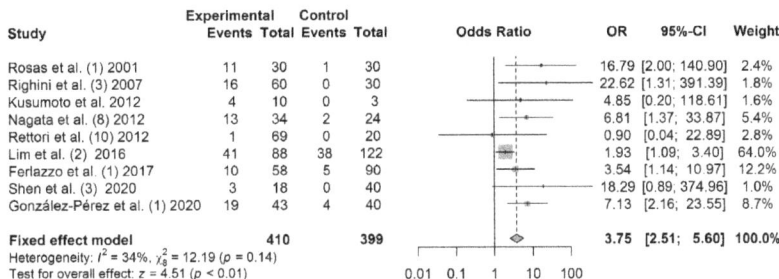

Figure 2. Forest plot for the association between *p16* promoter hypermethylation and HNC risk. The squares represent the ORs for individual studies. Bars represent the 95% CIs. The center of the diamond represents the summary effect size.

3.6. Association between MGMT Promoter Hypermethylation and HNC Risk

A total of 328 cases and 231 controls from 5 studies were included to estimate the effect of *MGMT* promoter hypermethylation on HNC risk. As shown in Figure 3, salivary *MGMT* promoter hypermethylation was associated with an increased HNC risk (OR: 5.72; 95% CI = 3.00–10.91). Visual analysis of the funnel plot revealed a symmetrical distribution of the studies (Egger's test, $p = 0.767$; Begg's test, $p = 0.624$), indicating no evidence of publication bias (Figure S3).

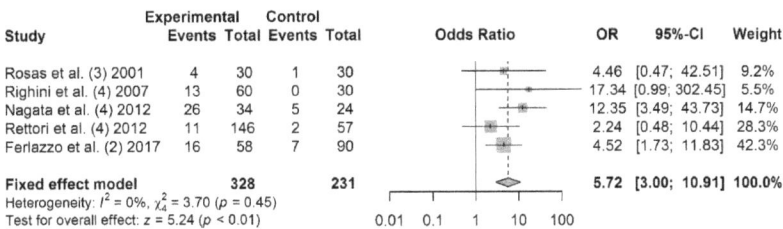

Figure 3. Forest plot for the association between *MGMT* promoter hypermethylation and HNC risk. The squares represent the ORs for individual studies. Bars represent the 95% CIs. The center of the diamond represents the summary effect size.

3.7. Association between DAPK Promoter Hypermethylation and HNC Risk

A total of 270 cases and 123 controls from 4 studies were included to estimate the effect of *DAPK* promoter hypermethylation on HNC risk. As shown in Figure 4, the rate of salivary *DAPK* promoter hypermethylation was significantly higher in HNC patients compared to controls (OR: 5.34; 95% CI = 2.18–13.10). Visual examination of the funnel plot revealed a symmetrical distribution of the studies (Begg's test, $p = 0.041$; Egger's test, $p = 0.187$;), indicating no evidence of publication bias (Figure S4).

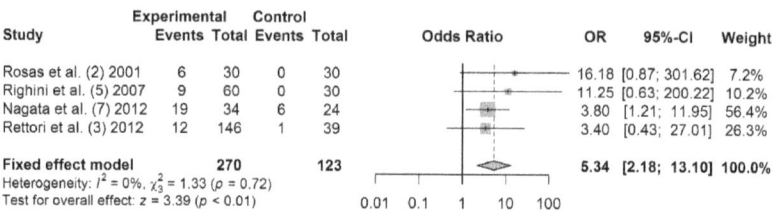

Figure 4. Forest plot for the association between *DAPK* promoter hypermethylation and HNC risk. The squares represent the ORs for individual studies. Bars represent the 95% CIs. The center of the diamond represents the summary effect size.

3.8. Association between TIMP3 Promoter Hypermethylation and HNC Risk

A total of 328 cases and 236 controls from 4 studies were included to estimate the effect of *TIMP3* promoter hypermethylation on HNC risk. As shown in Figure 5, a significant association was found between salivary *TIMP3* promoter hypermethylation and HNC risk (OR: 3.42; 95% CI = 1.99–5.88). Visual inspection of the funnel plot revealed a symmetrical distribution of the studies (Begg's test, $p = 0.174$; Egger's test, $p = 0.419$), indicating no evidence of publication bias (Figure S5).

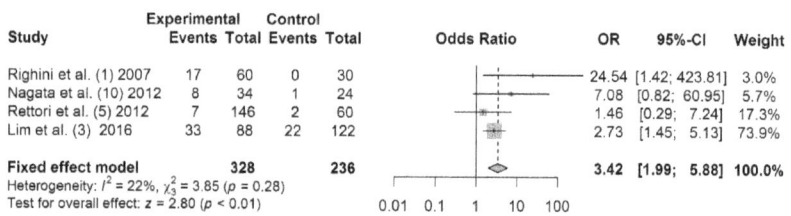

Figure 5. Forest plot for the association between TIMP3 promoter hypermethylation and HNC risk. The squares represent the ORs for individual studies. Bars represent the 95% CIs. The center of the diamond represents the summary effect size.

3.9. Association between RASSF1A Promoter Hypermethylation and HNC Risk

A total of 191 cases and 192 controls from 3 studies were included to estimate the effect of *RASSF1A* promoter hypermethylation on HNC risk. As shown in Figure 6, salivary *RASSF1A* promoter hypermethylation was associated with an increased HNC risk (OR: 7.69; 95% CI = 3.88–15.23). Visual examination of the funnel plot revealed a symmetrical distribution of the studies (Begg's test, $p = 0.601$; Egger's test, $p = 0.858$), indicating no evidence of publication bias (Figure S6).

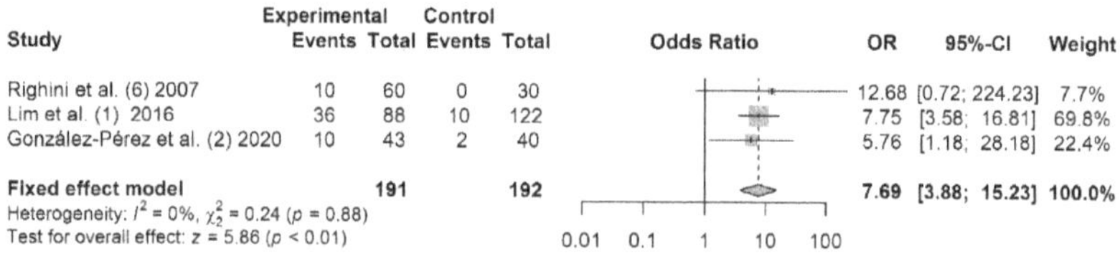

Figure 6. Forest plot for the association between *RASSF1A* promoter hypermethylation and the HNC risk. The squares represent the ORs for individual studies. Bars represent the 95% CIs. The center of the diamond represents the summary effect size.

3.10. Association between APC Promoter Hypermethylation and HNC Risk

A total of 156 cases and 74 controls from 3 studies were included to estimate the effect of *APC* promoter hypermethylation on HNC risk. As shown in Figure 7, salivary *APC* promoter hypermethylation was not significantly associated with HNC (OR: 2.15; 95% CI = 0.84–5.51). Visual examination of the funnel plot revealed no potential asymmetry (Begg's test, $p = 0.601$; Egger's test, $p = 0.609$), indicating no evidence of publication bias (Figure S7).

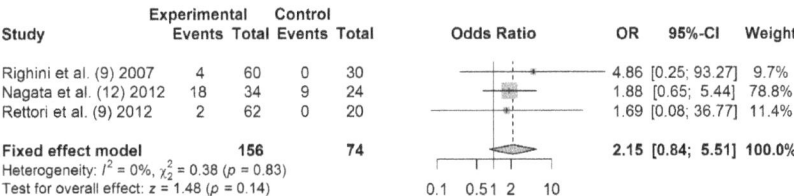

Figure 7. Forest plot for the association between *APC* promoter hypermethylation and HNC risk. The squares represent the ORs for individual studies. Bars represent the 95% CIs. The center of the diamond represents the summary effect size.

4. Discussion

Aberrant DNA hypermethylation has been recognized as an important epigenetic mechanism involved in head and neck carcinogenesis [1], suggesting its potential as a biomarker for evaluating cancer risk. Although prior studies have focused on the detection of promoter DNA hypermethylation in saliva from HNC patients [10,17], the evidence of a direct relationship is unclear and findings have been inconsistent.

To the best of our knowledge, this is the first meta-analysis evaluating the contribution of salivary promoter hypermethylation to HNC risk. The present comprehensive analysis included 18 studies comprising 4453 patients and 3233 controls. Overall, our results indicate that salivary promoter hypermethylation was significantly associated with an 8.34-fold increase in HNC risk.

As significant heterogeneity was observed among studies, meta-regression and subgroup analyses were carried out based on anatomic tumor location, sample type, sample size, DNA methylation method, and methylation gene profiling. The stratified analysis revealed that salivary DNA hypermethylation was associated with HNC risk in all subgroups. The association between salivary DNA promoter hypermethylation and HNC risk was stronger in oral rinses compared to saliva. This could be explained by the higher methylation proportion of oral exfoliated cells in oral rinse compared to saliva samples. Subgroup analysis of anatomic tumor location showed that the OR was higher in oral cancer and oropharyngeal cancer than overall HNC. These findings could be explained by the direct contact of saliva samples with tumors located in the oral cavity and oropharynx, which could result in an increased number of exfoliated tumoral cells during sample collection. Based on the methylation detection method subgroup, the frequency of salivary DNA promoter methylation was higher in MSP than in qMSP. This may be because MSP was the most commonly used technique for detecting aberrant DNA methylation in saliva samples (11 studies). In addition, the qualitative nature and lower specificity of MSP could lead to an overestimation of methylation data compared to qMSP methods [18]. However, quantitative approaches, such as qMSP or pyrosequencing, have shown better sensitivity than MSP [19]. With respect to sample size, a similar significant association was found between $n < 100$ and $n > 100$ subgroups. On the other hand, the gene profiling subgroup revealed that HNC risk was clearly higher when aberrant gene-specific DNA methylation was analyzed using gene panels rather than single gene analysis. This suggests that multiple tumor suppressor genes are epigenetically silenced in HNC pathogenesis, and, therefore, gene methylation panels should be used to better identify HNC risk.

We also explored the association between gene-specific promoter DNA methylation and HNC risk by analyzing the methylation frequency of genes reported in at least three studies. Thus, promoter hypermethylation of *p16*, *DAPK*, *TIMP3*, *MGMT*, and *RASSF1A* was significantly higher in HNC patients compared to controls, suggesting that the methylation of these tumor suppressor genes may play an important role in head and neck carcinogenesis. The *p16* gene acts as a negative cell cycle regulator that prevents the inactivation of retinoblastoma (Rb) protein by inhibiting the cyclin-dependent kinases and, therefore, cell cycle progression at G1/S phase [20]. Hypermethylation of *p16* promoter has been reported as a frequent epigenetic event in oral carcinogenesis [21,22]. In the present meta-analysis, methylation of *p16* promoter was significantly associated with a 3.75-fold increase in HNC risk, which is consistent with the study by Shi et al. (OR: 3.37) based on tissue and liquid biopsy methylation data [23]. In line with this, a more recent meta-analysis comprising 67 case-control studies reported an OR of 6.72. However, subgroup analysis in this study based on sample type revealed that OR was much higher in saliva (OR: 12.45) and blood (OR: 16.40) than in tissue (OR: 6.40) [24]. Overall, these findings indicate that hypermethylation of *p16* gene promoter in saliva could be a predictive biomarker for HNC risk. The *MGMT* gene is involved in the repair of O6-methylguanine in DNA sequences originating from the carcinogenic effects of alkylating agents [25]. The inactivation of *MGMT* promoter by aberrant hypermethylation has been associated with an increased frequency of GC > AT transition mutations in *TP53* and in *KRAS* oncogene, contributing to carcinogenesis and tumor progression [26,27]. In fact, our meta-analysis showed that methylation of *MGMT* promoter leads to a 5.72-fold increase in HNC risk. *DAPK* plays a critical role in the apoptotic process triggered by interferon-gamma (IFN-γ), tumor necrosis factor (TNF)-alpha, Fas ligand, and detachment from extracellular matrix [28]. Hypermethylation of *DAPK* gene promoter is a frequent alteration in HNC [29,30]. The results of the present meta-analysis show that individuals with salivary hypermethylation of *DAPK* gene promoter had a 5.34-fold higher HNC risk. A previous meta-analysis also showed that the frequency of *DAPK* promoter methylation was significantly higher in HNC vs. control groups (OR: 6.72) [31]. The *TIMP3* gene is a tissue inhibitor of matrix metalloproteinases, which acts as a potential anticancer agent by inducing apoptosis and inhibiting proliferation, angiogenesis, and metastasis [32]. The methylation of *TIMP3* promoter has been associated with HNC [33,34]. Interestingly, our meta-analysis revealed a significant association between salivary *TIMP3* promoter methylation and HNC with an OR of 3.42. The *RASSF1A* gene prevents tumorigenesis through multiple cellular process, such as cell cycle arrest, migration, microtubular stabilization, and apoptosis promotion [35]. Epigenetic inactivation of *RASSF1A* by hypermethylation has been observed in various cancers, including HNC [36]. Our data showed that methylation of *RASSF1A* promoter led to a 7.69-fold increase in HNC risk compared to the control group. In a previous study, Meng et al. evaluated the methylation prevalence of *RASSF1A* between cancerous tissues and controls, finding a significant association (OR: 2.93) between aberrant methylation of *RASSF1A* and HNC [37]. The *APC* gene acts as a negative regulator in the Wnt/beta-catenin signaling pathway and its dysfunction leads to increased β-catenin transcriptional activity, promoting the activation of downstream targets involved in tumorigenesis, such as cyclin D1 and Myc [38]. Hypermethylation of the *APC* promoter has been reported as a mechanism for *APC*-gene inactivation in oral carcinogenesis [39]. Our study did not reveal a significant association between salivary *APC* promoter hypermethylation and HNC, which could be explained by the low *APC*-gene methylation rates detected in saliva from HNC patients. Until now, few studies have reported *APC* hypermethylation in saliva from HNC patients [40–42]; however, this epigenetic alteration has been frequently observed in head and neck tumors [29,39,43,44]. It is important to note that hypermethylation of *p16, DAPK, TIMP3, MGMT,* and *RASSF1A* plays an important role in the carcinogenesis of various tumors, such as lung, breast, colorectal, renal, or gastric [45–50]. In line with this, several studies have focused on the association of cancer risk with the hypermethylation of these tumor suppressor genes [51–55], which highlights its potential for early diagnosis of the disease.

The present study has several strengths. It is the first meta-analysis highlighting the association between salivary DNA promoter hypermethylation and HNC. It explores the magnitude of the association both overall and by specific hypermethylated gene. In addition, it involved a comprehensive literature review without language restrictions. However, our study is not exempt from limitations. Firstly, all included research involved case-control retrospective studies, which could lead to selection bias. Some bias could also stem from the fact that cases and controls were not matched for demographic variables, such as age, sex, and lifestyle habits. Secondly, significant heterogeneity was found among studies. Despite performing subgroup analysis by anatomic tumor location, sample type, sample size, DNA methylation method, and methylation gene profiling, we were unable to elucidate the potential sources of this heterogeneity. Further subgroup analysis was hindered by the lack of original data regarding lifestyle habits or ethnicity. Thirdly, the association of salivary DNA promoter hypermethylation and clinicopathological variables (i.e., TNM stage, histological grade) was not explored due to insufficient data. Therefore, well-designed prospective clinical studies with large sample sizes are necessary to validate the results of this meta-analysis.

5. Conclusions

Overall, the findings from this meta-analysis showed that salivary DNA promoter hypermethylation was associated with HNC risk. Salivary hypermethylation of *p16*, *MGMT*, *DAPK*, *TIMP3*, and *RASSF1A* showed an important role in HNC development. Thus, saliva could be used as a potential source of epigenetic biomarkers for predicting HNC. The development of HNC screening programs based on the combination of these 5-methylated genes in saliva could be useful for identifying high-risk patients and for detecting cancer before the occurrence of initial clinical symptoms. The clinical implementation of this salivary panel would represent the beginning of precision medicine for HNC. To attain this, prospective and multicenter studies should be carried out in order to validate the present results.

Supplementary Materials: The following are available online at https://www.mdpi.com/article/10.3390/jpm11070606/s1, Figure S1: Funnel plot for studies (of 18 studies) on the association between salivary DNA hypermethylation and HNC, Figure S2: Funnel plot for studies (of 9 studies) on the association between salivary hypermethylation of *p16* gene promoter and HNC, Figure S3: Funnel plot for studies (of 5 studies) on the association between salivary hypermethylation of *MGMT* gene promoter and HNC, Figure S4: Funnel plot for studies (of 4 studies) on the association between salivary hypermethylation of *DAPK* gene promoter and HNC, Figure S5 Funnel plot for studies (of 4 studies) on the association between salivary hypermethylation of *TIMP3* gene promoter and HNC, Figure S6: Funnel plot for studies (of 3 studies) on the association between salivary hypermethylation of *RASSF1A* gene promoter and HNC, Figure S7: Funnel plot for studies (of 3 studies) on the association between salivary hypermethylation of *APC* gene promoter and HNC, Table S1: The Newcastle-Ottawa Scale (NOS) for assessing the quality of included studies, Table S2: Subgroup analysis of salivary DNA methylation for HNC detection based on different covariates.

Author Contributions: Conceptualization, Ó.R.-G. and M.M.S.-C.; methodology, Ó.R.-G. and M.M.S.-C.; software, C.M.-R. and Á.S.-B.; formal analysis, C.M.-R. and Á.S.-B.; investigation, Ó.R.-G., Á.D.-L. and M.M.S.-C.; data curation, Ó.R.-G., Á.D.-L. and M.M.S.-C.; Writing—Original draft preparation, Ó.R.-G. and M.M.S.-C.; Writing—Review and Editing, Á.D.-L. and R.L.-L.; visualization, Ó.R.-G. and M.A.S.; supervision, M.M.S.-C.; project administration, M.M.S.-C.; funding acquisition, M.M.S.-C. All authors have read and agreed to the published version of the manuscript.

Funding: This work was co-funded by the Instituto de Salud Carlos III (ISCIII) (PI20/01449) and the European Regional Development Fund (FEDER). ADL is funded by a contract "Juan Rodés" from ISCIII (JR17/00016).

Institutional Review Board Statement: Not applicable.

Informed Consent Statement: Not applicable.

Conflicts of Interest: R.L.-L. reports other from Nasasbiotech, during the conduct of the study; grants and personal fees from Roche, grants and personal fees from Merck, personal fees from AstraZeneca, personal fees from Bayer, personal fees and non-financial support from BMS, personal fees from Pharmamar, personal fees from Leo, outside the submitted work. The rest of the authors have nothing to disclose. The funders had no role in the design of the study; in the collection, analyses, or interpretation of data; in the writing of the manuscript, or in the decision to publish the results.

References

1. Castilho, R.M.; Squarize, C.H.; Almeida, L.O. Epigenetic modifications and head and neck cancer: Implications for tumor progression and resistance to therapy. *Int. J. Mol. Sci.* **2017**, *18*, 1506. [CrossRef] [PubMed]
2. Pfeifer, G.P. Defining driver DNA methylation changes in human cancer. *Int. J. Mol. Sci.* **2018**, *19*, 1166. [CrossRef]
3. Laird, P.W. The power and the promise of DNA methylation markers. *Nat. Rev. Cancer* **2003**, *3*, 253–266. [CrossRef]
4. Hulbert, A.; Jusue-Torres, I.; Stark, A.; Chen, C.; Rodgers, K.; Lee, B.; Griffin, C.; Yang, A.; Huang, P.; Wrangle, J.; et al. Early detection of lung cancer using DNA promoter hypermethylation in plasma and sputum. *Clin. Cancer Res.* **2017**, *23*, 1998–2005. [CrossRef]
5. Patai, Á.V.; Valcz, G.; Hollósi, P.; Kalmár, A.; Peterfia, B.; Wichmann, B.; Spisák, S.; Barták, B.K.; Leiszter, K.; Tóth, K.; et al. Comprehensive DNA methylation analysis reveals a common ten-gene methylation signature in colorectal adenomas and carcinomas. *PLoS ONE* **2015**, *10*, e0133836. [CrossRef]
6. Glodzik, D.; Bosch, A.; Hartman, J.; Aine, M.; Vallon-Christersson, J.; Reuterswärd, C.; Karlsson, A.; Mitra, S.; Niméus, E.; Holm, K.; et al. Comprehensive molecular comparison of BRCA1 hypermethylated and BRCA1 mutated triple negative breast cancers. *Nat. Commun.* **2020**, *11*, 1–15. [CrossRef]
7. Chou, J.L.; Huang, R.L.; Shay, J.; Chen, L.Y.; Lin, S.J.; Yan, P.S.; Chao, W.T.; Lai, Y.H.; Lai, Y.L.; Chao, T.K.; et al. Hypermethylation of the TGF-β target, ABCA1 is associated with poor prognosis in ovarian cancer patients. *Clin. Epigenet.* **2015**, *7*, 1. [CrossRef] [PubMed]
8. Gaździcka, J.; Gołąbek, K.; Strzelczyk, J.K.; Ostrowska, Z. Epigenetic modifications in head and neck cancer. *Biochem. Genet.* **2020**, *58*, 213–244. [CrossRef] [PubMed]
9. Schröck, A.; Leisse, A.; De Vos, L.; Gevensleben, H.; Dröge, F.; Franzen, A.; Wachendörfer, M.; Schröck, F.; Ellinger, J.; Teschke, M.; et al. Free-circulating methylated DNA in blood for diagnosis, staging, prognosis, and monitoring of head and neck squamous cell carcinoma patients: An observational prospective cohort study. *Clin. Chem.* **2017**, *63*, 1288–1296. [CrossRef] [PubMed]
10. Liyanage, C.; Wathupola, A.; Muraleetharan, S.; Perera, K.; Punyadeera, C.; Udagama, P. Promoter hypermethylation of tumor-suppressor genes $p16^{INK4a}$, *RASSF1A*, *TIMP3*, and *PCQAP/MED15* in salivary DNA as a quadruple biomarker panel for early detection of oral and oropharyngeal cancers. *Biomolecules* **2019**, *9*, 148. [CrossRef] [PubMed]
11. Sanchez-Cespedes, M.; Esteller, M.; Wu, L.; Nawroz-Danish, H.; Yoo, G.H.; Koch, W.M.; Jen, J.; Herman, J.G.; Sidransky, D. Gene promoter hypermethylation in tumors and serum of head and neck cancer patients. *Cancer Res.* **2000**, *60*, 892–895.
12. McInnes, M.; Moher, D.; Thombs, B.D.; McGrath, T.A.; Bossuyt, P.M.; Clifford, T.; Cohen, J.F.; Deeks, J.J.; Gatsonis, C.; Hooft, L.; et al. Preferred Reporting Items for a Systematic Review and Meta-analysis of Diagnostic Test Accuracy Studies. *JAMA* **2018**, *319*, 388–396. [CrossRef] [PubMed]
13. Rapado-González, Ó.; Martínez-Reglero, C.; Salgado-Barreira, Á.; Muinelo-Romay, L.; Muinelo-Lorenzo, J.; López-López, R.; Díaz-Lagares, Á.; Suárez-Cunqueiro, M. Salivary DNA methylation as an epigenetic biomarker for head and neck cancer. Part I: A diagnostic accuracy meta-analysis. *J. Pers. Med.* **2021**, *11*, 568. [CrossRef]
14. Wells, G.A.; Shea, B.; O'Connell, D.; Peterson, J.; Welch, V.; Losos, M.; Tugwell, P. The Newcastle-Ottawa Scale (NOS) for assessing the quality of nonrandomized studies in meta-analysis. Available online: http://www.ohri.ca/programs/clinical_epidemiology/oxford.asp (accessed on 16 December 2019).
15. Egger, M.; Smith, G.D.; Schneider, M.; Minder, C. Bias in meta-analysis detected by a simple, graphical test. *BMJ* **1997**, *315*, 629–634. [CrossRef]
16. Begg, C.B.; Mazumdar, M. Operating characteristics of a rank correlation test for publication bias. *Biometrics* **1994**, *50*, 1088. [CrossRef] [PubMed]
17. Guerrero-Preston, R.; Soudry, E.; Acero, J.; Orera, M.; Moreno-López, L.A.; Macía-Colón, G.; Jaffe, A.; Berdasco, M.; Ili, C.; Brebi-Mieville, P.; et al. NID2 and HOXA9 promoter hypermethylation as biomarkers for prevention and early detection in oral cavity squamous cell carcinoma tissues and saliva. *Cancer Prev. Res.* **2011**, *4*, 1061–1072. [CrossRef] [PubMed]
18. Claus, R.; Wilop, S.; Hielscher, T.; Sonnet, M.; Dahl, E.; Galm, O.; Jost, E.; Plass, C. A systematic comparison of quantitative high-resolution DNA methylation analysis and methylation-specific PCR. *Epigenetics* **2012**, *7*, 772–780. [CrossRef]
19. Lee, E.-S.; Issa, J.-P.; Roberts, D.B.; Williams, M.D.; Weber, R.S.; Kies, M.S.; El-Naggar, A.K. Quantitative promoter hypermethylation analysis of cancer-related genes in salivary gland carcinomas: Comparison with methylation-specific PCR technique and clinical significance. *Clin. Cancer Res.* **2008**, *14*, 2664–2672. [CrossRef]
20. Padhi, S.S.; Roy, S.; Kar, M.; Saha, A.; Roy, S.; Adhya, A.; Baisakh, M.; Banerjee, B. Role of CDKN2A/p16 expression in the prognostication of oral squamous cell carcinoma. *Oral Oncol.* **2017**, *73*, 27–35. [CrossRef]
21. Kulkarni, V.; Saranath, D. Concurrent hypermethylation of multiple regulatory genes in chewing tobacco associated oral squamous cell carcinomas and adjacent normal tissues. *Oral Oncol.* **2004**, *40*, 145–153. [CrossRef]

22. Su, P.F.; Huang, W.L.; Wu, H.T.; Wu, C.H.; Liu, T.Y.; Kao, S.Y. p16INK4A promoter hypermethylation is associated with invasiveness and prognosis of oral squamous cell carcinoma in an age-dependent manner. *Oral Oncol.* **2010**, *46*, 734–739. [CrossRef] [PubMed]
23. Shi, H.; Chen, X.; Lu, C.; Gu, C.; Jiang, H.; Meng, R.; Niu, X.; Huang, Y.; Lu, M. Association between P16INK4a promoter methylation and HNSCC: A meta-analysis of 21 published studies. *PLoS ONE* **2015**, *10*, e0122302. [CrossRef]
24. Zhou, C.; Shen, Z.; Ye, D.; Li, Q.; Deng, H.; Liu, H.; Li, J. The association and clinical significance of CDKN2A promoter methylation in head and neck squamous cell carcinoma: A meta-analysis. *Cell. Physiol. Biochem.* **2018**, *50*, 868–882. [CrossRef]
25. Zhao, J.J.; Li, H.Y.; Wang, D.; Yao, H.; Sun, D.W. Abnormal MGMT promoter methylation may contribute to the risk of esophageal cancer: A meta-analysis of cohort studies. *Tumor Biol.* **2014**, *35*, 10085–10093. [CrossRef] [PubMed]
26. Matsuda, S.; Mafune, A.; Kohda, N.; Hama, T.; Urashima, M. Associations among smoking, MGMT hypermethylation, TP53-mutations, and relapse in head and neck squamous cell carcinoma. *PLoS ONE* **2020**, *15*, e0231932. [CrossRef] [PubMed]
27. Zuo, C.; Ai, L.; Ratliff, P.; Suen, J.Y.; Hanna, E.; Brent, T.P.; Fan, C.Y. O6-methylguanine-DNA methyltransferase gene: Epigenetic silencing and prognostic value in head and neck squamous cell carcinoma. *Cancer Epidemiol. Biomark. Prev.* **2004**, *13*, 967–975.
28. Cohen, O.; Kimchi, A. DAP-kinase: From functional gene cloning to establishment of its role in apoptosis and cancer. *Cell Death Differ.* **2001**, *8*, 6–15. [CrossRef]
29. Šupić, G.; Kozomara, R.; Branković-Magić, M.; Jović, N.; Magić, Z. Gene hypermethylation in tumor tissue of advanced oral squamous cell carcinoma patients. *Oral Oncol.* **2009**, *45*, 1051–1057. [CrossRef]
30. Dammann, R.H.; Steinmann, K.; Sandner, A.; Schagdarsurengin, U. Frequent promoter hypermethylation of tumor-related genes in head and neck squamous cell carcinoma. *Oncol. Rep.* **2009**, *22*, 1519–1526. [CrossRef]
31. Cai, F.; Xiao, X.; Niu, X.; Zhong, Y. Association between promoter methylation of DAPK gene and HNSCC: A meta-analysis. *PLoS ONE* **2017**, *12*, e0173194. [CrossRef]
32. Su, C.W.; Lin, C.W.; Yang, W.E.; Yang, S.F. TIMP-3 as a therapeutic target for cancer. *Ther. Adv. Med. Oncol.* **2019**, *11*. [CrossRef]
33. Rettori, M.M.; De Carvalho, A.C.; Longo, A.L.B.; De Oliveira, C.Z.; Kowalski, L.P.; Carvalho, A.L.; Vettore, A.L. TIMP3 and CCNA1 hypermethylation in HNSCC is associated with an increased incidence of second primary tumors. *J. Transl. Med.* **2013**, *11*, 316. [CrossRef]
34. Sun, W.; Zaboli, D.; Wang, H.; Liu, Y.; Arnaoutakis, D.; Khan, T.; Khan, Z.; Koch, W.M.; Califano, J.A. Detection of TIMP3 promoter hypermethylation in salivary rinse as an independent predictor of local recurrence-free survival in head and neck cancer. *Clin. Cancer Res.* **2012**, *18*, 1082–1091. [CrossRef]
35. Dubois, F.; Bergot, E.; Zalcman, G.; Levallet, G. RASSF1A, puppeteer of cellular homeostasis, fights tumorigenesis, and metastasis—An updated review. *Cell Death Dis.* **2019**, *10*, 1–13. [CrossRef]
36. Raos, D.; Ulamec, M.; Bojanac, A.K.; Bulic-Jakus, F.; Jezek, D.; Sincic, N. Epigenetically inactivated RASSF1A as a tumor biomarker. *Bosn. J. Basic Med. Sci.* **2020**. [CrossRef]
37. Meng, R.W.; Li, Y.C.; Chen, X.; Huang, Y.X.; Shi, H.; Du, D.D.; Niu, X.; Lu, C.; Lu, M.X. Aberrant methylation of RASSF1A closely associated with HNSCC, a meta-Analysis. *Sci. Rep.* **2016**, *6*, 20756. [CrossRef]
38. Zhang, L.; Shay, J.W. Multiple roles of APC and its therapeutic implications in colorectal cancer. *J. Natl. Cancer Inst.* **2017**, *109*, 109. [CrossRef] [PubMed]
39. Uesugi, H.; Uzawa, K.; Kawasaki, K.; Shimada, K.; Moriya, T.; Tada, A.; Shiiba, M.; Tanzawa, H. Status of reduced expression and hypermethylation of the APC tumor suppressor gene in human oral squamous cell carcinoma. *Int. J. Mol. Med.* **2005**, *15*, 597–602. [CrossRef] [PubMed]
40. Rettori, M.M.; De Carvalho, A.C.; Longo, A.L.B.; De Oliveira, C.Z.; Kowalski, L.P.; Carvalho, A.; Vettore, A.L. Prognostic significance of TIMP3 hypermethylation in post-treatment salivary rinse from head and neck squamous cell carcinoma patients. *Carcinogenesis* **2012**, *34*, 20–27. [CrossRef] [PubMed]
41. Nagata, S.; Hamada, T.; Yamada, N.; Yokoyama, S.; Kitamoto, S.; Kanmura, Y.; Nomura, M.; Kamikawa, Y.; Yonezawa, S.; Sugihara, K. Aberrant DNA methylation of tumor-related genes in oral rinse. *Cancer* **2012**, *118*, 4298–4308. [CrossRef]
42. Righini, C.A.; De Fraipont, F.; Timsit, J.-F.; Faure, C.; Brambilla, E.; Reyt, E.; Favrot, M.-C. Tumor-specific methylation in saliva: A promising biomarker for early detection of head and neck cancer recurrence. *Clin. Cancer Res.* **2007**, *13*, 1179–1185. [CrossRef] [PubMed]
43. López, F.; Sampedro, T.; Llorente, J.L.; Dominguez, F.; Hermsen, M.; Suárez, C.; Álvarez-Marcos, C. Utility of MS-MLPA in DNA methylation profiling in primary laryngeal squamous cell carcinoma. *Oral Oncol.* **2014**, *50*, 291–297. [CrossRef]
44. Chen, K.; Sawhney, R.; Khan, M.; Benninger, M.S.; Hou, Z.; Sethi, S.; Stephen, J.K.; Worsham, M.J. Methylation of multiple genes as diagnostic and therapeutic markers in primary head and neck squamous cell carcinoma. *Arch. Otolaryngol. Head Neck Surg.* **2007**, *133*, 1131–1138. [CrossRef] [PubMed]
45. Seike, M.; Gemma, A.; Hosoya, Y.; Hemmi, S.; Taniguchi, Y.; Fukuda, Y.; Yamanaka, N.; Kudoh, S. Increase in the frequency of p16INK4 gene inactivation by hypermethylation in lung cancer during the process of metastasis and its relation to the status of p53. *Clin. Cancer Res.* **2000**, *6*, 4307–4313. [PubMed]
46. Wahab, A.H.A.; El-Mezayen, H.A.; Sharad, H.; Rahman, S.A. Promoter hypermethylation of RASSF1A, MGMT, and HIC-1 genes in benign and malignant colorectal tumors. *Tumor Biol.* **2011**, *32*, 845–852. [CrossRef] [PubMed]
47. Masson, D.; Rioux-Leclercq, N.; Fergelot, P.; Jouan, F.; Mottier, S.; Théoleyre, S.; Bach-Ngohou, K.; Patard, J.J.; Denis, M. Loss of expression of TIMP3 in clear cell renal cell carcinoma. *Eur. J. Cancer* **2010**, *46*, 1430–1437. [CrossRef]

48. Van Der Auwera, I.; Bovie, C.; Svensson, C.; Trinh, X.B.; Limame, R.; Van Dam, P.; Van Laere, S.J.; Van Marck, E.A.; Dirix, L.Y.; Vermeulen, P.B. Quantitative methylation profiling in tumor and matched morphologically normal tissues from breast cancer patients. *BMC Cancer* **2010**, *10*, 97. [CrossRef]
49. Krassenstein, R.; Sauter, E.; Dulaimi, E.; Battagli, C.; Ehya, H.; Klein-Szanto, A.; Cairns, P. Detection of breast cancer in nipple aspirate fluid by CpG island hypermethylation. *Clin. Cancer Res.* **2004**, *10*, 28–32. [CrossRef] [PubMed]
50. Mittag, F.; Kuester, D.; Vieth, M.; Peters, B.; Stolte, B.; Roessner, A.; Schneider-Stock, R. DAPK promotor methylation is an early event in colorectal carcinogenesis. *Cancer Lett.* **2006**, *240*, 69–75. [CrossRef]
51. Belinsky, S.A.; Klinge, D.M.; Dekker, J.D.; Smith, M.W.; Bocklage, T.J.; Gilliland, F.D.; Crowell, R.E.; Karp, D.D.; Stidley, C.A.; Picchi, M.A. Gene promoter methylation in plasma and sputum increases with lung cancer risk. *Clin. Cancer Res.* **2005**, *11*, 6505–6511. [CrossRef]
52. Shi, D.-T.; Han, M.; Gao, N.; Tian, W.; Chen, W. Association of RASSF1A promoter methylation with gastric cancer risk: A meta-analysis. *Tumor Biol.* **2013**, *35*, 943–948. [CrossRef] [PubMed]
53. Chen, R.; Zheng, Y.; Zhuo, L.; Wang, S. Association between MGMT promoter methylation and risk of breast and gynecologic cancers: A systematic review and meta-Analysis. *Sci. Rep.* **2017**, *7*, 12783. [CrossRef]
54. Cao, J.; Li, Z.; Yang, L.; Liu, C.; Luan, X. Association between tissue inhibitor of metalloproteinase-3 gene methylation and gastric cancer risk: A meta-Analysis. *Genet. Test. Mol. Biomark.* **2016**, *20*, 427–431. [CrossRef] [PubMed]
55. Tang, B.; Li, Y.; Qi, G.; Yuan, S.; Wang, Z.; Yu, S.; Li, B.; He, S. Clinicopathological significance of cdkn2a promoter hypermethylation frequency with pancreatic cancer. *Sci. Rep.* **2015**, *5*, srep13563. [CrossRef] [PubMed]

Article

Circulating p16-Positive and p16-Negative Tumor Cells Serve as Independent Prognostic Indicators of Survival in Patients with Head and Neck Squamous Cell Carcinomas

Pei-Hung Chang [1,2,3,†], Hung-Ming Wang [1,3,†], Yung-Chia Kuo [1,3,4,†], Li-Yu Lee [1,5], Chia-Jung Liao [1], Hsuan-Chih Kuo [1,4], Cheng-Lung Hsu [1,3], Chun-Ta Liao [1,6], Sanger Hung-Chi Lin [3], Pei-Wei Huang [1,3], Tyler Min-Hsien Wu [3,4,7] and Jason Chia-Hsun Hsieh [1,3,4,*]

1. College of Medicine, Chang Gung University, Taoyuan 33382, Taiwan; ph555chang@gmail.com (P.-H.C.); whm526@cgmh.org.tw (H.-M.W.); 8705024@cgmh.org.tw (Y.-C.K.); r22068@cgmh.org.tw (L.-Y.L.); L329735@ms49.hinet.net (C.-J.L.); hsuanchihkuo@gmail.com (H.-C.K.); hsu2221@adm.cgmh.org.tw (C.-L.H.); liaoct@cgmh.org.tw (C.-T.L.); freewind05@cgmh.org.tw (P.-W.H.)
2. Division of Hematology-Oncology, Department of Internal Medicine, Chang Gung Memorial Hospital at Keelung, Keelung 20448, Taiwan
3. Circulating Tumor Cell Lab., Division of Hematology-Oncology, Department of Internal Medicine, Chang Gung Memorial Hospital at Linkou, Taoyuan 33382, Taiwan; sangerhj@gmail.com (S.H.-C.L.); mhwu@mail.cgu.edu.tw (T.M.-H.W.)
4. Division of Hematology-Oncology, Department of Internal Medicine, New Taipei City Municipal TuCheng Hospital, New Taipei City 23600, Taiwan
5. Department of Pathology, Chang Gung Memorial Hospital at Linkou, Taoyuan 33382, Taiwan
6. Department of Otorhinolaryngology, Head and Neck Surgery, Linkou Chang Gung Memorial Hospital, Chang Gung University, Taoyuan 33382, Taiwan
7. Tissue Engineering and Microfluidic Biochip Lab., Graduate Institute of Biomedical Engineering, Chang Gung University, Taoyuan 33382, Taiwan
* Correspondence: wisdom5000@gmail.com; Tel.: +886-3281-200 (ext. 8825) or +886-3281-200 (ext. 2517); Fax: +886-3281-200 (ext. 2362)
† Chang, Wang, and Kuo contributed equally to this work.

Abstract: Background: Decisions regarding the staging, prognosis, and treatment of patients with head and neck squamous cell carcinomas (HNSCCs) are made after determining their p16 expression levels and human papillomavirus (HPV) infection status. Methods: We investigated the prognostic roles of p16-positive and p16-negative circulating tumor cells (CTCs) and their cell counts in HNSCC patients. We enrolled patients with locally advanced HNSCCs who received definitive concurrent chemoradiotherapy for final analysis. We performed CTC testing and p16 expression analysis before chemoradiotherapy. We analyzed the correlation between p16-positive and p16-negative CTCs and HPV genotyping, tissue p16 expression status, response to chemoradiotherapy, disease-free survival, and overall survival. Results: Forty-one patients who fulfilled the study criteria were prospectively enrolled for final analysis. The detection rates of p16-positive (>0 cells/mL blood) and p16-negative (≥3 cells/mL blood) CTCs were 51.2% (n = 21/41) and 70.7%, respectively. The best responses of chemoradiotherapy and the p16 positivity of CTCs are independent prognostic factors of disease progression, with hazard ratios of 1.738 (95% confidence interval (CI): 1.031–2.927), 5.497 (95% CI: 1.818–16.615), and 0.176 (95% CI: 0.056–0.554), respectively. The p16 positivity of CTCs was a prognostic factor for cancer death, with a hazard ratio of 0.294 (95% CI: 0.102–0.852). Conclusions: The p16-positive and p16-negative CTCs could predict outcomes in HNSCC patients receiving definitive chemoradiotherapy. This non-invasive CTC test could help stratify the risk and prognosis before chemoradiotherapy in clinical practice and enable us to perform de-intensifying therapies.

Keywords: circulating tumor cells; p16 expression; head and neck squamous cell carcinoma; HPV genotyping; biomarker; liquid biopsy

1. Introduction

Human papillomavirus (HPV)-associated head and neck squamous cell carcinoma (HNSCC) has been widely investigated and thought of as a critical biomarker in HNSCC [1,2]. HPV-positive HNSCC, especially with oropharyngeal-originated tumors [3,4], has a significantly better prognosis (50% reduction of death risk) compared with those without HPV infection [5,6]. Even though patients with HPV-associated HNSCC had a better prognosis, primary concurrent chemoradiotherapy (CCRT) remains one of the standards of care [7]. De-intensification of CCRT has been widely investigated in very recent years [8,9]. HPV-positive tumors can also be found in patients with cancers originating in the head and neck region other than the oropharynx, such as the paranasal sinus [10], the hypopharynx [11], the larynx [12], and the oral cavity [13]. Given the positive prognostic impact of HPV infection, the p16 expression or HPV genotyping status at cancer staging have been strongly suggested for patients presenting with neck squamous cell carcinoma, without identified primary sites [14,15]. The importance of p16 expression or HPV infection status in HNSCC is well established [16].

P16 expression by immunohistochemistry staining is much easier to perform than PCR for HPV infection; therefore, p16 expression was much more widely used in clinical practices considering the test's price and accessibility. However, some investigators reported that p16 expression does not always equal HPV infection [17]. HPV-DNA/RNA testing is still recommended for confirming p16 results, with increased specificity and diagnostic accuracy [17,18]. However, these tests are all tissue-based and require invasive procedures to obtain cancer tissue [17]. Sometimes, these invasive procedures caused unwanted complications, such as tumor bleeding [19]. However, there is no validated blood test for detecting HPV infection or evaluating p16 expression for diagnosis or monitoring, except for some exploratory or small-scale observational studies [20]. Only a few studies have reported technology detecting p16 expression levels in circulating tumor cells (CTCs) [21,22] or circulating tumor DNA [23,24] in plasma samples in HPV-associated HNSCC. The significance of p16 or HPV in peripheral blood remains unknown.

Therefore, we hypothesized the following: (1) that expressed p16 can be detected in CTCs; (2) that circulating p16-negative CTCs and p16-positive CTCs might have different effects on survival rates; (3) that p16 expression levels in the tissue and blood (CTCs) could be correlated with each other.

2. Materials and Methods

2.1. Patient Enrollment

Our prospective study was conducted in two medical centers, Chang Gung Memorial Hospital, Linkou, and Keelung, Taiwan. All patients provided written informed consent. The Institutional Review Board in Chang Gung Memorial Hospital approved the study protocols, with approval IDs 104-2620B, 103-7795B, and 201700867B0. Eligible patients with histologically- or pathologically- confirmed head and neck squamous cell carcinomas with p16 expression status were considered medically unfit for surgery or had surgically unresectable, locally advanced presentation (stage IIb–IV, American Joint Committee on Cancer [AJCC], 8th edition). In addition, patients with (1) age ≥ 20 years; (2) the ability to understand the protocol and provide informed consent out of their own free will; (3) primary HNSCC; and (4) adequate liver and renal function and white blood cell counts before undergoing anticancer therapies, especially chemoradiotherapy, were included. The exclusion criteria contained patients who (1) receive therapies except for CCRT, including curative surgery without CCRT, salvage surgery, or radiation alone; (2) refused to blood drawing in the protocol; (3) had rapidly worsened performance status to complete CCRT; or (4) had metachronous or synchronous double cancer. Physicians performed disease staging and management according to the standard treatment protocols detailed in institutional guidelines. Results were reported following the REMARK guidelines [25]. Examinations for the initial staging and response evaluation processes included magnetic resonance imaging and positron emission tomography. In accordance with standard treatment guidelines, concurrent chemoradiotherapy was scheduled and delivered by medical oncologists and radiation oncologists. In accordance with the guidelines of version 1.1 of the response evaluation criteria in solid tumors (RECIST), the treatment response was determined based on whether the patient exhibited complete remission (CR), partial response (PR), stable disease (SD), or progressive disease (PD). This was determined by the multidisciplinary head and neck cancer tumor board at Chang Gung Memorial Hospital.

2.2. Tissue Immunohistochemistry Staining for p16 Expression Analysis

Immunohistochemistry staining (IHC) was performed using a mouse monoclonal antibody against p16 (Roche E6H4™, catalog #725-4713) on a Ventana Benchmark LT automated immunostainer (Tucson, AZ, USA), per the standard protocol. Positive and negative controls were included routinely. A positive signal was defined as that obtained with nuclear and or cytoplasmic staining. If cells were stained via cytoplasmic staining alone, the result was considered negative. In this study, the positivity of tumor samples is defined by a result where $\geq 70\%$ of cells are stained via cytoplasmatic and nuclear staining in two medical centers [26]. In this study, we performed HPV genotyping to confirm the HPV infection status in p16-positive cancer tissues (Supplementary Table S1).

2.3. The Isolation and Identification of Circulating Tumor Cells via Microscopy

Blood samples (8 mL for each patient, including 4 mL for microscopy and the other 4 mL for flow cytometry) were drawn before anticancer therapies, including chemotherapy or radiotherapy. A CTC enrichment procedure was performed by red blood cell lysis (by mixing 155 mM NH4Cl, 14 mM NaHCO3, and 0.1 mM EDTA in a 10:1 ratio with whole blood samples) and CD45-positive leukocyte depletion, using EasySep Human CD45 Depletion kits (Cat. NO. 18259, STEMCELL Technologies Inc., Vancouver, BC, Canada), following the manufacturer's instructions. A previously detailed method for CTC enrichment and counting was used [27–29].

We further fixed CTCs, isolated from 4 mL of whole blood samples, using 4% paraform aldehyde (Cat. No. 15710, Electron Microscopy Sciences) for 10 min at 25 °C. Permeabilization was performed by treating cells with PBST (0.1% Triton X-100 in PBS) for 10 min at 25 °C. After washing cells with PBS, they were blocked with PBS containing 2% BSA and the HuFcR Binding Inhibitor (Cat. No. 14-9161-73, eBioscience, San Diego, CA, USA) for 30 min at room temperature. Before the antibody reaction, 0.0025% Trypan Blue (Cat. No. 15250061, Thermo Fisher Scientific, CA, USA) was added to block auto-fluorescence. The antibody reaction was allowed to occur upon the addition of anti-EpCAM antibody conjugated Alexa Fluor 488 (1:400, one hour, Cat. No. 5198, Cell Signaling, Danvers, MA, USA) and anti-p16 antibody conjugated Alexa Fluor 647 (Cat. No. ab199819, Abcam, Cambridge, UK). We used the Hoechst (Cat. No. 62249, Thermo Fisher Scientific, CA, USA) stain to stain the cell nucleus. Cell fluorescence images were captured using a fluorescence microscope (Zeiss Axioskop 2 Plus Fluorescence Microscope, Carl Zeiss Microscopy, LLC, United States; Leica TCS SP2 Confocal Laser Scanning Microscope, Leica Microsystems, Wetzlar, Germany). CTCs were defined as cells expressing $EpCAM^{pos}Hoechst^{pos}CD45^{neg}$ and were further divided into p16-positive or p16-negative status.

2.4. Analysis of p16 Expression in Circulating Tumor Cells via Flow Cytometry

To determine the status of p16 expression in CTCs, we first fixed cells enriched via RBC lysis and CD45 depletion, using Fix & Perm Cell Permeabilization Reagents (Cat. NO. GAS003, molecular probes by Life Technologies, Thermo Fisher Scientific, CA, USA). Then, we added an anti-EpCAM antibody conjugated PE (Cat. No. FAB960P-100, R & D Systems) and an anti-p16 antibody conjugated Alexa Fluor 647 (Cat. No. ab199819, Abcam) during fixation and permeabilization. A secondary antibody, i.e., goat anti-mouse IgG H&L conjugated Alexa Fluor 488 (Cat. No. ab150113, Abcam), was also added to exclude the residual CD45-positive leukocytes (labeled with the CD45 antibody cocktail provided in the CD45 depletion kit). Isotype control antibodies were used as the negative control.

After staining, cell samples were analyzed using a Flow Cytometer (CytoFLEXTM Flow Cytometer, Beckman Coulter, Inc., Pasadena, CA, USA).

Positive and negative controls for the analysis of p16 expression were carried out on circulating tumor cells.

As experimental controls for p16-positive cells, we used the HeLa (cervical cancer cells, ATCC® CRM-CCL-2TM) cells as positive controls in CTC samples; whereas, HCT116 (colorectal cancer cells, ATCC® CCL-247TM) and H1975 (non-small cell lung cancer cells, ATCC® CRL-5908™) cells were used as negative controls to analyze p16 expression. We cultivated these cells per the instructions provided by the American Type Culture Collection (ATCC). Briefly, HeLa, HCT116, and H1975 cells were maintained in DMEM medium (Cat. No. 11965092, GIBCO), McCoy's 5A medium (GIBCO), and RPMI-1640 medium (GIBCO), respectively, along with fetal bovine serum (Cat. NO. 10437028, GIBCO), while ensuring that the final concentration was 10%. Cells were cultured at 37 °C in a humidified incubator in a 5% CO_2 atmosphere.

2.5. Human Papillomavirus Genotyping

Human papillomavirus genotyping of cancer tissues was carried out via the Roche Cobas 4800 HPV test. Briefly, cancer cells were stored in 800 µL Cobas PCR Cell Collection Media (Cat. No. 05619637190, Roche Molecular Systems, Inc., Branchburg, CA, USA), and Roche Cobas X 480 instruments were used to purify the DNA. Real-time PCR was performed using the Roche Cobas 4800 HPV Test on Roche Cobas Z 480 analyzers. The assay was performed and validated by the Taipei Institute of pathology, Taiwan. In addition, CTCs, isolated after negative-selection processes, were sent for HPV genotyping. All the 41 patients had p16, evidenced by immunohistochemistry staining, and had p16 CTC, evidenced by flow cytometry and genotyping.

2.6. Statistical Analysis

The basic characteristics of enrolled patients are demonstrated using descriptive statistics. Progression-free survival (PFS) was calculated from the date of CTC sampling, seven days before systemic chemotherapy, to cancer-specific progression or recurrence after CCRT or death from any causes. Overall survival (OS) was defined as the period from the date of CTC sampling to death from any cause. We applied chi-square and Fisher's exact tests to determine the difference between the p16 expression status in the tissue and blood (CTCs). We also used Kaplan–Meier survival plots with the log-rank test to demonstrate the individual factors affecting survival. Patients who did not experience the event (disease progression or death) were defined to be censored in the analysis. After checking the assumptions of clinicopathological factors, we used the univariate and multivariate Cox proportional hazard regression models to identify the independent prognostic factors of PFS and OS. All potential predictor variables were analyzed in the multivariate analysis, including p16-positive and p16-negative CTC status. They are essential items in this research, although they are mutually exclusive. Statistical analysis was performed using SPSS for Windows (version 18, SPSS Inc., Chicago, IL, USA). A p-value of 0.05 was considered statistically significant.

3. Results

3.1. Patient Enrollment

A total of 76 subjects (including 16 healthy donors) were prospectively enrolled, and overall, 41 patients met all the treatment criteria and were analyzed at Chang Gung Memorial Hospital, Linkou, and at Keelung between August 2017 and August 2018. Figure 1 demonstrates the study flow and patient numbers at different stages of enrollment in this prospective study. Table 1 summarizes the characteristics of the entire population. A total of 28 (68.3%) patients had oropharyngeal cancer, while 13 (31.7%) patients had non-oropharyngeal cancer. These 13 patients were enrolled because they had initially presented with an unknown primary cancer or a huge confluent mass in the hypopharynx and oropharynx area. The median age of the cohort was 55 (37–74) years old, and 78% of the patients have relatively good performance status before the concurrent chemoradiotherapy. Totals of 19 (46.3%) and 22 (53.7%) patients tested positive and negative, respectively, for p16 expression, upon immunohistochemistry analysis. The stages of the enrolled patients were relatively advanced: 16 (39.0%) were stage IV patients. With a median follow-up time of 34.0 (3.0–44.9) months, 22 (53.7%) patients exhibited disease progression after concurrent chemoradiotherapy, and 14 (34.1%) patients died. The detection rates of p16-positive (>0 cells/mL blood) and p16-negative (≥3 cells/mL blood) CTCs were 51.2% ($n = 21/41$) and 70.7%, respectively. The cutoff values of CTCs (3 cells/mL) were the same as those used in previous studies [30,31].

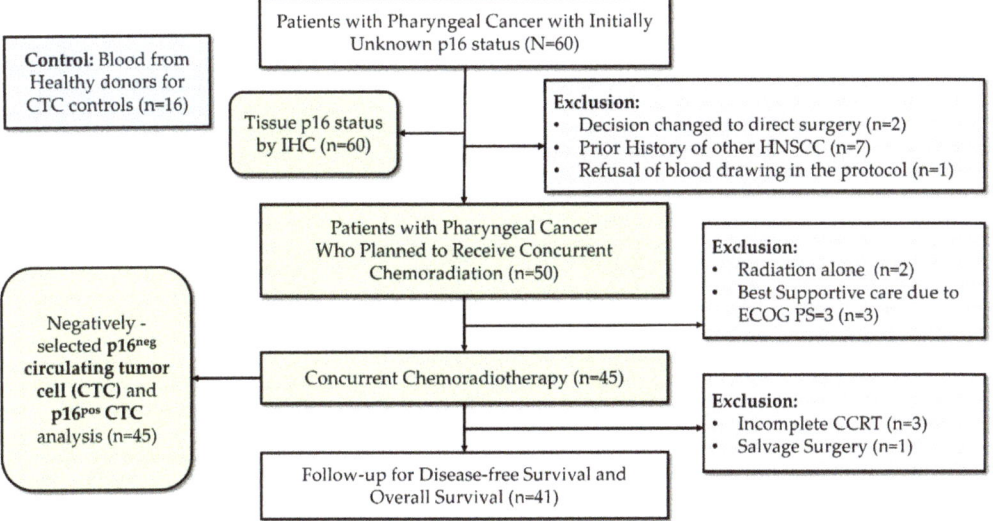

Figure 1. CONSORT algorithm.

Table 1. Patient characteristics ($n = 41$).

Characters	n	%
Age (median, range) in years		55 (37–74)
Sex		
Female	8	19.5%
Male	33	80.5%
Tumor type		
Oropharynx	28	68.3%
Non-oropharynx [a]	13	31.7%
ECOG PS		
0–1	32	78.0%
2	9	22.0%
Tumor stage (AJCC 8th edition) [b]		
II	20	48.8%
III	5	12.2%

Table 1. Cont.

Characters	n	%
IV	16	39.0%
T classification		
T1-2	28	68.3%
T3-4	13	31.7%
Lymph node Involvement		
Negative (N0)	11	26.8%
Positive (N1-3)	30	73.2%
p16 status by IHC staining		0.0%
Negative (0–70%)	22	53.7%
Positive (>70%)	19	46.3%
CCRT completion	41	100.0%
Disease Progression after CCRT		
No	19	46.3%
Yes	22	53.7%
Cancer-related death [c]		
No	27	65.9%
Yes	14	34.1%
Circulating tumor cells (CTCs) detection rate		
p16-positive CTCs (p16^{pos}EpCAM^{pos}Hoechst^{pos})	21	51.2%
p16-negative CTCs (p16^{neg}EpCAM^{pos}Hoechst^{pos}) [d]	20	48.8%

[a] Patients with non-oropharyngeal cancer included 10 hypopharyngeal cancers, and 3 cancers of unknown primary site. [b] The staging contained p16-positive and p16-negative tumors according to AJCC 8th edition staging system. [c] The cancer death was updated at a median follow-up time of 34.0 (range: 3.0–44.9) months. [d] Circulating tumor cell counts of ≥3 cells/mL was defined positive. Abbreviations: ECOG PS—Eastern Cooperative Oncology Group performance status; AJCC—The American Joint Committee on Cancer; IHC—immunohistochemistry; CCRT—concurrent chemoradiotherapy.

3.2. The Identification of p16-Positive Circulating Tumor Cells in Cancer Patients

To determine whether p16-positive CTCs could be detected in blood samples of pharyngeal cancer patients, we performed immunofluorescence staining. We used HCT116 (ATCC CCL-247)—the EpCAM^{pos}p16^{neg} human colon cancer cell line—for our analysis, while HeLa (ATCC CCL-2) cells—from the EpCAM^{neg}p16^{pos} human cervical cancer cell line—were used as control cells during immunofluorescence staining, as shown in Figure 2A,B. The remaining white blood cells in the CTC samples are shown in Figure 2C, while Figure 2D demonstrates the images of p16-positive CTCs, which were defined as p16^{pos}- or EpCAM^{pos}-nucleated cells. Otherwise, CTC without any p16 expression was be categorized as p16-negative CTCs, which have a threshold of 3 cells/mL as positive [31]. After CTC isolation, HPV genotyping was performed using commercial kits (Roche Cobas 4800 test), in accordance with the manufacturer's instructions. Experiments involving the spiking of human blood samples with HeLa cells enabled us to identify detection limits (10 cells/2 mL human blood), as demonstrated in Supplementary Table S1.

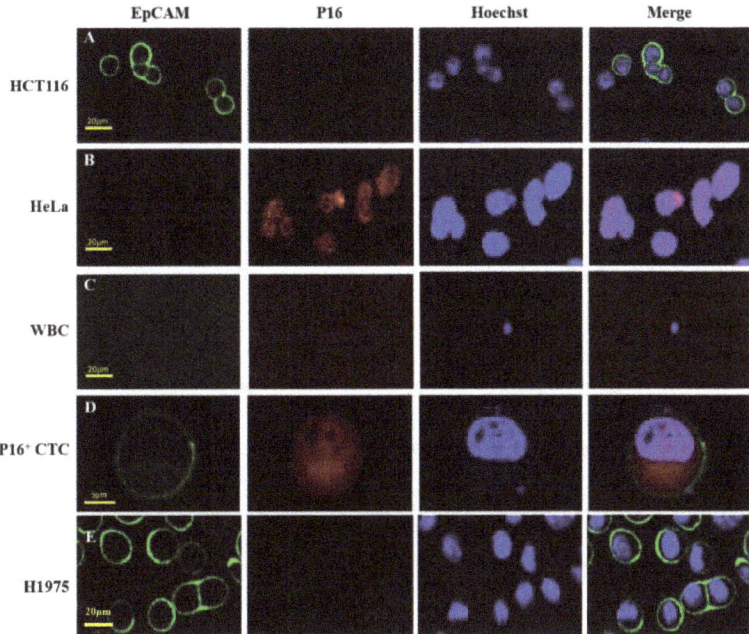

Figure 2. Demonstration of p16-positive circulating tumor cells identified in patients with oropharyngeal cancer. Immunofluorescence staining was used to identify cells in purified cells from blood samples. HCT116 (**A**) and HeLa (**B**) cells positively expressed EpCAM and p16, respectively. White blood cells (EpCAM^{neg}/P16^{neg}/Hoechst^{pos}) (**C**) and p16-positive circulating tumor cells (EpCAM^{pos}/P16^{pos}/Hoechst^{pos}) (**D**) were shown. H1975 cells (**E**) also serve as a positive control in this study. Abbreviations: CTC—circulating tumor cells.

We then applied flow cytometry-based CTC enumeration strategies and determined CTC counts after identifying the p16 expression status. In the present study, flow cytometry analysis was performed following the negative selection of CTCs, to analyze the p16-positive CTCs and p16-negative CTC in this cohort. First, HeLa (p16-positive) and H1975 (EpCAM-positive) cells were used to determine the staining conditions and set up the running template for flow cytometric analysis (Figure 3A–F). Then, tumor cell spike-in feasibility tests were carried out, as demonstrated in Figure 3G. Figure 3H,I show how p16-positive CTCs can be identified in a representative cancer patient's blood sample. The CTC detection ($\geq$1 cell/mL) rate was 70.7% (n = 29/41), and the p16 positivity rate of CTCs in the entire group was 51.2% (n = 21/41), irrespective of the tissue p16 status.

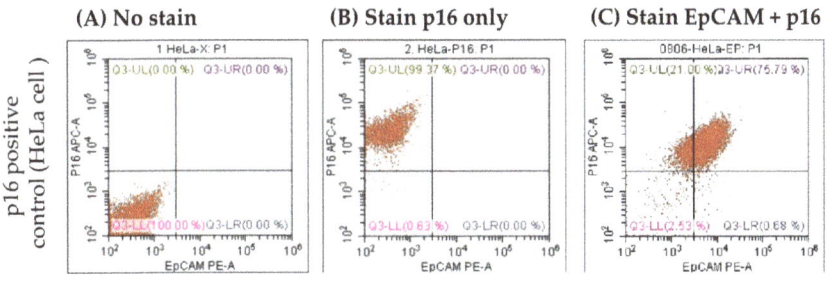

Figure 3. *Cont.*

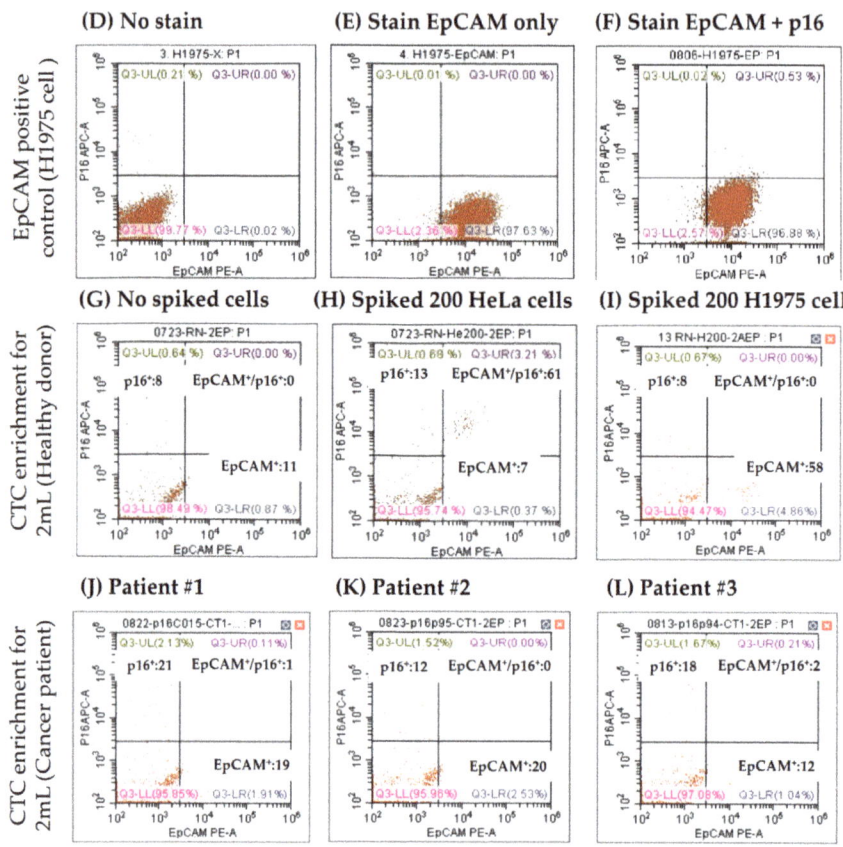

Figure 3. EpCAM+ and p16+ cell detection in blood samples. As standard controls, HeLa served as positive expression cells for flow cytometric analysis for p16 (**A–C**), and H1975 cells served as EpCAM expression (**D–F**). Accordingly, the protocol with controls can demonstrate p16 expression status in circulating tumor cells in one healthy individual by spiking different control cell lines (**G–I**) and three cancer patients (**J–L**).

3.3. HPV Genotyping of Cancer Tissues and Circulating Tumor Cells

To illustrate the concordance of p16 expression in CTCs and tissue, we compared p16 expression levels in the tissue and blood (CTC) via IHC and flow cytometry analyses. The results showed no statistical significance ($p = 0.155$, Table 2). The results of HPV genotyping and tissue p16 expression analysis of cancer tissues were analyzed further—we found that p16 positivity on CTCs was statistically related to p16 expression or a positive HPV genotype ($p < 0.019$, Table 2). However, the number of CTCs with HPV-positive genotypes was zero.

Table 2. Comparison between tissue p16 and blood p16 expressions.

	Tissue IHC p16 Negative	Tissue IHC p16 Positive	p-Value	Tissue IHC p16 Negative AND HPV Genotyping Negative	Tissue IHC p16 Positive OR HPV Genotyping Positive	p-Value
p16^{pos} CTC Negative	13	7	0.155	13	7	0.019 *
p16^{pos} CTC Positive	9	12		6	15	

* Fisher exact test was used for the statistical significance because numbers in some cells were less than 5.

3.4. Effects of CTCs and p16-Positive CTCs on Survival

We used Kaplan–Meier survival curves to compare the factors that might influence survival in this cohort (Figure 4). Figure 4A showed that patients with CTC count ≥3 cells/mL are associated with a short PFS ($p = 0.002$). Patients with p16-positive CTCs ($p = 0.012$, Figure 4B) who exhibited disease control after concurrent chemoradiotherapy ($p < 0.001$, Figure 4C) were associated with a prolonged PFS. However, tissue p16 expression was only marginally significant for the PFS ($p = 0.089$, Figure 4D). A prolonged OS was associated with patients with CTC counts <3 cells/mL ($p = 0.022$, Figure 4E), who exhibited p16 positivity of CTCs ($p = 0.017$, Figure 4F) and disease control after CCRT ($p = 0.003$, Figure 4G). Nevertheless, tissue p16 did not affect OS in this cohort ($p = 0.365$, Figure 4H).

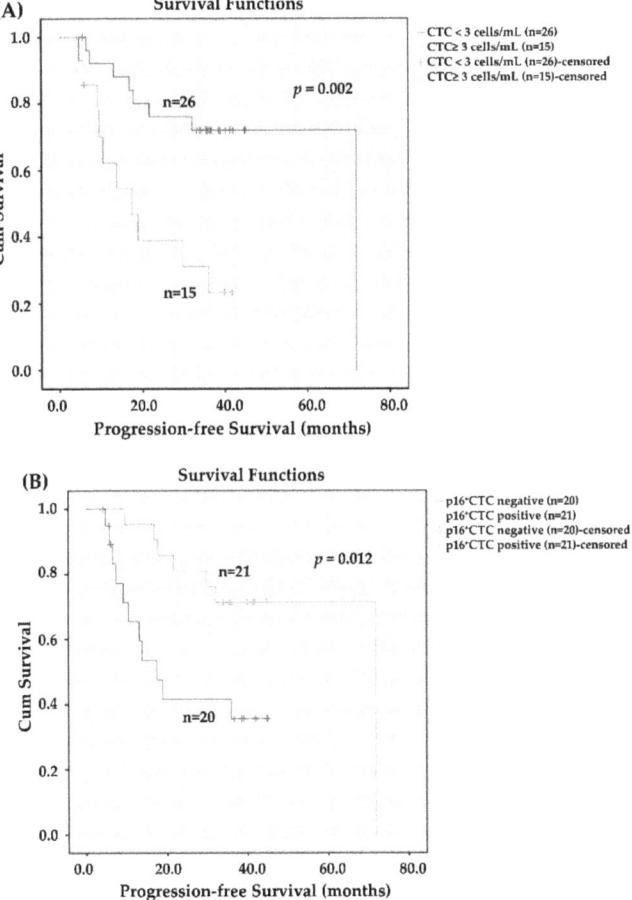

Figure 4. *Cont.*

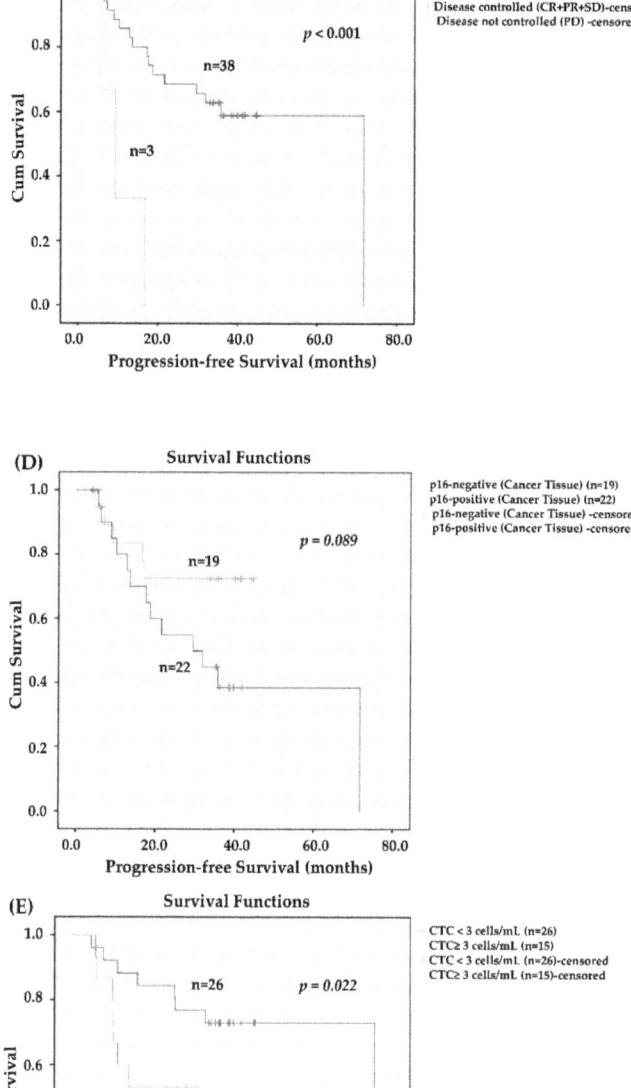

Figure 4. Cont.

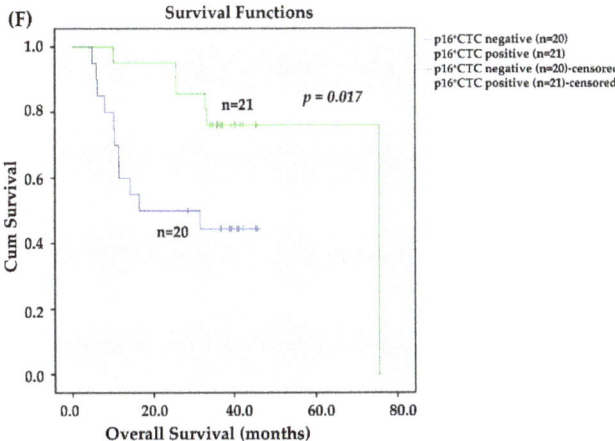
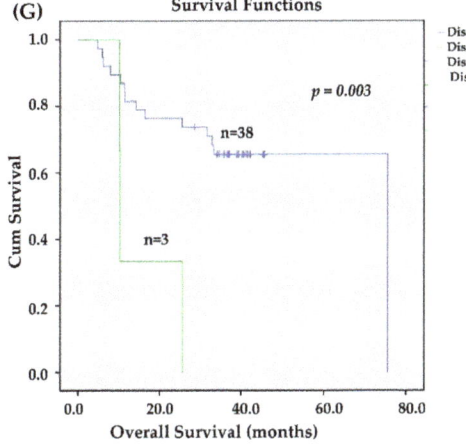

Figure 4. Cont.

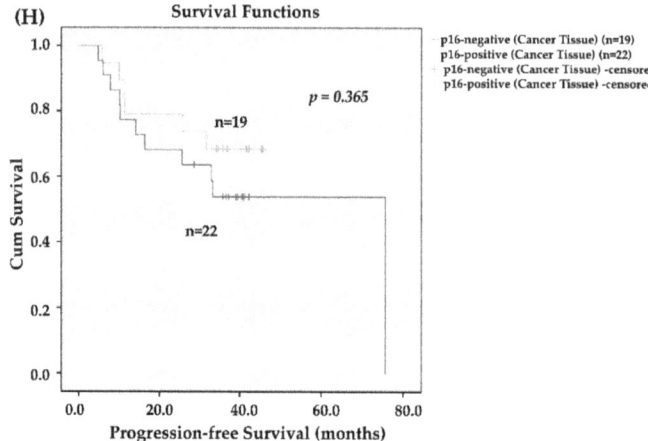

Figure 4. Kaplan–Meier Curves. In this cohort, patients with circulating tumor cell (CTC) numbers >=3 cells/mL had shown to negatively impact progression-free survival (PFS, $p = 0.002$) (**A**). Patients are associated with prolonged PFS, for those with p16-positive CTC ($p = 0.012$) (**B**), disease control after concurrent chemoradiotherapy was carried out (CCRT, $p < 0.001$) (**C**). However, tissue p16 expression is only marginally significant to PFS ($p = 0.089$) (**D**). For overall survival (OS), a prolonged OS is associated with patients harboring a CTC count of <3 cells/mL ($p = 0.022$) (**E**), p16 positivity of CTCs ($p = 0.017$) (**F**), and disease control after CCRT ($p = 0.003$) (**G**). Nevertheless, tissue p16 expression has no significant impact on OS in this cohort ($p = 0.365$) (**H**).

We analyzed all the factors in univariate and multivariate Cox regression models to identify independent prognostic factors. All factors involved in univariate analysis were analyzed during multivariate analysis using the forward LR model. Table 3 shows that the best response to CCRT, p16-negative CTC counts, and p16-positive CTCs were independent prognostic factors of disease progression, with hazard ratios (95% confidence interval) of 1.738 (95% confidence interval (CI): 1.031–2.927), 5.497 (95% CI: 1.818–16.615), and 0.176 (95% CI: 0.056–0.554), respectively. Only p16-positive CTCs (positive vs. negative) were found to be prognostic factors for cancer death, with a hazard ratio of 0.294 (95% CI: 0.102–0.852).

Table 3. Univariate and multivariate Cox regression analysis.

Factors	Progression-Free Survival						Overall Survival					
	Univariate Analysis			Multivariate Analysis			Univariate Analysis			Multivariate Analysis *		
	p	HR	(95% CI)	p	HR	(95% CI)	p	HR	(95% CI)	p	HR	(95% CI)
Age	0.135	0.957	(0.904–1.014)	0.003	0.909	(0.852–0.969)	0.338	0.973	(0.920–1.029)			
Sex (female vs. male)	0.262	2.328	(0.532–10.195)				0.371	1.967	(0.447–8.659)			
Tumor type (ORX vs. non-ORX)	0.924	0.95	(0.334–2.707)				0.415	1.524	(0.553–4.198)			
T classification (cT1-2 vs. cT3-4)	0.975	0.985	(0.374–2.592)				0.548	1.350	(0.506–3.600)			
N classification	0.421	1.214	(0.758–1.944)				0.968	0.991	(0.624–1.574)			
ECOG PS	0.171	0.377	(0.093–1.523)				0.248	0.439	(0.109–1.775)			
Best response of CCRT (Non-responders vs. responders)	0.104	1.535	(0.916–2.573)	0.038	1.738	1.031–2.927	0.114	1.499	(0.907–2.475)			
Tissue p16 IHC (Positive vs. negative)	0.099	0.415	(0.146–1.180)				0.369	0.629	(0.228–1.731)			
p16^{neg} CTC (Positive vs. negative)	0.005	4.029	(1.522–10.668)	0.003	5.497	1.818–16.615	0.029	3.037	(1.123–8.213)			
p16^{pos} CTC (Positive vs. negative)	0.018	0.300	(0.110–0.816)	0.003	0.176	0.056–0.554	0.024	0.294	(0.102–0.852)	0.024	0.294	(0.102–0.852)

* All factors in the univariate analysis were examined in the multivariate model. Abbreviations: CTC—circulating tumor cells; AJCC—The American Joint Cancer Committee; ECOG PS—Eastern Cooperative Oncology Group performance status; CCRT—concurrent chemoradiotherapy; PD—progressive disease; SD—stable disease; PR—partial response; CR—complete remission; IHC—immunohistochemistry; HR—hazard ratio; CI—confidence interval.

4. Discussion

This study found that p16-positive and p16-negative CTCs were uniquely correlated with survival in pharyngeal cancer patients. Our results have corroborated the results of previous studies, showing that it is feasible to detect p16-expressing CTCs [21,22]. Different methods were used to detect p16 (RT-qPCR assay [21,32,33] vs. protein expression [22]) in various studies to identify whether p16-expressing CTCs might improve risk discrimination in patients with early-stage oropharyngeal squamous cell carcinomas [21]. However, unlike the study findings by Dr. Economopoulou et al. (2019), a correlation between HPV16 E6/E7 expression in CTCs and a shorter PFS was identified in ten oropharyngeal cancer patients [21]. Hence, we hypothesized that p16-positive CTCs played a role similar to that of expressed p16 in tissues or the HPV genotyping process (Tables 2 and 3 and Figure 4). We hypothesized that the differences might be attributable to the relatively small case number (n = 10) and a short period of observation in that study. Compared with other studies' results, our study has provided relatively long-term follow-up outcomes with a median follow-up time of 34.0 months in pharyngeal cancer patients receiving CCRT. We concluded that (p16-negative) CTCs were correlated with a poor prognosis in patients with head and neck cancer: these findings were similar to those of previous studies [34–40].

The prognosis of patients with HPV-positive and HPV-negative head and neck cancers, especially oropharyngeal cancers, is notably different [41,42]. The p16 status of patients needs to be determined to enable clinicians to decide on the treatment plan [42,43]. Our study has provided evidence that CTC p16 positivity was independently associated with a prolonged PFS and OS. At the same time, p16-negative CTC counts were correlated with rapid disease progression after CCRT. In addition, our study has provided protocols for the identification and isolation of CTCs, and further analyzed the p16 status via a negative selection-based flow cytometric method.

We performed IHC staining and showed that the p16-positive CTC counts in flow cytometric analysis were not statistically related to the tissue p16 expression status (p = 0.155, Table 2) but were associated with both tissue p16 expression and HPV genotyping results (p = 0.019, Table 2). The main differences resulted from three cases with negative tissue IHC p16 expression but positive HPV genotyping—they were all in the p16-positive CTC group. It is well known that the discordance rate between tissue p16 expression and HPV infection by PCR could be up to 24–32% [42,44]. Some investigators have proposed that: (i) different HPV genotyping kits cannot fully identify all subtypes of HPV; (ii) the diagnostic efficacy of IHC staining p16 across countries was variable; (iii) p16 overexpression may be related to an Rb dysfunction, but Rb dysfunction may not be related to HPV infection; (iv) tumor heterogeneity or sampling bias [45,46] might cause a discrepancy between tissue p16 levels and HPV genotyping results [44]. In a meta-analysis involving 2963 patients, the IHC staining p16 expression level was more consistent with that observed during the in-situ hybridization test. It could prove to be prognostically more valuable in patients with pharyngeal cancer [45]. Our findings support that p16 expression analysis remains a cost-effective method for predicting HPV infections in daily clinical practice and makes it feasible to detect p16 expression in CTCs.

One of the most exciting findings of this study was that we found no positive HPV infection in CTC samples. We have several possible explanations for our findings. First, HPV-positive CTCs, and not p16-positive CTCs, might not intravasate into the bloodstream. Dok et al. (2017) have demonstrated that different dissemination patterns were observed in HPV-negative and HPV-positive HNSCCs because of the dual role of p16 [47]. Though p16 might impair angiogenesis, it promotes lymphatic vessel formation in patients with HPV-positive head and neck cancer [47], which might explain why patients with HPV-positive oropharyngeal cancer receive a good prognosis [48]. These findings might also explain why HPV-positive CTCs were rarely detected in the present study.

Taken together, p16-positive CTCs could provide a new risk stratification tool for diagnosis and enable us to monitor p16 expression and CTCs after curative therapy dynamically. With reference to the p16-positivity of CTCs, the de-escalation strategy in selected patients with baseline p16-positive CTCs could reduce the intensity of anticancer treatment and prevent the unnecessary physical, mental, and economic damage resulting from treatments. More importantly, a non-invasive test based on the p16 status in CTCs was able to predict the prognosis (PFS and OS) in pharyngeal cancer patients receiving CCRT. Therefore, the use of the non-invasive test could be an add-on prognostic strategy when the tissue specimen is unavailable or serial tests are required.

5. Limitations

The limitations of the study need to be addressed before our findings can be used in further studies or clinical practice. First, after excluding those unfit for final analysis, the sample size was relatively small ($n = 41$), which could explain why tissue p16 expression levels were not correlated to survival in the cohort. Second, the study included patients who were eventually diagnosed with hypopharyngeal cancer. This might result in some confusion during survival analysis because patients with p16-positive hypopharyngeal cancer have a different prognosis from those with p16-positive oropharyngeal cancer. The p16 expression level was found to be poorly correlated to HPV infections in patients with non-oropharyngeal cancer [12,49,50]. Although some investigators found that patients with HPV DNA and p16-positive hypopharyngeal cancer exhibited better clinical outcomes, as compared to those of patients with other types of HPV-unrelated hypopharyngeal cancers [51,52], the current consensus is that there is no clear correlation between p16 expression and survival in non-oropharyngeal cancer patients. Our findings show that, even if the data for some patients with hypopharyngeal cancer was mixed up, p16-positive CTCs and p16-negative CTCs could still play a positive role in prognosis. Third, the detection rate of HPV infection in CTCs was zero. This limitation needs to be investigated further via the analysis of HPV biology in the circulation of HNSCC patients.

6. Conclusions

The p16-positive and p16-negative CTCs could serve as prognostic markers for pharyngeal cancer patients receiving CCRT. A liquid biopsy might help clinicians to perform risk stratification before curative therapy and play a role in de-escalation trials in the future.

Supplementary Materials: The following are available online at https://www.mdpi.com/article/10.3390/jpm11111156/s1, Table S1: the limits of human papillomavirus genotyping detection on blood samples.

Author Contributions: Conceptualization, P.-H.C., H.-M.W., Y.-C.K. and J.C.-H.H.; methodology, L.-Y.L., C.-J.L., S.H-C Lin, H.-C.K., S.H.-C.L., T.M.-H.W. and C.-L.H.; software, C.-J.L. and J.C.-H.H.; validation, C.-T.L. and P.-W.H.; formal analysis, P.-H.C.; investigation, Y.-C.K.; resources, C.-T.L., S.H.-C.L. and C.-J.L.; data curation, P.-W.H.; writing—original draft preparation, P.-H.C.; writing—review and editing, H.-M.W.; funding acquisition, P.-H.C. and J.C.-H.H. All authors have read and agreed to the published version of the manuscript.

Funding: This research was funded by the Ministry of Science and Technology of Taiwan (MOST-109-2628-B-182A-001- to Jason CH Hsieh). Also, Chang Gung Memorial Hospital grants CMRPVVK0091 and CMRPVVL0021 (to Jason CH Hsieh), CMRPG2K0271 (to PH Chang).

Institutional Review Board Statement: The study was conducted according to the guidelines of the Declaration of Helsinki, and approved by the Institutional Review Board of Chang Gung Memorial Hospital (with approval ID of 104-2620B [date of approval:14-May-2015], approval ID of 103-7795B [date of approval:11 March 2015]; and approval ID of 201700867B0 [date of approval: 20 June 2017).

Informed Consent Statement: Written informed consent has been obtained from the patient(s) to publish this paper.

Data Availability Statement: All the data are available on demand.

Acknowledgments: The study team thank all the subjects who participated the trials.

Conflicts of Interest: The authors declare no conflict of interest. The funders had no role in the design of the study; in the collection, analyses, or interpretation of data; in the writing of the manuscript, or in the decision to publish the results.

Abbreviations

Full name	Abbreviations
head and neck squamous cell carcinomas	HNSCC
human papillomavirus	HPV
circulating tumor cells	CTCs
confidence interval	CI
concurrent chemoradiotherapy	CCRT
American Joint Committee on Cancer	AJCC
response evaluation criteria in solid tumors	RECIST
complete remission	CR
partial response	PR
Stable disease	SD
progressive disease	PD
Immunohistochemistry staining	IHC
Progression-free survival	PFS
Overall survival	OS
Eastern Cooperative Oncology Group performance status	ECOG PS

References

1. McKaig, R.G.; Baric, R.S.; Olshan, A.F. Human papillomavirus and head and neck cancer: Epidemiology and molecular biology. *Head Neck* **1998**, *20*, 250–265. [CrossRef]
2. Mork, J.; Lie, A.K.; Glattre, E.; Clark, S.; Hallmans, G.; Jellum, E.; Koskela, P.; Møller, B.; Pukkala, E.; Schiller, J.T. Human papillomavirus infection as a risk factor for squamous-cell carcinoma of the head and neck. *N. Engl. J. Med.* **2001**, *344*, 1125–1131. [CrossRef] [PubMed]
3. Nelson, H.H.; Pawlita, M.; Michaud, D.S.; McClean, M.; Langevin, S.M.; Eliot, M.N.; Kelsey, K.T. Immune Response to HPV16 E6 and E7 Proteins and Patient Outcomes in Head and Neck Cancer. *JAMA Oncol.* **2017**, *3*, 178–185. [CrossRef]
4. Ferreira, C.d.C.; Dufloth, R.; de Carvalho, A.C.; Reis, R.M.; Santana, I.; Carvalho, R.S.; Gama, R.R. Correlation of p16 immunohistochemistry with clinical and epidemiological features in oropharyngeal squamous-cell carcinoma. *PLoS ONE* **2021**, *16*, e0253418. [CrossRef]
5. Fakhry, C.; Gillison, M.L. Clinical implications of human papillomavirus in head and neck cancers. *J. Clin. Oncol.* **2006**, *24*, 2606–2611. [CrossRef] [PubMed]
6. Burbure, N.; Handorf, E.; Ridge, J.A.; Bauman, J.; Liu, J.C.; Giri, A.; Galloway, T.J. Prognostic significance of human papillomavirus status and treatment modality in hypopharyngeal cancer. *Head Neck* **2021**, *43*, 3042–3052. [CrossRef]
7. Saito, Y.; Hayashi, R.; Iida, Y.; Mizumachi, T.; Fujii, T.; Matsumoto, F.; Beppu, T.; Yoshida, M.; Shinomiya, H.; Kamiyama, R. Optimization of therapeutic strategy for p16-positive oropharyngeal squamous cell carcinoma: Multi-institutional observational study based on the national Head and Neck Cancer Registry of Japan. *Cancer* **2020**, *126*, 4177–4187. [CrossRef]
8. Patel, R.R.; Ludmir, E.B.; Augustyn, A.; Zaorsky, N.G.; Lehrer, E.J.; Ryali, R.; Trifiletti, D.M.; Adeberg, S.; Amini, A.; Verma, V. De-intensification of therapy in human papillomavirus associated oropharyngeal cancer: A systematic review of prospective trials. *Oral Oncol.* **2020**, *103*, 104608. [CrossRef]
9. Iganej, S.; Beard, B.W.; Chen, J.; Buchschacher, G.L., Jr.; Abdalla, I.A.; Thompson, L.D.; Bhattasali, O. Triweekly carboplatin as a potential de-intensification agent in concurrent chemoradiation for early-stage HPV-associated oropharyngeal cancer. *Oral Oncol.* **2019**, *97*, 18–22. [CrossRef]
10. Cohen, E.; Coviello, C.; Menaker, S.; Martinez-Duarte, E.; Gomez, C.; Lo, K.; Kerr, D.; Franzmann, E.; Leibowitz, J.; Sargi, Z. P16 and human papillomavirus in sinonasal squamous cell carcinoma. *Head Neck* **2020**, *42*, 2021–2029. [CrossRef]
11. Shinn, J.R.; Davis, S.J.; Lang-Kuhs, K.A.; Rohde, S.; Wang, X.; Liu, P.; Dupont, W.D.; Plummer Jr, D.; Thorstad, W.L.; Chernock, R.D. Oropharyngeal Squamous Cell Carcinoma With Discordant p16 and HPV mRNA Results: Incidence and Characterization in a Large, Contemporary United States Cohort. *Am. J. Surg. Pathol.* **2021**, *45*, 951–961. [PubMed]
12. Lifsics, A.; Groma, V.; Cistjakovs, M.; Skuja, S.; Deksnis, R.; Murovska, M. Identification of High-Risk Human Papillomavirus DNA, p16, and E6/E7 Oncoproteins in Laryngeal and Hypopharyngeal Squamous Cell Carcinomas. *Viruses* **2021**, *13*, 1008. [CrossRef] [PubMed]
13. Nauta, I.H.; Heideman, D.A.; Brink, A.; van der Steen, B.; Bloemena, E.; Koljenović, S.; Baatenburg de Jong, R.J.; Leemans, C.R.; Brakenhoff, R.H. The unveiled reality of human papillomavirus as risk factor for oral cavity squamous cell carcinoma. *Int. J. cancer* **2021**, *149*, 420–430. [CrossRef] [PubMed]

14. Pfister, D.G.; Spencer, S.; Adelstein, D.; Adkins, D.; Anzai, Y.; Brizel, D.M.; Bruce, J.Y.; Busse, P.M.; Caudell, J.J.; Cmelak, A.J.; et al. Head and Neck Cancers, Version 2.2020, NCCN Clinical Practice Guidelines in Oncology. *J. Natl. Compr. Cancer Netw.* **2020**, *18*, 873–898. [CrossRef] [PubMed]
15. Lydiatt, W.M.; Patel, S.G.; O'Sullivan, B.; Brandwein, M.S.; Ridge, J.A.; Migliacci, J.C.; Loomis, A.M.; Shah, J.P. Head and Neck cancers-major changes in the American Joint Committee on cancer eighth edition cancer staging manual. *CA Cancer J. Clin.* **2017**, *67*, 122–137. [CrossRef]
16. Bryant, A.K.; Sojourner, E.J.; Vitzthum, L.K.; Zakeri, K.; Shen, H.; Nguyen, C.; Murphy, J.D.; Califano, J.A.; Cohen, E.E.; Mell, L.K. Prognostic role of p16 in nonoropharyngeal head and neck cancer. *JNCI J. Natl. Cancer Inst.* **2018**, *110*, 1393–1399. [CrossRef]
17. Mariz, B.; Kowalski, L.P.; William, W.N., Jr.; de Castro, G., Jr.; Chaves, A.L.F.; Santos, M.; de Oliveira, T.B.; Araujo, A.L.D.; Normando, A.G.C.; Ribeiro, A.C.P.; et al. Global prevalence of human papillomavirus-driven oropharyngeal squamous cell carcinoma following the ASCO guidelines: A systematic review and meta-analysis. *Crit. Rev. Oncol. Hematol.* **2020**, *156*, 103116. [CrossRef]
18. Arsa, L.; Siripoon, T.; Trachu, N.; Foyhirun, S.; Pangpunyakulchai, D.; Sanpapant, S.; Jinawath, N.; Pattaranutaporn, P.; Jinawath, A.; Ngamphaiboon, N. Discrepancy in p16 expression in patients with HPV-associated head and neck squamous cell carcinoma in Thailand: Clinical characteristics and survival outcomes. *BMC Cancer* **2021**, *21*, 504. [CrossRef]
19. Spector, M.E.; Farlow, J.L.; Haring, C.T.; Brenner, J.C.; Birkeland, A.C. The potential for liquid biopsies in head and neck cancer. *Discov. Med.* **2018**, *25*, 251.
20. Balachandra, S.; Kusin, S.B.; Lee, R.; Blackwell, J.M.; Tiro, J.A.; Cowell, L.G.; Chiang, C.M.; Wu, S.Y.; Varma, S.; Rivera, E.L. Blood-based biomarkers of human papillomavirus–associated cancers: A systematic review and meta-analysis. *Cancer* **2021**, *127*, 850–864. [CrossRef]
21. Economopoulou, P.; Koutsodontis, G.; Avgeris, M.; Strati, A.; Kroupis, C.; Pateras, I.; Kirodimos, E.; Giotakis, E.; Kotsantis, I.; Maragoudakis, P. HPV16 E6/E 7 expression in circulating tumor cells in oropharyngeal squamous cell cancers: A pilot study. *PLoS ONE* **2019**, *14*, e0215984. [CrossRef]
22. Geraldine, O.; Wang, L.; Zlott, J.; Juwara, L.; Covey, J.M.; Beumer, J.H.; Cristea, M.C.; Newman, E.M.; Koehler, S.; Nieva, J.J. Intravenous 5-fluoro-2′-deoxycytidine administered with tetrahydrouridine increases the proportion of p16-expressing circulating tumor cells in patients with advanced solid tumors. *Cancer Chemother. Pharmacol.* **2020**, *85*, 979–993.
23. Lefevre, A.C.; Pallisgaard, N.; Kronborg, C.; Wind, K.L.; Krag, S.R.P.; Spindler, K.G. The Clinical Value of Measuring Circulating HPV DNA during Chemo-Radiotherapy in Squamous Cell Carcinoma of the Anus. *Cancers* **2021**, *13*, 2451. [CrossRef] [PubMed]
24. Asante, D.-B.; Calapre, L.; Ziman, M.; Meniawy, T.M.; Gray, E.S. Liquid biopsy in ovarian cancer using circulating tumor DNA and cells: Ready for prime time? *Cancer Lett.* **2020**, *468*, 59–71. [CrossRef] [PubMed]
25. McShane, L.M.; Altman, D.G.; Sauerbrei, W.; Taube, S.E.; Gion, M.; Clark, G.M. Reporting recommendations for tumor marker prognostic studies (REMARK). *J. Natl. Cancer Inst.* **2005**, *97*, 1180–1184. [CrossRef] [PubMed]
26. Larsen, C.G.; Gyldenløve, M.; Jensen, D.; Therkildsen, M.; Kiss, K.; Norrild, B.; Konge, L.; Von Buchwald, C. Correlation between human papillomavirus and p16 overexpression in oropharyngeal tumours: A systematic review. *Br. J. Cancer* **2014**, *110*, 1587–1594. [CrossRef] [PubMed]
27. Su, P.-J.; Wu, M.-H.; Wang, H.-M.; Lee, C.-L.; Huang, W.-K.; Wu, C.-E.; Chang, H.-K.; Chao, Y.-K.; Tseng, C.-K.; Chiu, T.-K. Circulating tumour cells as an independent prognostic factor in patients with advanced oesophageal squamous cell carcinoma undergoing chemoradiotherapy. *Sci. Rep.* **2016**, *6*, 1–9. [CrossRef]
28. Hsieh, J.C.; Lin, H.C.; Huang, C.Y.; Hsu, H.L.; Wu, T.M.; Lee, C.L.; Chen, M.C.; Wang, H.M.; Tseng, C.P. Prognostic value of circulating tumor cells with podoplanin expression in patients with locally advanced or metastatic head and neck squamous cell carcinoma. *Head Neck* **2015**, *37*, 1448–1455. [CrossRef]
29. Liao, C.J.; Hsieh, C.H.; Hung, F.C.; Wang, H.M.; Chou, W.P.; Wu, M.H. The Integration of a Three-Dimensional Spheroid Cell Culture Operation in a Circulating Tumor Cell (CTC) Isolation and Purification Process: A Preliminary Study of the Clinical Significance and Prognostic Role of the CTCs Isolated from the Blood Samples of Head and Neck Cancer Patients. *Cancers* **2019**, *11*, 783. [CrossRef]
30. Wu, C.Y.; Fu, J.Y.; Wu, C.F.; Hsieh, M.J.; Liu, Y.H.; Liu, H.P.; Hsieh, J.C.; Peng, Y.T. Malignancy Prediction Capacity and Possible Prediction Model of Circulating Tumor Cells for Suspicious Pulmonary Lesions. *J. Pers. Med.* **2021**, *11*, 444. [CrossRef]
31. Wu, C.-Y.; Lee, C.-L.; Wu, C.-F.; Fu, J.-Y.; Yang, C.-T.; Wen, C.-T.; Liu, Y.-H.; Liu, H.-P.; Hsieh, J.C.-H. Circulating Tumor Cells as a Tool of Minimal Residual Disease Can Predict Lung Cancer Recurrence: A longitudinal, Prospective Trial. *Diagnostics* **2020**, *10*, 144. [CrossRef]
32. Ikoma, D.; Ichikawa, D.; Ueda, Y.; Tani, N.; Tomita, H.; Sai, S.; Kikuchi, S.; Fujiwara, H.; Otsuji, E.; Yamagishi, H. Circulating tumor cells and aberrant methylation as tumor markers in patients with esophageal cancer. *Anticancer. Res.* **2007**, *27*, 535–539.
33. Zhang, X.; Li, H.; Yu, X.; Li, S.; Lei, Z.; Li, C.; Zhang, Q.; Han, Q.; Li, Y.; Zhang, K. Analysis of circulating tumor cells in ovarian cancer and their clinical value as a biomarker. *Cell. Physiol. Biochem.* **2018**, *48*, 1983–1994. [CrossRef]
34. Kulasinghe, A.; Kapeleris, J.; Kimberley, R.; Mattarollo, S.R.; Thompson, E.W.; Thiery, J.P.; Kenny, L.; O'Byrne, K.; Punyadeera, C. The prognostic significance of circulating tumor cells in head and neck and non-small-cell lung cancer. *Cancer Med.* **2018**, *7*, 5910–5919. [CrossRef] [PubMed]
35. Tada, H.; Takahashi, H.; Kuwabara-Yokobori, Y.; Shino, M.; Chikamatsu, K. Molecular profiling of circulating tumor cells predicts clinical outcome in head and neck squamous cell carcinoma. *Oral Oncol.* **2020**, *102*, 104558. [CrossRef] [PubMed]

36. Wang, H.M.; Wu, M.H.; Chang, P.H.; Lin, H.C.; Liao, C.D.; Wu, S.M.; Hung, T.M.; Lin, C.Y.; Chang, T.C.; Tzu-Tsen, Y. The change in circulating tumor cells before and during concurrent chemoradiotherapy is associated with survival in patients with locally advanced head and neck cancer. *Head Neck* **2019**, *41*, 2676–2687. [CrossRef] [PubMed]
37. Garrel, R.; Mazel, M.; Perriard, F.; Vinches, M.; Cayrefourcq, L.; Guigay, J.; Digue, L.; Aubry, K.; Alfonsi, M.; Delord, J.-P. Circulating tumor cells as a prognostic factor in recurrent or metastatic head and neck squamous cell carcinoma: The CIRCUTEC prospective study. *Clin. Chem.* **2019**, *65*, 1267–1275. [CrossRef]
38. Ng, S.P.; Bahig, H.; Wang, J.; Cardenas, C.E.; Lucci, A.; Hall, C.S.; Meas, S.; Sarli, V.N.; Yuan, Y.; Urbauer, D.L. Predicting treatment Response based on Dual assessment of magnetic resonance Imaging kinetics and Circulating Tumor cells in patients with Head and Neck cancer (PREDICT-HN): Matching 'liquid biopsy' and quantitative tumor modeling. *BMC Cancer* **2018**, *18*, 1–8. [CrossRef]
39. Cho, J.K.; Lee, G.J.; Kim, H.D.; Moon, U.Y.; Kim, M.J.; Kim, S.; Baek, K.H.; Jeong, H.S. Differential impact of circulating tumor cells on disease recurrence and survivals in patients with head and neck squamous cell carcinomas: An updated meta-analysis. *PLoS ONE* **2018**, *13*, e0203758. [CrossRef]
40. Hsieh, J.C.; Wang, H.M.; Wu, M.H.; Chang, K.P.; Chang, P.H.; Liao, C.T.; Liau, C.T. Review of emerging biomarkers in head and neck squamous cell carcinoma in the era of immunotherapy and targeted therapy. *Head Neck* **2019**, *41* (Suppl S1), 19–45. [CrossRef] [PubMed]
41. Reimers, N.; Kasper, H.U.; Weissenborn, S.J.; Stutzer, H.; Preuss, S.F.; Hoffmann, T.K.; Speel, E.J.; Dienes, H.P.; Pfister, H.J.; Guntinas-Lichius, O.; et al. Combined analysis of HPV-DNA, p16 and EGFR expression to predict prognosis in oropharyngeal cancer. *Int. J. Cancer* **2007**, *120*, 1731–1738. [CrossRef] [PubMed]
42. Lewis, J.S., Jr.; Thorstad, W.L.; Chernock, R.D.; Haughey, B.H.; Yip, J.H.; Zhang, Q.; El-Mofty, S.K. p16 positive oropharyngeal squamous cell carcinoma: An entity with a favorable prognosis regardless of tumor HPV status. *Am. J. Surg. Pathol.* **2010**, *34*. [CrossRef] [PubMed]
43. Bigelow, E.O.; Seiwert, T.Y.; Fakhry, C. Deintensification of treatment for human papillomavirus-related oropharyngeal cancer: Current state and future directions. *Oral Oncol.* **2020**, *105*, 104652. [CrossRef]
44. Singhi, A.D.; Westra, W.H. Comparison of human papillomavirus in situ hybridization and p16 immunohistochemistry in the detection of human papillomavirus-associated head and neck cancer based on a prospective clinical experience. *Cancer* **2010**, *116*, 2166–2173. [CrossRef] [PubMed]
45. Wang, H.; Zhang, Y.; Bai, W.; Wang, B.; Wei, J.; Ji, R.; Xin, Y.; Dong, L.; Jiang, X. Feasibility of Immunohistochemical p16 Staining in the Diagnosis of Human Papillomavirus Infection in Patients With Squamous Cell Carcinoma of the Head and Neck: A Systematic Review and Meta-Analysis. *Front. Oncol.* **2020**, *10*, 524928. [CrossRef] [PubMed]
46. Joseph, A.W.; D'Souza, G. Epidemiology of human papillomavirus-related head and neck cancer. *Otolaryngol. Clin. North. Am.* **2012**, *45*, 739–764. [CrossRef]
47. Dok, R.; Glorieux, M.; Holacka, K.; Bamps, M.; Nuyts, S. Dual role for p16 in the metastasis process of HPV positive head and neck cancers. *Mol. Cancer* **2017**, *16*, 113. [CrossRef]
48. Ang, K.K.; Harris, J.; Wheeler, R.; Weber, R.; Rosenthal, D.I.; Nguyen-Tan, P.F.; Westra, W.H.; Chung, C.H.; Jordan, R.C.; Lu, C.; et al. Human papillomavirus and survival of patients with oropharyngeal cancer. *N. Engl. J. Med.* **2010**, *363*, 24–35. [CrossRef]
49. Wendt, M.; Romanitan, M.; Nasman, A.; Dalianis, T.; Hammarstedt, L.; Marklund, L.; Ramqvist, T.; Munck-Wiklund, E. Presence of human papillomaviruses and p16 expression in hypopharyngeal cancer. *Head Neck* **2014**, *36*, 107–112. [CrossRef]
50. Meshman, J.; Wang, P.C.; Chin, R.; John, M.S.; Abemayor, E.; Bhuta, S.; Chen, A.M. Prognostic significance of p16 in squamous cell carcinoma of the larynx and hypopharynx. *Am. J. Otolaryngol.* **2017**, *38*, 31–37. [CrossRef]
51. Dalianis, T.; Grün, N.; Koch, J.; Vlastos, A.; Tertipis, N.; Nordfors, C.; Näsman, A.; Wendt, M.; Romanitan, M.; Bersani, C. Human papillomavirus DNA and p16INK4a expression in hypopharyngeal cancer and in relation to clinical outcome, in Stockholm, Sweden. *Oral Oncol.* **2015**, *51*, 857–861. [CrossRef] [PubMed]
52. Lassen, P.; Schou, M.; Overgaard, J.; Alsner, J. Correlation and prognostic impact of human papilloma virus and p16-expression in advanced hypopharynx and larynx cancer treated with definitive radiotherapy. *Acta Oncol.* **2021**, *60*, 646–648. [CrossRef] [PubMed]

Circulating Tumor Cell Detection during Neoadjuvant Chemotherapy to Predict Early Response in Locally Advanced Oropharyngeal Cancers: A Prospective Pilot Study

Arnaud Gauthier [1,2,†], Pierre Philouze [1,3,†], Alexandra Lauret [1], Gersende Alphonse [1,2], Céline Malesys [1], Dominique Ardail [1,2], Léa Payen [2], Philippe Céruse [1,3], Anne-Sophie Wozny [1,2,†] and Claire Rodriguez-Lafrasse [1,2,*,†]

1 Laboratory of Cellular and Molecular Radiobiology, UMR CNRS5822/IP2I, Lyon-Sud Medical School, Univ Lyon 1, Lyon University, 69921 Oullins, France; arnaud.gauthier@univ-lyon1.fr (A.G.); pierre.philouze@chu-lyon.fr (P.P.); alexandra.lauret@univ-lyon1.fr (A.L.); gersende.alphonse@univ-lyon1.fr (G.A.); celine.malesys@univ-lyon1.fr (C.M.); dominique.ardail@univ-lyon1.fr (D.A.); philippe.ceruse@chu-lyon.fr (P.C.); anne-sophie.wozny@univ-lyon1.fr (A.-S.W.)
2 Department of Biochemistry and Molecular Biology, Lyon-Sud Hospital, Hospices Civils de Lyon, 69310 Pierre-Bénite, France; lea.payen-gay@chu-lyon.fr
3 Department of OtoRhinoLaryngology Head and Neck Surgery, Croix-Rousse Hospital, Hospices Civils de Lyon, 69004 Lyon, France
* Correspondence: claire.rodriguez-lafrasse@univ-lyon1.fr; Tel.: +33-4-26-23-59-65
† These authors contributed equally to this work.

Abstract: Patients with locally advanced oropharyngeal carcinoma treated with neoadjuvant chemotherapy are reassessed both radiologically and clinically to adapt their treatment after the first cycle. However, some responders show early tumor progression after adjuvant radiotherapy. This cohort study evaluated circulating tumor cells (CTCs) from a population of locally advanced oropharyngeal carcinoma patients treated with docetaxel, cisplatin, and 5-fluorouracil (DCF) induction chemotherapy or DCF with a modified dose and fractioned administration. The counts and phenotypes of CTCs were assessed at baseline and at day 21 of treatment, after isolation using the RosetteSep^{TM} technique based on negative enrichment. At baseline, 6 out of 21 patients had CTCs (28.6%). On day 21, 5 out of 11 patients had CTCs (41.6%). There was no significant difference in the overall and progression-free survival between patients with or without CTCs at baseline ($p = 0.44$ and 0.78) or day 21 ($p = 0.88$ and 0.5). Out of the 11 patients tested at day 21, 4 had a positive variation of CTCs (33%). Patients with a positive variation of CTCs display a lower overall survival. Our findings suggest that the variation in the number of CTCs would be a better guide to the management of treatment, with possible early changes in treatment strategy.

Keywords: circulating tumor cells; predictive biomarker; HNSCC

1. Introduction

Head and neck squamous cell cancer (HNSCC) is the sixth most common cancer worldwide [1]. Current therapeutic strategies are multimodal and use either a combination of surgery followed by radiochemotherapy, neoadjuvant chemotherapy followed by radiotherapy, or radiochemotherapy, depending on the tumor location and stage. These strategies have not demonstrated any superiority to date, and locoregional recurrences and/or metastases lead to therapeutic failure with a less than 50% overall survival (OS) at 5 years. Moreover, the onset of metastasis within 12 months following diagnosis is responsible for nearly 88% of deaths [2].

Circulating tumor cells (CTCs) represent a heterogeneous population with wide plasticity and include epithelial cancer cells, cells in the process of epithelial-to-mesenchymal transition, mesenchymal cells, and cancer stem cells (CSCs). CSCs are demonstrated to be responsible for self-renewal and tumor growth in HNSCC [3,4]. In addition, the number of circulating CTCs is correlated with poor prognoses in lung, colorectal, prostate, and breast cancers [5–9].

In HNSCC, and in oropharyngeal carcinoma, few studies have explored the role of CTCs before, during, or after treatments. Despite CTCs having been found in 18% to 33% of HNSCC patients [10,11], their impact on progression-free survival (PFS) and OS remains to be established. Some studies have reported that the presence of CTCs correlates with a poor prognosis [11–13], i.e., lower PFS and OS; however, other studies did not find any correlation [10,14]. Therefore, further investigation is needed to determine if the identification and numeration of CTCs could help with the management of patients with oropharyngeal carcinoma.

Patients with locally advanced oropharyngeal carcinoma who are treated with neoadjuvant chemotherapy are currently reassessed both radiologically and clinically after the first cycle of chemotherapy. Patients with a response of more than a 50% response are referred for adjuvant radiotherapy with or without surgery, whereas patients with less than a 50% response or with tumor progression are directed to a palliative chemotherapeutic strategy.

Currently, except for HPV-driven oropharyngeal carcinoma [15,16], there are no biological markers to identify the response to neoadjuvant chemotherapy upon reassessment. Furthermore, some responders at clinical and radiological re-evaluation show early tumor progression after the end of adjuvant radiotherapy.

In this prospective pilot study, 21 patients with locally advanced oropharyngeal carcinoma treated with neoadjuvant chemotherapy were enrolled, and the evolution of CTCs during treatment was explored, both in terms of their cell number and morphological characteristics. Our primary objective was to define whether the number of CTCs before and after the first cycle of neoadjuvant chemotherapy could be a predictive biomarker of therapeutic response. The secondary objective was to determine whether a variation in the number of CTCs between the beginning and the end of the first cycle of neoadjuvant chemotherapy could be predictive of survival.

2. Materials and Methods

Study population and sample collection. Twenty-one patients displaying a histologically proven squamous cell carcinoma from the oropharynx were recruited between May 2016 and November 2018 from the Head and Neck Department at Croix-Rousse Hospital (Lyon, France). Tumors were not resectable. The HPV status was obtained by PCR on a tissue biopsy analyzed by the HPV DNA test Clinical® Array Human Papillomavirus Genomica (R-Biopharm, Lyon, France). All HPV-positive patients were p16 positive, except one who was p26 positive. According to our therapeutic protocol, the patients were treated with either neoadjuvant docetaxel, cisplatin, and 5-fluorouracil (DCF) or with docetaxel, cisplatin, 5-fluorouracil, with modified dose and fractioned administration (mDCF) chemotherapy. Some patients later received adjuvant radiotherapy. This study (NCT02714920) was conducted in compliance with French legislation and was approved by the local independent ethics committee in November 2015. Written consent was obtained from each patient. The patients were followed up for 24 months, and the last follow-up was conducted in November 2020. Blood samples were collected from every patient at baseline, i.e., before treatment. On day 21, blood samples were collected only from patients who received DCF chemotherapy due to the schedule of the chemotherapy administration. Blood from one patient who received mDCF chemotherapy was also collected on day 21. The patients' characteristics are summarized in Table 1.

Table 1. Patients' characteristics.

Patients' Characteristics	Number of Patients	n CTC+ at Baseline	p-Value
Age (years)			
<65	16 (76.2%)	5	1
>65	5 (23.8%)	1	
Gender			
Male	18 (85.7%)	5	1
Female	3 (14.3%)	1	
T stage			
T2	2 (9.5%)	0	0.57
T3	8 (38.1%)	3	
T4	11 (52.4%)	3	
N stage			
N0	3 (14.3%)	2	0.18
N+	18 (85.7%)	4	
Tobacco			
Exposed	19 (90.5%)	6	1
None	2 (9.5%)	0	
Alcohol			
Exposed	7 (33.3%)	3	0.29
None	14 (61.9%)	3	
HPV status			
Positive	5 (23.8%)	0	
Negative	13 (61.9%)	4	0.12
Unknown	3 (14.3%)	2	

n CTC+ at Baseline: number of patients with CTC at baseline.

Classification of patients and change in CTCs. Patients were clinically stratified into early responders or early nonresponders according to their clinical response at 4 months follow-up. The group of responders corresponded to clinical and radiological RECIST remission, while the nonresponders corresponded to disease progression. Changes in CTC number were classified into two categories, positive variation and no positive variation (i.e., stable and negative variation). An absence of CTCs at baseline compared with a presence of CTCs at day 21 was considered as a positive variation. The presence of CTCs at baseline compared with an absence of CTC at day 21 was considered as a negative variation. An absence of CTC at baseline and day 21 was considered as stable. A positive CTC count at baseline and day 21 with an increase in CTC was considered as a positive variation, whereas a decrease was considered as a negative variation.

Isolation of CTCs by RosetteSep™. Blood samples were collected in two EDTA tubes of 10 mL and centrifuged in a 50 mL tube at 1200× g for 10 min at room temperature. Plasma was then replaced by phosphate-buffered saline (PBS) at an equivalent volume without mixing. A small volume of residual plasma was left on the surface of the red blood cells to avoid collecting CTCs at the interface. The sample was then incubated with the RosetteSep reagent (Stemcell Technologies, Vancouver, Canada) [17] at 50 µL/mL for 30 min at room temperature under slight agitation. Thereafter, the cellular separation was achieved in SepMate 50 mL tubes containing 17 mL of Lymphoprep density gradient medium (Stemcell Technologies). Samples were centrifuged at 1200× g for 20 min at room temperature. The upper phase was transferred to a 50 mL tube and reconstituted to 50 mL with PBS with 2% fetal bovine serum (FBS). After centrifugation at 1200× g for 10 min at room temperature, the cell pellet was rinsed twice with 50 mL of PBS and 2% FBS. Finally, the enriched cellular suspension was resuspended in 2 mL of PreservCyt (Hologic, Marlborough, MA, USA) and transferred to a cryotube for storage at 4 °C until analysis.

Detection of CTCs. The cells stored in the PreservCyt were cytospined on a slide at 18 g for 4 min at room temperature. A droplet of blocking solution (100 µL; PBS with

0.1% bovine serum albumin (BSA), 1% FBS) was dropped onto the slide. After 30 min incubation at room temperature, blocking solution was replaced by 100 µL antibody solution (PBS, 0.1% BSA, 1% FBS), 1:100 anti-cytokeratin-FITC (Miltenyi Biotec, 130-080-101), 1:100 anti-CD44-APC (Miltenyi Biotec, 130-113-338), 1:100 anti-CD45-PE (Miltenyi Biotec, 130-110-632), 1:100 anti-N-Cadherin-Cy5 (Abcam, Cambridge, UK) and 0.1 mg/mL of 4′,6-diamidino-2-phenylindole (DAPI) (Sigma-Aldrich, St Louis, MO, USA). The slide was then maintained at 4 °C overnight. The next day, the solution containing antibodies was removed and 100 µL of 1:1000 AlexaFluor 594 anti-mouse antibody was added and incubated for one hour in the dark. Then, after four washes with 200 µL PBS followed by 5 min of drying at room temperature, the slide was mounted under a coverslip with Fluoromount (Sigma-Aldrich) and polymerized overnight at room temperature before analysis by fluorescence microscopy (Microscope Axio Imager Z2, Zeiss, Marly-Le-Roi, France; Metafer, MetaSystems, Altlussheim, Germany). CTCs were defined based on their morphology and specific staining. Expression of N-cadherin is associated with mesenchymal phenotype, cytokeratin is associated with epithelial phenotype, and CD44 is associated with HNSCC stem cell phenotype. Antibody specificity was validated on SQ20-CD44+ cells [18], a subpopulation of CSCs isolated from the HNSCC cell line, SQ20B, that expresses N-cadherin, cytokeratin, and CD44 (Figure 1A). Morphological studies enabled elimination of apoptotic bodies, cell debris, and neutrophilic polynuclear cells. Moreover, CTC is a cell with a round nucleus and a diameter around 20 µm without a real cut-off that can be defined. The use of anti-CD45 antibody specific for leukocytes enabled CD45-free cells to be the focus of our analysis. The combination of both evaluations allowed us to eliminate this population considered as false positive in contrast to the other cells considered as CTCs. CTCs could be positive for one marker and for DAPI and associated with the corresponding phenotype, or positive only for DAPI with an undefined phenotype. Representative images of immunostaining of CTCs are presented in Figure 1B and 1C. Two slides per patient per time point were analyzed. Results were reported as the number of CTCs identified per mL of whole blood. When more than three CTCs were aggregated, they were considered as a cluster, and each cell was counted.

Statistical analysis. Statistical analyses were performed using GraphPad Prism (v.8.4.2, GraphPad Software, San Diego, CA, USA). The association between CTC count and clinical characteristics described in Table 1 was evaluated using Fisher's exact test. PFS and OS were assessed in the groups stratified according to their clinical response to treatment at 24 months, and the association between changes in CTC count, treatment response, and prognosis were evaluated. Survival rates were assessed using the Kaplan–Meier method. The minimum level of significance was set at $p < 0.05$.

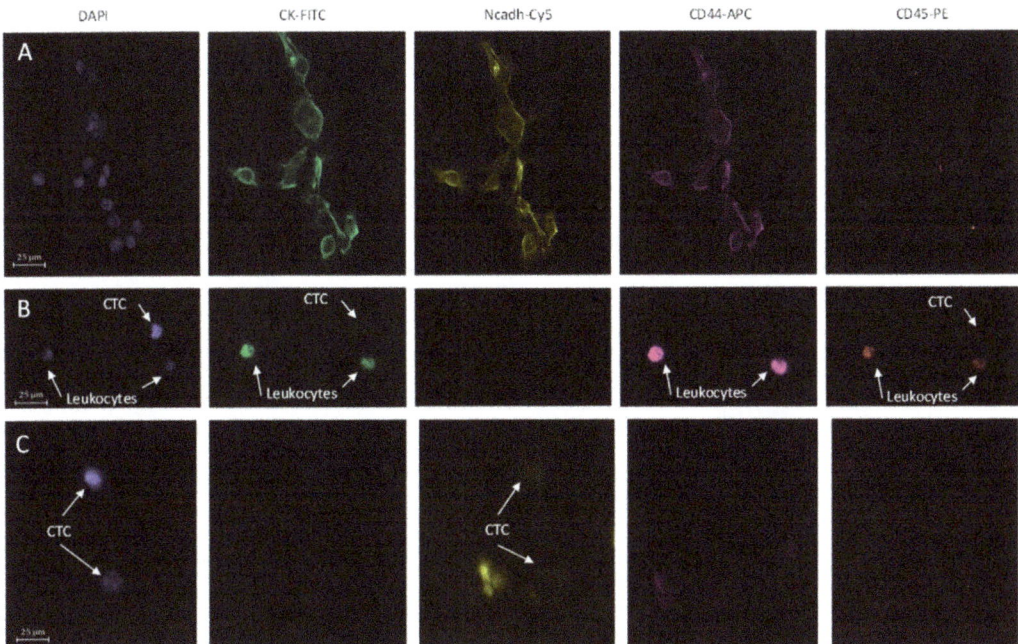

Figure 1. Immunofluorescence stainings (X20): (**A**) SQ20B-CD44+ cancer stem cells expressing cytokeratin, N-cadherin and CD44; (**B**) Representative CTC observed in patient #12, expressing cytokeratin at baseline; (**C**) Representative CTCs observed in patient #1, expressing N-Cadherin at baseline.

3. Results

3.1. Counting and Characterization of CTCs

Before any treatment, 6 out of 21 patients were found to have CTCs (28.6%) (Table 2). The minimum, maximum, and median CTC counts were 0.07 CTC/mL, 3.34 CTC/mL, and 0.22 CTC/mL, respectively. No significant associations were observed between the number of CTCs at baseline and clinical characteristics of the patients, including sex, age, clinical stage (tumor and nodes), tobacco use, alcohol intake, and human papillomavirus status (Table 1). On day 21, before the second course of DCF treatment, CTCs were collected from 11 patients who received DCF chemotherapy, as well as from 1 patient who received mDCF chemotherapy. Among the 12 patients from whom blood was collected at day 21, 5 (41.6%) had CTCs (Table 2).

The characterization of CTCs at baseline identified two patients with an epithelial CTC phenotype (cytokeratin expression), three patients with a mesenchymal CTC phenotype (N-cadherin), but no patients displaying stem cell CD44 expression. The six patients with CTCs also exhibited CTCs with undefined phenotypes. The characterization of CTCs at day 21 identified one patient with an epithelial CTC phenotype, two patients with a mesenchymal CTC phenotype, and one patient with a stem cell phenotype. The five patients with CTCs also exhibited CTCs with undefined phenotypes, including two patients with clusters (Table 2). No significant associations were observed between CTC phenotype at baseline or day 21 and OS or PFS (data not shown).

Table 2. Identification and characterization of CTC patients according to treatment protocol.

	Patients	Baseline (Number of Cells)					Day 21 (Number of Cells)					Variation
		Epithelial	Mesenchymal	Stem Cell	Undefined	CTC/mL	Epithelial	Mesenchymal	Stem Cell	Undefined	CTC/mL	
DCF	#1	0	1	0	2	0.315	0	0	0	0	0	-
	#2	0	1	0	2	0.255	1	1	0	20 *	1.505	+
	#3	0	0	0	1	0.190	0	0	0	6	0.315	+
	#4	0	0	0	0	0	0	0	0	0	0	=
	#5	0	0	0	0	0	0	3	1	15	1.190	+
	#6	0	0	0	0	0	0	0	0	0	0	=
	#7	0	0	0	0	0	0	0	0	0	0	=
	#14°	0	0	0	0	0	0	0	0	38 *	2.375	+
	#15°	0	0	0	3	0.880	0	0	0	1	0.055	-
	#16°	0	0	0	0	0	0	0	0	0	0	=
	#17°	0	0	0	0	0	0	0	0	0	0	=
	#18°	0	0	0	0	0	/	/	/	/	/	NA
mDCF	#8	0	0	0	0	0	/	/	/	/	/	NA
	#9	0	0	0	0	0	/	/	/	/	/	NA
	#10	0	0	0	0	0	0	0	0	0	0	=
	#11	0	0	0	0	0	/	/	/	/	/	NA
	#12	1	0	0	1	0.125	/	/	/	/	/	NA
	#13	0	0	0	0	0	/	/	/	/	/	NA
	#19°	0	0	0	0	0	/	/	/	/	/	NA
	#20°	1	1	0	8	3.335	/	/	/	/	/	NA
	#21°	0	0	0	0	0	/	/	/	/	/	NA

DCF: Docetaxel, Cisplatine, 5-Fluorouracil. mDCF: DCF modified (dose adapted). Variation: increase (+), decrease (-) or stable (=) variation in CTCs between baseline and day 21. *: cluster of CTCs. °: early nonresponder patient. /: unmeasured. NA: not applicable.

3.2. Association between the Presence of CTCs at Baseline or Day21 and the Survival Rate

Of the 21 patients, 8 were considered as early nonresponders at 4 months follow-up. The PFS was significantly lower for early nonresponders compared to responders ($p < 0.0001$, hazard ratio (HR) 30.4; confidence interval (CI), 6.6–139.3), and the OS was not statistically different ($p = 0.11$) (Figure 2A). Regarding CTCs, 4 early responders (patients #1, #2, #3, and #12) and 2 early nonresponders (patients #15 and #20) had CTCs at baseline. On day 21, 3 early responders (patient #2, #3, and #5) and 2 early nonresponders (patients #14 and #15) had CTCs (Table 2). Regardless of responder classification, there was no significant difference in the OS and PFS between patients with or without CTCs at baseline ($p = 0.44$ and 0.78, respectively) or day 21 ($p = 0.88$ and 0.5, respectively) (Figure 2B,C).

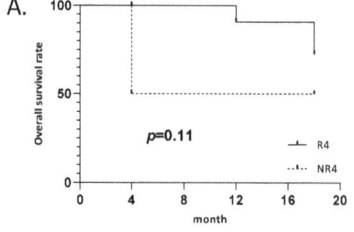 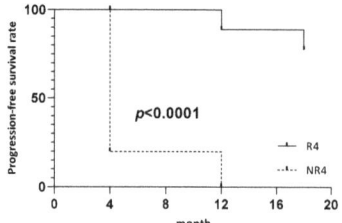

Figure 2. Cont.

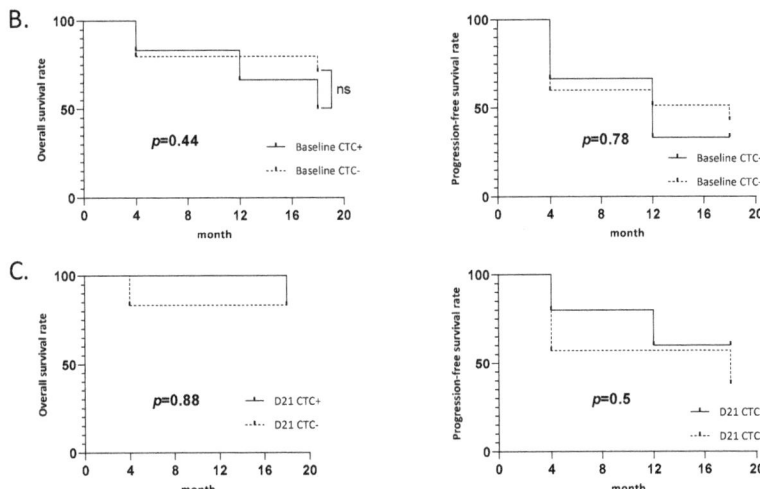

Figure 2. Survival curves: (**A**) Overall survival and progression-free survival of early responder. (R4: responder at 4 months post treatment, NR4: nonresponder at 4 months post treatment). (**B**) Overall survival and progression-free survival of patients depending on CTC at baseline. (**C**) Overall survival and progression-free survival of patients depending on CTC at day 21 (D21).

3.3. Variation in the Number of CTCs between Baseline and D21 and the Survival Rate

Of the 11 DCF patients who had a blood sample collected at day 21, 4 patients (33%) (patients #2, #3, #5, and #14) had a positive variation of CTCs (an increase in CTCs between baseline and day 21) and 2 had clusters at day 21 (patient #3, early responder and patient #14, early nonresponder). Two patients (16.7%) (patients #1 and #15) had a negative variation (decreased CTCs between baseline and day 21). There was no significant association between a positive variation in the CTC number and CTC phenotype or cluster. Despite the absence of a significant difference in the OS and PFS between patients with a positive variation and negative variation in CTCs (p = 0.48 and 0.75, respectively) (Figure 3), we observed a clear tendency in patients with a positive variation to have a lower OS.

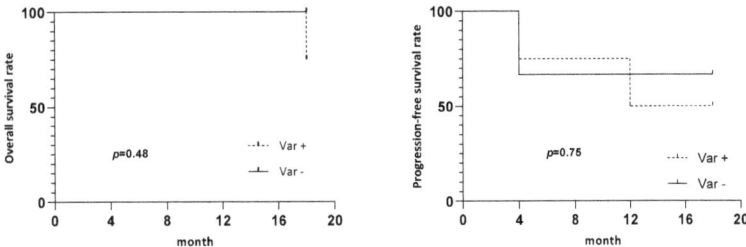

Figure 3. Survival curves. Overall survival and progression-free survival of patients depending on CTC's variation between baseline and day 21. Var +: positive variation of CTCs, Var -: no positive variation.

4. Discussion

We conducted a prospective pilot study to explore the potential role of CTCs during neoadjuvant chemotherapy to predict PFS and OS in patients with oropharyngeal cancer. We showed that there was no significant difference in PFS or OS between patients with and without CTCs at either baseline or day 21, but we observed a variation in the number of detected CTCs in some patients during the first 3 weeks of treatment. Despite the positive variation in CTCs, meaning an increase in CTCs during chemotherapy treatment, the change was not statistically significant because of the relatively low number of patients. Even so, a clear tendency to poor prognosis emerged from the results.

CTCs have already been evaluated in various cancers, and some studies showed that increased CTCs are correlated with a poorer prognosis [19,20]. A meta-analysis of CTCs in breast cancer indicated that CTC-positive patients (≥ 5 CTCs/7.5 mL) displayed an increased risk of both tumor progression and death [21]. In a recent review concerning lung cancer, non-small cell lung cancer, and small cell lung cancer, it was shown that an increase in CTCs correlated with a poor prognosis [22]. In a study of 216 patients with ovarian cancer, patients exhibiting ≥ 2 CTCs at baseline presented a decreased PFS and OS [23,24].

For HNSCC, few studies explored the role of CTCs before, during, and after treatment, and unfortunately, they are based on small cohorts and different methods for both isolation and counting CTCs. In a cohort of 73 patients with hypo- and oropharynx tumors, Buglione et al. demonstrated that a partial or complete response to chemotherapy was associated with the absence or disappearance of CTCs during treatment. In addition, a decrease in the number of CTCs or their absence during treatment also appeared to be associated with non-progressive disease. Unfortunately, in this study, the authors examined different anatomic subsites of cancer and different histopathological types of cancers such as squamous cell carcinoma or sinonasal undifferentiated carcinoma, which have different clinical outcomes [25]. Another study showed that the presence of CTCs expressing markers such as, cytokeratin, vimentin, EGFR, CD44, or N-Cadherin was correlated with a poor prognosis [26]. In a cohort of 25 patients with oropharynx cancer treated with neoadjuvant chemotherapy, Inhestern et al. showed that there was no correlation between the presence of CTCs and age, sex, tumor site, stage, or lymph node involvement. Furthermore, a high number of CTCs at baseline and after the treatment was proposed as a prognostic marker for OS [12], but an analysis of the correlation between the variations in the numbers of CTCs and PFS and OS was not addressed in this study.

The results concerning the number of CTCs often vary between studies due to the techniques used. The previously cited studies used CellSearch and flow cytometric assays based on a positive epithelial cell adhesion molecule (EpCAM) expression to isolate CTCs. Currently, only the CellSearch technique from Veridex has received approval from the Food and Drug Administration for clinical use in colorectal, lung, prostate, and breast cancers. The analysis is based on an immunological method that counts CD45-, cytokeratin+, and EpCAM+ cells. However, the EpCAM protein is an epithelial marker normally found in most carcinomas, but is weakly expressed in HNSCC tumors [27]. This explains why few experiments that used this device mention the presence of CTCs in patients with HNSCC.

Three isolation techniques were compared by Kulasinghe et al. in patients with advanced HNSCC: the CellSearch system, ScreenCell (microfiltration device), and RosetteSep (negative enrichment). They found that CellSearch detected CTCs in 8 out of 43 cases (18.6%), ScreenCell in 13 out of 28 cases (46.4%), and RosetteSep in 16 out of 25 cases (64.0%), the latter being able to also detect CTC clusters [28]. These results confirm that RosetteSep is an appropriate tool for the isolation of CTCs in HNSCC.

Concerning the kinetics of CTCs during treatment, the French multicenter CIRCUTEC study focused on patients with nonoperable or metastatic tumor relapse. Sixty-five patients treated with cetuximab chemotherapy were included. CTCs were isolated and detected by three methods: CellSearch, EPISPOT, and flow cytometry. Patients were tested at baseline and on days 7 and 21. Median PFS time was significantly lower in patients with increasing or stable CTC counts (36/54) from baseline to day 7 with EPISPOT and in patients with one CTC detected with a combination of 2 tests at day 7 [29]. For patients with curative intent, Wang et al. analyzed CTC counts before and during radiochemotherapy treatment in patients with locally advanced HNSCC. CTCs were detected using a negative selection strategy and a flow cytometry protocol. The positive variation in the number of CTCs correlated with lower PFS (and OS) [30].

Our results suggest the use of CTC kinetics during treatment is much more relevant than the detection of CTC levels alone. Our results are encouraging because it is important to develop predictive biomarkers for responses to neoadjuvant chemotherapy. Indeed, these treatments are associated with complications (e.g., hematological, renal, and auditory side effects). Thus, the early identification of nonresponding patients through an analysis of the variation in the number of CTCs may allow an early adjustment of the therapeutic strategy. This would improve survival while limiting the side effects of unnecessary treatments. Moreover, we observed that responding patients at 4 months after the end of treatment had significantly better OS and PFS. Four responder patients at 4 months (patients #2, #9, #12, and #20) showed tumor progression in the following weeks. Patient #2 was a clinical and radiological responder but showed tumor progression and died at 12 months. This patient had CTCs at baseline and a positive variation in CTCs on day 21. The variation in CTCs could not be assessed for the other three patients because they received mDCF treatment. Adding the CTC count to clinical and radiological investigations could help to earlier orientate the management of patients.

5. Conclusions

Our pilot study offers preliminary results that should be consolidated using a larger prospective study. The results suggest that the evaluation of variations in CTCs could be used as a predictive biomarker during treatment, particularly at the time of the morphological and clinical evaluations performed to assess the response to neoadjuvant chemotherapy. When patients show a good response to the clinical and morphological evaluation, they are referred to adjuvant radiotherapy. Unfortunately, some patients will show tumor progression after radiotherapy with worse survival. Thus, the study of the variation in the number of CTCs and their appearance or disappearance would be useful during treatment to better guide management decisions with possible early changes in strategy.

Author Contributions: Conceptualization, P.P., P.C. and C.R.-L.; methodology, P.P., P.C. and C.R.-L.; validation, P.P., P.C. and C.R.-L.; formal analysis, A.G.; investigation, P.P., P.C. and C.R.-L.; resources, P.C. and C.R.-L.; data curation, P.P., A.G., A.L., C.M., G.A., L.P.; writing—original draft preparation, A.G., P.P.; writing—review and editing, D.A., A.-S.W., C.R.-L.; visualization, P.C., C.R.-L.; supervision, P.C., C.R.-L., A.-S.W.; project administration, G.A., C.M., C.R.-L.; funding acquisition, C.R.-L. All authors have read and agreed to the published version of the manuscript.

Funding: This research was funded by Cancéropôle CLARA, Grenoble-Alpes-Métropole, Conseil Régional Rhône-Alpes, Conseil Général du Rhône, Ligue contre le cancer, section de Haute-Savoie and LABEX PRIMES (ANR-11-LABX-0063) of Université de Lyon, within the program "Investissements d'Avenir" (ANR-11-IDEX-0007) operated by the French National Research Agency (ANR).

Institutional Review Board Statement: The study was conducted in accordance with the Declaration of Helsinki and approved by the Ethics Committee of Hospices Civils de Lyon, 2015-47 (protocol code NCT02714920 and date of approval 16 December 2015).

Informed Consent Statement: Informed consent was obtained from all subjects involved in the study.

Data Availability Statement: Pr Rodriguez-Lafrasse had full access to all the data in the study and takes responsibility for the integrity of the data and the accuracy of the data analysis. The data that support the findings of this study are available upon reasonable request to the corresponding author.

Acknowledgments: We also thank the clinical research associates' team and especially Bénédicte Poumaroux and Stéphanie Vicente for their help in setting up the protocol and collecting the samples.

Conflicts of Interest: The authors declare no conflict of interest.

References

1. Ferlay, J.; Soerjomataram, I.; Dikshit, R.; Eser, S.; Mathers, C.; Rebelo, M.; Parkin, D.M.; Forman, D.; Bray, F. Cancer incidence and mortality worldwide: Sources, methods and major patterns in GLOBOCAN 2012. *Int. J. Cancer* **2015**, *136*, E359–E386. [CrossRef] [PubMed]
2. Takes, R.P.; Rinaldo, A.; Silver, C.E.; Haigentz, M.; Woolgar, J.A.; Triantafyllou, A.; Mondin, V.; Paccagnella, D.; de Bree, R.; Shaha, A.R.; et al. Distant metastases from head and neck squamous cell carcinoma. Part I. Basic aspects. *Oral Oncol.* **2012**, *48*, 775–779. [CrossRef] [PubMed]
3. Routray, S.; Mohanty, N. Cancer stem cells accountability in progression of head and neck squamous cell carcinoma: The most recent trends! *Mol. Biol. Int.* **2014**, *2014*, 375325. [CrossRef] [PubMed]
4. Yoshida, G.J.; Saya, H. Molecular pathology underlying the robustness of cancer stem cells. *Regen. Ther.* **2021**, *17*, 38–50. [CrossRef] [PubMed]
5. Lianidou, E.S.; Markou, A.; Strati, A. The Role of CTCs as Tumor Biomarkers. *Adv. Exp. Med. Biol.* **2015**, *867*, 341–367. [CrossRef]
6. Galletti, G.; Portella, L.; Tagawa, S.T.; Kirby, B.J.; Giannakakou, P.; Nanus, D.M. Circulating tumor cells in prostate cancer diagnosis and monitoring: An appraisal of clinical potential. *Mol. Diagn. Ther.* **2014**, *18*, 389–402. [CrossRef]
7. Giordano, A.; Egleston, B.L.; Hajage, D.; Bland, J.; Hortobagyi, G.N.; Reuben, J.M.; Pierga, J.-Y.; Cristofanilli, M.; Bidard, F.-C. Establishment and validation of circulating tumor cell-based prognostic nomograms in first-line metastatic breast cancer patients. *Clin. Cancer Res.* **2013**, *19*, 1596–1602. [CrossRef]
8. Tsai, W.-S.; Chen, J.-S.; Shao, H.-J.; Wu, J.-C.; Lai, J.-M.; Lu, S.-H.; Hung, T.-F.; Chiu, Y.-C.; You, J.-F.; Hsieh, P.-S.; et al. Circulating Tumor Cell Count Correlates with Colorectal Neoplasm Progression and Is a Prognostic Marker for Distant Metastasis in Non-Metastatic Patients. *Sci. Rep.* **2016**, *6*, 24517. [CrossRef]
9. Zou, K.; Yang, S.; Zheng, L.; Wang, S.; Xiong, B. Prognostic Role of the Circulating Tumor Cells Detected by Cytological Methods in Gastric Cancer: A Meta-Analysis. *Biomed. Res. Int.* **2016**, *2016*, 2765464. [CrossRef]
10. Bozec, A.; Ilie, M.; Dassonville, O.; Long, E.; Poissonnet, G.; Santini, J.; Chamorey, E.; Ettaiche, M.; Chauvière, D.; Peyrade, F.; et al. Significance of circulating tumor cell detection using the CellSearch system in patients with locally advanced head and neck squamous cell carcinoma. *Eur. Arch. Oto-Rhino-Laryngol.* **2013**, *270*, 2745–2749. [CrossRef]
11. Grisanti, S.; Almici, C.; Consoli, F.; Buglione, M.; Verardi, R.; Bolzoni-Villaret, A.; Bianchetti, A.; Ciccarese, C.; Mangoni, M.; Ferrari, L.; et al. Circulating tumor cells in patients with recurrent or metastatic head and neck carcinoma: Prognostic and predictive significance. *PLoS ONE* **2014**, *9*, e103918. [CrossRef] [PubMed]
12. Inhestern, J.; Oertel, K.; Stemmann, V.; Schmalenberg, H.; Dietz, A.; Rotter, N.; Veit, J.; Görner, M.; Sudhoff, H.; Junghanß, C.; et al. Prognostic Role of Circulating Tumor Cells during Induction Chemotherapy Followed by Curative Surgery Combined with Postoperative Radiotherapy in Patients with Locally Advanced Oral and Oropharyngeal Squamous Cell Cancer. *PLoS ONE* **2015**, *10*, e0132901. [CrossRef] [PubMed]
13. Jatana, K.R.; Balasubramanian, P.; Lang, J.C.; Yang, L.; Jatana, C.A.; White, E.; Agrawal, A.; Ozer, E.; Schuller, D.E.; Teknos, T.N.; et al. Significance of circulating tumor cells in patients with squamous cell carcinoma of the head and neck: Initial results. *Arch. Otolaryngol. Head Neck Surg.* **2010**, *136*, 1274–1279. [CrossRef]
14. Winter, S.C.; Stephenson, S.-A.; Subramaniam, S.K.; Paleri, V.; Ha, K.; Marnane, C.; Krishnan, S.; Rees, G. Long term survival following the detection of circulating tumour cells in head and neck squamous cell carcinoma. *BMC Cancer* **2009**, *9*, 424. [CrossRef] [PubMed]
15. Tang, K.D.; Vasani, S.; Taheri, T.; Walsh, L.J.; Hughes, B.G.M.; Kenny, L.; Punyadeera, C. An Occult HPV-Driven Oropharyngeal Squamous Cell Carcinoma Discovered Through a Saliva Test. *Front. Oncol.* **2020**, *10*, 408. [CrossRef] [PubMed]
16. Ekanayake Weeramange, C.; Liu, Z.; Hartel, G.; Li, Y.; Vasani, S.; Langton-Lockton, J.; Kenny, L.; Morris, L.; Frazer, I.; Tang, K.D.; et al. Salivary High-Risk Human Papillomavirus (HPV) DNA as a Biomarker for HPV-Driven Head and Neck Cancers. *J. Mol. Diagn.* **2021**, *23*, 1334–1342. [CrossRef] [PubMed]
17. Hodgkinson, C.L.; Morrow, C.J.; Li, Y.; Metcalf, R.L.; Rothwell, D.G.; Trapani, F.; Polanski, R.; Burt, D.J.; Simpson, K.L.; Morris, K.; et al. Tumorigenicity and genetic profiling of circulating tumor cells in small-cell lung cancer. *Nat. Med.* **2014**, *20*, 897–903. [CrossRef]
18. Moncharmont, C.; Guy, J.-B.; Wozny, A.-S.; Gilormini, M.; Battiston-Montagne, P.; Ardail, D.; Beuve, M.; Alphonse, G.; Simoëns, X.; Rancoule, C.; et al. Carbon ion irradiation withstands cancer stem cells' migration/invasion process in Head and Neck Squamous Cell Carcinoma (HNSCC). *Oncotarget* **2016**, *7*, 47738–47749. [CrossRef]
19. Kulasinghe, A.; Zhou, J.; Kenny, L.; Papautsky, I.; Punyadeera, C. Capture of Circulating Tumour Cell Clusters Using Straight Microfluidic Chips. *Cancers* **2019**, *11*, 89. [CrossRef]
20. Kulasinghe, A.; Schmidt, H.; Perry, C.; Whitfield, B.; Kenny, L.; Nelson, C.; Warkiani, M.E.; Punyadeera, C. A Collective Route to Head and Neck Cancer Metastasis. *Sci. Rep.* **2018**, *8*, 746. [CrossRef]
21. Moussavi-Harami, S.F.; Wisinski, K.B.; Beebe, D.J. Circulating Tumor Cells in Metastatic Breast Cancer: A Prognostic and Predictive Marker. *J. Patient Cent. Res. Rev.* **2014**, *1*, 85–92. [CrossRef] [PubMed]
22. Kapeleris, J.; Kulasinghe, A.; Warkiani, M.E.; Vela, I.; Kenny, L.; O'Byrne, K.; Punyadeera, C. The Prognostic Role of Circulating Tumor Cells (CTCs) in Lung Cancer. *Front. Oncol.* **2018**, *8*, 311. [CrossRef] [PubMed]

23. Poveda, A.; Kaye, S.B.; McCormack, R.; Wang, S.; Parekh, T.; Ricci, D.; Lebedinsky, C.A.; Tercero, J.C.; Zintl, P.; Monk, B.J. Circulating tumor cells predict progression free survival and overall survival in patients with relapsed/recurrent advanced ovarian cancer. *Gynecol. Oncol.* **2011**, *122*, 567–572. [CrossRef]
24. Zhang, X.; Li, H.; Yu, X.; Li, S.; Lei, Z.; Li, C.; Zhang, Q.; Han, Q.; Li, Y.; Zhang, K.; et al. Analysis of Circulating Tumor Cells in Ovarian Cancer and Their Clinical Value as a Biomarker. *Cell Physiol. Biochem.* **2018**, *48*, 1983–1994. [CrossRef] [PubMed]
25. Buglione, M.; Grisanti, S.; Almici, C.; Mangoni, M.; Polli, C.; Consoli, F.; Verardi, R.; Costa, L.; Paiar, F.; Pasinetti, N.; et al. Circulating Tumour Cells in locally advanced head and neck cancer: Preliminary report about their possible role in predicting response to non-surgical treatment and survival. *Eur. J. Cancer* **2012**, *48*, 3019–3026. [CrossRef] [PubMed]
26. Balasubramanian, P.; Lang, J.C.; Jatana, K.R.; Miller, B.; Ozer, E.; Old, M.; Schuller, D.E.; Agrawal, A.; Teknos, T.N.; Summers, T.A.; et al. Multiparameter analysis, including EMT markers, on negatively enriched blood samples from patients with squamous cell carcinoma of the head and neck. *PLoS ONE* **2012**, *7*, e42048. [CrossRef] [PubMed]
27. Kulasinghe, A.; Hughes, B.G.M.; Kenny, L.; Punyadeera, C. An update: Circulating tumor cells in head and neck cancer. *Expert Rev. Mol. Diagn.* **2019**, *19*, 1109–1115. [CrossRef] [PubMed]
28. Kulasinghe, A.; Kenny, L.; Perry, C.; Thiery, J.-P.; Jovanovic, L.; Vela, I.; Nelson, C.; Punyadeera, C. Impact of label-free technologies in head and neck cancer circulating tumour cells. *Oncotarget* **2016**, *7*, 71223–71234. [CrossRef]
29. Garrel, R.; Mazel, M.; Perriard, F.; Vinches, M.; Cayrefourcq, L.; Guigay, J.; Digue, L.; Aubry, K.; Alfonsi, M.; Delord, J.-P.; et al. Circulating Tumor Cells as a Prognostic Factor in Recurrent or Metastatic Head and Neck Squamous Cell Carcinoma: The CIRCUTEC Prospective Study. *Clin. Chem.* **2019**, *65*, 1267–1275. [CrossRef]
30. Wang, H.-M.; Wu, M.-H.; Chang, P.-H.; Lin, H.-C.; Liao, C.-D.; Wu, S.-M.; Hung, T.-M.; Lin, C.-Y.; Chang, T.-C.; Tzu-Tsen, Y.; et al. The change in circulating tumor cells before and during concurrent chemoradiotherapy is associated with survival in patients with locally advanced head and neck cancer. *Head Neck* **2019**, *41*, 2676–2687. [CrossRef]

Review

The IL6-like Cytokine Family: Role and Biomarker Potential in Breast Cancer

Carlos Martínez-Pérez [1,2,*], Charlene Kay [1,2], James Meehan [2], Mark Gray [2], J. Michael Dixon [1] and Arran K. Turnbull [1,2]

[1] Breast Cancer Now Edinburgh Research Team, MRC Institute of Genetics and Cancer, Western General Hospital, University of Edinburgh, Edinburgh EH4 2XU, UK; charlene.kay@ed.ac.uk (C.K.); mike.dixon@ed.ac.uk (J.M.D.); arran.turnbull@ed.ac.uk (A.K.T.)

[2] Translational Oncology Research Group, MRC Institute of Genetics and Cancer, Western General Hospital, University of Edinburgh, Edinburgh EH8 9YL, UK; james.meehan@ed.ac.uk (J.M.); mark.gray@ed.ac.uk (M.G.)

* Correspondence: carlos.martinez-perez@ed.ac.uk; Tel.: +0131-651-8694; Fax: +0131-651-8800

Abstract: IL6-like cytokines are a family of regulators with a complex, pleiotropic role in both the healthy organism, where they regulate immunity and homeostasis, and in different diseases, including cancer. Here we summarise how these cytokines exert their effect through the shared signal transducer IL6ST (gp130) and we review the extensive evidence on the role that different members of this family play in breast cancer. Additionally, we discuss how the different cytokines, their related receptors and downstream effectors, as well as specific polymorphisms in these molecules, can serve as predictive or prognostic biomarkers with the potential for clinical application in breast cancer. Lastly, we also discuss how our increasing understanding of this complex signalling axis presents promising opportunities for the development or repurposing of therapeutic strategies against cancer and, specifically, breast neoplasms.

Keywords: breast cancer; cytokine signalling; IL6ST; gp130; biomarkers; translational research

1. Introduction

Breast cancer (BC) is a heterogeneous disease comprising well-characterised molecular subtypes that differ in their underlying biology, response to treatments, and prognosis. As with all cancer types, biomarkers with prognostic and/or predictive power are essential tools in the clinical management of this disease, with the oestrogen receptor α (ER) and the human epidermal growth factor 2 receptor (HER2) being the foremost biomarkers in BC. Assessment of both receptors to help select patients likely to respond to endocrine and HER2-targeted therapies has been established in clinical practice for many decades and has considerably improved the prognosis and survival for patients with hormone-dependent and HER2-overexpressing BC [1,2].

Despite said advances, many challenges remain in the management of BC, particularly as it pertains to advanced disease. In order to meet these needs, extensive research efforts are devoted to gaining a better understanding of the underlying complexity of the disease, as well as to identifying and validating potential molecular markers that might enable better patient stratification and treatment selection [2–4]. Valuable markers are typically involved in and serve as surrogates for cancer-promoting mechanisms or biological processes known to be altered by disease. Discovery studies continue to identify novel biomarkers predictive of BC development and progression, which can be differentially-expressed or mutated proteins or genes, as well as other genomic markers such as microRNAs or long non-coding RNAs [5–8].

Here, we review the role of IL6-like cytokines in BC and summarise evidence on the role of members of this ligand family and their receptors as biomarkers, both based

on their expression levels or the presence of polymorphisms. Recent years have seen a wealth of evidence reported on this, given the central role of this signalling axis in many cancer-related processes. To our knowledge, this is the most comprehensive review to date on the role and biomarker potential of this cytokine family in breast neoplasms.

2. The IL6-like Cytokine Family

Cytokines are a superfamily of small polypeptide regulators involved in cell signalling and the regulation of health and disease. They are often subdivided into families according to their features [9–11]. As interleukin-6 (IL6) is the best characterised cytokine of its kind, the group of cytokines with similar structural features and signalling machinery is referred to as the IL6 or IL6-like family. This is also referred to as the gp130 family, as the central feature of this group of cytokines is the transmembrane signalling receptor glycoprotein 130, one or more molecules of which are found in all oligomeric signalling complexes. This signal transducer is also known as CD130, IL-6 receptor subunit β (IL6Rβ) or IL6 signal transducer (IL6ST, which is also its gene name).

Besides the eponymous IL6, other canonical members of this cytokine family are interleukin-11 (IL11), ciliary neurotrophic factor (CNTF), leukemia inhibitory factor (LIF), oncostatin M (OSM), cardiotrophin 1 (CT1), cardiotrophin-like cytokine (CLC) and neuropoietin (NPN). Interleukin-31 (IL31) is often described as a member of this family, although its signalling complex does not include gp130/IL6ST, but other related signalling and non-signalling receptors [12–14]. Other cytokines, such as interleukin-27 (IL27), interleukin-35 (IL35) and interleukin-37 (IL37), have been described by different authors as belonging to either the IL6 or IL12 cytokine families [15–20]. Indeed, phylogenetic analysis has shown a close relationship between both groups [21–23].

Cytokines act as extracellular ligands, binding transmembrane receptors with high affinity to form oligomeric protein complexes. These lead to the formation of gp130/IL6ST homo- or heterodimers (depending on the cytokine and its respective receptors), which trigger intracellular signalling (see graphical abstract). The diversity of ligand-receptor complexes that can be formed, together with signalling through a shared, ubiquitously expressed transducer [24] and interaction with varied downstream regulators, make this cytokine group a highly pleiotropic protein family, involved in a wide range of biological functions, both in vitro and in vivo [25].

IL6-like cytokines exhibit a long chain 'four-helix bundle' topology. This consists of four tightly packed α helices of 15–22 residues in length arranged in two pairs of antiparallel helices connected by three polypeptide loops [9,25]. Each ligand then associates with a specific set of receptors which can be classified as non-signalling or signalling. 'Non-signalling' receptors (also known as α receptors) are only required by some ligands (namely IL6, IL11, CNTF and CLC) and are involved in the formation of the signalling complex, but do not actively participate in intracellular signalling; their cytoplasmic regions determine intracellular distribution in polarised cells but lack signalling capacities [26]. 'Signalling' receptors (also known as β receptors) are required by all ligands, as they are transmembrane proteins whose cytoplasmic domains activate the signalling machinery; gp130/IL6ST is the signalling receptor common to all family members. Where both kinds of receptors are required, the association between the ligand and the non-signalling receptor is typically the limiting step for complex formation and the subsequent activation of downstream signalling, as the ligand can bind a non-signalling receptor with high affinity on its own, but only binds the signalling receptor when in the presence of said non-signalling receptor [27].

The receptors in this family are modular in form and present distinct structural motifs in their extracellular region (or ectodomain): a single immunoglobulin-like domain, a cytokine homology region and, in signalling receptors, a third element including several copies of the fibronectin type III-like domain [9] (see Figure 1 and next section). While all family members bind gp130/IL6ST, their differential affinity for other receptors to form their respective complexes is central to the complex specificity of signalling through this cytokine family. The receptors associated with each member of the cytokine family are summarised in Table 1. Cytokines in the IL6 family are characterised by the existence of 3 topologically discrete sites (I, II and III) that act as functional epitopes for interaction with their receptors. Mutagenesis studies have shown that the specificity of these sites is dictated by a small number of residues in close spatial proximity, some of which are conserved across members of the family [9,25].

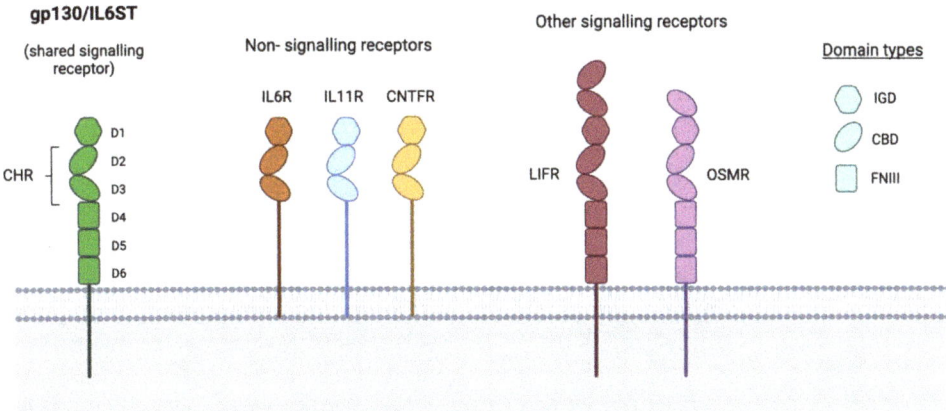

Figure 1. Structure of receptors in the IL6-like family. Receptors for cytokines in the IL6-like family present a modular structure with conserved motifs. Both signalling and non-signalling receptors include a single immunoglobin-like domain (IGD) and a cytokine-homology region (CHR), made up of cytokine-binding domains (CBD). Signalling receptors also present a membrane-proximal element including several copies of a fibronectin type III-like (FNIII) domain. The ectodomain of the shared signal transducer gp130/IL6ST consists of 6 domains, with the 3 membrane-distal ones (D1-D3) being essential for binding to the cytokine (and the non-signalling receptor, where this is required). Other signalling receptors, such as LIFR and OSMR, present larger ectodomains consisting of variations of this modular structures.

Table 1. Members of the IL6-like cytokine family and their respective receptors.

Cytokine	Site I: Non-Signalling: Receptor	Site II: Signalling Receptor	Site III: Signalling Receptor
IL6	IL6R (IL6Rα)	gp130/IL6ST	gp130/IL6ST
IL11	IL11R (IL11Rα)	gp130/IL6ST	gp130/IL6ST
CLC	CNTFR (CNTFRα)	gp130/IL6ST	LIFR (LIFRβ)
CNTF	CNTFR (CNTFRα)	gp130/IL6ST	LIFR (LIFRβ)
CT1	-	gp130/IL6ST	LIFR (LIFRβ)
LIF	-	gp130/IL6ST	LIFR (LIFRβ)
NPN	-	gp130/IL6ST	-
OSM	-	gp130/IL6ST	LIFR (LIFRβ) or OSMR (OSMRβ)

CLCF1, cardiotrophin-like cytokine; CNTF, ciliary neurotrophic factor; CNTFR, CNTF receptor α; CT1, cardiotrophin 1; IL6, interleukin-6; IL6R, IL6 receptor α; IL11, interleukin-11; IL11R, IL11 receptor α; IL27, interleukin-27; IL31, interleukin-31; gp130/IL6ST, glycoprotein 130, also known as IL6 signal transducer; LIF, leukemia inhibitory factor; LIFR, LIF receptor β; NPN, neuropoietin; OSM, oncostatin M; OSMR, OSM receptor β.

Site I, used only by some cytokines, is a binding site for non-signalling receptors only (e.g., IL6R for IL6 or IL11R for IL11). A recent study has reported on the different mechanisms for complex formation, evidencing the biological specificity for each ligand-receptor pair [28]. Consistently across all family members, site II is always the binding site for the shared receptor gp130/IL6ST. Site III is always used for association with a second signalling receptor, such as gp130/IL6ST, LIFR or OSMR, depending on the ligand. IL6 and IL11 have been shown to use sites II and III to bind different regions of the same gp130/IL6ST molecule [9]. Following this receptor recognition, higher order complexes are formed combining 2 ligands and their respective receptors, as described in the following section.

3. Soluble Receptors and Signalling Modes

As the prototypical and best-characterised member of its cytokine family, IL6 represents the best model to describe the complex signalling machinery observed in this family. Importantly for the scope of this review, IL6 also plays an important role in BC, so the description of its specific signalling partners and modes will be informative to the sections focusing on this disease. IL6 binds the non-signalling receptor IL6R at the cytokine's site I before the IL6-IL6R complex can bind the signalling receptor gp130/IL6ST using IL6's sites II and III. Two such complexes then dimerise to form a final ternary complex with a hexameric conformation and stoichiometry that includes two molecules each of IL6, IL6R and gp130/IL6ST [29–31] (see Figure 2). It is this complex that creates the gp130/IL6ST homodimer necessary to activate downstream signalling in the cytoplasm. While a tetrameric complex model (comprising one molecule each of IL6 and IL6R and two molecules of gp130/IL6ST) has also been proposed [32], the higher order hexameric conformation, similarly described in other IL6-like cytokines such as IL11 [33] and CNTF [34], has become the canonical model for complex formation and signal activation [35].

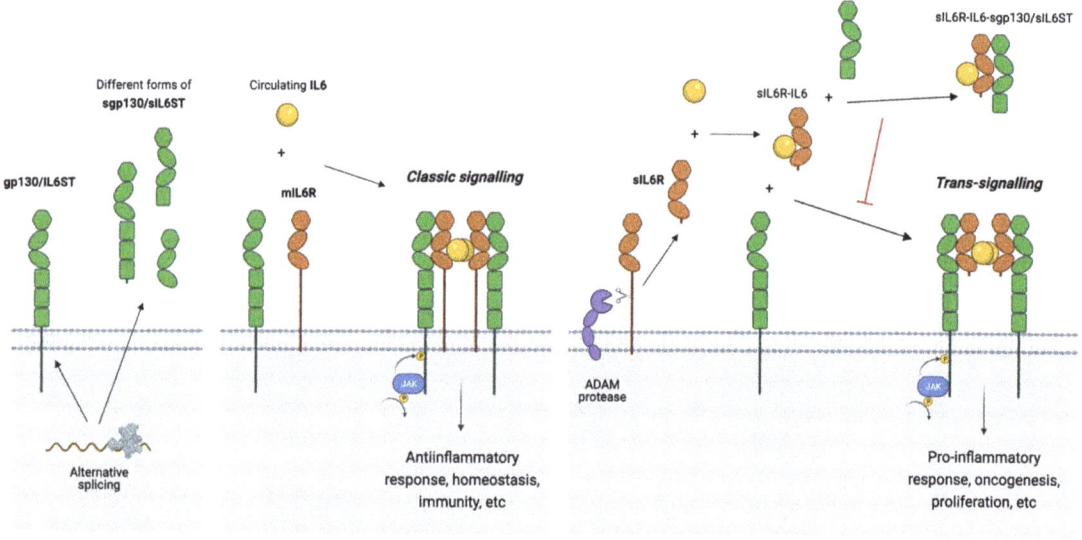

Figure 2. Signalling modes in IL6-like cytokine signalling. The shared signalling receptor gp130/IL6ST is ubiquitously expressed across all cell types in its full-length, membrane-bound form. Different soluble forms (sgp130/sIL6ST) are also produced, mainly through alternative splicing. Different cell types produce membrane-bound or soluble forms of non-signalling receptors such as IL6R (mIL6R or sIL6R, respectively). Receptor availability will determine what signalling mode is induced by a cytokine. In the classic signalling mode, IL6 forms a hexameric signalling complex by binding mIL6R and gp130/IL6ST. Alternatively, circulating sIL6R can act a cytokine agonist, capturing IL6 to trigger trans-signalling, associated with pro-inflammatory and pro-carcinogenic responses. In turn, sgp130/sIL6ST can act as a cytokine antagonist, sequestering the sIL6R-IL6 complex and inhibiting trans-signalling.

Two different modes of IL6 signalling have been described that are determined by the existence of two different forms of the IL6 receptor α (see Figure 2): classic IL6 signalling involves the full-length, membrane-bound form (mIL6R), while trans-signalling involves the soluble form (sIL6R), produced from mIL6R, mainly by ectodomain cleavage or shedding by a disintegrin and metalloproteinase domain-containing protein (ADAM10 and

ADAM17) [36,37] or, in a smaller proportion, by alternative splicing [38,39]. Recently, a third signalling mode referred to as trans-presentation has been described, by which IL6 binds mIL6R on the surface of a dendritic cell and the resulting complex is then presented to adjacent CD4+ T cells, leading to h17 cell differentiation [40]. However, this mechanism has not yet been observed in human models.

As gp130/IL6ST is ubiquitously expressed in all cell types [24], the form of IL6R available will determine whether IL6 elicits signalling through the classic or trans-signalling routes which, importantly, have been shown to have divergent functions (see Figure 2). Classic signalling is limited to mIL6R-expressing hepatocytes, leukocytes, and immune cells, and has been shown to control homeostasis and promote anti-inflammatory responses [41,42]. In contrast, evidence has shown that sIL6R is produced by a broad range of cell types, including malignant types such as BC cells, which produce sIL6R endogenously. sIL6R can circulate through the bloodstream where it binds up to 70% of the circulating IL6, thus increasing the cytokine's half-life and bioavailability and acting as a carrier for its delivery to gp130/IL6ST, available in the membrane of all cell types [43,44]. In this way, trans-signalling broadens the target cell repertoire of IL6, enabling response to the cytokine in cells lacking mIL6R. Depending on the levels of sIL6R produced, trans-signalling can take place as a paracrine action or at both local and systemic levels [45]. Trans-signalling has been linked to pro-inflammatory effects and the observed role of IL6 in chronic diseases and cancer [27,42,46,47]. Trans-signalling mechanisms have also been described for the IL6-like cytokines IL11 and CNTF through soluble forms of their respective non-signalling receptors, sIL11R and sCNTFR [48,49]. The soluble receptors sIL6R, sIL11R, and sCNFTR are considered agonists, since they act as ligand-binding receptors that enable cytokine presentation and complex formation [47].

The other essential receptor in all IL6 signalling is, obviously, gp130/IL6ST. While this is ubiquitously expressed, its role is also complicated by the existence of circulating forms. The extracellular portion of gp130/IL6ST consists of 6 domains (see Figure 1): 1 N-terminal immunoglobulin-like domain (IGD), 2 cytokine-binding domains (CBD) and 3 fibronectin type III-like (FNIII) domains. The 3 membrane-distal domains are essential for ligand recognition, since the 2 CBDs (D2-D3) form the cytokine homology region (CHR) and the IGD (D1) is also required for the receptor to be functionally responsive to the cytokine; the 3 membrane-proximal FNIII domains (D4-D6) provide the right spatial orientation to enable formation of the hexameric receptor complex and signal transduction [35,50,51]. At least 4 soluble forms of gp130/IL6ST (sgp130/sIL6ST) have been reported, which consist of the entire (D1-D6) or part (D1-D3 or D1-D4) of the ectodomain [39,42,52], often presenting stabilising glycosylations [53]. These soluble receptors, found at levels of up to 400ng/mL in the blood [54–56], are produced mainly through alternative splicing, although ectodomain shedding might also contribute to a very small proportion of their production [39,57,58].

Unlike soluble non-signalling receptors like sIL6R, sgp130/sIL6ST acts as a cytokine antagonist, competing with membrane-bound gp130/IL6ST to bind the circulating IL6-sIL6R complex and, thus, selectively blocking IL6 trans-signalling [59–61] (see Figure 2). All identified forms of sgp130/sIL6ST include the N-terminal cytokine-binding portion of the receptor. To date, there is no clear evidence of differential antagonistic abilities between the different known forms of sgp130/sIL6ST [42,62]. Evidence has shown that sgp130/sIL6ST can also inhibit IL11 trans-signalling [48,63]. Soluble forms of the signalling receptors OSMR and LIFR have also been reported [64,65], which act as antagonists for OSM and LIF signalling, respectively.

Given the opposing effects of sIL6R and sgp130/sIL6ST on IL6 signalling, and the fact that their plasma levels remain relatively stable (40–75 ng/mL for sIL6R [66] and 250–400 ng/mL for sgp130/sIL6ST [56,67]), these soluble forms of the receptors act as a buffer for circulating IL6. Plasma levels of this cytokine vary broadly by up to six orders of magnitude between health and disease and in response to different local and systemic processes [39]. Thus, this buffering mechanism might prevent unspecific overstimulation by IL6 trans-signalling unless systemic or local IL6 levels surpass a certain threshold.

Research has also reported cell type-specific expression patterns for the different existing forms of sgp130/sIL6ST, which might enable local fine-tuning of the antagonistic effect on IL6 trans-signalling [57].

4. Shared Cytokine Signalling: Pleiotropy, Redundancy and Specificity

Cytokine-driven dimerisation of gp130/IL6ST leads to signal transduction and activation of 3 major downstream pathways: the Janus-activated kinase—signal transducer and activator of transcription (JAK/STAT) pathway, the Ras-Raf mitogen-activated protein kinase (MAPK/MERK/ERK) signalling cascade, and the phosphoinositol-3 kinase—protein kinase B/Akt (PI3K/AKT) pathway. This versatile signalling cascade is initiated by tyrosine kinases in the JAK family, such as JAK1, JAK2 and TYK2, which can be found constitutively associated with the cytoplasmic region of gp130/IL6ST by a non-covalent bond. Dimerisation of gp130/IL6ST causes auto-phosphorylation and activation of JAK. One cascade can see JAK phosphorylating the signal transducer and activator of transcription 3 (STAT3), leading to its dimerisation and translocation to the nucleus, where it modulates proliferation and cell survival. JAK can also activate the SH2 domain-containing cytoplasmic protein tyrosine phosphatase (SHP2), which in turn activates the Ras/Raf pathway, leading to the hyperphosphorylation of mitogen activated protein kinases (MAPK) and triggering its increased serine/threonine kinase activity and complex downstream cascade, which includes various transcription factors linked to cell growth [68,69]. Thirdly, JAK can also activate the PI3K/AKT pathway. These signalling pathways are under regulation by a number of negative-feedback mechanisms, including temporal attenuation of the activity of SHP2 and the induction of the suppressor of cytokine signalling (SOCS) protein family [70].

These three main signalling pathways, with their own complex and pleiotropic effects, lead to the wide range of functions of IL6 and related cytokines in the healthy organism and in diseases, such as immune disorders and cancer. The tumour-promoting effects of these cytokines include both cancer cell-intrinsic processes, such as cell proliferation, differentiation, survival, invasion and metastasis, and extrinsic processes that affect the tumour microenvironment (TME), such as modulation of inflammation and angiogenesis [71,72]. Reliance on gp130/IL6ST as a shared signal transducer enables a certain level of functional redundancy across family members [68]. Despite relative selectivity in ligand-receptor recognition, structural similarities still allow for some level of receptor promiscuity, which can lead to crosstalk, where a cytokine associates with receptors other than their own with lower affinity. In vitro studies have previously reported non-canonical cytokine-receptor complexes such as OSM-LIFR [12] or CNTF-IL6R [73], which might widen a cytokine's target spectrum, enabling them to elicit effects normally associated with other ligands in the family.

Nevertheless, there is extensive evidence of significant functional specificity for different cytokines in vivo, with specific members exerting unique functions or the same cytokine being able to elicit different responses in different cell types [17]. How exactly this cytokine family circumvents its built-in redundancy to achieve specificity remains unclear, although a number of features in this family's complex signalling machinery are likely to contribute to this modulation. The expression patterns for the different cytokines are different across cell populations and can be modulated by the extracellular matrix, while levels or bioavailability of circulating cytokines might also vary. The same is true of the expression levels of different receptors which, as previously mentioned, are often the limiting factor in cytokine signalling and can determine the signalling mode triggered.

The complex formation process relies on a sophisticated network of interactions between each cytokine and the relevant non-signalling and signalling receptors. The complex extracellular portion of gp130/IL6ST enables additional functional complexity, as multiple domains and regions are involved in ligand recognition and activation [35]. This explains how, for instance, IL6 is only able to associate with gp130/IL6ST as part of a binary IL6-IL6R complex. Research has shown that different cytokines bind different specific residues in gp130/IL6ST [74–76]. These differences in the complex formation mechanism are likely to contribute to distinct changes in the intracellular portion of gp130/IL6ST. Differential target response to signalling thresholds across cell types, as well as a range of modifications and many potential regulatory mechanisms also likely contribute to the plasticity and specificity in gp130/IL6ST-mediated signalling [70]. Additionally, crosstalk with other pathways through shared signalling components and factors in the cytoplasm adds to the modulation of a cytokine's effect [77]. Authors have suggested that the tissue-specific effects of cytokines might be the result of signalling orchestration, where certain cell types can integrate the range of possibly opposing signals with interplaying mechanisms and factors for a balanced final response [78]. Further studies are needed to better elucidate the complex signalling machinery enabling functional specificity. In the meantime, this poses an interesting challenge in understanding how to tackle signalling of IL6 and other cytokines for therapeutic purposes, as effective agents would need to achieve a similar degree of specificity to enable targeting certain deleterious processes without compromising other essential activities (see Section 8).

5. The Role of the IL6-like Cytokine Family in BC

Among their broad range of pleiotropic functions, IL6-like cytokines are well-established as secretory factors contributing to many pro-carcinogenic changes, including disease progression or the development of treatment resistance, in a wide range of types of cancers [79–83]. This signalling axis is also involved in the regulation of homeostasis and other essential functions such as inflammation and immunity. In fact, the complex interaction between these functions and cancer has been thoroughly described [81,84,85]. For example, as pro-inflammatory signals regulated by these cytokines have been shown to the play a role in neoplastic aetiology and progression, and this could be exploited for therapeutic purposes. In the following paragraphs, we will focus on the role of these cytokines in BC, describing the activity and biomarker potential in this disease of different members of the IL6-like family.

5.1. IL6 in BC

5.1.1. Signalling Role in BC

As the prototypical pro-tumourigenic member of its cytokine family, IL6 has been shown to exert a wide range of pro-cancer effects, including promoting tumour initiation and progression, survival, invasion, metastasis and chemo-resistance. The evidence on the role of IL6 in cancer has been reviewed extensively [19,86,87]. Here we will summarise the important roles of IL6, its downstream effectors and pathways in BC, as well as the role of IL6 as a marker in this disease.

STAT3, highly active in more than 50% of BCs [88,89], has been described as a key signalling orchestrator of many of the cancer-promoting effects exerted by IL6, but also IL11, LIF and OSM [70,79]. Evidence has also shown that STAT3 enables cross-talk of the JAK/STAT pathway with the other gp130/IL6ST-dependent pathways, contributing to cancer-promoting effects of the MAPK/MEK/ERK and PI3K/AKT signalling pathways, such as chemo-resistance and epithelial-mesenchymal transition (EMT) [90,91].

Numerous studies have assessed the effect of IL6 on proliferation in vitro [86,92], with diverging conclusions: while most evidence has suggested that recombinant IL6 can inhibit proliferation in ER-positive (ER+) cell line models [93–98], this has been contested by other studies and some have also shown a divergent motility-promoting effect [96,97]. Indeed, some mechanistic studies have shown that STAT3 can induce cell cycle progression and inhibit various apoptotic genes [19,72,99,100], while PI3K/AKT signalling inhibits p53, Chk-1, and transcription of tumour suppressors, and induces cyclin D1, myc, and mTOR transcription [101,102], with these changes contributing to proliferation and cell survival. Conversely, other studies have reported that IL6 might induce apoptosis and help inhibit proliferation in MCF-7 cells [93,95,103]. Interestingly, one study on ER+ cells in 3D culture found that IL6 could promote proliferation [104], while another on xenograft models found that IL6 increased expression of EMT-related genes through STAT3 [105]. Similarly, a study found that the blockade of gp130/IL6ST signalling had different effects in vitro compared to in vivo: inhibition led to higher proliferation in cell line models, but reduced malignancy in mice models [106]. This evidence from more complex models suggests that the effect of this cytokine on proliferation might be dependent on the TME. Another likely determining factor is receptor expression in the different models studied, but most in vitro studies have not considered the potential role of membrane-bound or soluble receptors [86].

The effect of IL6 on proliferation is only one of its many roles in BC. Studies have shown that IL6 can exert a pro-metastatic effect by modulating genes related to EMT, a process essential to the metastatic process, via STAT3 [107,108]. Transition from a stationary to a motile phenotype is also aided by the downregulation of E-cadherin expression, leading to loss of adhesion [97,105,107,109]. IL6 can also induce pro-angiogenic effects, inducing VEGF expression in tumour-associated endothelial cells through STAT3 and MAPK [110], while the IL6 inflammatory loop can activate mechanisms linked to drug resistance [111–113].

5.1.2. Circulating IL6 Level as a Biomarker

Different sources of IL6 can induce a tumour-promoting effect in BC cells. Malignant cells are known to be a major source of IL6, with BC cells producing much higher levels of both the cytokines and its receptors than normal epithelial breast cells [114]. This endogenously-produced IL6 is used as a growth factor in an autocrine manner. In addition to production by cancer cells, tumour-associated macrophages (TAMs), helper T (Th) cells and tumour-associated fibroblasts have been shown to be primary sources of IL6 in the TME, suggesting these cell types enable paracrine IL6 signalling which can in turn contribute to oncogenesis and proliferation [104,115–117]. Interestingly, an IL6 activation signature revealed that pathway activity as measured in breast tumour samples correlated with circulating serum IL6 levels [118]. Both local autocrine and paracrine release are able to activate IL6 trans-signalling, thus contributing to the cytokine's tumour-promoting effects [119–122]. In turn, these circulating cytokines could also enable endocrine IL6 signalling in distant lesions as the disease spreads.

Although pre-clinical studies have produced diverging evidence on the effects of IL6 and the utility of its in-tumour levels as a marker remains unclear [123–126], the role of circulating serum IL6 levels as a negative prognosticator in BC has been well established [79,92]. Indeed, compared to healthy women, serum IL6 levels are significantly elevated in BC patients and correlate with the stage of disease [127,128]. Higher levels were also observed in patients with widely dispersed metastatic BC compared to single metastatic disease, in recurrent compared with non-recurrent disease and in progressive compared to stable disease [128–133]. Elevated serum levels were also associated with worse prognosis and survival, as well as reduced response to chemo- or endocrine therapy [128–131]. Multivariate analysis confirmed that IL6 is an independent negative prognosticator in metastatic BC [130]. A meta-analysis of previous studies also found that high IL6 expression is associated with poor overall survival [134]. Although correlation with other clinical factors or tumour characteristics was not identified, the analysis was limited by the heterogeneity across studies in the different sources for IL6 detection (tumour vs. serum). In short, despite diverging reports on IL6's role from in vitro studies, clinical evidence firmly supports that IL6 is involved in and a biomarker of BC development and progression. Despite the evidence on its prognostic and potentially predictive value of serum IL6 levels, prospective studies are needed before their assessment can be applied in BC detection and monitoring or to guide treatment selection.

Studies have reported changes in IL6 levels during treatment with taxane-based chemotherapy [135,136], but not with anthracycline-based chemotherapy [137,138] or endocrine therapy [139]. Some effort has gone into investigating a potential role of elevated cytokine level in the development of adverse events in patients receiving treatment for BC. Plasma IL6 levels were found to be associated with the development of fatigue in early-stage BC patients receiving radiotherapy [140,141], although other studies found no link between fatigue and IL6 levels in later stages [142,143]. Pre-treatment levels of IL6 were also higher in patients who went on to develop depression [144–146]. In long-term survival, plasma IL6 levels were associated with reduced cognitive function and poorer memory [147,148]. These findings suggest that anti-inflammatory therapies and, more specifically, agents inhibiting IL6 might help alleviate some of the side effects associated with anti-cancer treatment.

Researchers have also studied whether the systemic IL6 level could be an indicator of BC risk in healthy women, since evidence of non-steroidal anti-inflammatory treatment leading to reduced risk of developing cancer suggested that circulating pro-inflammatory factors such as this cytokine might be linked to a higher predisposition to breast neoplasms [149]. No correlation between IL6 and breast cancer risk was found in two prospective studies in older populations, although the studies were limited by a low predictive power [150,151].

Circulating levels of sIL6R have been found to be elevated (in comparison to healthy individuals) in patients with different cancer types, including myeloma, leukaemia, bladder, prostate and hepatocellular cancer, with higher levels associated with tumour grade, volume or disease spread [152–157]. While significantly higher sIL6R levels have also been reported in patients with BC [158,159], it has been suggested that larger sample cohorts need to be assessed before conclusions can be drawn regarding the prognostic potential of sIL6R levels in this cancer type [27].

5.2. Other IL6-like Cytokines in BC

Other cytokines and associated receptors have also been associated with breast disease. As previously discussed, there is a certain degree of functional redundancy through the shared signal transducer and the 3 main downstream pathways described. For instance, some studies have shown that both IL6 and OSM are capable of inducing tumour-promoting effects through STAT3, while IL6, IL11, LIF and OSM can all promote invasion and metastasis through the JAK/STAT3 and PI3K/AKT pathways [79,160–166]. On the other hand, different cytokines can also trigger diverging responses through modulation of common mediators and signalling pathways. For instance, IL6 induces numerous cancer-promoting changes through STAT3, while IL27 or OSM can induce an opposing effect through the related signal transducer STAT1, also activated as part of the JAK/STAT pathway [167,168]. As is the case for IL6, other cytokines might exert a complex range of possibly opposite effects, with evidence of anti- and pro-tumour effects. In the next paragraphs we summarise some of the evidence to date on the role and potential clinical implications of some of the other IL6-like cytokines in BC.

5.2.1. IL11

Having long been established as a haematopoietic growth factor, research in recent years has highlighted the potential pro-tumourigenic role of IL11 in epithelial cancers, with abundant evidence of its role in gastric cancer [169]. BC cell line studies have shown that, while only some models secreted IL11 [170], most expressed its specific receptor, IL11R [171,172]. In line with the typical signalling of cytokines in the IL6-like family, STAT3 is a central orchestrator of most of the known effects of IL11 in cancer, including promotion of cell growth and survival [173]. For example, studies on a triple-negative BC cell line model showed that blockade of IL11 led to increased response to chemotherapy [174]. The same researchers reported that higher IL11 levels were associated with poorer survival in BC patients treated with chemotherapy, suggesting this cytokine might play a similar role in vivo.

Characterisation of a range of cancer cell lines, including a BC model, found that IL11 and IL11R expression was induced by the development of hypoxia [175]. STAT3 has also been shown to contribute to the effects of the hypoxia-inducible factor-1α (HIF1α) transcription factor to promote angiogenesis, a key process in cancer progression and dissemination [176]. Animal studies using triple-negative human BC xenografts including IL11-overexpressing subclones have supported the role of IL11 in tumour growth, metastasis and angiogenesis [175,177].

The best-established role of IL11 in BC is in metastasis promotion. BC metastases to the bone are normally osteolytic, involving the activation of osteoclasts (bone-resorbing cells) that cause bone degradation and in turn enable tumour expansion. Recent animal studies have shown that IL11-driven activation of JAK/STAT signalling is an essential factor in promoting metastases-driven osteolysis [91]. Research has shown that different IL6-like cytokines are produced by osteoblasts (bone-forming cells) in physiological conditions and can be involved in normal bone remodelling [178–180]. Interestingly, IL11 is also expressed by bone marrow stromal cells and its secretion plays a central role in osteoclastogenesis [181]. The fact that IL11 is produced endogenously by stromal cells under physiological conditions and is essential to the formation and differentiation of bone-deforming cells (rather than just signalling for their activation like other related ligands) supports the notion that this cytokine plays a particularly important role in the bone microenvironment both during normal remodelling in healthy individuals and in the presence of colonising malignant cells [182]. Indeed, IL11 expression has been shown to correlate with risk of developing bone metastasis [183] and is higher in both tumour and serum samples from BC patients who presented these distant metastatic lesions [184].

Numerous other studies have investigated the clinical implications and potential prognostic value of IL11 expression. One study found that IL11 was elevated in tumour samples compared to matched normal breast tissue, regardless of subtypes and grade [172,185,186]. Another study showed higher levels of both IL11 and IL11R in clinical BC samples compared to normal breast tissue. Results also showed that IL11 was higher in tumours with node-positive status and poorer prognosis and that a higher level of the cytokine was linked to poorer survival [187]. Higher IL11 was also observed in patients who relapsed within 3–5 years compared to those who remained relapse-free [188].

Interestingly, a meta-analysis of 26 datasets from microarray studies found that higher IL11R expression was a positive prognosticator, associated with better survival in the lower-risk cohort of patients with negative node status [172]. Diverging from that earlier study, interrogation of multiple publicly-available datasets reported that IL11R was downregulated in most BC subtypes [172,189–192]. One notable exception is the mesenchymal stem cell-like subgroup of triple-negative BC. This subtype does express IL11R and presents an aggressive phenotype associated with poorer patient outcomes, suggesting the receptor might be a negative prognosticator [174]. The contrasting evidence on the role of IL11R as a biomarker in BC hints at the complex interaction between the receptor, its cytokine and other related factors, as well as at the fact that this signalling might vary across different BC subtypes.

5.2.2. LIF

Evidence exists of a range of effects of LIF in BC. The role of LIF signalling on proliferation is unclear, with studies using the same ER+ cell line model reporting both growth-promoting [193,194] and anti-proliferative effects [171,195,196] of LIF. This suggests different factors might affect the role of LIF in promoting or inhibiting proliferation and further studies are needed to better characterise its function in BC. However, the proliferative effect of LIF has been reported in models expressing low LIFR, suggesting this activity is likely not dependent on this receptor.

Although cell line studies have suggested that LIF can lead to increased migration and metastasis [164,165], its receptor LIFR has been shown to act as a breast tumour and metastasis suppressor through activation of the pro-dormancy activity of STAT3 [195,197–201]. In line with this, cell line models with low metastatic potential have been shown to express higher levels of LIFR and be responsive to LIF, whereas highly metastatic cells did not express the receptor and were unresponsive to the ligand [195]. Studies in ER+ cell line models have also shown that its knockdown leads to increased invasion, downregulation of dormancy genes and increased osteolytic bone destruction [195]. Overall, there is abundant evidence of a metastasis-supressing role of LIFR, suggesting a systemic pro-dormancy role in cancer cells disseminated to distant sites [195]. However, several cytokines (including LIF, OSM and CNTF) can recruit LIFR to initiate their signalling, so it remains unclear what ligands might drive this specific mechanism.

5.2.3. OSM

Several studies have reported growth-inhibitory effects of OSM in BC cell line models [196,202–205], as well as in normal human mammary epithelial cells [206]. Conversely, cell line studies have also suggested that OSM might exert a pro-tumorigenic and pro-metastatic effect through induction of detachment, invasiveness, bone dissemination and EMT [160,207–209]. Studies in animal models showed that OSM knockdown reduced the formation of metastases, with findings suggesting that autocrine and paracrine signalling might be linked to metastasis to bone and lungs, respectively [210,211]. Data from clinical samples showed that patients with higher OSM levels had decreased survival, further supporting this role in metastasis.

OSM can signal through both OSMR and LIFR [179,212], with evidence suggesting different downstream effects of OSM might be dependent on different signalling receptors [195,207]. Expression of OSMR has been shown to be associated with shorter recurrence-free and overall survival in BC patients [213], while evidence suggests a metastasis-suppressing role for LIFR. This supports the notion that OSM could have opposing effects depending on which receptor it binds to initiate its signalling, with OSMR potentially being involved in the cancer-promoting role of the ligand, while association with LIFR could be linked to its growth-inhibitory effect. Nevertheless, further work is needed before the complex signalling machinery and specific role of OSM in vivo can be fully elucidated.

5.3. IL6ST as a Biomarker in BC

As the cornerstone of all signalling by cytokines in the IL6-like family, the shared signal transducer gp130/IL6ST holds particular potential as both a therapeutic target and a candidate biomarker. We recently reviewed the extensive evidence of gp130/IL6ST as a promising predictor in BC [214]. In short, in recent years ten different independent studies based on the analysis of clinical BC samples have shown the value of gp130/IL6ST as a prognostic and/or predictive biomarker. Six studies reported that gp130/IL6ST serves as an independent marker and is, specifically, a positive prognostic marker in BC; its expression is significantly correlated with ER expression and better prognosis, and inversely correlated with adverse events such as invasion, metastasis and recurrence [214].

We also showed that IL6ST has been included in four different multifactor signatures (including the clinically-available EndoPredict assay), where gp130/IL6ST also served as a positive prognostic factor for ER+ BC [214]. These multigene signatures enable stratification of BC patients into prognostic groups with differing risks of recurrence and rates of response to different therapeutic strategies, so they could aid treatment selection. These findings suggest that inclusion of gp130/IL6ST, along with other molecular and clinicopathological factors, might provide further insight into the complex underlying biology of the disease and could, in turn, enable better patient stratification. As a result, these multifactor tools represent a promising avenue for the potential clinical translation of gp130/IL6ST's value as a biomarker in BC.

6. IL6-like Cytokines and Oestrogen Signalling

Given the essential role that oestrogen and its signalling play in the majority of BCs, extensive efforts have gone into assessing their potential interaction with cytokine signalling. Research has shown a complex association between the oestrogen receptor and IL6 signalling. ER+ cells have been shown to be more responsive to IL6 than hormone-independent cells. Interestingly, an in vivo study showed that IL6 could drive engraftment of xenografts derived from an oestrogen-dependent BC cell line in the absence of hormonal supplementation and this could be blocked using an IL6 inhibitor [118]. Although earlier evidence suggested IL6 expression might correlate with ER [215], several more recent studies have shown that ER-negative (ER-) cultured models produce higher levels of the cytokine [216–218]; thus, ER- cells might be exposed to constant autocrine IL6 signalling, rendering them less sensitive to fluctuations in exogenous IL6 levels when compared to ER+ cells [219]. Research has also assessed a potential link between ER status and IL6R expression, with diverging results: cell line studies have suggested that ER+ cells release mainly sIL6R, whereas ER- cells express mIL6R [95]; however, a more recent study of serum samples showed that patients with ER+ tumours had lower levels of sIL6R compared to ER- patients [220].

IL6 can activate oestrogen-generating enzymes in both the tumour and adjacent tissues [126,221–223], acting as a key modulator of the conversion of estrone to estradiol [223]. Thus, IL6 can lead to an increase in local and circulating levels of oestrogen, as well as oestrogen sulfate, which can remain in circulation longer and acts as a hormone reservoir. Interestingly, research has shown that the IL6-like cytokines IL11 and OSM can stimulate aromatase expression via binding of STAT1/3, supporting the notion that cytokines secreted locally by cells in the TME contribute to upregulation of aromatase activity [224–226].

A study of primary BC-derived cultures showed that IL6 can cause direct transcriptional activation of ER [227]. This supports the notion that, even if lower levels of IL6 are produced endogenously in ER+ tumours, the cytokine still contributes to the advancement of ER-driven disease. Interestingly, cell line studies have reported that ER can trans-repress the expression of IL6, suggesting a potential negative feedback loop. This could explain why ER- cells, where this negative feedback loop is inactive, produce more IL6 [228]. In turn, this cytokine up-regulation has been suggested as a contributing factor to the greater invasive and metastatic potential of ER- breast cancer [92,106,228].

In vitro studies on endometrium and decidua cells have shown that oestrogen and ER signalling might play a role in determining the balance between soluble and membrane-bound forms of gp130/IL6ST [229]. This suggests that hormonal regulation could also modulate expression and release of the different isoforms in BC, thus altering cytokine signalling and its downstream effects on the disease.

Researchers have also investigated the association between hormone receptor status and other cytokines and receptors in the IL6-like family. Studies have shown that expression of LIFR correlates with ER in clinical samples [230] and its level and function are also higher in ER+ cell line models [195]. Evidence has suggested that higher LIFR levels are associated with favourable biological features and better outcome and loss of LIFR favours bone metastasis [195,230]. In line with this, a recent study reported a metastasis-promoting mechanism in which ER is involved in down-regulation of LIFR expression, suggesting a potential negative feedback loop, similar to the one observed for IL6 [231].

Levels of OSM and its receptor OSMR have been shown to be inversely correlated with expression of ER and its target genes. Studies in ER+ cell lines also showed an antagonistic relation between ER and OSM and their associated signalling [213]. While the role of IL6-like cytokines in BC appears to be complex and the relation between ER status and the different cytokines in the family is heterogeneous, there is a common trend of evidence of ER-driven mechanisms of negative regulation of IL6, LIF and OSM signalling.

Lastly, the positive correlation between gp130/IL6ST expression and ER has been well established, as we recently reviewed [214] (also see Section 5.3). In short, evidence suggests that gp130/IL6ST might act as a robust surrogate marker of active oestrogen-related signalling. Importantly, its expression levels might provide an insight into the heterogenous underlying biology of ER+ BC, helping to stratify ER+ tumours into subsets that differ in their true level of hormone dependence and, consequently, their likelihood of response to endocrine therapy.

7. Polymorphisms in gp130/IL6ST-Dependent Signalling

Mutations leading to changes in the expression, activation or stability of molecules in the gp130/IL6ST signalling axis represent potential mechanisms for the development and progression of disease. Epigenetic alterations might also play an important role in aberrant activation of cytokine signalling in cancer. While the evidence to date has been reviewed extensively before [17,232], in this Section we will summarise the role polymorphisms in this protein family specifically as it pertains to BC.

7.1. Polymorphisms in IL6-like Cytokines

While there is no evidence of naturally-occurring gain-of-function mutations in IL6, numerous studies have reported loss-of-function mutations that might alter its expression or downstream signalling and are linked to a range of pathologies [17]. In BC, mutations have mainly been described in the promoter region of the IL6 gene. The best characterised SNP, -174G/C (rs1800795), has been shown to cause IL6 overexpression at least in part by enabling recognition by other transcription factors [233,234]. However, the potential clinical implications of this mutation are unclear. Some studies found the mutant GG genotype was a negative prognosticator associated with reduced disease-free survival (DFS) in ER+ patients receiving chemotherapy [235] and with increased risk of metastasis irrespective of ER status [236]. In contrast, other studies found that a wild-type CC genotype was associated with a more aggressive phenotype and worse overall survival [237] or with lymphovascular invasion [238]. Another group also reported that C-carrying cancers had a higher risk of early events; this association was observed in ER- tumours, particularly after radiotherapy, but also irrespective of ER status for chemotherapy-treated cancers [239]. Overall, researchers have hypothesised that this polymorphism might alter the effect of the TME on IL6 expression: while wild-type tumours increase production of the cytokine in response of inflammatory stimuli such as radio- or chemotherapy or in ER- disease, mutant tumours might produce more IL6 regardless of systemic changes [239]. Other studies have suggested that hormonal status might influence how genetic variation affects the role of IL6 in BC [240]. Another polymorphism in the IL6 promoter is the SNP -597G>A (rs1800797), which has been linked to worse DFS and higher risk of early events [238,239,241]. Diplotypes including both -174G>C and -597G>A have also been associated with worse DFS (IL6-174G/C (rs1800795) [235].

Numerous studies have assessed the effect of IL6 polymorphisms on BC risk with inconsistent conclusions. For example, while some studies found that the -174G>C SNP led to an increased risk of disease [240,242], other studies and meta-analyses found no association between this or other polymorphisms in IL6 or IL6R and BC risk [234,243,244]. Interestingly, a large study evaluating 16 genes for interleukins and their receptors, which included over 100 SNPs, found that associations of polymorphisms (including some in IL6 and IL6R) with risk differed depending on ethnic background [245]; this suggests that heterogeneity in patient cohorts across different studies might contribute to the diverging results seen to date. Further work is needed to better elucidate the effect of IL6 SNPs on BC risk and progression, which most likely depends on several interacting factors.

There is less knowledge on mutations in other IL6-like cytokines. SNPs in LIF, CNTF and OSM that lead to systemic cytokine deficiency have been linked to a broad range of pathologies [17]. However, there is little evidence of such aberrations contributing to cancer, although a recent study showed that polymorphisms in IL11 could affect susceptibility to gastric cancer [246].

7.2. Polymorphisms in Non-Signalling Receptors

Numerous polymorphisms in IL6R have been identified in humans, which largely alter the well-established role of IL6 in inflammation [247]. For example, the SNP p.D358A (rs2228145), affecting the site of cleavage by ADAM proteinases, leads to higher levels of sIL6R [248,249]. This has a systemic protective effect against inflammation that is translated into differential risks of inflammation-related conditions [249–251]. Another, rarer loss-of-function polymorphism in IL6R is linked with severe immune and inflammatory disorders [252].

Genotypic changes in IL6R can also affect BC, with a study finding that the rs11265608 SNP was associated with worse prognosis and reduced DFS [253]. Mutations in IL11R CNTFR have been linked to musculoskeletal alteration humans [254–257], but no aberrations in these receptors have been shown to affect BC.

7.3. Polymorphisms in Signalling Receptors

As the cornerstone of signalling by IL6-like cytokines, aberrations in the signal transducer gp130/IL6ST have the potential to affect many downstream pathways and effects. Introduction of a knock-in gp130/IL6ST mutation was shown to cause hyperactivation of STAT3 through the disruption of a negative feedback mechanism, leading to the promotion of the development of adenocarcinomas [258]. In an important first finding in humans, a study of inflammatory hepatocellular adenomas reported small in-frame deletions in gp130/IL6ST in 60% of samples [259]. These gain-of-function somatic aberrations in the ligand-binding domain of gp130/IL6ST caused changes in intracellular distribution of the receptor and constitutive ligand-independent activation of STAT3 [259,260], exemplifying a novel mechanism for overactivation of pro-tumourigenic signals observed in many tumours. The inflammatory phenotype of these benign cancerous lesions evidences the role of IL6-like cytokines in both inflammation and cancer.

Further work is needed to determine the role of these or other gp130/IL6ST aberrations in other epithelial cancers, including breast neoplasms. Interestingly, as binding of different cytokines involves specific residues in gp130/IL6ST, mutations might affect signalling by different ligands differently. For example, characterisation of a human mutation within a patient report showed that the N404Y mutation affected signal transduction of IL6, IL11, IL27 and OSM, but not LIF [261].

As for the other signalling receptors in the IL6-like cytokine family, there is no significant evidence of any activating mutations in OSMR or LIFR, although loss-of-function polymorphisms have been reported in patients with rare genetic disorders [262–264]. Polymorphisms in OSMR have been shown to be associated with increased risk and differential prognosis in different cancers [265–267], although no aberrations have been shown to play a role in BC.

7.4. Polymorphisms in Downstream Factors

Signalling can also be affected by polymorphisms in factors involved in the main pathways under gp130/IL6ST modulation, as has been extensively reviewed [232]. For example, both JAK2 mutations that lead to constitutive activation of JAK/STAT3 [268–271] and JAK2 fusion proteins [272–275] have been reported at high frequencies in patients with a range of pathologies. JAK polymorphisms have also been reported, albeit at a lower frequency, in solid tumours, including mutations in JAK1 and JAK3 in BC [271,276,277]. Activating mutations in these genes appear to function similarly, by blocking mechanisms of JAK activity autoinhibition [270,278]. Similar aberrations in the kinase TYK2, including activating mutations and gene fusions, have also been reported in myeloid disorders [279,280], but have not been observed in BC.

STAT3 mutations that lead to enhanced dimerisation and the constitutive activation of the JAK/STAT signalling cascade have been reported in leukemia [281,282] and lymphoma [283]. STAT3 polymorphisms have been detected in benign liver tumours [284], the same type of lesions shown to carry activating mutations in gp130/IL6ST [259], JAK1 and the kinase FRK [277], which also lead to constitutive activation of STAT3 [285].

Although mutations in STAT3 have not been reported in other solid tumours, there is extensive evidence of other aberrations causing downregulation of SOCS proteins or inhibition of SPH phosphatases that also lead to JAK/STAT dysregulation and STAT3 hyperactivity [286]. In BC, there is evidence of the expression and high signalling activity of STAT3 [89,287], which could be driven by aberrant JAK forms (reported in some clinical samples) or by other mutations affecting this signalling cascade. For instance, BC cell line models have been shown to carry activating mutations in Src kinases, which also modulate STAT3 and its effect on proliferation [288–290], and anti-STAT3 strategies has been shown to reduce tumour growth in animal studies [291].

These represent only some of the mechanisms that might affect the complex cytokine-driven signalling network in BC. While genotypic changes might vary between tumour types, it seems that alterations to the signalling cascade are a common feature across cancers and present a potential therapeutic strategy. Although this section has focused on polymorphisms in the main signalling axis JAK/STAT, aberrations with potential pro-tumourigenic effects also affect other pathways modulated by gp130/IL6ST, such as PI3K/AKT. For instance, in recent years much work has focused on the role and potential as therapeutic targets of PIK3CA genetic alterations, which are amongst the most common in BC [292,293].

8. Therapeutic Targeting of gp130/IL6ST Signalling

The involvement of IL6-like cytokines and their downstream signalling in many processes considered hallmarks of cancer has highlighted their potential as therapeutic targets. The pleiotropic role of these ligands means that blockade or inhibition of this signalling axis must be fine-tuned to prevent unwanted dysregulations. Despite this challenge, evidence shows that a high level of specificity can be achieved; for example, by recognising a critical residue in site III, the IL6 inhibitor olokizumab can hinder the interaction of gp130/IL6ST with the IL6-IL6R complex, but not other cytokines [294]. On the other hand, this pleiotropy also means that some drugs are already available that might be repurposed. A good example are anti-inflammatory agents that could also be used as anti-cancer treatments (see last paragraph in this section).

Therapeutic approaches to date have included monoclonal antibodies for direct blockade of a ligand or receptor, recombinant cytokine regimes or small-molecule agents that interfere with downstream signalling. A variety of agents are at different stages of pre-clinical and clinical development and some are already in use. While emerging drugs targeting IL6-like cytokines and their signalling have been reviewed recently [19,79,87,88,295,296], in this section we will summarise the main therapeutic approaches and highlight those agents with current or emerging applications in BC.

IL6-mediated signalling might be blocked through direct inhibition of IL6 or IL6R using monoclonal antibodies. Promising pre-clinical evidence led to numerous clinical trials of the anti-IL6 antibody siltuximab to treat multiple myeloma or solid tumours, but these largely reported a lack of efficacy [79,295]. Although other novel anti-IL6 antibodies are being assessed for the treatment of immune disorders [297,298] or COVID19 (NCT04348500), further work is needed to evaluate their potential against BC.

The anti-IL6R antibody tocilizumab is already used to treat inflammation-related disorders and is currently in clinical trials to treat different cancer types. Following evidence from BC cell line models [113], a phase I trial is currently underway to assess use of tocilizumab in combination with HER2-targeted therapy to treat trastuzumab-resistant HER2+ metastatic BC (NCT03135171). Other anti-IL6R antibodies such as sarilumab and NI-1201 are currently also in development [299,300].

The characterisation of IL6-dependent signalling has suggested that specific targeting of its trans-signalling route might be a good strategy to block this cytokine's pro-cancer effects without altering other important roles in homeostasis that normally rely on classic signalling [296]. Most drugs targeting IL6 or IL6R block both the classic and trans-signalling routes, although the emerging junctional epitope antibody VHH6, which binds the IL6-IL6R complex, has been shown to selectively inhibit trans-signalling [301]. The most common approach to achieve selective inhibition of trans-signalling has been the development of fusion proteins incorporating sgp130/sIL6ST, taking advantage of the natural antagonist role of the soluble form of the signal transducer. A prime example is olamkicept, a fusion product including the extracellular portion of the signal transducer (sgp130/sIL6ST) and the Fc portion (fragment crystallisable, a constant region of an immunoglobulin heavy chain) of a human IgG1 antibody [59]. This recombinant protein has been shown to exert a 10-fold greater inhibitory effect than sgp130/sIL6T and completely block IL6 trans-signalling both in vitro and in vivo [48,59,296,302]. Pre-clinical evidence has suggested the promise of olamkicept as an agent with potential to inhibit the role of IL6 in inflammation and cancer [79,302,303]. In theory, it could also block IL11 trans-signalling, although no evidence has been reported to date.

Another strategy involves the direct targeting of gp130/IL6ST with small-molecule inhibitors such as SC144, which has shown promise in preclinical ovarian cancer models [304]. A small-molecule inhibitor named LMT-28, has also been shown to directly target gp130/IL6ST but only inhibits the effects of IL6 and not those of other cytokines in the family [305]. This supports the notion that, as with olokizumab's targeting of IL6's site III, the specific epitopes targeted by an inhibitor can determine which ligand-receptor complexes are blocked from binding the signal transducer. This might greatly impact the selectiveness of a given inhibitor's effect on downstream signalling, so that finer blockade of the effects of a specific cytokine might be achieved [305]. Also in line with this, a recent study showed that targeted mutagenesis of different residues in CLC, which mediate interactions in the CNTF-CNTFR-LIFR-gp130/IL6ST signalling complex, can yield novel recombinant variants with distinct functions [306].

Of particular importance in the context of BC, the selective oestrogen receptor modulators (SERMs) raloxifene and bazedoxifene have been shown to inhibit the IL6-gp130/IL6ST interface [307]. Both agents are currently used to prevent and treat postmenopausal osteoporosis and raloxifene is also used to prevent BC in high-risk women [308,309]. Bazedoxifene has been shown to overcome hormone resistance in BC cells [310,311] and has a gp130/IL6ST-inhibiting effect in preclinical models of several cancer types [169,312–315]. Several BC clinical trials are currently underway: either in conjugation with oestrogens on benign proliferation or preinvasive breast lesions (NCT02729701, NCT02694809) or in combination with palbociclib to treat women with hormone receptor-positive breast tumours [316]. A recent study reported the development of novel bazedoxifene analogues designed to improve on the drug's affinity for and targeting of gp130/IL6ST [317]. Results showed a lead analogue selectively inhibited IL6-dependent activation of JAK2 and STAT3 and suppressed tumour progression both in vitro and in vivo in xenograft lung cancer models. This evidence supports the promise of repurposing bazedoxifene and, now, its improved analogues for specific inhibition of gp130/IL6ST signalling in cancer treatment. This is particularly interesting for the management of BC, where this anti-oestrogen is already in clinical development and could potentially exert a double inhibitory effect.

While therapies targeting IL6-like cytokines such as IL11, OSM and CNTF are at different stages of development for the treatment of a range of diseases [17], only the IL11R-targeted agent BMTP-11 is being developed for its potential use in cancer treatment. There is preclinical evidence of its effect on several cancer types [318,319] and a prostate cancer clinical trial is currently underway [320].

Drugs might also target factors central to signalling downstream of the cytokine signalling complex, an approach with the potential to modulate the effect of most cytokines in the family. Pre-clinical studies have shown that JAK inhibition can inhibit growth of a wide range of cancer types, including breast [321,322]. Several ongoing studies are assessing the potential repurposing of ruxolitinib for cancer treatment [323], including an early phase clinical trial using this agent in combination with HER2-targeted therapy in BC (NCT02066532).

The key role of STAT3 in cytokine-dependent signalling also makes it an attractive therapeutic target but the development of effective inhibitors has proven difficult, due partly to the diffuse localisation of STAT3 in the cell and the high level of homology that complicates specific targeting of STAT3 alone and not other STAT proteins [324]. Despite some authors having labelled STAT3 an "undruggable" factor [79,325], several agents have now been developed to block its expression or function that have shown promise in pre-clinical cancer studies [79,88,324,326]. Some of these have now gone into clinical trials, with some promising preliminary results from a phase I study of OPB-51602, an inhibitor targeting the SH2 domain [327]. An interesting recent study has reported a proteolysis-targeted chimera (PROTAC) that enables potent and specific STAT3 degradation, with results showing complete tumour regression in mouse models of blood cancers [328]. This represents a novel strategy with potential for its application in other cancer types.

As for evidence in BC, the inhibitors G-quartet and S3I-201 have been shown to block STAT3's ability to bind to DNA both in vitro and in vivo, with evidence of tumour regression in BC xenograft models [329,330]. Several drugs blocking STAT3 phosphorylation have also been shown to exert inhibitory effects in triple-negative cell line models [331–334]. However, further research is needed to better assess the therapeutic potential of any of these agents in BC.

As previously mentioned, efforts to develop agents to block gp130/IL6ST in cancer will benefit from a better understanding of the signalling machinery, including the structure of cytokines and receptors and the specific residues involved in the recognition of protein partners and triggering of distinct downstream effects [324]. Better biomarkers are also needed to help guide selection of treatment plans that including targeted agents, either alone or in combination with other therapies. Indeed, evidence suggests that both combination treatments and the use of repurposed agents might be particular promising strategies [19]. For example, anti-IL6 therapies are already commonly used to manage side effects caused by the cytokine release syndrome in patients treated with immunotherapy, which can lead to over-activation of the gp130/IL6ST signalling and increased levels of IL6 [335–340]. In addition to alleviating these adverse effects and enabling better treatment adherence (see also Section 5.1.2), pre-clinical evidence has suggested that this combination treatments might also lead to a greater overall anticancer effect [79,340]. In line with this, research has also shown that, besides its pro-cancer effects, active JAK/STAT signalling also suppresses antitumour immune responses within the TME, suggesting that inhibition of this pathway might lead to a dual anticancer effect through activation of local immunity and also that combination with immunotherapy might enhance treatment response. On the other hand, clinical evidence has shown that IL6 inhibitors can also lead to immune-related side effects, such as increased infections in patients receiving tocilizumab [341].

9. Conclusions

Amongst their many functions, IL6-like cytokines play important roles in breast cancer. Both the prototypical member IL6 and the shared receptor gp130/IL6ST have been established as biomarkers with significant clinical potential in this disease. Other cytokines and receptors might also hold potential as predictors, as do specific polymorphisms in these molecules that continue to be investigated.

Extensive research has led to a better characterisation of the structure and complex interaction between these cytokines and receptors, as well as a more detailed understanding of their intricate downstream signalling. These advances have shed light on the potential for therapeutic targeting of this signalling axis in cancer. Evidence suggests that inhibition of trans-signalling might be a particularly promising strategy. Although the pleiotropic function of these cytokines means that a high level of specificity is needed to achieve effective targeting, numerous novel or repurposed agents are currently at different phases of assessment for their use as single or combination treatments.

Further work is still needed to validate the role of some of these molecules as biomarkers and bring them closer to the clinic. Translation of this biomarker potential, which could help improve patient stratification and treatment selection, together with the potential application of the targeted agents currently under pre-clinical and clinical development, would represent a multi-pronged approach to exploit the central role of IL6-like cytokines in the management of cancer and, specifically, in breast neoplasms.

Author Contributions: C.M.-P. and A.K.T. conceived the review. All authors contributed to and agreed to the published version of the manuscript. All authors have read and agreed to the published version of the manuscript.

Funding: This research received no external funding.

Institutional Review Board Statement: Not applicable.

Informed Consent Statement: Not applicable.

Data Availability Statement: Not applicable.

Acknowledgments: Figures created with Biorender.com.

Conflicts of Interest: The authors declare no conflict of interest.

References

1. Duffy, M.J.; Harbeck, N.; Nap, M.; Molina, R.; Nicolini, A.; Senkus, E.; Cardoso, F. Clinical use of biomarkers in breast cancer: Updated guidelines from the European Group on Tumor Markers (EGTM). *Eur. J. Cancer* **2017**, *75*, 284–298. [CrossRef]
2. Martínez-Pérez, C.; Turnbull, A.K.; Dixon, J.M. The evolving role of receptors as predictive biomarkers for metastatic breast cancer. *Expert Rev. Anticancer. Ther.* **2018**, *19*, 121–138. [CrossRef] [PubMed]
3. Nagaraj, G.; Ma, C.X. Clinical Challenges in the Management of Hormone Receptor-Positive, Human Epidermal Growth Factor Receptor 2-Negative Metastatic Breast Cancer: A Literature Review. *Adv. Ther.* **2020**, *38*, 109–136. [CrossRef]
4. Lim, B.; Hortobagyi, G.N. Current challenges of metastatic breast cancer. *Cancer Metastasis Rev.* **2016**, *35*, 495–514. [CrossRef]
5. Wu, M.; Li, Q.; Wang, H. Identification of Novel Biomarkers Associated With the Prognosis and Potential Pathogenesis of Breast Cancer via Integrated Bioinformatics Analysis. *Technol. Cancer Res. Treat.* **2021**, *20*, 1–16. [CrossRef]
6. Falzone, L.; Grimaldi, M.; Celentano, E.; Augustin, L.S.A.; Libra, M. Identification of Modulated MicroRNAs Associated with Breast Cancer, Diet, and Physical Activity. *Cancers* **2020**, *12*, 2555. [CrossRef]
7. Zhang, K.; Luo, Z.; Zhang, Y.; Song, X.; Zhang, L.; Wu, L.; Liu, J. Long non-coding RNAs as novel biomarkers for breast cancer invasion and metastasis. *Oncol. Lett.* **2017**, *14*, 1895–1904. [CrossRef] [PubMed]
8. Walsh, M.F.; Nathanson, K.L.; Couch, F.J.; Offit, K. Genomic Biomarkers for Breast Cancer Risk. *Adv. Exp. Med. Biol.* **2016**, *882*, 1–32. [CrossRef] [PubMed]
9. Bravo, J.; Heath, J.K. New embo members' review: Receptor recognition by gp130 cytokines. *EMBO J.* **2000**, *19*, 2399–2411. [CrossRef] [PubMed]
10. Boulay, J.-L.; Paul, W.E. Hematopoietin sub-family classification based on size, gene organization and sequence homology. *Curr. Biol.* **1993**, *3*, 573–581. [CrossRef]
11. Sprang, S.R.; Bazan, J.F. Cytokine structural taxonomy and mechanisms of receptor engagement: Current opinion in structural biology 1993, 3:815–827. *Curr. Opin. Struct. Biol.* **1993**, *3*, 815–827. [CrossRef]
12. Garbers, C.; Rose-John, S. Dissecting Interleukin-6 Classic- and Trans-Signaling in Inflammation and Cancer. In *Methods in Molecular Biology*; Humana Press Inc.: New York, NY, USA, 2018; Volume 1725, pp. 127–140.
13. Hermanns, H.M. Oncostatin M and interleukin-31: Cytokines, receptors, signal transduction and physiology. *Cytokine Growth Factor Rev.* **2015**, *26*, 545–558. [CrossRef]
14. Ferretti, E.; Corcione, A.; Pistoia, V. The IL-31/IL-31 receptor axis: General features and role in tumor microenvironment. *J. Leukoc. Biol.* **2017**, *102*, 711–717. [CrossRef]
15. Rose-John, S. Interleukin-6 Family Cytokines. *Cold Spring Harb. Perspect. Biol.* **2017**, *10*, a028415. [CrossRef]
16. Kastelein, R.A.; Hunter, C.A.; Cua, D.J. Discovery and Biology of IL-23 and IL-27: Related but Functionally Distinct Regulators of Inflammation. *Annu. Rev. Immunol.* **2007**, *25*, 221–242. [CrossRef] [PubMed]
17. Murakami, M.; Kamimura, D.; Hirano, T. Pleiotropy and Specificity: Insights from the Interleukin 6 Family of Cytokines. *Immunity* **2019**, *50*, 812–831. [CrossRef]
18. Collison, L.W.; Workman, C.J.; Kuo, T.T.; Boyd, K.; Wang, Y.; Vignali, K.M.; Cross, R.; Sehy, D.; Blumberg, R.S.; Vignali, D.A.A. The inhibitory cytokine IL-35 contributes to regulatory T-cell function. *Nature* **2007**, *450*, 566–569. [CrossRef] [PubMed]
19. Jones, S.A.; Jenkins, B.J. Recent insights into targeting the IL-6 cytokine family in inflammatory diseases and cancer. *Nat. Rev. Immunol.* **2018**, *18*, 773–789. [CrossRef]
20. Sun, L.; He, C.; Nair, L.; Yeung, J.; Egwuagu, C.E. Interleukin 12 (IL-12) family cytokines: Role in immune pathogenesis and treatment of CNS autoimmune disease. *Cytokine* **2015**, *75*, 249–255. [CrossRef] [PubMed]
21. Huising, M.O.; Kruiswijk, C.P.; Flik, G. Phylogeny and evolution of class-I helical cytokines. *J. Endocrinol.* **2006**, *189*, 1–25. [CrossRef] [PubMed]
22. Brocker, C.; Thompson, D.; Matsumoto, A.; Nebert, D.W.; Vasiliou, V. Evolutionary divergence and functions of the human interleukin (IL) gene family. *Hum. Genom.* **2010**, *5*, 30–55. [CrossRef]
23. Jones, L.; Vignali, D.A.A. Molecular interactions within the IL-6/IL-12 cytokine/receptor superfamily. *Immunol. Res.* **2011**, *51*, 5–14. [CrossRef] [PubMed]

24. Silver, J.S.; Hunter, C.A. gp130 at the nexus of inflammation, autoimmunity, and cancer. *J. Leukoc. Biol.* **2010**, *88*, 1145–1156. [CrossRef]
25. Heinrich, P.C.; Behrmann, I.; Müller-Newen, G.; Schaper, F.; Graeve, L. Interleukin-6-type cytokine signalling through the gp130/Jak/STAT pathway. *Biochem. J.* **1998**, *334*, 297–314. [CrossRef] [PubMed]
26. Monhasery, N.; Moll, J.; Cuman, C.; Franke, M.; Lamertz, L.; Nitz, R.; Görg, B.; Häussinger, D.; Lokau, J.; Floss, D.M.; et al. Transcytosis of IL-11 and Apical Redirection of gp130 Is Mediated by IL-11α Receptor. *Cell Rep.* **2016**, *16*, 1067–1081. [CrossRef]
27. Knüpfer, H.; Preiss, R. Lack of Knowledge: Breast Cancer and the Soluble Interleukin-6 Receptor. *Breast Care* **2010**, *5*, 177–180. [CrossRef] [PubMed]
28. Metcalfe, R.D.; Putoczki, T.L.; Griffin, M.D.W. Structural Understanding of Interleukin 6 Family Cytokine Signaling and Targeted Therapies: Focus on Interleukin 11. *Front. Immunol.* **2020**, *11*, 1424. [CrossRef]
29. Ward, L.D.; Howlett, G.J.; Discolo, G.; Yasukawa, K.; Hammacher, A.; Moritz, R.L.; Simpson, R. High affinity interleukin-6 receptor is a hexameric complex consisting of two molecules each of interleukin-6, interleukin-6 receptor, and gp-130. *J. Biol. Chem.* **1994**, *269*, 23286–23289. [CrossRef]
30. Boulanger, M.J.; Chow, D.-C.; Brevnova, E.E.; Garcia, K.C. Hexameric Structure and Assembly of the Interleukin-6/IL-6 α-Receptor/gp130 Complex. *Science* **2003**, *300*, 2101–2104. [CrossRef]
31. Paonessa, G.; Graziani, R.; De Serio, A.; Savino, R.; Ciapponi, L.; Lahm, A.; Salvati, A.L.; Toniatti, C.; Ciliberto, G. Two distinct and independent sites on IL-6 trigger gp 130 dimer formation and signalling. *EMBO J.* **1995**, *14*, 1942–1951. [CrossRef]
32. Grotzinger, J.; Kernebeck, T.; Kallen, K.-J.; Rose-John, S. IL-6 Type Cytokine Receptor Complexes: Hexamer, Tetramer or Both. *Biol. Chem.* **1999**, *380*, 803–813. [CrossRef]
33. Barton, V.A.; Hall, M.A.; Hudson, K.R.; Heath, J.K. Interleukin-11 Signals through the Formation of a Hexameric Receptor Complex. *J. Biol. Chem.* **2000**, *275*, 36197–36203. [CrossRef] [PubMed]
34. De Serio, A.; Graziani, R.; Laufer, R.; Ciliberto, G.; Paonessa, G. In vitro Binding of Ciliary Neurotrophic Factor to its Receptors: Evidence for the Formation of an IL-6-type Hexameric Complex. *J. Mol. Biol.* **1995**, *254*, 795–800. [CrossRef]
35. Müller-Newen, G. The Cytokine Receptor gp130: Faithfully Promiscuous. *Sci. Signal.* **2003**, *2003*, pe40. [CrossRef]
36. Matthews, V.; Schuster, B.; Schütze, S.; Bussmeyer, I.; Ludwig, A.; Hundhausen, C.; Sadowski, T.; Saftig, P.; Hartmann, D.; Kallen, K.-J.; et al. Cellular Cholesterol Depletion Triggers Shedding of the Human Interleukin-6 Receptor by ADAM10 and ADAM17 (TACE). *J. Biol. Chem.* **2003**, *278*, 38829–38839. [CrossRef] [PubMed]
37. Mülberg, J.; Schooltink, H.; Stoyan, T.; Günther, M.; Graeve, L.; Buse, J.; Mackiewicz, A.; Heinrich, P.C.; Rose-John, S. The soluble interleukin-6 receptor is generated by shedding. *Eur. J. Immunol.* **1993**, *23*, 473–480. [CrossRef]
38. Lust, J.A.; Donovan, K.A.; Kline, M.P.; Greipp, P.R.; Kyle, R.A.; Maihle, N.J. Isolation of an mRNA encoding a soluble form of the human interleukin-6 receptor. *Cytokine* **1992**, *4*, 96–100. [CrossRef]
39. Rose-John, S. The soluble interleukin-6 receptor and related proteins. *Best Pract. Res. Clin. Endocrinol. Metab.* **2015**, *29*, 787–797. [CrossRef]
40. Heink, S.; Yogev, N.; Garbers, C.; Herwerth, M.; Aly, L.; Gasperi, C.; Husterer, V.; Croxford, A.L.; Möller-Hackbarth, K.; Bartsch, H.S.; et al. Trans-presentation of IL-6 by dendritic cells is required for the priming of pathogenic TH17 cells. *Nat. Immunol.* **2016**, *18*, 74–85. [CrossRef]
41. Chalaris, A.; Garbers, C.; Rabe, B.; Rose-John, S.; Scheller, J. The soluble Interleukin 6 receptor: Generation and role in inflammation and cancer. *Eur. J. Cell Biol.* **2011**, *90*, 484–494. [CrossRef]
42. Hunter, C.A.; Jones, S.A. IL-6 as a keystone cytokine in health and disease. *Nat. Immunol.* **2015**, *16*, 448–457. [CrossRef] [PubMed]
43. Singh, A.; Purohit, A.; Wang, D.Y.; Duncan, L.; Ghilchik, M.W.; Reed, M.J. IL-6sR: Release from mcf-7 breast cancer cells and role in regulating peripheral oestrogen synthesis. *J. Endocrinol.* **1995**, *147*, R9–R12. [CrossRef]
44. Gaillard, J.; Pugnière, M.; Tresca, J.; Mani, J.; Klein, B.; Brochier, J. Interleukin-6 receptor signaling. II. Bio-availability of interleukin-6 in serum. *Eur. Cytokine Netw.* **1999**, *10*, 337–343. [PubMed]
45. Peters, M.; Odenthal, M.; Schirmacher, P.; Blessing, M.; Fattori, E.; Ciliberto, G.; Buschenfelde, K.H.M.Z.; Rose-John, S. Soluble IL-6 receptor leads to a paracrine modulation of the IL-6-induced hepatic acute phase response in double transgenic mice. *J. Immunol.* **1997**, *159*, 1474–1481. [PubMed]
46. Becker, C.; Fantini, M.C.; Schramm, C.; Lehr, H.A.; Wirtz, S.; Nikolaev, A.; Burg, J.; Strand, S.; Kiesslich, R.; Huber, S.; et al. TGF-β Suppresses Tumor Progression in Colon Cancer by Inhibition of IL-6 trans-Signaling. *Immunity* **2004**, *21*, 491–501. [CrossRef]
47. Jones, S.A.; Scheller, J.; Rose-John, S. Therapeutic strategies for the clinical blockade of IL-6/gp130 signaling. *J. Clin. Investig.* **2011**, *121*, 3375–3383. [CrossRef]
48. Lokau, J.; Nitz, R.; Agthe, M.; Monhasery, N.; Aparicio-Siegmund, S.; Schumacher, N.; Wolf, J.; Möller-Hackbarth, K.; Waetzig, G.H.; Grötzinger, J.; et al. Proteolytic Cleavage Governs Interleukin-11 Trans-signaling. *Cell Rep.* **2016**, *14*, 1761–1773. [CrossRef] [PubMed]
49. Davis, S.; Aldrich, T.H.; Ip, N.Y.; Stahl, N.; Scherer, S.; Farruggella, T.; DiStefano, P.S.; Curtis, R.; Panayotatos, N.; Gascan, H.; et al. Released Form of CNTF Receptor α Component as a Soluble Mediator of CNTF Responses. *Science* **1993**, *259*, 1736–1739. [CrossRef]
50. Pflanz, S.; Kernebeck, T.; Giese, B.; Herrmann, A.; Pachta-Nick, M.; Stahl, J.; Wollmer, A.; Heinrich, P.C.; Müller-Newen, G.; Grötzinger, J. Signal transducer gp130: Biochemical characterization of the three membrane-proximal extracellular domains and evaluation of their oligomerization potential. *Biochem. J.* **2001**, *356*, 605–612. [CrossRef]

51. Xu, Y.; Kershaw, N.; Luo, C.S.; Soo, P.; Pocock, M.J.; Czabotar, P.; Hilton, D.; Nicola, N.; Garrett, T.P.J.; Zhang, J.-G. Crystal Structure of the Entire Ectodomain of gp130: Insights into the molecular assembly of the tall cytokine receptor complexes. *J. Biol. Chem.* **2010**, *285*, 21214–21218. [CrossRef] [PubMed]
52. Zhang, J.-G.; Zhang, Y.; Owczarek, C.M.; Ward, L.D.; Moritz, R.L.; Simpson, R.; Yasukawa, K.; Nicola, N. Identification and Characterization of Two Distinct Truncated Forms of gp130 and a Soluble Form of Leukemia Inhibitory Factor Receptor α-Chain in Normal Human Urine and Plasma. *J. Biol. Chem.* **1998**, *273*, 10798–10805. [CrossRef]
53. Waetzig, G.H.; Chalaris, A.; Rosenstiel, P.; Suthaus, J.; Holland, C.; Karl, N.; Uriarte, L.V.; Till, A.; Scheller, J.; Grötzinger, J.; et al. N-Linked Glycosylation Is Essential for the Stability but Not the Signaling Function of the Interleukin-6 Signal Transducer Glycoprotein 130. *J. Biol. Chem.* **2010**, *285*, 1781–1789. [CrossRef]
54. Diamant, M.; Rieneck, K.; Mechti, N.; Zhang, X.-G.; Svenson, M.; Bendtzen, K.; Klein, B. Cloning and expression of an alternatively spliced mRNA encoding a soluble form of the human interleukin-6 signal transducer gp1301. *FEBS Lett.* **1997**, *412*, 379–384. [CrossRef]
55. Montero-Julian, F.A.; Brailly, H.; Sautès, C.; Joyeux, I.; Dorval, T.; Mosseri, V.; Yasukawa, K.; Wijdenes, J.; Adler, A.; Gorin, I.; et al. Characterization of soluble gp130 released by melanoma cell lines: A polyvalent antagonist of cytokines from the interleukin 6 family. *Clin. Cancer Res.* **1997**, *3*, 1443–1451. [PubMed]
56. Narazaki, M.; Yasukawa, K.; Saito, T.; Ohsugi, Y.; Fukui, H.; Koishihara, Y.; Yancopoulos, G.; Taga, T.; Kishimoto, T. Soluble forms of the interleukin-6 signal-transducing receptor component gp130 in human serum possessing a potential to inhibit signals through membrane-anchored gp130. *Blood* **1993**, *82*, 1120–1126. [CrossRef]
57. Wolf, J.; Waetzig, G.H.; Chalaris, A.; Reinheimer, T.M.; Wege, H.; Rose-John, S.; Garbers, C. Different Soluble Forms of the Interleukin-6 Family Signal Transducer gp130 Fine-tune the Blockade of Interleukin-6 Trans-signaling. *J. Biol. Chem.* **2016**, *291*, 16186–16196. [CrossRef] [PubMed]
58. Müllberg, J.; Oberthür, W.; Lottspeich, F.; Mehl, E.; Dittrich, E.; Graeve, L.; Heinrich, P.C.; Rose-John, S. The soluble human IL-6 receptor. Mutational characterization of the proteolytic cleavage site. *J. Immunol.* **1994**, *152*, 4958–4968.
59. Jostock, T.; Müllberg, J.; Özbek, S.; Atreya, R.; Blinn, G.; Voltz, N.; Fischer, M.; Neurath, M.F.; Rose-John, S. Soluble gp130 is the natural inhibitor of soluble interleukin-6 receptor transsignaling responses. *JBIC J. Biol. Inorg. Chem.* **2001**, *268*, 160–167. [CrossRef]
60. Rabe, B.; Chalaris, A.; May, U.; Waetzig, G.H.; Seegert, D.; Williams, A.S.; Jones, S.A.; Rose-John, S.; Scheller, J. Transgenic blockade of interleukin 6 transsignaling abrogates inflammation. *Blood* **2008**, *111*, 1021–1028. [CrossRef]
61. Rose-John, S.; Heinrich, P.C. Soluble receptors for cytokines and growth factors: Generation and biological function. *Biochem. J.* **1994**, *300*, 281–290. [CrossRef]
62. Richards, P.J.; Nowell, M.A.; Horiuchi, S.; McLoughlin, R.M.; Fielding, C.A.; Grau, S.; Yamamoto, N.; Ehrmann, M.; Rose-John, S.; Williams, A.S.; et al. Functional characterization of a soluble gp130 isoform and its therapeutic capacity in an experimental model of inflammatory arthritis. *Arthritis Rheum.* **2006**, *54*, 1662–1672. [CrossRef]
63. Lamertz, L.; Rummel, F.; Polz, R.; Baran, P.; Hansen, S.; Waetzig, G.H.; Moll, J.M.; Floss, D.M.; Scheller, J. Soluble gp130 prevents interleukin-6 and interleukin-11 cluster signaling but not intracellular autocrine responses. *Sci. Signal.* **2018**, *11*, eaar7388. [CrossRef]
64. Diveu, C.; Venereau, E.; Froger, J.; Ravon, E.; Grimaud, L.; Rousseau, F.; Chevalier, S.; Gascan, H. Molecular and Functional Characterization of a Soluble Form of Oncostatin M/Interleukin-31 Shared Receptor. *J. Biol. Chem.* **2006**, *281*, 36673–36682. [CrossRef]
65. Heaney, M.L.; Golde, D.W. Soluble cytokine receptors. *Blood* **1996**, *87*, 847–857. [CrossRef]
66. Honda, M.; Yamamoto, S.; Cheng, M.; Yasukawa, K.; Suzuki, H.; Saito, T.; Osugi, Y.; Tokunaga, T.; Kishimoto, T. Human soluble IL-6 receptor: Its detection and enhanced release by HIV infection. *J. Immunol.* **1992**, *148*, 2175–2180. [PubMed]
67. Padberg, F.; Feneberg, W.; Schmidt, S.; Schwarz, M.; Körschenhausen, D.; Greenberg, B.D.; Nolde, T.; Müller, N.; Trapmann, H.; König, N.; et al. CSF and serum levels of soluble interleukin-6 receptors (sIL-6R and sgp130), but not of interleukin-6 are altered in multiple sclerosis. *J. Neuroimmunol.* **1999**, *99*, 218–223. [CrossRef]
68. Heinrich, P.C.; Behrmann, I.; Haan, S.; Hermanns, H.M.; Müller-Newen, G.; Schaper, F. Principles of interleukin (IL)-6-type cytokine signalling and its regulation. *Biochem. J.* **2003**, *374*, 1–20. [CrossRef] [PubMed]
69. Bousoik, E.; Aliabadi, H.M. "Do We Know Jack" About JAK? A Closer Look at JAK/STAT Signaling Pathway. *Front. Oncol.* **2018**, *8*, 287. [CrossRef] [PubMed]
70. Ernst, M.; Jenkins, B. Acquiring signalling specificity from the cytokine receptor gp130. *Trends Genet.* **2004**, *20*, 23–32. [CrossRef] [PubMed]
71. Schaper, F.; Rose-John, S. Interleukin-6: Biology, signaling and strategies of blockade. *Cytokine Growth Factor Rev.* **2015**, *26*, 475–487. [CrossRef]
72. Johnston, P.; Grandis, J.R. Stat3 signaling: Anticancer Strategies and Challenges. *Mol. Interv.* **2011**, *11*, 18–26. [CrossRef]
73. Schuster, B.; Kovaleva, M.; Sun, Y.; Regenhard, P.; Matthews, V.; Grötzinger, J.; Rose-John, S.; Kallen, K.-J. Signaling of Human Ciliary Neurotrophic Factor (CNTF) Revisited: The interleukin-6 receptor can serve as an α-receptor for CNTF. *J. Biol. Chem.* **2003**, *278*, 9528–9535. [CrossRef]

74. Chevalier, S.; Fourcin, M.; Robledo, O.; Wijdenes, J.; Pouplard-Barthelaix, A.; Gascan, H. Interleukin-6 Family of Cytokines Induced Activation of Different Functional Sites Expressed by gp130 Transducing Protein. *J. Biol. Chem.* **1996**, *271*, 14764–14772. [CrossRef]
75. Gu, Z.-J.; Wijdenes, J.; Zhang, X.-G.; Hallet, M.-M.; Clement, C.; Klein, B. Anti-gp130 transducer monoclonal antibodies specifically inhibiting ciliary neurotrophic factor, interleukin-6, interleukin-11, leukemia inhibitory factor or oncostatin M. *J. Immunol. Methods* **1996**, *190*, 21–27. [CrossRef] [PubMed]
76. Sommer, J.; Effenberger, T.; Volpi, E.; Waetzig, G.H.; Bernhardt, M.; Suthaus, J.; Garbers, C.; Rose-John, S.; Floss, D.M.; Scheller, J. Constitutively Active Mutant gp130 Receptor Protein from Inflammatory Hepatocellular Adenoma Is Inhibited by an Anti-gp130 Antibody That Specifically Neutralizes Interleukin 11 Signaling. *J. Biol. Chem.* **2012**, *287*, 13743–13751. [CrossRef]
77. Garbers, C.; Hermanns, H.; Schaper, F.; Müller-Newen, G.; Grötzinger, J.; Rose-John, S.; Scheller, J. Plasticity and cross-talk of interleukin 6-type cytokines. *Cytokine Growth Factor Rev* **2012**, *23*, 85–97. [CrossRef] [PubMed]
78. Hirano, T.; Matsuda, T.; Nakajima, K. Signal transduction through gp130 that is shared among the receptors for the interleukin 6 related cytokine subfamily. *Stem Cells* **1994**, *12*, 262–277. [CrossRef]
79. Johnson, D.E.; O'Keefe, R.A.; Grandis, J.R. Targeting the IL-6/JAK/STAT3 signalling axis in cancer. *Nat. Rev. Clin. Oncol.* **2018**, *15*, 234–248. [CrossRef] [PubMed]
80. Kumari, N.; Dwarakanath, B.S.; Das, A.; Bhatt, A.N. Role of interleukin-6 in cancer progression and therapeutic resistance. *Tumor Biol.* **2016**, *37*, 11553–11572. [CrossRef]
81. Candido, S.; Tomasello, B.M.R.; Lavoro, A.; Falzone, L.; Gattuso, G.; Libra, M. Novel Insights into Epigenetic Regulation of IL6 Pathway: In Silico Perspective on Inflammation and Cancer Relationship. *Int. J. Mol. Sci.* **2021**, *22*, 10172. [CrossRef]
82. Vainer, N.; Dehlendorff, C.; Johansen, J.S. Systematic literature review of IL-6 as a biomarker or treatment target in patients with gastric, bile duct, pancreatic and colorectal cancer. *Oncotarget* **2018**, *9*, 29820–29841. [CrossRef]
83. Dranoff, G. Cytokines in cancer pathogenesis and cancer therapy. *Nat. Rev. Cancer* **2004**, *4*, 11–22. [CrossRef] [PubMed]
84. Hirano, T. IL-6 in inflammation, autoimmunity and cancer. *Int. Immunol.* **2020**, *33*, 127–148. [CrossRef]
85. Browning, L.; Patel, M.R.; Horvath, E.B.; Tawara, K.; Jorcyk, C.L. IL-6 and ovarian cancer: Inflammatory cytokines in promotion of metastasis. *Cancer Manag. Res.* **2018**, *10*, 6685–6693. [CrossRef] [PubMed]
86. Omokehinde, T.; Johnson, R.W. GP130 Cytokines in Breast Cancer and Bone. *Cancers* **2020**, *12*, 326. [CrossRef]
87. Kaur, S.; Bansal, Y.; Kumar, R.; Bansal, G. A panoramic review of IL-6: Structure, pathophysiological roles and inhibitors. *Bioorganic Med. Chem.* **2020**, *28*, 115327. [CrossRef]
88. Masjedi, A.; Hashemi, V.; Hojjat-Farsangi, M.; Ghalamfarsa, G.; Azizi, G.; Yousefi, M.; Jadidi-Niaragh, F. The significant role of interleukin-6 and its signaling pathway in the immunopathogenesis and treatment of breast cancer. *Biomed. Pharmacother.* **2018**, *108*, 1415–1424. [CrossRef]
89. Barbieri, I.; Pensa, S.; Pannellini, T.; Quaglino, E.; Maritano, D.; Demaria, M.; Voster, A.; Turkson, J.; Cavallo, F.; Watson, C.J.; et al. Constitutively Active Stat3 Enhances Neu-Mediated Migration and Metastasis in Mammary Tumors via Upregulation of Cten. *Cancer Res.* **2010**, *70*, 2558–2567. [CrossRef] [PubMed]
90. Leslie, K.; Gao, S.P.; Berishaj, M.; Podsypanina, K.; Ho, H.; Ivashkiv, L.; Bromberg, J. Differential interleukin-6/Stat3 signaling as a function of cellular context mediates Ras-induced transformation. *Breast Cancer Res.* **2010**, *12*, R80. [CrossRef]
91. Liang, F.; Ren, C.; Wang, J.; Wang, S.; Yang, L.; Han, X.; Chen, Y.; Tong, G.; Yang, G. The crosstalk between STAT3 and p53/RAS signaling controls cancer cell metastasis and cisplatin resistance via the Slug/MAPK/PI3K/AKT-mediated regulation of EMT and autophagy. *Oncogenesis* **2019**, *8*, 59. [CrossRef] [PubMed]
92. Knüpfer, H.; Preiss, R. Significance of interleukin-6 (IL-6) in breast cancer (review). *Breast Cancer Res. Treat.* **2006**, *102*, 129–135. [CrossRef] [PubMed]
93. Danforth, D.N.; Sgagias, M.K. Interleukin-1α and Interleukin-6 Act Additively to Inhibit Growth of MCF-7 Breast Cancer Cells in Vitro. *Cancer Res.* **1993**, *53*.
94. Morinaga, Y.; Suzuki, H.; Takatsuki, F.; Akiyama, Y.; Taniyama, T.; Matsushima, K.; Onozaki, K. Contribution of IL-6 to the antiproliferative effect of IL-1 and tumor necrosis factor on tumor cell lines. *J. Immunol.* **1989**, *143*, 3538–35342. [PubMed]
95. Chiu, J.J.; Sgagias, M.K.; Cowan, K.H. Interleukin 6 acts as a paracrine growth factor in human mammary carcinoma cell lines. *Clin. Cancer Res.* **1996**, *2*, 215–221. [PubMed]
96. Tamm, I.; Cardinale, I.; Krueger, J.; Murphy, J.S.; May, L.T.; Sehgal, P.B. Interleukin 6 decreases cell-cell association and increases motility of ductal breast carcinoma cells. *J. Exp. Med.* **1989**, *170*, 1649–1669. [CrossRef] [PubMed]
97. Asgeirsson, K.S.; Olafsdottir, K.; Jonasson, J.G.; Ögmundsdóttir, H.M. The effects of il-6 on cell adhesion and e-cadherin expression in breast cancer. *Cytokine* **1998**, *10*, 720–728. [CrossRef] [PubMed]
98. Badache, A.; Hynes, N.E. Interleukin 6 inhibits proliferation and, in cooperation with an epidermal growth factor receptor autocrine loop, increases migration of T47D breast cancer cells. *Cancer Res.* **2001**, *61*, 383–391. [PubMed]
99. Wang, X.H.; Liu, B.R.; Qu, B.; Xing, H.; Gao, S.L.; Yin, J.M.; Cheng, Y.Q. Silencing STAT3 may inhibit cell growth through regulating signaling pathway, telomerase, cell cycle, apoptosis and angiogenesis in hepatocellular carcinoma: Potential uses for gene therapy. *Neoplasma* **2011**, *58*, 158–164. [CrossRef]
100. Zhang, X.; Zhang, J.; Wei, H.; Tian, Z. STAT3-decoy oligodeoxynucleotide inhibits the growth of human lung cancer via down-regulating its target genes. *Oncol. Rep.* **2007**, *17*, 1377–1382. [CrossRef] [PubMed]

101. Liu, W.; Zhou, Y.; Reske, S.N.; Shen, C. PTEN mutation: Many birds with one stone in tumorigenesis. *Anticancer. Res.* **2009**, *28*, 3613–3619.
102. Trotman, L.C.; Pandolfi, P.P. PTEN and p53: Who will get the upper hand. *Cancer Cell* **2003**, *3*, 97–99. [CrossRef]
103. Shen, W.; Zhou, J.-H.; Broussard, S.R.; Freund, G.G.; Dantzer, R.; Kelley, K.W. Proinflammatory cytokines block growth of breast cancer cells by impairing signals from a growth factor receptor. *Cancer Res.* **2002**, *62*, 4746–4756.
104. Studebaker, A.W.; Storci, G.; Werbeck, J.L.; Sansone, P.; Sasser, A.K.; Tavolari, S.; Huang, T.; Chan, M.; Marini, F.C.; Rosol, T.; et al. Fibroblasts Isolated from Common Sites of Breast Cancer Metastasis Enhance Cancer Cell Growth Rates and Invasiveness in an Interleukin-6–Dependent Manner. *Cancer Res.* **2008**, *68*, 9087–9095. [CrossRef]
105. Sullivan, N.J.; Sasser, A.K.; Axel, A.E.; Vesuna, F.; Raman, V.; Ramirez, N.; Oberyszyn, T.M.; Hall, B.M. Interleukin-6 induces an epithelial–mesenchymal transition phenotype in human breast cancer cells. *Oncogene* **2009**, *28*, 2940–2947. [CrossRef] [PubMed]
106. Selander, K.S.; Li, L.; Watson, L.; Merrell, M.; Dahmen, H.; Heinrich, P.C.; Mü Ller-Newen, G.; Harris, K.W. Inhibition of gp130 Signaling in Breast Cancer Blocks Constitutive Activation of Stat3 and Inhibits in vivo Malignancy. *Cancer Res.* **2004**, *64*, 6924–6933. [CrossRef] [PubMed]
107. Arihiro, K.; Oda, H.; Kaneko, M.; Inai, K. Cytokines facilitate chemotactic motility of breast carcinoma cells. *Breast Cancer* **2000**, *7*, 221–230. [CrossRef] [PubMed]
108. Lin, C.; Liao, W.; Jian, Y.; Peng, Y.; Zhang, X.; Ye, L.; Cui, Y.; Wang, B.; Wu, X.; Xiong, Z.; et al. CGI-99 promotes breast cancer metastasis via autocrine interleukin-6 signaling. *Oncogene* **2017**, *36*, 3695–3705. [CrossRef]
109. Verhasselt, B.; Van Damme, J.; Van Larebeke, N.; Put, W.; Bracke, M.; De Potter, C.; Mareel, M. Interleukin-1 is a motility factor for human breast carcinoma cells in vitro: Additive effect with interleukin-6. *Eur. J. Cell Biol.* **1992**, *59*, 449–457.
110. Yang, X.-M.; Wang, Y.-S.; Zhang, J.; Li, Y.; Xu, J.-F.; Zhu, J.; Zhao, W.; Chu, D.-K.; Wiedemann, P. Role of PI3K/Akt and MEK/ERK in Mediating Hypoxia-Induced Expression of HIF-1 and VEGF in Laser-Induced Rat Choroidal Neovascularization. *Investig. Opthalmol. Vis. Sci.* **2009**, *50*, 1873–1879. [CrossRef]
111. Conze, D.; Weiss, L.; Regen, P.S.; Bhushan, A.; Weaver, D.; Johnson, P.; Rincón, M. Autocrine production of interleukin 6 causes multidrug resistance in breast cancer cells. *Cancer Res.* **2001**, *61*, 8851–8858.
112. Haverty, A.A.; Harmey, J.H.; Redmond, H.; Bouchier-Hayes, D.J. Interleukin-6 Upregulates GP96 Expression in Breast Cancer. *J. Surg. Res.* **1997**, *69*, 145–149. [CrossRef]
113. Korkaya, H.; Kim, G.-I.; Davis, A.; Malik, F.; Henry, N.L.; Ithimakin, S.; Quraishi, A.A.; Tawakkol, N.; D'Angelo, R.; Paulson, A.; et al. Activation of an IL6 Inflammatory Loop Mediates Trastuzumab Resistance in HER2+ Breast Cancer by Expanding the Cancer Stem Cell Population. *Mol. Cell* **2012**, *47*, 570–584. [CrossRef] [PubMed]
114. Garcia-Tuñón, I.; Ricote, M.; Ruiz, A.; Fraile, B.; Paniagua, R.; Royuela, M. IL-6, its receptors and its relationship with bcl-2 and bax proteins in infiltrating and in situ human breast carcinoma. *Histopathology* **2005**, *47*, 82–89. [CrossRef]
115. Motallebnezhad, M.; Jadidi-Niaragh, F.; Qamsari, E.S.; Bagheri, S.; Gharibi, T.; Yousefi, M. The immunobiology of myeloid-derived suppressor cells in cancer. *Tumor Biol.* **2015**, *37*, 1387–1406. [CrossRef]
116. Erez, N.; Glanz, S.; Raz, Y.; Avivi, C.; Barshack, I. Cancer Associated Fibroblasts express pro-inflammatory factors in human breast and ovarian tumors. *Biochem. Biophys. Res. Commun.* **2013**, *437*, 397–402. [CrossRef]
117. Lieblein, J.C.; Ball, S.; Hutzen, B.; Sasser, A.K.; Lin, H.-J.; Huang, T.H.; Hall, B.M.; Lin, J. STAT3 can be activated through paracrine signaling in breast epithelial cells. *BMC Cancer* **2008**, *8*, 302–314. [CrossRef] [PubMed]
118. Sasser, A.K.; Casneuf, T.; Axel, A.E.; King, P.; Alvarez, J.D.; Werbeck, J.L.; Verhulst, T.; Verstraeten, K.; Hall, B.M. Interleukin-6 is a potential therapeutic target in interleukin-6 dependent, estrogen receptor-α-positive breast cancer. *Breast Cancer Targets Ther.* **2016**, *8*, 13–27. [CrossRef]
119. Crichton, M.B.; Nichols, J.E.; Zhao, Y.; Bulun, S.E.; Simpson, E.R. Expression of transcripts of interleukin-6 and related cytokines by human breast tumors, breast cancer cells, and adipose stromal cells. *Mol. Cell. Endocrinol.* **1996**, *118*, 215–220. [CrossRef]
120. Grivennikov, S.; Karin, M. Autocrine IL-6 Signaling: A Key Event in Tumorigenesis? *Cancer Cell* **2008**, *13*, 7–9. [CrossRef]
121. Fisher, D.T.; Appenheimer, M.M.; Evans, S.S. The two faces of IL-6 in the tumor microenvironment. *Semin. Immunol.* **2014**, *26*, 38–47. [CrossRef] [PubMed]
122. Lederle, W.; Depner, S.; Schnur, S.; Obermueller, E.; Catone, N.; Just, A.; Fusenig, N.E.; Mueller, M.M. IL-6 promotes malignant growth of skin SCCs by regulating a network of autocrine and paracrine cytokines. *Int. J. Cancer* **2010**, *128*, 2803–2814. [CrossRef]
123. Green, A.R.; Green, V.L.; White, M.C.; Speirs, V. Expression of cytokine messenger RNA in normal and neoplastic human breast tissue: Identification of interleukin-8 as a potential regulatory factor in breast tumours. *Int. J. Cancer* **1997**, *72*, 937–941. [CrossRef] [PubMed]
124. Basolo, F.; Conaldi, P.G.; Fiore, L.; Calvo, S.; Toniolo, A. Normal breast epithelial cells produce interleukins 6 and 8 together with tumor-necrosis factor: Defective il6 expression in mammary carcinoma. *Int. J. Cancer* **1993**, *55*, 926–930. [CrossRef] [PubMed]
125. Karczewska, A.; Nawrocki, S.; Breborowicz, D.; Filas, V.; Mackiewicz, A. Expression of interleukin-6, interleukin-6 receptor, and glycoprotein 130 correlates with good prognoses for patients with breast carcinoma. *Cancer* **2000**, *88*, 2061–2071. [PubMed]
126. Purohit, A.; Ghilchik, M.W.; Walker, M.M.; Duncan, L.; Wang, D.Y.; Singh, A.; Reed, M.J. Aromatase activity and interleukin-6 production by normal and malignant breast tissues. *J. Clin. Endocrinol. Metab.* **1995**, *80*, 3052–3058. [CrossRef] [PubMed]
127. Kozłowski, L.; Zakrzewska, I.; Tokajuk, P.; Wojtukiewicz, M. Concentration of interleukin-6 (IL-6), interleukin-8 (IL-8) and interleukin-10 (IL-10) in blood serum of breast cancer patients. *Rocz. Akad. Med. Bialymst.* **2003**, *48*, 82–84. [PubMed]

128. Zhang, G.J.; Adachi, I. Serum interleukin-6 levels correlate to tumor progression and prognosis in metastatic breast carcinoma. *Anticancer. Res.* **1999**, *19*, 1427–1432. [PubMed]
129. Bozcuk, H.; Uslu, G.; Samur, M.; Yıldız, M.; Özben, T.; Özdoğan, M.; Artaç, M.; Altunbaş, H.; Akan, I.; Savaş, B. Tumour necrosis factor-alpha, interleukin-6, and fasting serum insulin correlate with clinical outcome in metastatic breast cancer patients treated with chemotherapy. *Cytokine* **2004**, *27*, 58–65. [CrossRef]
130. Salgado, R.; Junius, S.; Benoy, I.; Van Dam, P.; Vermeulen, P.; Van Marck, E.; Huget, P.; Dirix, L.Y. Circulating interleukin-6 predicts survival in patients with metastatic breast cancer. *Int. J. Cancer* **2002**, *103*, 642–646. [CrossRef]
131. Bachelot, T.; Ray-Coquard, I.; Ménétrier-Caux, C.; Rastkha, M.; Duc, A.; Blay, J.-Y. Prognostic value of serum levels of interleukin 6 and of serum and plasma levels of vascular endothelial growth factor in hormone-refractory metastatic breast cancer patients. *Br. J. Cancer* **2003**, *88*, 1721–1726. [CrossRef]
132. Nishimura, R.; Nagao, K.; Miyayama, H.; Matsuda, M.; Baba, K.; Matsuoka, Y.; Yamashita, H.; Fukuda, M.; Mizumoto, T.; Hamamoto, R. An Analysis of Serum Interleukin-6 Levels to Predict Benefits of Medroxyprogesterone Acetate in Advanced or Recurrent Breast Cancer. *Oncology* **2000**, *59*, 166–173. [CrossRef] [PubMed]
133. Yokoe, T.; Lino, Y.; Morishita, Y. Trends of IL-6 and IL-8 levels in patients with recurrent breast cancer: Preliminary report. *Breast Cancer* **2000**, *7*, 187–190. [CrossRef] [PubMed]
134. Lin, S.; Gan, Z.; Han, K.; Yao, Y.; Min, D. Interleukin-6 as a Prognostic Marker for Breast Cancer: A Meta-analysis. *Tumori J.* **2015**, *101*, 535–541. [CrossRef]
135. Pusztai, L.; Mendoza, T.R.; Reuben, J.M.; Martinez, M.M.; Willey, J.S.; Lara, J.; Syed, A.; Fritsche, H.A.; Bruera, E.; Booser, D.; et al. Changes in plasma levels of inflammatory cytokines in response to paclitaxel chemotherapy. *Cytokine* **2003**, *25*, 94–102. [CrossRef]
136. Tsavaris, N.; Kosmas, C.; Vadiaka, M.; Kanelopoulos, P.; Boulamatsis, D. Immune changes in patients with advanced breast cancer undergoing chemotherapy with taxanes. *Br. J. Cancer* **2002**, *87*, 21–27. [CrossRef]
137. Mills, P.J.; Ancoli-Israel, S.; Parker, B.; Natarajan, L.; Hong, S.; Jain, S.; Sadler, G.R.; von Känel, R. Predictors of inflammation in response to anthracycline-based chemotherapy for breast cancer. *Brain Behav. Immun.* **2008**, *22*, 98–104. [CrossRef]
138. Kang, D.-H.; Weaver, M.T.; Park, N.-J.; Smith, B.; McArdle, T.; Carpenter, J. Significant Impairment in Immune Recovery After Cancer Treatment. *Nurs. Res.* **2009**, *58*, 105–114. [CrossRef]
139. Oner-Iyidogan, Y.; Oner, P.; Kocak, H.; Lama, A.; Gurdol, F.; Bekpınar, S.; Unur, N.; Özbek-Kır, Z. Evaluation of leukocyte arylsulphatase a, serum interleukin-6 and urinary heparan sulphate following tamoxifen therapy in breast cancer. *Pharmacol. Res.* **2005**, *52*, 340–345. [CrossRef] [PubMed]
140. Saligan, L.N.; Kim, H.S. A systematic review of the association between immunogenomic markers and cancer-related fatigue. *Brain, Behav. Immun.* **2012**, *26*, 830–848. [CrossRef]
141. Bower, J.E.; Ganz, P.A.; Tao, M.L.; Hu, W.; Belin, T.R.; Sepah, S.; Cole, S.; Aziz, N. Inflammatory Biomarkers and Fatigue during Radiation Therapy for Breast and Prostate Cancer. *Clin. Cancer Res.* **2009**, *15*, 5534–5540. [CrossRef] [PubMed]
142. Cameron, B.A.; Bennett, B.; Li, H.; Boyle, F.; Desouza, P.; Wilcken, N.; Friedlander, M.; Goldstein, D.; Lloyd, A.R. Post-cancer fatigue is not associated with immune activation or altered cytokine production. *Ann. Oncol.* **2012**, *23*, 2890–2895. [CrossRef]
143. Orre, I.J.; Reinertsen, K.V.; Aukrust, P.; Dahl, A.A.; Fosså, S.D.; Ueland, T.; Murison, R. Higher levels of fatigue are associated with higher CRP levels in disease-free breast cancer survivors. *J. Psychosom. Res.* **2011**, *71*, 136–141. [CrossRef] [PubMed]
144. Soygur, H.; Palaoglu, O.; Akarsu, E.S.; Cankurtaran, E.S.; Ozalp, E.; Turhan, L.; Ayhan, I.H. Interleukin-6 levels and HPA axis activation in breast cancer patients with major depressive disorder. *Prog. Neuro-Psychopharmacol. Biol. Psychiatry* **2007**, *31*, 1242–1247. [CrossRef]
145. Musselman, D.L.; Miller, A.H.; Porter, M.R.; Manatunga, A.; Gao, F.; Penna, S.; Pearce, B.D.; Landry, J.; Glover, S.; McDaniel, J.S.; et al. Higher Than Normal Plasma Interleukin-6 Concentrations in Cancer Patients With Depression: Preliminary Findings. *Am. J. Psychiatry* **2001**, *158*, 1252–1257. [CrossRef]
146. Jehn, C.F.; Flath, B.; Strux, A.; Krebs, M.; Possinger, K.; Pezzutto, A.; Lüftner, D. Influence of age, performance status, cancer activity, and IL-6 on anxiety and depression in patients with metastatic breast cancer. *Breast Cancer Res. Treat.* **2012**, *136*, 789–794. [CrossRef] [PubMed]
147. Janelsins, M.C.; Mustian, K.M.; Palesh, O.G.; Mohile, S.G.; Peppone, L.J.; Sprod, L.K.; Heckler, C.E.; Roscoe, J.A.; Katz, A.W.; Williams, J.P.; et al. Differential expression of cytokines in breast cancer patients receiving different chemotherapies: Implications for cognitive impairment research. *Support. Care Cancer* **2011**, *20*, 831–839. [CrossRef] [PubMed]
148. Kesler, S.; Janelsins, M.; Koovakkattu, D.; Palesh, O.; Mustian, K.; Morrow, G.; Dhabhar, F.S. Reduced hippocampal volume and verbal memory performance associated with interleukin-6 and tumor necrosis factor-alpha levels in chemotherapy-treated breast cancer survivors. *Brain, Behav. Immun.* **2013**, *30*, S109–S116. [CrossRef] [PubMed]
149. Hudis, C.A.; Subbaramaiah, K.; Morris, P.G.; Dannenberg, A.J. Breast Cancer Risk Reduction: No Pain, No Gain? *J. Clin. Oncol.* **2012**, *30*, 3436–3438. [CrossRef] [PubMed]
150. Heikkila, K.; Harris, R.; Lowe, G.; Rumley, A.; Yarnell, J.; Gallacher, J.; Ben-Shlomo, Y.; Ebrahim, S.; Lawlor, D.A. Associations of circulating C-reactive protein and interleukin-6 with cancer risk: Findings from two prospective cohorts and a meta-analysis. *Cancer Causes Control.* **2008**, *20*, 15–26. [CrossRef] [PubMed]
151. Il'yasova, D.; Colbert, L.H.; Harris, T.B.; Newman, A.B.; Bauer, D.C.; Satterfield, S.; Kritchevsky, S.B. Circulating Levels of Inflammatory Markers and Cancer Risk in the Health Aging and Body Composition Cohort. *Cancer Epidemiol. Biomark. Prev.* **2005**, *14*, 2413–2418. [CrossRef]

152. Andrews, B.; Shariat, S.F.; Kim, J.-H.; Wheeler, T.M.; Slawin, K.M.; Lerner, S.P. Preoperative plasma levels of interleukin-6 and its soluble receptor predict disease recurrence and survival of patients with bladder cancer. *J. Urol.* **2002**, *167*, 1475–1481. [CrossRef]
153. Shariat, S.F.; Kattan, M.; Traxel, E.; Andrews, B.; Zhu, K.; Wheeler, T.M.; Slawin, K.M. Association of Pre- and Postoperative Plasma Levels of Transforming Growth Factor β1 and Interleukin 6 and Its Soluble Receptor with Prostate Cancer Progression. *Clin. Cancer Res.* **2004**, *10*, 1992–1999. [CrossRef] [PubMed]
154. Alexandrakis, M.; Passam, F.; Boula, A.; Christophoridou, A.; Aloizos, G.; Roussou, P.; Kyriakou, D. Relationship between circulating serum soluble interleukin-6 receptor and the angiogenic cytokines basic fibroblast growth factor and vascular endothelial growth factor in multiple myeloma. *Ann. Hematol.* **2003**, *82*, 19–23. [CrossRef] [PubMed]
155. Stasi, R.; Brunetti, M.; Parma, A.; Di Giulio, C.; Terzoli, E.; Pagano, A. The prognostic value of soluble interleukin-6 receptor in patients with multiple myeloma. *Cancer* **1998**, *82*, 1860–1866. [PubMed]
156. Robak, T.; Wierzbowska, A.; Błasińska-Morawiec, M.; Korycka, A.; Blonski, J.Z. Serum Levels of IL-6 Type Cytokines and Soluble IL-6 Receptors in Active B-Cell Chronic Lymphocytic Leukemia and in Cladribine Induced Remission. *Mediat. Inflamm.* **1999**, *8*, 277–286. [CrossRef]
157. Soresi, L.G.M.; Antona, A.M.F.; Alessandro, G.M. Interleukin-6 and its soluble receptor in patients with liver cirrhosis and hepatocellular carcinoma. *World J. Gastroenterol.* **2006**, *12*, 2563–2568. [CrossRef] [PubMed]
158. Jabłońska, E.; Kiluk, M.; Markiewicz, W.; Piotrowski, L.; Grabowska, Z.; Jabłoński, J. TNF-alpha, IL-6 and their soluble receptor serum levels and secretion by neutrophils in cancer patients. *Arch. Immunol. Ther. Exp.* **2001**, *49*, 63–69.
159. Kovacs, E. Investigation of interleukin-6 (IL-6), soluble IL-6 receptor (sIL-6R) and soluble gp130 (sgp130) in sera of cancer patients. *Biomed. Pharmacother.* **2001**, *55*, 391–396. [CrossRef]
160. Tawara, K.; Scott, H.; Emathinger, J.; Wolf, C.; Lajoie, D.; Hedeen, D.; Bond, L.; Montgomery, P.; Jorcyk, C. HIGH expression of OSM and IL-6 are associated with decreased breast cancer survival: Synergistic induction of IL-6 secretion by OSM and IL-1β. *Oncotarget* **2019**, *10*, 2068–2085. [CrossRef]
161. Lapeire, L.; Hendrix, A.; Lambein, K.; Van Bockstal, M.R.; Braems, G.; Broecke, R.V.D.; Limame, R.; Mestdagh, P.; Vandesompele, J.; Vanhove, C.; et al. Cancer-Associated Adipose Tissue Promotes Breast Cancer Progression by Paracrine Oncostatin M and Jak/STAT3 Signaling. *Cancer Res.* **2014**, *74*, 6806–6819. [CrossRef]
162. Tawara, K.; Scott, H.; Emathinger, J.; Ide, A.; Fox, R.; Greiner, D.; LaJoie, D.; Hedeen, D.; Nandakumar, M.; Oler, A.J.; et al. Co-Expression of VEGF and IL-6 Family Cytokines is Associated with Decreased Survival in HER2 Negative Breast Cancer Patients: Subtype-Specific IL-6 Family Cytokine-Mediated VEGF Secretion. *Transl. Oncol.* **2018**, *12*, 245–255. [CrossRef]
163. Winship, A.; Van Sinderen, M.; Donoghue, J.; Rainczuk, K.; Dimitriadis, E. Targeting Interleukin-11 Receptor-α Impairs Human Endometrial Cancer Cell Proliferation and Invasion In Vitro and Reduces Tumor Growth and Metastasis In Vivo. *Mol. Cancer Ther.* **2016**, *15*, 720–730. [CrossRef]
164. Li, X.; Yang, Q.; Yu, H.; Wu, L.; Zhao, Y.; Zhang, C.; Yue, X.; Liu, Z.; Wu, H.; Haffty, B.G.; et al. LIF promotes tumorigenesis and metastasis of breast cancer through the AKT-mTOR pathway. *Oncotarget* **2014**, *5*, 788–801. [CrossRef] [PubMed]
165. Yue, X.; Zhao, Y.; Zhang, C.; Li, J.; Liu, Z.; Liu, J.; Hu, W. Leukemia inhibitory factor promotes EMT through STAT3-dependent miR-21 induction. *Oncotarget* **2015**, *7*, 3777–3790. [CrossRef]
166. Junk, D.J.; Bryson, B.; Smigiel, J.M.; Parameswaran, N.; Bartel, C.A.; Jackson, M.W. Oncostatin M promotes cancer cell plasticity through cooperative STAT3-SMAD3 signaling. *Oncogene* **2017**, *36*, 4001–4013. [CrossRef]
167. Cocco, C.; Giuliani, N.; DI Carlo, E.; Ognio, E.; Storti, P.; Abeltino, M.; Sorrentino, C.; Ponzoni, M.; Ribatti, D.; Airoldi, I. Interleukin-27 Acts as Multifunctional Antitumor Agent in Multiple Myeloma. *Clin. Cancer Res.* **2010**, *16*, 4188–4197. [CrossRef] [PubMed]
168. Pan, C.-M.; Wang, M.-L.; Chiou, S.-H.; Chen, H.-Y.; Wu, C.-W. Oncostatin M suppresses metastasis of lung adenocarcinoma by inhibiting SLUG expression through coordination of STATs and PIASs signalings. *Oncotarget* **2016**, *7*, 60395–60406. [CrossRef]
169. Thilakasiri, P.; Huynh, J.; Poh, A.; Tan, C.W.; Nero, T.; Tran, K.; Parslow, A.C.; Afshar-Sterle, S.; Baloyan, D.; Hannan, N.J.; et al. Repurposing the selective estrogen receptor modulator bazedoxifene to suppress gastrointestinal cancer growth. *EMBO Mol. Med.* **2019**, *11*, e9539. [CrossRef] [PubMed]
170. Lacroix, M.; Siwek, B.; Marie, P.J.; Body, J.J. Production and regulation of interleukin-11 by breast cancer cells. *Cancer Lett.* **1998**, *127*, 29–35. [CrossRef]
171. Douglas, A.M.; Goss, G.A.; Sutherland, R.L.; Hilton, D.J.; Berndt, M.C.; Nicola, N.A.; Begley, C.G. Expression and function of members of the cytokine receptor superfamily on breast cancer cells. *Oncogene* **1997**, *14*, 661–669. [CrossRef] [PubMed]
172. Johnstone, C.N.; Chand, A.; Putoczki, T.L.; Ernst, M. Emerging roles for IL-11 signaling in cancer development and progression: Focus on breast cancer. *Cytokine Growth Factor Rev.* **2015**, *26*, 489–498. [CrossRef]
173. Putoczki, T.; Wilson, N.; Edwards, K.; McKenzie, B.; Greten, F.; Ernst, M. Interleukin-11 is the dominant IL-6 family cytokine during gastrointestinal 20umourigenesis. *Cytokine* **2013**, *63*, 290. [CrossRef]
174. Bockhorn, J.; Dalton, R.; Nwachukwu, C.; Huang, S.; Prat, A.; Yee, K.; Chang, Y.-F.; Huo, D.; Wen, Y.; Swanson, K.E.; et al. MicroRNA-30c inhibits human breast tumour chemotherapy resistance by regulating TWF1 and IL-11. *Nat. Commun.* **2013**, *4*, 1–14. [CrossRef]
175. Onnis, B.; Fer, N.; Rapisarda, A.; Perez, V.S.; Melillo, G. Autocrine production of IL-11 mediates tumorigenicity in hypoxic cancer cells. *J. Clin. Investig.* **2013**, *123*, 1615–1629. [CrossRef]

176. Jung, J.E.; Lee, H.-G.; Cho, I.-H.; Chung, D.H.; Yoon, S.-H.; Yang, Y.M.; Lee, J.W.; Choi, S.; Park, J.-W.; Ye, S.-K.; et al. STAT3 is a potential modulator of HIF-1-mediated VEGF expression in human renal carcinoma cells. *FASEB J.* **2005**, *19*, 1296–1298. [CrossRef]
177. Marusyk, A.; Tabassum, D.P.; Altrock, P.; Almendro, V.; Michor, F.; Polyak, K. Non-cell-autonomous driving of tumour growth supports sub-clonal heterogeneity. *Nature* **2014**, *514*, 54–58. [CrossRef] [PubMed]
178. Bellido, T.; Borba, V.Z.C.; Roberson, P.; Manolagas, S.C. Activation of the Janus Kinase/STAT (Signal Transducer and Activator of Transcription) Signal Transduction Pathway by Interleukin-6-Type Cytokines Promotes Osteoblast Differentiation*. *Endocrinology* **1997**, *138*, 3666–3676. [CrossRef]
179. Walker, E.C.; McGregor, N.E.; Poulton, I.J.; Solano, M.; Pompolo, S.; Fernandes, T.J.; Constable, M.J.; Nicholson, G.; Zhang, J.-G.; Nicola, N.; et al. Oncostatin M promotes bone formation independently of resorption when signaling through leukemia inhibitory factor receptor in mice. *J. Clin. Investig.* **2010**, *120*, 582–592. [CrossRef] [PubMed]
180. Walker, E.C.; McGregor, N.E.; Poulton, I.J.; Pompolo, S.; Allan, E.H.; Quinn, J.M.; Gillespie, M.T.; Martin, T.J.; Sims, N.A. Cardiotrophin-1 Is an Osteoclast-Derived Stimulus of Bone Formation Required for Normal Bone Remodeling. *J. Bone Miner. Res.* **2008**, *23*, 2025–2032. [CrossRef]
181. Le Pape, F.; Vargas, G.; Clézardin, P. The role of osteoclasts in breast cancer bone metastasis. *J. Bone Oncol.* **2016**, *5*, 93–95. [CrossRef] [PubMed]
182. Girasole, G.; Passeri, G.; Jilka, R.L.; Manolagas, S.C. Interleukin-11: A new cytokine critical for osteoclast development. *J. Clin. Investig.* **1994**, *93*, 1516–1524. [CrossRef]
183. Sotiriou, C.; Lacroix, M.; Lespagnard, L.; Larsimont, D.; Paesmans, M.; Body, J.-J. Interleukins-6 and -11 expression in primary breast cancer and subsequent development of bone metastases. *Cancer Lett.* **2001**, *169*, 87–95. [CrossRef] [PubMed]
184. Ren, L.; Wang, X.; Dong, Z.; Liu, J.; Zhang, S. Bone metastasis from breast cancer involves elevated IL-11 expression and the gp130/STAT3 pathway. *Med Oncol.* **2013**, *30*, 1–9. [CrossRef]
185. Glück, S.; Ross, J.S.; Royce, M.; McKenna, E.F.; Perou, C.; Avisar, E.; Wu, L. TP53 genomics predict higher clinical and pathologic tumor response in operable early-stage breast cancer treated with docetaxel-capecitabine ± trastuzumab. *Breast Cancer Res. Treat.* **2011**, *132*, 781–791. [CrossRef] [PubMed]
186. Finak, G.; Bertos, N.; Pepin, F.; Sadekova, S.; Souleimanova, M.; Zhao, H.; Chen, H.; Omeroglu, G.; Meterissian, S.; Omeroglu, A.; et al. Stromal gene expression predicts clinical outcome in breast cancer. *Nat. Med.* **2008**, *14*, 518–527. [CrossRef]
187. Hanavadi, S.; Martin, T.A.; Watkins, G.; Mansel, R.E.; Jiang, W.G. Expression of Interleukin 11 and Its Receptor and Their Prognostic Value in Human Breast Cancer. *Ann. Surg. Oncol.* **2006**, *13*, 802–808. [CrossRef] [PubMed]
188. Desmedt, C.; Piette, F.; Loi, S.; Wang, Y.; Lallemand, F.; Haibe-Kains, B.; Viale, G.; Delorenzi, M.; Zhang, Y.; D'Assignies, M.S.; et al. Strong Time Dependence of the 76-Gene Prognostic Signature for Node-Negative Breast Cancer Patients in the TRANSBIG Multicenter Independent Validation Series. *Clin. Cancer Res.* **2007**, *13*, 3207–3214. [CrossRef]
189. Koboldt, D.C.; Fulton, R.S.; McLellan, M.D.; Schmidt, H.; Kalicki-Veizer, J.; McMichael, J.F.; Fulton, L.L.; Dooling, D.J.; Ding, L.; Mardis, E.R.; et al. Comprehensive molecular portraits of human breast tumours. *Nature* **2012**, *490*, 61–70. [CrossRef]
190. Curtis, C.; Shah, S.P.; Chin, S.-F.; Turashvili, G.; Rueda, O.M.; Dunning, M.; Speed, D.; Lynch, A.; Samarajiwa, S.; Yuan, Y.; et al. The genomic and transcriptomic architecture of 2,000 breast tumours reveals novel subgroups. *Nature* **2012**, *486*, 346–352. [CrossRef]
191. Albert, R.K.; Connett, J.; Bailey, W.C.; Casaburi, R.; Cooper, J.A.D.; Criner, G.J.; Curtis, J.; Dransfield, M.T.; Han, M.K.; Lazarus, S.C.; et al. Azithromycin for Prevention of Exacerbations of COPD. *N. Engl. J. Med.* **2011**, *365*, 689–698. [CrossRef]
192. Abdollahi, A.; Hahnfeldt, P.; Maercker, C.; Gröne, H.-J.; Debus, J.; Ansorge, W.; Folkman, J.; Hlatky, L.; Huber, P.E. Endostatin's Antiangiogenic Signaling Network. *Mol. Cell* **2004**, *13*, 649–663. [CrossRef]
193. Estrov, Z.; Samal, B.; Lapushin, R.; Kellokumpu-Lehtinen, P.; Sahin, A.A.; Kurzrock, R.; Talpaz, M.; Aggarwal, B.B. Leukemia Inhibitory Factor Binds to Human Breast Cancer Cells and Stimulates Their Proliferation. *J. Interf. Cytokine Res.* **1995**, *15*, 905–913. [CrossRef]
194. Kellokumpu-Lehtinen, P.; Talpaz, M.; Harris, D.; Van, Q.; Kurzrock, R.; Estrov, Z. Leukemia-inhibitory factor stimulates breast, kidney and prostate cancer cell proliferation by paracrine and autocrine pathways. *Int. J. Cancer* **1996**, *66*, 515–519. [CrossRef]
195. Johnson, R.W.; Finger, E.C.; Olcina, M.M.; Vilalta, M.; Aguilera, T.; Miao, Y.; Merkel, A.; Johnson, J.R.; Sterling, J.A.; Wu, J.Y.; et al. Induction of LIFR confers a dormancy phenotype in breast cancer cells disseminated to the bone marrow. *Nat. Cell Biol.* **2016**, *18*, 1078–1089. [CrossRef] [PubMed]
196. Douglas, A.M.; Grant, S.L.; Goss, G.A.; Clousion, D.R.; Sulhirland, R.L.; Beflly, C.G. Oncostatin M induces the differentiation of breast cancer cells. *Int. J. Cancer* **1998**, *75*, 64–73. [CrossRef]
197. Chen, D.; Sun, Y.; Wei, Y.; Zhang, P.; Rezaeian, A.H.; Teruya-Feldstein, J.; Gupta, S.; Liang, H.; Lin, H.-K.; Hung, M.-C.; et al. LIFR is a breast cancer metastasis suppressor upstream of the Hippo-YAP pathway and a prognostic marker. *Nat. Med.* **2012**, *18*, 1511–1517. [CrossRef] [PubMed]
198. Zeng, H.; Qu, J.; Jin, N.; Xu, J.; Lin, C.; Chen, Y.; Yang, X.; He, X.; Tang, S.; Lan, X.; et al. Feedback Activation of Leukemia Inhibitory Factor Receptor Limits Response to Histone Deacetylase Inhibitors in Breast Cancer. *Cancer Cell* **2016**, *30*, 459–473. [CrossRef]
199. Kim, R.S.; Avivar-Valderas, A.; Estrada, Y.; Bragado, P.; Sosa, M.S.; Aguirre-Ghiso, J.A.; Segall, J.E. Dormancy Signatures and Metastasis in Estrogen Receptor Positive and Negative Breast Cancer. *PLoS ONE* **2012**, *7*, e35569. [CrossRef]

200. Wang, X.-J.; Qiao, Y.; Xiao, M.M.; Wang, L.; Chen, J.; Lv, W.; Xu, L.; Li, Y.; Wang, Y.; Tan, M.-D.; et al. Opposing Roles of Acetylation and Phosphorylation in LIFR-Dependent Self-Renewal Growth Signaling in Mouse Embryonic Stem Cells. *Cell Rep.* **2017**, *18*, 933–946. [CrossRef] [PubMed]
201. Iorns, E.; Ward, T.M.; Dean, S.; Jegg, A.; Thomas, D.; Murugaesu, N.; Sims, D.; Mitsopoulos, C.; Fenwick, K.; Kozarewa, I.; et al. Whole genome in vivo RNAi screening identifies the leukemia inhibitory factor receptor as a novel breast tumor suppressor. *Breast Cancer Res. Treat.* **2012**, *135*, 79–91. [CrossRef] [PubMed]
202. Grant, S.L.; Douglas, A.M.; Goss, G.A.; Begley, C.G. Oncostatin M and Leukemia Inhibitory Factor Regulate the Growth of Normal Human Breast Epithelial Cells. *Growth Factors* **2001**, *19*, 153–162. [CrossRef]
203. Franken, N.A.P.; Rodermond, H.M.; Stap, J.; Haveman, J.; Van Bree, C. Clonogenic assay of cells in vitro. *Nat. Protoc.* **2006**, *1*, 2315–2319. [CrossRef] [PubMed]
204. Liu, J.; Spence, M.J.; Wallace, P.M.; Forcier, K.; Hellström, I.; Vestal, R.E. Oncostatin M-specific receptor mediates inhibition of breast cancer cell growth and down-regulation of the c-myc proto-oncogene. *Cell Growth Differ.* **1997**, *8*, 667–676.
205. Li, C.; Ahlborn, T.E.; Kraemer, F.; Liu, J. Oncostatin M–induced growth inhibition and morphological changes of MDA-MB231 breast cancer cells are abolished by blocking the MEK/ERK signaling pathway. *Breast Cancer Res. Treat.* **2001**, *66*, 111–121. [CrossRef]
206. Liu, J.; Hadjokas, N.; Mosley, B.; Estrov, Z.; Spence, M.J.; Vestal, R.E. Oncostatin m-specific receptor expression and function in regulating cell proliferation of normal and malignant mammary epithelial cells. *Cytokine* **1998**, *10*, 295–302. [CrossRef]
207. Underhill-Day, N.; Heath, J. Oncostatin M (OSM) Cytostasis of Breast Tumor Cells: Characterization of an OSM Receptor β–Specific Kernel. *Cancer Res.* **2006**, *66*, 10891–10901. [CrossRef] [PubMed]
208. Jorcyk, C.; Holzer, R.; Ryan, R. Oncostatin M induces cell detachment and enhances the metastatic capacity of T-47D human breast carcinoma cells. *Cytokine* **2006**, *33*, 323–336. [CrossRef]
209. Omokehinde, T.; Jotte, A.; Johnson, R.W. gp130 Cytokines Activate Novel Signaling Pathways and Alter Bone Dissemination in ER + Breast Cancer Cells. *J. Bone Miner. Res.* **2021**, *12*, 326. [CrossRef]
210. Bolin, C.; Tawara, K.; Sutherland, C.; Redshaw, J.; Aranda, P.; Moselhy, J.; Anderson, R.; Jorcyk, C.L. Oncostatin M Promotes Mammary Tumor Metastasis to Bone and Osteolytic Bone Degradation. *Genes Cancer* **2012**, *3*, 117–130. [CrossRef] [PubMed]
211. Tawara, K.; Bolin, C.; Koncinsky, J.; Kadaba, S.; Covert, H.; Sutherland, C.; Bond, L.; Kronz, J.; Garbow, J.R.; Jorcyk, C.L. OSM potentiates preintravasation events, increases CTC counts, and promotes breast cancer metastasis to the lung. *Breast Cancer Res.* **2018**, *20*, 53. [CrossRef]
212. Mosley, B.; De Imus, C.; Friend, D.; Boiani, N.; Thoma, B.; Park, L.S.; Cosman, D. Dual Oncostatin M (OSM) Receptors. Cloning and characterization of an alternative signaling subunit conferring OSM-specific receptor activation. *J. Biol. Chem.* **1996**, *271*, 32635–32643. [CrossRef]
213. West, N.; Murphy, L.C.; Watson, P.H. Oncostatin M suppresses oestrogen receptor-α expression and is associated with poor outcome in human breast cancer. *Endocrine-Related Cancer* **2012**, *19*, 181–195. [CrossRef]
214. Martínez-Pérez, C.; Leung, J.; Kay, C.; Meehan, J.; Gray, M.; Dixon, J.; Turnbull, A. The Signal Transducer IL6ST (gp130) as a Predictive and Prognostic Biomarker in Breast Cancer. *J. Pers. Med.* **2021**, *11*, 618. [CrossRef] [PubMed]
215. Fontanini, G.; Campani, D.; Roncella, M.; Cecchetti, D.; Calvo, S.; Toniolo, A.; Basolo, F. Expression of Interleukin 6 (IL-6) Correlates with Oestrogen Receptor in Human Breast Carcinoma. *Brit J Cancer* **1999**, *80*, 579–584. [CrossRef]
216. Hartman, Z.C.; Yang, X.-Y.; Glass, O.; Lei, G.; Osada, T.; Dave, S.S.; Morse, M.A.; Clay, T.M.; Lyerly, H. HER2 Overexpression Elicits a Proinflammatory IL-6 Autocrine Signaling Loop That Is Critical for Tumorigenesis. *Cancer Res.* **2011**, *71*, 4380–4391. [CrossRef]
217. Hartman, Z.C.; Poage, G.M.; Hollander, P.D.; Tsimelzon, A.; Hill, J.; Panupinthu, N.; Zhang, Y.; Mazumdar, A.; Hilsenbeck, S.G.; Mills, G.B.; et al. Growth of Triple-Negative Breast Cancer Cells Relies upon Coordinate Autocrine Expression of the Proinflammatory Cytokines IL-6 and IL-8. *Cancer Res.* **2013**, *73*, 3470–3480. [CrossRef] [PubMed]
218. Chavey, C.; Bibeau, F.; Gourgou-Bourgade, S.; Burlinchon, S.; Boissière, F.; Laune, D.; Roques, S.; Lazennec, G. Oestrogen receptor negative breast cancers exhibit high cytokine content. *Breast Cancer Res.* **2007**, *9*, R15. [CrossRef]
219. Dethlefsen, C.; Højfeldt, G.; Hojman, P. The role of intratumoral and systemic IL-6 in breast cancer. *Breast Cancer Res. Treat.* **2013**, *138*, 657–664. [CrossRef]
220. Won, H.S.; Kim, Y.A.; Lee, J.S.; Jeon, E.K.; An, H.J.; Sun, D.S.; Ko, Y.H.; Kim, J.S. Soluble Interleukin-6 Receptor is a Prognostic Marker for Relapse-Free Survival in Estrogen Receptor-Positive Breast Cancer. *Cancer Investig.* **2013**, *31*, 516–521. [CrossRef]
221. Singh, A.; Purohit, A.; Ghilchik, M.W.; Reed, M.J. The regulation of aromatase activity in breast fibroblasts: The role of interleukin-6 and prostaglandin E 2. *Endocr. Relat. Cancer* **1999**, *6*, 139–147. [CrossRef] [PubMed]
222. Irahara, N.; Miyoshi, Y.; Taguchi, T.; Tamaki, Y.; Noguchi, S. Quantitative analysis ofaromatasemRNA expression derived from various promoters (I.4, I.3, PII and I.7) and its association with expression ofTNF-α,IL-6andCOX-2mRNAs in human breast cancer. *Int. J. Cancer* **2005**, *118*, 1915–1921. [CrossRef]
223. Purohit, A.; Newman, S.P.; Reed, M.J. The role of cytokines in regulating estrogen synthesis: Implications for the etiology of breast cancer. *Breast Cancer Res.* **2002**, *4*, 65–69. [CrossRef]
224. Simpson, E.R.; Michael, M.D.; Agarwal, V.R.; Hinshelwood, M.M.; Bulun, S.E.; Zhao, Y. Expression of the CYP19 (aromatase) gene: An unusual case of alternative promoter usage. *FASEB J.* **1997**, *11*, 29–36. [CrossRef]

225. De Miguel, F.; Lee, S.O.; Onate, S.A.; Gao, A.C. Stat3 enhances transactivation of steroid hormone receptors. *Nucl. Recept.* **2003**, *1*, 3. [CrossRef]
226. Zhang, Z.; Jones, S.A.; Hagood, J.S.; Fuentes, N.L.; Fuller, G.M. STAT3 Acts as a Co-activator of Glucocorticoid Receptor Signaling. *J. Biol. Chem.* **1997**, *272*, 30607–30610. [CrossRef]
227. Speirs, V.; Kerin, M.J.J.; Walton, D.S.S.; Newton, C.J.J.; Desai, S.B.B.; Atkin, S.L. Direct activation of oestrogen receptor-alpha by interleukin-6 in primary cultures of breast cancer epithelial cells. *Br. J. Cancer* **2000**, *82*, 1312–1316. [CrossRef] [PubMed]
228. Bhat-Nakshatri, P.; Campbell, R.A.; Patel, N.M.; Newton, T.R.; King, A.J.; Marshall, M.S.; Ali, S.; Nakshatri, H. Tumour necrosis factor and PI3-kinase control oestrogen receptor alpha protein level and its transrepression function. *Br. J. Cancer* **2004**, *90*, 853–859. [CrossRef]
229. Classen-Linke, I.; Müller-Newen, G.; Heinrich, P.C.; Beier, H.M.; Von Rango, U. The cytokine receptor gp130 and its soluble form are under hormonal control in human endometrium and decidua. *Mol. Hum. Reprod.* **2004**, *10*, 495–504. [CrossRef]
230. Dhingra, K.; Sahin, A.; Emami, K.; Hortobagyi, G.N.; Estrov, Z. Expression of leukemia inhibitory factor and its receptor in breast cancer: A potential autocrine and paracrine growth regulatory mechanism. *Breast Cancer Res. Treat.* **1998**, *48*, 165–174. [CrossRef] [PubMed]
231. Li, Y.; Zhang, H.; Zhao, Y.; Wang, C.; Cheng, Z.; Tang, L.; Gao, Y.; Liu, F.; Li, J.; Li, Y.; et al. A mandatory role of nuclear PAK4-LIFR axis in breast-to-bone metastasis of ERα-positive breast cancer cells. *Oncogene* **2018**, *38*, 808–821. [CrossRef] [PubMed]
232. Lokau, J.; Garbers, C. Activating mutations of the gp130/JAK/STAT pathway in human diseases. In *Advances in Protein Chemistry and Structural Biology*; Academic Press Inc.: Cambridge, MA, USA, 2019; Volume 116, pp. 283–309, ISBN 9780128155615.
233. Saha, A.; Bairwa, N.K.; Ranjan, A.; Gupta, V.; Bamezai, R. Two novel somatic mutations in the human interleukin 6 promoter region in a patient with sporadic breast cancer. *Eur. J. Immunogenet.* **2003**, *30*, 397–400. [CrossRef]
234. Yu, K.-D.; Di, G.-H.; Fan, L.; Chen, A.-X.; Yang, C.; Shao, Z.-M. Lack of an association between a functional polymorphism in the interleukin-6 gene promoter and breast cancer risk: A meta-analysis involving 25,703 subjects. *Breast Cancer Res. Treat.* **2009**, *122*, 483–488. [CrossRef]
235. DeMichele, A.; Gray, R.; Horn, M.; Chen, J.; Aplenc, R.; Vaughan, W.P.; Tallman, M.S. Host Genetic Variants in the Interleukin-6 Promoter Predict Poor Outcome in Patients with Estrogen Receptor-Positive, Node-Positive Breast Cancer. *Cancer Res.* **2009**, *69*, 4184–4191. [CrossRef] [PubMed]
236. Abana, C.O.; Bingham, B.S.; Cho, J.H.; Graves, A.J.; Koyama, T.; Pilarski, R.T.; Chakravarthy, A.B.; Xia, F. IL-6 variant is associated with metastasis in breast cancer patients. *PLoS ONE* **2017**, *12*, e0181725. [CrossRef]
237. Iacopetta, B.; Grieu, F.; Joseph, D. The −174 G/C gene polymorphism in interleukin-6 is associated with an aggressive breast cancer phenotype. *Br. J. Cancer* **2004**, *90*, 419–422. [CrossRef] [PubMed]
238. Sa-Nguanraksa, D.; Suntiparpluacha, M.; Kulprom, A.; Kummalue, T.; Chuangsuwanich, T.; Avirutnan, P.; O-Charoenrat, P. Association of estrogen receptor alpha and interleukin 6 polymorphisms with lymphovascular invasion, extranodal extension, and lower disease-free survival in thai breast cancer patients. *Asian Pacific J. Cancer Prev.* **2016**, *17*, 2935–2940.
239. Markkula, A.; Simonsson, M.; Ingvar, C.; Rose, C.; Jernström, H. IL6 genotype, tumour ER-status, and treatment predicted disease-free survival in a prospective breast cancer cohort. *BMC Cancer* **2014**, *14*, 759. [CrossRef]
240. Slattery, M.L.; Curtin, K.; Baumgartner, R.; Sweeney, C.; Byers, T.; Giuliano, A.R.; Baumgartner, K.B.; Wolff, R.R. IL6, Aspirin, Nonsteroidal Anti-inflammatory Drugs, and Breast Cancer Risk in Women Living in the Southwestern United States. *Cancer Epidemiology Biomarkers Prev.* **2007**, *16*, 747–755. [CrossRef] [PubMed]
241. Snoussi, K.; Strosberg, A.D.; Bouaouina, N.; Ahmed, S.B.; Chouchane, L. Genetic variation in pro-inflammatory cytokines (interleukin-1β, interleukin-1α and interleukin-6) associated with the aggressive forms, survival, and relapse prediction of breast carcinoma. *Eur. Cytokine Netw.* **2005**, *16*, 253–260.
242. Hefler, L.A.; Grimm, C.; Lantzsch, T.; Lampe, D.; Leodolter, S.; Koelbl, H.; Heinze, G.; Reinthaller, A.; Tong-Cacsire, D.; Tempfer, C.; et al. Interleukin-1 and Interleukin-6 Gene Polymorphisms and the Risk of Breast Cancer in Caucasian Women. *Clin. Cancer Res.* **2005**, *11*, 5718–5721. [CrossRef]
243. Madeleine, M.M.; Johnson, L.G.; Malkki, M.; Resler, A.J.; Petersdorf, E.W.; McKnight, B.; Malone, K.E. Genetic variation in proinflammatory cytokines IL6, IL6R, TNF-region, and TNFRSF1A and risk of breast cancer. *Breast Cancer Res. Treat.* **2011**, *129*, 887–899. [CrossRef]
244. Balasubramanian, S.P.; Azmy, I.A.F.; Higham, S.E.; Wilson, A.G.; Cross, S.S.; Cox, A.; Brown, N.J.; Reed, M.W. Interleukin gene polymorphisms and breast cancer: A case control study and systematic literature review. *BMC Cancer* **2006**, *6*, 188.
245. Slattery, M.L.; Herrick, J.S.; Torres-Mejia, G.; John, E.M.; Giuliano, A.R.; Hines, L.M.; Stern, M.C.; Baumgartner, K.B.; Presson, A.P.; Wolff, R.K. Genetic variants in interleukin genes are associated with breast cancer risk and survival in a genetically admixed population: The Breast Cancer Health Disparities Study. *Carcinogenesis* **2014**, *35*, 1750–1759. [CrossRef] [PubMed]
246. Liao, C.; Hu, S.; Zheng, Z.; Tong, H. Contribution of interaction between genetic variants of interleukin-11 and *Helicobacter pylori* infection to the susceptibility of gastric cancer. *OncoTargets Ther.* **2019**, *12*, 7459–7466. [CrossRef]
247. Kim, L.H.; Lee, H.-S.; Kim, Y.J.; Jung, J.H.; Kim, J.Y.; Park, B.L.; Shin, H.D. Identification of novel SNPs in the interleukin 6 receptor gene (IL6R). *Hum. Mutat.* **2003**, *21*, 450–451. [CrossRef]
248. Garbers, C.; Monhasery, N.; Aparicio-Siegmund, S.; Lokau, J.; Baran, P.; Nowell, M.A.; Jones, S.A.; Rose-John, S.; Scheller, J. The interleukin-6 receptor Asp358Ala single nucleotide polymorphism rs2228145 confers increased proteolytic conversion rates by ADAM proteases. *Biochim. Biophys. Acta Mol. Basis Dis.* **2014**, *1842*, 1485–1494. [CrossRef]

249. Ferreira, R.C.; Freitag, D.F.; Cutler, A.; Howson, J.; Rainbow, D.B.; Smyth, D.; Kaptoge, S.; Clarke, P.; Boreham, C.; Coulson, R.M.; et al. Functional IL6R 358Ala Allele Impairs Classical IL-6 Receptor Signaling and Influences Risk of Diverse Inflammatory Diseases. *PLoS Genet.* **2013**, *9*, e1003444. [CrossRef]
250. Sarwar, N.; Butterworth, A.S.; Hung, J.; Mcquillan, B.M. Interleukin-6 receptor pathways in coronary heart disease: A collaborative meta-analysis of 82 studies. *Lancet* **2012**, *379*, 1205–1213. [CrossRef]
251. Esparza-Gordillo, J.; Schaarschmidt, H.; Liang, L.; Cookson, W.; Bauerfeind, A.; Lee-Kirsch, M.-A.; Nemat, K.; Henderson, J.; Paternoster, L.; Harper, J.I.; et al. A functional IL-6 receptor (IL6R) variant is a risk factor for persistent atopic dermatitis. *J. Allergy Clin. Immunol.* **2013**, *132*, 371–377. [CrossRef] [PubMed]
252. Spencer, S.; Bal, S.K.; Egner, W.; Allen, H.L.; Raza, S.I.; Ma, C.A.; Gürel, M.; Zhang, Y.; Sun, G.; Sabroe, R.A.; et al. Loss of the interleukin-6 receptor causes immunodeficiency, atopy, and abnormal inflammatory responses. *J. Exp. Med.* **2019**, *216*, 1986–1998. [CrossRef]
253. Choi, J.; Song, N.; Han, S.; Chung, S.; Sung, H.; Lee, J.-Y.; Jung, S.J.; Park, S.K.; Yoo, K.-Y.; Han, W.; et al. The Associations between Immunity-Related Genes and Breast Cancer Prognosis in Korean Women. *PLoS ONE* **2014**, *9*, e103593. [CrossRef]
254. Metcalfe, R.D.; Aizel, K.; Zlatic, C.O.; Nguyen, P.M.; Morton, C.; Lio, D.S.-S.; Cheng, H.-C.; Dobson, R.C.J.; Parker, M.; Gooley, P.R.; et al. The structure of the extracellular domains of human interleukin 11α receptor reveals mechanisms of cytokine engagement. *J. Biol. Chem.* **2020**, *295*, 8285–8301. [CrossRef]
255. Brischoux-Boucher, E.; Trimouille, A.; Baujat, G.; Goldenberg, A.; Schaefer, E.; Guichard, B.; Hannequin, P.; Paternoster, G.; Baer, S.; Cabrol, C.; et al. IL11RA-related Crouzon-like autosomal recessive craniosynostosis in 10 new patients: Resemblances and differences. *Clin. Genet.* **2018**, *94*, 373–380. [CrossRef] [PubMed]
256. Keupp, K.; Li, Y.; Vargel, I.; Hoischen, A.; Richardson, R.; Neveling, K.; Alanay, Y.; Uz, E.; Elcioğlu, N.; Rachwalski, M.; et al. Mutations in the interleukin receptor IL 11 RA cause autosomal recessive Crouzon-like craniosynostosis. *Mol. Genet. Genom. Med.* **2013**, *1*, 223–237. [CrossRef]
257. De Mars, G.; Windelinckx, A.; Beunen, G.; Delecluse, C.; Lefevre, J.; Thomis, M.A.I. Polymorphisms in the CNTF and CNTF receptor genes are associated with muscle strength in men and women. *J. Appl. Physiol.* **2007**, *102*, 1824–1831. [CrossRef] [PubMed]
258. Jenkins, B.; Grail, D.; Nheu, T.; Najdovska, M.; Wang, B.; Waring, P.; Inglese, M.; McLoughlin, R.; Jones, S.A.; Topley, N.; et al. Hyperactivation of Stat3 in gp130 mutant mice promotes gastric hyperproliferation and desensitizes TGF-β signaling. *Nat. Med.* **2005**, *11*, 845–852. [CrossRef]
259. Rebouissou, S.; Amessou, M.; Couchy, G.; Poussin, K.; Imbeaud, S.; Pilati, C.; Izard, T.; Balabaud, C.; Bioulac-Sage, P.; Zucman-Rossi, J. Frequent in-frame somatic deletions activate gp130 in inflammatory hepatocellular tumours. *Nature* **2008**, *457*, 200–204. [CrossRef]
260. Schmidt-Arras, D.; Müller, M.; Stevanovic, M.; Horn, S.; Schütt, A.; Bergmann, J.; Wilkens, R.; Lickert, A.; Rose-John, S. Oncogenic deletion mutants of gp130 signal from intracellular compartments. *J. Cell Sci.* **2013**, *127*, 341–353. [CrossRef]
261. Schwerd, T.; Twigg, S.R.; Aschenbrenner, D.; Manrique, S.; Miller, K.A.; Taylor, I.B.; Capitani, M.; McGowan, S.J.; Sweeney, E.; Weber, A.; et al. A biallelic mutation in IL6ST encoding the GP130 co-receptor causes immunodeficiency and craniosynostosis. *J. Exp. Med.* **2017**, *214*, 2547–2562. [CrossRef]
262. Arita, K.; South, A.P.; Hans-Filho, G.; Sakuma, T.H.; Lai-Cheong, J.; Clements, S.; Odashiro, M.; Odashiro, D.N.; Hans-Neto, G.; Hans, N.R.; et al. Oncostatin M Receptor-β Mutations Underlie Familial Primary Localized Cutaneous Amyloidosis. *Am. J. Hum. Genet.* **2008**, *82*, 73–80. [CrossRef] [PubMed]
263. Mikelonis, D.; Jorcyk, C.L.; Tawara, K.; Oxford, J.T. Stüve-Wiedemann syndrome: LIFR and associated cytokines in clinical course and etiology. *Orphanet J. Rare Dis.* **2014**, *9*, 1–11.
264. Kosfeld, A.; Brand, F.; Weiss, A.-C.; Kreuzer, M.; Goerk, M.; Martens, H.; Schubert, S.; Schäfer, A.-K.; Riehmer, V.; Hennies, I.; et al. Mutations in the leukemia inhibitory factor receptor (LIFR) gene and Lifr deficiency cause urinary tract malformations. *Hum. Mol. Genet.* **2017**, *26*, 1716–1731. [CrossRef]
265. Deng, S.; He, S.Y.; Zhao, P.; Zhang, P. The role of oncostatin M receptor gene polymorphisms in bladder cancer. *World J. Surg. Oncol.* **2019**, *17*, 1–9. [CrossRef]
266. Zhong, Y.; Li, J.; Zhang, Y.; Qiu, W.; Luo, Y. The polymorphisms of oncostatin M receptor gene associated with increased risk of lung cancer. *Int. J. Clin. Exp. Med.* **2018**, *11*, 12421–12428.
267. Hong, I.K.; Eun, Y.G.; Chung, D.H.; Kwon, K.H.; Kim, D.Y. Association of the Oncostatin M Receptor Gene Polymorphisms with Papillary Thyroid Cancer in the Korean Population. *Clin. Exp. Otorhinolaryngol.* **2011**, *4*, 193–198. [CrossRef]
268. Senkevitch, E.; Durum, S. The promise of Janus kinase inhibitors in the treatment of hematological malignancies. *Cytokine* **2016**, *98*, 33–41. [CrossRef] [PubMed]
269. Kralovics, R.; Passamonti, F.; Buser, A.S.; Teo, S.-S.; Tiedt, R.; Passweg, J.R.; Tichelli, A.; Cazzola, M.; Skoda, R.C. A Gain-of-Function Mutation of JAK2 in Myeloproliferative Disorders. *N. Engl. J. Med.* **2005**, *352*, 1779–1790. [CrossRef]
270. Flex, E.; Petrangeli, V.; Stella, L.; Chiaretti, S.; Hornakova, T.; Knoops, L.; Ariola, C.; Fodale, V.; Clappier, E.; Paoloni, F.; et al. Somatically acquired JAK1 mutations in adult acute lymphoblastic leukemia. *J. Exp. Med.* **2008**, *205*, 751–758. [CrossRef]
271. Jeong, E.G.; Kim, M.S.; Nam, H.K.; Min, C.K.; Lee, S.; Chung, Y.J.; Yoo, N.J.; Lee, S.H. Somatic Mutations of JAK1 and JAK3 in Acute Leukemias and Solid Cancers. *Clin. Cancer Res.* **2008**, *14*, 3716–3721. [CrossRef]

272. Nebral, K.; Denk, D.M.; Attarbaschi, A.; Konig, M.; Mann, G.E.; Haas, O.A.; Strehl, S. Incidence and diversity of PAX5 fusion genes in childhood acute lymphoblastic leukemia. *Leukemia* **2008**, *23*, 134–143. [CrossRef]
273. Lacronique, V.; Boureux, A.; Della Valle, V.; Poirel, H.; Quang, C.T.; Mauchauffé, M.; Berthou, C.; Lessard, M.; Berger, R.; Ghysdael, J.; et al. A TEL-JAK2 Fusion Protein with Constitutive Kinase Activity in Human Leukemia. *Science* **1997**, *278*, 1309–1312. [CrossRef] [PubMed]
274. Reiter, A.; Walz, C.; Watmore, A.; Schoch, C.; Blau, I.; Schlegelberger, B.; Berger, U.; Telford, N.; Aruliah, S.; Yin, J.A.; et al. The t(8;9)(p22;p24) Is a Recurrent Abnormality in Chronic and Acute Leukemia that Fuses PCM1 to JAK2. *Cancer Res.* **2005**, *65*, 2662–2667. [CrossRef]
275. Poitras, J.L.; Cin, P.D.; Aster, J.C.; DeAngelo, D.J.; Morton, C.C. NovelSSBP2-JAK2fusion gene resulting from a t(5;9)(q14.1;p24.1) in pre-B acute lymphocytic leukemia. *Genes, Chromosom. Cancer* **2008**, *47*, 884–889. [CrossRef]
276. Kan, Z.; Zheng, H.; Liu, X.; Li, S.; Barber, T.D.; Gong, Z.; Gao, H.; Hao, K.; Willard, M.D.; Xu, J.; et al. Whole-genome sequencing identifies recurrent mutations in hepatocellular carcinoma. *Genome Res.* **2013**, *23*, 1422–1433. [CrossRef] [PubMed]
277. Pilati, C.; Letouzé, E.; Nault, J.-C.; Imbeaud, S.; Boulai, A.; Calderaro, J.; Poussin, K.; Franconi, A.; Couchy, G.; Morcrette, G.; et al. Genomic Profiling of Hepatocellular Adenomas Reveals Recurrent FRK-Activating Mutations and the Mechanisms of Malignant Transformation. *Cancer Cell* **2014**, *25*, 428–441. [CrossRef]
278. Lupardus, P.J.; Ultsch, M.; Wallweber, H.; Kohli, P.B.; Johnson, A.R.; Eigenbrot, C. Structure of the pseudokinase-kinase domains from protein kinase TYK2 reveals a mechanism for Janus kinase (JAK) autoinhibition. *Proc. Natl. Acad. Sci. USA* **2014**, *111*, 8025–8030. [CrossRef]
279. Waanders, E.; Scheijen, B.; Jongmans, M.C.J.; Venselaar, H.; Van Reijmersdal, S.V.; Van Dijk, A.H.A.; Pastorczak, A.; Weren, R.D.A.; Van Der Schoot, C.E.; Van De Vorst, J.M.; et al. Germline activating TYK2 mutations in pediatric patients with two primary acute lymphoblastic leukemia occurrences. *Leukemia* **2016**, *31*, 821–828. [CrossRef]
280. Velusamy, T.; Kiel, M.J.; Sahasrabuddhe, A.A.; Rolland, D.; Dixon, C.A.; Bailey, N.G.; Betz, B.L.; Brown, N.A.; Hristov, A.C.; Wilcox, R.A.; et al. A novel recurrent NPM1-TYK2 gene fusion in cutaneous CD30-positive lymphoproliferative disorders. *Blood* **2014**, *124*, 3768–3771. [CrossRef]
281. Koskela, H.L.M.; Eldfors, S.; Ellonen, P.; Van Adrichem, A.J.; Kuusanmäki, H.; Andersson, E.; Lagström, S.; Clemente, M.J.; Olson, T.; Jalkanen, S.E.; et al. SomaticSTAT3Mutations in Large Granular Lymphocytic Leukemia. *N. Engl. J. Med.* **2012**, *366*, 1905–1913. [CrossRef] [PubMed]
282. Rajala, H.L.M.; Olson, T.; Clemente, M.J.; Lagström, S.; Ellonen, P.; Lundan, T.; Hamm, D.E.; Zaman, S.A.U.; Marti, J.M.L.; Andersson, E.I.; et al. The analysis of clonal diversity and therapy responses using STAT3 mutations as a molecular marker in large granular lymphocytic leukemia. *Haematologica* **2014**, *100*, 91–99. [CrossRef]
283. Sim, S.H.; Kim, S.; Kim, T.M.; Jeon, Y.K.; Nam, S.J.; Ahn, Y.-O.; Keam, B.; Park, H.H.; Kim, D.-W.; Kim, C.W.; et al. Novel JAK3-Activating Mutations in Extranodal NK/T-Cell Lymphoma, Nasal Type. *Am. J. Pathol.* **2017**, *187*, 980–986. [CrossRef] [PubMed]
284. Pilati, C.; Amessou, M.; Bihl, M.P.; Balabaud, C.; Van Nhieu, J.T.; Paradis, V.; Nault, J.C.; Izard, T.; Bioulac-Sage, P.; Couchy, G.; et al. Somatic mutations activating STAT3 in human inflammatory hepatocellular adenomas. *J. Exp. Med.* **2011**, *208*, 1359–1366. [CrossRef] [PubMed]
285. Pilati, C.; Zucman-Rossi, J. Mutations leading to constitutive active gp130/JAK1/STAT3 pathway. *Cytokine Growth Factor Rev.* **2015**, *26*, 499–506. [CrossRef] [PubMed]
286. Buchert, M.; Burns, C.J.; Ernst, M. Targeting JAK kinase in solid tumors: Emerging opportunities and challenges. *Oncogene* **2016**, *35*, 939–951. [CrossRef]
287. Sonnenblick, A.; Shriki, A.; Galun, E.; Axelrod, J.H.; Daum, H.; Rottenberg, Y.; Hamburger, T.; Mali, B.; Peretz, T. Tissue microarray-based study of patients with lymph node-positive breast cancer shows tyrosine phosphorylation of signal transducer and activator of transcription 3 (tyrosine705-STAT3) is a marker of good prognosis. *Clin. Transl. Oncol.* **2012**, *14*, 232–236. [CrossRef]
288. Egan, C.; Pang, A.; Durda, D.; Cheng, H.-C.; Wang, J.H.; Fujita, D.J. Activation of Src in human breast tumor cell lines: Elevated levels of phosphotyrosine phosphatase activity that preferentially recognizes the Src carboxy terminal negative regulatory tyrosine 530. *Oncogene* **1999**, *18*, 1227–1237. [CrossRef] [PubMed]
289. Schwarz, L.; Fox, E.M.; Balko, J.M.; Garrett, J.T.; Kuba, M.G.; Estrada, M.V.; González-Angulo, A.M.; Mills, G.B.; Red-Brewer, M.; Mayer, I.A.; et al. LYN-activating mutations mediate antiestrogen resistance in estrogen receptor–positive breast cancer. *J. Clin. Investig.* **2014**, *124*, 5490–5502. [CrossRef]
290. Garcia, R.; Bowman, T.L.; Niu, G.; Yu, H.; Minton, S.; Muro-Cacho, C.A.; Cox, C.E.; Falcone, R.; Fairclough, R.; Parsons, S.; et al. Constitutive activation of Stat3 by the Src and JAK tyrosine kinases participates in growth regulation of human breast carcinoma cells. *Oncogene* **2001**, *20*, 2499–2513. [CrossRef] [PubMed]
291. Sun, Z.; Yao, Z.; Liu, S.; Tang, H.; Yan, X. An oligonucleotide decoy for Stat3 activates the immune response of macrophages to breast cancer. *Immunobiology* **2006**, *211*, 199–209. [CrossRef]
292. Mosele, F.; Stefanovska, B.; Lusque, A.; Dien, A.T.; Garberis, I.; Droin, N.; Le Tourneau, C.; Sablin, M.-P.; Lacroix, L.; Enrico, D.; et al. Outcome and molecular landscape of patients with PIK3CA-mutated metastatic breast cancer. *Ann. Oncol.* **2020**, *31*, 377–386. [CrossRef]

293. Freitag, C.E.; Mei, P.; Wei, L.; Parwani, A.V.; Li, Z. Genetic alterations and their association with clinicopathologic characteristics in advanced breast carcinomas: Focusing on clinically actionable genetic alterations. *Hum. Pathol.* **2020**, *102*, 94–103. [CrossRef]
294. Shaw, S.; Bourne, T.; Meier, C.; Carrington, B.; Gelinas, R.; Henry, A.; Popplewell, A.; Adams, R.; Baker, T.; Rapecki, S.; et al. Discovery and characterization of olokizumab: A humanized antibody targeting interleukin-6 and neutralizing gp130-signaling. *mAbs* **2014**, *6*, 773–781. [CrossRef]
295. Heo, T.-H.; Wahler, J.; Suh, N. Potential therapeutic implications of IL-6/IL-6R/gp130-targeting agents in breast cancer. *Oncotarget* **2016**, *7*, 15460–15473. [CrossRef]
296. Rose-John, S. Therapeutic targeting of IL-6 trans-signaling. *Cytokine* **2021**, *144*, 155577. [CrossRef]
297. Nasonov, E.; Fatenejad, S.; Feist, E.; Ivanova, M.; Korneva, E.; Krechikova, D.G.; Maslyanskiy, A.L.; Samsonov, M.; Stoilov, R.; Zonova, E.V.; et al. Olokizumab, a monoclonal antibody against interleukin 6, in combination with methotrexate in patients with rheumatoid arthritis inadequately controlled by methotrexate: Efficacy and safety results of a randomised controlled phase III study. *Ann. Rheum. Dis.* **2021**. [CrossRef]
298. Eskandary, F.; Dürr, M.; Budde, K.; Doberer, K.; Reindl-Schwaighofer, R.; Waiser, J.; Wahrmann, M.; Regele, H.; Spittler, A.; Lachmann, N.; et al. Clazakizumab in late antibody-mediated rejection: Study protocol of a randomized controlled pilot trial. *Trials* **2019**, *20*, 1–13. [CrossRef]
299. Lacroix, M.; Rousseau, F.; Guilhot, F.; Malinge, P.; Magistrelli, G.; Herren, S.; Jones, S.A.; Jones, G.; Scheller, J.; Lissilaa, R.; et al. Novel Insights into Interleukin 6 (IL-6) Cis- and Trans-signaling Pathways by Differentially Manipulating the Assembly of the IL-6 Signaling Complex. *J. Biol. Chem.* **2015**, *290*, 26943–26953. [CrossRef]
300. Genovese, M.C.; Fleischmann, R.; Kivitz, A.; Lee, E.-B.; Van Hoogstraten, H.; Kimura, T.; John, G.S.; Mangan, E.K.; Burmester, G.R. Efficacy and safety of sarilumab in combination with csDMARDs or as monotherapy in subpopulations of patients with moderately to severely active rheumatoid arthritis in three phase III randomized, controlled studies. *Arthritis Res.* **2020**, *22*, 1–17. [CrossRef]
301. Adams, R.; Burnley, R.J.; Valenzano, C.R.; Qureshi, O.; Doyle, C.; Lumb, S.; Lopez, M.D.C.; Griffin, R.; McMillan, D.; Taylor, R.D.; et al. Discovery of a junctional epitope antibody that stabilizes IL-6 and gp80 protein: Protein interaction and modulates its downstream signaling. *Sci. Rep.* **2017**, *7*, 37716. [CrossRef]
302. Scheller, J.; Chalaris, A.; Schmidt-Arras, D.; Rose-John, S. The pro- and anti-inflammatory properties of the cytokine interleukin-6. *Biochim. Biophys. Mol. Cell Res.* **2011**, *1813*, 878–888.
303. Rose-John, S. The Soluble Interleukin 6 Receptor: Advanced Therapeutic Options in Inflammation. *Clin. Pharmacol. Ther.* **2017**, *102*, 591–598. [CrossRef]
304. Xu, S.; Grande, F.; Garofalo, A.; Neamati, N.; Gu, D.; Liu, H.; Su, G.H.; Zhang, X.; Chin-Sinex, H.; Hanenberg, H.; et al. Discovery of a Novel Orally Active Small-Molecule gp130 Inhibitor for the Treatment of Ovarian Cancer. *Mol. Cancer Ther.* **2013**, *12*, 937–949. [CrossRef]
305. Hong, S.-S.; Choi, J.H.; Lee, S.Y.; Park, Y.-H.; Park, K.-Y.; Lee, J.Y.; Kim, J.; Gajulapati, V.; Goo, J.-I.; Singh, S.; et al. A Novel Small-Molecule Inhibitor Targeting the IL-6 Receptor β Subunit, Glycoprotein 130. *J. Immunol.* **2015**, *195*, 237–245. [CrossRef]
306. Kim, J.W.; Marquez, C.; Sperberg, R.A.P.; Wu, J.; Bae, W.G.; Huang, P.-S.; Sweet-Cordero, E.A.; Cochran, J.R. Engineering a potent receptor superagonist or antagonist from a novel IL-6 family cytokine ligand. *Proc. Natl. Acad. Sci. USA* **2020**, *117*, 14110–14118. [CrossRef] [PubMed]
307. Li, H.; Xiao, H.; Lin, L.; Jou, D.; Kumari, V.; Lin, J.; Li, C. Drug Design Targeting Protein–Protein Interactions (PPIs) Using Multiple Ligand Simultaneous Docking (MLSD) and Drug Repositioning: Discovery of Raloxifene and Bazedoxifene as Novel Inhibitors of IL-6/GP130 Interface. *J. Med. Chem.* **2014**, *57*, 632–641. [CrossRef]
308. Barrett-Connor, E.; Mosca, L.; Collins, P.; Geiger, M.J.; Grady, D.; Kornitzer, M.; McNabb, M.A.; Wenger, N.K. Effects of Raloxifene on Cardiovascular Events and Breast Cancer in Postmenopausal Women. *N. Engl. J. Med.* **2006**, *355*, 125–137. [CrossRef]
309. Gennari, L.; Merlotti, D.; De Paola, V.; Martini, G.; Nuti, R. Bazedoxifene for the prevention of postmenopausal osteoporosis. *Ther. Clin. Risk Manag.* **2008**, *4*, 1229–1242. [CrossRef]
310. Fanning, S.W.; Jeselsohn, R.; Dharmarajan, V.; Mayne, C.G.; Karimi, M.; Buchwalter, G.; Houtman, R.; Toy, W.; Fowler, C.E.; Laine, M.; et al. The SERM/SERD Bazedoxifene Disrupts ESR1 Helix 12 to Overcome Acquired Hormone Resistance in Breast Cancer Cells. *Elife* **2018**, *7*, 1–26. [CrossRef]
311. Wardell, S.E.; Nelson, E.; Chao, C.A.; McDonnell, D.P. Bazedoxifene Exhibits Antiestrogenic Activity in Animal Models of Tamoxifen-Resistant Breast Cancer: Implications for Treatment of Advanced Disease. *Clin. Cancer Res.* **2013**, *19*, 2420–2431. [CrossRef]
312. Wu, X.; Cao, Y.; Xiao, H.; Li, C.; Lin, J. Bazedoxifene as a Novel GP130 Inhibitor for Pancreatic Cancer Therapy. *Mol. Cancer Ther.* **2016**, *15*, 2609–2619. [CrossRef]
313. Ma, H.; Yan, D.; Wang, Y.; Shi, W.; Liu, T.; Zhao, C.; Huo, S.; Duan, J.; Tao, J.; Zhai, M.; et al. Bazedoxifene exhibits growth suppressive activity by targeting interleukin-6/glycoprotein 130/signal transducer and activator of transcription 3 signaling in hepatocellular carcinoma. *Cancer Sci.* **2019**, *110*, 950–961. [CrossRef]
314. Xiao, H.; Bid, H.K.; Chen, X.; Wu, X.; Wei, J.; Bian, Y.; Zhao, C.; Li, H.; Li, C.; Lin, J. Repositioning Bazedoxifene as a novel IL-6/GP130 signaling antagonist for human rhabdomyosarcoma therapy. *PLoS ONE* **2017**, *12*, e0180297. [CrossRef]
315. Wei, J.; Ma, L.; Lai, Y.-H.; Zhang, R.; Li, H.; Li, C.; Lin, J. Bazedoxifene as a novel GP130 inhibitor for Colon Cancer therapy. *J. Exp. Clin. Cancer Res.* **2019**, *38*, 1–13. [CrossRef]

316. Jeselsohn, R.; Guo, H.; Rees, R.; Barry, W.T.; Barlett, C.H.; Tung, N.M.; Krop, I.E.; Brown, M.; Winer, E.P. Abstract PD1-05: Results from the phase Ib/II clinical trial of bazedoxifene and palbociclib in hormone receptor positive metastatic breast cancer. *Cancer Res.* **2019**, *79*, PD1-05.
317. Song, D.; Yu, W.; Ren, Y.; Zhu, J.; Wan, C.; Cai, G.; Guo, J.; Zhang, W.; Kong, L. Discovery of bazedoxifene analogues targeting glycoprotein 130. *Eur. J. Med. Chem.* **2020**, *199*, 112375. [CrossRef] [PubMed]
318. Lewis, V.O.; Devarajan, E.; Cardó-Vila, M.; Thomas, D.G.; Kleinerman, E.S.; Marchiò, S.; Sidman, R.L.; Pasqualini, R.; Arap, W. BMTP-11 is active in preclinical models of human osteosarcoma and a candidate targeted drug for clinical translation. *Proc. Natl. Acad. Sci. USA* **2017**, *114*, 8065–8070. [CrossRef]
319. Cardó-Vila, M.; Marchio, S.; Sato, M.; Staquicini, F.I.; Smith, T.L.; Bronk, J.K.; Yin, G.; Zurita, A.J.; Sun, M.; Behrens, C.; et al. Interleukin-11 Receptor Is a Candidate Target for Ligand-Directed Therapy in Lung Cancer: Analysis of Clinical Samples and BMTP-11 Preclinical Activity. *Am. J. Pathol.* **2016**, *186*, 2162–2170. [CrossRef] [PubMed]
320. Pasqualini, R.; Millikan, R.E.; Christianson, D.R.; Cardó-Vila, M.; Driessen, W.H.P.; Giordano, R.J.; Hajitou, A.; Hoang, A.G.; Wen, S.; Barnhart, K.F.; et al. Targeting the interleukin-11 receptor α in metastatic prostate cancer: A first-in-man study. *Cancer* **2015**, *121*, 2411–2421. [CrossRef]
321. Xin, H.; Herrmann, A.; Reckamp, K.; Zhang, W.; Pal, S.; Hedvat, M.; Zhang, C.; Liang, W.; Scuto, A.; Weng, S.; et al. Antiangiogenic and Antimetastatic Activity of JAK Inhibitor AZD1480. *Cancer Res.* **2011**, *71*, 6601–6610. [CrossRef]
322. Hedvat, M.; Huszar, D.; Herrmann, A.; Gozgit, J.M.; Schroeder, A.; Sheehy, A.; Buettner, R.; Proia, D.; Kowolik, C.M.; Xin, H.; et al. The JAK2 Inhibitor AZD1480 Potently Blocks Stat3 Signaling and Oncogenesis in Solid Tumors. *Cancer Cell* **2009**, *16*, 487–497. [CrossRef]
323. Tavallai, M.; Booth, L.; Roberts, J.L.; Poklepovic, A.; Dent, P. Rationally Repurposing Ruxolitinib (Jakafi®) as a Solid Tumor Therapeutic. *Front. Oncol.* **2016**, *6*, 142. [CrossRef]
324. Wong, A.L.A.; Hirpara, J.L.; Pervaiz, S.; Eu, J.-Q.; Sethi, G.; Goh, B.-C. Expert Opinion on Investigational Drugs Do STAT3 inhibitors have potential in the future for cancer therapy. *Expert Opin. Investig. Drugs* **2017**, *26*, 883–887. [CrossRef] [PubMed]
325. Zhang, J.T.; Liu, J.Y. Drugging the "undruggable" DNA-binding domain of STAT3. *Oncotarget* **2016**, *7*, 66324–66325. [CrossRef] [PubMed]
326. Yue, P.; Turkson, J. Targeting STAT3 in cancer: How successful are we. *Expert Opin. Investig. Drugs* **2008**, *18*, 45–56. [CrossRef]
327. Wong, A.L.; Soo, R.A.; Tan, D.S.; Lee, S.C.; Lim, J.S.; Marban, P.C.; Kong, L.R.; Lee, Y.J.; Wang, L.Z.; Thuya, W.L.; et al. Phase I and biomarker study of OPB-51602, a novel signal transducer and activator of transcription (STAT) 3 inhibitor, in patients with refractory solid malignancies. *Ann. Oncol.* **2015**, *26*, 998–1005. [CrossRef]
328. Bai, L.; Zhou, H.; Xu, R.; Zhao, Y.; Chinnaswamy, K.; McEachern, D.; Chen, J.; Yang, C.-Y.; Liu, Z.; Wang, M.; et al. A Potent and Selective Small-Molecule Degrader of STAT3 Achieves Complete Tumor Regression In Vivo. *Cancer Cell* **2019**, *36*, 498–511.e17. [CrossRef] [PubMed]
329. Siddiquee, K.; Zhang, S.; Guida, W.C.; Blaskovich, M.A.; Greedy, B.; Lawrence, H.R.; Yip, M.L.R.; Jove, R.; McLaughlin, M.M.; Lawrence, N.J.; et al. Selective chemical probe inhibitor of Stat3, identified through structure-based virtual screening, induces antitumor activity. *Proc. Natl. Acad. Sci. USA* **2007**, *104*, 7391–7396. [CrossRef]
330. Jing, N.; Li, Y.; Xiong, W.; Sha, W.; Jing, L.; Tweardy, D.J. G-Quartet Oligonucleotides: A new class of signal transducer and activator of transcription 3 inhibitors that suppresses growth of prostate and breast tumors through induction of apoptosis. *Cancer Res.* **2004**, *64*, 6603–6609. [CrossRef]
331. Ishdorj, G.; Johnston, J.B.; Gibson, S. Inhibition of Constitutive Activation of STAT3 by Curcurbitacin-I (JSI-124) Sensitized Human B-Leukemia Cells to Apoptosis. *Mol. Cancer Ther.* **2010**, *9*, 3302–3314. [CrossRef] [PubMed]
332. Yu, X.; He, L.; Cao, P.; Yu, Q. Eriocalyxin B Inhibits STAT3 Signaling by Covalently Targeting STAT3 and Blocking Phosphorylation and Activation of STAT3. *PLoS ONE* **2015**, *10*, e0128406. [CrossRef]
333. Zinzalla, G.; Haque, M.R.; Basu, B.P.; Anderson, J.; Kaye, S.L.; Haider, S.; Hasan, F.; Antonow, D.; Essex, S.; Rahman, K.M.; et al. A novel small-molecule inhibitor of IL-6 signalling. *Bioorganic Med. Chem. Lett.* **2010**, *20*, 7029–7032. [CrossRef]
334. Schust, J.; Sperl, B.; Hollis, A.; Mayer, T.; Berg, T. Stattic: A Small-Molecule Inhibitor of STAT3 Activation and Dimerization. *Chem. Biol.* **2006**, *13*, 1235–1242. [CrossRef] [PubMed]
335. Uemura, M.; Trinh, V.A.; Haymaker, C.; Jackson, N.; Kim, D.W.; Allison, J.P.; Sharma, P.; Vence, L.; Bernatchez, C.; Hwu, P.; et al. Selective inhibition of autoimmune exacerbation while preserving the anti-tumor clinical benefit using IL-6 blockade in a patient with advanced melanoma and Crohn's disease: A case report. *J. Hematol. Oncol.* **2016**, *9*, 81. [CrossRef] [PubMed]
336. Kim, S.T.; Tayar, J.; Trinh, V.A.; Suarez-Almazor, M.; Garcia, S.; Hwu, P.; Johnson, D.H.; Uemura, M.; Diab, A. Successful treatment of arthritis induced by checkpoint inhibitors with tocilizumab: A case series. *Ann. Rheum. Dis.* **2017**, *76*, 2061–2064. [CrossRef]
337. Rotz, S.J.; Leino, D.; Szabo, S.; Mangino, J.L.; Turpin, B.K.; Pressey, J.G. Severe cytokine release syndrome in a patient receiving PD-1-directed therapy. *Pediatr. Blood Cancer* **2017**, *64*, e26642. [CrossRef] [PubMed]
338. Bertucci, F.; Boudin, L.; Finetti, P.; Van Berckelaer, C.; Van Dam, P.; Dirix, L.; Viens, P.; Gonçalves, A.; Ueno, N.T.; Van Laere, S.; et al. Immune landscape of inflammatory breast cancer suggests vulnerability to immune checkpoint inhibitors. *OncoImmunology* **2021**, *10*, 1929724. [CrossRef] [PubMed]
339. Christofi, T.; Baritaki, S.; Falzone, L.; Libra, M.; Zaravinos, A. Current Perspectives in Cancer Immunotherapy. *Cancers* **2019**, *11*, 1472. [CrossRef] [PubMed]

340. Jin, K.; Pandey, N.B.; Popel, A.S. Simultaneous blockade of IL-6 and CCL5 signaling for synergistic inhibition of triple-negative breast cancer growth and metastasis. *Breast Cancer Res.* **2018**, *20*, 1–10. [CrossRef]
341. Smolen, J.S.; Schoels, M.M.; Nishimoto, N.; Breedveld, F.C.; Burmester, G.R.; Dougados, M.; Emery, P.; Ferraccioli, G.; Gabay, C.; Gibofsky, A.; et al. Consensus statement on blocking the effects of interleukin-6 and in particular by interleukin-6 receptor inhibition in rheumatoid arthritis and other inflammatory conditions. *Ann. Rheum. Dis.* **2013**, *72*, 482–492. [CrossRef] [PubMed]

Review

The Signal Transducer IL6ST (gp130) as a Predictive and Prognostic Biomarker in Breast Cancer

Carlos Martínez-Pérez [1,2,*], Jess Leung [1], Charlene Kay [1,2], James Meehan [2], Mark Gray [2], J Michael Dixon [1] and Arran K Turnbull [1,2]

1. Breast Cancer Now Edinburgh Research Team, MRC Institute of Genetics and Cancer, Western General Hospital, University of Edinburgh, Edinburgh EH4 2XU, UK; s1446279@sms.ed.ac.uk (J.L.); charlene.kay@ed.ac.uk (C.K.); mike.dixon@ed.ac.uk (J.M.D.); arran.turnbull@ed.ac.uk (A.K.T.)
2. Translational Oncology Research Group, MRC Institute of Genetics and Cancer, Western General Hospital, University of Edinburgh, Edinburgh EH4 2XU, UK; james.meehan@ed.ac.uk (J.M.); mark.gray@ed.ac.uk (M.G.)
* Correspondence: carlos.martinez-perez@ed.ac.uk

Abstract: Novel biomarkers are needed to continue to improve breast cancer clinical management and outcome. IL6-like cytokines, whose pleiotropic functions include roles in many hallmarks of malignancy, rely on the signal transducer IL6ST (gp130) for all their signalling. To date, 10 separate independent studies based on the analysis of clinical breast cancer samples have identified IL6ST as a predictor. Consistent findings suggest that IL6ST is a positive prognostic factor and is associated with ER status. Interestingly, these studies include 4 multigene signatures (EndoPredict, EER4, IRSN-23 and 42GC) that incorporate IL6ST to predict risk of recurrence or outcome from endocrine or chemotherapy. Here we review the existing evidence on the promising predictive and prognostic value of IL6ST. We also discuss how this potential could be further translated into clinical practice beyond the EndoPredict tool, which is already available in the clinic. The most promising route to further exploit IL6ST's promising predicting power will likely be through additional hybrid multifactor signatures that allow for more robust stratification of ER+ breast tumours into discrete groups with distinct outcomes, thus enabling greater refinement of the treatment-selection process.

Keywords: breast cancer; predictive tools; prognostic tools; translational research; IL6ST; gp130; cytokine signalling

1. Background: The Essential Role of Biomarkers in Breast Cancer

Breast cancer (BC) is a heterogeneous disease comprising well-characterised molecular subtypes that differ in their oncogenic drivers, pathogenesis and prognosis. Clinical management and outcome have improved considerably over time, in part due to the identification and clinical application of biological markers (or biomarkers), which have been defined as "characteristics that can be objectively measured and evaluated as indicators of certain normal biological processes, pathogenic processes, or pharmacologic responses to a therapeutic intervention" [1]. Biomarkers can be classified as prognostic, when they indicate the likelihood of an event such as disease recurrence or progression, or predictive, when they indicate the likelihood of response or resistance to a given treatment [2]. They can be clinical or histopathological factors, such as patient or tumour characteristics, or molecular markers, such as the expression level of a certain protein or gene or the presence or frequency of a genomic event (e.g., a mutation).

Molecular biomarkers are often molecules playing a role in processes such as disease progression or treatment response. Thus, they may act as surrogates for the activity of a given driver and provide insight into the complex underlying tumour biology. A biomarker might be utilised qualitatively or quantitatively, as a continuous variable or with discrete cut-offs, alone or in combination with other markers in the form of multifactor tests or signatures. In their different capacities, biomarkers are highly valuable in disease detection, staging, monitoring or prognosis estimation and they can guide the treatment selection and decision-making process in the management of many cancers, including breast [3].

The foremost examples of BC biomarkers are the oestrogen receptor α (ERα or ER) and the human epidermal growth factor receptor 2 (HER2), which indicate differences in prognosis and predict responsiveness to endocrine and HER2-targeted therapies, respectively. Assessment of ER and HER2 status has long been mandatory for all new BC diagnoses to help guide treatment selection [4] and has considerably improved prognosis and survival for patients with hormone-dependent and HER2-overexpressing BC. Importantly, the role of ER and HER2 as biomarkers continues to evolve, with growing evidence on different genomic aberrations contributing to the development of treatment resistance [5]. Research on ER mutations in particular has been extensive, with their prevalence and clinical implications being assessed in several retrospective and currently-ongoing prospective trials, making their translation into clinical practice in the near future a strong possibility [6,7].

Research over the last two decades has led to the identification of numerous other molecular biomarkers. These include proteins referred to as cancer antigens, such as CA15-3, CA19-9, CA27-29, the carcinoembryonic antigen (CEA) or mucin-like carcinoma antigen (MCA), which can be measured in patient serum to enable early detection and prognostic assessment [8–11]. In addition to well-established genomic markers such as BRCA1/2 mutations [12], translational studies continue to describe aberrations and rearrangements that could serve as prognostic factors or actionable targets [13]. Recent studies have also highlighted microRNAs as molecules with an emerging potential as biomarkers due to their complex regulatory role in breast cancer [14–16].

Many challenges still remain in the clinical management of BC. Evidence suggests that the current diagnostic tools and available biomarkers fail to sufficiently discriminate the underlying heterogeneity of the disease. Both basic and translational research continue to add to our understanding of the complex and evolving biology, shedding light on the pathways and mechanisms involved in phenomena such as the development of acquired resistance to treatment. Biomarker discovery studies can identify promising candidates with prognostic or predictive value which will be essential to continue to improve BC management and outcome. Here we will review the evidence on one molecule in particular, the interleukin-6 signal transducer (IL6ST), which has emerged as a novel and exciting BC biomarker in recent years.

2. The IL6-Like Cytokine Family and Its Signalling in Breast Cancer

Interleukin-6 (IL6) is the best characterised cytokine of a class that also includes interleukin-11 (IL11), interleukin-31 (IL31), ciliary neurotrophic factor (CNTF), leukemia inhibitory factor (LIF), oncostatin M (OSM), cardiotrophin 1 (CT1), cardiotrophin-like cytokine (CLC) and neuropoietin (NPN). This group of cytokines, with similar structural and functional features, are normally referred to as the IL6 or IL6-like family [17,18]. They are also known as the gp130 family, after the shared transmembrane signalling receptor glycoprotein 130, which acts as a signal transducer in all signalling by this cytokine family. Each oligomeric signalling complex includes one or more gp130 molecules, depending on the cytokine. This signal transducer is also known as CD130, IL-6 receptor subunit β (IL6Rβ) or IL6 signal transducer (IL6ST, which is also its gene name). For naming consistency, in this review we will refer to this cytokine group as the IL6-like family and to the signal transducer as IL6ST.

Indeed, the common dependence on IL6ST for signalling is the defining characteristic of this cytokine family. The signal transducer is ubiquitously expressed in all cell types [19] and has been shown to be essential for survival in knockout in vivo studies in mice [20]. IL6-like cytokines act as extracellular ligands that bind the membrane-bound IL6ST and different non-signalling receptors with high affinity (see Figure 1 for diagram). This leads to the formation of signalling complexes including IL6ST homo- or heterodimers (depending on the cytokine). The cytoplasmic portions of the dimerised transducers then trigger intracellular signalling primarily through tyrosine kinases in the JAK family, such as JAK1 and JAK2, which are constitutively associated with IL6ST. JAK1/2 dimerisation and autophosphorylation lead to signalling through 3 major pathways: (i) the Janus-activated kinase – signal transducer and activator of transcription (JAK/STAT) pathway, (ii) the Ras-Raf mitogen-activated protein kinase (MAPK/MERK/ERK) signalling cascade, and (iii) the phosphoinositol-3 kinase – protein kinase B/Akt pathway (PI3K/AKT).

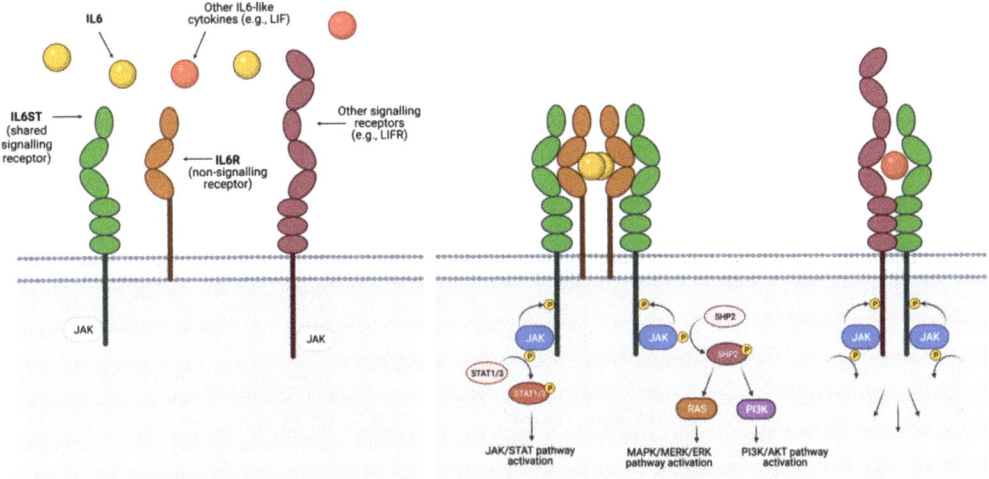

Figure 1. Summary of signalling by cytokines in the IL6-like family. Cytokines bind membrane-bound receptors with similar modular structures to form signalling complexes including 2 signalling receptors, of which at least 1 is always the shared signal transducer IL6ST. Dimerisation of these receptors leads to the activation of tyrosine kinases bound to their cytoplasmic sections, which in turn trigger a signalling cascade that can activate 3 pathways with known roles in breast cancer: JAK/STAT, MAPK/MERK/ERK and PI3K/AKT. Signalling complexes are different for each cytokine in the family. For example, IL6 is recruited by the non-signalling (lacking cytoplasmic domains) receptor IL6 receptor α (IL6R), leading to the formation of a hexameric signalling complex including an IL6ST homodimer. Other cytokines in the family form complexes comprising heterodimers (with IL6ST and a different signalling receptor) with or without the need of a non-signalling receptor, depending on the cytokine. Members of the IL6-like family can exert specific functions due to variations in ligand and receptor concentrations and in the activity of modulating signals across different cell and tissue types.

Through the sophisticated signalling machinery downstream of IL6ST, subject to complex modulation by a wide range of regulatory mechanisms, interacting factors and cross-talking pathways, IL6-like cytokines are among the most pleiotropic protein families in the human body. They have been shown to play important roles in homeostasis, immunity, inflammation and disease pathogenesis, including a well-established role in numerous cancer types [21]. This includes breast neoplasms, where they are involved in many of the hallmarks of cancer development and progression. As the prototypical member of this cytokine family, the role of IL6 in particular has been extensively studied [22–24]: although in vitro studies have reported both pro- and anti-tumourigenic effects, the role of IL6 as a negative prognosticator in BC is firmly established [25,26], with circulating serum levels in patients correlating with disease stage and higher levels being associated with worse prognosis and survival and poorer response to chemo- and endocrine therapy [27–30]. Other IL6-like cytokines have been shown to play important roles in BC, including IL11 and LIF, which can promote migration and metastasis [22,23,31].

Although signalling through a shared transducer can entail some redundancy in the roles of different IL6-like cytokines, there is also extensive evidence of functional specificity for the different ligands in vivo: specific cytokines can exert unique functions, which can result from a balance of distinct, often contrasting effects; additionally, one cytokine might elicit different responses in different cell types [32–34]. This balance of redundancy and specificity is an inherent trait of this cytokine family and is likely made possible by differences in the expression patterns of different ligands and receptors across varying tissues and cell types [32,35].

The involvement of IL6-like cytokines in many BC-related processes has highlighted their promise not only as biomarkers, but also as therapeutic targets [36]; the signal transducer itself, its ligands, co-receptors or downstream interacting factors could be modulated using either novel agents or re-purposed similarly-targeted drugs already used in the clinic for the management of other pathologies. While some such agents are currently in pre-clinical or clinical testing for their use in BC [23–25,36,37], targeting of such a complex and pleiotropic signalling axis might prove difficult, as effective inhibition will need to be fine-tuned to achieve sufficient specificity. The central signalling role in BC also suggests potential for the identification of novel biomarkers, as already established for serum levels of IL6. As the central transducer of this family, IL6ST expression could be an indicator of overall signalling activity in this cytokine class and has been identified as a potential predictor in several biomarker discovery studies. The next sections will summarise the evidence to date on the role of IL6ST as a biomarker in BC, which has led to its incorporation into several molecular signatures with prognostic and predictive value.

3. IL6ST as an Independent Predictor in BC

To date, ten independent studies based on the analysis of clinical samples by different research groups have reported IL6ST as a predictor with potential clinical utility in BC (see Table 1). Six of these studies assessed the role of the signal transducer as an independent biomarker, showing an association between IL6ST expression and prognosis in BC.

In their study of primary breast carcinomas, Karczewska et al. found that 5-year rates of both overall (OS) and disease-free survival (DFS) were significantly higher in the IL6ST-positive (IL6ST+) compared to IL6ST-negative groups (90% vs. 9% and 88% vs. 0%, respectively) [38]. Similar trends were observed for IL6 and its non-signalling receptor α (IL6R), although the survival differences were more marked in relation to IL6ST expression. Indeed, univariate analysis found significant differences in OS and DFS associated with IL6ST status ($p < 0.0001$). Subgroup analysis showed IL6ST was independent from other well-established prognostic factors, while multivariate analysis found that IL6ST expression was the strongest positive prognostic factor. The researchers concluded that IL6ST expression was associated with earlier stages of BC but, in advanced stages, its active expression correlated with better prognosis. This study also showed that IL6ST expression was negatively correlated with both lymph node status and tumour size. These findings were consistent with a more recent study by Klahan et al., which found that IL6ST expression was significantly downregulated in breast tumours with lympho-vascular invasion ($p = 0.037$) [39].

In a study of triple-negative (negative status for ER, HER2 and progesterone receptor (PR)) BC (TNBC) across 3 independent sample cohorts, Mathe et al. found that IL6ST was one of only 4 genes that were differentially expressed between normal and BC tissues and which also differed in expression between TNBC and ER-positive (ER+) BC subtypes [40]. They showed that IL6ST expression was lower in TNBC than in the ER+ group, but also that higher IL6ST levels were significantly associated ($p < 0.05$) with better OS in TNBC patients. Subsequent validation on a larger cohort of publicly-available cases also showed that higher IL6ST expression was associated with significantly increased relapse-free survival [41].

In their assessment of cases from 2 large publicly-available BC datasets, Fertig et al. found that IL6ST was significantly overexpressed ($p = 2 \times 10 - 16$) in tumours classified as luminal A or luminal B intrinsic subtypes (characterised by ER+/PR+ status) [42], consistent with previous reports of lower levels in TNBC. Survival analysis showed a trend towards longer survival in IL6ST-expressing luminal A tumours ($p = 0.06$) but not in other subtypes (Table 1).

Table 1. Summary of studies reporting on the role of IL6ST as a biomarker in breast cancer, including study cohorts and main findings. Studies are listed in chronological order of the original publication. See Table 2 for further description of the multifactor signatures. All the described associations achieved statistical significance (at least $p < 0.05$).

Original Publication	Study Type	Study Cohorts	Associations Reported	Main Predictive or Prognostic Value
Karczewska et al. (2000) [38]	Independent biomarker	75 PBCs who received surgery +/− adjuvant therapy.	IL6ST expression strongly correlates with earlier disease stages. In advanced stages, IL6ST expression is associated with better prognosis and higher OS and DFS rates. IL6ST negatively correlates with lymph node status and tumour size. IL6ST is independent from other well established clinicopathological factors.	IL6ST is a positive prognostic factor.
Tozlu et al. (2006) [43]	Independent biomarker	PBCs who received surgery (+ ET for ER+): - 12 in screening set. - 36 in validation set.	IL6ST is a perfect discriminator of ER+ status.	IL6ST is predictive for ER status and likely endocrine responsiveness.
Filipits et al. (2011) [44]	Molecular signatures: **EP and EPclin**	Original cohorts of ER+/HER2− BCs treated with ET: - 964 in training set. - 2948 in validation sets [44–47]. ER+/HER2− BCs chemotherapy study [48]: - 2630 in ET alone arm. - 1116 in ET + chemotherapy arm.	EP and EPclin scores (linked to lower IL6ST expression) are continuous predictors of the risk of distant recurrence. EPclin is also prognostic for disease recurrence in patients who received chemotherapy, regardless of menopausal status. Patients with higher EPclin score derive benefit from the addition of chemotherapy to ET.	EP and EPclin stratify into risk groups that are prognostic for risk of distant recurrence at 5, 10 and 15 years in ER+/HER2− patients. EPclin is also prognostic for LRFS. EPclin high-risk group is predictive for chemotherapy benefit in pre- and postmenopausal ER+/HER2− patients.
Sota et al. (2014) [49]	Molecular signature: **IRSN-23**	PBCs who received NAC: - 58 in training set. - 59 in validation set. - 901 in external validation set (publicly-available data).	Higher IL6ST is associated with lack of pCR from NAC. IRSN-23 classifies into Gp-R and Gp-NR groups, with differential response to NAC.	IRSN-23 signature stratifies into groups predictive of response to NAC, regardless of BC subtype of chemotherapy regimen.
Andres et al. (2014) [50]	Independent biomarker	Tumour marker analysis: - 98 male BCs (publicly-available data). - 18,366 female BCs (publicly-available data). Gene expression analysis validation: - 12 male BCs. - 233 female BCs.	IL6ST expression is significantly elevated in male BCs compared to female malignancies. IL6ST correlates with ER expression.	

Table 1. *Conts.*

Original Publication	Study Type	Study Cohorts	Associations Reported	Main Predictive or Prognostic Value
Mathe et al. (2015) [40]	Independent biomarker	Screening set: - 33 TNBCs; 17/33 with matched normal tissue, 15/33 with lymph node metastases. Validation sets: - 16 TNBCs; 4/16 with matched normal tissue - 255 non-TNBC. - Independent validation sets [41]: - 255 (publicly-available data) TNBCs. - 148 TNBCs.	IL6ST expression is associated with longer survival. IL6ST expression is lower in TNBC than ER+ tumours.	IL6ST is prognostic for OS and RFS in TNBC.
Fertig et al. (2015) [42]	Independent biomarker	638 + 897 PBCs from publicly-available sets.	IL6ST expression is higher in luminal tumours (ER+/PR+) than in other BC subtypes. Positive trend towards longer survival in IL6ST+ luminal A tumours.	
Turnbull et al. (2015) [51]	Molecular signatures: **EER4, EA2 and EA2clin**	EER4 cohort of ER+ postmenopausal IBCs treated with NET & ET: - 73 training set. - 44 validation set. EA/EA2clin study cohort of ER+ IBCs treated with NET & ET [52,53]: - 186 postmenopausal. - 51 premenopausal.	IL6ST alone is an independent predictor of response to AIs. EER4 predicts response to AIs with greater accuracy and also predict RFS and BCSS. EA2 and EA2clin predict outcome from adjuvant ET with greater accuracy and also predict RFS and BCSS. EA2 also predicts outcome in premenopausal women. EA2clin predicts treatment response regardless of ET regimen.	IL6ST is an independent predictive marker for AI response in ER+/HER2- patients. EER4 further improves on this predictive ability. Models are prognostic of outcome (RFS, BCSS) from adjuvant ET response, regardless of menopausal status or ET regimen in ER+/HER2- patients.
Klahan et al. (2017) [39]	Independent biomarker	108 pretreated IBCs: - 79 LVI+ - 29 LVI-	IL6ST correlates with LVI in samples without lymph node metastasis and perineural invasion.	
Tsunashima et al. (2018) [54]	Molecular signature: 42GC	ER+ BCs treated with ET who recurred: - 177 training set (from publicly-available sets); 84 LR, 93 NLR. - 201 validation set; 137 LR, 84 NLR.	Higher IL6ST is associated with lower risk of early recurrence but higher risk of late recurrence. 42GC classified into LR and NLR groups, with differential risk of recurrence over time. could predict late recurrence	42GC stratifies into prognostic groups for risk of early and late recurrence in ER+ BC intervals.

42GC, 42-gene classifier; **AI**, aromatase inhibitor; **BC**, breast cancer; **BCSS**, BC-specific survival; **DFS**, disease-free survival; **EA2**, EndoAdjuvant 2; **EA2clin**, EndoAdjuvant 2 clinical; **EER4**, Edinburgh EndoResponse 4; **EP**, EndoPredict; **EPclin**, EndoPredict clinical; **ER**, oestrogen receptor; **ET**, endocrine therapy; **Gp-NR**, genomically-predicted non-responders; **Gp-R**, genomically-predicted responders; **HER2**, human epidermal growth factor receptor 2; **IBC**, invasive BC; **IRSN-23**, immune-related 23-gene signature for NAC; **LN+**: lymph node positive status; **LR**, late recurrence-like; **LRFR**, local recurrence-free survival; **LVI**, lympho-vascular invasion; **NAC**, neoadjuvant chemotherapy; **NET**, neoadjuvant ET; **NLR**, non-late recurrence-like; **OS**, overall survival; **PBC**, primary BC; **pCR**, pathological complete response; **PR**, progesterone receptor; **RFS**, recurrence-free survival; **TNBC**, triple negative BC.

Consistent with previous observations, Tozlu et al. also showed that the expression of IL6ST and ER were significantly associated ($p = 1.4 \times 10^{-6}$) and positive expression of the signal transducer was highly predictive of ER+ status, perfectly discriminating between ER+ and ER- tumours (area under the receiver operating characteristic curve = 1) [43]. Andres et al. also reported that IL6ST expression was associated with ER+ status ($p < 0.05$), in addition to finding that it was upregulated in male breast tumours compared to those from female patients ($p < 0.05$) [50].

4. Molecular Signatures Incorporating IL6ST

The most relevant work in the literature comes from studies that developed molecular signatures including IL6ST and which showed prognostic or predictive power. This section describes the development to date of these signatures (also summarised in Table 1 and Table 2 and Figures 2 and 3), likely to be the best avenues for the clinical application of IL6ST as a biomarker in BC.

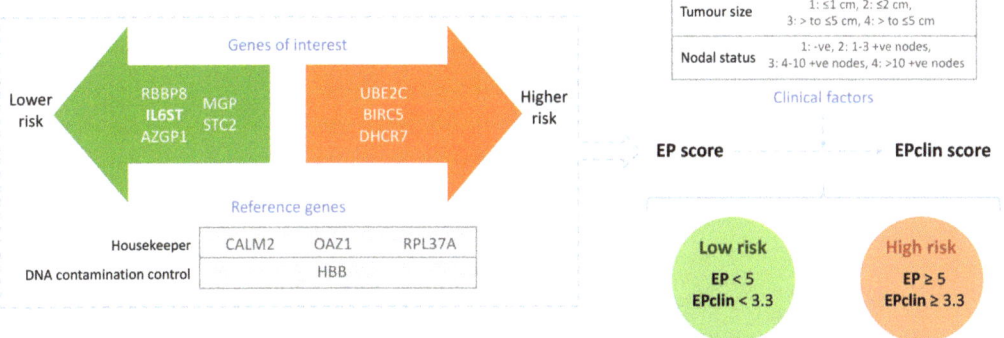

Figure 2. Summary of the markers included in the EndoPredict (EP) molecular signature and the EPclin hybrid signature, which combines EP with clinical factors. Both continuous scores allow for stratification into discrete risk groups with differential rates of distance recurrence.

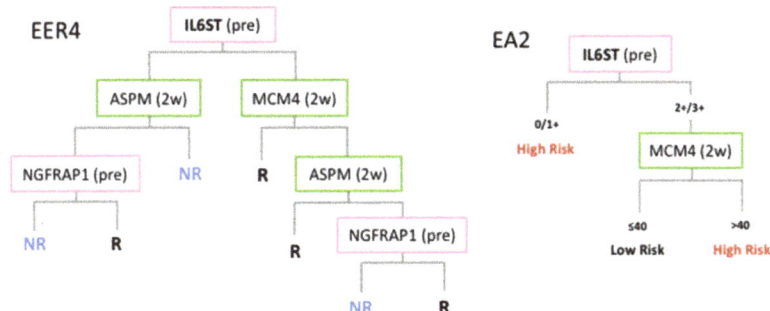

Figure 3. *Conts.*

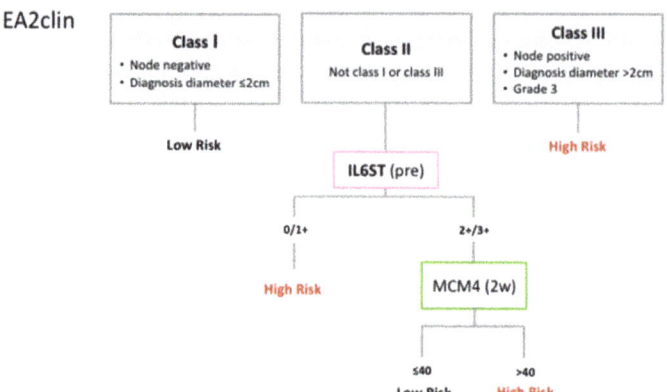

Figure 3. Summary of the markers included in the different predictive models developed in Edinburgh: the 4-gene classifier Edinburgh EndoResponse 4 (EER4) incorporates the expression level of 2 genes at pretreatment (pre) and 2 genes after 2 weeks of neoadjuvant endocrine therapy (2w); this was simplified into the EndoAdjuvant2 (EA2) signature, which uses IHC assessment of the 2 main classifiers to stratify cases into discrete risk groups; EndoAdjuvant2 clinical (EA2clin) combined EA2 with clinical factors to produce a more accurate hybrid model.

4.1. EndoPredict and EPclin Scores for Prediction of Risk of Distant Recurrence

In 2011, Filipits et al. presented a prognostic signature named EndoPredict (EP) that predicted the likelihood of distant recurrence (DR) at 5 and 10-years in patients with ER+/HER2- BC treated with endocrine therapy (ET) alone [44]. This molecular classifier was based on the assessment of the expression level of 8 cancer-related genes (3 linked to proliferation and 5 linked to ER signalling, including IL6ST) and 3 reference genes using reverse transcription quantitative polymerase chain reaction (RT-qPCR). Higher IL6ST expression led to a lower EP score and, consequently, lower associated risk of recurrence and better prognosis. This molecular score was combined with lymph node status and tumour size to provide a hybrid score named EPclin (see Figure 2 for diagram).

Initial independent validation showed that the continuous EP score was an independent predictor of DR in multivariate analysis and also provided additional prognostic information. EPclin was able to stratify patients into low (score < 3.3) and high-risk (score ≥ 3.3) groups with significantly different 10-year rates of DR (4 vs. 22–28%, $p < 0.001$), outperforming conventional clinicopathological parameters [44]. Further analysis of one validation cohort also showed that the two risk groups exhibited statistically significant different rates of local recurrence-free survival (LRFS) at 10 years, but also concluded that EPclin was not useful to help tailor local therapy [55]. Another validation study showed that EP was also an independent prognostic parameter in both pre- and postmenopausal patients who received chemotherapy, although it could not predict differences in efficacy between drug regimens [45]. Another study showed that EP was significantly associated with distant metastasis, with higher expression of the module of genes linked to ER signalling in particular contributing to reduced risk [46]. The multigene classifier was subsequently revised, adding 1 control gene for a final 12-gene molecular assay [56]. Subsequent studies validated the performance characteristics and robustness of the test, supporting its reliability for decentralised molecular assessment of luminal breast tumours [56–58].

Further work assessed the potential of EP to predict benefit from the addition of adjuvant chemotherapy to ET in both pre- and postmenopausal ER+/HER2- BC patients [48]. EPclin was highly prognostic for 10-year DR in both patients who received ET alone and in those that received it in combination with chemotherapy ($p < 0.0001$ for both groups). Results also showed that 10-year DR risk was significantly lower among patients with a high EPclin score who received chemotherapy, but no differences were found between the treatment groups for patients with low EPclin score. This suggested that a high EPclin score can predict benefit from chemotherapy in ER+/HER2- BC patients and could be used to guide treatment selection.

A recent study reassessed the prognostic power of the assay in the original validation cohorts including longer clinical follow-up to assess distant recurrence-free rates at 10 and 15 years [47]. Results showed that the EPclin score also had significant prognostic value in predicting 15-year DR, irrespective of nodal status. Additionally, they suggested that this score could help guide treatment selection: a low EPclin score may help identify patients with reduced risk of recurrence who could safely forgo adjuvant chemotherapy at diagnosis (particularly any low-risk patients with nodal involvement who would be likely to receive chemotherapy without added benefit) or extended ET at the 5-year mark.

4.2. Immune-Related 23-Gene Signature for Prediction of Response to Neoadjuvant Chemotherapy

Sota et al. constructed a signature based on gene expression microarray analysis which included 23 probes (for 19 genes, with IL6ST being represented by 3 probes) to predict the response to neoadjuvant chemotherapy (NAC) in BC patients [49]. The immune-related 23-gene signature for NAC (IRSN-23) classified patients into 2 groups, the genomically-predicted responders (Gp-R) and non-responders (Gp-NR). The Gp-R group had significantly higher rates of pathological complete response (pCR) after NAC in both the internal (38 vs. 0%, $p = 1.04 \times 10^{-6}$) and external validation (40 vs. 11%, $p = 4.98 \times 10^{-23}$) sets. This study did not select patients based on ER status and the results showed that IRSN-23 held prognostic power regardless of the patients' receptor status or chemotherapy regimen. Importantly, IL6ST was the most statistically significant marker of poorer response to NAC in the signature, with its higher expression being associated with non-pCR ($p < 0.005$). This is consistent with the previous study showing that patients with lower EP and EPclin scores (and, thus, higher IL6ST expression) derived no benefit from NAC [41].

4.3. Edinburgh EndoResponse4, EndoAdjuvant2 and EA2clin for Prediction of Response to and Outcome from Adjuvant Endocrine Therapy

Work from our group has led to the development of tools for the prediction of response to ET in postmenopausal ER+ BC patients who received neoadjuvant endocrine therapy (NET). The Edinburgh EndoResponse4 (EER4) predictive model is a 4-gene classifier incorporating the expression level of 2 genes (including IL6ST) before treatment and another 2 genes after 2 weeks of NET to classify patients into discrete responder (R) and non-responder (NR) groups [51] (see Figure 3 for diagram). IL6ST+ status alone could predict good clinical response to aromatase inhibitors (AI) with high accuracy (85%). This was further improved by EER4, which included IL6ST as its primary classifier, in both the training (96%) and independent validation (91%) sets. EER4 was also shown to significantly predict recurrence-free (RFS) ($p = 0.029$) and BC-specific survival (BCSS) ($p = 0.009$). We also showed that this 4-marker test could be performed using qPCR or immunohistochemistry (IHC).

Subsequent work has continued to revise this model. EndoAdjuvant2 (EA2) consisted of an improved tool incorporating IHC-based assessment of 2 markers at different timepoints: IL6ST at diagnosis and the proliferation-related MCM4 at 2 weeks on-treatment [52]. EA2 clinical (EA2clin) is a hybrid tool combining EA2 with clinical factors, namely node status, tumour size and grade, also included in the Nottingham Prognostic Index (NPI) tool [59] (see Figure 3 for diagram). Interestingly, EA2 (but not EA2clin) was shown to accurately predict outcome from adjuvant ET in both postmenopausal ($p = 0.001$ and $p = 0.016$ for RFS and BCSS, respectively) and premenopausal women ($p = 0.002$ and $p = 0.016$ for RFS and BCSS, respectively). EA2clin showed the best performance in postmenopausal patients, outperforming both EA2 and NPI and accurately predicting outcome ($p < 0.001$ for both RFS and BCSS) regardless of the type of adjuvant ET received [53].

4.4. 42-Gene Classifier for Prediction Risk of Late Recurrence

The team that developed IRSN-23 also sought to generate a molecular assay for prediction of recurrence in ER+ BC treated with ET alone [54]. Tsunashima et al. constructed a 42-gene classifier (42GC) including 42 probes (37 genes, including IL6ST represented by 5 probes) identified from gene expression microarray data. This signature was used to classify patients into the late-recurrence-like (LR) and non-late recurrence-like (NLR) groups. IL6ST was the most statistically significant marker in the 42GC signature ($p < 0.005$), with the LR group presenting higher expression of the signal transducer.

Results showed that the prognosis of the 2 groups identified was different and varied over time. The LR group showed significantly higher rates of late recurrence (5–15 years) and significantly lower rates of early recurrence (0–5 years) when compared to NLR in both the training ($p = 0.006$ and $p = 1.6 \times 10^{-13}$, respectively) and validation ($p = 0.02$ and $p = 5.7 \times 10^{-5}$, respectively) sets. Based on the previously established link between IL6ST expression and response to ET [51], the researchers hypothesised that the higher IL6ST expression in the LR group suggested these patients would benefit from extended ET (Table 2).

5. Discussion

In the search for novel candidate biomarkers to continue to improve the management and outcome of BC, IL6ST has emerged as a signal transducer with potential value as a predictor. We sought to review studies to date based on patient samples and data, rather than pre-clinical studies, in order to focus on results with greater clinical relevance and, thus, more likely translation into practice. We identified ten independent studies to date reporting IL6ST as a prognostic or predictive BC biomarker, either alone or as part of a multi-marker signature. Overall, these studies analysed samples and/or data from over 30,000 patients including both prospective processing of tissue samples (n > 9000) and analysis of publicly-available data (n > 25,000). Here we have reviewed and summarised this research, from which several trends have emerged.

Firstly, IL6ST seems to be a positive prognostic marker, with its higher expression being associated with better prognosis and survival rates both as an independent marker [38,40–42] and when the signal transducer is incorporated into a multi-factor signature [44,47,51]. The signal transducer has also been shown to be significantly associated with a number of other biomarkers. One prominent association reported in numerous studies across the literature is the correlation between IL6ST expression and ER+ status [40,42,43,50]. IL6ST levels have also been shown to negatively correlate with tumour size [38] and grade [51], as well as with nodal [38] or lymphovascular invasion [39].

The importance of this signal transducer in disease is well established, given its role as the signalling cornerstone for IL6-like cytokines, whose pleiotropic functions include regulation of cellular processes linked to the hallmarks of BC [22]. Interestingly, some authors had previously suggested IL6ST might instead correlate with malignancy, given its higher expression in infiltrating cancers compared with in situ or benign lesions [60]. This observation would also be in line with the fact that IL6, whose activity is dependent on IL6ST expression, has been shown to correlate with poorer prognosis in BC. Nevertheless, the complexity of the IL6ST signalling axis and the many cross-talking pathways modulating its downstream effects prevent a straight-forward description of the biological and clinical significance of this signal transducer. Indeed, the literature summarised here provides consistent evidence of IL6ST as a positive prognostic biomarker. This also includes through its association with other markers, which suggests IL6ST expression is linked to a lower risk of invasion, metastasis and recurrence and, thus, to better prognosis. IL6ST expression has also been shown to be higher in luminal tumours, which are characterised by a better clinical prognosis than other BC subtypes.

Table 2. Summary of molecular signatures incorporating IL6ST. The markers included in each model are listed, as well as its prognostic or predictive value. See Figures 2 and 3 for further description of the hybrid multifactor signatures.

Original Publication	Signature	Biomarkers Incorporated in the Signature	Clinical Significance
Filipits et al. (2011) [44]	EndoPredict	Low risk-associated (surrogates for ER signalling/cell differentiation): RBBP8, **IL6ST**, AZGP1, MGP, STC2 High risk-associated (surrogates for proliferation/cell cycle): BIRC5, UBE2C, DHCR7 Housekeeper genes: CALM2, OAZ1, RPL37A Control gene: HBB	- Stratifies into prognostic groups for risk of distant recurrence in ER+/HER2- BC patients - Predictive for benefit from the addition of chemotherapy in the high-risk group in ER+/HER2- patients
	EPclin	Clinical factors: Lymph node status, tumour size Molecular factors: EndoPredict genes	
Sota et al. (2014) [49]	IRSN-23	LR-associated: **IL6ST** (5 probes), NPY1R, ELOVL5, ASAH1 (2 probes), ALDH6A1, SYBU, RAB5C, PTP4A2, HSPA2, SLC7A8 ADRA2A, MYCBP, CX3CR1, ERCC1, DNAJA3, NINJ1, C4orf43, IFI35, ZNF688, SNX1, CREBL2, HPN, NME3, PDHB, NKX3-1, DEXI, GSTM3, LCMT1 Non-pCR-associated: **IL6ST** (3 probes), CX3CR1, ZEB1 (2 probes), SEMA3C, HFE, EDA pCR-associated: CARD9, IDO1, CXCL9, PNP, CXCL11 (2 probes), CEBPB, CD83, CD1D, CTSC, CXCL10, IGHG1, VEGFA, CR2	- Stratifies into groups predictive for response to NAC.
Turnbull et al. (2016) [51]	EER4	Pretreatment levels: **IL6ST**, NGFRAP1 2-week levels: ASPM, MCM4	- Predictive for response to AIs in postmenopausal ER+/HER2- BC patients. - Prognostic for long term outcome (RFS and BCSS) in postmenopausal ER+/HER2- BC patients treated with AIs.
	EA2	Pretreatment levels: **IL6ST** 2-week levels: MCM4	- Prognostic for long term outcome (RFS and BCSS) in ER+/HER2- BC patients treated with ET, regardless of menopausal status.
	EA2clin	Clinical factors: Lymph node involvement, tumour size and tumour grade Molecular factors: Pretreatment level: **IL6ST** 2-week level: MCM4	- Prognostic for long term outcome (RFS and BCSS) in ER+/HER2- BC patients treated with ET, regardless of ET regimen.
Tsunashima et al. (2018) [54]	42GC	NLR-associated: KLF7, STS, RALA, SMURF2, OXTR, ABCC10, ASAP2, CALB2, OPA1 LR-associated: **IL6ST** (5 probes), NPY1R, ELOVL5, ASAH1 (2 probes), ALDH6A1, SYBU, RAB5C, PTP4A2, HSPA2, SLC7A8 ADRA2A, MYCBP, CX3CR1, ERCC1, DNAJA3, NINJ1, C4orf43, IFI35, ZNF688, SNX1, CREBL2, HPN, NME3, PDHB, NKX3-1, DEXI, GSTM3, LCMT1	- Stratifies into prognostic groups for risk of early and late recurrence in ER+ BC.

42GC, 42-gene classifier; AI, aromatase inhibitor; BC, breast cancer; BCSS, BC-specific survival; EA2, EndoAdjuvant 2; EA2clin, EndoAdjuvant 2 clinical; EER4, Edinburgh EndoResponse 4; ER, oestrogen receptor; HER2, human epidermal growth factor receptor 2; EA2, EndoAdjuvant2; EA2clin, EndoAdjuvant2 clinical; ET, endocrine therapy; IRSN-23, immune-related 23-gene signature for NAC; NAC, neoadjuvant chemotherapy; pCR, pathological complete response; RFS, recurrence-free survival.

Multigene signatures including IL6ST have demonstrated prognostic value. Specifically, EP/EPclin and 42GC have shown that IL6ST expression is associated with differences in recurrence rates. The prognostic signature EPclin is already a well-established molecular assay, having been validated and reviewed with longer follow-up, and is currently commercially-available from Myriad Genetics. Expert panels in the USA and Europe have endorsed the use of EPclin to help guide treatment selection for patients with ER+/HER2-, node-negative early BC when the indication for adjuvant therapy is uncertain [61–63]. Most recently, the American National Comprehensive Cancer Network endorsed its use for prognostic purposes [64], while guidelines from the UK's National Institute of Health and Care Excellence state that EPclin may be used for patients who had an intermediate risk of DR in other tools such as NPI [65]. The extent of the use of EPclin will vary between countries depending on each territory's recommendations and health system. While it is relatively early to assess its adoption into practice in most countries, a recent prospective assessment estimated 63% cost-effectiveness for EP (versus usual care) within the Canadian health system [66], although it should be mentioned that this study also reported greater probability of cost-effectiveness for other clinically-available gene expression profiling tests. Some research has sought to assess the potential effect on the treatment decision-making process: two retrospective studies found that EPclin would lead to changes in therapy recommendation, either escalation or de-escalation, in ~35% of cases [67,68]; another study assessing how physicians' level of experience affected the decision-making process found that EPclin could be particularly beneficial to help less experienced physicians prevent over or undertreatment [69].

Interestingly, EPclin and 42GC research reported some contrasting findings. Evidence from EPclin studies was consistent with previous research [70–74] in showing that recurrence risk trends (low vs. high-risk) were consistent across time; as it pertains to IL6ST, patients with higher expression (i.e., lower score) showed decreased rates of both early and later distant recurrences. In contrast, 42GC results suggested the risk of recurrence might change overtime; thus, higher IL6ST expression was associated with the LR group of patients, with lower risk of early recurrence but higher risk of later recurrence. This could be interpreted as being in line with the described correlation between IL6ST and ER, as ER+ BC has been shown to sustain risk of recurrence over a longer period of time post-treatment than other subtypes [75].

The 42GC study used a distinct approach that likely contributed to these diverging findings, as the team specifically focused on the biological differences between malignancies that lead to early and late recurrences in their study design and supervised analysis. This differentiation in 42GC would mean a more complex prognostic role and patient stratification, compared with the other recurrence-predicting tool EPclin, in which IL6ST was very clearly a positive prognostic marker whose higher expression was linked to lower rates of distant recurrence and better prognosis [44]. Interestingly, researchers also drew different conclusions from their findings. EPclin researchers interpreted their evidence as indicative that patients in the high-IL6ST/low-risk group may be able to safely forgo extended ET [47], while the 42GC researchers hypothesised that the higher IL6ST expression in the LR group suggested that these patients would benefit from extended ET, based on the previously established link between IL6ST expression and response to ET [51]. Despite these diverging evidence and conclusions, EPclin benefits from extensive validation and its already-established clinical use.

Other molecular signatures incorporating IL6ST have also been shown to hold predictive power, with potential to help guide the selection of endocrine and chemotherapy. While the pretreatment level of IL6ST alone was shown to be a good predictor of response to AIs, this predictive ability was further improved in the EER4 model, which incorporates IL6ST as its main classifier [51]. The revised tools EA2 and EA2clin have shown great accuracy and robustness in prediction of outcomes from treatment with adjuvant ET across several validation cohorts and, importantly, regardless of menopausal status and type of ET. These tools are also advantageous in that, unlike other molecular tests such as EP, they are based on IHC assessment and, thus, could be easily implemented in local laboratories. They also enable discrete risk stratification, making its interpretation for potential clinical application more straight-forward than continuous scores.

Evidence has also shown that IL6ST expression is predictive of a lack of response to chemotherapy. In the IRSN-23 signature, designed specifically to predict response to NAC, higher IL6ST expression was linked to a lack of pCR. In line with this, a recent study showed that ER+ BC patients with higher IL6ST expression and, consequently, lower EPclin scores did not benefit from the addition of chemotherapy to ET. While these results diverge from previous evidence in that IL6ST acts as a negative predictor, they are consistent with the fact that different BC subtypes, as well as some subsets within the ER+ BC population, will respond differently to chemotherapy. Indeed, while the IRSN-23 study did not select patients according to hormone receptor status, results showed that the Gp-NR group was significantly enriched for luminal breast tumours ($p < 0.005$), which would typically show less response to chemotherapy [76,77].

Finally, IL6ST might hold particular promise as a biomarker in ER+ disease. The link between IL6-like cytokines and oestrogen-related signalling in BC is already well documented [26,37] and, as summarised here, numerous studies have reported a correlation between both biomarkers. Evidence suggests that, in addition to its prognostic role, IL6ST might be a robust surrogate marker of active oestrogen signalling and, consequently, responsiveness to ET. Indeed, we have shown that IL6ST can identify subsets of breast lesions with active ER-dependent signalling within larger ER+ populations [78–80]. This suggests that the predictive value of IL6ST might partly emerge from the biomarker's ability to discriminate the complex underlying biology of hormone-dependent disease, possibly enabling a finer stratification than histological assessment of ER status alone currently allows. In this way, IL6ST might serve as a marker to identify those ER+ tumours that are more likely to respond to readily-available endocrine therapy.

6. Conclusions

In recent years, IL6ST has emerged as a biomarker with prognostic and predictive value in BC. Although the complex role of IL6ST signalling in the disease might prevent the description of a simple mechanism behind this predictive value, there is extensive evidence that expression of this signal transducer is a positive prognostic factor in BC.

While current research efforts are investigating the potential of targeting the IL6ST signalling axis as a therapeutic approach, studies to date support the notion that the best route for exploiting IL6ST as biomarker in the clinical setting will be as part of multifactor hybrid signature and likely within the ER+ subset of the disease.

In this way, tools incorporating IL6ST could enable patient stratification into discrete groups that more accurately reflect the underlying biology of the disease and, consequently, better predict prognosis and the likelihood of treatment response. As with any tools of this type, successful clinical translation will necessitate prospective studies to both corroborate the prognostic and predictive ability of IL6ST and its related signatures, and to help define any potential clinical guidelines, particularly on whether lower-risk patients might be able to safely forgo neoadjuvant or extended therapy. Overall, with sufficient validation, tools incorporating IL6ST as a molecular biomarker could improve the management of BC by helping to make a better, more targeted use of the therapeutic strategies already available in the clinic.

Author Contributions: C.M.-P., J.L. and A.K.T. conceived the review. C.M.-P., J.L. and C.K. conducted an extensive review of the literature. All authors contributed to and approved the final article. All authors have read and agreed to the published version of the manuscript.

Funding: This research received no external funding.

Institutional Review Board Statement: Not applicable.

Informed Consent Statement: Not applicable.

Data Availability Statement: Not applicable.

Conflicts of Interest: The authors declare no conflict of interest.

References

1. Atkinson, A.J.; Colburn, W.A.; DeGruttola, V.G.; DeMets, D.L.; Downing, G.J.; Hoth, D.F.; Oates, J.A.; Peck, C.C.; Schooley, R.T.; Spilker, B.A.; et al. Biomarkers and surrogate endpoints: Preferred definitions and conceptual framework. *Clin. Pharmacol. Ther.* **2001**, *69*, 89–95.
2. FDA-NIH Biomarker Working Group. *BEST (Biomarkers, EndpointS, and Other Tools)*; National Institute of Health: Bethesda, MD, USA, 2017.
3. Nicolini, A.; Carpi, A.; Rossi, G. Cytokines in breast cancer. *Cytokine Growth Factor Rev.* **2006**, *17*, 325–337. [CrossRef] [PubMed]
4. Duffy, M.J.; Harbeck, N.; Nap, M.; Molina, R.; Nicolini, A.; Senkus, E.; Cardoso, F. Clinical use of biomarkers in breast cancer: Updated guidelines from the European Group on Tumor Markers (EGTM). *Eur. J. Cancer* **2017**, *75*, 284–298. [CrossRef]
5. Martínez-Pérez, C.; Turnbull, A.K.; Dixon, J.M. The evolving role of receptors as predictive biomarkers for metastatic breast cancer. *Expert Rev. Anticancer Ther.* **2019**, *19*, 121–138. [CrossRef]
6. Reinert, T.; Saad, E.D.; Barrios, C.H.; Bines, J. Clinical Implications of ESR1 Mutations in Hormone Receptor-Positive Advanced Breast Cancer. *Front. Oncol.* **2017**, *7*, 26. [CrossRef]
7. Angus, L.; Beije, N.; Jager, A.; Martens, J.W.M.; Sleijfer, S. ESR1 mutations: Moving towards guiding treatment decision-making in metastatic breast cancer patients. *Cancer Treat. Rev.* **2017**, *52*, 33–40. [CrossRef] [PubMed]
8. Shao, Y.; Sun, X.; He, Y.; Liu, C.; Liu, H. Elevated levels of serum tumor markers CEA and CA15-3 are prognostic parameters for different molecular subtypes of breast cancer. *PLoS ONE* **2015**, *10*, e0133830. [CrossRef] [PubMed]
9. Kazarian, A.; Blyuss, O.; Metodieva, G.; Gentry-Maharaj, A.; Ryan, A.; Kiseleva, E.M.; Prytomanova, O.M.; Jacobs, I.J.; Widschwendter, M.; Menon, U.; et al. Testing breast cancer serum biomarkers for early detection and prognosis in pre-diagnosis samples. *Br. J. Cancer* **2017**, *116*, 501–508. [CrossRef]
10. Kabel, A.M. Tumor markers of breast cancer: New prospectives. *J. Oncol. Sci.* **2017**, *3*, 5–11. [CrossRef]
11. Gaughran, G.; Aggarwal, N.; Shadbolt, B.; Stuart-Harris, R. The utility of the tumor markers CA15.3, CEA, CA-125 and CA19.9 in metastatic breast cancer. *Breast Cancer Manag.* **2020**, *9*, BMT50. [CrossRef]
12. Walsh, M.F.; Nathanson, K.L.; Couch, F.J.; Offit, K. Genomic biomarkers for breast cancer risk. In *Advances in Experimental Medicine and Biology*; Springer New York LLC: New York, NY, USA, 2016; Volume 882, pp. 1–32.
13. Akcakanat, A.; Zheng, X.; Cruz Pico, C.X.; Kim, T.-B.; Chen, K.; Korkut, A.; Sahin, A.; Holla, V.; Tarco, E.; Singh, G.; et al. Genomic, Transcriptomic, and Proteomic Profiling of Metastatic Breast Cancer. *Clin. Cancer Res.* **2021**, *27*, 3243–3252. [CrossRef]
14. Falzone, L.; Grimaldi, M.; Celentano, E.; Augustin, L.S.A.; Libra, M. Identification of Modulated MicroRNAs Associated with Breast Cancer, Diet, and Physical Activity. *Cancers* **2020**, *12*, 2555. [CrossRef] [PubMed]
15. Loh, H.Y.; Norman, B.P.; Lai, K.S.; Rahman, N.M.A.N.A.; Alitheen, N.B.M.; Osman, M.A. The regulatory role of microRNAs in breast cancer. *Int. J. Mol. Sci.* **2019**, *20*, 4940. [CrossRef] [PubMed]
16. Abolghasemi, M.; Tehrani, S.S.; Yousefi, T.; Karimian, A.; Mahmoodpoor, A.; Ghamari, A.; Jadidi-Niaragh, F.; Yousefi, M.; Kafil, H.S.; Bastami, M.; et al. MicroRNAs in breast cancer: Roles, functions, and mechanism of actions. *J. Cell. Physiol.* **2020**, *235*, 5008–5029. [CrossRef] [PubMed]
17. Rose-John, S. Interleukin-6 family cytokines. *Cold Spring Harb. Perspect. Biol.* **2018**, *10*, a028415. [CrossRef] [PubMed]
18. Heinrich, P.C.; Behrmann, I.; Haan, S.; Hermanns, H.M.; Schaper, F. Principles of interleukin (IL)-6-type cytokine signalling and its regulation. *Biochem. J* **2003**, *374*, 1–20. [CrossRef]
19. Silver, J.S.; Hunter, C.A. gp130 at the nexus of inflammation, autoimmunity, and cancer. *J. Leukoc. Biol.* **2010**, *88*, 1145–1156. [CrossRef] [PubMed]
20. Yoshida, K.; Taga, T.; Saito, M.; Suematsu, S.; Kumanogoh, A.; Tanaka, T.; Fujiwara, H.; Hirata, M.; Yamagami, T.; Nakahata, T.; et al. Targeted disruption of gp130, a common signal transducer for the interleukin 6 family of cytokines, leads to myocardial and hematological disorders. *Proc. Natl. Acad. Sci. USA* **1996**, *93*, 407–411. [CrossRef]
21. Bravo, J.; Heath, J. Receptor recognition by gp130 cytokines. *EMBO J.* **2000**, *19*, 2399–2411. [CrossRef] [PubMed]
22. Omokehinde, T.; Johnson, R.W. GP130 Cytokines in Breast Cancer and Bone. *Cancers* **2020**, *12*, 326. [CrossRef]
23. Jones, S.A.; Jenkins, B.J. Recent insights into targeting the IL-6 cytokine family in inflammatory diseases and cancer. *Nat. Rev. Immunol.* **2018**, *18*, 773–789. [CrossRef]
24. Kaur, S.; Bansal, Y.; Kumar, R.; Bansal, G. A panoramic review of IL-6: Structure, pathophysiological roles and inhibitors. *Bioorganic Med. Chem.* **2020**, *28*, 115327. [CrossRef] [PubMed]
25. Johnson, D.E.; O'keefe, R.A.; Grandis, J.R. Targeting the IL-6/JAK/STAT3 signalling axis in cancer. *Nat. Rev. Clin. Oncol.* **2018**, *15*, 234–248. [CrossRef] [PubMed]
26. Knüpfer, H.; Preiß, R. Significance of interleukin-6 (IL-6) in breast cancer (review). *Breast Cancer Res. Treat.* **2007**, *102*, 129–135. [CrossRef]
27. Taher, M.Y.; Davies, D.M.; Maher, J. The role of the interleukin (IL)-6/IL-6 receptor axis in cancer. *Biochem. Soc. Trans.* **2018**, *46*, 1449–1462. [CrossRef]

28. Nishimura, R.; Nagao, K.; Miyayama, H.; Matsuda, M.; Baba, K.; Matsuoka, Y.; Yamashita, H.; Fukuda, M.; Mizumoto, T.; Hamamoto, R. An Analysis of Serum Interleukin-6 Levels to Predict Benefits of Medroxyprogesterone Acetate in Advanced or Recurrent Breast Cancer. *Oncology* **2000**, *59*, 166–173. [CrossRef]
29. Yokoe, T.; Lino, Y.; Morishita, Y. Trends of IL-6 and IL-8 levels in patients with recurrent breast cancer: Preliminary report. *Breast Cancer* **2000**, *7*, 187–190. [CrossRef] [PubMed]
30. Bachelot, T.; Ray-Coquard, I.; Menetrier-Caux, C.; Rastkha, M.; Duc, A.; Blay, J.Y. Prognostic value of serum levels of interleukin 6 and of serum and plasma levels of vascular endothelial growth factor in hormone-refractory metastatic breast cancer patients. *Br. J. Cancer* **2003**, *88*, 1721–1726. [CrossRef] [PubMed]
31. Johnstone, C.N.; Chand, A.; Putoczki, T.L.; Ernst, M. Emerging roles for IL-11 signaling in cancer development and progression: Focus on breast cancer. *Cytokine Growth Factor Rev.* **2015**, *26*, 489–498. [CrossRef]
32. Murakami, M.; Kamimura, D.; Hirano, T. Pleiotropy and Specificity: Insights from the Interleukin 6 Family of Cytokines. *Immunity* **2019**, *50*, 812–831. [CrossRef]
33. Hirano, T.; Matsuda, T.; Nakajima, K. Signal transduction through gp130 that is shared among the receptors for the interleukin 6 related cytokine subfamily. *Stem Cells* **1994**, *12*, 262–277. [CrossRef]
34. Heinrich, P.C.; Behrmann, I.; Müller-Newen, G.; Schaper, F.; Graeve, L. Interleukin-6-type cytokine signalling through the gp130/Jak/STAT pathway. *Biochem. J.* **1998**, *334*, 297–314. [CrossRef]
35. Ernst, M.; Jenkins, B.J. Acquiring signalling specificity from the cytokine receptor gp130. *Trends Genet.* **2004**, *20*, 23–32. [CrossRef]
36. Heo, T.H.T.-H.; Wahler, J.; Suh, N. Potential therapeutic implications of IL-6/IL-6R/gp130-targeting agents in breast cancer. *Oncotarget* **2016**, *7*, 15460–15473. [CrossRef]
37. Masjedi, A.; Hashemi, V.; Hojjat-Farsangi, M.; Ghalamfarsa, G.; Azizi, G.; Yousefi, M.; Jadidi-Niaragh, F. The significant role of interleukin-6 and its signaling pathway in the immunopathogenesis and treatment of breast cancer. *Biomed. Pharmacother.* **2018**, *108*, 1415–1424. [CrossRef]
38. Karczewska, A.; Nawrocki, S.; Bręborowicz, D.; Filas, V.; Mackiewicz, A. Expression of interleukin-6, interleukin-6 receptor, and glycoprotein 130 correlates with good prognoses for patients with breast carcinoma. *Cancer* **2000**, *88*, 2061–2071. [CrossRef]
39. Klahan, S.; Wong, H.S.C.; Tu, S.H.; Chou, W.H.; Zhang, Y.F.; Ho, T.F.; Liu, C.Y.; Yih, S.Y.; Lu, H.F.; Chen, S.C.C.; et al. Identification of genes and pathways related to lymphovascular invasion in breast cancer patients: A bioinformatics analysis of gene expression profiles. *Tumor Biol.* **2017**, *39*. [CrossRef]
40. Mathe, A.; Wong-Brown, M.; Morten, B.; Forbes, J.F.; Braye, S.G.; Avery-Kiejda, K.A.; Scott, R.J. Novel genes associated with lymph node metastasis in triple negative breast cancer. *Sci. Rep.* **2015**, *5*, 15832. [CrossRef]
41. Pariyar, M.; Scott, R.; Avery-Kiejda, K. OR20 Validation of Four Triple Negative Breast Cancer–Specific Genes and their Association with Prognosis. *Asia-Pac. J. Clin. Oncol.* **2017**, *13* (Suppl. 5), 7–18.
42. Fertig, E.J.; Lee, E.; Pandey, N.B.; Popel, A.S. Analysis of gene expression of secreted factors associated with breast cancer metastases in breast cancer subtypes. *Sci. Rep.* **2015**, *5*, 12133. [CrossRef] [PubMed]
43. Tozlu, S.; Girault, I.; Vacher, S.; Vendrell, J.; Andrieu, C.; Spyratos, F.; Cohen, P.; Lidereau, R.; Bieche, I. Identification of novel genes that co-cluster with estrogen receptor alpha in breast tumor biopsy specimens, using a large-scale real-time reverse transcription-PCR approach. *Endocr. Relat. Cancer* **2006**, *13*, 1109–1120. [CrossRef]
44. Filipits, M.; Rudas, M.; Jakesz, R.; Dubsky, P.; Fitzal, F.; Singer, C.F.; Dietze, O.; Greil, R.; Jelen, A.; Sevelda, P.; et al. A new molecular predictor of distant recurrence in ER-positive, HER2-negative breast cancer adds independent information to conventional clinical risk factors. *Clin. Cancer Res.* **2011**, *17*, 6012–6020. [CrossRef]
45. Martin, M.; Brase, J.C.; Calvo, L.; Krappmann, K.; Ruiz-Borrego, M.; Fisch, K.; Ruiz, A.; Weber, K.E.; Munarriz, B.; Petry, C.; et al. Clinical validation of the EndoPredict test in node-positive, chemotherapy-treated ER+/HER2- breast cancer patients: Results from the GEICAM 9906 trial. *Breast Cancer Res.* **2014**, *16*, R38. [CrossRef]
46. Dubsky, P.; Brase, J.C.; Jakesz, R.; Rudas, M.; Singer, C.F.; Greil, R.; Dietze, O.; Luisser, I.; Klug, E.; Sedivy, R.; et al. The EndoPredict score provides prognostic information on late distant metastases in ER+/HER2-breast cancer patients. *Br. J. Cancer* **2013**, *109*, 2959–2964. [CrossRef]
47. Filipits, M.; Dubsky, P.; Rudas, M.; Greil, R.; Balic, M.; Bago-Horvath, Z.; Singer, C.; Hlauschek, D.; Brown, K.; Bernhisel, R.; et al. Prediction of distant recurrence using EndoPredict among women with ER+, HER2-node-positive and node-negative breast cancer treated with endocrine therapy only. *Clin. Cancer Res.* **2019**, *25*, 3865–3872. [CrossRef]
48. Sestak, I.; Martín, M.; Dubsky, P.; Kronenwett, R.; Rojo, F.; Cuzick, J.; Filipits, M.; Ruiz, A.; Gradishar, W.; Soliman, H.; et al. Prediction of chemotherapy benefit by EndoPredict in patients with breast cancer who received adjuvant endocrine therapy plus chemotherapy or endocrine therapy alone. *Breast Cancer Res. Treat.* **2019**, *176*, 377–386. [CrossRef]
49. Sota, Y.; Naoi, Y.; Tsunashima, R.; Kagara, N.; Shimazu, K.; Maruyama, N.; Shimomura, A.; Shimoda, M.; Kishi, K.; Baba, Y.; et al. Construction of novel immune-related signature for prediction of pathological complete response to neoadjuvant chemotherapy in human breast cancer. *Ann. Oncol.* **2014**, *25*, 100–106. [CrossRef]
50. Andres, S.A.; Smolenkova, I.A.; Wittliff, J.L. Gender-associated expression of tumor markers and a small gene set in breast carcinoma. *Breast* **2014**, *23*, 226–233. [CrossRef]
51. Turnbull, A.K.; Arthur, L.M.; Renshaw, L.; Larionov, A.A.; Kay, C.; Dunbier, A.K.; Thomas, J.S.; Dowsett, M.; Sims, A.H.; Dixon, J.M. Accurate Prediction and Validation of Response to Endocrine Therapy in Breast Cancer. *J. Clin. Oncol.* **2015**, *33*, 2270–2278. [CrossRef]

52. Turnbull, A.; Lee, Y.; Pearce, D.; Martinez-Perez, C.; Uddin, S.; Webb, H.; Fernando, A.; Thomas, J.; Renshaw, L.; Sims, A.; et al. A test utilising diagnostic and on-treatment biomarkers to improve prediction of response to endocrine therapy in breast cancer. *J. Clin. Oncol.* **2016**, *34*, 555. [CrossRef]
53. Turnbull, A.; Fernando, A.; Renshaw, L.; Keys, J.; Thomas, J.; Sims, A. EA2Clin: A novel immunohistochemical prognostic and predictive test for patients with estrogen receptor-Positive breast cancer. *Cancer Res.* **2018**, *78*, P6-09-27.
54. Tsunashima, R.; Naoi, Y.; Shimazu, K.; Kagara, N.; Shimoda, M.; Tanei, T.; Miyake, T.; Kim, S.J.; Noguchi, S. Construction of a novel multi-gene assay (42-gene classifier) for prediction of late recurrence in ER-positive breast cancer patients. *Breast Cancer Res. Treat.* **2018**, *171*, 33–41. [CrossRef]
55. Fitzal, F.; Filipits, M.; Rudas, M.; Greil, R.; Dietze, O.; Samonigg, H.; Lax, S.; Herz, W.; Dubsky, P.; Bartsch, R.; et al. The genomic expression test EndoPredict is a prognostic tool for identifying risk of local recurrence in postmenopausal endocrine receptor-positive, her2neu-negative breast cancer patients randomised within the prospective ABCSG 8 trial. *Br. J. Cancer* **2015**, *112*, 1405–1410. [CrossRef] [PubMed]
56. Warf, M.B.; Rajamani, S.; Krappmann, K.; Doedt, J.; Cassiano, J.; Brown, K.; Reid, J.E.; Kronenwett, R.; Roa, B.B. Analytical validation of a 12-gene molecular test for the prediction of distant recurrence in breast cancer. *Futur. Sci. OA* **2017**, *3*, FSO221. [CrossRef] [PubMed]
57. Kronenwett, R.; Bohmann, K.; Prinzler, J.; Sinn, B.V.; Haufe, F.; Roth, C.; Averdick, M.; Ropers, T.; Windbergs, C.; Brase, J.C.; et al. Decentral gene expression analysis: Analytical validation of the Endopredict genomic multianalyte breast cancer prognosis test. *BMC Cancer* **2012**, *12*, 456. [CrossRef] [PubMed]
58. Denkert, C.; Kronenwett, R.; Schlake, W.; Bohmann, K.; Penzel, R.; Weber, K.E.; Höfler, H.; Lehmann, U.; Schirmacher, P.; Specht, K.; et al. Decentral gene expression analysis for ER+/Her2− breast cancer: Results of a proficiency testing program for the EndoPredict assay. *Virchows Arch.* **2012**, *460*, 251–259. [CrossRef]
59. Haybittle, J.L.; Blamey, R.W.; Elston, C.W.; Johnson, J.; Doyle, P.J.; Campbell, F.C.; Nicholson, R.I.; Griffiths, K. A prognostic index in primary breast cancer. *Br. J. Cancer* **1982**, *45*, 361–366. [CrossRef]
60. García-Tuñón, I.; Ricote, M.; Ruiz, A.; Fraile, B.; Paniagua, R.; Royuela, M. OSM, LIF, Its Receptors, and Its Relationship with the Malignance in Human Breast Carcinoma (In Situ and in Infiltrative). *Cancer Investig.* **2008**, *26*, 222–229. [CrossRef]
61. Senkus, E.; Kyriakides, S.; Ohno, S.; Penault-Llorca, F.; Poortmans, P.; Rutgers, E.; Zackrisson, S.; Cardoso, F. Primary breast cancer: ESMO Clinical Practice Guidelines for diagnosis, treatment and follow-up. *Ann. Oncol.* **2015**, *26*, v8–v30. [CrossRef]
62. Krop, I.; Ismaila, N.; Andre, F.; Bast, R.C.; Barlow, W.; Collyar, D.E.; Hammond, M.E.; Kuderer, N.M.; Liu, M.C.; Mennel, R.G.; et al. Use of biomarkers to guide decisions on adjuvant systemic therapy for women with early-stage invasive breast cancer: American society of clinical oncology clinical practice guideline focused update. *J. Clin. Oncol.* **2017**, *35*, 2838–2847. [CrossRef]
63. Vieira, A.F.; Schmitt, F. An update on breast cancer multigene prognostic tests-emergent clinical biomarkers. *Front. Med.* **2018**, *5*, 248. [CrossRef] [PubMed]
64. NCCN. *2019 Clinical Practice Guidelines in Oncology: Breast Cancer*; National Comprehensive Cancer Network: Plymouth Meeting, PA, USA, 2019.
65. NICE. *Tumour Profiling Tests to Guide Adjuvant Chemotherapy Decisions in Early Breast Cancer*; National Institute of Health and Care Excellence: London, UK, 2018.
66. Ontario Health (Quality). *Gene Expression Profiling Tests for Early-Stage Invasive Breast Cancer: A Health Technology Assessment*; Ontario Health (Quality): Toronto, ON, Canada, 2020; Volume 20.
67. Almstedt, K.; Mendoza, S.; Otto, M.; Battista, M.J.; Steetskamp, J.; Heimes, A.S.; Krajnak, S.; Poplawski, A.; Gerhold-Ay, A.; Hasenburg, A.; et al. EndoPredict® in early hormone receptor-positive, HER2-negative breast cancer. *Breast Cancer Res. Treat.* **2020**, *182*, 137–146. [CrossRef]
68. Müller, B.M.; Keil, E.; Lehmann, A.; Winzer, K.-J.; Richter-Ehrenstein, C.; Prinzler, J.; Bangemann, N.; Reles, A.; Stadie, S.; Schoenegg, W.; et al. The EndoPredict Gene-Expression Assay in Clinical Practice—Performance and Impact on Clinical Decisions. *PLoS ONE* **2013**, *8*, e68252. [CrossRef]
69. Thangarajah, F.; Eichler, C.; Fromme, J.; Malter, W.; Caroline Radosa, J.; Ludwig, S.; Puppe, J.; Paepke, S.; Warm, M.; Caroline Radosa JuliaRadosa, J.; et al. The impact of EndoPredict® on decision making with increasing oncological work experience: Can overtreatment be avoided? *Arch. Gynecol. Obstet.* **2019**, *299*, 1437–1442. [CrossRef] [PubMed]
70. Buus, R.; Sestak, I.; Kronenwett, R.; Denkert, C.; Dubsky, P.; Krappmann, K.; Scheer, M.; Petry, C.; Cuzick, J.; Dowsett, M. Comparison of EndoPredict and EPclin with Oncotype DX recurrence score for prediction of risk of distant recurrence after endocrine therapy. *J. Natl. Cancer Inst.* **2016**, *108*, djw149. [CrossRef]
71. Zhang, Y.; Schnabel, C.A.; Schroeder, B.E.; Jerevall, P.L.; Jankowitz, R.C.; Fornander, T.; Stål, O.; Brufsky, A.M.; Sgroi, D.; Erlander, M.G. Breast cancer index identifies early-stage estrogen receptor-positive breast cancer patients at risk for early- and late-distant recurrence. *Clin. Cancer Res.* **2013**, *19*, 4196–4205. [CrossRef]
72. Sgroi, D.C.; Sestak, I.; Cuzick, J.; Zhang, Y.; Schnabel, C.A.; Schroeder, B.; Erlander, M.G.; Dunbier, A.; Sidhu, K.; Lopez-Knowles, E.; et al. Prediction of late distant recurrence in patients with oestrogen-receptor-positive breast cancer: A prospective comparison of the breast-cancer index (BCI) assay, 21-gene recurrence score, and IHC4 in the TransATAC study population. *Lancet Oncol.* **2013**, *14*, 1067–1076. [CrossRef]

73. Schroeder, B.; Zhang, Y.; Stål, O.; Fornander, T.; Brufsky, A.; Sgroi, D.C.; Schnabel, C.A. Risk stratification with Breast Cancer Index for late distant recurrence in patients with clinically low-risk (T1N0) estrogen receptor-positive breast cancer. *NPJ Breast Cancer* **2017**, *3*, 1–3. [CrossRef] [PubMed]
74. Filipits, M.; Nielsen, T.O.; Rudas, M.; Greil, R.; Stoger, H.; Jakesz, R.; Bago-Horvath, Z.; Dietze, O.; Regitnig, P.; Gruber-Rossipal, C.; et al. The PAM50 risk-of-recurrence score predicts risk for late distant recurrence after endocrine therapy in postmenopausal women with endocrine-responsive early breast cancer. *Clin. Cancer Res.* **2014**, *20*, 1298–1305. [CrossRef] [PubMed]
75. Pan, H.; Gray, R.; Braybrooke, J.; Davies, C.; Taylor, C.; McGale, P.; Peto, R.; Pritchard, K.I.; Bergh, J.; Dowsett, M.; et al. 20-Year Risks of Breast-Cancer Recurrence after Stopping Endocrine Therapy at 5 Years. *N. Engl. J. Med.* **2017**, *377*, 1836–1846. [CrossRef]
76. Barrios, C.H.; Sampaio, C.; Vinholes, J.; Caponero, R. What is the role of chemotherapy in estrogen receptor-positive, advanced breast cancer? *Ann. Oncol.* **2009**, *20*, 1157–1162. [CrossRef]
77. Herr, D.; Wischnewsky, M.; Joukhadar, R.; Chow, O.; Janni, W.; Leinert, E.; Fink, V.; Stüber, T.; Curtaz, C.; Kreienberg, R.; et al. Does chemotherapy improve survival in patients with nodal positive luminal A breast cancer? A retrospective Multicenter Study. *PLoS ONE* **2019**, *14*, e0218434. [CrossRef]
78. Turnbull, A.; Webber, V.; McStay, D.; Arthur, L.; Martinez-Perez, C.; Fernando, A.; Renshaw, L.; Keys, J.; Clarke, R.; Sims, A. Predicting benefit from HER2-targeted therapies in patients with ER+/HER2+ breast cancer. *Cancer Res.* **2019**, *79*, P3-10-26.
79. Turnbull, A.K.; Webber, V.; McStay, D.; Arthur, L.M.; Martinez-Perez, C.; Meehan, J.; Gray, M.; Kay, C.; Renshaw, L.; Keys, J.; et al. Abstract P1-18-07: Can some ER+/HER2+ patients be safely spared from treatment with chemotherapy plus herceptin? *Cancer Res.* **2020**, *80*, P1-18-07. [CrossRef]
80. Martinez-Perez, C.; Kay, C.; Swan, R.; Ekatah, G.E.; Arthur, L.M.; Meehan, J.; Gray, M.; Sims, A.H.; Oikonomidou, O.; Turnbull, A.K.; et al. Abstract P6-16-04: IL6ST, a biomarker of endocrine therapy response, has potential in identifying a subgroup of women with ER+ DCIS who are more likely to benefit from adjuvant endocrine therapy. *Cancer Res.* **2020**, *80*, P6-16-04.

Article

A Novel Approach for the Discovery of Biomarkers of Radiotherapy Response in Breast Cancer

James Meehan [1,*,†], Mark Gray [2,†], Carlos Martínez-Pérez [1,3], Charlene Kay [1,3], Jimi C. Wills [4,5], Ian H. Kunkler [4], J. Michael Dixon [3] and Arran K. Turnbull [1,3]

1. Translational Oncology Research Group, Institute of Genetics and Cancer, Western General Hospital, University of Edinburgh, Edinburgh EH4 2XU, UK; carlos.martinez-perez@ed.ac.uk (C.M.-P.); charlene.kay@ed.ac.uk (C.K.); a.turnbull@ed.ac.uk (A.K.T.)
2. The Royal (Dick) School of Veterinary Studies and Roslin Institute, University of Edinburgh, Midlothian EH25 9RG, UK; mark.gray@ed.ac.uk
3. Breast Cancer Now Edinburgh Research Team, Institute of Genetics and Cancer, Western General Hospital, University of Edinburgh, Edinburgh EH4 2XU, UK; mike.dixon@ed.ac.uk
4. Cancer Research UK Edinburgh Centre, Institute of Genetics and Cancer, University of Edinburgh, Edinburgh EH4 2XU, UK; jimi@firefinch.io (J.C.W.); iankunkler@yahoo.com (I.H.K.)
5. Firefinch Software Ltd., Edinburgh EH12 9DQ, UK
* Correspondence: james.meehan@ed.ac.uk
† These authors contributed equally to this work.

Abstract: Radiotherapy (RT) is an important treatment modality for the local control of breast cancer (BC). Unfortunately, not all patients that receive RT will obtain a therapeutic benefit, as cancer cells that either possess intrinsic radioresistance or develop resistance during treatment can reduce its efficacy. For RT treatment regimens to become personalised, there is a need to identify biomarkers that can predict and/or monitor a tumour's response to radiation. Here we describe a novel method to identify such biomarkers. Liquid chromatography-mass spectrometry (LC-MS) was used on conditioned media (CM) samples from a radiosensitive oestrogen receptor positive (ER$^+$) BC cell line (MCF-7) to identify cancer-secreted biomarkers which reflected a response to radiation. A total of 33 radiation-induced secreted proteins that had higher (up to 12-fold) secretion levels at 24 h post-2 Gy radiation were identified. Secretomic results were combined with whole-transcriptome gene expression experiments, using both radiosensitive and radioresistant cells, to identify a signature related to intrinsic radiosensitivity. Gene expression analysis assessing the levels of the 33 proteins showed that 5 (YBX3, EIF4EBP2, DKK1, GNPNAT1 and TK1) had higher expression levels in the radiosensitive cells compared to their radioresistant derivatives; 3 of these proteins (DKK1, GNPNAT1 and TK1) underwent in-lab and initial clinical validation. Western blot analysis using CM samples from cell lines confirmed a significant increase in the release of each candidate biomarker from radiosensitive cells 24 h after treatment with a 2 Gy dose of radiation; no significant increase in secretion was observed in the radioresistant cells after radiation. Immunohistochemistry showed that higher intracellular protein levels of the biomarkers were associated with greater radiosensitivity. Intracellular levels were further assessed in pre-treatment biopsy tissues from patients diagnosed with ER$^+$ BC that were subsequently treated with breast-conserving surgery and RT. High DKK1 and GNPNAT1 intracellular levels were associated with significantly increased recurrence-free survival times, indicating that these two candidate biomarkers have the potential to predict sensitivity to RT. We suggest that the methods highlighted in this study could be utilised for the identification of biomarkers that may have a potential clinical role in personalising and optimising RT dosing regimens, whilst limiting the administration of RT to patients who are unlikely to benefit.

Keywords: breast cancer; radiotherapy; radiosensitivity biomarkers; secretome; radioresistance

Citation: Meehan, J.; Gray, M.; Martínez-Pérez, C.; Kay, C.; Wills, J.C.; Kunkler, I.H.; Dixon, J.M.; Turnbull, A.K. A Novel Approach for the Discovery of Biomarkers of Radiotherapy Response in Breast Cancer. *J. Pers. Med.* **2021**, *11*, 796. https://doi.org/10.3390/jpm11080796

Academic Editor: Cynthia Aristei

Received: 29 July 2021
Accepted: 11 August 2021
Published: 14 August 2021

Publisher's Note: MDPI stays neutral with regard to jurisdictional claims in published maps and institutional affiliations.

Copyright: © 2021 by the authors. Licensee MDPI, Basel, Switzerland. This article is an open access article distributed under the terms and conditions of the Creative Commons Attribution (CC BY) license (https://creativecommons.org/licenses/by/4.0/).

1. Introduction

Radiotherapy (RT), initially utilised for cancer treatment in the 1890s [1], still has a crucial role in the multidisciplinary management of breast cancer (BC) today, in spite of many advances in both surgery and systemic therapy. Studies have shown that RT can benefit up to 83% of BC patients [2] and that whole-breast RT following breast-conserving surgery provides local control and survival rates comparable to mastectomy [3–5]. Unfortunately, not all BC patients obtain a therapeutic benefit from RT; although overall five-year BC survival rates after RT are ~80%, it has been estimated that local recurrences or metastatic disease will develop in 30% of these patients, the majority of whom will die within 5 years [6]. In BC and other solid tumours, the clinical effects of RT are also only observed near the end or after the treatment course has been completed; as such, patients who do not respond to RT (due to either innate [7] or acquired radioresistance [8]), will initially go undetected. This delay in identifying non-responding cancers exposes patients to the risk of acquiring RT-induced side effects for no therapeutic gain [9], allows tumour progression, impacts long-term survival and delays the delivery of alternate, more effective treatments [10].

The precision medicine initiative is a concept that is increasingly being implemented into BC clinical practices. It can be defined as the prevention, examination and treatment of disease, while also considering individual variability [11]. Molecular classification systems, based on gene expression signatures of BC tissue, are currently being used to classify these cancers into specific subtypes that can predict prognosis and treatment response [12–16]. While these tools have led to improvements in the systemic treatment of BC patients, the incorporation of RT into the precision medicine initiative is lagging behind such achievements [17]. To improve BC patient outcomes and allow RT to become fully integrated into the precision medicine initiative, we need to identify biomarkers that can not only predict RT response before the initiation of treatment but also allow the evaluation of a tumour's response to RT during treatment [18]. These biomarkers could enable personalised RT treatment regimens to be given to individuals on the basis of individual risk and tumour biology and also allow the identification of patients who are unlikely to benefit from RT.

In response to this unmet clinical need, studies have attempted to produce radiation sensitivity gene signatures that can predict tumour radiation response and identify those resistant to conventional RT regimens [19–22]. Unfortunately, as of yet, none of these gene signatures have been sufficiently validated for clinical use. Rather than using tissue-based biomarkers, another method that could be used to personalise RT is the detection and/or measurement of tumour secreted biomarkers. Several secretomic studies have used conditioned media (CM, spent media harvested from cultured cells) from BC cell lines cultured in vitro in an attempt to detect clinically relevant biomarkers [23–27]. While these secretomic studies have distinguished novel biomarkers of aggressive phenotypes [23,25] or biomarkers that act as predictors of chemotherapy response [26], no study has yet explored the immediate impact of radiation on the secretome of cancer cells as a means of evaluating radiation response and/or determining radiosensitivity [28].

We have previously developed and characterised radioresistant (RR) cells derived from oestrogen receptor positive (ER$^+$) BC cell lines [29]. In-depth genotypic, phenotypic and functional characterisation identified several important mechanisms (including EMT, reduced proliferation, metabolic changes and activation of PI3K, AKT and WNT signalling) that may contribute to the development of radioresistance. In this current study, we utilised these RR models, along with their parental cells, to describe a novel method for the identification of gene, intracellular protein and secreted protein biomarkers that can be used to provide prognostic and/or predictive information on a tumour's response to RT. Utilising secretomic data obtained through liquid chromatography-mass spectrometry (LC-MS) with a radiosensitive ER$^+$ BC cell line (MCF-7), we characterised the cancer secretome and identified cancer-secreted biomarkers whose release reflected an acute radiation response. In addition, we combined the secretomic results with data from whole-transcriptome gene expression experiments, using both radiosensitive and resistant cells, to identify a signature related to intrinsic radiosensitivity. Candidate secreted and intracellular biomarkers were then successfully validated in-lab using cell lines, BC xenograft tumours and patient tissue samples (Figure 1). We suggest that our methods can be utilised for the identification of biomarkers that could have a clinical role in personalising RT dosing regimens, thus optimising treatment and limiting the administration of RT to patients who are unlikely to benefit.

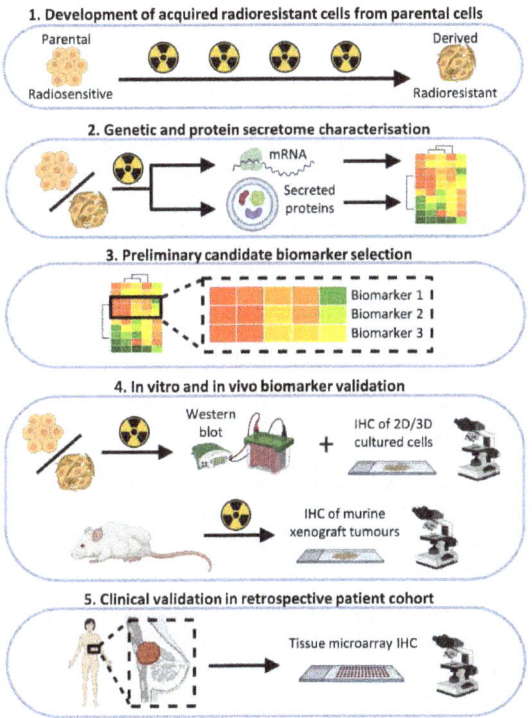

Figure 1. Biomarker discovery pipeline. Outline of the methods used to identify and validate biomarkers of BC RT response. Figure created with Biorender.com.

2. Materials and Methods

2.1. Cell Culture

Unless indicated otherwise, cell culture reagents were acquired from Gibco Thermo Fisher Scientific (Loughborough, England). MCF-7 and ZR-751 BC cell lines were cultured in Dulbecco's Modified Eagle's Medium (DMEM) supplemented with 10% foetal calf serum, 50 U mL^{-1} penicillin and 50 mg mL^{-1} streptomycin. Cells were incubated at 37 °C in a humidified atmosphere with 5% CO_2. These cell lines were obtained from the American Type Culture Collection (LGC Standards, Teddington, England). Cells were authenticated by short tandem repeat profiling carried out at Public Health England (Porton Down, Salisbury, England). Spinner flasks (Cellcontrol Spinner Flask, Integra, Zizers, Switzerland), placed onto a magnetic stirrer platform (Cellspin, Integra, Zizers, Switzerland), were used to produce multicellular tumour spheroids (MTS) from single cell suspensions. MTS were allowed to form over 7 days in normal incubation conditions before use.

2.2. Irradiation of Cells and Development of Radioresistant Cell Lines

Radioresistant (MCF-7 RR and ZR-751 RR) cells were established from their parental cell lines within our lab, as described previously [29]. Briefly, parental cell lines were treated with weekly doses of radiation using a Faxitron cabinet X-ray system 43855D (Faxitron X-ray Corporation, Lincolnshire, IL, USA). After a starting dose of 2 Gy, the radiation doses were increased by 0.5 Gy per week over a three-month period. Cells were subsequently maintained with additional weekly doses of 5 Gy after the development of radioresistance.

2.3. Cell Irradiation and Secretome Sample Preparation

Cells were seeded into six well plates to achieve ~40–50% confluency at 24 h. Cells were washed three times with PBS before 2 mL of serum-free media (SFM) was added. The cells were serum-starved for 2 h. Cells were then exposed to radiation and the CM was harvested at appropriate time points. Secretome samples underwent processing for LC-MS or western blot (WB) analysis immediately following collection. Following CM harvesting, cells were routinely trypsinised and counted using a haemocytometer with trypan blue exclusion (Sigma-Aldrich, Gillingham, England).

CM samples were centrifuged at $3000\times g$ for 15 min at 4 °C to remove dead cells and large debris. Proteins were concentrated from the supernatant using the Amicon Ultra-0.5 Centrifugal Filter Unit with Ultracel-3 membrane (Merck Millipore, Livingston, Scotland) as per the manufacturer's protocol. Briefly, 500 µL of the CM was added to the Amicon Ultra filter device and the sample was centrifuged at $14,000\times g$ for 30 min at 4 °C. The filter was removed and placed upside down into a new 1.5 mL microcentrifuge tube. The sample was centrifuged at $1000\times g$ for 2 min at 4 °C to elute the concentrated protein. The ultrafiltrate was then stored at −80 °C.

2.4. Liquid Chromatography-Mass Spectrometry and Secretome Analysis

In-solution digests of secretomic samples were performed for LC-MS analysis. Protein concentrations of the CM samples were ascertained using a bicinchoninic acid assay (Sigma-Aldrich, Gillingham, England). 50 µg of protein was added to 100 mM tris/2 M urea/10 mM DTT and heated for 30 min at 50 °C; this was performed in 96 well plates with silicon lids. 55 mM iodoacetamide was then added and incubated in darkness for 30 min at room temperature. After this, trypsin (1:100 dilution) was added and incubation was performed overnight at room temperature. Of this peptide solution, 10 µg was inserted into an activated (20 µL methanol) and equilibrated (100 µL 0.1% trifluoroacetic acid (TFA)) C18 StAGE tip; washing was performed with 100 µL of 0.1% TFA. The bound peptides were eluted into Protein LoBind tubes with 20 µL of 80% acetonitrile (ACN) and 0.1% TFA solution. The samples were concentrated to volumes <4 µL using a vacuum concentrator. Final sample volumes were adjusted to 6 µL using 0.1% TFA. Online LC was performed using a Dionex RSLC Nano. After the C18 clean-up, 5 µg of the peptide solution was injected onto a C18 packed emitter and eluted over a gradient of 2–80% ACN for 2 h with 0.1% TFA. Eluted peptides were ionised at +2 kV and data-dependent analysis was carried out on a Thermo Q-Exactive Plus. MS1 was obtained with resolution 70,000 and mz range 300–1650 and the top 12 ions were chosen for fragmentation with a normalised collision energy of 26 and an exclusion window of 30 sec. MS2 was collected with a resolution of 17,500. The AGC targets for MS1 and MS2 were 3×10^6 and 5×10^4, respectively. All spectra were obtained with 1 microscan without lockmass.

Data were analysed using MaxQuant in conjunction with uniport fasta database with matching between runs. Prior to the analysis, all data were $\log_2$ transformed. For fold change analysis, data were normalised to untreated controls at each time point using R (Bioconductor) software and packages [30]. Venn diagrams were generated using jvenn [31]. Heatmap and cluster analyses were performed using TM4 MeV (multiple experiment viewer) software [32]. Heatmap clustering was carried out using Pearson correlation with average linkage. Protein interaction networks of candidate biomarkers were generated using the STRING protein interaction database [33] and Markov clustering algorithms [34].

All secretomic datasets generated and/or analysed within this study are available on the PRoteomics IDEntifications Database (PRIDE) [35,36]; these can be found with the PRIDE project accession number PXD027572.

2.5. Lactate Dehydrogenase Assay

Lactate dehydrogenase (LDH) levels within the CM used for secretome analysis were analysed to confirm the absence of cell death after radiation treatment. LDH levels were measured using the CyQUANT LDH Cytotoxicity Assay Kit (Invitrogen, Inchinnan, Scotland) as per the manufacturer's protocol. Briefly, 50 µL of CM was transferred to a 96-well plate, along with 50 µL of the reaction mixture. The plates were incubated at room temperature for 30 min. 50 µL of stop solution was then added to the wells and absorbance was measured at 490 nm and 680 nm using a Spark 20M multimode reader (Tecan, Männedorf, Switzerland).

2.6. RNA Extraction and Whole-Transcriptome Gene Expression Analysis

Cells were seeded into 75 mm plates (3×10^6 cells/plate). Following 24 h of incubation, cells were serum-starved for 2 h (providing the same experimental conditions as for CM collection) and then exposed to radiation. Pellets containing up to 10,000,000 cells were collected by trypsinisation at 0, 2 and 8 h post-radiation, snap-frozen on dry ice and stored at −70 °C. RNA was extracted from the cells with the RNeasy Mini Kit using QIAshredder technology (Qiagen, Manchester, England). Spin technology was used to purify total RNA from the cells, as per the manufacturer's protocol. RNA was quantified and examined for contaminants using the NanoDrop™ Spectrophotometer ND1000 and the Qubit RNA IQ Assay (Thermo Fischer Scientific, Loughborough, England). RNA quality was assessed by producing RNA integrity numbers (RIN) for each of the samples using the Agilent Bioanalyzer (Agilent Technologies Ltd., Stockport, England); each sample had RIN values above 9.7 (Supplementary Table S1). The ZR-751 2 h 2 Gy sample failed in sequencing and was removed from further analysis. Lexogen QuantSeq 3′ FWD sequencing technology produced full genome expression read-counts on an Illumina flow cell; these were scanned using the Illumina HiScanSQ system (Edinburgh Clinical Research Facility, University of Edinburgh, Scotland). Next-generation sequencing reads were generated towards the poly(A) tail with read 1 directly reflecting the mRNA sequence. The FASTQ files were pre-processed with the BlueBee high-performance next generation sequencing analysis software; this uses poly(A) tail trimming and alignment to the Genome Reference Consortium Human genome build 38 reference genome using the Spliced Transcripts Alignment to a Reference (STAR) algorithm [37].

Filtering was carried out on the data, removing all genes that had fewer than five reads per sample in at least 90% of samples. Overall, 17,243 genes were mapped to human Ensembl gene identifiers. Data were $\log_2$ transformed and quantile normalised in R (Bioconductor) software and packages [30] before any analysis was carried out. Heatmap and cluster analyses were performed with the TM4 MeV (multiple experiment viewer) software [32]. Heatmap clustering was implemented using Pearson correlation with average linkage. Correction for batch effects was performed to integrate gene expression data produced in this study with public datasets; this was carried out using the ComBat package in R, as described previously [38,39]. Gene enrichment analysis was performed in DAVID Functional Annotation Bioinformatics Microarray Analysis [40] and also using the KEGG [41] and Reactome [42,43] databases. Differential gene expression analysis was performed using ranked products with a false discovery rate of 0.01. All gene transcriptomic datasets generated and/or analysed within this study are available in the NCBI's Gene Expression Omnibus [44]; these can be found with the GEO Series accession number GSE120798.

2.7. Protein Isolation and Detection

Whole-cell lysates were procured as previously described [45], with protein concentrations ascertained using a bicinchoninic acid assay. Sodium dodecyl sulphate (SDS) polyacrylamide gel electrophoresis was used to separate proteins. After separation, proteins were transferred to Immobilon-P transfer membranes (Merck Millipore, Livingston, Scotland). Membranes were incubated in LI-COR Odyssey blocking buffer solution (1:1 with PBS) for 1 h at room temperature. The membranes were then incubated overnight at 4 °C with primary antibodies DKK1 (abcam ab93017), GNPNAT1 (abcam ab234981) and TK1 (abcam ab76495). IRDye 800CW and IRDye 680LT fluorescently labelled secondary antibodies (LI-COR, Bioscience, Cambridge, England), diluted in LI-COR Odyssey blocking buffer solution, were used to bind to the primary antibodies. An LI-COR Odyssey Imager was used to detect the presence of signals from the bound secondary antibodies.

2.8. Murine Xenograft Experiments

As part of a complementary study, radiation-treated mouse xenograft tissue was available for analysis. These in vivo murine studies were undertaken under a UK Home Office Project Licence in accordance with the Animals (Scientific Procedures) Act 1986. All experiments received approval from the University of Edinburgh Animal Welfare and Ethical Review Board. The recommended guidelines for the welfare and use of animals in research were followed. CD-1 immunodeficient female nude mice (Charles River Laboratories, Tranent, Scotland) ≥ 8 weeks of age were allowed at least a seven-day period of acclimatisation to a sterile, pathogen-free environment with ad libitum access to food and water. Mice were housed in groups of five in individually ventilated cages in a barrier environment.

Approximately 5×10^8 MCF-7 and MCF-7 RR cells were grown routinely and resuspended in individual aliquots of 0.5 mL of SFM and 0.5 mL of Matrigel Matrix (Corning, Ewloe, Wales). Under gaseous isoflurane anaesthesia, each mouse received a 0.72 mg 17B-Oestradiol pellet (60-day release, Innovative Research of America, Sarasota, FL, USA) implanted subcutaneously in the dorsum using a 10 G trocar. 0.1 mL of either the MCF-7 or MCF-7 RR cell suspension was injected bilaterally into subcutaneous flank tissue using a 22 G needle connected to a 1 mL syringe. Once stock tumours had grown to ~1.0 cm in length mice were euthanised by cervical dislocation. In a sterile cabinet, xenograft tumours were harvested and placed into DMEM with no additives and sectioned into fragments ~1–2 mm in length. Implantation of tumour fragments into experimental mice was performed under gaseous isoflurane anaesthesia using a 12 G trocar. Each mouse received a 0.72 mg 17B-Oestradiol pellet as previously described and one tumour fragment was injected into the subcutaneous tissue of the flank. Mice were monitored for the development of xenograft tumours which occurred within 6–8 weeks post-implantation. Once the tumours had grown to ~1.0 cm in length, they were radiated. Mice were euthanised 24 h post-radiation and the tumours harvested ($n = 5$). Control tumours were left untreated and harvested at the same time ($n = 5$).

2.9. Human Breast Tissue Experiments

To investigate whether candidate biomarkers could predict response to RT we identified ER$^+$ positive breast cancer patients within a unique series of patient-derived BC tissues known as the Edinburgh Breast-Conserving Series (BCS) [46]. The Edinburgh BCS comprises a fully documented consecutive cohort of 1812 patients treated by breast conservation surgery, axillary node sampling or clearance and whole breast radiotherapy between 1981 and 1998. Over the study period, patients were managed by a specialist multidisciplinary team of surgeons, radiologists, pathologists and oncologists. Patients were those considered suitable for breast-conserving therapy and were T1 or T2 (<30 mm), N0 or N1 and M0 based on conventional TNM staging. Post-operative breast radiotherapy was given over 4–5 weeks at a dose of 45 Gy in 20–25 fractions. Notably, 12.7% of patients received no additional adjuvant therapy (chemotherapy or endocrine therapy) and of those

37% were ER-rich tumours ($n = 80$). It is these cases which were selected for analysis in this study. Clinicopathological data were available, including patient age, lymph node status, ER and PR status, tumour size and grade (see Supplementary Table S2). To generate tissue microarrays (TMAs) from these patients, formalin-fixed paraffin-embedded tissue blocks were initially created from patient-derived surgical excision specimens. These blocks were analysed by a pathologist to identify tumour regions. TMAs were then constructed in triplicate with representative cores (diameter ~700 µm) taken from three different random areas of the tumour. Each of the triplicates was then placed into three different TMA blocks. For use in our study, these three blocks were stained independently to assess intracellular protein levels of candidate biomarkers. The staining results of the three matched cores were then averaged. Following TMA processing, between 74 and 78 cases with intact triplicate samples were available for analysis. Recurrence-free survival data were available with a median follow-up of 12.7 years. Ethical approval for the study was granted under the Lothian NRS BioResource approval number 20/ES/0061.

2.10. Immunohistochemistry

Immunohistochemistry (IHC) was performed on formalin-fixed human TMAs, MTS and murine xenograft tumours, in addition to methanol-fixed cells cultured in Lab-Tek II chamber slides (Thermo Fisher Scientific, Loughborough, England). Formalin-fixed samples were deparaffinised and rehydrated, after which antigen retrieval was performed. 3% H_2O_2 (Dako, Ely, England) was used to block endogenous peroxidase activity. All samples were incubated with Total Protein Block (Dako, Ely, England) for 1 h at room temperature. Primary antibodies DKK1 (abcam ab93017), GNPNAT1 (abcam ab234981) and TK1 (abcam ab76495) were incubated for 1 h at room temperature. One drop of Envision labelled polymer (Dako, Ely, England) was added to each sample for 30 min, after which DAB and substrate buffer (Dako, Ely, England) was applied for 10 min. Haematoxylin was used to counterstain the tissues, after which the slides were dehydrated and mounted with coverslips using a DXP mountant (Sigma-Aldrich, Gillingham, England).

IHC scoring of the Breast-Conserving Series TMAs was performed independently by two researchers. The scoring system used depended on the staining pattern observed. If staining intensity was consistent within a sample for a candidate biomarker (DKK1 and GNPNAT1), the scores given ranged from 0 (no staining), 1+ (weak staining), 2+ (moderate staining) and 3+ (strong staining). If staining intensity varied within a sample (TK1), then each sample was given a score that was dependent on the staining intensity (0, 1+, 2+ or 3+) combined with the percentage of cells with that intensity of staining, providing a final score ranging from 0–300.

2.11. Statistical Analysis

One-way ANOVA, with Holm–Šídák multiple comparisons tests, was used to check for differences in secretion levels of candidate biomarkers within a cell line in the western blot CM experiments. Two-way ANOVA tests were performed to assess for differences in intracellular levels of candidate biomarkers between parental and RR cell lines in the western blot experiments using whole-cell lysate samples. For the Kaplan–Meier analysis of recurrence-free survival data in relation to candidate biomarker expression levels, the p-value was derived from log-rank (Mantel-cox) tests. The p-values ≤ 0.05 were deemed statistically significant. Graphs and statistical analysis were generated with GraphPad Prism 9 for Windows (GraphPad Software Ltd., San Diego, CA, USA).

3. Results

3.1. Characterisation of the MCF-7 Basal Secretome

Initial analysis was performed using CM samples procured from untreated MCF-7 cells 24 h after serum starvation to characterise the basal secretome before irradiation. The total number of proteins identified in the untreated secretome was 808. A cut-off of 2 was used to enable a functional analysis to be performed for the identification of key enriched pathways. Using this cut-off value, 318 proteins were detected within the CM of untreated MCF-7 cells; of these, 231 were shown to interact with one another. These secreted proteins were predominately involved in metabolic pathways, immune and cytokine signalling and cell cycle regulation (Figure 2A). The majority of these proteins have been reported/predicted to be secreted in exosomes/microvesicles or are released directly; only 37 had an unknown method of secretion (Figure 2B).

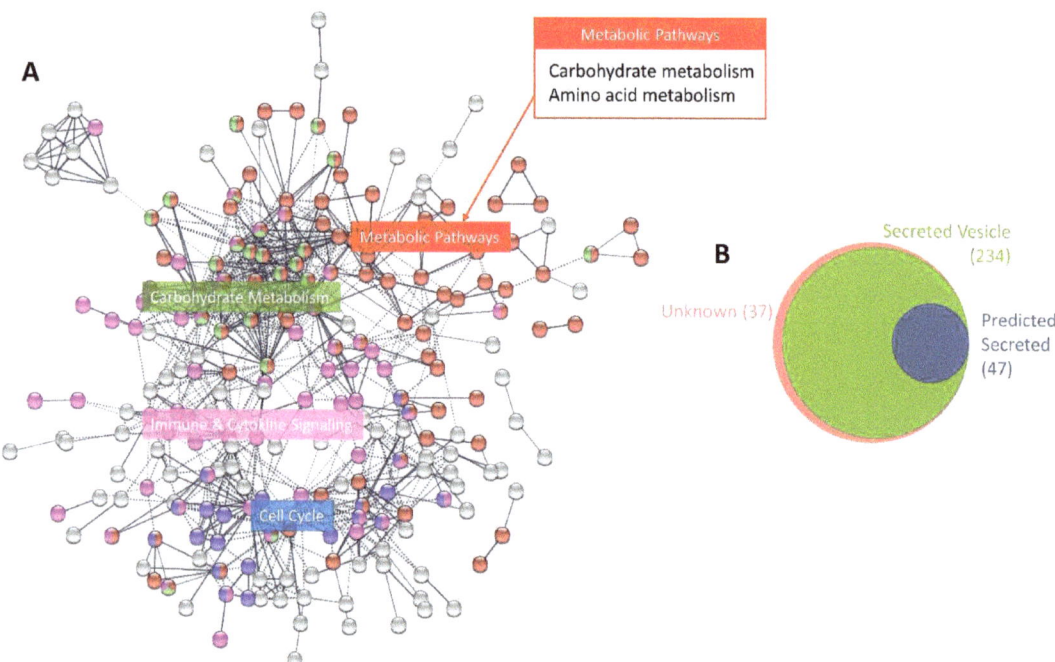

Figure 2. Characterisation of the MCF-7 basal secretome. (**A**) Functional protein association network showing the subset of 231 secreted proteins with known interactions from the total 318 proteins identified in the untreated basal secretome (after cut-offs were applied). Graph produced in STRING based on co-expression with high-confidence interaction score (0.7), clustered using the Markov Clustering algorithm. Significantly enriched pathways from the KEGG [41] and Reactome [42,43] databases are highlighted and labelled (lists of proteins in each pathway are provided in Supplementary Table S3). (**B**) Venn diagram showing proportions of proteins identified in the basal secretome and their reported/predicted method of secretion; (pink) unknown, (green) secreted in exosomes/microvesicles (ExoCarta [47] and Vesiclepedia [48]), and (blue) directly secreted (Human Protein Atlas [49], SignalP [50], Phobius [51] and SPOCTOPUS [52]).

3.2. Characterisation of the MCF-7 Radiation-Induced Secretome

Following characterisation of the MCF-7 untreated secretome, we wished to identify differentially secreted proteins in response to radiation. To achieve this, MCF-7 cells were treated with a single dose of 2 Gy and CM samples were obtained up to 24 h post-radiation. To ensure that radiation treatment was not causing significant cell death, cell counts (using trypan blue exclusion) and LDH quantification (using CM from these cells) were performed. Results demonstrated no difference in total cell numbers or LDH levels between untreated and radiation treated groups at 24 h (Supplementary Figure S1).

The total number of proteins detected in the CM 24 h after 2 Gy was 552. A total of 159 proteins were identified which exhibited at least a 50% increase in secretion levels following 2 Gy of radiation compared with 24 h untreated controls. As in the basal secretome, some of the secreted proteins were involved in immune and cytokine signalling and metabolism, whereas proteins involved in translation, spliceosome, RNA processing, protein metabolism and the proteasome were found only in the secretome of irradiated MCF-7 cells (Figure 3A). While there was some overlap between the secretomes of untreated and treated cells, the majority of the proteins isolated in the irradiated secretome were not found in the basal CM (Figure 3B). Like the basal secretome, most of the proteins identified in the radiation secretome were reported/predicted to be secreted (Figure 3C).

Analysis was performed to assess differences in the enriched pathways identified in the 24 h treated secretome across earlier time points. Secretion levels, relative to untreated controls at each time point, were assessed following 2 Gy of radiation at 1, 2, 4, 8 and 24 h (Figure 4). Results showed that the pathways enriched in the secretome at 24 h were also identified at the earlier time points, but secretion levels of the proteins were highest at 24 h. These results provided justification for focusing on the 24 h time point for biomarker discovery.

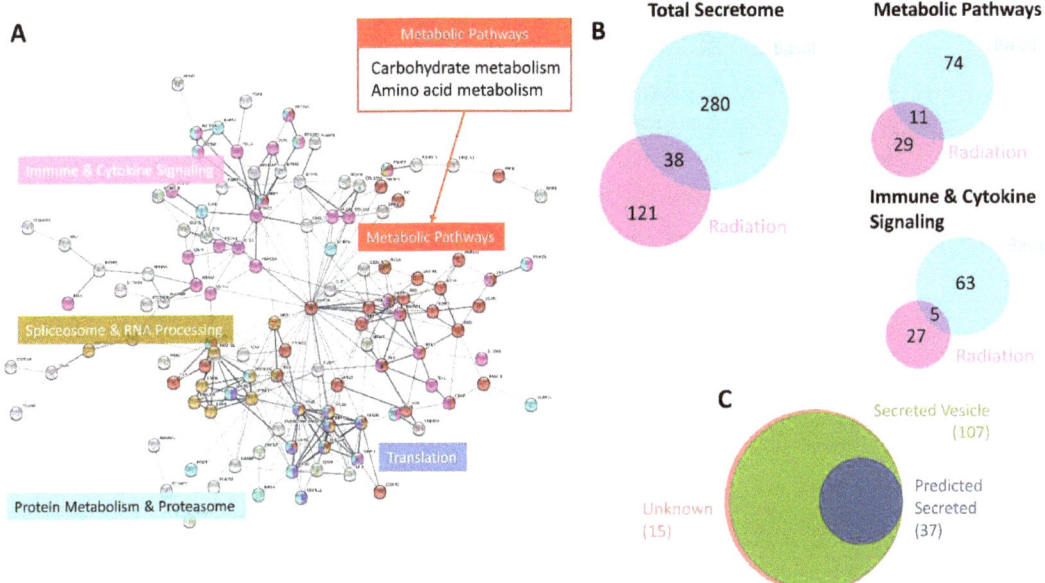

Figure 3. Characterisation of radiation-induced MCF-7 secretome. (A) Functional protein association network of the subset of 120 proteins with known interactions from the total number of 159 secreted proteins at 24 h with at least a 50% increase in secretion level following 2 Gy of radiation compared with 24 h untreated controls. Graph produced in STRING based on co-expression and reported STRING interactions with high-confidence interaction score (0.7), clustered using the Markov Clustering algorithm. Significantly enriched pathways from the KEGG [41] and Reactome [42,43] databases are highlighted and labelled (lists of proteins in each pathway are provided in Supplementary Table S4). (B) Venn diagrams showing the overlap in secreted proteins between the basal secretome and the radiation-induced secretome in respect of all secreted proteins and enriched pathways in both secretome profiles. (C) Venn diagram showing proportions of proteins identified in the radiation-induced secretome and their reported/predicted method of secretion; (pink) unknown, (green) secreted in exosomes/microvesicles (ExoCarta [47] and Vesiclepedia [48]), and (blue) directly secreted (Human Protein Atlas [49], SignalP [50], Phobius [51] and SPOCTOPUS [52]).

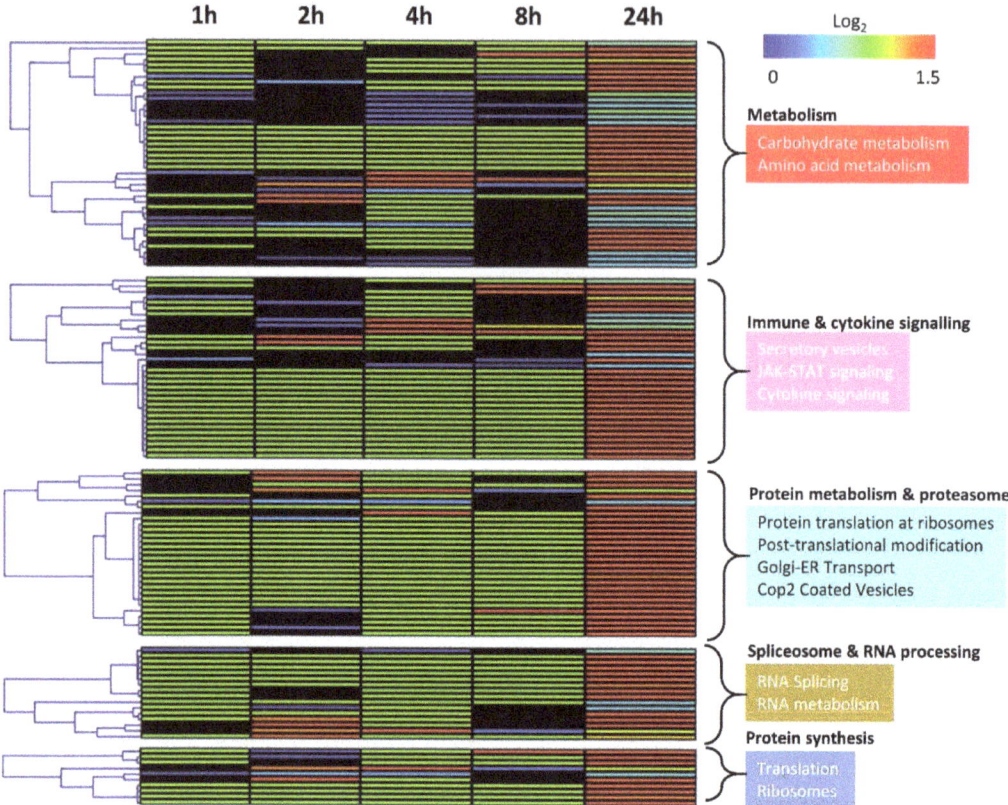

Figure 4. Comparison of secreted protein level by enriched pathways across all timepoints. Heatmap is based on log$_2$ secretion levels following 2 Gy of radiation at 1, 2, 4, 8 and 24 h compared to untreated controls at each timepoint in respect of pathways enriched in the radiation-induced secretome. Functional enrichment was performed in STRING using the KEGG [41] and Reactome [42,43] databases. Clustering of proteins is based on Pearson correlation with average linkage. Heatmap colours denote log$_2$ change in secretion level compared to untreated controls at each time point as denoted by the colour bar.

3.3. Gene Expression Changes Associated with Response to Radiation in Parental Radiosensitive and Derived Radioresistant MCF-7 Cells

Global gene expression analysis was carried out to identify differences between the parental radiosensitive MCF-7 cells and their RR derivatives at 2 and 8 h post-radiation, time points that have previously been used to assess differences in DNA damage response pathways between radiosensitive and RR cells [29,53]. Within the MCF-7 radiosensitive cells, a 2 Gy radiation dose led to the upregulation of genes involved in DNA damage repair, apoptosis and cell cycle arrest; whereas genes involved in cell cycle, gene splicing and transcription were downregulated. The radiation response of the MCF-7 RR cells was different from that of the radiosensitive cells, with an overall reduction in gene expression changes being observed (Figure 5). Similar results were observed within the ZR-751 parental and RR cell lines (Supplementary Figure S2).

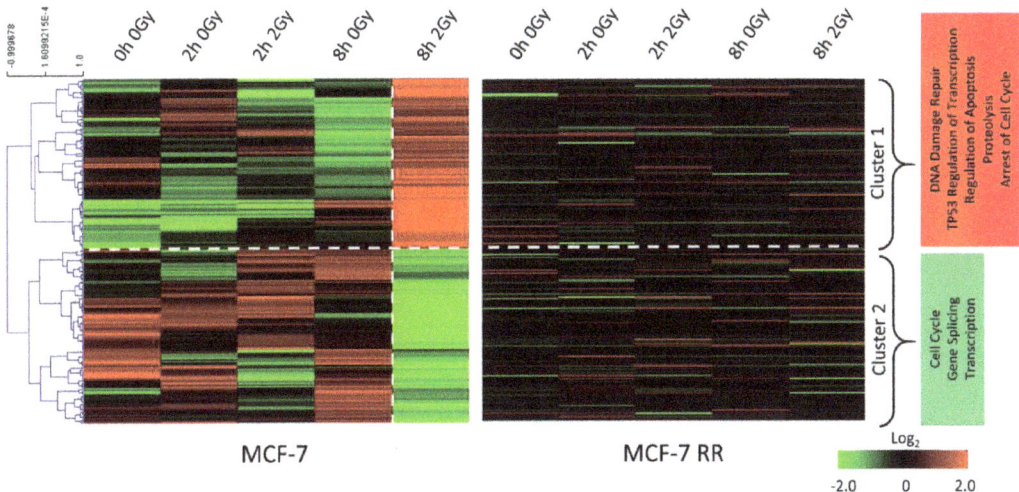

Figure 5. MCF7 and MCF-7 RR gene expression changes associated with response to radiation. Heatmaps reflect $\log_2$ mean-centred gene expression changes with clustering based on Pearson correlation with average linkage (red = higher expression, black = no change, green = lower expression). Radiosensitive MCF-7 parental cells and their RR derivatives are shown in adjacent heatmaps. For each cell line, untreated baseline controls at 0 h are shown along with both the treated (2 Gy radiation) and untreated controls at 2 h and 8 h. The genes shown are the most differentially expressed in sensitive parental MCF-7 cells, with the largest gene expression differences seen between the untreated controls and the 2 Gy treated cells at 8 h.

3.4. MCF-7 Candidate Biomarker Selection

From the 159 proteins which exhibited at least a 50% increase in secretion at 24 h following 2 Gy of radiation, cluster analysis identified 33 proteins that had significantly increased secretion levels (up to 12-fold) at all radiation doses tested (Figure 6A). While a small number of these proteins exhibited increased or decreased secretion levels compared to untreated controls at earlier time points, the secretion levels of the majority of the proteins did not change (Figure 6B). From these 33 proteins, we identified those which were known to be secreted and those which belonged to the previously identified enriched pathways; we hypothesised that it might be these biomarkers that play a role in RT response. Gene expression analysis assessing the levels of these 33 proteins in both MCF-7 and MCF-7 RR cells showed that 5 of the 33 proteins had higher levels of expression in the radiosensitive compared to the RR cells (Figure 6C); similar results were observed within the ZR-751 parental and RR cell lines (Supplementary Figure S3). We chose to focus on these 5 proteins (DKK1, EIF4EBP2, GNPNAT1, TK1 and YBX3) as our candidate biomarkers.

3.5. Candidate Biomarker Expression and Intrinsic Sensitivity to Radiation

As these five candidate biomarkers had higher inherent gene expression levels within the radiosensitive cells compared to their acquired RR derivatives, we further investigated whether these biomarkers might be linked to intrinsic radiosensitivity. SF2 values (a commonly used experimental indicator of cellular radiosensitivity) of parental and derived RR cells determined within our lab [29] were combined with SF2 values of a panel of ER$^+$ BC cell lines ascertained by others in the literature [19,54–57]. Cell lines with SF2 values <0.4 and >0.4 were classed as radiosensitive and RR, respectively (this threshold has been previously used to define radiosensitivity and radioresistance [58]). Gene expression levels of our five biomarkers were observed to be higher in the more radiosensitive cell lines than in RR models (Figure 7). These results suggest that our candidate biomarkers may be associated with intrinsic radiosensitivity.

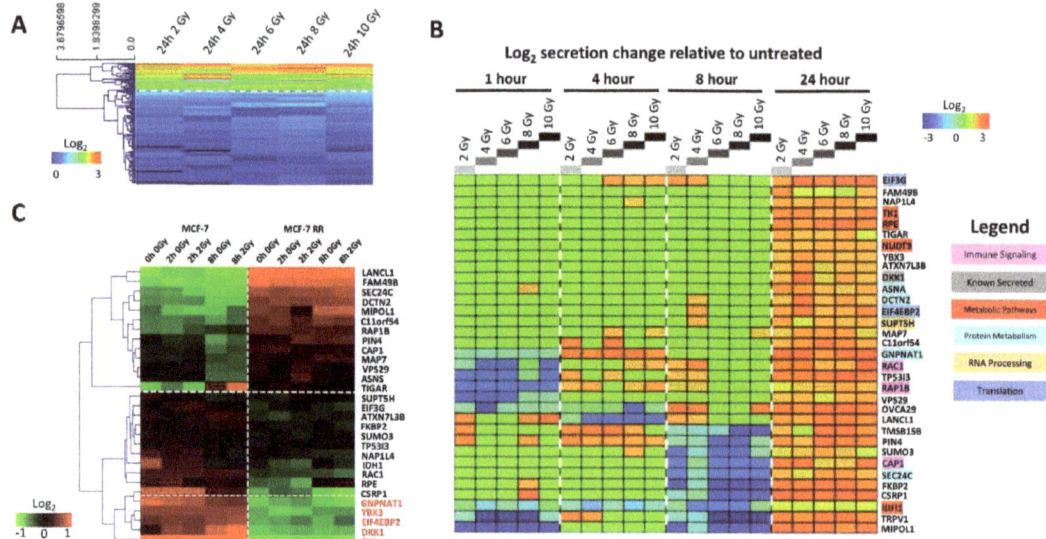

Figure 6. Candidate biomarker selection. (**A**) Protein secretion heatmap showing the $\log_2$ secretion level of all 159 proteins identified from the radiation-induced secretome across all doses of radiation (2, 4, 6, 8 and 10 Gy) at 24 h. Cluster analysis, performed using Pearson correlation with average linkage, gave rise to two clusters. The upper cluster was found to contain 33 proteins with significantly higher levels of secretion in response to radiation across all doses at 24 h. Heatmap colours indicate $\log_2$ secretion level as denoted by the colour bar. (**B**) Protein secretion heatmap showing the $\log_2$ secretion level changes of the 33 proteins from the upper cluster in Figure 3A across all timepoints and radiation doses, normalised to untreated controls at each timepoint. Heatmap colours indicate $\log_2$ secretion level changes compared to untreated controls at each timepoint as denoted by the colour bar (red = higher expression, green = no change, blue = lower expression). Proteins belonging to pathways found to be enriched in the radiation-induced secretome are highlighted according to the legend. (**C**) Heatmap of $\log_2$ mean-centred gene expression data from both untreated controls and radiation treated MCF-7 and MCF-7 RR cells, comparing the expression levels of the 33 secreted proteins at the gene level. Clustering was performed using Pearson correlation with average linkage (red = higher expression, black = no change, green = lower expression).

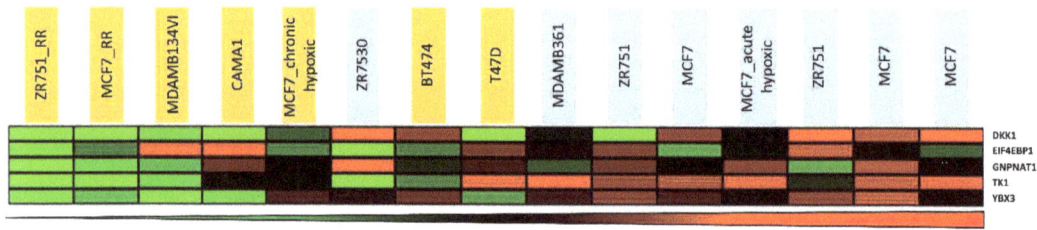

Figure 7. Candidate biomarker expression and intrinsic sensitivity to radiation. Mean-centred gene expression heatmap (red = higher expression, black = no change, green = lower expression) showing the levels of genes encoding the 5 lead candidate biomarkers, ranked left to right by highest mean expression, across a panel of ER[+] BC cell lines from a public dataset (GSE50811). SF2 values of parental and derived RR cells determined within our lab [29] were combined with SF2 values of a panel of ER[+] BC cell lines ascertained by others in the literature [19,54–57]. Cell lines with SF2 values <0.4 and >0.4 were classed as radiosensitive and RR, respectively [58]. The intrinsic radiosensitivity of individual cell lines is indicated by highlighted colour (blue = sensitive, yellow = resistant).

3.6. In Vitro and In Vivo Validation of Candidate Biomarkers

To validate the secretomic results and further investigate the potential use of these proteins as biomarkers of radiosensitivity, the secreted and intracellular protein levels of our candidate biomarkers were assessed through WB and IHC, respectively, using both parental radiosensitive and derived RR cell lines. While we initially set out to validate all five candidate biomarkers, we were unable to find suitable antibodies for two of the proteins (EIF4EBP2 and YBX3); we therefore focused on validating DKK1, GNPNAT1 and TK1.

WB analysis was performed using CM samples to assess secreted protein levels from MCF-7 parental and RR cell lines 24 h after the cells had received a single radiation dose of 2 Gy (Figure 8A). Compared to untreated controls, the secretion levels of DKK1, GNPNAT1 and TK1 were significantly increased in MCF-7 cells 24 h after irradiation. In comparison, biomarker levels in the CM samples from untreated and radiation-treated MCF-7 RR cells remained low. Increased levels of secretion of our candidate biomarkers after irradiation was also observed in radiosensitive ZR-751 cells, with no increase in secretion detected in ZR-751 RR cells (Supplementary Figure S4).

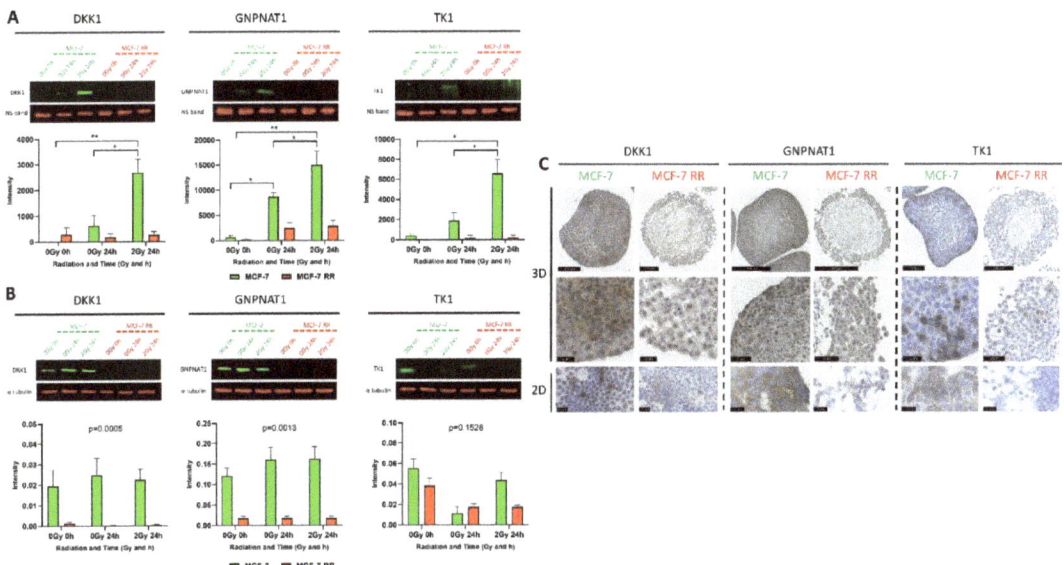

Figure 8. In vitro validation of lead candidate biomarkers. (**A**) WB analysis assessing the secretion levels of lead candidate biomarkers in MCF-7 and MCF-7 RR cell lines using CM samples obtained up to 24 h following 2 Gy of radiation. NS is a non-specific band used to confirm equal loading (One-way ANOVA with Holm–Šídák multiple comparisons test; data expressed as mean ± SEM, n = 3, * $p \leq 0.05$, ** $p \leq 0.01$). (**B**) WB analysis assessing the intracellular levels of lead candidate biomarkers in whole-cell lysates of MCF-7 and MCF-7 RR cell lines obtained up to 24 h following 2 Gy of radiation (Two-way ANOVA; data expressed as mean ± SEM, n = 3). (**C**) IHC assessing the intracellular levels of the lead candidate biomarkers in MCF-7 and MCF-7 RR cells cultured in 2D and 3D environments.

Intracellular expression levels of the candidate biomarkers were assessed in both 2D and 3D culture conditions. WB analysis of whole cell lysates of cells cultured in 2D showed that the protein expression levels of DKK1 and GNPNAT1 were significantly higher in the radiosensitive parental MCF-7 cells compared to the RR cells (Figure 8B). Both the 2D ICC and 3D IHC indicated that the parental MCF-7 cells had higher basal levels of the three candidate biomarkers compared to the RR cells (Figure 8C). Similar results were also observed with the ZR-751 radiosensitive and RR cell lines (Supplementary Figure S4).

We further assessed the link between the intracellular levels of these biomarkers and radiosensitivity using mouse xenograft tumours consisting of either MCF-7 parental or MCF-7 RR cells. IHC was performed on these mouse xenograft tumours, which were harvested 24 h post-radiation. Results showed that, while there was no increase in intracellular protein expression levels 24 h after radiation, the intracellular basal levels of the biomarkers were higher in the parental tumours compared to the RR tumours (Figure 9).

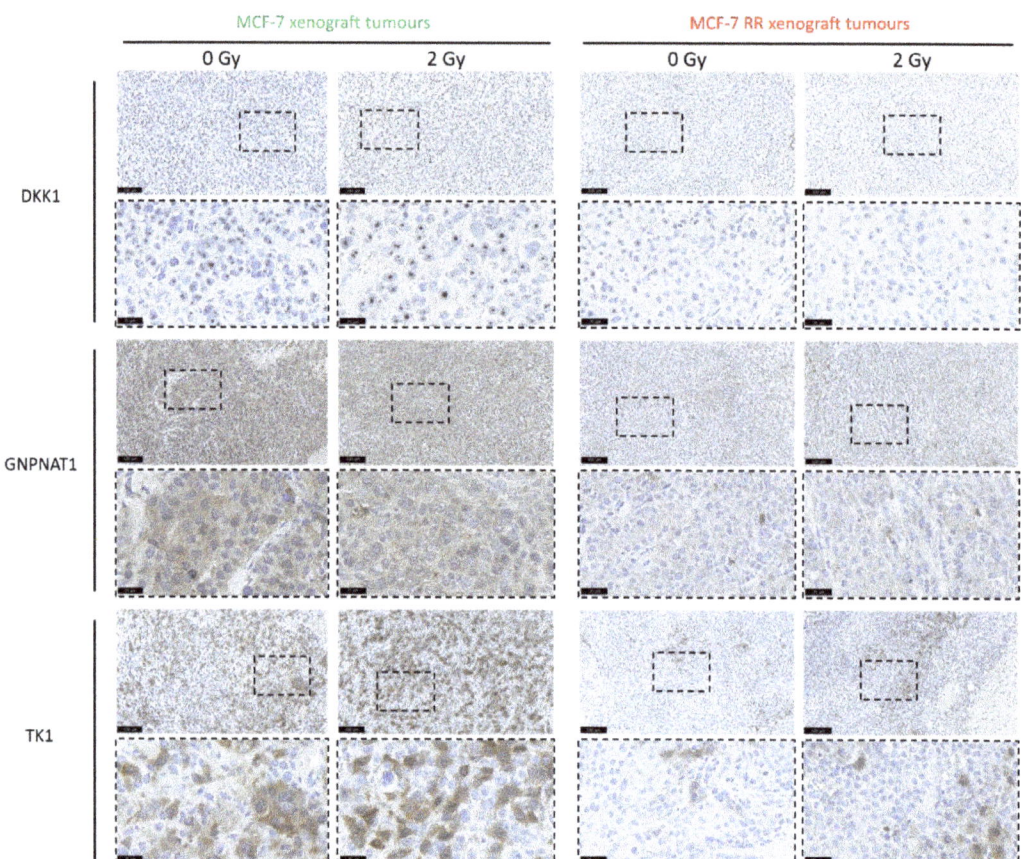

Figure 9. In vivo intracellular levels of lead candidate biomarkers. IHC assessing the intracellular levels of the lead candidate biomarkers in mouse xenograft tumours harvested 24 h after radiation. Representative images taken from five MCF-7 and five MCF-7 RR xenograft tumours.

3.7. Validation in a Retrospective Patient Cohort

Previous gene and protein expression analysis indicated that intracellular levels of the candidate biomarkers may be linked with radiosensitivity. Further investigation into whether these candidate biomarkers could predict response to RT was carried out. To do this, we performed IHC to assess the intracellular levels of the three candidate biomarkers using pre-treatment biopsy tissues from ER$^+$ BC patients identified in the Breast-Conserving Series. We hypothesised that patients exhibiting higher levels of our candidate biomarkers would have a better response to RT compared to those with lower levels. High intracellular levels of both DKK1 (Figure 10Ci) and GNPNAT1 (Figure 10Cii) were associated with significantly increased recurrence-free survival (DKK1, $p = 0.014$; GNPNAT1, $p = 0.022$), indicating that these two candidate biomarkers have the potential to predict sensitivity to

RT. No significant differences in recurrence-free survival were observed in those patients with either low or high intracellular TK1 levels (Figure 10Ciii). High magnification images of the TMA samples are presented in Supplementary Figures S5–S7.

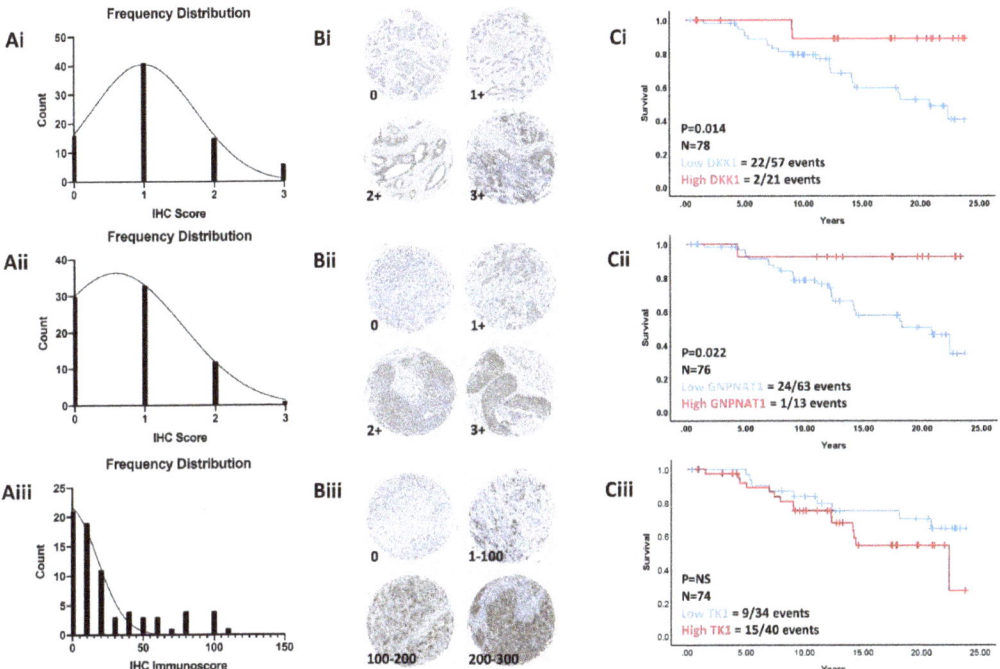

Figure 10. Validation in a retrospective patient cohort. (**A**) Frequency distribution histograms with gaussian regression curves fitted showing distribution of IHC grading histoscores of DKK1 (Ai) and GNPNAT1 (Aii), along with the distribution of IHC grading immunoscores for TK1 (Aiii), across a cohort ($n = 78$) of post-menopausal ER$^+$ BC patients treated with surgery and adjuvant RT alone. (**B**) Representative images of IHC staining for DKK1 (Bi), GNPNAT1 (Bii) and TK1 (Biii). (**C**) Kaplan–Meier analysis of recurrence-free survival in relation to DKK1 (Ci), GNPNAT1 (Cii) and TK1 (Ciii) biomarker expression in the patient cohort. Median follow-up is 12.3 years. p-value derived from log-rank (Mantel-cox) test.

4. Discussion

RT is a frequently used curative and palliative treatment for BC. However, for some patients intrinsic and acquired radioresistance can substantially limit the efficacy of RT, ultimately leading to local recurrence, disease progression and/or metastasis. While some studies have investigated tissue-based gene signatures as a way of predicting tumour radiation response [19–21], others appreciate the advantages of using blood-based biomarkers as they can be detected less invasively pre-, post- and during treatment; this can allow a patient to be continually monitored. Various clinical studies have explored the utilisation of blood-based biomarkers, such as carbohydrate antigen 15-3 and carcinoembryonic antigen for primary cancer diagnosis and metastatic disease detection [59–64], while the association between serum human epidermal growth factor receptor 2 (HER2) levels and tumour HER2 status has also been studied [65–68]. Pre-clinical studies typically focus on the cancer secretome for the identification of secreted biomarkers. Several secretomic studies have used it to identify biomarkers of aggressive phenotypes or predictors of chemotherapeutic response [23,25,26]. Previous work has also identified secreted biomarkers related to radiosensitivity. One study examined the secretome of BC cells 6 days after treatment with a single dose of 10 Gy, showing that the secretion of cyclophilin A was related to intrinsic radiosensitivity [27]. While this study demonstrated that protein secretion can increase following radiation, and that secreted proteins can relate to radiosensitivity, acute cancer secretome changes after radiation treatment were not assessed. It is these early changes that could potentially be more useful in a clinical setting. As a result of increased clinical interest in the use of blood-based biomarkers to evaluate pre- and on-treatment RT response [11], along with the potential of tissue-based biomarkers to predict tumour radiosensitivity, our study aimed to develop a novel method to identify both secreted and intracellular biomarkers of RT response.

The ER$^+$ MCF-7 cell line was chosen as the initial model for biomarker discovery, as it is a well-characterised cell line that has been used in many previous secretomic studies [25,26,69–73]. The first stage of our study involved the acquisition of CM samples from MCF-7 cells for LC-MS. For this, we used the CM of cells cultured in SFM, as serum bovine proteins can dilute the cancer secretome and hinder the identification of secreted proteins due to the close sequence homology of cattle proteins to many human proteins [74]. Even though the effect of serum starvation on cancer cells is disputed [75–78], studies have demonstrated that culturing cells in SFM does not significantly alter the composition of secreted proteins [79,80] and that cell death is minimised under appropriate culture conditions [25,69,81]. Researchers have recommended that optimal incubation times and cell numbers are needed to diminish the cytosolic protein contamination that arises from cell death. Incubating cells with SFM for up to 30 h, with less than 70% cell confluency, are considered optimal conditions for the acquisition of secretome samples; these culture conditions were followed in all of our experiments. A washing step was also carried out in our study before incubating the cells in SFM; previous studies have demonstrated that washing reduces the contamination of CM with serum proteins and also increases the quantity of secreted proteins isolated, without having any effect on cell growth or viability [82].

All CM samples underwent centrifugation to reduce contamination by dead cells and debris, with concentration performed to enrich secreted proteins. This approach has been successfully used previously [83,84] and is necessary because secreted proteins are generally present in low abundance [85]. Control secretome samples were also acquired at each time point to account for the potential effects of serum starvation. To confirm that radiation was not having an effect on cell number or causing significant cell death at 24 h post-treatment, we performed cell counts and LDH assays. LDH is an intracellular enzyme involved in metabolism, if present in the CM it indicates that plasma membrane rupture and cell death has occurred [86]. Our results showed no significant differences in viable cell numbers or LDH levels between the controls and radiation-treated samples. This suggests that radiation-induced changes in secreted protein levels would be a result of changes in secretion processes rather than altered proliferation rates or radiation-induced cell lysis. Our results are in accordance with other secretomic studies that have demonstrated the absence of any significant levels of cell death up to 24 h after treatment with 10 Gy [67–69].

Our secretome sample preparation method likely led to the co-collection of directly secreted proteins and those secreted through exosome/microvesicle pathways. Using databases such as ExoCarta and Vesipedia we identified that a proportion of our identified secreted proteins had been previously identified within exosomes/microvesicles. Interestingly, exosomal structural proteins were not present within our samples. One possible explanation for this is that exosomes and microvesicles can differ in the composition of their structural proteins including ALIX, TSG101, CD81, CD63 and CD9 [87]. It may be that the primary method of secretion for the proteins we identified using ExoCarta and Vesipedia (which do not differentiate between exosomes and microvesicles) is via microvesicles or even direct secretion rather than in exosomes. Indeed, our current work is focused on answering this important question by repeating our proteomic analysis of secreted samples after applying specific methods to isolate exosomes, microvesicles and directly secreted proteins.

Our secretomic analysis initially focused on CM samples obtained 24 h after irradiation. Cancer patients are typically treated with daily radiation fractions; therefore, the measurement of biomarkers at 24 h after the first dose of fractionated RT might be appropriate in clinical practice. In theory, biomarker levels could be analysed just before daily treatment, that is, 24 h after a patient's preceding dose. Initial analysis characterised the MCF-7 untreated basal secretome. The number of proteins isolated and the key enriched pathways in which they function (metabolism, carbohydrate metabolism, immune and cytokine signalling and cell cycle regulation) were in agreement with previous studies using various tumour types, including BC cell lines [24,88]. The majority of the proteins detected in the secretome 24 h after radiation differed from those of the basal secretome, specifically those involved in translation, spliceosome and RNA processing, protein metabolism and the proteasome. Proteins involved in some of these pathways have previously been shown to be secreted from BC cells 6 days after a 10 Gy radiation dose [27].

In order to identify the most suitable candidate biomarkers to be taken forward for validation, we wanted to identify biomarkers that exhibited a straightforward secretion profile, whereby levels were minimal at earlier time points, then demonstrated a large increase at 24 h, as this might potentially increase the probability of successful validation. Of the proteins that had been identified in the radiation-induced secretome, 33 proteins were found to have significantly increased secretion levels (up to 12-fold) at all radiation doses tested at 24 h, with low secretion at earlier time points.

Further analysis of these 33 proteins focused on their gene expression levels within the MCF-7 radiosensitive and RR cell lines. Initial comparative analysis of the two cell lines showed differences in their gene expression patterns in response to 2 Gy treatment, with radiosensitive cells exhibiting up-regulation of genes involved in DNA damage repair pathways and arrest of the cell cycle, and down-regulation of genes involved in the cell cycle. Similar gene expression changes have been found in other studies using the MCF-7 cell line [89] and patient samples [90]. These recognised radiation-induced gene expression changes did not occur in the RR cells. DNA damage repair pathways play a crucial role in the response of cells to radiation; previous studies have also shown there to be differences in the expression of DNA damage related genes between radiosensitive and RR cell lines [53]. Given the differences in response to radiation, we proposed that any of our 33 secretomic candidate biomarkers that were differentially expressed between the sensitive and resistant cell lines could hold value as biomarkers of RT response or acquired radioresistance. Gene expression analysis assessing the 33 proteins showed that DKK1, EIF4EBP2, GNPNAT1, TK1 and YBX3 had higher expression levels in the radiosensitive cells. Further evidence of a relationship between the gene expression levels of these 5 candidate biomarkers and radiosensitivity was shown in a panel of ER$^+$ cells, with the more radiosensitive cells expressing higher levels of the candidate biomarkers. Validation experiments focusing on DKK1, GNPNAT1 and TK1 showed that these biomarkers were secreted in response to radiation treatment, but only in radiosensitive cells. These results were recapitulated in a second ER$^+$ cell line (ZR-751). Results from the in vitro and in vivo experiments indicated that intracellular protein levels of these three biomarkers may also be associated with radiosensitivity. Further evidence of the biomarkers potential to predict RT response was seen through assessing intracellular protein expression levels using samples from the Breast-Conserving Series. Here, survival analysis identified that patients with higher intracellular DKK1 and GNPNAT1 expression levels were associated with significantly increased recurrence-free survival.

Prior studies have linked our three lead candidate biomarkers with cancer. DKK1 is a soluble antagonist of Wnt/β-catenin signalling [91]. Previous work has suggested that Wnt signalling and DKK1 are involved in bone metastasis [92] and that DKK1 can stimulate osteoclast activity and inhibit the production and differentiation of osteoblasts. Inhibition of the effects of Wnt on the bone can help generate a microenvironment that allows tumours to expand [93]. DKK1's role in stimulating osteolytic metastases has been established in investigations of multiple myloma-associated bone disease [94,95], with differing studies also supporting the role of DKK1 in BC bone metastasis. Serum concentrations of DKK1 have also been shown to be increased in BC patients; moreover, patients with bone metastases were shown to have significantly increased serum DKK1 levels when compared to non-metastatic BC patients [96]. Elevated serum DKK1 concentrations have also been correlated with more advanced disease stage and grade of BC, along with shorter recurrence-free and overall survival times [97]. A further study demonstrated that although DKK1 was present in 70% of BC tissues, it could be identified in all patients using serum samples [96]. Altogether, these studies show that DKK1 is a promising intracellular and secreted biomarker for assessing BC prognosis.

GNPNAT1 is an enzyme involved in the hexosamine biosynthetic pathway (HBP). The HBP produces UDP-N-acetylglucosamine (UDP-GlcNAc), which is thought to be an essential nutrient sensor [98]. UDP-GlcNAc itself is used as substrate in glycosylation reactions; these post-translational changes are highly altered in tumour cells and can regulate the function of proteins involved in various tumour-associated processes such as gene regulation, metabolism, cell signalling and epithelial-to-mesenchymal-transition [98]. GNPNAT1 expression has been linked with prognosis in prostate cancer; higher expression levels have been associated with a lower risk of biochemical recurrence [99], whereas lower levels are typically seen in advanced, castrate-resistant prostate cancer when compared to localised disease [100]. Studies have demonstrated that GNPNAT1 is upregulated in lung adenocarcinoma tissues compared to normal tissues [101,102], with Liu et al. concluding that this protein may have potential as a prognostic biomarker [101]. Our results indicate that GNPNAT1 may additionally have a role to play in BC. This is in line with other recent studies which have demonstrated that elevated GNPNAT1 gene expression levels are present in BC tissue samples [103].

TK1 is involved in cell cycle regulation through the production of thymidine monophosphate, an essential requirement for DNA replication [104,105]. TK1 has been identified in extracellular vesicles from numerous cancer types [106–109]. Some studies have suggested that it could be used as a proliferation biomarker [110] with both diagnostic and prognostic potential [104,111]. In BC, increased intracellular TK1 expression has been correlated with disease grade and stage [112], with serum levels having been investigated for monitoring treatment responses [113] and for predicting the risk of developing distant and/or regional recurrence post-surgery [114].

In BC, RT is traditionally carried out in the adjuvant setting, after breast-conserving surgery and sometimes after mastectomy to eliminate any residual cancer cells left behind after surgery. While our results are promising, there are potential limitations to their translatability to the clinic. A potential issue is that there could be differences in secreted biomarker levels when RT is given neoadjuvantly to shrink in situ cancers compared with levels seen after post-operative adjuvant RT dealing with residual tumour cells. However, RT does also have a role in the management of BC in the neoadjuvant setting, where it can be combined with chemotherapy in patients with locally advanced cancer [115–121]. Neoadjuvant RT alone has been used for the treatment of BCs that are unsuitable for primary conservative surgery [122]. There is also increasing interest in the use of neoadjuvant accelerated partial breast irradiation alone to help reduce treatment-related morbidities associated with external beam irradiation [123,124]. Recent work has additionally shown that neoadjuvant RT alone may significantly increase disease-free survival without decreasing overall survival in patients with early-stage BC; these results were most evident for ER$^+$ BC patients [125]. As our study used ER$^+$ BC cell lines, our results may be of particular utility to early-stage patients suffering from this BC subtype. Recent work has also shown that neoadjuvant RT alone, followed by radical surgery, is a feasible treatment option and is associated with good long-term locoregional control [126]. Therefore, while pre-surgical RT is not currently the standard treatment option for patients, neoadjuvant RT has the potential to challenge the current treatment paradigm. This BC treatment strategy will ultimately require biomarkers, such as ours, that can predict and monitor RT response.

Although previous studies have shown that each of our candidate biomarkers is secreted from BC cells [106], with some of them linked to BC prognosis, ours is the first study to describe a link between the intracellular/secreted levels of these biomarkers and radiosensitivity. Whilst our initial model for secreted biomarker discovery was only performed using the MCF-7 cell line, our secretomic results have been comprehensively validated using two different ER$^+$ cell lines. Although these results are promising, additional work is now needed to assess whether these biomarkers can be detected in blood samples using animal models. Following on from our successful use of BC xenograft tumours and patient tissues from the Breast-Conserving Series, we will now look to investigate the biomarker's ability to predict radiosensitivity in larger patient cohorts. Furthermore, experiments will be needed to investigate the mechanisms of biomarker secretion and elucidate what roles these biomarkers play in cellular radiosensitivity. Although our study is particularly focused on BC, it is possible that the biomarkers we have identified are not BC-specific but may be more generic measures of tumour radiosensitivity. The methods we have used to identify biomarkers of radiation response are equally applicable to other solid tumours; future studies could therefore utilise our validated methods for biomarker discovery in other cancer types.

5. Conclusions

For clinicians to be able to deliver biologically adapted, personalised RT for BC patients they must be able to stratify patients based on individual tumour radiosensitivity before commencing treatment. Clinicians should also be able to monitor RT responses during treatment. To begin to address these clinical needs we developed an integrated secretomic and transcriptomic approach using both radiosensitive and RR cell lines to identify biomarkers of radiation sensitivity and response. To our knowledge, we are the first to report the use of secretomic experiments to identify radiation-induced BC secreted biomarkers that are released within 24 h of treatment. Furthermore, we showed that differential biomarker secretion, gene expression and intracellular protein levels can indicate cellular radiosensitivity. Initial validation using clinical samples also suggested that two of our selected candidate biomarkers have the potential to predict RT outcomes in ER$^+$ BC patients. For any of these intracellular/secreted candidate biomarkers to be used in the clinic, further research will have to prove their validity and demonstrate their ability to improve outcomes or refine patient selection for RT. The incorporation of individual biomarkers and/or signatures with advanced radiation delivery techniques, already available in the clinic, would enable the development of a precision medicine platform that could significantly improve the efficacy of RT in the treatment of BC patients.

Supplementary Materials: The following are available online at https://www.mdpi.com/article/10.3390/jpm11080796/s1. Supplementary Figure S1. Cell numbers and LDH cytotoxicity assays at 24 h post-radiation treatment. (**A**) Cell counts using trypan blue exclusion were performed with MCF-7 and ZR-751 parental and RR cell lines to confirm that no changes in proliferation or cell death were occurring after treatment with a single dose of up to 10 Gy radiation (one-way ANOVA with the Holm–Šídák multiple comparisons test, comparing only values within each cell line; data expressed as mean ± SEM, $n = 3$). (**B**) LDH cytotoxicity assays were performed with MCF-7 and ZR-751 parental and RR cell lines to confirm that no cell death was occurring after treatment with 2 Gy of radiation (unpaired t-test performed on the control and treated cells for each cell line; data expressed as mean ± SEM, $n = 3$). Supplementary Figure S2. ZR-Z51 and ZR-751 RR gene expression changes associated with response to radiation. Heatmaps reflect log$_2$ mean-centred gene expression changes with clustering based on Pearson correlation with average linkage (red = higher expression, black = no change, green = lower expression). Radiosensitive ZR-751 parental cells and their RR derivatives are shown in adjacent heatmaps. For each cell line, untreated baseline controls at 0 h are shown along with both treated (2 Gy radiation) and untreated controls at 2 h and 8 h. The 2 h 2 Gy ZR-751 sample failed in sequencing and was removed from further analysis. The genes shown are the most differentially expressed in sensitive parental MCF-7 cells, with the largest gene expression differences seen between the untreated controls and the 2 Gy treated cells at 8 h. Supplementary Figure S3. Gene expression levels of the 33 candidate biomarkers within ZR-751 parental and RR cell lines. Heatmap of $\log_2$ mean-centred gene expression data from both untreated controls and radiation-treated parental ZR-751 and ZR-751 RR cells. Clustering was performed using Pearson correlation with average linkage (red = higher expression, black = no change, green = lower expression). Supplementary Figure S4. In-lab validation of lead candidate biomarkers in the ZR-751 cell line. (**A**) WB analysis assessing the secretion levels of lead candidate biomarkers in ZR-751 and ZR-751 RR cell lines using CM samples obtained up to 24 h following 2 Gy of radiation. NS is a non-specific band used to confirm equal loading (One-way ANOVA with Holm–Šídák multiple comparisons test; data expressed as mean ± SEM, $n = 3$, * $p \leq 0.05$, *** $p \leq 0.001$, **** $p \leq 0.0001$). (**B**) WB analysis assessing the intracellular levels of lead candidate biomarkers in whole-cell lysates of ZR-751 and ZR-751 RR cell lines obtained up to 24 h following 2 Gy of radiation (Two-way ANOVA; data expressed as mean ± SEM, $n = 3$). (**C**) IHC assessing the intracellular levels of the lead candidate biomarkers in ZR-751 and ZR-751 RR cells cultured in 2D and 3D environments. Supplementary Figure S5: High magnification TMA images stained for DKK1. Images are taken from those tissues presented in Figure 10. Supplementary Figure S6: High magnification TMA images stained for GNPNAT1. Images are taken from those tissues presented in Figure 10. Supplementary Figure S7: High magnification TMA images stained for TK1. Images are taken from those tissues presented in Figure 10. Supplementary Table S1. RNA quality of the samples used for gene expression analysis. RNA integrity numbers (RIN) for the gene expression analysis samples. Supplementary Table S2. Clinicopathological data from 80 patients within the Breast-Conserving Series were used to investigate whether the candidate biomarkers could predict response to RT. Supplementary Table S3. List of proteins identified in each pathway from the untreated MCF-7 cell secretome. In total, 318 proteins were identified in the untreated basal MCF-7 secretome. Proteins involved in the significantly enriched pathways identified from the KEGG and Reactome databases are shown. Supplementary Table S4. List of proteins identified that exhibited at least a 50% increase in secretion level following 2 Gy of radiation compared with 24 h untreated controls. A proportion of the secreted proteins were involved in immune signalling, metabolism, translation, RNA processing and the proteasome.

Author Contributions: J.M., M.G., I.H.K. and A.K.T. designed and conceptualised the study. A.K.T. and J.M.D. secured funding for this research. J.M., M.G., C.M.-P., C.K. performed, analysed and interpreted the laboratory work. A.K.T. performed the bioinformatic and statistical analysis. J.M. and M.G. wrote the manuscript. J.M., M.G. and A.K.T. created the figures. Mass spectrometry was performed by J.C.W. C.M.-P. and J.C.W. uploaded data to GEO and PRIDE respectively. Critical revisions were made by all authors. All authors have read and agreed to the published version of the manuscript.

Funding: This work was supported by funding from the Breast Cancer Institute Fund (Edinburgh and Lothians Health Foundation), from the Chief Scientist Office (Scottish Government Health Directorates, CGA/19/35) and from the Welcome Trust (Multiuser Equipment Grant, 208402/Z/17/Z).

Institutional Review Board Statement: Animal studies were undertaken under a UK Home Office Project License (70/9016, approved July 2016) in accordance with the Animals (Scientific Procedures) Act 1986 and with approval from the University of Edinburgh Animal Welfare and Ethical Review Board. The recommended guidelines for the welfare and use of animals in research were followed. Ethical approval for use of tissue microarray samples from the Breast-Conserving Series was granted under the Lothian NRS BioResource approval number 20/ES/0061.

Informed Consent Statement: Not applicable.

Data Availability Statement: The gene expression datasets generated and/or analysed during the current study are available in the NCBI's Gene Expression Omnibus [44]; accessible with GEO Series accession number GSE120798. All secretomic data are available on the PRoteomics IDEntifications Database [35,36]; accessible with the project accession number PXD027572.

Acknowledgments: The authors gratefully acknowledge the teams at the Genetics Core Lab (Edinburgh Clinical Research Facility, University of Edinburgh) and Mass Spectrometry Facility (Institute of Genetics and Cancer, University of Edinburgh) for their expertise and assistance in the generation of transcriptomic and LC-MS data, respectively. C. Cunningham, M Sobel and T Piper at the Institute of Genetics and Cancer, University of Edinburgh cut the archived Breast-Conserving Series tissue microarray blocks. We would also like to thank the Breast-Conserving Series group of researchers including Ian H Kunkler, Gillian R Kerr, Jeremy S Thomas, Wilma JL Jack, John MS Bartlett, Hans C Pedersen, David A Cameron and Udi Chetty for the donation of the tissue microarray samples.

Conflicts of Interest: The authors declare that the research was conducted in the absence of any commercial or financial relationships that could be viewed as a potential conflict of interest.

References

1. Connell, P.P.; Hellman, S. Advances in radiotherapy and implications for the next century: A historical perspective. *Cancer Res.* **2009**, *69*, 383–392. [CrossRef]
2. Delaney, G.; Jacob, S.; Featherstone, C.; Barton, M. The role of radiotherapy in cancer treatment: Estimating optimal utilization from a review of evidence-based clinical guidelines. *Cancer Interdiscip. Int. J. Am. Cancer Soc.* **2005**, *104*, 1129–1137. [CrossRef]
3. Onitilo, A.A.; Engel, J.M.; Stankowski, R.V.; Doi, S.A. Survival comparisons for breast conserving surgery and mastectomy revisited: Community experience and the role of radiation therapy. *Clin. Med. Res.* **2015**, *13*, 65–73. [CrossRef]
4. Cao, J.; Olson, R.; Tyldesley, S. Comparison of recurrence and survival rates after breast-conserving therapy and mastectomy in young women with breast cancer. *Curr. Oncol.* **2013**, *20*, 593–601. [CrossRef] [PubMed]
5. Poortmans, P. Evidence based radiation oncology: Breast cancer. *Radiother. Oncol.* **2007**, *84*, 84–101. [CrossRef] [PubMed]
6. Allemani, C.; Sant, M.; Weir, H.K.; Richardson, L.C.; Baili, P.; Storm, H.; Siesling, S.; Torrella-Ramos, A.; Voogd, A.C.; Aareleid, T. Breast cancer survival in the US and Europe: A CONCORD high-resolution study. *Int. J. Cancer* **2013**, *132*, 1170–1181. [CrossRef] [PubMed]
7. Kim, J.-K.; Jeon, H.-Y.; Kim, H. The molecular mechanisms underlying the therapeutic resistance of cancer stem cells. *Arch. Pharmacal Res.* **2015**, *38*, 389–401. [CrossRef] [PubMed]
8. Chang, L.; Graham, P.H.; Ni, J.; Hao, J.; Bucci, J.; Cozzi, P.J.; Li, Y. Targeting PI3K/Akt/mTOR signaling pathway in the treatment of prostate cancer radioresistance. *Crit. Rev. Oncol. Hematol.* **2015**, *96*, 507–517. [CrossRef]
9. Shapiro, C.L.; Recht, A. Side effects of adjuvant treatment of breast cancer. *N. Engl. J. Med.* **2001**, *344*, 1997–2008. [CrossRef] [PubMed]
10. Nix, P.; Cawkwell, L.; Patmore, H.; Greenman, J.; Stafford, N. Bcl-2 expression predicts radiotherapy failure in laryngeal cancer. *Br. J. Cancer* **2005**, *92*, 2185–2189. [CrossRef]
11. Meehan, J.; Gray, M.; Martínez-Pérez, C.; Kay, C.; Pang, L.Y.; Fraser, J.A.; Poole, A.V.; Kunkler, I.H.; Langdon, S.P.; Argyle, D.; et al. Precision Medicine and the Role of Biomarkers of Radiotherapy Response in Breast Cancer. *Front. Oncol.* **2020**, *10*, 628. [CrossRef] [PubMed]
12. Paik, S.; Shak, S.; Tang, G.; Kim, C.; Baker, J.; Cronin, M.; Baehner, F.L.; Walker, M.G.; Watson, D.; Park, T. A multigene assay to predict recurrence of tamoxifen-treated, node-negative breast cancer. *N. Engl. J. Med.* **2004**, *351*, 2817–2826. [CrossRef] [PubMed]
13. Tutt, A.; Wang, A.; Rowland, C.; Gillett, C.; Lau, K.; Chew, K.; Dai, H.; Kwok, S.; Ryder, K.; Shu, H. Risk estimation of distant metastasis in node-negative, estrogen receptor-positive breast cancer patients using an RT-PCR based prognostic expression signature. *BMC Cancer* **2008**, *8*, 339. [CrossRef] [PubMed]
14. Van De Vijver, M.J.; He, Y.D.; Van't Veer, L.J.; Dai, H.; Hart, A.A.; Voskuil, D.W.; Schreiber, G.J.; Peterse, J.L.; Roberts, C.; Marton, M.J. A gene-expression signature as a predictor of survival in breast cancer. *N. Engl. J. Med.* **2002**, *347*, 1999–2009. [CrossRef] [PubMed]
15. Parker, J.S.; Mullins, M.; Cheang, M.C.; Leung, S.; Voduc, D.; Vickery, T.; Davies, S.; Fauron, C.; He, X.; Hu, Z. Supervised risk predictor of breast cancer based on intrinsic subtypes. *J. Clin. Oncol.* **2009**, *27*, 1160. [CrossRef]
16. Ellis, M.J.; Suman, V.J.; Hoog, J.; Lin, L.; Snider, J.; Prat, A.; Parker, J.S.; Luo, J.; DeSchryver, K.; Allred, D.C. Randomized phase II neoadjuvant comparison between letrozole, anastrozole, and exemestane for postmenopausal women with estrogen receptor–rich stage 2 to 3 breast cancer: Clinical and biomarker outcomes and predictive value of the baseline PAM50-based intrinsic subtype—ACOSOG Z1031. *J. Clin. Oncol.* **2011**, *29*, 2342.
17. Hall, W.A.; Bergom, C.; Thompson, R.F.; Baschnagel, A.M.; Vijayakumar, S.; Willers, H.; Li, X.A.; Schultz, C.J.; Wilson, G.D.; West, C.M. Precision oncology and genomically guided radiation therapy: A report from the American Society for radiation oncology/American association of physicists in medicine/national cancer institute precision medicine conference. *Int. J. Radiat. Oncol. Biol. Phys.* **2018**, *101*, 274–284. [CrossRef] [PubMed]
18. Bernier, J. Precision medicine for early breast cancer radiotherapy: Opening up new horizons? *Crit. Rev. Oncol. Hematol.* **2017**, *113*, 79–82. [CrossRef]
19. Speers, C.; Zhao, S.; Liu, M.; Bartelink, H.; Pierce, L.J.; Feng, F.Y. Development and Validation of a Novel Radiosensitivity Signature in Human Breast Cancer. *Clin. Cancer Res.* **2015**, *21*, 3667–3677. [CrossRef]
20. Forrest, A.P.; Stewart, H.J.; Everington, D.; Prescott, R.J.; McArdle, C.S.; Harnett, A.N.; Smith, D.C.; George, W.D. Scottish Cancer Trials Breast Group. Randomised controlled trial of conservation therapy for breast cancer: 6-year analysis of the Scottish trial. *Lancet* **1996**, *348*, 708–713. [CrossRef]
21. Sjöström, M.; Chang, S.L.; Fishbane, N.; Davicioni, E.; Zhao, S.G.; Hartman, L.; Holmberg, E.; Feng, F.Y.; Speers, C.W.; Pierce, L.J.; et al. Clinicogenomic Radiotherapy Classifier Predicting the Need for Intensified Locoregional Treatment After Breast-Conserving Surgery for Early-Stage Breast Cancer. *J. Clin. Oncol.* **2019**, *37*, 3340–3349. [CrossRef] [PubMed]
22. Eschrich, S.A.; Fulp, W.J.; Pawitan, Y.; Foekens, J.A.; Smid, M.; Martens, J.W.M.; Echevarria, M.; Kamath, V.; Lee, J.-H.; Harris, E.E.; et al. Validation of a Radiosensitivity Molecular Signature in Breast Cancer. *Clin. Cancer Res.* **2012**, *18*, 5134–5143. [CrossRef]

23. Mbeunkui, F.; Metge, B.J.; Shevde, L.A.; Pannell, L.K. Identification of Differentially Secreted Biomarkers Using LC-MS/MS in Isogenic Cell Lines Representing a Progression of Breast Cancer. *J. Proteome Res.* **2007**, *6*, 2993–3002. [CrossRef] [PubMed]
24. Liang, X.; Huuskonen, J.; Hajivandi, M.; Manzanedo, R.; Predki, P.; Amshey, J.R.; Pope, R.M. Identification and quantification of proteins differentially secreted by a pair of normal and malignant breast-cancer cell lines. *Proteomics* **2009**, *9*, 182–193. [CrossRef]
25. Lai, T.-C.; Chou, H.-C.; Chen, Y.-W.; Lee, T.-R.; Chan, H.-T.; Shen, H.-H.; Lee, W.-T.; Lin, S.-T.; Lu, Y.-C.; Wu, C.-L.; et al. Secretomic and Proteomic Analysis of Potential Breast Cancer Markers by Two-Dimensional Differential Gel Electrophoresis. *J. Proteome Res.* **2010**, *9*, 1302–1322. [CrossRef] [PubMed]
26. Yao, L.; Zhang, Y.; Chen, K.; Hu, X.; Xu, L.X. Discovery of IL-18 As a Novel Secreted Protein Contributing to Doxorubicin Resistance by Comparative Secretome Analysis of MCF-7 and MCF-7/Dox (Secretome Comparison of MCF-7 and MCF-7/Dox). *PLoS ONE* **2011**, *6*, 1–13. [CrossRef] [PubMed]
27. Chevalier, F.; Depagne, J.; Hem, S.; Chevillard, S.; Bensimon, J.; Bertrand, P.; Lebeau, J. Accumulation of cyclophilin A isoforms in conditioned medium of irradiated breast cancer cells. *Proteomics* **2012**, *12*, 1756–1766. [CrossRef]
28. Forker, L.-J.; Choudhury, A.; Kiltie, A. Biomarkers of tumour radiosensitivity and predicting benefit from radiotherapy. *Clin. Oncol.* **2015**, *27*, 561–569. [CrossRef] [PubMed]
29. Gray, M.; Turnbull, A.K.; Ward, C.; Meehan, J.; Martinez-Perez, C.; Bonello, M.; Pang, L.Y.; Langdon, S.P.; Kunkler, I.H.; Murray, A.; et al. Development and characterisation of acquired radioresistant breast cancer cell lines. *Radiat. Oncol.* **2019**, *14*, 64–83. [CrossRef]
30. Gentleman, R.C.; Carey, V.J.; Bates, D.M.; Bolstad, B.; Dettling, M.; Dudoit, S.; Ellis, B.; Gautier, L.; Ge, Y.; Gentry, J. Bioconductor: Open software development for computational biology and bioinformatics. *Genome Biol.* **2004**, *5*, 80–96. [CrossRef] [PubMed]
31. Bardou, P.; Mariette, J.; Escudié, F.; Djemiel, C.; Klopp, C. jvenn: An interactive Venn diagram viewer. *BMC Bioinform.* **2014**, *15*, 286–293. [CrossRef] [PubMed]
32. Howe, E.; Holton, K.; Nair, S.; Schlauch, D.; Sinha, R.; Quackenbush, J. MeV: MultiExperiment Viewer. In *Biomedical Informatics for Cancer Research*; Ochs, M.F., Casagrande, J.T., Davuluri, R.V., Eds.; Springer: Boston, MA, USA, 2010; pp. 267–277.
33. Szklarczyk, D.; Morris, J.H.; Cook, H.; Kuhn, M.; Wyder, S.; Simonovic, M.; Santos, A.; Doncheva, N.T.; Roth, A.; Bork, P.; et al. The STRING database in 2017: Quality-controlled protein-protein association networks, made broadly accessible. *Nucleic Acids Res.* **2017**, *45*, 362–368. [CrossRef] [PubMed]
34. Vlasblom, J.; Wodak, S.J. Markov clustering versus affinity propagation for the partitioning of protein interaction graphs. *BMC Bioinform.* **2009**, *10*, 99–113. [CrossRef]
35. Martens, L.; Hermjakob, H.; Jones, P.; Adamski, M.; Taylor, C.; States, D.; Gevaert, K.; Vandekerckhove, J.; Apweiler, R. PRIDE: The proteomics identifications database. *Proteomics* **2005**, *5*, 3537–3545. [CrossRef] [PubMed]
36. Perez-Riverol, Y.; Csordas, A.; Bai, J.; Bernal-Llinares, M.; Hewapathirana, S.; Kundu, D.J.; Inuganti, A.; Griss, J.; Mayer, G.; Eisenacher, M. The PRIDE database and related tools and resources in 2019: Improving support for quantification data. *Nucleic Acids Res.* **2019**, *47*, D442–D450. [CrossRef] [PubMed]
37. Dobin, A.; Davis, C.A.; Schlesinger, F.; Drenkow, J.; Zaleski, C.; Jha, S.; Batut, P.; Chaisson, M.; Gingeras, T.R. STAR: Ultrafast universal RNA-seq aligner. *Bioinformatics* **2013**, *29*, 15–21. [CrossRef]
38. Leek, J.T.; Johnson, W.E.; Parker, H.S.; Jaffe, A.E.; Storey, J.D. The sva package for removing batch effects and other unwanted variation in high-throughput experiments. *Bioinformatics* **2012**, *28*, 882–883. [CrossRef] [PubMed]
39. Turnbull, A.K.; Kitchen, R.R.; Larionov, A.A.; Renshaw, L.; Dixon, J.M.; Sims, A.H. Direct integration of intensity-level data from Affymetrix and Illumina microarrays improves statistical power for robust reanalysis. *BMC Med. Genom.* **2012**, *5*, 5–35. [CrossRef]
40. Sherman, B.T.; Tan, Q.; Collins, J.R.; Alvord, W.G.; Roayaei, J.; Stephens, R.; Baseler, M.W.; Lane, H.C.; Lempicki, R.A. The DAVID Gene Functional Classification Tool: A novel biological module-centric algorithm to functionally analyze large gene lists. *Genome Biol.* **2007**, *8*, 183.
41. Ogata, H.; Goto, S.; Sato, K.; Fujibuchi, W.; Bono, H.; Kanehisa, M. KEGG: Kyoto Encyclopedia of Genes and Genomes. *Nucleic Acids Res.* **1999**, *27*, 29–34. [CrossRef] [PubMed]
42. Croft, D.; O'kelly, G.; Wu, G.; Haw, R.; Gillespie, M.; Matthews, L.; Caudy, M.; Garapati, P.; Gopinath, G.; Jassal, B. Reactome: A database of reactions, pathways and biological processes. *Nucleic Acids Res.* **2010**, *39* (Suppl. 1), D691–D697. [CrossRef] [PubMed]
43. Fabregat, A.; Jupe, S.; Matthews, L.; Sidiropoulos, K.; Gillespie, M.; Garapati, P.; Haw, R.; Jassal, B.; Korninger, F.; May, B. The reactome pathway knowledgebase. *Nucleic Acids Res.* **2018**, *46*, D649–D655. [CrossRef] [PubMed]
44. Edgar, R.; Domrachev, M.; Lash, A.E. Gene Expression Omnibus: NCBI gene expression and hybridization array data repository. *Nucleic Acids Res.* **2002**, *30*, 207–210. [CrossRef]
45. Meehan, J.; Ward, C.; Turnbull, A.; Bukowski-Wills, J.; Finch, A.J.; Jarman, E.J.; Xintaropoulou, C.; Martinez-Perez, C.; Gray, M.; Pearson, M. Inhibition of pH regulation as a therapeutic strategy in hypoxic human breast cancer cells. *Oncotarget* **2017**, *8*, 42857–42875. [CrossRef]
46. Kunkler, I.H.; Kerr, G.R.; Thomas, J.S.; Jack, W.J.L.; Bartlett, J.M.S.; Pedersen, H.C.; Cameron, D.A.; Dixon, J.M.; Chetty, U. Impact of Screening and Risk Factors for Local Recurrence and Survival After Conservative Surgery and Radiotherapy for Early Breast Cancer: Results From a Large Series With Long-Term Follow-Up. *Int. J. Radiat. Oncol. Biol. Phys.* **2012**, *83*, 829–838. [CrossRef] [PubMed]
47. Mathivanan, S.; Fahner, C.J.; Reid, G.E.; Simpson, R.J. ExoCarta 2012: Database of exosomal proteins, RNA and lipids. *Nucleic Acids Res.* **2012**, *40*, D1241–D1244. [CrossRef] [PubMed]

48. Pathan, M.; Fonseka, P.; Chitti, S.V.; Kang, T.; Sanwlani, R.; Van Deun, J.; Hendrix, A.; Mathivanan, S. Vesiclepedia 2019: A compendium of RNA, proteins, lipids and metabolites in extracellular vesicles. *Nucleic Acids Res.* **2019**, *47*, D516–D519. [CrossRef] [PubMed]
49. Pontén, F.; Schwenk, J.M.; Asplund, A.; Edqvist, P.H.D. The Human Protein Atlas as a proteomic resource for biomarker discovery. *J. Intern. Med.* **2011**, *270*, 428–446. [CrossRef]
50. Armenteros, J.J.A.; Tsirigos, K.D.; Sønderby, C.K.; Petersen, T.N.; Winther, O.; Brunak, S.; von Heijne, G.; Nielsen, H. SignalP 5.0 improves signal peptide predictions using deep neural networks. *Nat. Biotechnol.* **2019**, *37*, 420–423. [CrossRef]
51. Käll, L.; Krogh, A.; Sonnhammer, E.L. Advantages of combined transmembrane topology and signal peptide prediction—The Phobius web server. *Nucleic Acids Res.* **2007**, *35* (Suppl. 2), W429–W432. [CrossRef] [PubMed]
52. Viklund, H.; Bernsel, A.; Skwark, M.; Elofsson, A. SPOCTOPUS: A combined predictor of signal peptides and membrane protein topology. *Bioinformatics* **2008**, *24*, 2928–2929. [CrossRef]
53. Kim, K.H.; Yoo, H.Y.; Joo, K.M.; Jung, Y.; Jin, J.; Kim, Y.; Yoon, S.J.; Choi, S.H.; Seol, H.J.; Park, W.-Y.; et al. Time-course analysis of DNA damage response-related genes after in vitro radiation in H460 and H1229 lung cancer cell lines. *Exp. Mol. Med.* **2011**, *43*, 419–426. [CrossRef] [PubMed]
54. Torres-Roca, J.F.; Eschrich, S.; Zhao, H.; Bloom, G.; Sung, J.; McCarthy, S.; Cantor, A.B.; Scuto, A.; Li, C.; Zhang, S.; et al. Prediction of Radiation Sensitivity Using a Gene Expression Classifier. *Cancer Res.* **2005**, *65*, 7169–7176. [CrossRef]
55. Choi, C.; Park, S.; Cho, W.K.; Choi, D.H. Cyclin D1 is Associated with Radiosensitivity of Triple-Negative Breast Cancer Cells to Proton Beam Irradiation. *Int. J. Mol. Sci.* **2019**, *20*, 4943. [CrossRef]
56. Zhang, C.; Girard, L.; Das, A.; Chen, S.; Zheng, G.; Song, K. Nonlinear Quantitative Radiation Sensitivity Prediction Model Based on NCI-60 Cancer Cell Lines. *Sci. World J.* **2014**, *2014*, 903602. [CrossRef]
57. Amundson, S.A.; Do, K.T.; Vinikoor, L.C.; Lee, R.A.; Koch-Paiz, C.A.; Ahn, J.; Reimers, M.; Chen, Y.; Scudiero, D.A.; Weinstein, J.N. Integrating global gene expression and radiation survival parameters across the 60 cell lines of the National Cancer Institute Anticancer Drug Screen. *Cancer Res.* **2008**, *68*, 415–424. [CrossRef] [PubMed]
58. Björk-Eriksson, T.; West, C.; Karlsson, E.; Mercke, C. Tumor radiosensitivity (SF2) is a prognostic factor for local control in head and neck cancers. *Int. J. Radiat. Oncol. Biol. Phys.* **2000**, *46*, 13–19. [CrossRef]
59. Lamerz, R.; Stieber, P.; Fateh-Moghadam, A. Serum marker combinations in human breast cancer. *In Vivo* **1993**, *7*, 607–613. [PubMed]
60. Dnistrian, A.M.; Schwartz, M.K.; Greenberg, E.J.; Smith, C.A.; Schwartz, D.C. Evaluation of CA M26, CA M29, CA 15-3 and CEA as circulating tumor markers in breast cancer patients. *Tumor Biol.* **1991**, *12*, 82–90. [CrossRef] [PubMed]
61. Ebeling, F.G.; Stieber, P.; Untch, M.; Nagel, D.; Konecny, G.E.; Schmitt, U.M.; Fateh-Moghadam, A.; Seidel, D. Serum CEA and CA 15-3 as prognostic factors in primary breast cancer. *Br. J. Cancer* **2002**, *86*, 1217–1222. [CrossRef] [PubMed]
62. Stieber, P.; Nagel, D.; Ritzke, C.; Rössler, N.; Kirsch, C.; Eiermann, W.; Fateh-Moghadam, A. Significance of bone alkaline phosphatase, CA 15-3 and CEA in the detection of bone metastases during the follow-up of patients suffering from breast carcinoma. *Clin. Chem. Lab. Med.* **1992**, *30*, 809–814. [CrossRef]
63. Vizcarra, E.; Lluch, A.; Cibrian, R.; Jarque, F.; Garcia-Conde, J. CA15. 3, CEA and TPA tumor markers in the early diagnosis of breast cancer relapse. *Oncology* **1994**, *51*, 491–496. [CrossRef] [PubMed]
64. Duffy, M.J.; Evoy, D.; McDermott, E.W. CA 15-3: Uses and limitation as a biomarker for breast cancer. *Clin. Chim. Acta* **2010**, *411*, 1869–1874. [CrossRef]
65. Ludovini, V.; Gori, S.; Colozza, M.; Pistola, L.; Rulli, E.; Floriani, I.; Pacifico, E.; Tofanetti, F.R.; Sidoni, A.; Basurto, C.; et al. Evaluation of serum HER2 extracellular domain in early breast cancer patients: Correlation with clinicopathological parameters and survival. *Ann. Oncol.* **2008**, *19*, 883–890. [CrossRef] [PubMed]
66. Molina, R.; Augé, J.M.; Escudero, J.M.; Filella, X.; Zanon, G.; Pahisa, J.; Farrus, B.; Muñoz, M.; Velasco, M. Evaluation of tumor markers (HER-2/neu oncoprotein, CEA, and CA 15.3) in patients with locoregional breast cancer: Prognostic value. *Tumor Biol.* **2010**, *31*, 171–180. [CrossRef] [PubMed]
67. Asgeirsson, K.S.; Agrawal, A.; Allen, C.; Hitch, A.; Ellis, I.O.; Chapman, C.; Cheung, K.L.; Robertson, J.F. Serum epidermal growth factor receptor and HER2 expression in primary and metastatic breast cancer patients. *Breast Cancer Res.* **2007**, *9*, R75. [CrossRef] [PubMed]
68. Leyland-Jones, B.; Smith, B.R. Serum HER2 testing in patients with HER2-positive breast cancer: The death knell tolls. *Lancet Oncol.* **2011**, *12*, 286–295. [CrossRef]
69. Villarreal, L.; Méndez, O.; Salvans, C.; Gregori, J.; Baselga, J.; Villanueva, J. Unconventional secretion is a major contributor of cancer cell line secretomes. *Mol. Cell. Proteom. MCP* **2013**, *12*, 1046–1060. [CrossRef] [PubMed]
70. Ziegler, Y.S.; Moresco, J.J.; Yates, J.R., III; Nardulli, A.M. Integration of Breast Cancer Secretomes with Clinical Data Elucidates Potential Serum Markers for Disease Detection, Diagnosis, and Prognosis. *PLoS ONE* **2016**, *11*, e0158296. [CrossRef]
71. Shin, J.; Kim, G.; Lee, J.W.; Lee, J.E.; Kim, Y.S.; Yu, J.H.; Lee, S.T.; Ahn, S.H.; Kim, H.; Lee, C. Identification of ganglioside GM2 activator playing a role in cancer cell migration through proteomic analysis of breast cancer secretomes. *Cancer Sci.* **2016**, *107*, 828–835. [CrossRef]
72. Blache, U.; Horton, E.R.; Xia, T.; Schoof, E.M.; Blicher, L.H.; Schönenberger, A.; Snedeker, J.G.; Martin, I.; Erler, J.T.; Ehrbar, M. Mesenchymal stromal cell activation by breast cancer secretomes in bioengineered 3D microenvironments. *Life Sci. Alliance* **2019**, *2*, e201900304. [CrossRef]

73. Guo, W.; Li, H.; Zhu, Y.; Lan, L.; Yang, S.; Drukker, K.; Morris, E.A.; Burnside, E.S.; Whitman, G.J.; Giger, M.L. Prediction of clinical phenotypes in invasive breast carcinomas from the integration of radiomics and genomics data. *J. Med. Imaging* **2015**, *2*, 041007. [CrossRef]
74. Lin, Q.; Tan, H.T.; Lim, H.S.R.; Chung, M.C. Sieving through the cancer secretome. *Biochim. Biophys. Acta BBA Proteins Proteom.* **2013**, *1834*, 2360–2371. [CrossRef]
75. Cooper, S. Reappraisal of serum starvation, the restriction point, G0, and G1 phase arrest points. *FASEB J.* **2003**, *17*, 333–340. [CrossRef]
76. Shin, J.-S.; Hong, S.-W.; Lee, S.-L.O.; Kim, T.-H.; Park, I.-C.; An, S.-K.; Lee, W.-K.; Lim, J.-S.; Kim, K.-I.; Yang, Y. Serum starvation induces G1 arrest through suppression of Skp2-CDK2 and CDK4 in SK-OV-3 cells. *Int. J. Oncol.* **2008**, *32*, 435–439. [CrossRef] [PubMed]
77. Hasan, N.M.; Adams, G.E.; Joiner, M.C. Effect of serum starvation on expression and phosphorylation of PKCα and p53 in V79 cells: Implications for cell death. *Int. J. Cancer* **1999**, *80*, 400–405. [CrossRef]
78. Zander, L.; Bemark, M. Identification of genes deregulated during serum-free medium adaptation of a Burkitt's lymphoma cell line. *Cell Prolif.* **2008**, *41*, 136–155. [CrossRef]
79. Yamaguchi, N.; Yamamura, Y.; Koyama, K.; Ohtsuji, E.; Imanishi, J.; Ashihara, T. Characterization of new human pancreatic cancer cell lines which propagate in a protein-free chemically defined medium. *Cancer Res.* **1990**, *50*, 7008–7014. [PubMed]
80. Inoue, Y.; Kawamoto, S.; Shoji, M.; Hashizume, S.; Teruya, K.; Katakura, Y.; Shirahata, S. Properties of ras-amplified recombinant BHK-21 cells in protein-free culture. *Cytotechnology* **2000**, *33*, 21–26. [CrossRef] [PubMed]
81. Mbeunkui, F.; Fodstad, O.; Pannell, L.K. Secretory protein enrichment and analysis: An optimized approach applied on cancer cell lines using 2D LC− MS/MS. *J. Proteome Res.* **2006**, *5*, 899–906. [CrossRef]
82. Pellitteri-Hahn, M.; Warren, M.; Didier, D.; Winkler, E.; Mirza, S.; Greene, A.; Olivier, M. Improved mass spectrometric proteomic profiling of the secretome of rat vascular endothelial cells. *J. Proteome Res.* **2006**, *5*, 2861–2864. [CrossRef] [PubMed]
83. Yamashita, R.; Fujiwara, Y.; Ikari, K.; Hamada, K.; Otomo, A.; Yasuda, K.; Noda, M.; Kaburagi, Y. Extracellular proteome of human hepatoma cell, HepG2 analyzed using two-dimensional liquid chromatography coupled with tandem mass spectrometry. *Mol. Cell. Biochem.* **2007**, *298*, 83–92. [CrossRef] [PubMed]
84. Pardo, M.; García, Á.; Antrobus, R.; Blanco, M.J.; Dwek, R.A.; Zitzmann, N. Biomarker Discovery from Uveal Melanoma Secretomes: Identification of gp100 and Cathepsin D in Patient Serum. *J. Proteome Res.* **2007**, *6*, 2802–2811. [CrossRef] [PubMed]
85. Brandi, J.; Manfredi, M.; Speziali, G.; Gosetti, F.; Marengo, E.; Cecconi, D. Proteomic approaches to decipher cancer cell secretome. *Semin. Cell Dev. Biol.* **2018**, *78*, 93–101. [CrossRef] [PubMed]
86. Kumar, P.; Nagarajan, A.; Uchil, P.D. Analysis of Cell Viability by the Lactate Dehydrogenase Assay. *Cold Spring Harb. Protoc.* **2018**, *2018*. [CrossRef] [PubMed]
87. Willms, E.; Johansson, H.J.; Mäger, I.; Lee, Y.; Blomberg, K.E.M.; Sadik, M.; Alaarg, A.; Smith, C.I.E.; Lehtiö, J.; El Andaloussi, S.; et al. Cells release subpopulations of exosomes with distinct molecular and biological properties. *Sci. Rep.* **2016**, *6*, 22519. [CrossRef] [PubMed]
88. Wu, C.C.; Hsu, C.W.; Chen, C.D.; Yu, C.J.; Chang, K.P.; Tai, D.I.; Liu, H.P.; Su, W.H.; Chang, Y.S.; Yu, J.S. Candidate serological biomarkers for cancer identified from the secretomes of 23 cancer cell lines and the human protein atlas. *Mol. Cell. Proteom. MCP* **2010**, *9*, 1100–1117. [CrossRef]
89. Tsai, M.H.; Cook, J.A.; Chandramouli, G.V.; DeGraff, W.; Yan, H.; Zhao, S.; Coleman, C.N.; Mitchell, J.B.; Chuang, E.Y. Gene expression profiling of breast, prostate, and glioma cells following single versus fractionated doses of radiation. *Cancer Res.* **2007**, *67*, 3845–3852. [CrossRef]
90. Bosma, S.C.J.; Hoogstraat, M.; van der Leij, F.; de Maaker, M.; Wesseling, J.; Lips, E.; Loo, C.E.; Rutgers, E.J.; Elkhuizen, P.H.M.; Bartelink, H.; et al. Response to Preoperative Radiation Therapy in Relation to Gene Expression Patterns in Breast Cancer Patients. *Int. J. Radiat. Oncol. Biol. Phys.* **2020**, *106*, 174–181. [CrossRef]
91. Mao, B.; Wu, W.; Davidson, G.; Marhold, J.; Li, M.; Mechler, B.M.; Delius, H.; Hoppe, D.; Stannek, P.; Walter, C.; et al. Kremen proteins are Dickkopf receptors that regulate Wnt/beta-catenin signalling. *Nature* **2002**, *417*, 664–667. [CrossRef]
92. Mariz, K.; Ingolf, J.B.; Daniel, H.; Teresa, N.J.; Erich-Franz, S. The Wnt inhibitor dickkopf-1: A link between breast cancer and bone metastases. *Clin. Exp. Metastasis* **2015**, *32*, 857–866. [CrossRef]
93. Pinzone, J.J.; Hall, B.M.; Thudi, N.K.; Vonau, M.; Qiang, Y.-W.; Rosol, T.J.; Shaughnessy, J.D., Jr. The role of Dickkopf-1 in bone development, homeostasis, and disease. *Blood* **2009**, *113*, 517–525. [CrossRef] [PubMed]
94. Gunn, W.G.; Conley, A.; Deininger, L.; Olson, S.D.; Prockop, D.J.; Gregory, C.A. A crosstalk between myeloma cells and marrow stromal cells stimulates production of DKK1 and interleukin-6: A potential role in the development of lytic bone disease and tumor progression in multiple myeloma. *Stem Cells* **2006**, *24*, 986–991. [CrossRef]
95. Heath, D.J.; Chantry, A.D.; Buckle, C.H.; Coulton, L.; Shaughnessy, J.D., Jr.; Evans, H.R.; Snowden, J.A.; Stover, D.R.; Vanderkerken, K.; Croucher, P.I. Inhibiting Dickkopf-1 (Dkk1) removes suppression of bone formation and prevents the development of osteolytic bone disease in multiple myeloma. *J. Bone Miner. Res.* **2009**, *24*, 425–436. [CrossRef] [PubMed]
96. Kasoha, M.; Bohle, R.M.; Seibold, A.; Gerlinger, C.; Juhasz-Böss, I.; Solomayer, E.F. Dickkopf-1 (Dkk1) protein expression in breast cancer with special reference to bone metastases. *Clin. Exp. Metastasis* **2018**, *35*, 763–775. [CrossRef]
97. Zhou, S.-J.; Zhuo, S.-R.; Yang, X.-Q.; Qin, C.-X.; Wang, Z.-L. Serum Dickkopf-1 expression level positively correlates with a poor prognosis in breast cancer. *Diagn. Pathol.* **2014**, *9*, 161. [CrossRef]

98. Akella, N.M.; Ciraku, L.; Reginato, M.J. Fueling the fire: Emerging role of the hexosamine biosynthetic pathway in cancer. *BMC Biol.* **2019**, *17*, 52. [CrossRef] [PubMed]
99. Chu, J.; Li, N.; Gai, W. Identification of genes that predict the biochemical recurrence of prostate cancer. *Oncol. Lett.* **2018**, *16*, 3447–3452. [CrossRef] [PubMed]
100. Kaushik, A.K.; Shojaie, A.; Panzitt, K.; Sonavane, R.; Venghatakrishnan, H.; Manikkam, M.; Zaslavsky, A.; Putluri, V.; Vasu, V.T.; Zhang, Y.; et al. Inhibition of the hexosamine biosynthetic pathway promotes castration-resistant prostate cancer. *Nat. Commun.* **2016**, *7*, 11612. [CrossRef]
101. Liu, W.; Jiang, K.; Wang, J.; Mei, T.; Zhao, M.; Huang, D. Upregulation of GNPNAT1 Predicts Poor Prognosis and Correlates With Immune Infiltration in Lung Adenocarcinoma. *Front. Mol. Biosci.* **2021**, *8*, 605754. [CrossRef]
102. Zheng, X.; Li, Y.; Ma, C.; Zhang, J.; Zhang, Y.; Fu, Z.; Luo, H. Independent Prognostic Potential of GNPNAT1 in Lung Adenocarcinoma. *Biomed. Res. Int.* **2020**, *2020*, 8851437. [CrossRef]
103. Chokchaitaweesuk, C.; Kobayashi, T.; Izumikawa, T.; Itano, N. Enhanced hexosamine metabolism drives metabolic and signaling networks involving hyaluronan production and O-GlcNAcylation to exacerbate breast cancer. *Cell Death Dis.* **2019**, *10*, 803. [CrossRef] [PubMed]
104. Aufderklamm, S.; Todenhöfer, T.; Gakis, G.; Kruck, S.; Hennenlotter, J.; Stenzl, A.; Schwentner, C. Thymidine kinase and cancer monitoring. *Cancer Lett.* **2012**, *316*, 6–10. [CrossRef]
105. Eriksson, S.; Munch-Petersen, B.; Johansson, K.; Ecklund, H. Structure and function of cellular deoxyribonucleoside kinases. *Cell. Mol. Life Sci. CMLS* **2002**, *59*, 1327–1346. [CrossRef] [PubMed]
106. Hurwitz, S.N.; Rider, M.A.; Bundy, J.L.; Liu, X.; Singh, R.K.; Meckes, D.G. Proteomic profiling of NCI-60 extracellular vesicles uncovers common protein cargo and cancer type-specific biomarkers. *Oncotarget* **2016**, *7*, 86999–87015. [CrossRef] [PubMed]
107. Hong, B.S.; Cho, J.H.; Kim, H.; Choi, E.J.; Rho, S.; Kim, J.; Kim, J.H.; Choi, D.S.; Kim, Y.K.; Hwang, D.; et al. Colorectal cancer cell-derived microvesicles are enriched in cell cycle-related mRNAs that promote proliferation of endothelial cells. *BMC Genom.* **2009**, *10*, 556–578. [CrossRef] [PubMed]
108. Skog, J.; Wurdinger, T.; van Rijn, S.; Meijer, D.H.; Gainche, L.; Sena-Esteves, M.; Curry, W.T., Jr.; Carter, B.S.; Krichevsky, A.M.; Breakefield, X.O. Glioblastoma microvesicles transport RNA and proteins that promote tumour growth and provide diagnostic biomarkers. *Nat. Cell Biol.* **2008**, *10*, 1470–1476. [CrossRef] [PubMed]
109. Sinha, A.; Ignatchenko, V.; Ignatchenko, A.; Mejia-Guerrero, S.; Kislinger, T. In-depth proteomic analyses of ovarian cancer cell line exosomes reveals differential enrichment of functional categories compared to the NCI 60 proteome. *Biochem. Biophys. Res. Commun.* **2014**, *445*, 694–701. [CrossRef] [PubMed]
110. Topolcan, O.; Holubec, L., Jr. The role of thymidine kinase in cancer diseases. *Expert Opin. Med. Diagn.* **2008**, *2*, 129–141. [CrossRef]
111. Li, H.; Lei, D.; Wang, X.; Skog, S.; He, Q. Serum thymidine kinase 1 is a prognostic and monitoring factor in patients with non-small cell lung cancer. *Oncol. Rep.* **2005**, *13*, 145–149. [CrossRef] [PubMed]
112. Mao, Y.; Wu, J.; Wang, N.; He, L.; Wu, C.; He, Q.; Skog, S. A Comparative Study: Immunohistochemical Detection of Cytosolic Thymidine Kinase and Proliferating Cell Nuclear Antigen in Breast Cancer. *Cancer Investig.* **2002**, *20*, 922–931. [CrossRef] [PubMed]
113. He, Q.; Zou, L.; Zhang, P.; Lui, J.; Skog, S.; Fornander, T. The clinical significance of thymidine kinase 1 measurement in serum of breast cancer patients using anti-TK1 antibody. *Int. J. Biol. Markers* **2000**, *15*, 139–146. [CrossRef] [PubMed]
114. He, Q.; Fornander, T.; Johansson, H.; Johansson, U.; Hu, G.Z.; Rutqvist, L.-E.; Skog, S. Thymidine kinase 1 in serum predicts increased risk of distant or loco-regional recurrence following surgery in patients with early breast cancer. *Anticancer Res.* **2006**, *26*, 4753–4759.
115. Lerouge, D.; Touboul, E.; Lefranc, J.P.; Genestie, C.; Moureau-Zabotto, L.; Blondon, J. Combined chemotherapy and preoperative irradiation for locally advanced noninflammatory breast cancer: Updated results in a series of 120 patients. *Int. J. Radiat. Oncol. Biol. Phys.* **2004**, *59*, 1062–1073. [CrossRef]
116. Semiglazov, V.F.; Topuzov, E.E.; Bavli, J.L.; Moiseyenko, V.M.; Ivanova, O.A.; Seleznev, I.K.; Orlov, A.A.; Barash, N.Y.; Golubeva, O.M.; Chepic, O.F. Primary (neoadjuvant) chemotherapy and radiotherapy compared with primary radiotherapy alone in stage IIb-IIIa breast cancer. *Ann. Oncol.* **1994**, *5*, 591–595. [CrossRef]
117. Bondiau, P.Y.; Courdi, A.; Bahadoran, P.; Chamorey, E.; Queille-Roussel, C.; Lallement, M.; Birtwisle-Peyrottes, I.; Chapellier, C.; Pacquelet-Cheli, S.; Ferrero, J.M. Phase 1 clinical trial of stereotactic body radiation therapy concomitant with neoadjuvant chemotherapy for breast cancer. *Int. J. Radiat. Oncol. Biol. Phys.* **2013**, *85*, 1193–1199. [CrossRef] [PubMed]
118. Sousa, C.; Cruz, M.; Neto, A.; Pereira, K.; Peixoto, M.; Bastos, J.; Henriques, M.; Roda, D.; Marques, R.; Miranda, C.; et al. Neoadjuvant radiotherapy in the approach of locally advanced breast cancer. *ESMO Open* **2020**, *4* (Suppl. 2), e000640. [CrossRef]
119. Hughes, K.; Neoh, D. Neoadjuvant Radiotherapy: Changing the Treatment Sequence to Allow Immediate Free Autologous Breast Reconstruction. *J. Reconstr. Microsurg.* **2018**, *34*, 624–631. [PubMed]
120. Singh, P.; Hoffman, K.; Schaverien, M.V.; Krause, K.J.; Butler, C.; Smith, B.D.; Kuerer, H.M. Neoadjuvant Radiotherapy to Facilitate Immediate Breast Reconstruction: A Systematic Review and Current Clinical Trials. *Ann. Surg. Oncol.* **2019**, *26*, 3312–3320. [CrossRef] [PubMed]

121. Pazos, M.; Corradini, S.; Dian, D.; von Bodungen, V.; Ditsch, N.; Wuerstlein, R.; Schönecker, S.; Harbeck, N.; Scheithauer, H.; Belka, C. Neoadjuvant radiotherapy followed by mastectomy and immediate breast reconstruction: An alternative treatment option for locally advanced breast cancer. *Strahlenther. Onkol.* **2017**, *193*, 324–331. [CrossRef] [PubMed]
122. Calitchi, E.; Kirova, Y.M.; Otmezguine, Y.; Feuilhade, F.; Piedbois, Y.; Le Bourgeois, J.P. Long-term results of neoadjuvant radiation therapy for breast cancer. *Int. J. Cancer* **2001**, *96*, 253–259. [CrossRef] [PubMed]
123. Blitzblau, R.C.; Arya, R.; Yoo, S.; Baker, J.A.; Chang, Z.; Palta, M.; Duffy, E.; Horton, J.K. A phase 1 trial of preoperative partial breast radiation therapy: Patient selection, target delineation, and dose delivery. *Pract. Radiat. Oncol.* **2015**, *5*, e513–e520. [CrossRef] [PubMed]
124. Horton, J.K.; Blitzblau, R.C.; Yoo, S.; Geradts, J.; Chang, Z.; Baker, J.A.; Georgiade, G.S.; Chen, W.; Siamakpour-Reihani, S.; Wang, C.; et al. Preoperative Single-Fraction Partial Breast Radiation Therapy: A Novel Phase 1, Dose-Escalation Protocol With Radiation Response Biomarkers. *Int. J. Radiat. Oncol. Biol. Phys.* **2015**, *92*, 846–855. [CrossRef] [PubMed]
125. Poleszczuk, J.; Luddy, K.; Chen, L.; Lee, J.K.; Harrison, L.B.; Czerniecki, B.J.; Soliman, H.; Enderling, H. Neoadjuvant radiotherapy of early-stage breast cancer and long-term disease-free survival. *Breast Cancer Res.* **2017**, *19*, 75. [CrossRef] [PubMed]
126. Riet, F.G.; Fayard, F.; Arriagada, R.; Santos, M.A.; Bourgier, C.; Ferchiou, M.; Heymann, S.; Delaloge, S.; Mazouni, C.; Dunant, A.; et al. Preoperative radiotherapy in breast cancer patients: 32 years of follow-up. *Eur. J. Cancer* **2017**, *76*, 45–51. [CrossRef] [PubMed]

Article

Exosomes Derived from Radioresistant Breast Cancer Cells Promote Therapeutic Resistance in Naïve Recipient Cells

Chantell Payton, Lisa Y. Pang *, Mark Gray and David J. Argyle

Roslin Institute, The Royal (Dick) School of Veterinary Studies, The University of Edinburgh, Edinburgh EH25 9RG, UK; chantell.payton@ed.ac.uk (C.P.); mark.gray@ed.ac.uk (M.G.); david.argyle@ed.ac.uk (D.J.A.)
* Correspondence: Lisa.pang@ed.ac.uk; Tel.: +44-(0)131-651-9100

Abstract: Radiation resistance is a significant challenge in the treatment of breast cancer in humans. Human breast cancer is commonly treated with surgery and adjuvant chemotherapy/radiotherapy, but recurrence and metastasis upon the development of therapy resistance results in treatment failure. Exosomes are extracellular vesicles secreted by most cell types and contain biologically active cargo that, when transferred to recipient cells, can influence the cells' genome and proteome. We propose that exosomes secreted by radioresistant (RR) cells may be able to disseminate the RR phenotype throughout the tumour. Here, we isolated exosomes from the human breast cancer cell line, MDA-MB-231, and the canine mammary carcinoma cell line, REM134, and their RR counterparts to investigate the effects of exosomes derived from RR cells on non-RR recipient cells. Canine mammary cancer cells lines have previously been shown to be excellent translational models of human breast cancer. This is consistent with our current data showing that exosomes derived from RR cells can increase cell viability and colony formation in naïve recipient cells and increase chemotherapy and radiotherapy resistance, in both species. These results are consistent in cancer stem cell and non-cancer stem cell populations. Significantly, exosomes derived from RR cells increased the tumoursphere-forming ability of recipient cells compared to exosomes derived from non-RR cells. Our results show that exosomes are potential mediators of radiation resistance that could be therapeutically targeted.

Keywords: breast cancer; exosomes; chemoresistance; radioresistance; comparative oncology; One Health

1. Introduction

Breast cancer is the most common female malignancy and the leading cause of cancer-related deaths in women [1,2]. Similarly, naturally occurring canine mammary tumours are the most common cause of death in intact female dogs and have been proposed as a comparative model of the human disease [3]. Canine mammary tumours have a similar genetic predisposition, histopathology, disease progression and clinical outcome to the human disease. Human breast cancer is commonly classified into molecularly distinct subtypes: normal breast-like, HER2+, luminal A, luminal B and triple negative. These subtypes differ in clinical outcomes, patient survival and treatment strategy. However, there is gene expression heterogeneity within these subtypes and breast cancer can be considered as a spectrum of diseases. Kumar et al., 2012 [4] utilised microarray technology to highlight a 163-gene expression signature associated with prognosis, highlighting that, in the context of gene expression, this disease is highly heterogenous and individualised. Assessing the global gene expression and proteomic profiles of each individual patient and applying that information to a database of available treatment options may be more successful, in terms of survival rates, than following a rigid treatment plan based on tumour subtype [5]. This method of patient-specific therapy assignment would be more efficient in terms of time, expense and patient side effects and may be applicable in both human and veterinary medicine.

The emergence of resistance to key modalities, including chemotherapy and radiotherapy, and the subsequent re-initiation of tumour growth and relapse represent a significant clinical problem, often with limited treatment options and increased mortality. Understanding the underlying molecular mechanisms driving therapy resistance could help to identify potential biomarkers to track the emergence of resistance and novel therapeutic targets.

Tumours comprise a heterogenous mix of cell populations including cancer stem cells (CSCs) and non-CSCs, which make up the bulk of the tumour. CSCs are long-lived cells that drive tumourigenesis as they can self-renew and differentiate into other cellular subtypes. Breast CSCs are inherently resistant to conventional chemotherapy and radiotherapy [6]. Therefore, the relative size of a CSC pool within a tumour may influence the intrinsic radioresistance of that tumour. Radiation treatment will eliminate the majority of cancer cells; however, CSCs will survive and be able to re-initiate tumour growth and tumour cell repopulation leading to patient relapse [7]. The development of acquired therapy resistance can also occur due to selective pressures imposed by cancer therapies that can result in advantageous mutations in newly forming cancer cells and lead to increased survival by, for example, the activation of epithelial-to-mesenchymal transition (EMT), enhanced DNA damage repair and enhanced elimination of cytotoxic content from within the cancer cell [8] including the active chemotherapeutic agents or the reactive oxygen species produced during radiotherapy treatment [9].

Exosomes have been implicated in the acquisition of therapy resistance [10,11]. Exosomes are nanovesicles secreted from most living cells. They have a size range between of 30–150 nm in diameter, and they contain a biologically active cargo consisting of nucleic acids, miRNAs, proteins and lipids, encapsulated within their double membrane [12]. The outer surface of the membrane contains integrins, tetraspanins and cell signalling receptors [13]. The content of exosomes is reflective of the parental cell from which it is derived, and under non-diseased states, the role of exosomes is to mediate cell-to-cell communication [14,15]. As the formation of exosomes within the parental cell results in the incorporation of the contents of the parental cell, the exosome cargo can reflect the development and progression of the diseased state of the parental cell. Further research has shown that the active content of exosomes can result in phenotypic and genotypic changes in recipient cells. In cancer, exosomal transfer can occur between developing cancer cells, and between cancer cells and stromal cells, and can have a range of functions, for example, developing cancer cells can communicate via exosomes to programme stromal cells to provide nourishment in the form of amino acids and carbon [16–18]. As well as programming surrounding stromal cells to provide a nurturing environment for cancer cells, exosomes can also promote metastasis and mediate organotropism [19,20].

Exosomes have been shown to play a pivotal role in therapy resistance in humans [21,22], but the role of exosomes in canine therapy resistance has not yet been studied. Exosomes derived from human breast cancer cells have been shown to shuttle chemotherapeutic agents out of the cell [23], and chemotherapy-resistant breast cancer cells can transfer p-glycoprotein protein pumps to chemotherapy-sensitive breast cancer cells to allow the active removal of the chemotherapeutic agents [24]. However, the role of exosomes in the development of radiotherapy resistance in breast cancer cells and the CSC population is poorly understood, and the mechanisms by which exosomes can mediate chemoresistance cannot be directly applied to the development of radioresistance. We hypothesise that exosomes derived from radioresistant (RR) cells can disseminate the RR phenotype to non-RR cancer cells. In this study, we isolated exosomes from the human breast cancer cell line, MDA-MB-231, and the canine mammary carcinoma cell line, REM134, and their RR counterparts to investigate the effects of exosomes derived from RR cells on non-RR recipient cells. Our data show that exosomes derived from RR cells, compared to exosomes derived from non-RR cells, can increase cell viability and colony formation in recipient cells and increase chemotherapy and radiotherapy resistance. These results are consistent in CSC and non-CSC populations. Our results show that exosomes are potential mediators of RR that could be therapeutically targeted. Future research could focus on the profiling the exosomal cargo to identify emerging markers of radioresistance. These biomarkers could be monitored throughout treatment to optimise patient-specific treatment plans for anticancer interventions.

2. Materials and Methods

2.1. Cell Culture

The cell lines used in this study were the human breast cancer cell line, MDA-MB-231, and the canine mammary carcinoma cell line, REM134 [25]. Radioresistant MDA-MB-231 and REM134 cell lines were gifted by Dr. Mark Gray [26]. RR cell lines were established over several weeks by gradually irradiating the non-RR parental cell lines with increasing doses of Gray (Gy). MDA-MB-231 cells were cultured in Dulbecco's modified Eagle's medium (DMEM) + 1 g/L D-glucose, L-glutamine + pyruvate (Gibco Life Technologies, Invitrogen, UK). REM134 cells were grown in DMEM + 4.5 g/L D-glucose, L-glutamine—pyruvate (Gibco Life Technologies, Invitrogen, UK). All cell culture media were supplemented with 10% exosome-depleted FBS and 1% penicillin and streptomycin. Cells were maintained at 37 °C in 5% CO_2 in a humidified incubator. FBS was depleted of exosomes by ultracentrifugation in an SW32 Ti rotor (Beckman Coulter, IN, USA) at $12,000\times g$ for 18 h at 4 °C.

2.2. Radiation Treatment

To maintain the RR phenotype, RR cells were irradiated with 12 Gy every 3–4 weeks. Briefly, cells were grown until 70% confluence and, after standard trypsinisation, were resuspended as single cells in 10 mL of the appropriate media and immediately irradiated in the gamma cell irradiator (Gammacell 1000 Elite, Best Theratronics, Ottawa, ON, Canada) in 50 mL falcon tubes. After irradiation, cells were transferred into a T75 flask and maintained as previously described.

2.3. Exosome Isolation

Cells were seeded in T175 flasks and grown until 70% confluence. Cells were washed in PBS, and all media were replaced with 10 mL of exosome-free DMEM and incubated for 24 h. The medium was removed and centrifuged at $500\times g$ for 10 min to remove cell debris. The supernatant was then filtered through a 0.22 µm filter and ultracentrifuged at $120,000\times g$ for 90 min at 4 °C in an SW41 ultracentrifuge rotor (Beckman Coulter, IN, USA) with swing buckets. The supernatant was removed, and the exosome pellet was resuspended in 1 mL filtered PBS and stored at -70 °C until further use.

2.4. Exosome Quantification

Exosomes were lysed with RIPA buffer (50 mM Tris pH 6.8, 150 mM NaCl, 1 mM EDTA, 1% NP40) by adding 3:1 volume of RIPA buffer to the exosome sample and mixing thoroughly. The samples were incubated for 30 min on ice and then centrifuged at $13,000\times g$ for 5 min at 4 °C. Supernatants were transferred to 1.5 mL Eppendorf tubes and stored at -70 °C. The protein concentration of samples was determined by a Bradford assay. BSA standards at 0.1, 0.25, 0.5, 1, 2, 5 and 10 mg/mL were used as controls. Then, 1 µL of BSA standards were added to designated wells of a 96-well plate in duplicate, and 1 µL of protein samples were loaded in triplicate. Following this, 200 µL of Bradford reagent (Bio-Rad Laboratories, Watford, UK) was added to each well and mixed by pipetting. The plate was incubated at room temperature for 2 min. Absorbance at 595 nm was determined using the Victor3 plate reader (Perkin Elmer, Beaconsfield, UK) and the relative protein concentration of the samples was determined by comparing them to the BSA standards.

2.5. Transmission Electron Microscopy

Freshly isolated exosomes in 10 µL PBS were added in a 1:1 ratio with 2% paraformaldehyde and immediately processed for transmission electron microscopy (TEM). Briefly, 5 µL of sample was placed on formvar-coated grids and incubated for 20 min at room temperature. Grids were washed in 100 µL of PBS plus 50 µL of 1% glutaraldehyde for 5 min and then incubated with 100 µL of ddH$_2$O for 2 min. Wash steps were repeated eight times in total. After washing, 50 µL of 1% uranyl-oxalate solution (pH 7) was added to the grid for 5 min, then 50 µL of methyl cellulose-UA was added for 10 min on ice. The excess fluid was removed by blotting, and the grids were air dried for 5 to 10 min. Samples were viewed on a JEM-1400 Plus TEM (Jeol, Welwyn, UK) operating at 80 kV. Representative images were collected on an OneView camera (Gatan, Pleasanton, CA, USA). These experiments were carried out at King's Buildings at The University of Edinburgh.

2.6. Nanoparticle Tracking Analysis

Exosomes were analysed by nanoparticle tracking analysis (NanoSight LM10, Malvern Panalytical, Malvern, UK) to determine the size range and distribution. Briefly, 1 mL of diluted exosome sample (1:50–1:100) was loaded on to the NanoSight machine, and particle concentration was determined and diluted in the range of 4×10^8–12×10^8 particles/mL. Parameters were set at a detection rate of 15,000 particles per minute for capture settings, and the smallest vesicle size was set at 30 nm, with analysis performed by NanoSight software version 2.3 (Malvern Panalytical, Malvern, UK). The rate at which exosomes were produced per cell per hour was calculated by dividing the total number of exosomes by the total number of cells after exosome harvesting and then dividing by the number of hours over which the sample was collected.

2.7. Exosome Treatment

For exosome treatment, cells were seeded depending on cell type and experimental conditions. Generally, exosomes were added at a concentration of 50 µg/mL. To determine this concentration, 10 µL of isolated exosomes were lysed, and their protein concentration was quantified as in Section 2.4. From that concentration, we calculated the volume of isolated exosomes required to make up a solution at 50 µg/mL in exosome-free media. All exosome solutions were made up fresh prior to treatment. Controls were generated with PBS vehicle instead of exosomes.

2.8. Cell Viability Assay

Cells were seeded in 96-well plates at 500 cells per well. Exosomes were added at the indicated concentrations 24 h after seeding. Cell viability was determined 72 h post-treatment using the CellTiter-Glo® Luminescent Cell Viability Assay (Promega, Hampshire, UK) according to the manufacturer's instructions. Luminescence was measured by a Victor3 multilabel plate reader (Perkin Elmer, Beaconsfield, UK). Data were averaged and normalised against the average signal of the PBS control samples.

2.9. Colony Fromation Assay

MDA-MB-231 and MDA-MB-231 RR cell lines were trypsinised and seeded as single cells at 50 cells per well in a 6-well plate. REM134 and REM134 RR were trypsinised and seeded as single cells at 1000 cells per well in a 6-well plate. Immediately after seeding, either PBS (vehicle control), 50 µg/mL exosomes derived from non-RR cells or exosomes derived from RR cells were added to the appropriate well. All plates were incubated as previously described until colonies formed in the vehicle control (approximately 10 days). To stain the colonies, each well was washed with 5 mL PBS and then incubated with 5 mL of 100% methanol for 5 min at room temperature. The methanol was removed, and plates were air dried. Colonies were then stained with a Giemsa stain (20% Giemsa stain (Sigma-Aldrich, Gillingham, UK) plus 80% ddH$_2$O) for 20 min at room temperature. After staining, the plates were then washed twice with water and air dried. All colonies were counted and normalised to the control.

In experiments to determine the effect of exosomes derived from RR cells on the colony-forming ability after treatment with radiation, cells were seeded at 20,000 cells in 1 mL of medium in a 12-well plate and incubated for 24 h with either PBS, 50 µg/mL exosomes derived from non-RR cells or 50 µg/mL exosomes derived from RR cells. Cells were then seeded as single cells as described above. In addition, MDA-MB-231 CSCs and MDA-MB-231 RR CSCs were seeded at 750 cells in 3 mL media, and REM134 CSCs and REM134 RR CSCs were seeded at 1000 cells in 3 mL media. Single cells were immediately irradiated at either 0, 2.5 or 5 Gy. Colonies were allowed to form and were processed as described above.

2.10. Chemosensitivity Assays

MDA-MB-231 and MDA-MB-231 RR cells were seeded at 500 cells/50 µL per well in a 96-well plate. REM134 and REM134 RR cells were seeded at 1000 cells/50 µL per well in a 96-well plate. CSCs were seeded at 1000 cells/50 µL per well. Cells were incubated for 24 h before treating with 25 µL exosomes (50 µg/mL). Cells were then treated 12 h later with a dose titration of doxorubicin at the indicated concentrations in 25 µL. Cell viability was determined 72 h post-treatment with doxorubicin as described above.

2.11. Tumoursphere-Forming Assay

Cells were seeded at 20,000 cells/mL in 1 mL of exosome-free FBS DMEM media in 12-well plates and treated with either PBS, 50 µg/mL of exosomes derived from non-RR cells or 50 µg/mL of exosomes derived from RR cells and incubated for 24 h. Following incubation, MDA-MB-231 and MDA-MB-231 RR cells were seeded as single cells at 3000 cells per well, and REM134 and REM134 RR were seeded at 6000 cells per well, in 3 mL N2 media in 6-well low-attachment plates (Corning, Flintshire, UK). All samples were triplicated. N2 media was supplemented every 48 h with human EFG and human FGF at 10 ng/mL (Peptrotech, London, UK). Sphere formation was monitored for 7 days. Tumourspheres over 50 µm in diameter were counted in five random fields of vision using an Axiovert 40 CFL microscope (Zeiss, Hallbergmoos, Germany) with images taken at 5× and 10× magnification and size measurements recorded by Axiovision software version 4.7.2 (Zeiss, Hallbergmoos, Germany).

2.12. Migration Assay

Cells were seeded at 20,000 cells in 1 mL media per well in a 12-well plate and treated with either PBS or corresponding exosomes derived from either non-RR or RR cells at the indicated concentration and incubated for 24 h. Cells were then seeded into Ibidi® (Munich, Germany) chamber slides according to the manufacturer's instructions. Briefly, cells were trypsinised and seeded at varying concentrations: MDA-MB-231 at 4.5×10^5/well; MDA-MB-232 RR at 4.75×10^5/well; REM134 at 3.45×10^5/well; and REM134 RR at 3.75×10^5/well and incubated until confluent. Once confluent, each insert was removed to leave a gap. Then, 1 mL of media was added to each well and the width of the gap was measured at six points using the Axiovert 40 CFL microscope with an AxioCAM HRm camera (Zeiss, Hallbergmoos, Germany) and pictures were taken at 5× magnification at set time points until the gap was closed. The migration distance was recorded at stated time points with measurements by Axiovision software version 4.7.2. Percentage migration was calculated as $(A-B)/B$, with A being the size of the gap at 0 h, and B being the gap at the designated time point.

2.13. Statistical Analysis

Data were analysed for normality using the Anderson–Darling normality test and the appropriate parametric/non-parametric test was chosen to determine statistical significance. All statistical analyses were performed using Minitab 19 software, with statistical significance being defined as $p \leq 0.05$.

3. Results

3.1. Isolation of Exosomes from Canine and Human Breast Cancer Cell Lines and Their Derived RR Counterparts

Radioresistant cell lines MDA-MB-231 RR and REM134 RR were derived by exposing parental cells to increasing doses of radiation every week up until there was limited cell death at 8 Gy [27]. RR cells are morphologically distinct from non-RR parental cells: RR cells have extended cytoplasmic extensions and a spindle-like morphology (Figure 1A(ii,iv)) compared to non-RR cells (Figure 1A(i,ii)). Exosomes were isolated from all cell lines by ultracentrifugation and visualised using TEM. All exosomes exhibited the characteristic "cup shape" morphology [21] (Figure 1B(i–iv)) and expected size distribution as analysed by nanoparticle tracking analysis (NTA) (Figure 1C(i–iv)). NTA was also used to calculate the rate of exosome production per cell per hour and showed that RR cells produced more exosomes than non-RR cells. REM 134 RR cells and MDA-MB-231 RR produced approximately sixfold and threefold more exosomes than their non-RR counterparts, respectively (Figure 1D).

3.2. Exosomes Isolated from RR Cells Increased the Survival of Recipient Cells Compared to Exosomes Isolated from Non-RR Cells

To determine the effect of exosomes on cell viability, cells were seeded in 96-well plates, incubated for 24 h and then treated with exosome dilutions of 10, 20, 30, 50 and 75 µg/mL. Cell viability was determined 72 h after treatment. Our data show that exosomes derived from RR cell lines resulted in a significant increase in cell viability, which appeared to be dose dependant, resulting in an increase in cell viability from 100% to 150% (Figure 2A). To compliment the cell viability assay, we also performed colony formation assays. Single cells were immediately treated with either 50 or 100 µg/mL of the corresponding exosomes and incubated until colonies were visible. Exosomes derived from MDA-MB-231 RR and REM134 RR cell lines resulted in a significant increase in the number of colonies compared to both PBS control and exosomes derived from non-RR exosomes (Figure 2B). Based on these results, we selected 50 µg/mL of exosomes to be used in further experiments.

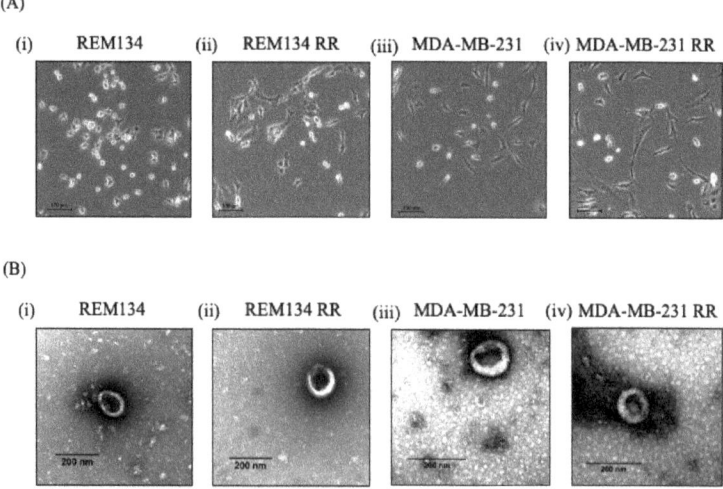

Figure 1. Cont.

(C)

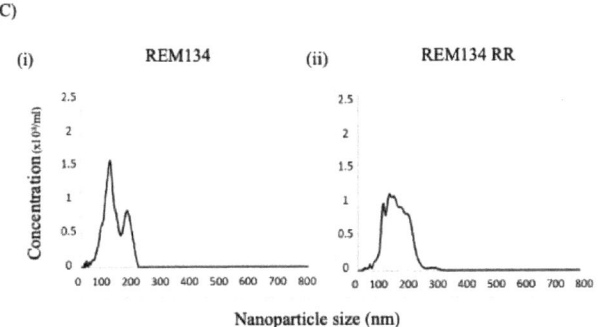

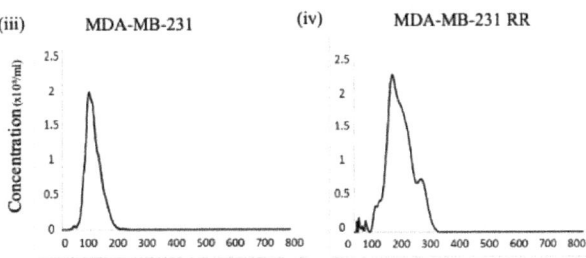

(D)

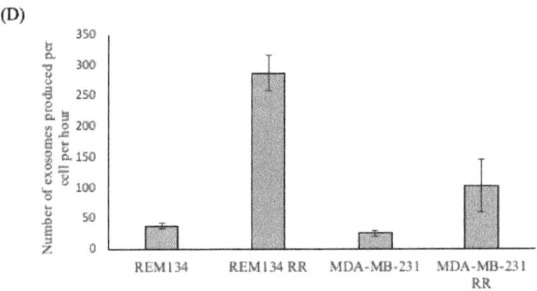

Figure 1. Isolation of exosomes from canine and human breast cancer cell lines and their derived isogenic RR counterparts. (**A**) Cell morphology of (**i**) REM134, (**ii**) REM134 RR, (**iii**) MDA-MB-231 and (**iv**) MDA-MB-231 RR cells. Scale bar represents 100 μm. (**B**) Visualisation, using TEM, of exosomes isolated from (**i**) REM134, (**ii**) REM134 RR, (**iii**) MDA-MB-231 and (**iv**) MDA-MB-231 RR cells. Scale bar represents 200 nm. Characterisation of exosomes using NTA to measure (**C**) particle distribution from (**i**) REM134 cells, (**ii**) REM134 RR, (**iii**) MDA-MB-231 and (**iv**) MDA-MB-231 RR and (**D**) rate of exosome production per cell per hour. Data are representative of three independent experiments.

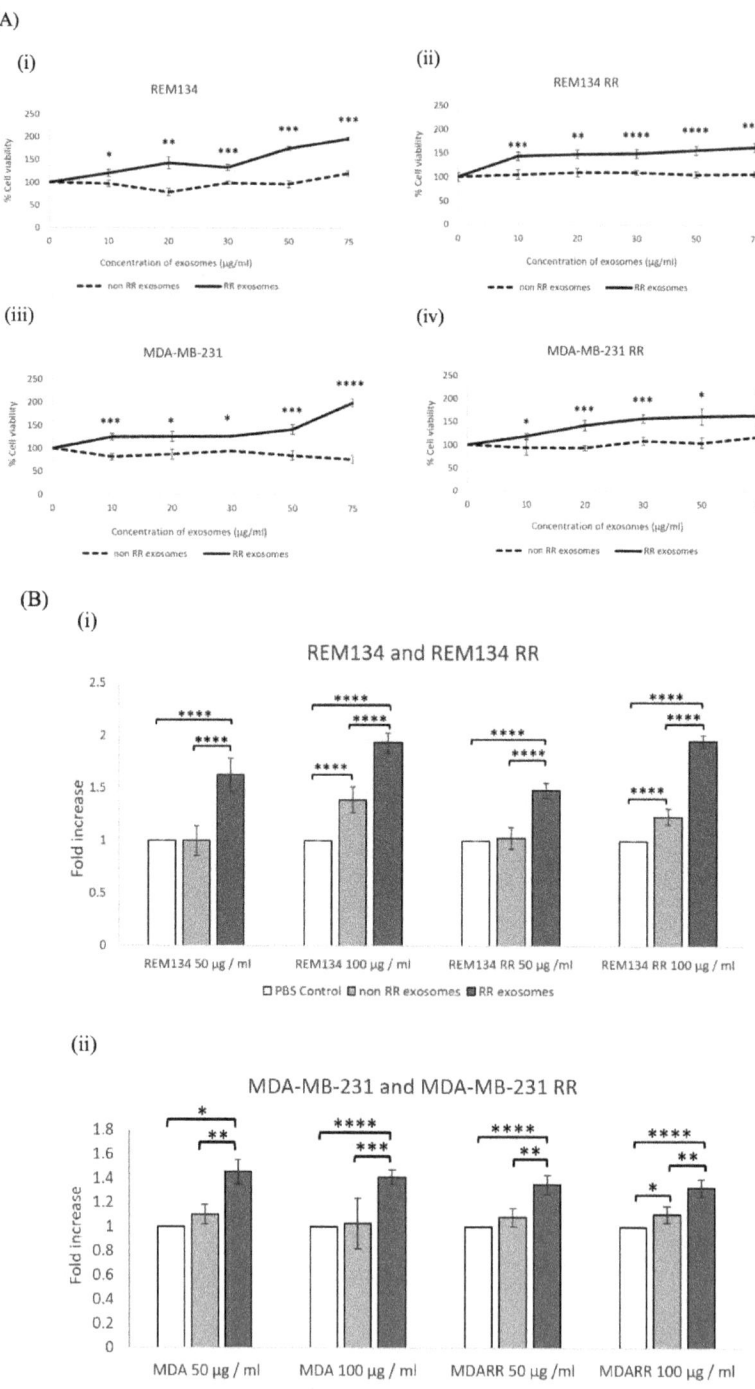

Figure 2. Exosomes isolated from RR cell lines increased the survival of recipient cells. Analysis of (**A**) cell viability and (**B**) colony-forming ability were assayed after (**i**) REM134, (**ii**) REM134 RR, (**iii**) MDA-MB-231 and (**iv**) MDA-MB-231 RR cells

were treated with the indicated dose of exosomes isolated from either corresponding non-RR or RR cells. All results are relative to the appropriate PBS control. Three repeats were performed and analysed by a two-sample t test. Error bars indicate ±SD. * $p \leq 0.05$, ** $p \leq 0.01$; *** $p \leq 0.001$, **** $p \leq 0.00001$.

3.3. Exosomes Isolated from RR Cells Enhanced the Migration Potential of Recipient Cells

To investigate the effect of exosomes derived from RR cells on the migration potential of REM134 and MDA-MB-231 cells and their RR derivatives, we utilised a 2D scratch assay. Here, cells were incubated with 50 µg/mL of exosomes for 24 h to allow for exosome uptake prior to seeding into a chamber cell with an ibidi insert. Removal of the insert created a defined wound in the cell monolayer. Closure of the wound was measured at the indicated time points until the wound was fully closed (Figure 3). The vehicle control showed that RR cells migrate inherently faster than non-RR cells: non-RR REM134 cells closed the wound 56 h after injury (Figure 3(Ai)) compared to RR REM134 cells, which closed the wound 24 h after injury (Figure 3(Bi)). Similar results, albeit less striking, were obtained for the MDA-MB-231 cell line, whereby non-RR cells closed the wound at 28 h (Figure 3(Ci)) compared to RR cells, which closed the wound at 24 h after injury (Figure 3(Cii)). Exosomes derived from both non-RR and RR cells enhanced the migration potential of recipient cells; however, this effect was more prominent in cells treated with RR exosomes. In non-RR REM134 cells treated with exosomes isolated from non-RR cells, the wound closed at 52 h compared to 48 h for those treated with exosomes derived from RR cells (Figure 3(Ci)). These results were significantly different compared to the control and between treatment groups, such as at 24 h ($p = 0.0000$) for the effect of exosomes derived from RR cells when compared to the control and exosomes derived from non-RR cells. In RR REM134 cells treated with exosomes isolated from non-RR cells, the wound closed at 12 h compared to 8 h for those treated with exosomes derived from RR cells (Figure 3(Cii)). These results were significantly different compared to the control and between treatment groups, such as at 8 h ($p = 0.0000$) for the effect of exosomes derived from RR cells when compared to the control and exosomes derived from non-RR cells. The human cell line showed similar results, in non-RR MDA-MB-231 cells treated with exosomes isolated from non-RR cells, the wound closed at 24 h compared to 12 h for those treated with exosomes derived from RR cells (Figure 3(Ciii)). These results were significantly different compared to the control and between treatment groups, for example at 8 h ($p = 0.0000$) for the effect of exosomes derived from RR cells when compared to the control and exosomes derived from non-RR cells. In RR MDA-MB-231 cells treated with exosomes isolated from non-RR cells, the wound closed at 12 h compared to 8 h for those treated with exosomes derived from RR cells (Figure 3(Ciii)). These results were significantly different compared to the control and between treatment groups such as at the time point of 8 h ($p = 0.0000$) for the effect of exosomes derived from RR cells when compared to the control and exosomes derived from non-RR cells.

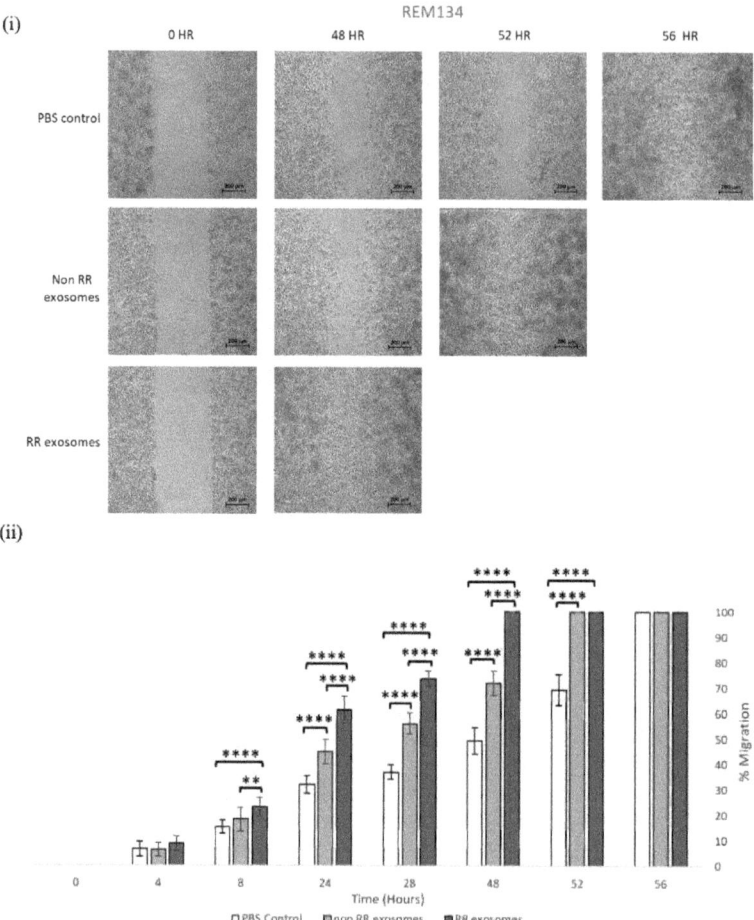

Figure 3. Cont.

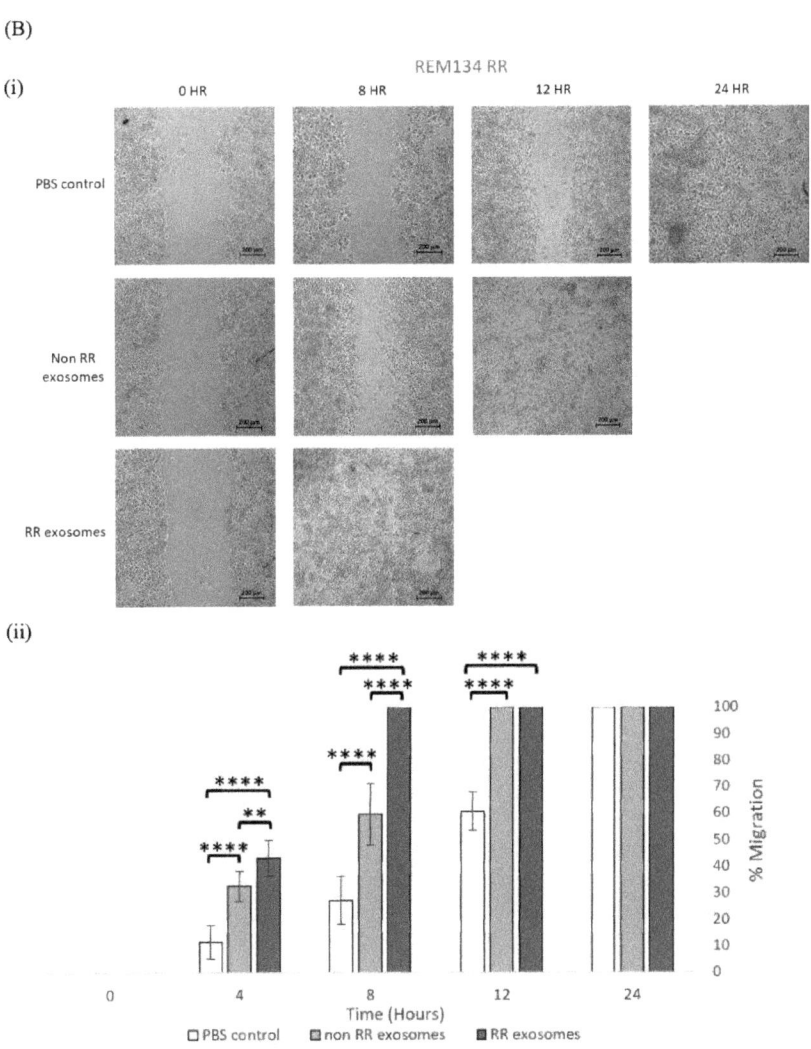

Figure 3. *Cont.*

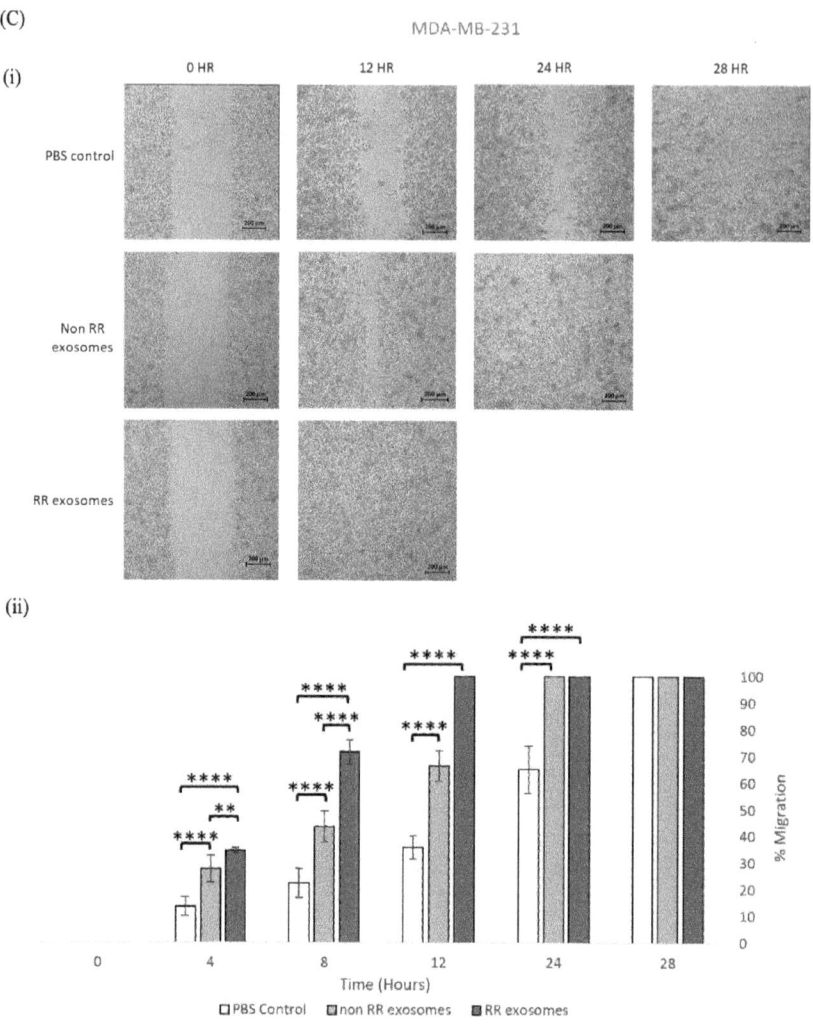

Figure 3. Cont.

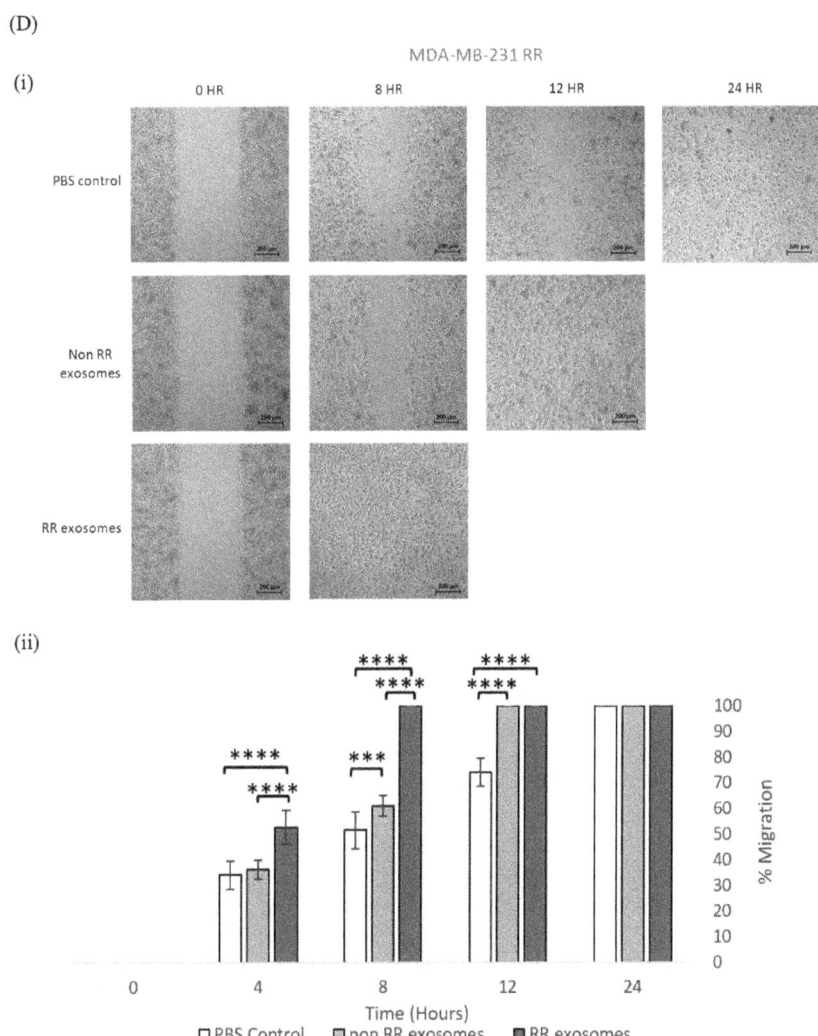

Figure 3. RR cells migrate faster than non-RR cells, and the migration potential in all cell types was enhanced after treatment with exosomes isolated from RR cell lines. Migration potential was assayed by an in vitro wound-healing assay in (**A**) REM134, (**B**) REM134 RR, (**C**) MDA-MB-231 and (**D**) MDA-MB-231 RR cells. The indicated cell line was treated with either PBS or exosomes isolated from either non-RR (50 μg/mL) or RR corresponding cells (50 μg/mL). (**i**) Light microscopy images of cell migration at the indicated time points are shown. (**ii**) Graphical representation of relative migration compared to the PBS control at the indicated time points. Three biological repeats were performed, and a two-sample t test was used for the analysis of data. Error bars indicate ±SD. ** $p \leq 0.01$; *** $p \leq 0.001$, **** $p \leq 0.00001$.

3.4. Recipient Cells of Exosomes Isolated from Estalished RR Cells Were More Resistant to Chemotherapy and Irradiation Compared to Those Treated with Exosomes from Non-RR Cells

Adjuvant chemotherapy and radiotherapy are commonly used modalities to treat breast cancer in both humans and dogs [22]. Doxorubicin is a common chemotherapeutic used in the treatment of mammary carcinomas [23,24]. To determine the effect of exosomes on the sensitivity of recipient cells to doxorubicin, cells were treated with 50 μg/mL of exosomes isolated from either RR or non-RR cells and incubated for 24 h prior to treatment with the indicated dose titration of doxorubicin. Cell viability was determined 72 h post-treatment (Figure 4A). Exosomes isolated from REM134 RR cells resulted in a significant increase in cell viability of both types of recipient cells, REM134 RR (Figure 4(Ai)) and REM134 non-RR (Figure 4(Aii)) compared to exosomes isolated from non-RR cells and PBS controls, such as at 0.001 µM ($p < 0.00001$) in both the REM134 and the REM134 RR cell line. The exosomes derived from the non-RR MDA-MB-231 cell line did not result in a significant increase in percentage cell viability when compared to the PBS control when added to the MDA-MB-231 cell line (Figure 4(Aiii)), except in the MDA-MB-231 RR cell line (Figure 4(Aiv)) at the concentration of 0.001 µM ($p < 0.01$).

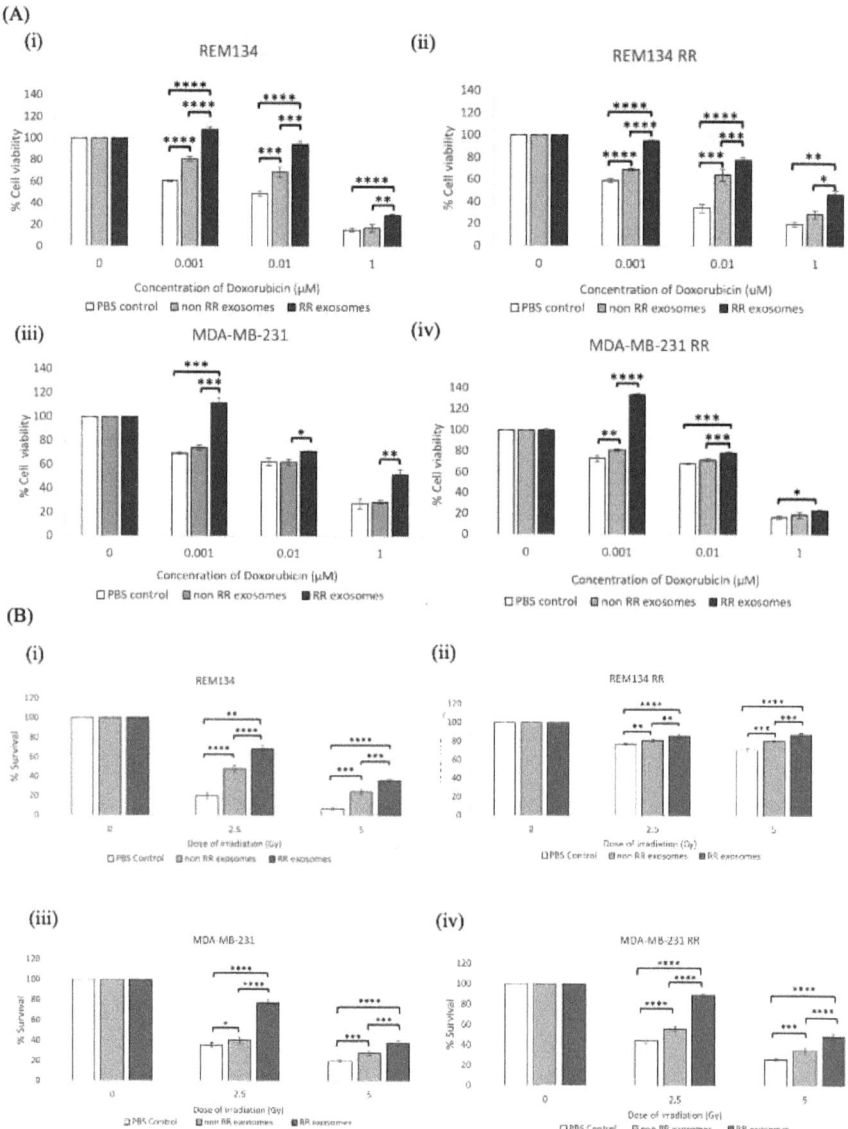

Figure 4. Exosomes isolated from RR cells increased the resistance of recipient cells to doxorubicin and ionising radiation. (**A**) Chemosensitivity to increasing doses of doxorubicin was determined for (**i**) REM134, (**ii**) REM134 RR, (**iii**) MDA-MB-231

and (**iv**) MDA-MB-231 RR cells. Cells were seeded for 24 h with exosomes (50 µg/mL) isolated from either non-RR or RR corresponding cell lines prior to treatment with the indicated dose of doxorubicin. Cell viability was assayed 72 h after doxorubicin treatment. (**B**) Colony-forming ability after treatment with 0, 2.5 or 5 Gy was determined for (**i**) REM134, (**ii**) REM134 RR, (**iii**) MDA-MB-231 and (**iv**) MDA-MB-231 RR cells. All cell lines were pretreated with exosomes (50 µg/mL) isolated from either non-RR or RR corresponding cell lines for 24 h prior to irradiation. Three repeats were performed, and significance was determined by a two-sample t test. Error bars indicate ±SD. * $p \leq 0.05$, ** $p \leq 0.01$; *** $p \leq 0.001$, **** $p \leq 0.00001$.

To assay the effect of exosomes isolated from RR cells on the resistance of recipient cells to radiotherapy, we utilised a colony formation assay to assess cell survival and clonogenic growth. Here, non-RR or RR cells were incubated with 50 µg/mL of exosomes isolated from either non-RR or RR cells prior to seeding as single cells at a low density and immediately irradiating at the indicated doses. The number of colonies were counted after 10 days. Exosomes isolated from RR cells significantly increased the colony-forming ability of recipient cells after irradiation at 2.5 and 5 Gy compared to exosomes isolated from non-RR cells or the PBS vehicle control (Figure 4B). This effect was more striking in the non-RR cells treated with exosomes isolated from RR cells in both canine (Figure 4(Bi)) and human (Figure 4(Biii)) cell lines, compared to RR cells treated with exosomes isolated from RR cell lines (Figure 4B(ii,iv)). Our results show that exosomes derived from the RR breast cancer cell lines can alter the phenotype of recipient cells and enhance their resistance to doxorubicin and irradiation.

3.5. Exosomes Isolated from RR Cells Can Alter the Phenotype of CSCs

CSCs are inherently more resistant to conventional cancer therapies than surrounding bulk (non-CSC) cancer cells. To determine the effect of exosomes isolated from RR cells on recipient CSCs, we enriched for CSCs using an established tumoursphere assay from all cell lines [28]. CSCs were pre-incubated with exosomes isolated from either RR, non-RR cells or PBS control for 24 h prior to treatment with the indicated dose titration of doxorubicin. Cell viability was assayed 72 h later. Our results show that exosomes isolated from RR cells significantly increased the percentage of cell viability for all recipient CSCs when compared to exosomes isolated from non-RR cells or the PBS vehicle control (Figure 5A). These results were consistent regardless of RR status and both in REM134 cell lines (Figure 5A(i,ii)) and in MDA-MB-231 cell lines (Figure 5A(iii,iv)). We also noted that PBS-treated RR CSCs were inherently more resistant to doxorubicin at all indicated doses than non-RR CSCs, and this was consistent in both cell lines (Figure 5A).

To investigate the effect of exosomes isolated from RR cells on recipient CSCs after radiotherapy, we assayed their colony-forming ability after irradiation. CSCs were pretreated with exosomes for 24 h prior to seeding as single cells at a low density and then immediately irradiated at 0, 2.5 and 5 Gy. The number of colonies were counted after approximately 10 days. Exosomes isolated from the REM134 RR and MDA-MB-231 RR cell lines significantly increased the number of colonies formed and, therefore, the radioresistance of all recipient CSCs compared to treatment with exosomes derived from non-RR cells or the PBS control (Figure 5B). To a much lesser extent, recipient cells treated with exosomes isolated from non-RR cell lines produced relatively more colonies after irradiation treatment compared to the PBS control. This was statistically significant in both non-RR REM134 CSCs ($p < 0.031$ at 2.5 Gy and $p < 0.003$ at 5 Gy) and RR REM134 CSCs ($p < 0.00001$) (Figure 5B(i,ii)) and for non-RR MDA-MB-231 CSCs at 2.5 Gy ($p < 0.05$) (Figure 5(Biii)). Significantly, our results show that exosomes derived from RR cells can change the radioresistance potential of recipient CSCs.

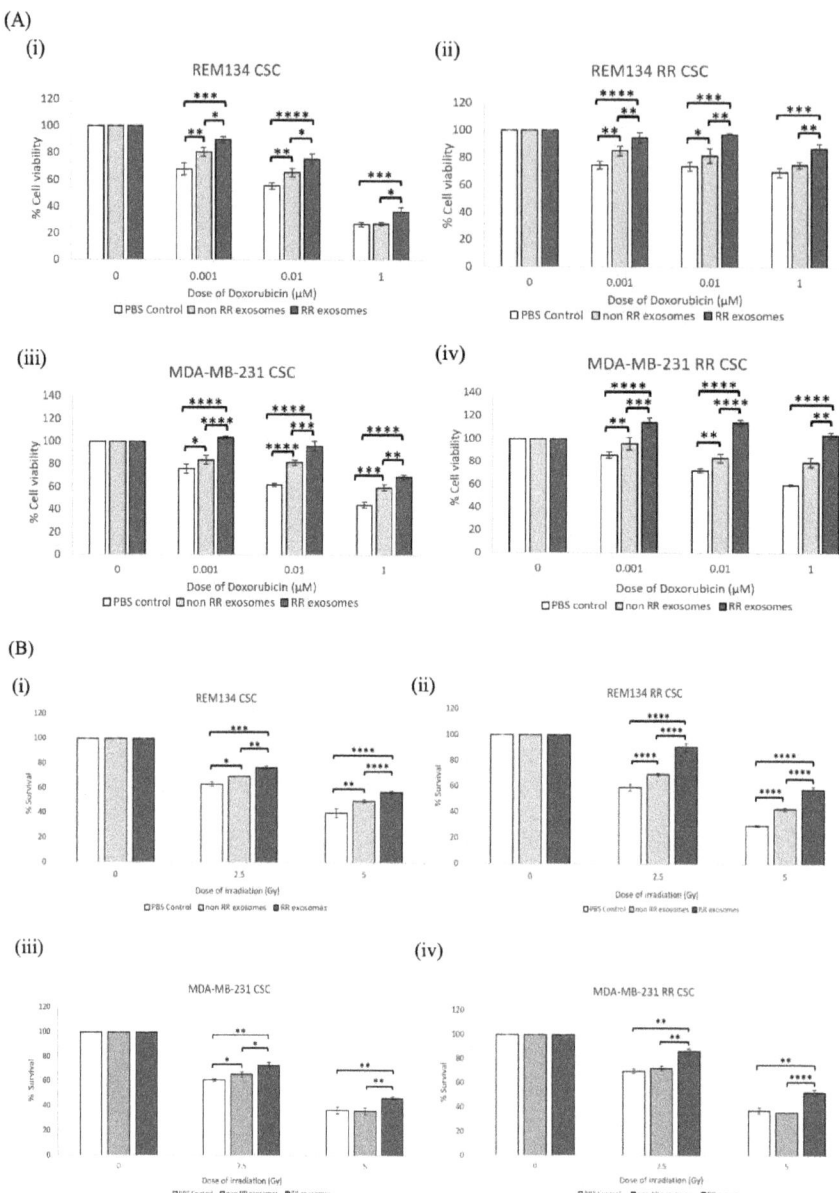

Figure 5. Exosomes isolated from RR cells can alter the resistant phenotype of CSCs. (**A**) Chemosensitivity to increasing doses of doxorubicin was determined for (**i**) REM134, (**ii**) REM134 RR, (**iii**) MDA-MB-231 and (**iv**) MDA-MB-231 RR CSCs. CSCs were pretreated for 24 h with 50 μg/mL exosomes isolated from either non-RR or RR corresponding cell lines prior to treatment with the indicated dose of doxorubicin. Cell viability was assayed 72 h after doxorubicin treatment. (**B**) Colony-forming ability after treatment with 0, 2.5 or 5 Gy was determined for (**i**) REM134, (**ii**) REM134 RR, (**iii**) MDA-MB-231 and (**iv**) MDA-MB-231 RR CSCs. All CSCs were pretreated with 50 μg/mL of exosomes isolated from either non-RR or RR corresponding cell lines for 24 h prior to irradiation. Three repeats were performed, and data were analysed by a two-sample t test. Error bars indicate ±SD. * $p \leq 0.05$, ** $p \leq 0.01$; *** $p \leq 0.001$, **** $p \leq 0.00001$.

3.6. Exosomes Derived from RR Cells Increased the Size of the CSC Pool

To observe the effect of exosomes isolated from RR cells on the tumoursphere-forming ability of recipient cells, REM134, REM134 RR, MDA-MB-231 and MDA-MB-231 RR cells were incubated with 50 µg/mL of exosomes isolated from the indicated cell lines for 24 h, cells were then seeded into low-attachment plates with N2 media to allow the formation of 3D tumourspheres. REM134 and REM134 RR tumourspheres were counted after 5 days. MDA-MB-231 and MDA-MB-231 RR tumourspheres were counted after 17 days. Our results showed that exosomes isolated from both non-RR cell lines (MDA-MB-231 and REM134) and RR cell lines (MDA-MB-231 RR and REM134 RR) significantly increased tumoursphere-forming capacity, both in the number of tumourspheres formed and in the relative size of individual tumourspheres (Figure 6A(i,iv)). Recipient cells of exosomes isolated from non-RR cells produced approximately twice as many tumourspheres compared to the PBS control. This was consistent in all cell lines (Figure 6B(i,iv)). REM134 and REM134 RR recipient cells treated with exosomes isolated from RR cells produced a 3-fold and 4.5-fold increase in tumoursphere formation compared to PBS control, respectively (Figure 6B(i,ii)). Both MDA-MB-231 and MDA-MB-231 RR recipient cells treated with exosomes isolated from RR cells produced approximately 2.5-fold increase in tumoursphere formation compared to PBS control (Figure 6B(iii,iv)). Recipient cells treated with exosomes isolated from RR cells produced significantly larger tumourspheres compared to those receiving exosomes isolated from non-RR cells or the PBS control (Figure 6C(i,iv)). Interestingly, recipient cells treated with exosomes isolated from non-RR cells produced significantly larger tumourspheres compared to the PBS control (Figure 6C(i,iv)). Together, our results indicate that exosomes derived from RR cell types can significantly increase the tumoursphere-forming ability of recipient cells and enhance the overall survival of CSCs, indicating that exosomes derived from RR cell lines may increase the size and hardiness of the CSC pool, and this may drive treatment failure in a clinical setting.

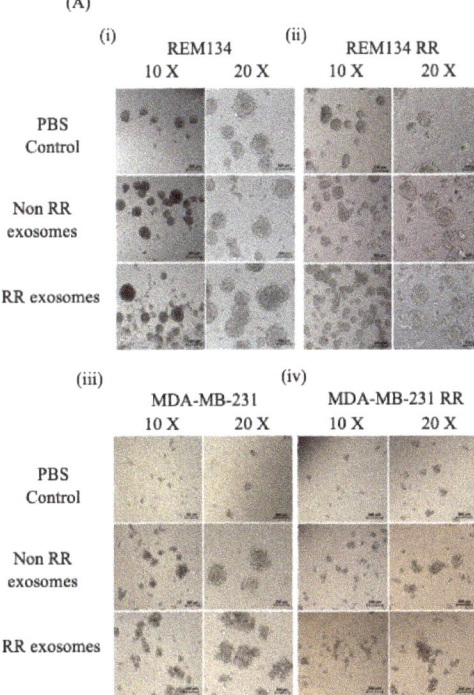

Figure 6. *Cont.*

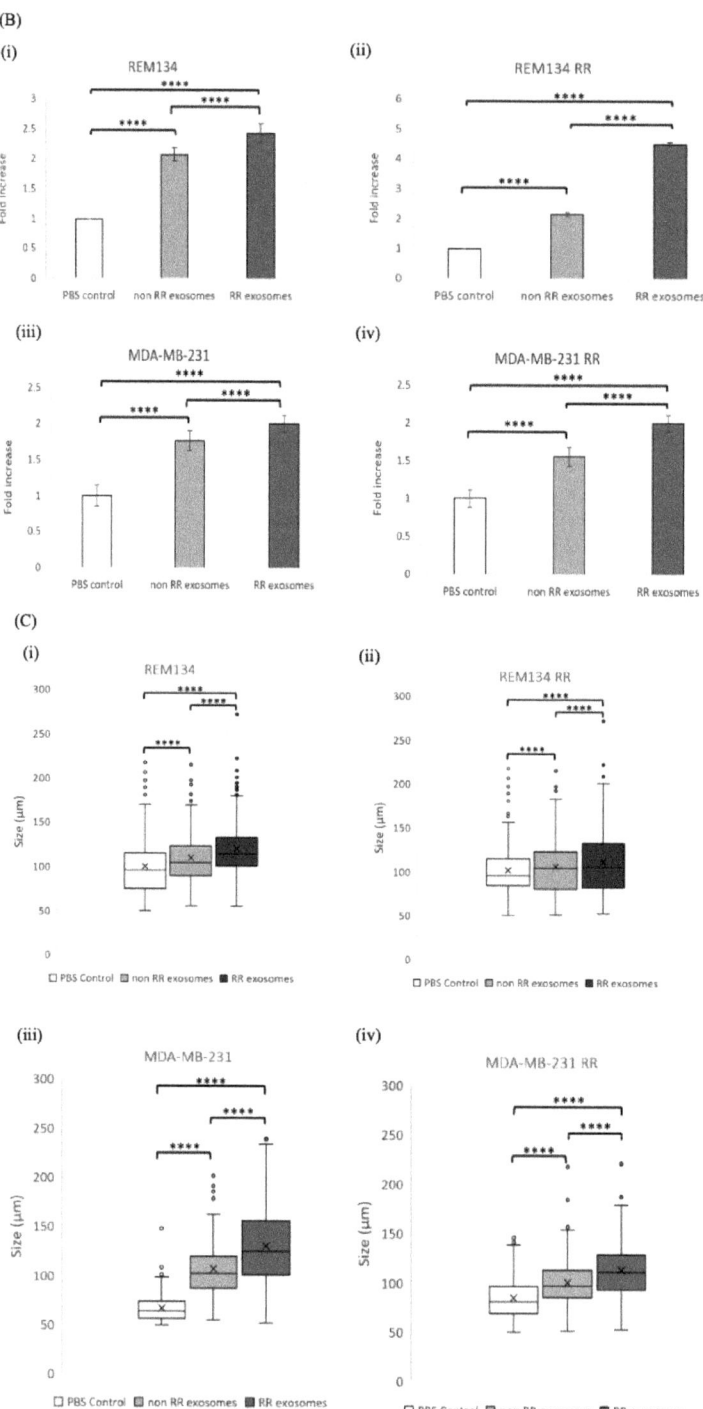

Figure 6. Exosomes isolated from RR cells enhanced sphere-forming ability. Spheres were characterised by (**A**) cell morphology, (**B**) number of spheres and (**C**) size of spheres. (**i**) REM134, (**ii**) REM134 RR, (**iii**) MDA-MB-231 and (**iv**) MDA-MB-231 RR cells

were treated with 50 µg/mL exosomes isolated from either non-RR or RR corresponding cell lines for 24 h prior to setting up the sphere assay. REM134 and REM134 RR spheres were grown for 7 days, and MDA-MB-231 and MDA-MB-231 RR spheres were grown for 17 days prior to analysis. Three repeats were performed, data were analysed by a two-sample t test and size data was analysed by a Wilcoxon signed rank test. Error bars indicate ±SD. **** $p \leq 0.00001$.

4. Discussion

Radiotherapy treatment is critical in the management of human breast cancers, with up to 94% of invasive breast cancer patients receiving radiotherapy treatment plans after surgery in conjugation with chemotherapy [29]. Despite progress made in the precision delivery of radiation and personalised radiotherapy schedules, the development of radioresistance in clinical settings is a significant clinical challenge, which ultimately leads to relapse and metastasis [27]. The tumour microenvironment plays an important role, driving tumour progression and therapeutic response. Exosomes are small extracellular vesicles, containing a large array of active biomolecules that are secreted by different cells into the extracellular matrix of the tumour microenvironment. They are then internalised by recipient cells and then release their content to mediate gene expression and protein activity [12]. Cellular stresses, including radiation and hypoxia, affect exosome secretion, composition, abundance and potential binding to recipient cells [28–31]. Previous studies have shown that radiation can enhance the release of exosomes and change their molecular composition and that exosomes are capable of transferring radiation-induced effects to non-irradiated cancer cells, therefore, potentially mediating radiation bystander effects [32,33]. Most of these reports have mainly focused on pre- and postradiation changes in exosomal proteins and miRNAs rather than on the mechanisms involved in these changes or their effect on biological functions [30,34,35]. In these studies, exosomes are usually harvested between 1 and 96 h after irradiation treatment [35]. In general, there is a lack of radioresistant model systems to facilitate elucidating the mechanisms underlying the development of radioresistance. In our lab, we previously developed and extensively characterised novel in vitro radioresistant cell lines from human breast cancer (MCF-7, ZR-751 and MDA-MB-231) and canine mammary carcinoma (REM-134) cell lines [26,36]. We found that the radioresistance phenotype was maintained long term, even in the absence of radiation exposure, and concluded that the acquisition of radioresistance was not transient [26]. In this study, we utilised these radioresistant model systems to show that exosomes derived from established RR breast cancer cell lines are capable of changing the phenotype of non-RR recipient cells and inducing radioresistance within 24 h of uptake. Our data suggest that radioresistance is transmittable via exosomes and that, once acquired and established, radioresistance could potentially spread throughout a tumour and beyond. This may be reflective of the observation that any factor affecting the phenotype of a donor cell likely affects the molecular composition of the exosome released by that cell. Our results are consistent with previous studies that investigated the functional role of exosomes in the response of exosomes to radiation exposure. These studies showed that exosomes secreted from head and neck cancer cells within 24 h of irradiation increased the proliferation, survival and migration potential of both non-irradiated and irradiated recipient cells [31,37]. Similarly, exosomes isolated from irradiated glioblastoma cells enhanced the migration phenotype of recipient cells, and molecular profiling revealed an abundance of molecules important for cell migration [38]. However, in these studies, as well as our study, conditioned media collected from irradiated cells prior to exosome isolation were not used as a positive control to confirm that exosomes can mediate this effect within the context of a more complex secretome including other extracellular vesicles.

To date, no studies have mapped changes in exosome composition through the process of acquiring radioresistance. In future studies, we aim to utilise our panel of established RR cell lines to compare the cargo of exosomes derived from RR cells and non-RR cells. Current knowledge in radiation-induced changes in exosome cargo is limited and refers mainly to proteomic changes. There are several studies showing that exosomes derived from irradiated cells can increase the levels of proteins involved in transcription and translation, chaperones, ubiquitin-related proteins and proteosome components and downregulate the proteins associated with response to stress, immunity, cell adhesion and immunity [31,35]. Future research should also focus on the minutiae of exosome uptake and processing to determine what drives the selective uptake of exosomes derived from radioresistant cells/cancer stem cell populations, as it would be beneficial to identify the fate of exosomes derived from radioresistant cells once they are internalised by recipient cells. Do all recipient cells take up exosomes equivalently? Or are subsets of cells primed to take up exosomes secreted by irradiated cells? Can we block this interaction using either small-molecule compound inhibitors or neutralising antibodies? Do all recipient cells respond the same once donor exosomes have been taken up? These are interesting questions that warrant further investigation.

The use of exosomes as a minimally invasive platform for evaluating the circulating biomarkers of a multitude of physiological and pathological processes (including cancer, pregnancy disorders, cardiovascular diseases and immune responses) is gaining traction. Exosomes exist in almost all body fluids and are very stable as they are encapsulated by lipid bilayers, this enhances the clinical applicability of exosomes. Exosomes and their cargo are also representative of parental cells and contain more biological information than cell-free DNA or conventional serum-based biomarkers. Within the context of solid cancers, although solid biopsy is still the gold standard for pathological diagnosis and basis for treatment, the use of serum-based exosomes as biomarkers of cancer has been demonstrated in gliomas [39–41], liver cancers [42,43], endometrial cancer [44] and gastrointestinal cancers [45,46]. Exosomes in urine have also been investigated for their possible use in the diagnosis and prognostication of prostate cancer [47,48]. As the production of exosomes and their composition is altered by radiation treatment, exosomes could potentially be used as non-invasive diagnostic markers for radiosensitivity and to monitor the emergence of radioresistance.

Breast cancers are highly heterogeneous and contain a small subset of CSCs. CSCs are inherently more resistant to radiation treatment that non-CSCs and more likely to survive treatment and re-initiate tumour growth [27]. Here, we show that exosomes isolated from RR breast cancer cells have similar effects on both CSCs and non-CSCs, notably conferring resistance to radiation. Interestingly, exosomes isolated from both RR and non-RR cells increased the sphere-forming ability of recipient cells, but this was enhanced by the former significantly more, indicating that exosomes isolated from RR breast cancer cells may increase the size of the CSC pool. We also showed that exosomes isolated from RR cells increased the migratory ability of recipient cells, indicating that that these exosomes activate an EMT, which is associated with cellular plasticity and the acquisition of CSC characteristics [49]. Although, we have shown that exosomes isolated from RR breast cancer cells confer a radioresistance phenotype on recipient cells and that recipient cells have enhanced sphere-forming ability, we have not unequivocally shown that the increased radioresistance is due to an increased proportion of inherently resistant CSCs. Further studies will focus on confirming whether recipient cells of exosomes isolated from RR cells activate an EMT and whether this process is the predominant underlying molecular mechanism driving emerging radiation resistance in naïve cells.

In this study, we compared human and canine breast cancer cells as canine mammary cancer is considered as an excellent translational model of human breast cancer. Naturally occurring mammary tumours are the most frequently diagnosed cancer in bitches, and these tumours represent 50% of all canine tumours, of which 50% are malignant [50]. The main treatment option for dogs is surgery alone due to a lack of receptor status evaluation or molecular subtype classification. Previously, in our lab, we compared the RR REM-134 cell line with a panel of RR human cell lines to investigate the mechanisms of acquired radioresistance and identified a number of similarities including the expression of epithelial and mesenchymal genes and WNT, PI3K and MAPK pathway activation [26]. Here, we demonstrate that exosomes isolated from human and canine RR cell lines have similar functional effects on recipient cells and that the process of potentiating exosome-mediated radioresistance is comparable in humans and dogs. We believe that a "One Health" approach is crucial to unpick tumourigenesis and to develop future treatment strategies that will benefit both species.

5. Conclusions

Our study provides compelling evidence that exosomes can serve as an effective communication tool in the development of radioresistance and can confer pro-survival signals and promote the radioresistant phenotype to non-radioresistant cells. This study indicates a functional role for exosomes within our models in the dissemination of aggressive cancer characteristics. Further studies are required to map the cargo of exosomes derived from RR cells and to identify and validate potential therapeutic targets to halt the perpetuation of acquired radioresistance throughout a tumour.

Author Contributions: Conceptualisation, L.Y.P.; methodology, L.Y.P.; software, C.P.; validation, C.P.; formal analysis, C.P.; investigation, C.P.; resources, D.J.A.; data curation, C.P. and L.Y.P.; writing—original draft preparation, L.Y.P. and C.P.; writing—review and editing, C.P., L.Y.P., M.G. and D.J.A.; supervision, L.Y.P. and D.J.A.; project administration, L.Y.P.; funding acquisition, D.J.A. All authors have read and agreed to the published version of the manuscript.

Funding: This study was funded by a University of Edinburgh scholarship. The TEM facility is supported by a Wellcome Trust Multiuser Equipment Grant (WT104915MA).

Institutional Review Board Statement: Not applicable.

Informed Consent Statement: Not applicable.

Data Availability Statement: Not applicable.

Acknowledgments: We would like to thank Rhona Muirhead for technical support.

Conflicts of Interest: The authors declare no conflict of interest.

References

1. Siegel, R.L.; Miller, K.D.; Jemal, A. Cancer statistics, 2015. *CA Cancer J. Clin.* **2015**, *65*, 5–29. [CrossRef] [PubMed]
2. Sung, H.; Ferlay, J.; Siegel, R.L.; Laversanne, M.; Soerjomataram, I.; Jemal, A.; Bray, F. Global cancer statistics 2020: Globocan estimates of incidence and mortality worldwide for 36 cancers in 185 countries. *CA Cancer J. Clin.* **2021**, *71*, 209–249. [CrossRef]
3. Gardner, H.L.; Fenger, J.M.; London, C.A. Dogs as a model for cancer. *Annu. Rev. Anim. Biosci.* **2016**, *4*, 199–222. [CrossRef]
4. Kumar, R.; Sharma, A.; Tiwari, R.K. Application of microarray in breast cancer: An overview. *J. Pharm. Bioallied Sci.* **2012**, *4*, 21–26. [CrossRef] [PubMed]
5. Pizarro, F.; Hernandez, A. Optimization of radiotherapy fractionation schedules based on radiobiological functions. *Br. J. Radiol.* **2017**, *90*, 20170400. [CrossRef]
6. Pang, L.Y.; Argyle, D.J. The evolving cancer stem cell paradigm: Implications in veterinary oncology. *Vet. J.* **2015**, *205*, 154–160. [CrossRef]
7. Phi, L.T.H.; Sari, I.N.; Yang, Y.G.; Lee, S.H.; Jun, N.; Kim, K.S.; Lee, Y.K.; Kwon, H.Y. Cancer stem cells (cscs) in drug resistance and their therapeutic implications in cancer treatment. *Stem Cells Int.* **2018**, *2018*, 5416923. [CrossRef]
8. Chen, W.X.; Liu, X.M.; Lv, M.M.; Chen, L.; Zhao, J.H.; Zhong, S.L.; Ji, M.H.; Hu, Q.; Luo, Z.; Wu, J.Z.; et al. Exosomes from drug-resistant breast cancer cells transmit chemoresistance by a horizontal transfer of micrornas. *PLoS ONE* **2014**, *9*, e95240. [CrossRef]
9. Yarana, C.; St Clair, D.K. Chemotherapy-induced tissue injury: An insight into the role of extracellular vesicles-mediated oxidative stress responses. *Antioxidants* **2017**, *6*, 75. [CrossRef] [PubMed]
10. Ni, J.; Bucci, J.; Malouf, D.; Knox, M.; Graham, P.; Li, Y. Exosomes in cancer radioresistance. *Front. Oncol.* **2019**, *9*, 869. [CrossRef]
11. Wang, X.; Zhou, Y.; Ding, K. Roles of exosomes in cancer chemotherapy resistance, progression, metastasis and immunity, and their clinical applications (review). *Int. J. Oncol.* **2021**, *59*, 44. [CrossRef] [PubMed]
12. Gurung, S.; Perocheau, D.; Touramanidou, L.; Baruteau, J. The exosome journey: From biogenesis to uptake and intracellular signalling. *Cell Commun. Signal.* **2021**, *19*, 47. [CrossRef]
13. Wei, H.; Chen, Q.; Lin, L.; Sha, C.; Li, T.; Liu, Y.; Yin, X.; Xu, Y.; Chen, L.; Gao, W.; et al. Regulation of exosome production and cargo sorting. *Int. J. Biol. Sci.* **2021**, *17*, 163–177. [CrossRef]
14. Muller, L.; Muller-Haegele, S.; Mitsuhashi, M.; Gooding, W.; Okada, H.; Whiteside, T.L. Exosomes isolated from plasma of glioma patients enrolled in a vaccination trial reflect antitumor immune activity and might predict survival. *Oncoimmunology* **2015**, *4*, e1008347. [CrossRef]

15. Hong, C.S.; Muller, L.; Whiteside, T.L.; Boyiadzis, M. Plasma exosomes as markers of therapeutic response in patients with acute myeloid leukemia. *Front. Immunol.* **2014**, *5*, 160. [CrossRef]
16. Zhao, H.; Yang, L.; Baddour, J.; Achreja, A.; Bernard, V.; Moss, T.; Marini, J.C.; Tudawe, T.; Seviour, E.G.; San Lucas, F.A.; et al. Tumor microenvironment derived exosomes pleiotropically modulate cancer cell metabolism. *eLife* **2016**, *5*, e10250. [CrossRef] [PubMed]
17. Yang, L.; Achreja, A.; Yeung, T.L.; Mangala, L.S.; Jiang, D.; Han, C.; Baddour, J.; Marini, J.C.; Ni, J.; Nakahara, R.; et al. Targeting stromal glutamine synthetase in tumors disrupts tumor microenvironment-regulated cancer cell growth. *Cell Metab.* **2016**, *24*, 685–700. [CrossRef]
18. Sousa, C.M.; Biancur, D.E.; Wang, X.; Halbrook, C.J.; Sherman, M.H.; Zhang, L.; Kremer, D.; Hwang, R.F.; Witkiewicz, A.K.; Ying, H.; et al. Pancreatic stellate cells support tumour metabolism through autophagic alanine secretion. *Nature* **2016**, *536*, 479–483. [CrossRef] [PubMed]
19. Xing, F.; Liu, Y.; Wu, S.Y.; Wu, K.; Sharma, S.; Mo, Y.Y.; Feng, J.; Sanders, S.; Jin, G.; Singh, R.; et al. Loss of xist in breast cancer activates msn-c-met and reprograms microglia via exosomal mirna to promote brain metastasis. *Cancer Res.* **2018**, *78*, 4316–4330. [CrossRef] [PubMed]
20. Hoshino, A.; Costa-Silva, B.; Shen, T.L.; Rodrigues, G.; Hashimoto, A.; Tesic Mark, M.; Molina, H.; Kohsaka, S.; Di Giannatale, A.; Ceder, S.; et al. Tumour exosome integrins determine organotropic metastasis. *Nature* **2015**, *527*, 329–335. [CrossRef]
21. Thery, C.; Amigorena, S.; Raposo, G.; Clayton, A. Isolation and characterization of exosomes from cell culture supernatants and biological fluids. *Curr. Protoc. Cell Biol.* **2006**, *3*, 3–22. [CrossRef] [PubMed]
22. Rossi, F.; Sabattini, S.; Vascellari, M.; Marconato, L. The impact of toceranib, piroxicam and thalidomide with or without hypofractionated radiation therapy on clinical outcome in dogs with inflammatory mammary carcinoma. *Vet. Comp. Oncol.* **2018**, *16*, 497–504. [CrossRef]
23. Zambrano-Estrada, X.; Landaverde-Quiroz, B.; Duenas-Bocanegra, A.A.; De Paz-Campos, M.A.; Hernandez-Alberto, G.; Solorio-Perusquia, B.; Trejo-Mandujano, M.; Perez-Guerrero, L.; Delgado-Gonzalez, E.; Anguiano, B.; et al. Molecular iodine/doxorubicin neoadjuvant treatment impair invasive capacity and attenuate side effect in canine mammary cancer. *BMC Vet. Res.* **2018**, *14*, 87. [CrossRef]
24. Zhao, Y.; Alakhova, D.Y.; Zhao, X.S.; Band, V.; Batrakova, E.V.; Kabanov, A.V. Eradication of cancer stem cells in triple negative breast cancer using doxorubicin/pluronic polymeric micelles. *Nanomed-Nanotechnology* **2020**, *24*, 102124. [CrossRef] [PubMed]
25. Else, R.W.; Norval, M.; Neill, W.A. The characteristics of a canine mammary-carcinoma cell-line, rem-134. *Brit. J. Cancer* **1982**, *46*, 675–681. [CrossRef] [PubMed]
26. Gray, M.; Turnbull, A.K.; Ward, C.; Meehan, J.; Martinez-Perez, C.; Bonello, M.; Pang, L.Y.; Langdon, S.P.; Kunkler, I.H.; Murray, A.; et al. Development and characterisation of acquired radioresistant breast cancer cell lines. *Radiat. Oncol.* **2019**, *14*, 64. [CrossRef] [PubMed]
27. Liu, Y.F.; Yang, M.; Luo, J.J.; Zhou, H.M. Radiotherapy targeting cancer stem cells "awakens" them to induce tumour relapse and metastasis in oral cancer. *Int. J. Oral Sci.* **2020**, *12*, 19. [CrossRef] [PubMed]
28. Augimeri, G.; La Camera, G.; Gelsomino, L.; Giordano, C.; Panza, S.; Sisci, D.; Morelli, C.; Gyorffy, B.; Bonofiglio, D.; Ando, S.; et al. Evidence for enhanced exosome production in aromatase inhibitor-resistant breast cancer cells. *Int. J. Mol. Sci.* **2020**, *21*, 5841. [CrossRef] [PubMed]
29. Hazawa, M.; Tomiyama, K.; Saotome-Nakamura, A.; Obara, C.; Yasuda, T.; Gotoh, T.; Tanaka, I.; Yakumaru, H.; Ishihara, H.; Tajima, K. Radiation increases the cellular uptake of exosomes through cd29/cd81 complex formation. *Biochem. Bioph. Res. Commun.* **2014**, *446*, 1165–1171. [CrossRef] [PubMed]
30. Abramowicz, A.; Wojakowska, A.; Marczak, L.; Lysek-Gladysinska, M.; Smolarz, M.; Story, M.D.; Polanska, J.; Widlak, P.; Pietrowska, M. Ionizing radiation affects the composition of the proteome of extracellular vesicles released by head-and-neck cancer cells in vitro. *J. Radiat. Res.* **2019**, *60*, 289–297. [CrossRef]
31. Mutschelknaus, L.; Azimzadeh, O.; Heider, T.; Winkler, K.; Vetter, M.; Kell, R.; Tapio, S.; Merl-Pham, J.; Huber, S.M.; Edalat, L.; et al. Radiation alters the cargo of exosomes released from squamous head and neck cancer cells to promote migration of recipient cells. *Sci. Rep.* **2017**, *7*, 12423. [CrossRef]
32. Du, Y.; Du, S.F.; Liu, L.; Gan, F.H.; Jiang, X.G.; Wangrao, K.J.; Lyu, P.; Gong, P.; Yao, Y. Radiation-induced bystander effect can be transmitted through exosomes using mirnas as effector molecules. *Radiat. Res.* **2020**, *194*, 89–100. [CrossRef] [PubMed]
33. Cagatay, S.T.; Mayah, A.; Mancuso, M.; Giardullo, P.; Pazzaglia, S.; Saran, A.; Daniel, A.; Traynor, D.; Meade, A.D.; Lyng, F.; et al. Phenotypic and functional characteristics of exosomes derived from irradiated mouse organs and their role in the mechanisms driving non-targeted effects. *Int. J. Mol. Sci.* **2020**, *21*, 8389. [CrossRef] [PubMed]
34. Widlak, P.; Pietrowska, M.; Polanska, J.; Rutkowski, T.; Jelonek, K.; Kalinowska-Herok, M.; Gdowicz-Klosok, A.; Wygoda, A.; Tarnawski, R.; Skladowski, K. Radiotherapy-related changes in serum proteome patterns of head and neck cancer patients; the effect of low and medium doses of radiation delivered to large volumes of normal tissue. *J. Transl. Med.* **2013**, *11*, 299. [CrossRef] [PubMed]
35. Jelonek, K.; Wojakowska, A.; Marczak, L.; Muer, A.; Tinhofer-Keilholz, I.; Lysek-Gladysinska, M.; Widlak, P.; Pietrowska, M. Ionizing radiation affects protein composition of exosomes secreted in vitro from head and neck squamous cell carcinoma. *Acta Biochim. Pol.* **2015**, *62*, 265–272. [CrossRef]

36. Gray, M.; Turnbull, A.K.; Meehan, J.; Martinez-Perez, C.; Kay, C.; Pang, L.Y.; Argyle, D.J. Comparative analysis of the development of acquired radioresistance in canine and human mammary cancer cell lines. *Front. Vet. Sci.* **2020**, *7*, 439. [CrossRef] [PubMed]
37. Mutschelknaus, L.; Peters, C.; Winkler, K.; Yentrapalli, R.; Heider, T.; Atkinson, M.J.; Moertl, S. Exosomes derived from squamous head and neck cancer promote cell survival after ionizing radiation. *PLoS ONE* **2016**, *11*, e0152213. [CrossRef] [PubMed]
38. Arscott, W.T.; Tandle, A.T.; Zhao, S.; Shabason, J.E.; Gordon, I.K.; Schlaff, C.D.; Zhang, G.; Tofilon, P.J.; Camphausen, K.A. Ionizing radiation and glioblastoma exosomes: Implications in tumor biology and cell migration. *Transl. Oncol.* **2013**, *6*, 638–648. [CrossRef]
39. Ebrahimkhani, S.; Vafaee, F.; Hallal, S.; Wei, H.; Lee, M.Y.T.; Young, P.E.; Satgunaseelan, L.; Beadnall, H.; Barnett, M.H.; Shivalingam, B.; et al. Deep sequencing of circulating exosomal microrna allows non-invasive glioblastoma diagnosis. *NPJ Precis. Oncol.* **2018**, *2*, 28. [CrossRef] [PubMed]
40. Lan, F.M.; Qing, Q.; Pan, Q.; Hu, M.; Yu, H.M.; Yue, X. Serum exosomal mir-301a as a potential diagnostic and prognostic biomarker for human glioma. *Cell Oncol.* **2018**, *41*, 25–33. [CrossRef]
41. Santangelo, A.; Imbruce, P.; Gardenghi, B.; Belli, L.; Agushi, R.; Tamanini, A.; Munari, S.; Bossi, A.M.; Scambi, I.; Benati, D.; et al. A microrna signature from serum exosomes of patients with glioma as complementary diagnostic biomarker. *J. Neuro-Oncol.* **2018**, *136*, 51–62. [CrossRef] [PubMed]
42. Sun, L.; Liu, X.; Pan, B.; Hu, X.; Zhu, Y.; Su, Y.; Guo, Z.; Zhang, G.; Xu, M.; Xu, X.; et al. Serum exosomal mir-122 as a potential diagnostic and prognostic biomarker of colorectal cancer with liver metastasis. *J. Cancer* **2020**, *11*, 630–637. [CrossRef] [PubMed]
43. Cui, Y.; Xu, H.F.; Liu, M.Y.; Xu, Y.J.; He, J.C.; Zhou, Y.; Cang, S.D. Mechanism of exosomal microrna-224 in development of hepatocellular carcinoma and its diagnostic and prognostic value. *World J. Gastroenterol.* **2019**, *25*, 1890–1898. [CrossRef] [PubMed]
44. Fan, X.; Zou, X.; Liu, C.; Cheng, W.; Zhang, S.; Geng, X.; Zhu, W. Microrna expression profile in serum reveals novel diagnostic biomarkers for endometrial cancer. *Biosci. Rep.* **2021**, *41*, 41. [CrossRef]
45. Shi, Y.T.; Wang, Z.H.; Zhu, X.J.; Chen, L.; Ma, Y.L.; Wang, J.Y.; Yang, X.Z.; Liu, Z. Exosomal mir-1246 in serum as a potential biomarker for early diagnosis of gastric cancer. *Int. J. Clin. Oncol.* **2020**, *25*, 89–99. [CrossRef] [PubMed]
46. Tang, S.L.; Cheng, J.N.; Yao, Y.F.; Lou, C.J.; Wang, L.; Huang, X.Y.; Zhang, Y.Q. Combination of four serum exosomal mirnas as novel diagnostic biomarkers for early-stage gastric cancer. *Front. Genet.* **2020**, *11*, 237. [CrossRef] [PubMed]
47. Li, Z.; Li, L.X.; Diao, Y.J.; Wang, J.; Ye, Y.; Hao, X.K. Identification of urinary exosomal mirnas for the non-invasive diagnosis of prostate cancer. *Cancer Manag. Res.* **2021**, *13*, 25–35. [CrossRef] [PubMed]
48. Shin, S.; Park, Y.H.; Jung, S.H.; Jang, S.H.; Kim, M.Y.; Lee, J.Y.; Chung, Y.J. Urinary exosome microrna signatures as a noninvasive prognostic biomarker for prostate cancer. *Npj Genom. Med.* **2021**, *6*, 45. [CrossRef] [PubMed]
49. Weidenfeld, K.; Schif-Zuck, S.; Abu-Tayeh, H.; Kang, K.; Kessler, O.; Weissmann, M.; Neufeld, G.; Barkan, D. Dormant tumor cells expressing loxl2 acquire a stem-like phenotype mediating their transition to proliferative growth. *Oncotarget* **2016**, *7*, 71362–71377. [CrossRef]
50. Salas, Y.; Marquez, A.; Diaz, D.; Romero, L. Epidemiological study of mammary tumors in female dogs diagnosed during the period 2002-2012: A growing animal health problem. *PLoS ONE* **2015**, *10*, e0127381. [CrossRef] [PubMed]

Article

Associations between the Levels of Estradiol-, Progesterone-, and Testosterone-Sensitive MiRNAs and Main Clinicopathologic Features of Breast Cancer

Tatiana Kalinina [1,*], Vladislav Kononchuk [1,2], Efim Alekseenok [1], Grigory Abdullin [3], Sergey Sidorov [4], Vladimir Ovchinnikov [5] and Lyudmila Gulyaeva [1,6]

[1] Federal Research Center of Fundamental and Translational Medicine, Timakova Str. 2/12, 630117 Novosibirsk, Russia; kononchuk@niimbb.ru (V.K.); alekseenok@niimbb.ru (E.A.); gulyaeva@niimbb.ru (L.G.)
[2] Meshalkin National Medical Research Center, Ministry of Health of the Russian Federation, Rechkunovskaya Str. 15, 630055 Novosibirsk, Russia
[3] Novosibirsk Regional Oncological Dispensary, Plakhotnogo Str., 2, 630000 Novosibirsk, Russia; grabdullin@mail.ru
[4] Department of Breast Pathology, Novosibirsk Municipal Publicly Funded Healthcare Institution Municipal Clinical Hospital #1, Zalessky Str. 6, 630047 Novosibirsk, Russia; svsidorov@yandex.ru
[5] Institute of Cytology and Genetics of Siberian Branch of the Russian Academy of Sciences, Prospekt Lavrentyeva 10, 630090 Novosibirsk, Russia; tdkaliki@gmail.com
[6] Institute for Medicine and Psychology, Novosibirsk State University, Pirogova Str. 2, 630090 Novosibirsk, Russia
* Correspondence: kalinina@niimbb.ru

Abstract: Despite the existing advances in the diagnosis and treatment of breast cancer (BC), the search for markers associated with the clinicopathological features of BC is still in demand. MiRNAs (miRs) have potential as markers, since a change in the miRNA expression profile accompanies the initiation and progression of malignant diseases. The receptors for estrogen, androgen, and progesterone (ER, AR, and PR) play an important role in breast carcinogenesis. Therefore, to search for miRNAs that may function as markers in BC, using bioinformatic analysis and the literature data, we selected 13 miRNAs whose promoter regions contain binding sites for ER or AR, or putative binding sites for ER, AR, and PR. We quantified their expression in MCF-7 cells treated with estradiol, progesterone, or testosterone. The levels of miRNAs sensitive to one or more of these hormones were quantified in BC samples ($n = 196$). We discovered that high expression levels of miR-190b in breast tumor tissue indicate a positive ER status, and miR-423 and miR-200b levels differ between patients with and without HER2 amplification. The miR-193b, -423, -190a, -324, and -200b levels were associated with tumor size or lymph node status in BC patients, but the presence of these associations depended on the status and expression level of ER, PR, HER2, and Ki-67. We also found that miR-21 expression depends on HER2 expression in ER- and/or PR-positive BC. The levels of miRNA were significantly different between HER2 0 and HER2 1+ tumors ($p = 0.027$), and between HER2 0 and HER2 2+, 3+ tumors ($p = 0.005$).

Keywords: microRNA; breast cancer; biomarker; lymph node metastasis; hormone-dependent carcinogenesis

1. Introduction

The choice of treatment for breast cancer (BC) depends on the expression level of the estrogen and progesterone receptors (ER and PR), HER2 (a receptor for epidermal growth factor), and Ki-67. The levels of these proteins in tumor tissues are determined by immunohistochemistry (IHC). However, IHC has several disadvantages, such as type and duration of tissue fixation, the choice of antibody, and the experience of the pathologist, all of which affect the reproducibility of the results [1]. Axillary lymph node (ALN) status is another important factor in the diagnosis of BC that predicts disease-free survival and overall survival. Conventional methods for diagnosing lymph node metastases (LNM) at the preoperative stage, such as ultrasound, mammography, and magnetic resonance imaging (MRI), have relatively low accuracy and sensitivity [2]. In recent years, sentinel lymph node biopsy (SLNB) has been used to diagnose lymph node status. However, not all medical facilities have the ability to conduct SLNB. In addition, the false-negative rate of the procedure is estimated between 4.6% and 16.7%, which also restricts the popularity of this method [3]. Therefore, in many regions of Russia, axillary lymph node dissection (ALND) is performed to diagnose metastatic lesions of the lymph nodes. Both SLNB and especially ALND can lead to a number of complications. Thus, despite the existing advances in the diagnosis and treatment of BC, the search for new markers associated with the clinicopathological features of BC is still relevant. MiRNAs (miRs) have great potential as markers, since a change in the miRNA expression profile accompanies the initiation and progression of malignant diseases.

ER, PR, and androgen receptor (AR) play important roles in breast carcinogenesis, targeting the regulators of cell cycle, signaling, differentiation, and apoptosis [4–6]. MiRNAs can also be targets of these receptors [7,8]. Therefore, to search for miRNA-markers of breast carcinogenesis using bioinformatics analysis and the literature data, we selected some miRNAs potentially regulated by ER, AR, or PR. We examined their levels in MCF-7 cells treated with estradiol, progesterone, and testosterone. The levels of those miRNAs that significantly responded to hormone treatment were analyzed in BC samples (n = 196).

2. Materials and Methods

2.1. Cell Culture

MCF-7 cells were obtained from the Russian Cell Culture Collection (St. Petersburg (Russia) branch of the ETCS). Cells were cultivated in Iscove's modified Dulbecco's medium (IMDM; Gibco BRL Co., Gaithersburg, MD, USA) with 10% of FBS (Gibco BRL Co., Gaithersburg, MD, USA), 2 mM L-alanyl-L-glutamine (Gibco BRL Co., Gaithersburg, MD, USA), 250 mg/mL amphotericin B, and 100 U/mL penicillin/streptomycin (Gibco BRL Co., Gaithersburg, MD, USA). The cells were maintained in a 5% CO_2 incubator at 37 °C. The medium was refreshed every 2–3 days, and the cells were passaged when 65–80% confluent. The absence of mycoplasma contamination was verified by conventional PCR assays. At 48 h prior to the addition of hormones, the culture medium was changed to phenol red-free IMDM (Gibco BRL Co., Gaithersburg, MD, USA) to eliminate the weak estrogen-agonistic activity of phenol red. Estradiol, progesterone, and testosterone (Sigma-Aldrich, St. Louis, MO, USA) were dissolved in dimethyl sulfoxide (DMSO; Sigma-Aldrich, USA), and then the solutions were diluted with the culture medium so that the final DMSO concentration was 0.1% (v/v). Cells treated with 0.1% DMSO were used as a control. The cells were treated for 6, 24, or 48 h.

2.2. Tissue Samples

A total of 196 pairs of BC tissue samples and samples of normal adjacent tissue from female patients who had not received preoperative pharmacotherapy, were collected between 2017 and 2020 at Novosibirsk municipal publicly-funded healthcare institution Municipal Clinical Hospital #1 and Novosibirsk Regional Oncological Dispensary. Tissue samples were placed in an RNAlater™ Stabilization Solution (Invitrogen™, Carlsbad, CA, USA) and kept at −20 °C until experiments were performed. Clinicopathologic information was obtained by reviewing medical records and reports on results of immunohistochemical assays. The following variables were determined: the T stage, N stage; IHC scores on ER, PR, HER2, and Ki-67 (Table 1). For cases with HER2 IHC-score 2+, the determination of the final HER2 status was carried out using FISH. All patients recruited into the study had grade 2 (G2) tumors. Breast cancer subtypes were categorized according to the St. Gallen Expert Consensus as follows [9]: luminal A (ER^+ and/or PR^+, $HER2^-$, and Ki-67 < 14%), luminal B HER2-negative (ER^+ and/or PR^+, $HER2^-$, and Ki-67 $\geq$ 14%) luminal B HER2-positive

(ER$^+$ and/or PR$^+$, HER2$^-$), HER2-positive (ER$^-$, PR$^-$, and HER2$^+$), and triple-negative (ER$^-$, PR$^-$, and HER2$^-$).

Table 1. Characteristics of the breast tumors under study.

Characteristics		ER- and/or PR-Positive (*n* = 156)	ER- and PR-Negative (*n* = 40)
Age (mean and range, year)		61 (27–90)	55 (38–76)
T stage	T1	71	18
	T2	81	20
	T3	2	1
	T4	2	1
N stage	N0	103	27
	N1	37	7
	N2	11	6
	N3	5	-
ER score	0–2	3	40
	3–5	7	-
	6–8	146	-
PR score	0–2	25	40
	3–5	34	-
	6–8	97	-
HER2 score	0	72	18
	1	47	5
	2–3	37	17

Estrogen and progesterone receptors (ER and PR). ER and PR were graded by the Allred scoring method [10].

2.3. MicroRNA Isolation

Total miRNA was extracted from human tissue by a previously published protocol [11].

2.4. MiRNA Reverse Transcription and Real-Time PCR (RT-PCR)

Relative expression levels for miRNAs were measured by real-time reverse transcription-PCR. A reverse-transcription reaction was carried out using stem-loop primers [12] and the RT-M-MuLV-RH kit (Biolabmix Ltd., Novosibirsk, Russia). Real-time PCR was performed with TaqMan probes and the BioMaster UDG HS-qPCR (2×) kit (Biolabmix Ltd.). To detect PCR products, a CFX96™ Detection System (Bio-Rad Laboratories, Hercules, CA, USA) was applied. Small nuclear RNAs U44 and U48 were used to normalize the data.

Primers for the reverse transcription were as follows: hsa-miR-190a-5p, 5'-GTCGTATCCAGTGCAGGGTCCGAGGTATTCGCACTGGATACGACACCTAATA-3'; hsa-miR-190b-5p, 5'-GTCGTATCCAGTGCAGGGTCCGAGGTATTCGCACTGGATACGACAACCCAA-3'; hsa-miR-23a-3p, 5'- GTCGTATCCAGTGCAGGGTCCGAGGTATTCGCACTGGATACGACGGAAATC -3'; hsa-miR-27a-3p, 5'-GTCGTATCCAGTGCAGGGTCCGAGGTATTCGCACTGGATACGACTGCTCACA -3'; hsa-miR-193b-3p, 5'-GTCGTATCCAGTGCAGGGTCCGAGGTATTCGCACTGGATACGACAGCGGGAC-3'; hsa-miR-324-5p, 5'-GTCGTATCCAGTGCAGGGTCCGAGGTATTCGCACTGGATACGACACACCAAT-3'; hsa-miR-423-3p, 5'-GTCGTATCCAGTGCAGGGTCCGAGGTATTCGCACTGGATACGACACTGAGGG-3'; hsa-miR-200b-3p, 5'-GTCGTATCCAGTGCAGGGTCCGAGGTATTCGCACTGGATACGACGTCATCAT-3'; hsa-miR-21-5p, 5'-GTCGTATCCAGTGCAGGGTCCGAGGTATTCGCACTGGATACGACTCAACATC-3'; hsa-miR-126-3p, 5'- GTCGTATCCAGTGCAGGGTCCGAGGTATTCGCACTGGATACGACCGCATTAT -3'; hsa-miR-378a-3p, 5'- GTCGTATCCAGTGCAGGGTCCGAGGTATTCGCACTGGATACGACGCCTTCT -3'; hsa-miR-149-5p, 5'- GTCGTATCCAGTGCAGGGTCCGAGGTATTCGCACTGGATACGACGGGAGTGA -3'; hsa-miR-342-3p, 5'- GTCGTATCCAGTGCAGGGTCCGAGGTATTCGCACTGGATACGACACGGGTG -3';

U44, 5′-GTCGTATCCAGTGCAGGGTCCGAGGTATTCGCACTGGATACGACAGTCAGTT-3′;
U48, 5′-GTCGTATCCAGTGCAGGGTCCGAGGTATTCGCACTGGATACGAGACGGTCAG-3′.

The following specific oligonucleotides were employed for RT-PCR: hsa-miR-190a-5p, (forward primer) 5′-GCCGCTGATATGTTTGATA-3′, (probe) 5′-(R6G)-TTCGCACTGGATACGACACCTAATA-(BHQ1)-3′; hsa-miR-190b-5p, (forward primer) 5′-GCCGCTGATATGTTTGATA-3′, (probe) 5′-(R6G)-TTCGCACTGGATACGACAACCCAA-(BHQ1)-3′; hsa-miR-23a-3p, (forward primer) 5′-GCCGCATCACATTGCCAGG-3′, (probe) 5′-(R6G)-TTCGCACTGGATACGACGGAAATC-(BHQ1)-3′; hsa-miR-27a-3p, (forward primer) 5′-GCCGCTTCACAGTGGCTAA-3′, (probe) 5′-(R6G)-TTCGCACTGGATACGACGCGGAAC-(BHQ1)-3′; hsa-miR-193b-3p, (forward primer) 5′-GCCGCAACTGGCCCTCAAA-3′, (probe) 5′-(R6G)-TTCGCACTGGATACGACAGCGGGAC-(BHQ1)-3′; hsa-miR-324-5p, (forward primer) 5′-CCCGCATCCCCTAGGGC-3′, (probe) 5′-(R6G)-TTCGCACTGGATACGACACACCAAT-(BHQ1)-3′; hsa-miR-423-3p, (forward primer) 5′-GCCGAGCTCGGTCTGAGGC-3′, (probe) 5′-(R6G)-TTCGCACTGGATACGACACTGAGG-(BHQ1)-3′; hsa-miR-200b-3p, (forward primer) 5′-GCCGCTAATACTGCCTGGTA-3′, (probe) 5′-(R6G)-TTCGCACTGGATACGACGTCATCAT-(BHQ1)-3′; hsa-miR-21-5p, (forward primer) 5′-GCCGCTAGCTTATCAGACT-3′, (probe) 5′-(R6G)-TTCGCACTGGATACGACTCAACATC-(BHQ1)-3′; hsa-miR-126-3p, (forward primer) 5′-GCCGCTCGTACCGTGAGTA-3′, (probe) 5′-(R6G)-TTCGCACTGGATACGACCGCATTAT-(BHQ1)-3′; hsa-miR-378a-3p, (forward primer) 5′-GCCGCACTGGACTTGGAGTC-3′, (probe) 5′-(R6G)-TTCGCACTGGATACGACGCCTTCT-(BHQ1)-3′; hsa-miR-149-5p, (forward primer) 5′-GCCGTCTGGCTCCGTGTCT-3′, (probe) 5′-(R6G)-TTCGCACTGGATACGACGGGAGTGA-(BHQ1)-3′; hsa-miR-342-3p, (forward primer) 5′-GCCGCTCTCACACAGAAATCG-3′, (probe) 5′-(R6G)-TTCGCACTGGATACGACACGGGTGC-(BHQ1)-3′; U44, (forward primer) 5′-GCCGCTCTTAATTAGCTCT-3′, (probe) 5′-(R6G)-TTCGCACTGGATACGACAGTCAGTT-(BHQ1)-3′; U48, (forward primer) 5′-CCCTGAGTGTGTCGCTGATG-3′, (probe) 5′-(R6G)-TTCGCACTGGATACGAGACGGTCAG-(BHQ1)-3′. A similar type of reverse primer targeting the stem-loop region in the synthesized cDNAs was 5′-AGTGCAGGGTCCGAGGTA-3′. Each sample was analyzed in triplicate. Relative expression level was assessed based on threshold cycle (Ct) values taking into account PCR efficacy (E) for both the analyzed and reference RNAs.

2.5. Bioinformatics Analysis

The list of miRNAs containing ER binding sites in their promoter regions was previously published [13]. To search for miRNAs potentially regulated by AR or PR receptors, putative miRNA promoter regions were extracted from the human genome (hg38) 10,000 nucleotides upstream from the start of a precursor miRNA sequence according to MirGeneDB [14]. AR and PR binding sites were searched in these regions using position weight matrices (MA0007.2, MA0113.3) from Jaspar (http://jaspar.genereg.net/, accessed on 12 December 2017) [15] (sequences of binding sites for these receptors are the same) using Biostrings (R Bioconductor package) [16]. We additionally performed a search for binding sites in the promoter regions of rat miRNAs (using MA0007.1 and MA0113.1). The putative promoter regions were extracted from rat genome (Rnor_6.0), 10,000 nucleotides upstream from the start of a precursor miRNA sequence according to miRBase v21 [17]. For further research, miRNAs were selected that have high expression in breast tissues according to the Human miRNA tissue atlas (https://ccb-web.cs.uni-saarland.de/tissueatlas/, accessed on 17 October 2021) [18].

2.6. Statistical Analysis

STATISTICA software (version 12; TIBCO Software Inc., Palo Alto, CA, USA) was used for statistical data analysis and plotting. Data are presented as median values. The Shapiro–Wilk test was used to check data normality. Since the distribution was not normal in some groups, the statistical analysis was carried out using the non-parametric Mann–Whitney U test. Data with $p < 0.05$ were regarded as statistically significant.

3. Results

3.1. Selection of Estradiol-, Progesterone-, Testosterone-Sensitive miRNAs

The list of microRNAs potentially regulated by ER was published earlier [13]. MiRNAs, whose promoter regions contain sequences corresponding to the AR and PR binding sites, were searched using Biostrings. For the study, we selected miRNAs with high expression in breast tissues. We were interested in miRNAs containing ER binding sites in promoter shared by the three species (human, rat, and mouse), and AR/PR binding sites in promoter in humans and rats (Table 2). Interest in such miRNAs is due, firstly, to the possibility of further studies of the regulation of their expression in vivo, and secondly, because the regulation of miRNA by ER, PR, and AR in different species indicates an essential role of such miRNAs in the signaling pathways of receptors. Thus, we chose miR-21, miR-190b, miR-200b, miR-23a, miR-27a, miR-342, miR-190a, miR-378a, miR-324, miR-423, and miR-149 for analysis. We also took miR-193b and miR-126 into study, since their targets are ER and PR.

Table 2. The miRNAs potentially regulated by ER, PR, and AR.

miRNA	ESR1 and ESR2 Binding Sites in Promoter According to ChipSeq Data (Homo Sapiens) [13]	ESR1 and ESR2 Binding Sites in Promoter According to Position Weight Matrix (Homo Sapiens) [13]	ESR1 and ESR2 Binding Sites in Promoter According to Position Weight Matrix (Mus Musculus and Rattus Norvegicus) [13]	AR/PR Binding Sites in Promoter According to Position Weight Matrix (Homo Sapiens)	AR/PR Binding Sites in Promoter According to Position Weight Matrix (Rattus Norvegicus)	Comments
hsa-mir-21	+	+	only mouse	+	+	It was demonstrated that androgen induced AR binding to the miR-21 promoter; MiR-21 expression was induced by R1881 in LNCaP and LAPC-4 cells [19]. Mibolerone inhibited basal expression of miR-21 in MCF-7 breast cancer cells [20]. Estradiol inhibited miR-21 expression in MCF-7 cells [21].
hsa-mir-190b	+	+	+	+	-	MiR-190b is the highest up-regulated miRNA in ER⁺ breast cancers compared to ER⁻ tumors. Did not observe an increase of miR-190b expression levels in MCF-7 or in T-47D treated by estradiol (1 nM for MCF-7 and 10 nM for T-47D, 6 h, 18 h, and 4 days) [22].
hsa-mir-200a/ hsa-mir-200b/ hsa-mir-429	+	+	+	-	-	MiR-200b showed the highest fold change under the influence of R1881 among androgen-sensitive miRNAs (PC3-AR cells) [23]. MiR-200b expression in MCF-7 cells decreased after 6 h of incubation with 10 nM estradiol [24].
hsa-mir-23a/ hsa-mir-24-2/ hsa-mir-27a	+	-	+	-	-	AR is able to associate transiently with the miR-23a/27a/24-2 promoter in response to androgen to initiate cluster transcription. The highest-fold change was observed for miR-27a and miR-23a (LNCaP cells treated with mibolerone) [25]. Estrogen induced miR-23a expression in SNU-387 cells [26].
hsa-mir-342	+	-	+	+	-	MiR-342 expression is positively correlated with ERα mRNA expression in human BC [27].
hsa-mir-190a	-	-	-	+	+	The promoter region of miR-190a contains half of an estrogen response element. ERα binds directly to this promoter [28]. Androgen inhibits miR-190a expression through direct binding to the half-site of ARE in miR-190a promoter (LNCaP cells) [29].
hsa-mir-378a	-	+	+	+	+	
hsa-mir-324	-	+	+	+	+	
hsa-mir-423	-	+	only rat	+	+	
hsa-mir-149	-	+	only rat	+	+	

Table 2. *Cont.*

miRNA	ESR1 and ESR2 Binding Sites in Promoter According to ChipSeq Data (Homo Sapiens) [13]	ESR1 and ESR2 Binding Sites in Promoter According to Position Weight Matrix (Homo Sapiens) [13]	ESR1 and ESR2 Binding Sites in Promoter According to Position Weight Matrix (Mus Musculus and Rattus Norvegicus) [13]	AR/PR Binding Sites in Promoter According to Position Weight Matrix (Homo Sapiens)	AR/PR Binding Sites in Promoter According to Position Weight Matrix (Rattus Norvegicus)	Comments
hsa-mir-365b	+	+	only rat	+	-	
hsa-mir-574	+	-	only mouse	+	-	
hsa-mir-30a	+	-	only mouse	+	-	AR does not target the miR-30a promoter; AR activating signal may indirectly downregulate miR-30a (MDA-MB-453 cells) [30].
hsa-mir-10a	+	+	+	-	-	
hsa-mir-483	+	+	+	-	-	
hsa-let-7a-3/ hsa-let-7b	+	-	+	-	-	
hsa-mir-196a-2	+	+	-	-	-	MiR-196a expression is regulated by the estrogen receptor [31].
hsa-mir-33b	+	-	-	-	-	
hsa-miR-193b	-	+	+	+	-	Targets ER [32].
hsa-miR-126	-	+	+	+	-	Targets PR (regulation confirmed using mouse mammary epithelial cells) [33].

A plus signifies the presence of binding site in promoter region of miRNA according to the analysis performed. The list did not include miRNAs with low expression in breast tissues.

The relative levels of selected miRNAs were determined in MCF-7 cells treated with estradiol (E2), progesterone (P4), or testosterone by RT-PCR (Table 3). Treatment of cells with E2 led to a significant change in the expression of miR-190b, miR-200b, and miR-193b. The miR-190b level decreased in cells treated with 100 nM E2 for 6 h, but increased 1.4-fold under the influence of both doses of the hormone after 24 h of incubation. Expression of miR-200b decreased in cells treated with 100 nM E2 for 6 h, and expression of miR-193b increased 1.3-fold after 48 h of incubation of cells with 100 nM E2. For miR-200b, the decrease in its level in MCF-7 cells after 6 h of treatment with E2 was also reported earlier (Table 2).

In cells treated with testosterone, the expression of miR-27a, miR-190a, miR-200b, miR-21, miR-423, miR-193b, and miR-324 significantly changed. After 6 h of incubation with 100 nM testosterone, the levels of miR-27a and miR-21 decreased (1.3- and 1.6-fold, respectively). However, after 48 h, the level of miR-21 increased 1.4-fold under the influence of both doses of testosterone. In addition, in cells treated with 10 or 100 nM testosterone, the levels of miR-190a, miR-200b, miR-423, and miR-193b increased. The level of miR-324 increased only in cells treated with a high dose of testosterone.

Treatment of MCF-7 with P4 led to changes in the levels of miR-190b, miR-190a, miR-21, and miR-324. The level of miR-190b increased 1.3-fold in cells treated with 100 nM P4 for 48 h. The expression of miR-190a decreased by 1.3-times under the action of both doses of P4 after 24 h of incubation. The level of miR-21 increased 1.4-fold after 48 h of incubation of cells with 100 nM P4. Finally, miR-324 expression was significantly increased 1.6- and 2.2-fold in cells treated with low and high doses of P4, respectively, after 6 h of incubation.

Thus, we identified hormone-sensitive miRNAs: miR-190b, miR-193b, miR-324, miR-190a, miR-200b, miR-21, miR-423, and miR-27a.

Table 3. Relative miRNA levels in MCF-7 cells treated with estradiol, testosterone, or progesterone.

miRNA	Time, h	Relative Level of miRNA					
		Estradiol		Testosterone		Progesterone	
		10 nM	100 nM	10 nM	100 nM	10 nM	100 nM
miR-23a	6	1.12	1.08	0.96	0.86	0.84	0.84
	24	0.89	0.95	0.88	1.00	0.92	0.88
	48	0.89	0.85	1.02	0.94	1.11	0.92
miR-27a	6	1.12	0.88	0.92	0.76 *	0.89	0.93
	24	1.07	1.10	1.09	1.08	0.93	0.92
	48	0.92	0.95	1.13	1.09	1.01	1.16
miR-190b	6	0.97	0.75 **	0.96	0.92	0.93	1.10
	24	1.35 *	1.37 **	1.13	1.09	1.02	1.03
	48	1.12	1.09	1.05	1.15	1.14	1.29 *
miR-190a	6	1.05	1.01	0.91	0.88	1.01	0.90
	24	1.09	1.02	0.90	0.99	0.79	0.75 *
	48	1.08	1.17	1.24 *	1.38 *	0.90	0.96
miR-200b	6	1.01	0.79 **	1.00	0.88	1.02	0.98
	24	0.98	1.01	1.10	0.99	0.91	0.96
	48	1.01	1.11	1.24 *	1.32 **	0.95	1.07
miR-21	6	1.03	0.98	0.92	0.62 **	1.17	1.36 *
	24	0.97	0.98	0.89	0.91	0.92	0.90
	48	1.07	1.08	1.36 **	1.40 **	1.11	1.08
miR-126	6	1.01	1.03	0.90	1.05	0.87	0.95
	24	0.94	0.88	1.06	0.94	1.01	1.12
	48	1.13	1.11	1.18	0.99	0.93	1.07
miR-378	6	1.09	1.00	0.88	0.94	0.90	1.08
	24	1.05	1.05	1.11	0.97	0.95	1.04
	48	1.00	1.17	1.15	1.19	0.94	1.12
miR-423	6	0.98	0.90	0.89	0.93	0.91	1.09
	24	1.04	1.06	0.99	0.87	0.96	0.99
	48	0.97	0.99	1.50 **	1.42 **	0.90	0.97
miR-149	6	1.25	1.01	0.95	0.91	0.96	1.06
	24	0.98	0.94	0.99	0.95	0.90	0.94
	48	0.89	1.05	1.05	0.89	1.02	1.18
miR-193b	6	1.07	1.00	0.93	0.95	0.92	1.07
	24	1.02	1.10	1.05	1.08	1.00	0.89
	48	1.24	1.30 **	1.30 **	1.40 **	1.07	1.08
miR-324	6	1.35 *	1.21	0.91	0.95	1.61 **	2.20 **
	24	0.99	1.06	1.06	1.05	1.05	1.04
	48	1.08	1.07	1.15	1.33 **	0.94	1.09

Table 3. Cont.

miRNA	Time, h	Relative Level of miRNA					
		Estradiol		Testosterone		Progesterone	
		10 nM	100 nM	10 nM	100 nM	10 nM	100 nM
miR-342	6	0.83	0.96	0.89	1.10	0.92	1.12
	24	0.96	1.02	1.15	1.10	0.89	0.94
	48	1.07	1.08	1.17	1.19	0.90	1.18

Control cells were treated with vehicle (DMSO). Each value represents the mean of four independent experiments; the results are normalized to the control. The statistical significance of differences in miRNA expression in MCF-7 cells treated with compounds was calculated using the Student's t-test. * $p < 0.05$ as compared with the control; ** $p < 0.01$ as compared with the control.

3.2. Analysis of the Hormone-Sensitive MiRNAs Expression in Breast Cancer

The relative levels of identified miRNAs were determined in 196 pairs of tumors and healthy tissues by RT-PCR. The amount of miR-190a, miR-27a, miR-193b, miR-324, and miR-423 was reduced in BC tissues compared to normal tissues (Figure 1). In contrast, miR-190b and miR-21 levels were increased in BC.

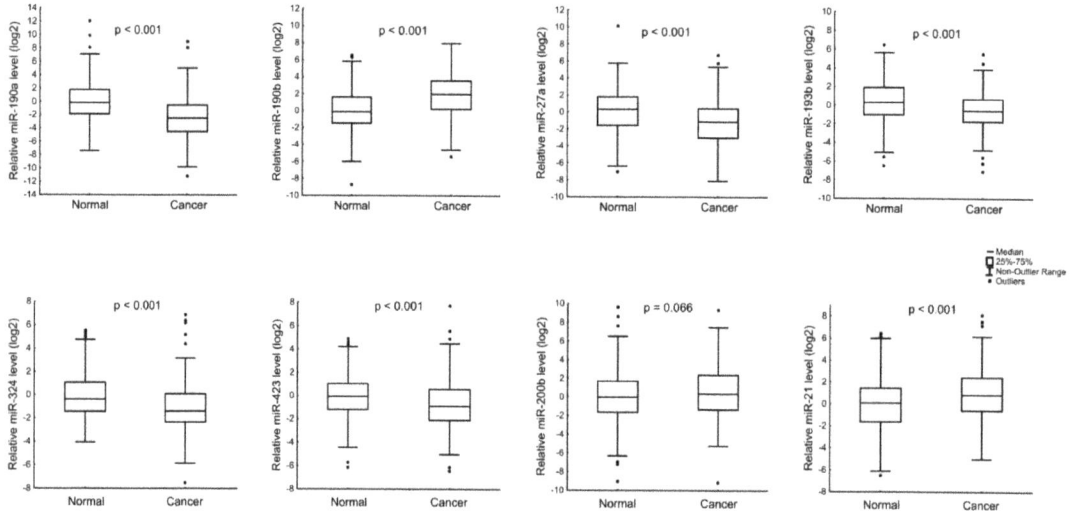

Figure 1. The comparison of miRNA expression between normal and cancerous tissue groups. The Y axis presents the expression level ($\log_2 2^{-\Delta\Delta Ct}$).

We investigated whether the expression of an identified miRNA depends on the status of ER, PR, HER2, Ki-67 index, or age. We observed that the amount of miR-190b and miR-21 in tissues depended on the ER and PR status, and the expression of miR-423 and miR-200b depended on the HER2 status (Table 4). The relative level of miR-190b was significantly higher in ER+ and/or PR+ tumors, and the level of miR-21 was significantly higher in ER− and PR− tumors. The expression levels of miR-423 and miR-200b were significantly higher in the tumors of patients with HER2-amplified cancer than in tumors with HER2 0 and HER2 1+ expression scores (according to IHC). Furthermore, the amount of miR-423 and miR-200b was associated with Ki-67 index. The levels of these miRNAs were higher in tumors with high Ki-67 ($\geq 14\%$).

Expression of miR-27a and miR-21 was lower in the tumors of patients older than 50 years compared to tumors of younger patients. MiR-193b was found to be associated with the status of the lymph nodes—the amount of miRNA was lower in the tumor tissues of patients with LNM.

Table 4. Associations between the amounts of miR-190a, miR-190b, miR-27a, miR-193b, miR-324, miR-423, miR-200b, or miR-21 in tissue samples from BC patients and ER, PR, HER2 status, Ki-67 index, age, or LN status.

Characteristics		n	Relative Level * of miRNA and p-Value							
			miR-190a	p-Value	miR-190b	p-Value	miR-27a	p-Value	miR-193b	p-Value
ER and PR status	ER+ and/or PR+	156	0.10	0.165	4.84	**<0.001**	0.33	0.443	0.58	0.168
	ER− and PR−	40	0.33		1.13		0.44		0.36	
HER2 status	HER2+	52	0.10	0.728	3.86	0.985	0.34	0.388	0.72	0.193
	HER2−	144	0.12		3.76		0.39		0.47	
Ki-67 index (%)	<14	65	0.11	0.753	3.89	0.941	0.49	0.394	0.49	0.567
	≥14	131	0.11		3.64		0.35		0.53	
Age	≤50	48	0.21	0.107	2.79	0.119	0.48	**0.045**	0.56	0.861
	>50	148	0.11		4.26		0.33		0.52	
N stage	N0	130	0.17	0.103	3.47	0.592	0.37	0.834	0.65	**0.022**
	N1-N3	66	0.09		4.21		0.42		0.37	
			miR-324	p-Value	miR-423	p-Value	miR-200b	p-Value	miR-21	p-Value
ER and PR status	ER+ and/or PR+	156	0.48	0.129	0.65	0.800	1.77	0.676	1.78	**0.004**
	ER− and PR−	40	0.68		0.74		1.53		3.48	
HER2 status	HER2+	52	0.60	0.180	0.80	**0.004**	2.99	**0.024**	1.62	0.342
	HER2−	144	0.47		0.55		1.46		1.95	
Ki-67 index (%)	<14	65	0.39	0.257	0.56	**0.038**	1.26	**0.030**	2.02	0.947
	≥14	131	0.60		0.71		2.04		1.88	
Age	≤50	48	0.60	0.103	0.69	0.958	1.68	0.598	3.52	**0.003**
	>50	148	0.48		0.65		1.50		1.68	
N stage	N0	130	0.55	0.210	0.68	0.068	1.51	0.491	1.84	0.445
	N1-N3	66	0.48		0.52		1.78		2.13	

* Median of differences in miRNA levels between BC tissue and normal adjacent tissue (control) samples; the results were normalized to the control. Significant differences are highlighted in bold.

3.3. Expression of MiR-190a, MiR-190b, MiR-27a, MiR-193b, MiR-324, MiR-423, MiR-200b, and MiR-21 in Relation to Clinicopathologic Features of ER- and/or PR-Positive BC

Next, we evaluated the relation between the expression of miRNAs and the clinicopathologic features of tumors with positive ER and PR status (Table 5). We also analyzed whether there is a relation between miRNA counts and tumor characteristics within specific BC subtypes. In the analysis, we separately considered the group of patients with the HER2 1+ expression score. HER2 expression is higher in HER2 1+ BC compared to HER2 0 tumors [34]. As previously shown, HER2 1+ and HER2 0 tumors differ in the expression profile of a number of genes, and clinically, HER2-low (i.e., 1+ and lack of *ERBB2* amplification) BC shows more ALN involvement compared to HER2 0 BC [35]. We have also previously demonstrated that the relation of miRNA levels with tumor characteristics can be different for these variants of tumors [36].

For miRNA-190a, a nearly significant tendency towards a decrease in its level was observed in the tissues of patients with LNM compared to cases without LNM. Detailed analysis revealed that for tumors with Ki-67 < 14%, the decrease in the level of miR-190a in the presence of metastases was significant (Figure 2A). The miR-190a relative level was decreased in tumor tissues of patients with high PR levels compared to those with PR IHC scores of 0–5. However, this association with the PR level was not observed in tumors with a Ki-67 < 14% (Figure 2B). Additionally, the amount of miRNA was decreased in the

BC tissues of patients over 50 years old compared to younger patients with the luminal B HER2-amplified BC subtype (Figure 2C).

Table 5. Association of miR-190a, miR-190b, miR-27a, miR-193b, miR-324, miR-423, miR-200b, and miR-21 expression levels with clinicopathologic characteristics of ER- and/or PR-positive BC.

Characteristics		n	Relative Level * of miRNA and p-Value							
			miR-190a	p-Value	miR-190b	p-Value	miR-27a	p-Value	miR-193b	p-Value
			ER⁺ and/or PR⁺							
T stage	T1	71	0.11	0.783	4.80	0.995	0.30	0.248	0.43	0.070
	T2–T4	85	0.09		4.93		0.46		0.74	
N stage	N0	103	0.13	0.058	4.57	0.477	0.37	0.391	0.69	**0.013**
	N1–N3	53	0.08		5.41		0.52		0.38	
Ki-67 index (%)	<M **	81	0.17	0.071	4.59	0.231	0.47	0.261	0.56	0.316
	≥M **	75	0.09		5.42		0.32		0.66	
ER score	6–8	146	0.09	0.475	5.04	0.078	0.32	0.113	0.54	**0.013**
	0–5	10	0.19		2.12		0.52		1.74	
PR score	6–8	97	0.09	**0.004**	4.71	0.840	0.29	**0.048**	0.49	0.258
	0–5	59	0.21		6.41		0.49		0.81	
Age	≤50	30	0.12	0.226	5.41	0.893	0.48	0.227	0.90	0.360
	>50	126	0.10		4.80		0.31		0.55	
			miR-324	p-Value	miR-423	p-Value	miR-200b	p-Value	miR-21	p-Value
			ER⁺ and/or PR⁺							
T stage	T1	71	0.43	0.216	0.59	0.290	2.09	0.109	1.68	0.247
	T2–T4	85	0.50		0.67		1.36		1.81	
N stage	N0	103	0.53	0.313	0.68	**0.039**	1.64	0.832	1.74	0.802
	N1–N3	53	0.46		0.49		1.77		1.95	
Ki-67 index (%)	<M **	81	0.44	0.694	0.57	0.055	1.53	0.253	1.84	0.494
	≥M **	75	0.49		0.68		2.04		1.69	
ER score	6–8	146	0.48	0.467	0.62	0.756	1.88	**0.006**	1.69	0.106
	0–5	10	0.89		0.88		0.28		2.09	
PR score	6–8	97	0.45	0.211	0.63	0.600	1.85	0.858	1.76	0.336
	0–5	59	0.61		0.65		1.50		1.83	
Age	≤50	30	0.57	0.232	0.71	0.741	1.72	0.729	2.28	0.085
	>50	126	0.46		0.62		1.73		1.65	

* Median of relative differences in miRNA amounts between breast tumors and paired samples of normal adjoining (control) tissue; the results were normalized to the control. ** M—median value. For ER- and/or PR-positive BC, median = 18. Significant differences are highlighted in bold.

For miR-190b, there was a nearly significant tendency to an increase in its level in tumor tissues of patients with LNM in HER2-non-expressing BC (HER2 0) (Figure 3A). In HER2-expressing BC (HER2 score 1+, 2+, 3+), the miRNA level was increased in the tissues of patients with Ki-67 levels above the median compared to cases with lower Ki-67 (Figure 3B).

For miR-27a, we noted the presence of an association with the level of PR expression. Detailed analysis showed that in the luminal B subtypes, the miRNA level was reduced in tumor tissues of patients with a high PR level (6–8 IHC score) compared to tissues of patients with a lower level of PR (Figure 4).

For miR-193b, an association was found with the presence of LNM, the level of ER expression, and we also observed a tendency for its level to increase with increasing tumor size. The relation with tumor size was found to be significant for HER2-non-expressing tumors (Figure 5A). A lower level of miR-193b in tumor tissues of patients with LNM was observed in all cases, except for tumors with HER2 1+ expression (Figure 5B). Since most cases with low ER expression belong to the luminal B HER2 0 variant, we separately assessed the association of miRNA with ER and PR expression in this BC type (Figure 5C).

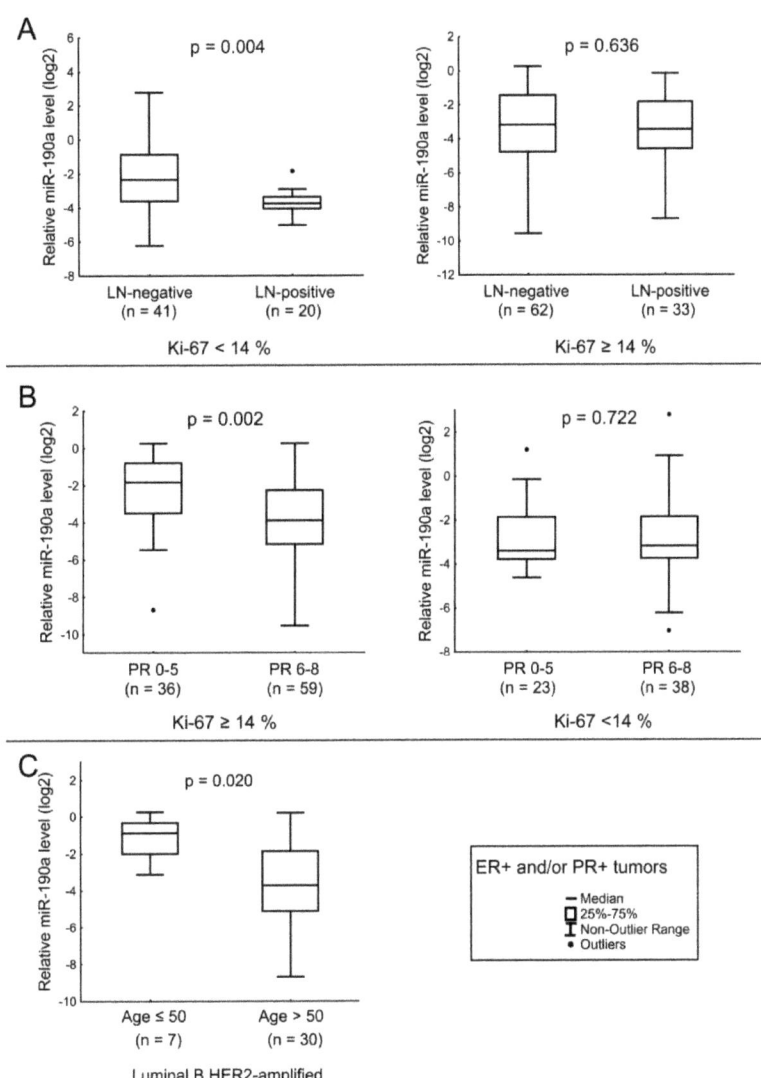

Figure 2. The comparison of miR-190a expression between different cancerous tissue groups: (**A**) relation of miR-190a level with the LNM in tumors with Ki-67 < 14% (left), relation of miR-190a level with the LNM in tumors with Ki-67 ≥ 14% (right); (**B**) relation of miR-190a level with the PR expression level in tumors with Ki-67 ≥ 14% (left), relation of miR-190a level with the PR expression level in tumors with Ki-67 < 14% (right); (**C**) relation of miR-190a level with age of patients with luminal B HER2-aplified BC. The Y axis presents the expression level ($\log_2 2^{-\Delta\Delta Ct}$); the results were normalized to the control (normal tissue).

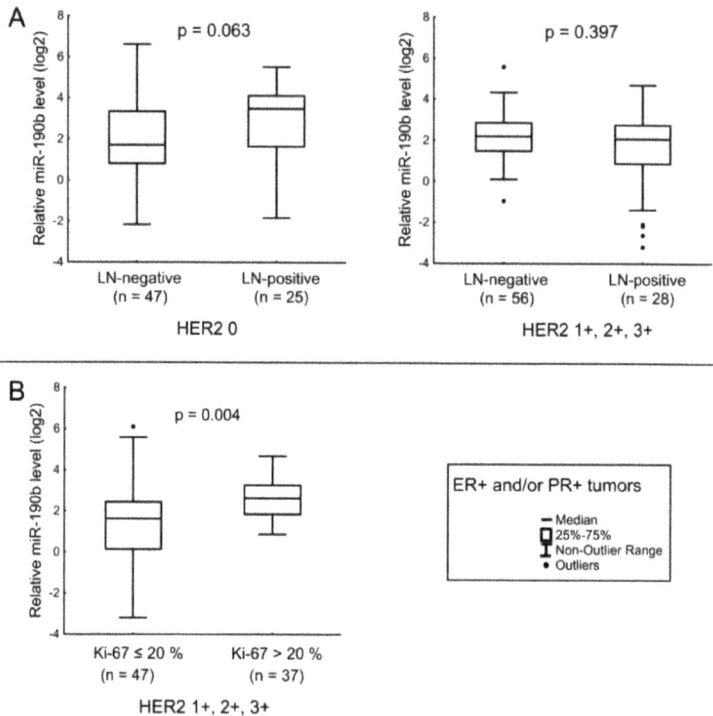

Figure 3. The comparison of miR-190b expression between different cancerous tissue groups: (**A**) relation of miR-190b level with the LNM in tumors with HER2 IHC score 0 (left), relation of miR-190b level with the LNM in tumors with HER2 IHC score 1+, 2+, 3+ (right); (**B**) relation of miR-190b level with Ki-67 in tissues of patients with luminal HER2-expressing BC (20% was the median value of Ki-67 index). The Y axis presents the expression level ($\log_2 2^{-\Delta\Delta Ct}$); the results were normalized to the control (normal tissue).

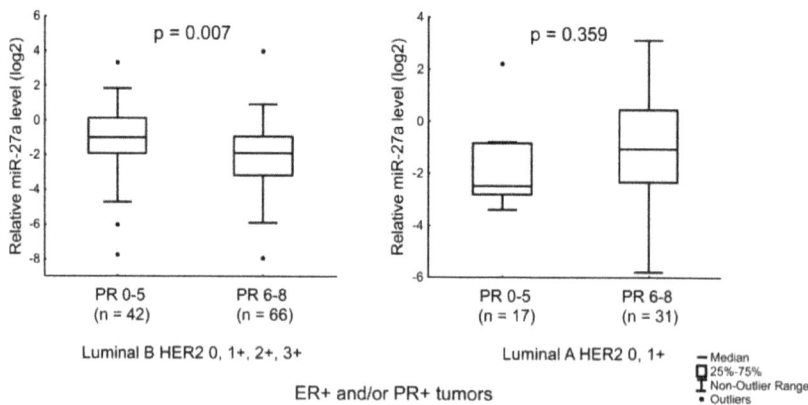

Figure 4. Relation of miR-27a level with PR expression level in luminal B tumors (left) and relation of miR-27a level with PR expression level in luminal A tumors (right). The Y axis presents the expression level ($\log_2 2^{-\Delta\Delta Ct}$); the results were normalized to the control (normal tissue).

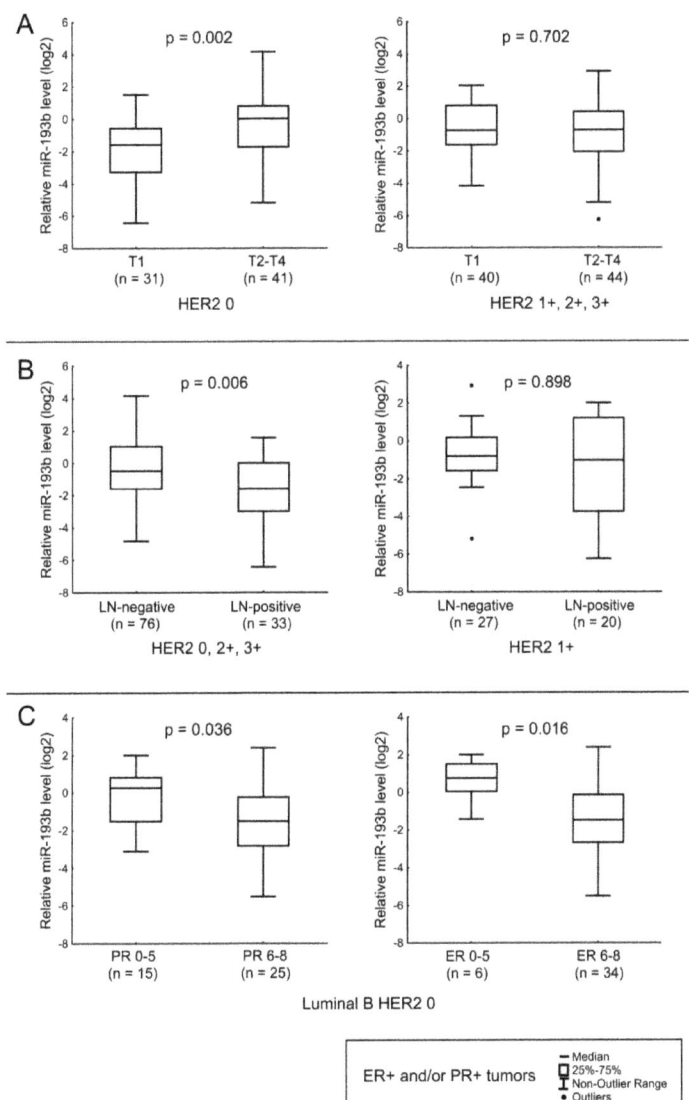

Figure 5. The comparison of miR-193b expression between different cancerous tissue groups: (**A**) relation of miR-193b level with the T stage in tumors with HER2 IHC score 0 (left), relation of miR-193b level with the T stage in tumors with HER2 IHC score 1+, 2+, 3+ (right); (**B**) relation of miR-193b level with LNM in tumors with HER2 IHC score 0, 2+, 3+ (left), relation of miR-193b level with LNM in tumors with HER2 IHC score 1+ (right); (**C**) relation of miR-193b level with PR and ER expression levels in luminal B HER2 0 BC. The Y axis presents the expression level ($\log_2 2^{-\Delta\Delta Ct}$); the results were normalized to the control (normal tissue).

MiR-324 levels have been found to be associated with tumor size in luminal B HER2-non-amplified BC (Figure 6A). Furthermore, the amount of miRNA was lower in the tumors of patients over 50 years old compared to younger patients in cases of the disease with Ki-67 < 14% (Figure 6B).

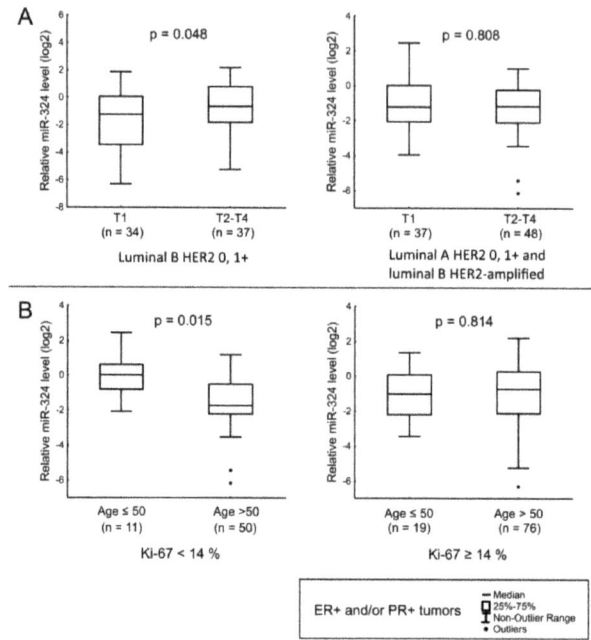

Figure 6. The comparison of miR-324 expression between different cancerous tissue groups: (**A**) relation of miR-324 level with the T stage in luminal B HER2-non-amplified tumors (left), relation of miR-324 level with the T stage in luminal A and luminal B HER2-amplified tumors (right); (**B**) relation of miR-324 level with age of patients in tumors with Ki-67 < 14% (left), relation of miR-324 level with age of patients in tumors with Ki-67 ≥ 14% (right). The Y axis presents the expression level ($\log_2 2^{-\Delta\Delta Ct}$); the results were normalized to the control (normal tissue).

We observed that the level of miR-423 is reduced in tumor tissues of patients with LNM. This decrease was not observed only for patients with luminal B HER2 0 BC (Figure 7).

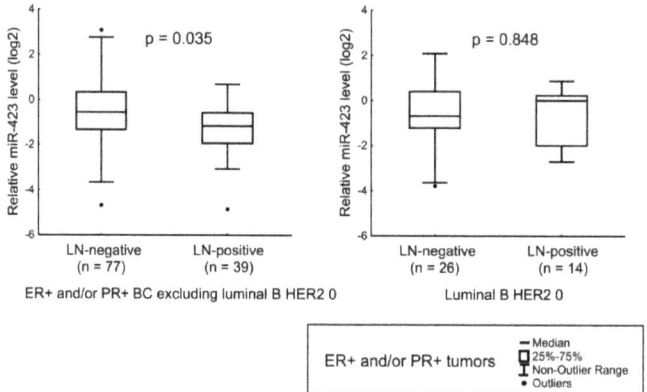

Figure 7. Relation of miR-423 level with LNM in luminal A, luminal B HER2-amplified, and luminal B HER2 1+ tumors (left); relation of miR-423 level with LNM in luminal B HER2-non-expressing tumors (right). The Y axis presents the expression level ($\log_2 2^{-\Delta\Delta Ct}$); the results were normalized to the control (normal tissue).

In patients with tumors with Ki-67 < 14%, the miR-200b level was lower in cases with tumors > 2 cm (Figure 8A). Additionally, when analyzing the general sample of patients with ER$^+$ and/or PR$^+$ it was found that the level of miR-200b is associated with the level of ER. In Figure 8B, we showed the association of the miRNA level with ER expression only for patients with luminal B HER2 0 BC.

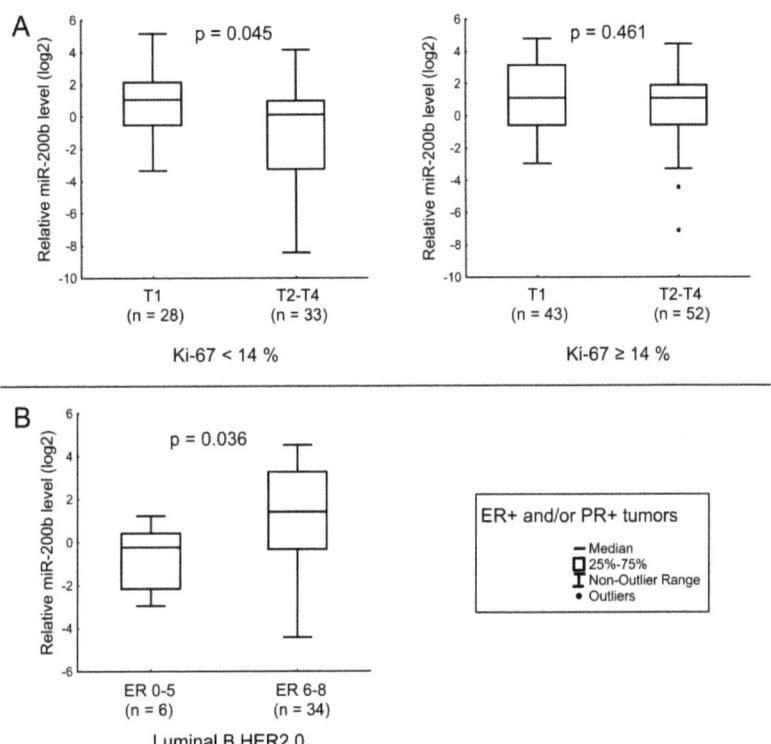

Figure 8. The comparison of miR-200b expression between different cancerous tissue groups: (A) relation of miR-200b level with the T stage in tumors with Ki-67 < 14% (left), relation of miR-200b level with the T stage in tumors with Ki-67 ≥ 14% (right); (B) relation of miR-200b level with ER expression level in luminal B HER2-non-expressing tumors. The Y axis presents the expression level ($\log_2 2^{-\Delta\Delta Ct}$); the results were normalized to the control (normal tissue).

For miR-21, there was a tendency to an increase in its level in cases with LNM for patients with ER$^+$ and/or PR$^+$ HER2-expressing tumors with Ki-67 ≥ 14% (Figure 9A). In luminal B HER2-non-expressing tumors, the amount of miR-21 was significantly lower in the BC tissues of patients with PR IHC scores of 3–8 than in patients with IHC scores of 0–2 (Figure 9B).

We also analyzed whether there is a relation between the HER2 level in the tumor, determined using IHC, and the expression levels of the studied miRNAs. We found that in tumors with Ki-67 < 14%, miR-324 levels were significantly higher in cases where HER2 amplification was confirmed, compared to cases without HER2 amplification ($p = 0.019$).

The amount of miR-21 in tumor tissue was significantly different in patients with a HER2 level estimated at a 0 score according to the IHC, from the amount of miRNA in patients with HER2-amplified BC or HER2 1+ BC (Figure 10).

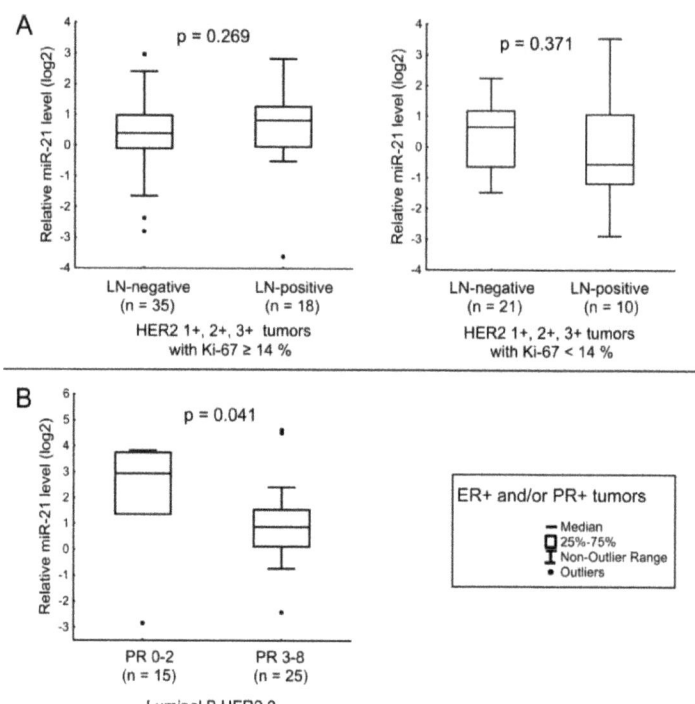

Figure 9. The comparison of miR-21 expression between different cancerous tissue groups: (**A**) relation of miR-21 level with LNM in HER2-expressing tumors with Ki-67 ≥ 14% (left), relation of miR-21 level with LNM in HER2-expressing tumors with Ki-67 < 14% (right); (**B**) relation of miR-21 level with PR expression level in luminal B HER2-non-expressing tumors. The Y axis presents the expression level ($\log_2 2^{-\Delta\Delta Ct}$); the results were normalized to the control (normal tissue).

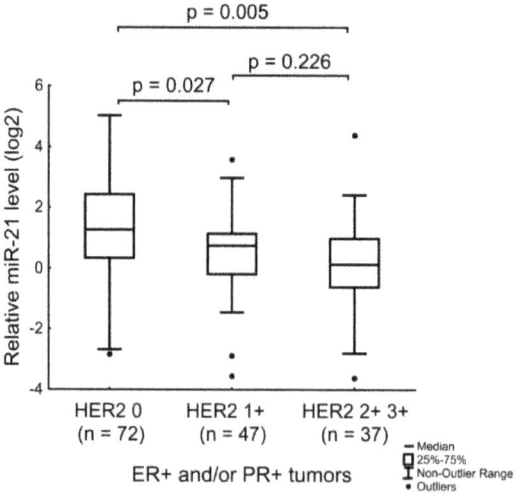

Figure 10. The relation of the miR-21 level with the HER2 expression level in ER-positive and/or PR-positive tumors. The Y axis presents the expression level ($\log_2 2^{-\Delta\Delta Ct}$); the results were normalized to the control (normal tissue).

Since an association with the level of HER2 expression was found for miR-21, we analyzed its level in the BC subtypes (Figure 11). We found that miR-21 level in luminal A HER2 0 tumors was significantly different from the level of this miRNA in luminal B HER2+ BC and luminal tumors with HER2 1+ expression levels. The expression of miRNA-21 in the tissues of the luminal B HER2 0 BC was significantly different only in comparison with its level in the tissues of the luminal HER2-amplified BC. However, we found that in ER- and PR-negative tumors, the highest level of miR-21 was in HER2-expressing tumors. In addition, the levels of miRNA were significantly differentiated between luminal B HER2-amplified cancer and ER- and PR-negative HER2-amplified BC, and between luminal HER2 1+ cancer and HER2-amplified BC.

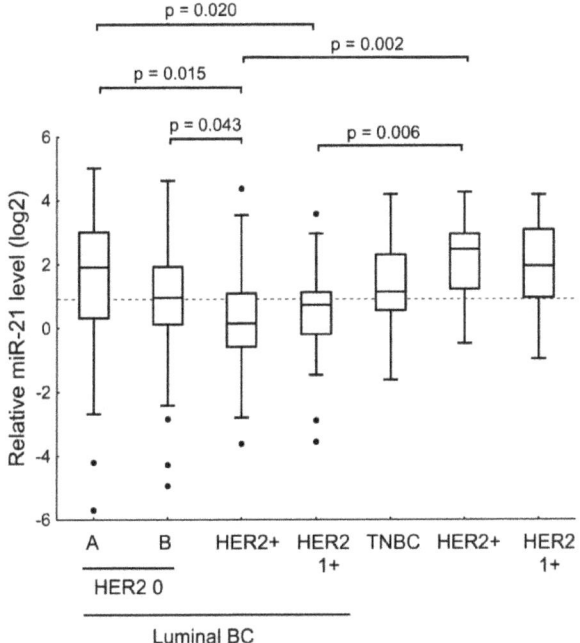

Figure 11. The comparison of miR-21 expression between different BC subtypes. HER2+ = HER2 amplification. Tumors with HER2 IHC scores of 1+ were treated as a separate group. The Y axis presents the expression level ($\log_2 2^{-\Delta\Delta Ct}$); the results were normalized to the control (normal tissue).

3.4. Expression of MiR-190a, MiR-190b, MiR-27a, MiR-193b, MiR-324, MiR-423, MiR-200b, and MiR-21 in Relation to Clinicopathologic Features of ER- and PR-Negative BC

Next, we analyzed the relation between the levels of the studied miRNAs and the characteristics of tumors in patients with ER- and PR-negative BC (Table 6). We found that miR-200b levels are significantly increased in BC tissues from patients with tumors > 2 cm. The amount of miR-21 in tumor tissue depended on the age of the patients and was significantly increased in patients under the age of 50.

We found that miR-190a was associated with a Ki-67 index in patients with triple-negative BC (TNBC): miR-190a was reduced in tumors with Ki-67 $\geq$ 75% (Figure 12A). In HER2-amplified BC, miR-190b was reduced in tumors with high Ki-67 (> 32%) (Figure 12B). Finally, miR-27a was found to be associated with the age of patients with HER2-expressing BC (Figure 12C).

Table 6. Association of miR-190a, miR-190b, miR-27a, miR-193b, miR-324, miR-423, miR-200b, and miR-21 expression with clinicopathologic characteristics of ER- and PR-negative BC.

Characteristics		n	Relative Level * of miRNA and p-Value							
			miR-190a	p-Value	miR-190b	p-Value	miR-27a	p-Value	miR-193b	p-Value
T stage	T1	18	0.33	0.372	0.90	0.629	0.40	0.240	0.57	0.176
	T2–T4	22	0.37		1.26		0.57		0.35	
N stage	N0	27	0.37	0.824	1.07	0.835	0.39	0.183	0.44	0.725
	N1–N3	13	0.24		1.14		0.52		0.35	
Ki-67 index (%)	≤M **	19	0.39	0.162	1.45	0.191	0.45	0.779	0.47	0.272
	>M **	21	0.08		1.02		0.44		0.35	
Age	≤50	18	0.34	0.228	0.64	0.275	0.55	0.091	0.37	0.617
	>50	22	0.26		1.24		0.39		0.43	
			miR-324	p-Value	miR-423	p-Value	miR-200b	p-Value	miR-21	p-Value
T stage	T1	18	0.65	0.987	0.80	0.729	1.11	**0.036**	3.87	0.703
	T2–T4	26	0.75		0.51		2.04		3.47	
N stage	N0	27	0.69	0.391	0.70	0.969	1.33	0.349	3.33	0.101
	N1–N3	13	0.62		0.80		2.04		6.68	
Ki-67 index (%)	≤M **	19	0.89	0.101	0.79	0.546	1.34	0.706	3.48	0.582
	>M **	21	0.56		0.49		2.07		3.47	
Age	≤50	18	0.65	0.565	0.55	0.617	1.78	0.266	6.15	**0.019**
	>50	22	0.68		0.80		1.32		1.95	

* Medians of relative differences in miRNA levels between breast tumors and paired samples of normal adjoining (control) tissue; the results were normalized to the control. ** M—median value. For ER- and PR-negative BC, median = 40. Significant differences are highlighted in bold.

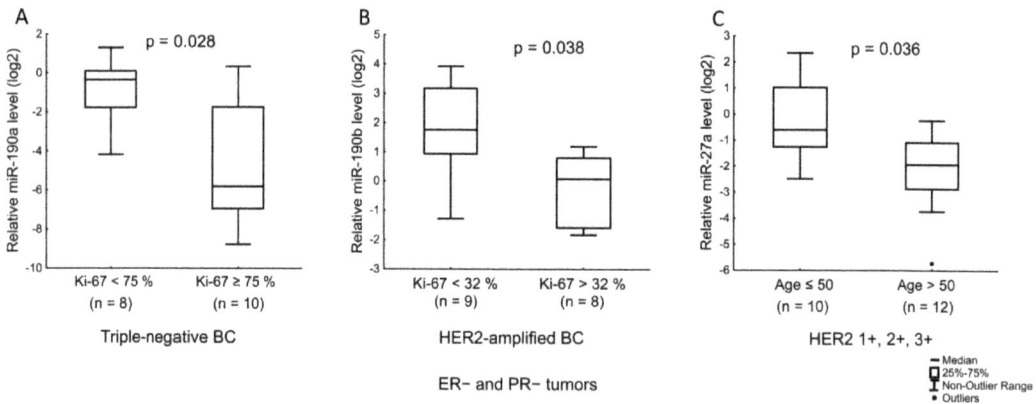

Figure 12. The comparison of miR-190a, miR-190b, and miR-27a expression between different cancerous tissue groups: (**A**) relation of miR-190a level with Ki-67 index in TNBC (75% was the median value of Ki-67); (**B**) relation of miR-190b level with Ki-67 in HER2-amplified tumors (32% was the median value of Ki-67); (**C**) relation of miR-27a level with age of patients with HER2-expressing tumors. The Y axis presents the expression level ($\log_2 2^{-\Delta\Delta Ct}$); the results were normalized to the control (normal tissue).

4. Discussion

The most common cancer in women is BC. In the past decade, significant progress has been made in the diagnosis and treatment of this disease. However, there are still some problems in diagnosing the disease. For example, it was previously noted that there may be a discrepancy between the results of the HER2 expression assessment by IHC and the results of fluorescence in situ hybridization, and the level of this discrepancy is higher in small laboratories [1]. In a recent study, it was shown that the concordance rates between the results of the IHC analysis of samples obtained during biopsy and the results of the IHC analysis of samples obtained during surgery for HER2 and Ki-67 are 84.8% and 83.5%, respectively [37]. In general, this is a good agreement, but it indicates that there may be an insufficient treatment or, conversely, an over-treatment of patients at the preoperative stage. Another problem is that the accuracy and sensitivity of the currently used imaging tests for determining the status of ALN in the preoperative phase are relatively low [2]. Thus, the search for markers that can subsequently serve to clarify the diagnosis or identify metastases is still in demand.

It is known that steroid receptors ER, PR, and AR play an important role in the initiation and progression of BC. Therefore, here, in order to search for markers associated with the clinicopathological features of tumors, we identified miRNAs sensitive to estradiol, progesterone, or testosterone, and analyzed their level in 196 pairs of tumors and healthy breast tissues.

For the study, we selected 13 miRNAs that are highly expressed in breast cells, and whose promoter regions contain sequences corresponding to the ER, PR, and AR binding sites. Of the selected miRNAs, the expression of eight were significantly altered in ER-, PR-, and AR-positive MCF-7 cells under the influence of one or more compounds (Table 7). We further analyzed the levels of identified hormone-sensitive miRNAs in BC samples.

Table 7. The observed changes in the expression of miRNAs potentially regulated by ER, PR, or AR in MCF-7 cells treated with estradiol, progesterone, or testosterone.

miRNA	Regulation of miRNA Expression According to Bioinformatic Analysis	Observed Changes		
		MCF-7		
		Estradiol	Testosterone	Progesterone
miR-27a	ER *, AR *	-	down (6 h)	-
miR-190b	ER *, AR, PR	down (6 h) up (24 h)	-	up (48 h)
miR-190a	ER *, AR, PR	-	up (48 h)	down (24 h)
miR-200b	ER *, AR **	down (6 h)	up (48 h)	-
miR-21	ER *, AR *, PR	-	down (6 h) up (48 h)	up (6 h)
miR-423	ER, AR, PR	-	up (48 h)	-
miR-193b	ER, PR, AR	up (48 h)	up (48 h)	-
miR-324	ER, PR, AR	up (6 h)	up (48 h)	up (6 h)

* The presence of a binding site was previously confirmed by chromatin immunoprecipitation or reporter assay.
** The binding site has not been identified, but the sensitivity of miRNA to androgen has been previously demonstrated.

In the analysis, we divided patients into groups depending on ER and PR status, HER2 status, or Ki-67 index. We separately analyzed a group of patients with a HER2 1+ IHC score; as previously shown, biologically, HER2 1+ tumors are significantly different from HER2 0 tumors [34]. In our study, there were no patients with HER2 2+ tumors with a lack of HER2 amplification. The main identified associations between miRNA expression levels and tumor characteristics are presented in Table 8.

We found that the expression of miR-190b and miR-21 depends on the status of ER and PR. The differences between miR-190b levels in tumor tissues with ER IHC scores of 6–8, and in tumors with ER IHC scores of 0–5, were also close to significant. Previously, it was demonstrated that the level of miR-190b is significantly increased in ER-positive breast tumors [22]. Thus, the result of our study is consistent with the previously obtained data. However, the same study reported no change in miR-190b expression in MCF-7 treated with estradiol (incubation with 1 nM E2 for 6, 18, and 96 h). Here, we detected an increase in the miR-190b level in MCF-7 cells treated with 10 nM and 100 nM E2 for 24 h. Thus, given the presence of an ER binding site in the promoter of this miRNA, according to the ChIP-seq data, it can be expected that a high level of miR-190b in ER$^+$ and/or PR$^+$ breast tumors is due to the increased expression and activity of ER.

Table 8. The identified relation between miRNAs levels and tumor characteristics.

miRNA	Associated Tumor Characteristics
miR-27a	Age of the patients and the level of PR expression.
miR-190b	ER receptor status. Ki-67 index in ER$^+$ and/or PR$^+$ HER2-expressing, and Ki-67 index in ER$^-$ and PR$^-$ HER2-amplified tumors.
miR-190a	Expression level of PR. LN status in ER$^+$ and/or PR$^+$ tumors with low Ki-67. Ki-67 index in triple-negative tumors.
miR-200b	HER2 status and Ki-67 index. Tumor size in ER$^+$ and/or PR$^+$ tumors with low Ki-67. Level of ER expression in Luminal B tumors not expressing HER2.
miR-21	ER receptor status and age of the patients. Expression level of HER2.
miR-423	HER2 status and Ki-67 index. LN status in luminal A, luminal B HER2-amplified, and luminal B HER2 1+ tumors.
miR-193b	LN status in ER$^+$ and/or PR$^+$ tumors (except HER2 1+ cases). Tumor size in ER$^+$ and/or PR$^+$ HER2-non-expressing tumors. Expression level of ER in Luminal B HER2 0 tumors.
miR-324	Tumor size in Luminal B HER2-non-amplified tumors. Age of the patients and HER2 status in ER$^+$ and/or PR$^+$ tumors with low Ki-67.

As for miR-21, earlier it was shown in smaller samples that its level is higher in ER-positive BC [38,39]. According to our data, the expression of miRNA is higher in tumors with a negative status of ER and PR. However, the observed dependence may be related to HER2. So, we observed the lowest level of miR-21 expression in ER$^+$ and/or PR$^+$ HER2-amplified and ER$^+$ and/or PR$^+$ HER2 1+ tumors. At the same time, the amount of miR-21 was high in ER- and PR-negative HER2-amplified and ER- and PR-negative HER2 1+ tumors. The miR-21 level in ER- and PR-negative HER2-amplified tumors was significantly higher than in ER$^+$ and/or PR$^+$ HER2-amplified BC.

In addition to miR-21, miR-423 and miR-200b were associated with the HER2 status. The levels of these miRNAs were higher in HER2-amplified tumors. The association with HER2 was also observed for miR-324, but only in tumors with Ki-67 < 14%. We found no association with ER and PR status for miR-200b and miR-193b; however, the levels of these miRNAs in ER$^+$ and/or PR$^+$ tumors with ER IHC scores of 0–5 differed significantly from miR-200b and miR-193b levels in tumors with higher ER expression. MiR-200b level was lower in tumors with ER expressions estimated at 0–5 IHC scores. According to previous studies, AR expression correlates with ER status [40]. We have also previously shown that in luminal B HER2 0 BC, AR level is higher for tumors with ER IHC scores of 6–8 compared to tumors with ER IHC scores of 0–5 [11]. Considering that miR-200b is an AR-regulated miRNA, the observed decrease in its levels in tumors with lower ER levels may be due to lower AR expression.

For miR-190a, we observed a decrease in expression in MCF-7 cells treated with progesterone. In ER$^+$ and/or PR$^+$ BC with high Ki-67 ($\geq$ 14%), the miRNA level was significantly lower in tumor tissues with high PR expression (scores of 6–8) than in tissues of patients with lower receptor expression. Thus, a significant decrease in miR-190a expression in ER$^+$ and/or PR$^+$ BC can be caused by increased PR expression.

To test whether the identified miRNAs could be diagnostic markers in BC, we also analyzed the relation between their expression levels and tumor size, the presence of metastatic lymph node lesions, and the Ki-67 index. The levels of miR-193b, miR-324, miR-423, and miR-200b were associated with tumor size in ER$^+$ and/or PR$^+$ BC. MiR-193b level was higher in tissues of patients with tumors > 2 cm in HER2 0 BC; miR-324—in luminal B HER2-non-amplified; miR-423—in luminal A. MiR-200b expression was reduced in tumors > 2 cm compared to smaller tumors in BC with Ki-67 < 14%. Thus, the association of the miRNA level with tumor size depended on Ki-67 and HER2 expression.

The level of miR-190a was significantly reduced in the tissues of patients with LNM compared with cases without metastases in ER+ and/or PR+ BC with Ki-67 < 14%. That is, in tumors with Ki-67 > 14%, the miR-190a level depended on the PR level, while in tumors with Ki-67 < 14%, the level of miR-190a was associated with the LNM. The level of miR-193b was also lower in the tissues of patients with LNM in all ER+ and/or PR+ BC, except for BC with HER2 1+ status. Lower levels of miR-423 in tumor tissues of patients with LNM was observed in all luminal BCs excluding luminal B HER2 0. The most significant association with metastasis was for miR-193b and miR-190a. It was previously demonstrated that a decrease in miR-193b level can lead to an increase in cellular invasion in MDA-MB-231 and MDA-MB-435 cells, and miR-190 suppresses BC metastasis both in vitro and in vivo [41,42]. Thus, our results are consistent with previously obtained data. However, since we found that the miR-193b level in HER2-non-expressing tumors is associated not only with metastases, but also with tumor size, we assume that this miRNA is a more useful marker for predicting metastasis in luminal HER2-amplified cancer.

We have also previously shown miR-21 levels in HER2-expressing BC may be associated with the presence of LNM [11]. Here, when using a larger sample, we observed only a tendency to an increase in the miR-21 level in the tissues of patients with LNM compared to cases without, in HER2-expressing BC with Ki-67 $\geq$ 14%. In HER2-expressing BC with Ki-67 < 14%, we observed a tendency towards a decrease in the level of miR-21 in the tissues of patients with LNM. Further research is required on the relation between miR-21 and LNM in BC.

No association between miRNA levels and the presence of LNM was found for ER- and PR-negative tumors.

For miR-423, miR-200b, and miR-190b, we also found an association with Ki-67 in ER+ and/or PR+ BC. The levels of miR-423, miR-200b were higher in tissues with high Ki-67. The miR-190b level was also higher in BC tissues with Ki-67 values above the median in HER2-expressing tumors. However, in ER- and PR-negative HER2-amplified BC, miRNA expression, on the contrary, was lower in patients with Ki-67 above the median value. It is possible that this miRNA plays different roles in ER- and/or PR-positive BC and ER- and PR-negative BC. In addition, in TNBC, the miR-190a level was significantly lower in tissues with high Ki-67 ($\geq$ 75%).

The levels of miR-27a and miR-21 were lower in the tissues of patients over the age of 50 compared to younger patients. Lower levels in tissues of patients over 50 years old were also observed for miR-190a—in luminal B HER2-amplified BC—and for miR-324—in ER+ and/or PR+ BC with Ki-67 < 14%. Menopause typically occurs between the ages of 49 and 52 years [43]. Therefore, the observed differences in the levels of miRNA in patients may be associated with hormonal changes after 50 years old.

Thus, we found that the expression of miR-190a, -190b, -27a, -193b, -324, -423, -200b, and -21, whose promoter regions contain binding sites for ER or AR, or putative binding sites for AR and PR, changes in MCF-7 cells when cells are treated with estradiol, progesterone, or testosterone. We have shown that the levels of these miRNAs may be associated with tumor size or the presence of lymph node metastases in breast cancer patients, but the presence of this association depends on the status and expression level of ER, PR, HER2, and Ki-67. As a result of our study we can draw the following conclusions: high expression levels of miRNA-190b in breast tumor tissue indicate a positive ER status; the assessment of miR-21, miR-423, and miR-200b levels could be used to confirm the amplification of HER2; and the levels of miR-190a (in ER+ and/or PR+ BC with Ki-67 < 14%), miR-193b (in luminal B HER2-amplified), and miR-423 (in luminal A, luminal B HER2 1+, and luminal B HER2-amplified tumors) can potentially be used as markers to predict the absence or presence of metastases in the lymph nodes.

Author Contributions: Conceptualization, T.K. and V.K.; methodology, T.K. and V.K.; validation, T.K. and V.K.; formal analysis, T.K. and V.O.; investigation, T.K., V.K., E.A. and G.A.; resources, L.G., S.S., G.A. and V.O.; data curation, T.K.; writing—original draft preparation, T.K.; writing—review and editing, L.G.; visualization, T.K.; supervision, L.G.; project administration, T.K. and V.K.; funding acquisition, L.G. All authors have read and agreed to the published version of the manuscript.

Funding: This research was funded by the Russian Science Foundation, grant number 19-15-00319.

Institutional Review Board Statement: The study was conducted according to the guidelines of the Declaration of Helsinki and approved by the Bioethics Committee of the Institute of Molecular Biology and Biophysics (protocol No 2/2017).

Informed Consent Statement: Informed consent was obtained from all subjects involved in the study.

Data Availability Statement: The data presented in this study are available on request from the corresponding author.

Acknowledgments: The work was performed using the equipment of the Center for Collective Use "Proteomic Analysis", supported by funding from the Ministry of Science and Higher Education of the Russian Federation (agreement No. 075-15-2021-691).

Conflicts of Interest: The authors declare no conflict of interest.

References

1. Gown, A.M. Current issues in ER and HER2 testing by IHC in breast cancer. *Mod. Pathol.* **2008**, *21*, S8–S15. [CrossRef] [PubMed]
2. Li, J.; Downs, B.M.; Cope, L.M.; Fackler, M.J.; Zhang, X.; Song, C.G.; VandenBussche, C.; Zhang, K.; Han, Y.; Liu, Y.; et al. Automated and rapid detection of cancer in suspicious axillary lymph nodes in patients with breast cancer. *NPJ Breast Cancer* **2021**, *7*, 89. [CrossRef]
3. Li, H.; Jun, Z.; Zhi-Cheng, G.; Xiang, Q. Factors that affect the false negative rate of sentinel lymph node mapping with methylene blue dye alone in breast cancer. *J. Int. Med. Res.* **2019**, *47*, 4841–4853. [CrossRef]
4. Shah, R.; Rosso, K.; Nathanson, S.D. Pathogenesis, prevention, diagnosis and treatment of breast cancer. *World J. Clin. Oncol.* **2014**, *5*, 283–298. [CrossRef] [PubMed]
5. Saha, S.; Dey, S.; Nath, S. Steroid hormone receptors: Links with cell cycle machinery and breast cancer progression. *Front. Oncol.* **2021**, *11*, 620214. [CrossRef]
6. Amaral, J.D.; Solá, S.; Steer, C.J.; Rodrigues, C.M. Role of nuclear steroid receptors in apoptosis. *Curr. Med. Chem.* **2009**, *16*, 3886–3902. [CrossRef] [PubMed]
7. Jansen, J.; Greither, T.; Behre, H.M. Androgen-regulated microRNAs (AndroMiRs) as novel players in adipogenesis. *Int. J. Mol. Sci.* **2019**, *20*, 5767. [CrossRef] [PubMed]
8. Klinge, C.M. Estrogen Regulation of MicroRNA Expression. *Curr Genomics.* **2009**, *10*, 169–183. [CrossRef]
9. Goldhirsch, A.; Wood, W.C.; Coates, A.S.; Gelber, R.D.; Thürlimann, B.; Senn, H.J.; Panel members. Strategies for subtypes–dealing with the diversity of breast cancer: Highlights of the St. Gallen International Expert Consensus on the Primary Therapy of Early Breast Cancer 2011. *Ann. Oncol.* **2011**, *22*, 1736–1747. [CrossRef] [PubMed]
10. Harvey, J.M.; Clark, G.M.; Osborne, C.K.; Allred, D.C. Estrogen receptor status by immunohistochemistry is superior to the ligand-binding assay for predicting response to adjuvant endocrine therapy in breast cancer. *J. Clin. Oncol.* **1999**, *17*, 1474–1481. [CrossRef]
11. Kalinina, T.S.; Kononchuk, V.V.; Yakovleva, A.K.; Alekseenok, E.Y.; Sidorov, S.V.; Gulyaeva, L.F. Association between lymph node status and expression levels of androgen receptor, miR-185, miR-205, and miR-21 in breast cancer subtypes. *Int. J. Breast Cancer* **2020**, *2020*, 3259393. [CrossRef]
12. Chen, C.; Ridzon, D.A.; Broomer, A.J.; Zhou, Z.; Lee, D.H.; Nguyen, J.T.; Barbisin, M.; Xu, N.L.; Mahuvakar, V.R.; Andersen, M.R.; et al. Real-time quantification of microRNAs by stem-loop RT-PCR. *Nucleic Acids Res.* **2005**, *33*, e179. [CrossRef]
13. Ovchinnikov, V.Y.; Antonets, D.V.; Gulyaeva, L.F. The search of CAR, AhR, ESRs binding sites in promoters of intronic and intergenic microRNAs. *J. Bioinform. Comput. Biol.* **2018**, *16*, 1750029. [CrossRef] [PubMed]
14. Fromm, B.; Billipp, T.; Peck, L.E.; Johansen, M.; Tarver, J.E.; King, B.L.; Newcomb, J.M.; Sempere, L.F.; Flatmark, K.; Hovig, E.; et al. A uniform system for the annotation of vertebrate microRNA genes and the evolution of the human microRNAome. *Annu. Rev. Genet.* **2015**, *49*, 213–242. [CrossRef]
15. Mathelier, A.; Fornes, O.; Arenillas, D.J.; Chen, C.Y.; Denay, G.; Lee, J.; Shi, W.; Shyr, C.; Tan, G.; Worsley-Hunt, R.; et al. JASPAR 2016: A major expansion and update of the open-access database of transcription factor binding profiles. *Nucleic Acids Res.* **2016**, *44*, D110–D115. [CrossRef] [PubMed]
16. Pagès, H.; Aboyoun, P.; Gentleman, R.; DebRoy, S. Biostrings: String Objects Representing Biological Sequences, and Matching Algorithms. R Package Version 2.44.2. 2017. Available online: https://bioconductor.org/packages/release/bioc/html/Biostrings.html (accessed on 25 October 2021).

17. Kozomara, A.; Griffiths-Jones, S. miRBase: Integrating microRNA annotation and deep-sequencing data. *Nucleic Acids Res.* **2011**, *39*, D152–D157. [CrossRef] [PubMed]
18. Ludwig, N.; Leidinger, P.; Becker, K.; Backes, C.; Fehlmann, T.; Pallasch, C.; Rheinheimer, S.; Meder, B.; Stähler, C.; Meese, E.; et al. Distribution of miRNA expression across human tissues. *Nucleic Acids Res.* **2016**, *44*, 3865–3877. [CrossRef]
19. Ribas, J.; Ni, X.; Haffner, M.; Wentzel, E.A.; Salmasi, A.H.; Chowdhury, W.H.; Kudrolli, T.A.; Yegnasubramanian, S.; Luo, J.; Rodriguez, R.; et al. miR-21: An androgen receptor-regulated microRNA that promotes hormone-dependent and hormone-independent prostate cancer growth. *Cancer Res.* **2009**, *69*, 7165–7169. [CrossRef]
20. Casaburi, I.; Cesario, M.G.; Donà, A.; Rizza, P.; Aquila, S.; Avena, P.; Lanzino, M.; Pellegrino, M.; Vivacqua, A.; Tucci, P.; et al. Androgens downregulate miR-21 expression in breast cancer cells underlining the protective role of androgen receptor. *Oncotarget* **2016**, *7*, 12651–12661. [CrossRef]
21. Wickramasinghe, N.S.; Manavalan, T.T.; Dougherty, S.M.; Riggs, K.A.; Li, Y.; Klinge, C.M. Estradiol downregulates miR-21 expression and increases miR-21 target gene expression in MCF-7 breast cancer cells. *Nucleic Acids Res.* **2009**, *37*, 2584–2595. [CrossRef]
22. Cizeron-Clairac, G.; Lallemand, F.; Vacher, S.; Lidereau, R.; Bieche, I.; Callens, C. MiR-190b, the highest up-regulated miRNA in ERα-positive compared to ERα-negative breast tumors, a new biomarker in breast cancers? *BMC Cancer* **2015**, *15*, 499. [CrossRef] [PubMed]
23. Williams, L.V.; Veliceasa, D.; Vinokour, E.; Volpert, O.V. miR-200b inhibits prostate cancer EMT, growth and metastasis. *PLoS ONE* **2013**, *8*, e83991. [CrossRef]
24. Manavalan, T.T.; Teng, Y.; Litchfield, L.M.; Muluhngwi, P.; Al-Rayyan, N.; Klinge, C.M. Reduced expression of miR-200 family members contributes to antiestrogen resistance in LY2 human breast cancer cells. *PLoS ONE* **2013**, *8*, e62334.
25. Fletcher, C.E.; Dart, D.A.; Sita-Lumsden, A.; Cheng, H.; Rennie, P.S.; Bevan, C.L. Androgen-regulated processing of the oncomir miR-27a, which targets Prohibitin in prostate cancer. *Hum. Mol. Genet.* **2012**, *21*, 3112–3127. [CrossRef] [PubMed]
26. Huang, F.Y.; Wong, D.K.; Seto, W.K.; Lai, C.L.; Yuen, M.F. Estradiol induces apoptosis via activation of miRNA-23a and p53: Implication for gender difference in liver cancer development. *Oncotarget* **2015**, *6*, 34941–34952. [CrossRef]
27. He, Y.J.; Wu, J.Z.; Ji, M.H.; Ma, T.; Qiao, E.Q.; Ma, R.; Tang, J.H. miR-342 is associated with estrogen receptor-α expression and response to tamoxifen in breast cancer. *Exp. Ther. Med.* **2013**, *5*, 813–818. [CrossRef] [PubMed]
28. Chu, H.W.; Cheng, C.W.; Chou, W.C.; Hu, L.Y.; Wang, H.W.; Hsiung, C.N.; Hsu, H.M.; Wu, P.E.; Hou, M.F.; Shen, C.Y.; et al. A novel estrogen receptor-microRNA 190a-PAR-1-pathway regulates breast cancer progression, a finding initially suggested by genome-wide analysis of loci associated with lymph-node metastasis. *Hum. Mol. Genet.* **2014**, *23*, 355–367. [CrossRef] [PubMed]
29. Xu, S.; Wang, T.; Song, W.; Jiang, T.; Zhang, F.; Yin, Y.; Jiang, S.W.; Wu, K.; Yu, Z.; Wang, C.; et al. The inhibitory effects of AR/miR-190a/YB-1 negative feedback loop on prostate cancer and underlying mechanism. *Sci. Rep.* **2015**, *5*, 13528. [CrossRef]
30. Lyu, S.; Liu, H.; Liu, X.; Liu, S.; Wang, Y.; Yu, Q.; Niu, Y. Interrelation of androgen receptor and miR-30a and miR-30a function in ER-, PR-, AR+ MDA-MB-453 breast cancer cells. *Oncol. Lett.* **2017**, *14*, 4930–4936. [CrossRef]
31. Milevskiy, M.J.G.; Gujral, U.; Del Lama Marques, C.; Stone, A.; Northwood, K.; Burke, L.J.; Gee, J.M.W.; Nephew, K.; Clark, S.; Brown, M.A. MicroRNA-196a is regulated by ER and is a prognostic biomarker in ER+ breast cancer. *Br. J. Cancer* **2019**, *120*, 621–632. [CrossRef] [PubMed]
32. Leivonen, S.K.; Mäkelä, R.; Ostling, P.; Kohonen, P.; Haapa-Paananen, S.; Kleivi, K.; Enerly, E.; Aakula, A.; Hellström, K.; Sahlberg, N.; et al. Protein lysate microarray analysis to identify microRNAs regulating estrogen receptor signaling in breast cancer cell lines. *Oncogene* **2009**, *28*, 3926–3936. [CrossRef]
33. Cui, W.; Li, Q.; Feng, L.; Ding, W. MiR-126-3p regulates progesterone receptors and involves development and lactation of mouse mammary gland. *Mol. Cell. Biochem.* **2011**, *355*, 17–25. [CrossRef]
34. Lambein, K.; Van Bockstal, M.; Vandemaele, L.; Geenen, S.; Rottiers, I.; Nuyts, A.; Matthys, B.; Praet, M.; Denys, H.; Libbrecht, L. Distinguishing score 0 from score 1+ in HER2 immunohistochemistry-negative breast cancer: Clinical and pathobiological relevance. *Am. J. Clin. Pathol.* **2013**, *140*, 561–566. [CrossRef] [PubMed]
35. Schettini, F.; Chic, N.; Brasó-Maristany, F.; Paré, L.; Pascual, T.; Conte, B.; Martínez-Sáez, O.; Adamo, B.; Vidal, M.; Barnadas, E.; et al. Clinical, pathological, and PAM50 gene expression features of HER2-low breast cancer. *NPJ Breast Cancer* **2021**, *7*, 1. [CrossRef] [PubMed]
36. Kalinina, T.; Kononchuk, V.; Alekseenok, E.; Obukhova, D.; Sidorov, S.; Strunkin, D.; Gulyaeva, L. Expression of estrogen receptor- and progesterone receptor-regulating microRNAs in breast cancer. *Genes (Basel)* **2021**, *12*, 582. [CrossRef]
37. You, K.; Park, S.; Ryu, J.M.; Kim, I.; Lee, S.K.; Yu, J.; Kim, S.W.; Nam, S.J.; Lee, J.E. Comparison of core needle biopsy and surgical specimens in determining intrinsic biological subtypes of breast cancer with immunohistochemistry. *J. Breast Cancer* **2017**, *20*, 297–303. [CrossRef] [PubMed]
38. Mattie, M.D.; Benz, C.C.; Bowers, J.; Sensinger, K.; Wong, L.; Scott, G.K.; Fedele, V.; Ginzinger, D.; Getts, R.; Haqq, C. Optimized high-throughput microRNA expression profiling provides novel biomarker assessment of clinical prostate and breast cancer biopsies. *Mol. Cancer* **2006**, *5*, 24. [CrossRef] [PubMed]
39. Petrović, N.; Mandušić, V.; Dimitrijević, B.; Roganović, J.; Lukić, S.; Todorović, L.; Stanojević, B. Higher miR-21 expression in invasive breast carcinomas is associated with positive estrogen and progesterone receptor status in patients from Serbia. *Med. Oncol.* **2014**, *31*, 977. [CrossRef]

40. Yu, Q.; Niu, Y.; Liu, N.; Zhang, J.Z.; Liu, T.J.; Zhang, R.J.; Wang, S.L.; Ding, X.M.; Xiao, X.Q. Expression of androgen receptor in breast cancer and its significance as a prognostic factor. *Ann. Oncol.* **2011**, *22*, 1288–1294. [CrossRef] [PubMed]
41. Li, X.F.; Yan, P.J.; Shao, Z.M. Downregulation of miR-193b contributes to enhance urokinase-type plasminogen activator (uPA) expression and tumor progression and invasion in human breast cancer. *Oncogene* **2009**, *28*, 3937–3948. [CrossRef]
42. Yu, Y.; Luo, W.; Yang, Z.J.; Chi, J.R.; Li, Y.R.; Ding, Y.; Ge, J.; Wang, X.; Cao, X.C. miR-190 suppresses breast cancer metastasis by regulation of TGF-β-induced epithelial-mesenchymal transition. *Mol. Cancer* **2018**, *17*, 70. [CrossRef] [PubMed]
43. Zhu, D.; Chung, H.F.; Dobson, A.J.; Pandeya, N.; Giles, G.G.; Bruinsma, F.; Brunner, E.J.; Kuh, D.; Hardy, R.; Avis, N.E.; et al. Age at natural menopause and risk of incident cardiovascular disease: A pooled analysis of individual patient data. *Lancet Public Health* **2019**, *4*, e553–e564. [CrossRef]

Article

GBP5 Serves as a Potential Marker to Predict a Favorable Response in Triple-Negative Breast Cancer Patients Receiving a Taxane-Based Chemotherapy

Shun-Wen Cheng [1], Po-Chih Chen [2,3,4], Tzong-Rong Ger [1], Hui-Wen Chiu [5,6,7,*] and Yuan-Feng Lin [5,8,*]

[1] Department of Biomedical Engineering, Chung Yuan Christian University, Taoyuan City 32023, Taiwan; g9975606@cycu.edu.tw (S.-W.C.); sunbow@cycu.org.tw (T.-R.G.)
[2] Neurology Department, Shuang-Ho Hospital, Taipei Medical University, New Taipei City 235, Taiwan; d620100001@tmu.edu.tw
[3] Taipei Neuroscience Institute, Taipei Medical University, New Taipei City 23561, Taiwan
[4] Department of Neurology, School of Medicine, College of Medicine, Taipei Medical University, Taipei 11031, Taiwan
[5] Graduate Institute of Clinical Medicine, College of Medicine, Taipei Medical University, Taipei 11031, Taiwan
[6] Department of Medical Research, Shuang Ho Hospital, Taipei Medical University, New Taipei City 23561, Taiwan
[7] TMU Research Center of Urology and Kidney, Taipei Medical University, Taipei 11031, Taiwan
[8] Cell Physiology and Molecular Image Research Center, Wan Fang Hospital, Taipei Medical University, Taipei 11031, Taiwan
* Correspondence: leu3@tmu.edu.tw (H.-W.C.); d001089012@tmu.edu.tw (Y.-F.L.); Tel.: +886-2-22490088 (ext. 8884) (H.-W.C.); +886-2-2736-1661 (ext. 3106) (Y.-F.L.); Fax: +886-2-2739-0500 (H.-W.C. & Y.-F.L.)

Abstract: Pre-operative (neoadjuvant) or post-operative (adjuvant) taxane-based chemotherapy is still commonly used to treat patients with triple-negative breast cancer (TNBC). However, there are still no effective biomarkers used to predict the responsiveness and efficacy of taxane-based chemotherapy in TNBC patients. Here we find that guanylate-binding protein 5 (GBP5), compared to other GBPs, exhibits the strongest prognostic significance in predicting TNBC recurrence and progression. Whereas GBP5 upregulation showed no prognostic significance in non-TNBC patients, a higher GBP5 level predicted a favorable recurrence and progression-free condition in the TNBC cohort. Moreover, we found that GBP5 expression negatively correlated with the 50% inhibitory concentration (IC_{50}) of paclitaxel in a panel of TNBC cell lines. The gene knockdown of GBP5 increased the IC_{50} of paclitaxel in the tested TNBC cells. In TNBC patients receiving neoadjuvant or adjuvant chemotherapy, a higher GBP5 level strongly predicted a good responsiveness. Computational simulation by the Gene Set Enrichment Analysis program and cell-based assays demonstrated that GBP5 probably enhances the cytotoxic effectiveness of paclitaxel via activating the Akt/mTOR signaling axis and suppressing autophagy formation in TNBC cells. These findings suggest that GBP5 could be a good biomarker to predict a favorable outcome in TNBC patients who decide to receive a taxane-based neoadjuvant or adjuvant therapy.

Keywords: triple-negative breast cancer; taxane; chemotherapy; GBP5; Akt/mTOR; autophagy

1. Introduction

Triple-negative breast cancer (TNBC) is a subset of breast cancer that does not express the estrogen receptor (ER), progesterone receptor (PR), and human epidermal growth factor receptor-2 (HER2) [1], and accounts for approximately 20% of breast cancers [2]. TNBC is most aggressive subtype of breast cancers, with a high metastatic ability and lack of specific targeted therapeutics [3]. It has been shown that TNBC patients with the BRCA mutation, higher levels of tumor-infiltrated lymphocytes and p53 abnormalities have a greater pathological complete response (pCR) rate to anthracycline and taxane regimens [4–6]. More recently, TNBCs were further classified into six different molecular subtypes—basal-like 1 (BL1), basal-like 2 (BL2), immunomodulatory (IM), mesenchymal stem like (MSL) and luminal androgen receptor (LAR), with a different pathological complete response rate to the standard neoadjuvant regimens include anthracyclines, taxanes, and cyclophosphamide [7]. This classification demonstrated that TNBCs are a heterogeneous group, which explains the lack of survival benefit for experimental drugs tested in several clinical trials. Therefore, identifying useful markers to predict the therapeutic responsiveness in TNBC subtypes is urgently needed in terms of precision oncology.

Guanylate-binding protein 5 (GBP5) has been known as part of the family of interferon-gamma (IFN-γ)-inducible GTPases and is involved in many cellular functions, including inflammasome activation [8] and innate immunity against microbial pathogens [9–12]. The human GBP family is composed of seven different members (GBP1-7) [13]. In addition to their immunomodulatory functions, a recent report showed that GBP1 upregulation predicts poor prognosis and is probably associated with the mechanism for erlotinib resistance in lung adenocarcinoma [14]. Moreover, GBP1 knockout by the CRISR/Cas9 tool dramatically suppressed the metastatic potential of prostate cancer cells [15]. In ER-negative breast cancer patients with brain metastasis, GBP1 was up-regulated by the stimulation of T lymphocytes, which promoted the ability of breast cancer cells to cross the blood–brain barrier [16]. GBP1 has also been proposed as a potential drug target for treating TNBC with elevated EGFR expression [17]. On the other hand, GBP2 appeared to correlate with favorable prognosis in breast cancer and indicate an efficient T cell response [18]. The methylation of GBP2 promoter was found in TNBC and associated with the malignant evolution of breast cancer [19]. Nevertheless, the prognostic significance of GBP5 and its roles in TNBC development remain largely unknown.

This study thus attempted to estimate the prognostic significance of GBP5 in TNBC patients with systemic chemotherapy. Our results showed that GBP5 upregulation strongly predicts a favorable recurrence and progression-free survival rate in TNBC patients. Particularly, GBP5 upregulation was significantly associated with a pCR rate in breast cancer patients receiving docetaxel/paclitaxel-based neoadjuvant therapy. Cell-based experiments revealed that GBP5 expression is negatively correlated with the 50% inhibitory concentration of paclitaxel in a panel of tested TNBC cell lines. Moreover, our results showed that GBP5 upregulation probably activates the Akt/mTOR pathway and suppresses autophagy formation in the paclitaxel-sensitive TNBC cells. These findings suggest a potential prognostic value of GBP5 in predicting the therapeutic effectiveness of taxane-based regimens in pre and post-operative settings for TNBC patients.

2. Materials and Methods

2.1. Clinical and Molecular Data for Breast Cancer Patients

The transcriptional profile generated by RNAseq (polyA þ Illumina HiSeq, Illumina, CA, USA) analysis of the TCGA breast cancer cohort was also downloaded from the UCSC Xena website (UCSC Xena. Available online: http://xena.ucsc.edu/welcome-to-ucsc-xena/, accessed on 1 February 2021). Microarray results with accession numbers GSE36133, GSE21997 and GSE32646 and the related clinical data were obtained from the Gene Expression Omnibus (GEO) database on the NCBI website and Kaplan–Meier Plotter website (http://kmplot.com/analysis/index.php?p=service&cancer=breast, accessed on 1 February 2021). The raw intensities in the .CEL files were normalized by robust multichip analysis (RMA), and fold-change analysis was performed using GeneSpring GX11 (Agilent Technologies, Santa Clara, CA, USA). Relative mRNA expression levels were normalized by the median of all samples and presented as $\log_2$ values. The gene lists of detected gene sets were obtained from the Molecular Signature Database (https://www.gsea-msigdb.org/gsea/msigdb, accessed on 1 February 2021).

2.2. Cell Lines and Cell Culture Condition

TNBC cell lines MDA-MB-231 and MDA-MB-468 were cultured in Leibovitz's (L-15) medium (Gibco Life Technologies, Grand Island, NY, USA), supplemented with 10% fetal bovine serum (FBS, Invitrogen, Thermo Fisher Scientific, Waltham, MA, USA), and incubated at 37 °C with free gas exchange with atmospheric air. TNBC cell lines HCC2157, HCC38, HCC1143 and HCC1937 were cultured in RPMI-1640 medium (Gibco Life Technologies, Thermo Fisher Scientific, Waltham, MA, USA) with 10% FBS and incubated at 37 °C with 5% CO2. TNBC cell line Hs578T and embryonic kidney cell line 293T were cultured in DMEM with 10% FBS and incubated at 37 °C with 5% CO2. Human non-malignant mammary epithelial cell lines H184B5F5/M10 and MCF10A were cultivated in Alpha-Minimum essential medium supplemented with 10% FBS and DMEM/F-12 medium supplemented with 5% horse serum, 20 ng/mL epithelium growth factor, 0.5 mg/mL Hydrocortisone, 100 ng/mL cholera toxin, and 10 µg/mL insulin, respectively. All cell lines, except H184B5F5/M10 from Bioresource Collection and Research Center (BCRC) in Taiwan, were obtained from American Type Culture Collection (ATCC). All cells were routinely authenticated on the basis of short tandem repeat (STR) analysis, morphologic and growth characteristics and mycoplasma detection.

2.3. Reverse Transcription PCR (RT-PCR)

The total RNA of detected cells was extracted by using TRIzol extraction kit (Invitrogen, Thermo Fisher Scientific, Waltham, MA, USA). The extracted total RNA (5 µg) were treated with M-MLV reverse transcriptase (Invitrogen) and then amplified by PCR protocol with a Taq-polymerase (Protech, Taipei, Taiwan) using paired primers (for GBP5, forward-GCCATTACGCAACCTGTAGTTGTG and reverse-CATTGTGCAGTAGGTCGATAGCAC; for PD-L1, forward-GCTGCACTTCAGATCACAGATGTG and reverse- GTGTTGATTCTCAGTGTGCTGGTC; for GAPDH, forward-AGGTCGGAGTCAACGGATTTG and reverse-GTGATGGCATGGACTGTGGTC).

2.4. MTT Assay

Cells (1×10^5/mL) were cultivated in a 96-well culture plate. At the endpoint of the designated treatments, 10 µL of MTT (3-(4,5-dimethylthiazol-2-yl)-2,5-diphenyltetrazolium bromide) (Molecular Probe, Invitrogen, CA, USA) stock solution was added into each well. The conversion of MTT to formazan by viable cells was performed at 37 °C for another 4 h. Then, to solubilize the formazan precipitates, 100 µL of DMSO solution was added into each well. The levels of formazan were measured by optical density at 540 nm using an ELISA reader in order to estimate cell survival rates.

2.5. Lentivirus-Driven shRNA Infection

Non-silencing and GBP5 shRNA clones (TRCN0000158813 (sh1): CCGGGCCATAATCTCTTCATTCAGACTCGAGTCTGAATGAAGAGATTATGGCTTTTTTG; TRCN0000159924 (sh2): CCGGCAAGGTAGTGATCAAAGAGTTCTCGAGAACTCTTTGATCACTACCTTGTTTTTTG) with a puromycin selection marker were obtained from the National RNAi Core Facility Platform in Taiwan. Lentiviruses were produced by transfecting 293T cells with the shRNA-expressing vector and pMDG/p△8.91 constructs using a calcium phosphate transfection kit (Invitrogen). After incubation for 48–72 h, the media containing lentiviral particles were collected. Cells with 50% confluence grown on six-well plates were cultivated in fresh media containing 5 µg/mL polybrene (SantaCruz, Dallas, TX, USA) prior to infection overnight with a lentiviral particle-driven control or candidate gene shRNA at 2–10 multiplicity of infection (MOI). Cells were further cultivated in the presence of puromycin (10 µg/mL) for 24 h in order to select cells stably expressing the control or candidate gene shRNA. RT-PCR analysis was used to confirm the efficiency of gene knockdown.

2.6. Western Blotting Analysis

Aliquots of total protein (20–100 μg) from designated experiments and TD-PM10315 TOOLS Pre-Stained Protein Marker (10–315 kDa) (BIOTOOLS Co., Ltd., Taipei, Taiwan) were subjected to SDS-PAGE and then transferred to PVDF membranes. The membranes were then incubated with blocking buffer (5% nonfat milk in TBS containing 0.1% Tween-20) for 2 hours at room temperature prior to incubation with primary antibodies against GBP5, (GeneTex, GTX118635, 1;1000), phosphorylated Akt (Thr308) (Taiclone, #tcea12931, 1:500), Akt (Cell Signaling, #4685, 1:1000), phosphorylated mTOR (Cell Signaling, #2971, 1:1000), mTOR (Cell Signaling, #2983, 1:1000), p62 (Mblintl, #PM045, 1:1000) ATG5 (Cell Signaling, #12994, 1:1000), Beclin-1 (Cell Signaling, #3738, 1:1000), LC3-I/II (Cell Signaling, #4108, 1:1000) or GAPDH (AbFrontier, #LF-PA0212, 1:5000) overnight at 4 °C. After excessive washes, the membranes were incubated with peroxidase-labeled species-specific secondary antibodies for another hour at room temperature. Immunoreactive bands were finally visualized by an enhanced chemiluminescence system (Amersham Bioscience, GE Healthcare, Billerica, MA, USA).

2.7. Statistical Analysis

SPSS 17.0 software (Informer Technologies, Roseau, Dominica) was used to analyze statistical significance. Paired *t*-test was utilized to compare GBP5 gene expression in the TNBC tissues. Pearson's correlation test was performed to estimate the association among mRNA levels of GBP5, IC50 of paclitaxel/doxorubicin and PI3K_AKT_MTOR/MTORC1 gene sets in the detected samples. Kaplan–Meier analysis and log-rank test were used to evaluate survival probabilities. Student's *t*-test was used to estimate the statistical significance of GBP5 gene expression in clinical samples. The non-parametric Mann–Whitney U was used to analyze the non-parametric data. p values < 0.05 in all analyses were considered statistically significant.

3. Results

We first dissected the gene expression status of GBP1, GBP2, GBP3, GBP4, GBP5 and GBP6 in TNBC cohorts stratified into the low and high-risk groups at a minimized log-rank p value of Kaplan–Meier analysis, a method determining the optimal cut point in continuous gene expression [20]. In comparison with other GBPs, GBP5 upregulation showed a great correlation with a favorable recurrence-free survival (RFS) rate in TNBC patients from the K–M Plotter (Figure 1A) and progression-free survival (PFS) condition in TNBC patients from the TCGA database (Figure 1B). According to the definition of National Cancer Institute (NCI, https://www.cancer.gov/, accessed on 1 February 2021), RFS and PFS associate with the length of time after primary treatment for a cancer ends that the patient survives without any signs of that cancer and lives with that cancer, but it does not get worse. Both survival conditions could reflect the therapeutic effectiveness in TNBC patients. Moreover, Cox regression test demonstrated that a higher GBP5 level in TNBC patients refers to a favorable hazard ratio, lower than that of other GBPs, under a recurrence and progression-free survival condition for the K–M Plotter and TCGA cohorts, respectively (Figure 1C). Similar views were also found in the other Kaplan–Meier analyses (Figure S1A,B) and Cox regression (Figure S1C) test using overall survival condition. Whereas GBP5 did not show a prognostic significance in the unclassified, ER-positive, non-TNBC population, GBP5 upregulation served as a potential biomarker, predicting a good outcome in TNBC patients under the conditions of recurrence- and progression-free survival probabilities (Figure 2A,B).

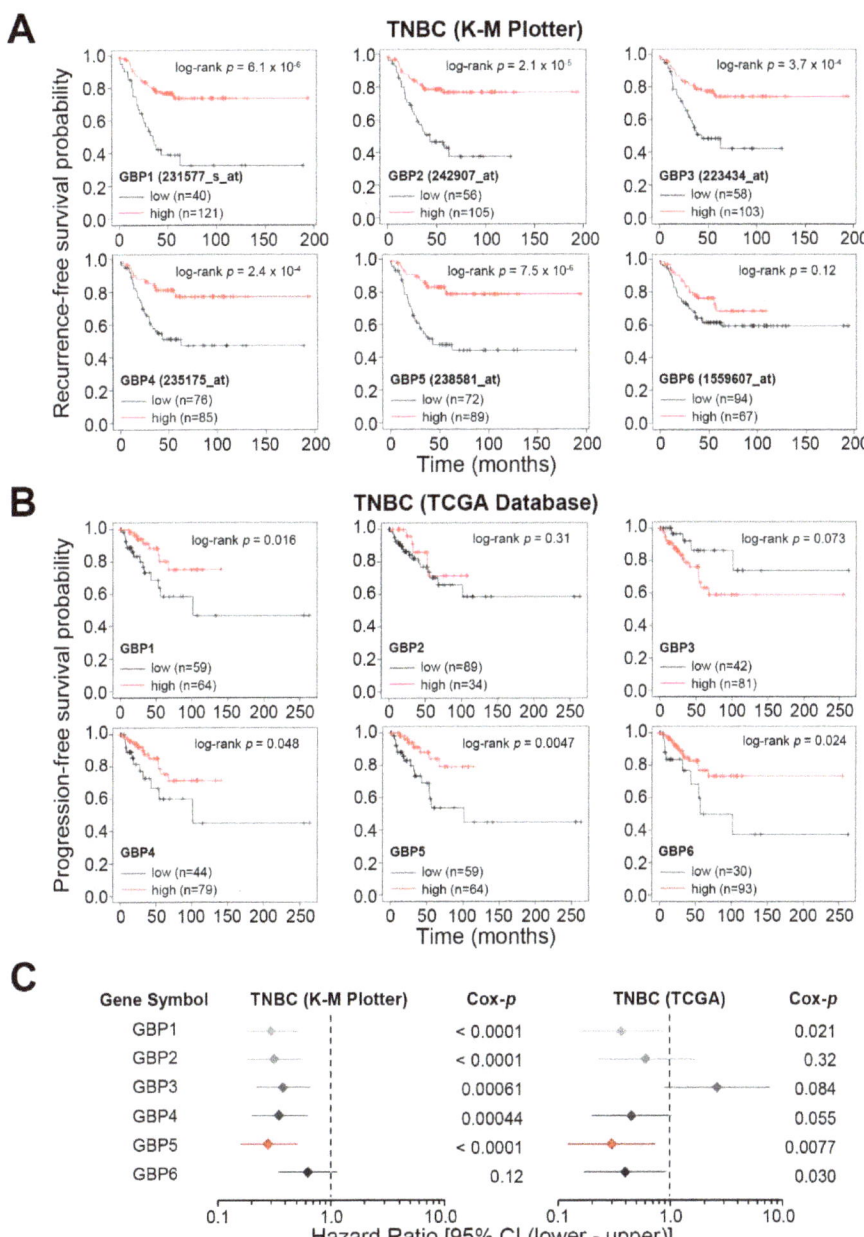

Figure 1. Guanylate-binding protein 5 (GBP5) upregulation predicts a good prognosis in triple-negative breast cancer (TNBC). (**A** and **B**) Kaplan–Meier analyses for GBP1, GBP2, GBP3, GBP4, GBP5 and GBP6 gene expression using recurrence-free survival condition against TNBC patients from K–M Plotter (**A**) and progression-free survival condition against TNBC patients from TCGA database (**B**) under a minimized p value. (**C**) Forest plot for the hazard ratio at a 95% confidence interval (CI), derived from Cox regression test using univariate mode for GBP1, GBP2, GBP3, GBP4, GBP5 and GBP6 against TNBC cohorts shown in A and B.

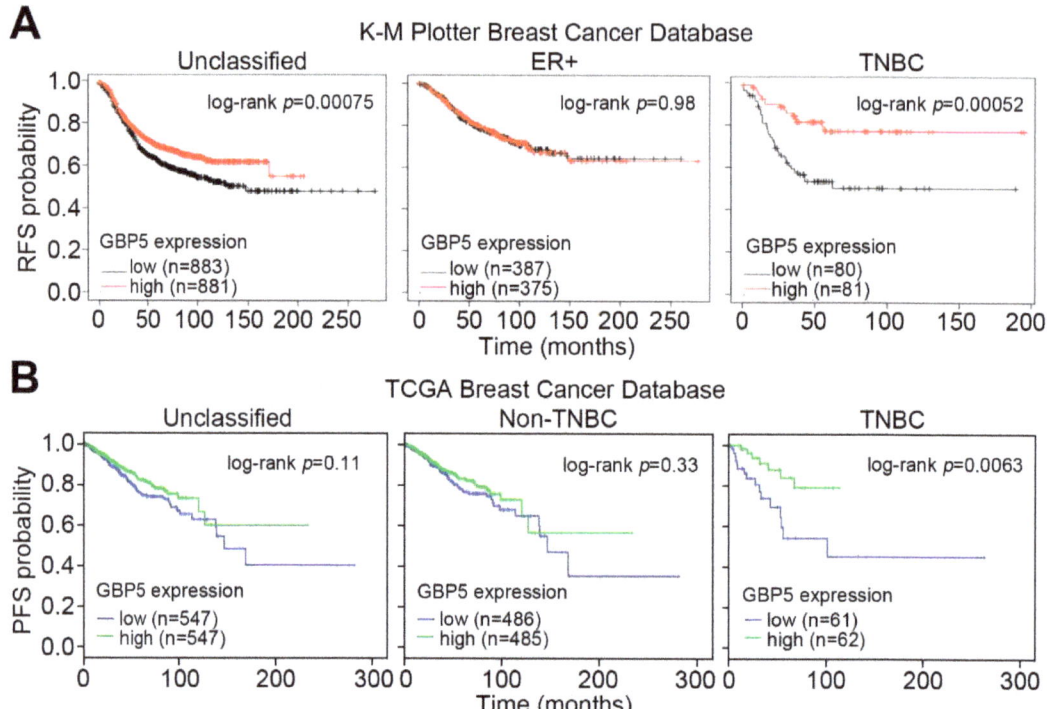

Figure 2. The prognostic significance of GBP5 is dominant for TNBC cohorts. (**A**,**B**) Kaplan–Meier analyses for GBP5 transcripts using recurrence-free for K–M Plotter cohort (**A**) and progression-free for TCGA cohort (**B**) survival conditions against the unclassified (**left**), ER+ or non-TNBC (**middle**), and TNBC (**right**) patients that were stratified by the media of GBP5 mRNA levels.

We next examined the endogenous mRNA levels of GBP5 in a panel of normal mammary epithelial cell lines H184B5F5/M10 and MCF10A, and TNBC cell lines HCC2157, HCC38, HCC1143, HCC1937, Hs578T, MDA-MB-231 and MDA-MB-468. The data showed that GBP5 mRNA levels in HCC38, HCC1143, Hs578T and MDA-MB-231 cells are much higher than that of HCC2157, HCC1937 and MDA-MB-468 cells, as well as non-malignant H184B5F5/M10 and MCF10A cells (Figure 3A). A similar outcome was also found in the microarray results from GSE36133 dataset for the GBP5 mRNA levels in HCC2157, HCC38, HCC1143, HCC1937, Hs578T, MDA-MB-231 and MDA-MB-468 cells (Figure 3B). WhileGBP5 expression was negatively correlated with the 50% of inhibitory concentration (IC_{50}) for paclitaxel (Figure 3C), GBP mRNA levels appeared to be positively correlated with the IC_{50} for doxorubicin (Figure 3D) in those TNBC cell lines. The gene knockdown of GBP5 (Figure 3E,F) by shRNA clone 2 (sh2) which has been validated to suppress GBP5 expression in the previous report [21] predominantly desensitizes MDA-MB-231 and Hs578T cells to the paclitaxel treatment as shown by an increased IC_{50} from 0.33 µM to over 1 µM and 0.00037 µM to 0.016 µM, respectively (Figure 3G,H).

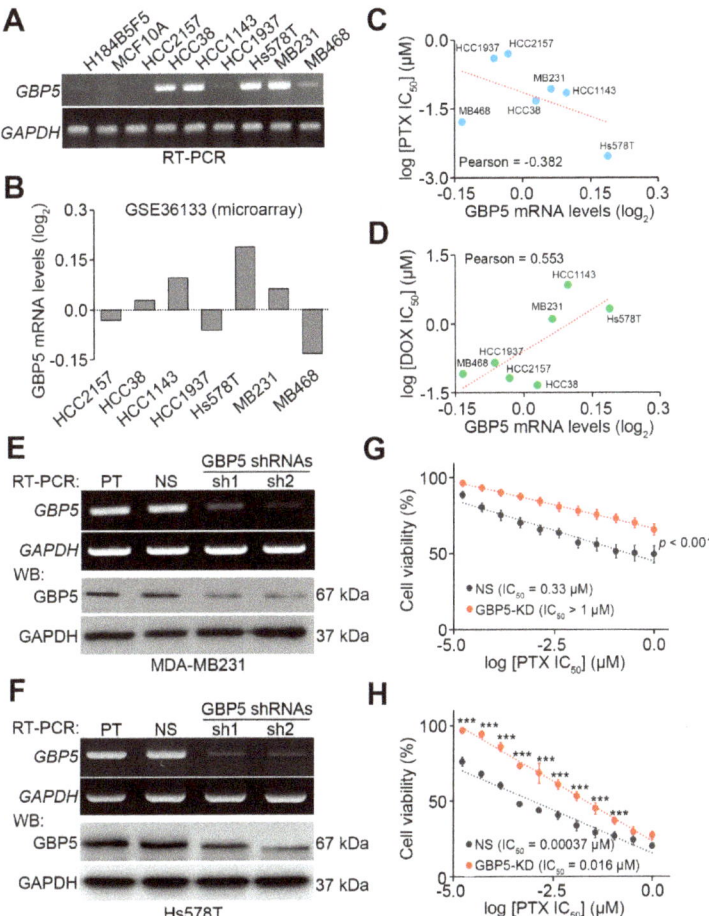

Figure 3. GBP5 knockdown desensitizes TNBC cells to paclitaxel treatment. (**A**) The mRNA levels of GBP5 and GAPDH detected by RT-PCR in a panel of normal mammary epithelial cell lines H184B5F5/M10 and MCF10A, and TNBC cell lines HCC2157, HCC38, HCC1143, HCC1937, Hs578T, MDA-MB231 (MB231) and MDA-MB468 (MB468). (**B**) GBP5 mRNA levels in the indicated TNBC cell lines from GSE36133 dataset. (**C**,**D**) Scatter plots for the correlation of GBP5 mRNA levels with paclitaxel (PTX, **C**) and doxorubicin (DOX, **D**) IC_{50} concentrations in the tested TNBC cells lines. Statistical significance was analyzed by Pearson correlation test. (**E**,**F**) The mRNA and protein levels of GBP5 and GAPDH detected by RT-PCR and Western blot (WB) analyses, respectively, in parental (PT) MDA-MB231 (**E**)/Hs578T (**F**) cells and MDA-MB231/Hs578T cells stably transfected non-silencing (NS) control or 2 independent GBP5 shRNA clones. In **A**, **E** and **F**, GAPDH was used as an internal control of experiments. (**G**,**H**) Dot plot for cell viability determined from non-silencing control and GBP5-knockdown (GBP5-KD), using sh2 clone, MDA-MB231 (**G**)/Hs578T (**H**) cells. Non-parametric Mann–Whitney test was used to estimate the statistical significances. The symbol "***" denotes $p < 0.001$.

While a higher GBP5 level was probably correlated with no complete response in breast cancer patients received doxorubicin neoadjuvant therapy, GBP5 upregulation appeared to significantly ($p = 0.031$) predict pathologic complete response in breast cancer patients receiving docetaxel neoadjuvant therapy (Figure 4A). Accordingly, in breast cancer cohort received paclitaxel neoadjuvant therapy, GBP5 upregulation significantly ($p < 0.001$)

referred to a pathologic complete response (Figure 4B). In the TNBC cohort receiving adjuvant chemotherapy, GBP5 upregulation was robustly correlated with a favorable recurrence-free survival condition (Figure 4C).

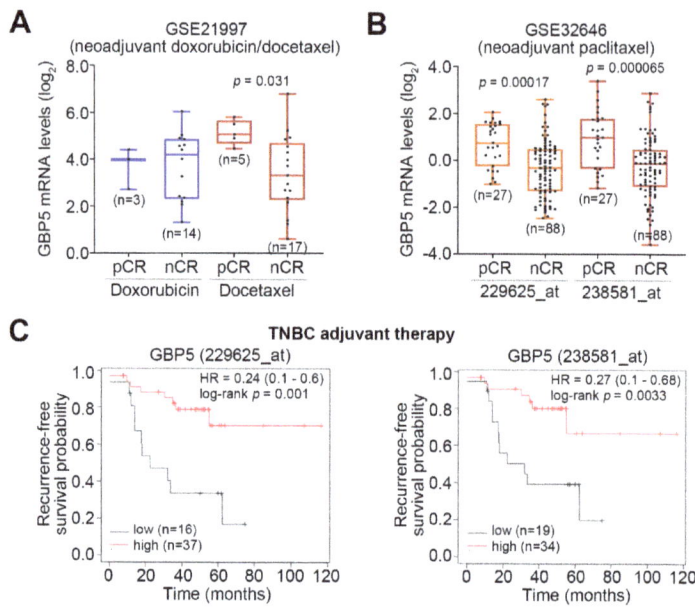

Figure 4. GBP5 upregulation predicts a good responsiveness to the taxol treatment in TNBC patients. (**A,B**) Box plots for the GBP5 mRNA levels in breast cancer patients that were recorded to be pathologic complete response (pCR) or no complete response (nCR) after neoadjuvant doxorubicin or docetaxel therapy from GSE21997 dataset (**A**) and after neoadjuvant paclitaxel therapy from GSE32646 dataset (**B**). In B, 229625_at and 238581_at denote the probe identifiers of GBP5 in the commercial microarray. Student's t-test was used to analyze the statistical significance. (**C**) Kaplan–Meier analyses using recurrence-free survival condition for GBP5 mRNA levels detected by two probes in K–M Plotter against TNBC patients receiving adjuvant chemotherapy.

To understand the possible mechanism by which GBP5 upregulation enhances the taxane sensitivity of TNBC, we next performed a computational simulation by using Gene Set Enrichment Analysis (GSEA) program. To obtain a GBP5-related signature, we first performed Spearman's Correlation tests against the co-expression of GBP5 with other somatic genes determined by the RNA-sequencing tool in TNBC samples from the TCGA database. Then, the ranked Spearman's coefficient p values was used as a GBP5-related signature for the further GSEA simulation (Figure 5A). GSEA results revealed that the GBP5 signature positively correlates with the mRNA levels of gene sets for the PI3K_AKT_MTOR and MTORC1 pathways in TNBC (Figure 5B,D). Western blot analyses revealed that GBP5 knockdown, via its two independent shRNA clones, dramatically suppresses the protein levels of phosphorylated Akt and mTOR, but elevates the protein levels of molecules, p62, ATG5, Beclin1 and LC3-II, related to autophagy formation in MDA-MB-231 and Hs578T cells (Figure 5E and Figure S2). The massive accumulation of LC3-II in the GBP5-silencd MDA-MB-231 cells treated with chloroquine indicate a generation of autophagic flux after GBP5 knockdown (Figure S3). Moreover, the pre-treatment with autophagy inhibitor 3-methyladenin (3-MA) dramatically restored the paclitaxel sensitivity of GBP5-sliencing MDA-MB-231 cells (Figure 5F).

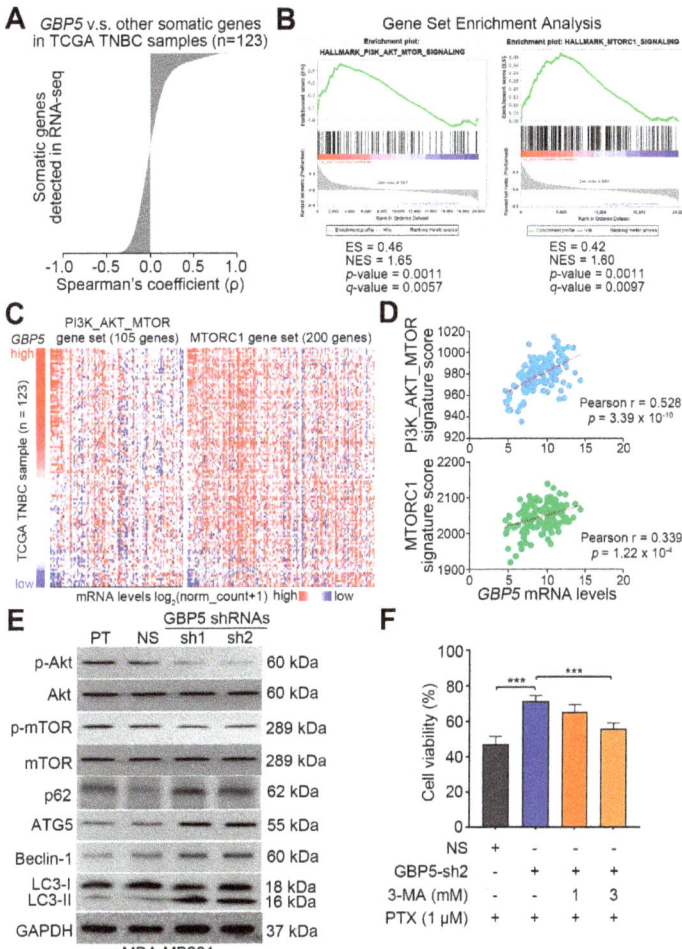

Figure 5. GBP5 activates Akt/mTOR signaling axis and inhibits autophagy activity to the paclitaxel-sensitive TNBC cells. (**A**) A histogram for the Spearman's coefficient (p) values derived from the Spearman correlation test against the co-expression of GBP5 with other somatic genes detected by RNA-sequencing method in 123 TNBC samples deposited in TCGA database. (**B**) The enrichment score (ES) derived from the correlation of GBP5 signature with the PI3K_AKT_MTOR (left) and MTORC1 (right) gene sets was plotted as the green curve. The parameters of enrichment score (NES), nominal p value and false discovery rate q value are shown as insets. (**C**) Heatmap for the transcriptional profiling of GBP5 and PI3K_AKT_MTOR (left)/MTORC1 (right) gene sets detected by RNA-sequencing tool in TNBC sample from TCGA database. (**D**) Scatchard plot for the expression of GBP5 and PI3K_AKT_MTOR (upper)/MTORC1 (lower) gene sets in the TNBC samples from TCGA database. (**E**) Western blot analyses for the protein levels of phosphorylated Akt (p-Akt), Akt, p-mTOR, mTOR, p62, ATG5, Beclin-1, LC3-I/II and GAPDH in the indicated cell variants of MDA-MB231 cells. (**F**) A histogram for the cell viability (percentages relative to untreated groups) in the non-silencing control MDA-MB231 cells and GBP5-silencing MDA-MB231 cells pretreated without or with autophagy inhibitor 3-Methyladenine (3-MA) at 1 and 3 mM prior to the treatment with paclitaxel (PTX) at 1 μM for 72 h. Non-parametric Mann–Whitney U test was used to estimate statistical significance. The symbol "***" denotes $p < 0.001$.

We further performed Kaplan–Meier analyses using minimize p value approach for determining the mRNA levels of PI3K_AKT_MTOR gene set in TNBC patients of TCGA database stratified into low and high-risk groups under progression-free survival condition. The data showed that a higher mRNA level of the PI3K_AKT_MTOR gene set refers to a good progression-free survival condition in TNBC patients (Figure 6A). Importantly, the signature of combining high-level GBP5 and PI3K_AKT_MTOR gene set predicted a prolonged time interval for cancer progression in TNBC patients from the TCGA database (Figure 6B). Collectively, we proposed that GBP5 upregulation probably enhances the activity of Akt/mTOR signaling cascades and suppresses autophagy formation in the paclitaxel-sensitive TNBC cells (Figure 6C).

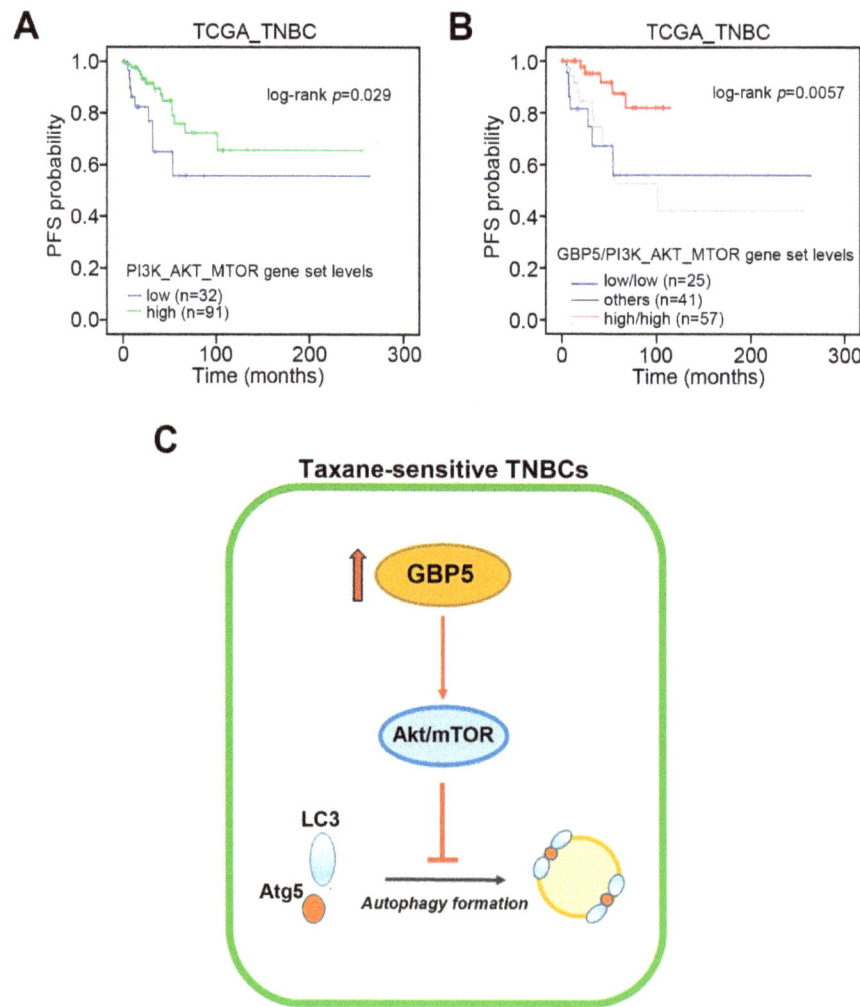

Figure 6. The signature of combining higher levels of GBP5 and PI3K_AKT_MTOR gene set correlates with a favorable progression-free condition in TNBC patients. (**A**,**B**) Kaplan–Meier analyses for the expression of PI3K_AKT_MTOR gene set without (**A**) or with (**B**) the combination of GBP5 expression using progression-free survival condition under a minimized log-rang p value against TCGA TNBC patients. (**C**) A possible mechanism for the GBP5-enhanced taxane sensitivity in TNBC.

4. Discussion

TNBC remains the breast cancer subtype with the poorest prognosis. Although transcriptional profiling has identified six different TNBC subtypes with sensitivity to therapies, the heterogeneous nature of TNBCs may point to the difficulty in the management of this breast subtype [22]. Therefore, systemic chemotherapy remains the major regimen for treating TNBC in current clinics, even though several targeted agents have been investigated in clinical trials without demonstrating a clear survival benefit [23]. Here, we show that GBP5 upregulation correlates with pathological complete response in TNBC patients who received docetaxel and paclitaxel neoadjuvanttherapy and a favorable recurrence-free survival condition in TNBC patients receiving post-operative systemic chemotherapy. In TNBC cell lines, GBP5 expression was appeared to highly correlate with cellular sensitivity to the cytotoxicity of paclitaxel. Robustly, GBP5 knockdown rendered the tested TNBC cells resistant to paclitaxel treatment. These findings not only highlight a critical role of GBP5 in regulating cellular responsiveness to paclitaxel but also provide GBP5 as a potential marker to predict the great therapeutic effectiveness of paclitaxel on TNBC patients.

Targeting the PI3K/Akt/mTOR signaling axis has been considered to be a promising therapy for the TNBC subtypes, including basal-like 2 (BL2), luminal androgen receptor (LAR), mesenchymal stem-like (MSL) and mesenchymal (M) [7]. The BL2 subtype has been identified to frequently overexpress growth factor receptors, such as epidermal growth factor receptor (EGFR), IGF1R, and myoepithelial markers and commonly exhibit the poorest response to chemotherapy in comparison with other TNBC subtypes [24]. Both MSL and M subtypes were found to highly associate with epithelial–mesenchymal transition and cell motility and frequently harbor a PI3KCA-activating mutations, which provides a therapeutic opportunity for the PI3K/mTOR inhibitor [7]. Although the LAR subtype expresses androgen receptors with sensitivity to an AR antagonist such as bicalutamide, TNBC patients with LAR tumors, compared to other TNBC subtypes, showed a decreased recurrence-free survival time [23]. Furthermore, the PI3K/AKT pathway plays a key role in tumorigenesis and metabolism, survival and proliferation in cancer cells. Previous research has shown that AKT activation by phosphorylation is a good predictor for paclitaxel treatment but a negative predictor for anthracycline-based chemotherapy in breast cancer [25]. Here, we find that the MDA-MB-231 cell line, as well as Hs578T, has been classified as an MSL subtype [23] and expresses enriched GBP5 levels. Moreover, the gene knockdown of GBP5 reduced cellular sensitivity to paclitaxel treatment and suppressed the activity of the Akt/mTOR pathway in MBA-MB-231 cells. These findings not only confirm the need for the Akt/mTOR pathway for the biologic functions of the MSL subtype, but also provide a predictive value of GBP5 for the therapeutic effectiveness of mTOR inhibitor on TNBC patients with MSL subtype.

Basal-like 1 (BL1) subtype has been identified to highly express cell-cycle and DNA-damage-response genes that suggest their great sensitivity to DNA-damaging agents such as platinum [23] and achieve a higher pCR rate in systemic chemotherapy compared to other subtype [24]. In this study, excepting MDA-MB-231 and Hs578T cells, other TNBC cell lines, HCC2157, HCC38, HCC1143, HCC1937 and MDA-MB-468, are of the BL1 subtype and express the endogenous GBP5 transcript at different levels. HCC38 and HCC1143 cells exhibiting higher GBP5 levels displayed a great sensitivity to paclitaxel treatment compared to HCC2157 and HCC1937 cells, which harbor a lower GBP5 expression. Conversely, the endogenous mRNA levels of GBP5 in these TNBC cell lines with BL1 characters appeared to be negatively correlated with the cellular sensitivity to doxorubicin treatment. Despite its lack of significance, breast cancer patients with tumors expressing a higher GBP5 transcript did not show a complete response to doxorubicin neoadjuvant therapy. Therefore, GBP5 may also serve as a potential marker to predict the therapeutic efficacy of DNA-damaging agents in TNBC patients with the BL1 subtype, even though this type shows a great pCR rate after systemic chemotherapy.

Compared to other breast cancer subtypes, TNBC was found to have the highest count of tumor-infiltrated lymphocytes (TILs) [26,27], indicating immune modulation as the new treatment paradigm in TNBC. Although the role of GBP5 in modulating the immune responses between TNBC and TIL needs to be further explored, it has been identified as an interferon-responsive effector [10,28] and was found to promote the activation of NLRP3-dependent inflammatory responses [8]. Cytotoxic drugs have been found to be capable of modifying the tumor microenvironment, thereby inducing dendritic cell activation and cytotoxic T cells [29–31], which support the concept that the immunotherapeutic effectiveness may be amplified by chemotherapy [32]. Besides this, it has been found that the induction of the inflammation-related pathway promotes metastatic progression in breast cancer [33,34]. NF-κB is recognized as a key transcription factor in regulating inflammation-related gene expression [35], as well as PD-L1 expression in lung cancer [36], thereby enhancing the metastatic potentials of TNBC [37–39]. Therefore, further experiments are needed to explore the role of GBP5-induced activation of NLRP3-dependent inflammatory response in the immunomodulatory capacity of TNBC after systemic chemotherapy.

5. Conclusions

Collectively, molecular subtyping provides a new era of precisely managing TNBC patients who decide to receive systemic chemotherapy, or who are probably sensitive to targeted therapies, e.g., Akt/mTOR inhibitors. Although the prognostic significance of p-Akt and p-mTOR in TNBC patients receiving pre- or post-operative chemotherapy remains controversial according to previous reports [23,40–42], in this study, the signature of combining low-level GBP5 with either a high- or low-level transcript of the PI3K/AKT/MOTR geneset predicted a poorer progression-free survival condition in TNBC patients. These findings suggest that GBP5 may serve as a useful biomarker to predict the therapeutic effectiveness of taxane-based chemotherapy on TNBC subtypes. Even in the BL1 subtype, which is highly sensitive to DNA-damaging agents, e.g., doxorubicin, GBP5 expression is able to distinguish an insensitive population. Importantly, this is the first documentation showing that GBP5 shows prognostic significance and is capable of regulating the activity of Akt/mTOR axis and autophagy formation in TNBC.

Supplementary Materials: The following are available online at https://www.mdpi.com/2075-4426/11/3/197/s1, Figure S1: Prognostic significance for GBPs against TNBC patients derived from K-M Plotter and TCGA cohorts under overall survival condition, Figure S2: Western blot analyses for the protein levels of phosphorylated Akt (p-Akt), Akt, p-mTOR, mTOR, p62, ATG5, Beclin-1, LC3-I/II and GAPDH in the indicated cell variants of Hs578T cells, Figure S3: Western blot analyses for the protein levels of LC3-I/II and GAPDH in the parental/non-silencing control MDA-MB231 cells and GBP5-silencing MDA-MB-231 cells without (untreated, UT) or with chloroquine (CQ) treatment at 20 μM for 24 h, Figure S4: Uncut blots for Figure 3E,F, Figure S5: Uncut blots for Figure 5E, Figure S6: Uncut blots for Figures S2 and S3.

Author Contributions: Conception and design: S.-W.C., P.-C.C., T.-R.G., H.-W.C. and Y.-F.L. Acquisition of data: S.-W.C., P.-C.C., and H.-W.C. Analysis and interpretation of data: S.-W.C., T.-R.G., H.-W.C. and Y.-F.L. Writing, review and/or revision of the manuscript: S.-W.C., T.-R.G., H.-W.C. and Y.-F.L. All authors have read and agreed to the published version of the manuscript.

Funding: This study was supported by the Ministry of Science and Technology, Taiwan (MOST 108-2320-B-038-017-MY3 to Yuan-Feng Lin; MOST 109-2314-B-038-078-MY3 to Hui-Wen Chiu).

Institutional Review Board Statement: Not applicable.

Informed Consent Statement: Not applicable.

Data Availability Statement: Publicly available datasets GSE21997 and GSE32646 were analyzed in this study and can be found here: https://www.ncbi.nlm.nih.gov/geo/, accessed on 1 February 2021.

Conflicts of Interest: The authors declare no conflict of interest.

References

1. Perou, C.M.; Sorlie, T.; Eisen, M.B.; van de Rijn, M.; Jeffrey, S.S.; Rees, C.A.; Pollack, J.R.; Ross, D.T.; Johnsen, H.; Akslen, L.A.; et al. Molecular portraits of human breast tumours. *Nature* **2000**, *406*, 747–752. [CrossRef]

2. Schmadeka, R.; Harmon, B.E.; Singh, M. Triple-negative breast carcinoma: Current and emerging concepts. *Am. J. Clin. Pathol.* **2014**, *141*, 462–477. [CrossRef]
3. Lee, K.L.; Chen, G.; Chen, T.Y.; Kuo, Y.C.; Su, Y.K. Effects of Cancer Stem Cells in Triple-Negative Breast Cancer and Brain Metastasis: Challenges and Solutions. *Cancers* **2020**, *12*, 2122. [CrossRef]
4. Salgado, R.; Denkert, C.; Demaria, S.; Sirtaine, N.; Klauschen, F.; Pruneri, G.; Wienert, S.; Van den Eynden, G.; Baehner, F.L.; Penault-Llorca, F.; et al. The evaluation of tumor-infiltrating lymphocytes (TILs) in breast cancer: Recommendations by an International TILs Working Group 2014. *Ann. Oncol.* **2015**, *26*, 259–271. [CrossRef]
5. Guarneri, V.; Barbieri, E.; Piacentini, F.; Giovannelli, S.; Ficarra, G.; Frassoldati, A.; Maiorana, A.; D'Amico, R.; Conte, P. Predictive and prognostic role of p53 according to tumor phenotype in breast cancer patients treated with preoperative chemotherapy: A single-institution analysis. *Int. J. Biol. Markers* **2010**, *25*, 104–111. [CrossRef]
6. Wang, C.; Zhang, J.; Wang, Y.; Ouyang, T.; Li, J.; Wang, T.; Fan, Z.; Fan, T.; Lin, B.; Xie, Y. Prevalence of BRCA1 mutations and responses to neoadjuvant chemotherapy among BRCA1 carriers and non-carriers with triple-negative breast cancer. *Ann. Oncol.* **2015**, *26*, 523–528. [CrossRef]
7. Omarini, C.; Guaitoli, G.; Pipitone, S.; Moscetti, L.; Cortesi, L.; Cascinu, S.; Piacentini, F. Neoadjuvant treatments in triple-negative breast cancer patients: Where we are now and where we are going. *Cancer Manag. Res.* **2018**, *10*, 91–103. [CrossRef]
8. Shenoy, A.R.; Wellington, D.A.; Kumar, P.; Kassa, H.; Booth, C.J.; Cresswell, P.; MacMicking, J.D. GBP5 promotes NLRP3 inflammasome assembly and immunity in mammals. *Science* **2012**, *336*, 481–485. [CrossRef]
9. Hotter, D.; Sauter, D.; Kirchhoff, F. Guanylate binding protein 5: Impairing virion infectivity by targeting retroviral envelope glycoproteins. *Small GTPases* **2017**, *8*, 31–37. [CrossRef]
10. Li, Z.; Qu, X.; Liu, X.; Huan, C.; Wang, H.; Zhao, Z.; Yang, X.; Hua, S.; Zhang, W. GBP5 Is an Interferon-Induced Inhibitor of Respiratory Syncytial Virus. *J. Virol.* **2020**, *94*, e01407-20. [CrossRef] [PubMed]
11. Matta, S.K.; Patten, K.; Wang, Q.; Kim, B.H.; MacMicking, J.D.; Sibley, L.D. NADPH Oxidase and Guanylate Binding Protein 5 Restrict Survival of Avirulent Type III Strains of Toxoplasma gondii in Naive Macrophages. *mBio* **2018**, *9*, e01393-18. [CrossRef]
12. Koltes, J.E.; Fritz-Waters, E.; Eisley, C.J.; Choi, I.; Bao, H.; Kommadath, A.; Serao, N.V.; Boddicker, N.J.; Abrams, S.M.; Schroyen, M.; et al. Identification of a putative quantitative trait nucleotide in guanylate binding protein 5 for host response to PRRS virus infection. *BMC Genom.* **2015**, *16*, 412. [CrossRef] [PubMed]
13. Tripal, P.; Bauer, M.; Naschberger, E.; Mortinger, T.; Hohenadl, C.; Cornali, E.; Thurau, M.; Sturzl, M. Unique features of different members of the human guanylate-binding protein family. *J. Interferon Cytokine Res.* **2007**, *27*, 44–52. [CrossRef] [PubMed]
14. Cheng, L.; Gou, L.; Wei, T.; Zhang, J. GBP1 promotes erlotinib resistance via PGK1activated EMT signaling in nonsmall cell lung cancer. *Int. J. Oncol.* **2020**, *57*, 858–870. [CrossRef]
15. Zhao, J.; Li, X.; Liu, L.; Cao, J.; Goscinski, M.A.; Fan, H.; Li, H.; Suo, Z. Oncogenic Role of Guanylate Binding Protein 1 in Human Prostate Cancer. *Front. Oncol.* **2019**, *9*, 1494. [CrossRef] [PubMed]
16. Mustafa, D.A.M.; Pedrosa, R.M.S.M.; Smid, M.; van der Weiden, M.; de Weerd, V.; Nigg, A.L.; Berrevoets, C.; Zeneyedpour, L.; Priego, N.; Valiente, M.; et al. T lymphocytes facilitate brain metastasis of breast cancer by inducing Guanylate-Binding Protein 1 expression. *Acta Neuropathol.* **2018**, *135*, 581–599. [CrossRef]
17. Quintero, M.; Adamoski, D.; Reis, L.M.D.; Ascencao, C.F.R.; Oliveira, K.R.S.; Goncalves, K.A.; Dias, M.M.; Carazzolle, M.F.; Dias, S.M.G. Guanylate-binding protein-1 is a potential new therapeutic target for triple-negative breast cancer. *BMC Cancer* **2017**, *17*, 727. [CrossRef]
18. Godoy, P.; Cadenas, C.; Hellwig, B.; Marchan, R.; Stewart, J.; Reif, R.; Lohr, M.; Gehrmann, M.; Rahnenfuhrer, J.; Schmidt, M.; et al. Interferon-inducible guanylate binding protein (GBP2) is associated with better prognosis in breast cancer and indicates an efficient T cell response. *Breast Cancer* **2014**, *21*, 491–499. [CrossRef]
19. Rahvar, F.; Salimi, M.; Mozdarani, H. Plasma GBP2 promoter methylation is associated with advanced stages in breast cancer. *Genet. Mol. Biol.* **2020**, *43*, e20190230. [CrossRef]
20. Budczies, J.; Klauschen, F.; Sinn, B.V.; Győrffy, B.; Schmitt, W.D.; Darb-Esfahani, S.; Denkert, C. Cutoff Finder: A comprehensive and straightforward Web application enabling rapid biomarker cutoff optimization. *PLoS ONE* **2012**, *7*, e51862. [CrossRef]
21. Qin, A.; Lai, D.H.; Liu, Q.; Huang, W.; Wu, Y.P.; Chen, X.; Yan, S.; Xia, H.; Hide, G.; Lun, Z.R.; et al. Guanylate-binding protein 1 (GBP1) contributes to the immunity of human mesenchymal stromal cells against Toxoplasma gondii. *Proc. Natl. Acad. Sci. USA* **2017**, *114*, 1365–1370. [CrossRef] [PubMed]
22. Kennedy, R.D.; Quinn, J.E.; Johnston, P.G.; Harkin, D.P. BRCA1: Mechanisms of inactivation and implications for management of patients. *Lancet* **2002**, *360*, 1007–1014. [CrossRef]
23. Lehmann, B.D.; Bauer, J.A.; Chen, X.; Sanders, M.E.; Chakravarthy, A.B.; Shyr, Y.; Pietenpol, J.A. Identification of human triple-negative breast cancer subtypes and preclinical models for selection of targeted therapies. *J. Clin. Investig.* **2011**, *121*, 2750–2767. [CrossRef] [PubMed]
24. Gluz, O.; Nitz, U.; Liedtke, C.; Christgen, M.; Grischke, E.M.; Forstbauer, H.; Braun, M.; Warm, M.; Hackmann, J.; Uleer, C.; et al. Comparison of Neoadjuvant Nab-Paclitaxel+Carboplatin vs Nab-Paclitaxel+Gemcitabine in Triple-Negative Breast Cancer: Randomized WSG-ADAPT-TN Trial Results. *J. Natl. Cancer Inst.* **2018**, *110*, 628–637. [CrossRef]
25. Yang, S.X.; Polley, E.; Lipkowitz, S. New insights on PI3K/AKT pathway alterations and clinical outcomes in breast cancer. *Cancer Treat. Rev.* **2016**, *45*, 87–96. [CrossRef]

26. Disis, M.L.; Stanton, S.E. Triple-negative breast cancer: Immune modulation as the new treatment paradigm. *Am. Soc. Clin. Oncol. Educ. Book.* **2015**, e25–e30. [CrossRef]
27. Garcia-Teijido, P.; Cabal, M.L.; Fernandez, I.P.; Perez, Y.F. Tumor-Infiltrating Lymphocytes in Triple Negative Breast Cancer: The Future of Immune Targeting. *Clin. Med. Insights Oncol.* **2016**, *10*, 31–39. [CrossRef]
28. Feng, J.; Cao, Z.; Wang, L.; Wan, Y.; Peng, N.; Wang, Q.; Chen, X.; Zhou, Y.; Zhu, Y. Inducible GBP5 Mediates the Antiviral Response via Interferon-Related Pathways during Influenza A Virus Infection. *J. Innate Immun.* **2017**, *9*, 419–435. [CrossRef]
29. Apetoh, L.; Ghiringhelli, F.; Tesniere, A.; Obeid, M.; Ortiz, C.; Criollo, A.; Mignot, G.; Maiuri, M.C.; Ullrich, E.; Saulnier, P.; et al. Toll-like receptor 4-dependent contribution of the immune system to anticancer chemotherapy and radiotherapy. *Nat. Med.* **2007**, *13*, 1050–1059. [CrossRef] [PubMed]
30. Schreiber, R.D.; Old, L.J.; Smyth, M.J. Cancer immunoediting: Integrating immunity's roles in cancer suppression and promotion. *Science* **2011**, *331*, 1565–1570. [CrossRef]
31. Andre, F.; Dieci, M.V.; Dubsky, P.; Sotiriou, C.; Curigliano, G.; Denkert, C.; Loi, S. Molecular pathways: Involvement of immune pathways in the therapeutic response and outcome in breast cancer. *Clin. Cancer Res.* **2013**, *19*, 28–33. [CrossRef]
32. Dieci, M.V.; Criscitiello, C.; Goubar, A.; Viale, G.; Conte, P.; Guarneri, V.; Ficarra, G.; Mathieu, M.C.; Delaloge, S.; Curigliano, G.; et al. Prognostic value of tumor-infiltrating lymphocytes on residual disease after primary chemotherapy for triple-negative breast cancer: A retrospective multicenter study. *Ann. Oncol.* **2014**, *25*, 611–618. [CrossRef] [PubMed]
33. Ershaid, N.; Sharon, Y.; Doron, H.; Raz, Y.; Shani, O.; Cohen, N.; Monteran, L.; Leider-Trejo, L.; Ben-Shmuel, A.; Yassin, M.; et al. NLRP3 inflammasome in fibroblasts links tissue damage with inflammation in breast cancer progression and metastasis. *Nat. Commun.* **2019**, *10*, 4375. [CrossRef]
34. Wellenstein, M.D.; Coffelt, S.B.; Duits, D.E.M.; van Miltenburg, M.H.; Slagter, M.; de Rink, I.; Henneman, L.; Kas, S.M.; Prekovic, S.; Hau, C.S.; et al. Loss of p53 triggers WNT-dependent systemic inflammation to drive breast cancer metastasis. *Nature* **2019**, *572*, 538–542. [CrossRef] [PubMed]
35. Ilchovska, D.D.; Barrow, D.M. An Overview of the NF-kB mechanism of pathophysiology in rheumatoid arthritis, investigation of the NF-kB ligand RANKL and related nutritional interventions. *Autoimmun. Rev.* **2020**, *20*, 102741. [CrossRef] [PubMed]
36. Asgarova, A.; Asgarov, K.; Godet, Y.; Peixoto, P.; Nadaradjane, A.; Boyer-Guittaut, M.; Galaine, J.; Guenat, D.; Mougey, V.; Perrard, J.; et al. PD-L1 expression is regulated by both DNA methylation and NF-kB during EMT signaling in non-small cell lung carcinoma. *Oncoimmunology* **2018**, *7*, e1423170. [CrossRef]
37. Ma, F.; Zu, X.; Liu, K.; Bode, A.M.; Dong, Z.; Liu, Z.; Kim, D.J. Knockdown of Pyruvate Kinase M Inhibits Cell Growth and Migration by Reducing NF-kB Activity in Triple-Negative Breast Cancer Cells. *Mol. Cells* **2019**, *42*, 628–636.
38. Arora, R.; Yates, C.; Gary, B.D.; McClellan, S.; Tan, M.; Xi, Y.; Reed, E.; Piazza, G.A.; Owen, L.B.; Dean-Colomb, W. Panepoxydone targets NF-kB and FOXM1 to inhibit proliferation, induce apoptosis and reverse epithelial to mesenchymal transition in breast cancer. *PLoS ONE* **2014**, *9*, e98370. [CrossRef]
39. Rajendran, P.; Ben, A.R.; Al-Saeedi, F.J.; Elsayed, M.M.; Islam, M.; Al-Ramadan, S.Y. Thidiazuron decreases epithelial-mesenchymal transition activity through the NF-kB and PI3K/AKT signalling pathways in breast cancer. *J. Cell. Mol. Med.* **2020**, *24*, 14525–14538. [CrossRef]
40. Khan, M.A.; Jain, V.K.; Rizwanullah, M.; Ahmad, J.; Jain, K. PI3K/AKT/mTOR pathway inhibitors in triple-negative breast cancer: A review on drug discovery and future challenges. *Drug Discov. Today* **2019**, *24*, 2181–2191. [CrossRef]
41. Pascual, J.; Turner, N.C. Targeting the PI3-kinase pathway in triple-negative breast cancer. *Ann. Oncol.* **2019**, *30*, 1051–1060. [CrossRef] [PubMed]
42. Ueng, S.H.; Chen, S.C.; Chang, Y.S.; Hsueh, S.; Lin, Y.C.; Chien, H.P.; Lo, Y.F.; Shen, S.C.; Hsueh, C. Phosphorylated mTOR expression correlates with poor outcome in early-stage triple negative breast carcinomas. *Int. J. Clin. Exp. Pathol.* **2012**, *5*, 806–813. [PubMed]

Article

Identification of Novel Biomarkers and Candidate Drug in Ovarian Cancer

Chia-Jung Li [1,2], Li-Te Lin [1,2,3], Pei-Yi Chu [4,5,6,7], An-Jen Chiang [1], Hsiao-Wen Tsai [1,2], Yi-Han Chiu [8], Mei-Shu Huang [1], Zhi-Hong Wen [9] and Kuan-Hao Tsui [1,2,3,10,11,12,*]

1. Department of Obstetrics and Gynaecology, Kaohsiung Veterans General Hospital, Kaohsiung 813, Taiwan; nigel6761@gmail.com (C.-J.L.); litelin1982@gmail.com (L.-T.L.); ajchiang490111@gmail.com (A.-J.C.); drtsai0627@gmail.com (H.-W.T.); m1221226@gmail.com (M.-S.H.)
2. Institute of Biopharmaceutical Sciences, National Sun Yat-sen University, Kaohsiung 804, Taiwan
3. Department of Obstetrics and Gynaecology, National Yang-Ming University School of Medicine, Taipei 112, Taiwan
4. School of Medicine, College of Medicine, Fu Jen Catholic University, New Taipei 242, Taiwan; chu.peiyi@msa.hinet.net
5. Department of Pathology, Show Chwan Memorial Hospital, Changhua 500, Taiwan
6. Department of Health Food, Chung Chou University of Science and Technology, Changhua 510, Taiwan
7. National Institute of Cancer Research, National Health Research Institutes, Tainan 704, Taiwan
8. Department of Microbiology, Soochow University, Taipei 111, Taiwan; chiuyiham@scu.edu.tw
9. Department of Marine Biotechnology and Resources, National Sun Yat-sen University, Kaohsiung 804, Taiwan; wzh@mail.nsysu.edu.tw
10. Department of Obstetrics and Gynecology, Taipei Veterans General Hospital, Taipei 112, Taiwan
11. Department of Pharmacy and Master Program, College of Pharmacy and Health Care, Tajen University, Pingtung County 907, Taiwan
12. Department of Medicine, Tri-Service General Hospital, National Defense Medical Center, Taipei 114, Taiwan
* Correspondence: khtsui60@gmail.com; Tel.: +886-7-342-2121

Citation: Li, C.-J.; Lin, L.-T.; Chu, P.-Y.; Chiang, A.-J.; Tsai, H.-W.; Chiu, Y.-H.; Huang, M.-S.; Wen, Z.-H.; Tsui, K.-H. Identification of Novel Biomarkers and Candidate Drug in Ovarian Cancer. *J. Pers. Med.* **2021**, *11*, 316. https://doi.org/10.3390/jpm11040316

Academic Editor: James Meehan

Received: 9 March 2021
Accepted: 15 April 2021
Published: 19 April 2021

Publisher's Note: MDPI stays neutral with regard to jurisdictional claims in published maps and institutional affiliations.

Copyright: © 2021 by the authors. Licensee MDPI, Basel, Switzerland. This article is an open access article distributed under the terms and conditions of the Creative Commons Attribution (CC BY) license (https://creativecommons.org/licenses/by/4.0/).

Abstract: This paper investigates the expression of the CREB1 gene in ovarian cancer (OV) by deeply excavating the gene information in the multiple databases and the mechanism thereof. In short, we found that the expression of the CREB1 gene in ovarian cancer tissue was significantly higher than that of normal ovarian tissue. Kaplan–Meier survival analysis showed that the overall survival was significantly shorter in patients with high expression of the CREB1 gene than those in patients with low expression of the CREB1 gene, and the prognosis of patients with low expression of the CREB1 gene was better. The CREB1 gene may play a role in the occurrence and development of ovarian cancer by regulating the process of protein. Based on differentially expressed genes, 20 small-molecule drugs that potentially target CREB1 with abnormal expression in OV were obtained from the CMap database. Among these compounds, we found that naloxone has the greatest therapeutic value for OV. The high expression of the CREB1 gene may be an indicator of poor prognosis in ovarian cancer patients. Targeting CREB1 may be a potential tool for the diagnosis and treatment of OV.

Keywords: ovarian cancer; bioinformatics; CREB1; drug perturbation

1. Introduction

Ovarian cancer is a common malignant tumor in gynecology, with the highest mortality rate and the second highest incidence rate of gynecologic malignancies, and the 5-year survival rate is only 25–30% [1]. Current studies have found that epigenetic modifications play an important role in the occurrence and development of ovarian cancer [2]. Therefore, a comprehensive study on the pathogenesis of ovarian cancer and the establishment of effective prevention and treatment programs are the urgent issues that we should address now. Nowadays, systematic analysis based on gene microarray technology using bioinformatics methods to study tumor-related genes and their regulatory mechanisms is one of the main research tools in functional genomics.

CREB1 is a nuclear protein in eukaryotic cells and a nuclear factor that regulates transcription. It is composed of 341 amino acid residues and has a molecular weight of 43 KDa. It belongs to the CREB/ATF subgroup of the leucine zipper family of transcription factors [3]. CREB is activated by phosphorylation, forming homodimers or heterodimers, and it is regulated by cofactors to recognize and bind the cAMP response element (CRE) in the target gene promoter, promoting the transcriptional expression of the gene and participating in tumor proliferation, differentiation, and metastasis [4]. Extracellular signals interact with receptors on the cell membrane to phosphorylate and activate CREB1 via signaling pathways such as PKA, PKC, PKB, and ERK [5]. It further regulates the cell cycle, promotes cell proliferation, inhibits apoptosis, etc. Several studies have shown that target genes regulated by CREB1 play an important role in cell proliferation, differentiation, survival, and cell cycle regulation. Its overexpression contributes to cell survival and proliferation and has an important role in the development of several tumors [6–9].

In this study, we first performed bioinformatics analysis to investigate the prognostic and therapeutic impact of CREB1 on OV. We identified human ovarian cancer tissue microarray and different stages of ovarian cancer cells for analysis of CREB1 protein and mRNA levels. Finally, pharmacogenomics was used to predict potential drugs. Since naloxone has been approved for clinical treatment of lung cancer [10], our results may support the use of CREB1 gene status as an ovarian cancer biomarker and precision treatment of OV patients with naloxone.

2. Materials and Methods

2.1. Cells and Cell Culture

Ovarian cancer cell lines OC-117-VGH cells (BCRC#60601, Hsinchu, Taiwan), OC-117-VGH cells (BCRC#60602), OCPC-2-VGH (BCRC#60603), OC-3-VGH (BCRC#60599), TOV-21G (BCRC#60407), and NIH-OVCAR-3 (BCRC#60551) were used and cultured in DMEM/F12 supplemented with 1.5 g/L sodium bicarbonate and 10% fetal bovine serum (Themo Fisher Scientific, Waltham, MA, USA) in a humidified atmosphere of 95% air and 5% CO_2 at 37 °C.

2.2. RNA Extraction and Real-Time PCR

The total RNA was extracted with the EasyPrep Total RNA Kit (BIOTOOLS Co., Ltd., Taipei, Taiwan). A total of 1 µg of RNA was reverse-transcribed with a ToolScript MMLV RT kit (BIOTOOLS Co., Ltd.) for cDNA synthesis. Real-time PCR was carried out using a StepOnePlusTM system (Applied Biosystems, Foster City, CA, USA) with TOOLS 2X SYBR qPCR Mix (BIOTOOLS Co., Ltd.). The expression levels of all the genes in cells were normalized to the internal control RNU6-1 gene. All the samples with a coefficient of variation for Ct value > 1% were retested.

2.3. Tissue microarrays (TMA) and Immunohistochemistry (IHC) Analysis

Tissue array slides (CJ2) containing human ovarian cancer, metastatic, and normal tissues were purchased from SuperBioChips Laboratories (Seoul, Republic of Korea). For immunohistochemistry (IHC), assays and scoring methods were performed as described [11]. The slides were treated with anti-CREB1 antibody (1:100, A11063, ABclonal, Boston, MA, USA). All glass slides were digitized with an Motic Easyscan Digital Slide Scanner (Motic Hong Kong Limited, Hong Kong, China) at ×40 (0.26 µm/pixel) with high precision (High precision autofocus). Motic Easyscan whole-slide images were viewed with DSAssistant and EasyScanner software at Li-Tzung Pathology Laboratory (Kaohsiung, Taiwan).

2.4. Multi-Omics Analysis

TumorMap is an integrated genomics portal for visual and exploratory analysis that biologists and bioinformaticians can use to query a rich set of cancer genomics data.

The intuitive and interactive layout helps to identify cancer subtypes based on common molecular activities in a set of tumor samples [12].

The gene mutations and co-expression of CREB1 were computed and analyzed using CBio-Cancer Genomics Portal (cBioPortal) databases. Searching the term "CREB1" enabled the acquisition of the full mutation distribution across all tumor and non-tumor tissues. We analyzed the expression of CREB1 in 9736 tumors and 8587 normal tissues using this tool [13].

Gene Expression Profiling Interactive Analysis (GEPIA) is an interactive network database that can be linked and analyzed with other databases (TCGA and GTEx). Using GEPIA, we analyzed 9736 tumors and 8587 normal tissues [13].

Metascape contains fully integrated data from multiple databases such as GO, KEGG, UniProt, and DrugBank. With metascape, it is possible to perform a complete pathway enrichment and biological process annotation, gene-related protein network analysis, and drug analysis [14].

We used Reactome to compare a pathway to its homolog in another species, and the protein–compound interactions from external databases were used to confirm our findings [15–17]. The data contained 51,745 PPIs among 10,177 human proteins. After filtering the PPI data for proteins encoded by genes having transcriptome data in TCGA datasets, a network was reconstructed with 34,604 PPIs among 8322 proteins.

The CMap database collects drug-induced gene expression profiles from human cancer cell lines, which can be used to compare the similarity and dissimilarity between the inputted DEGs and drug-induced gene expression [18].

2.5. Statistical Analyses

All data were presented as mean ± standard deviation (SD) or case number (%). The correlation between the clinicopathological parameters and the four gene expressions was analyzed using the chi-square or Fisher exact tests for categorical variables, and paired-sample t-test for continuous variables, using the GraphPad Prism 8.0 (GraphPad Software, San Diego, CA, USA). The Spearman rank correlation test was used to analyze the correlation results of expression of the four biomolecules. In this study, the endpoints were overall survival (OS) and disease-free survival (DFS). The results of univariable analysis of the variables and survival data were performed using the Kaplan-Meier method with the log-rank test. The relationship between the variables and survival data was analyzed via Cox's proportional hazards regression model. Statistical significance was defined as a p-value < 0.05.

3. Results

3.1. TumorMap and Integrated Cluster Identify Significant Features Distinguishing OV among PanCancer-33 Tumors

First, we explored whether CREB1 is involved in multiple types of cancer and physiological functions. We investigated the sample subgroups revealed by the integration of the multi-omics platform. We categorized the inverse significance of the similarity of the data to create an integrated graph with equal contributions from seven different data platforms. Each group represents a different type of physiological function. Several known connections between tumor types can be seen on the result graph. We found that CREB1 is widely expressed in reproductive system or breast disease and urinary system disease (Figure 1a). To identify a molecular signature-based classification, we conducted an integrated tumor map and cluster identify analysis of 9,759 tumor samples from PanCancer-33 cancers (Figure 1b) for which Gyn disease (Figure 1c), mutation (Figure 1d), and methylation (Figure 1e), and a smaller set of protein expression profiles were available. The integrated map separates OV patients into distinct cytogenetic subgroups, which are characterized by differential cancer types. The OV-like tumors are further characterized by mutations and methylation in CREB1. Therefore, we judged the maps to be biologically relevant.

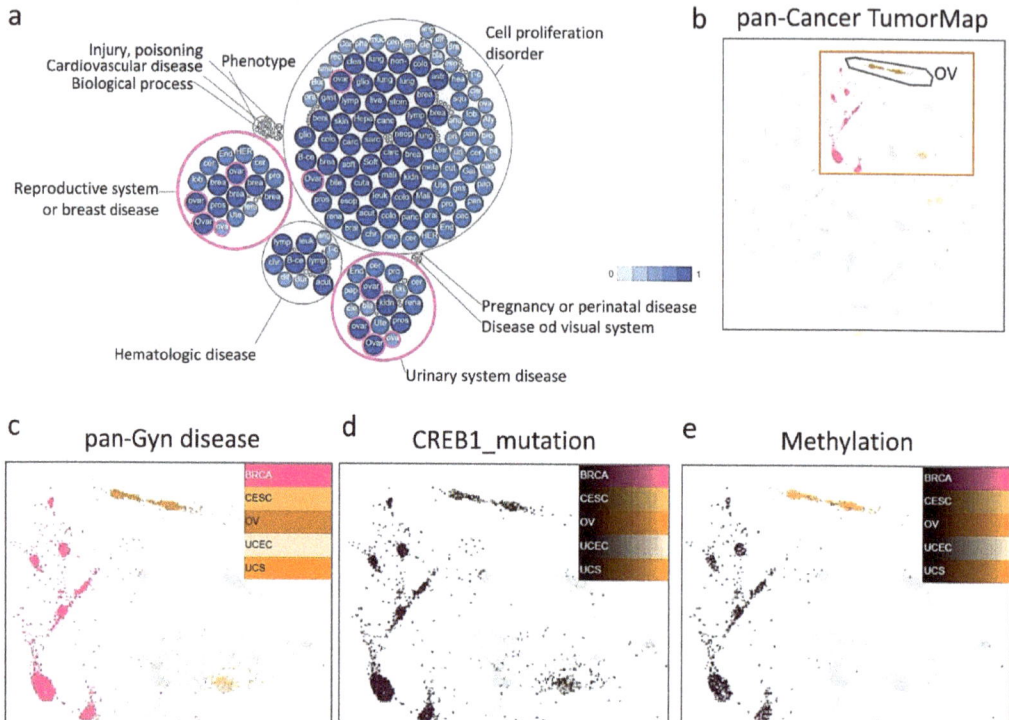

Figure 1. TumorMap and integrated cluster of gynecologic cancers from PanCancer-33 Analysis. (**a**) The distribution of CREB1 in various cancers and physiological functions. (**b**) TumorMap analysis visualizing close mapping of BRCA, CESC, OV, UCEC, and UCS among 28 PanCancer-33 islands. (**c**) Higher resolution view of TumorMap islands and distribution of Gyn cancers from five sites. (**d**) CREB1 mutation status showing the majority of mutation BRCA and OV map around a distinct island. (**e**) Methylation of Gyn cancers. Each spot in the map represents a sample. The colors of the sample spots represent attributes as described for each panel. BRCA: Breast invasive carcinoma, CESC: cervical squamous cell carcinoma and endocervical adenocarcinoma, OV: ovarian serous cystadenocarcinoma, UCEC: uterine corpus endometrial carcinoma, UCS: uterine carcinosarcoma.

3.2. CREB1 Gene Mutation Predicts A Poorer Disease-Free Survival in OV Patients

CREB1 expression is frequently found in OV. Indeed, we mined "TCGA, PanCancer Atlas" data via the cBioPortal website for the genetic alterations of CREB1 gene. Among the 32 tumor types we used as dataset, the expression levels of these hub genes varied from 0.17% to 2.08%, and the mutation frequency of each hub gene was shown in Figure 1a. This pan-cancer analysis also indicated that CREB1 gene alterations occurred most frequently in OV, compared with other cancer types (Figure 2a). The alterations for the CREB1 gene was calculated to be between 0% and 2.1% in the examined ALL samples. Genetic mutations of CREB1 were 2.1% (Figure 2b). From the diagram of CREB1 gene and the encoded protein, mutations occurred more frequently in the kinase inducible domain (KID) that is responsible for heteromerization and transactivation (Figure 2c).

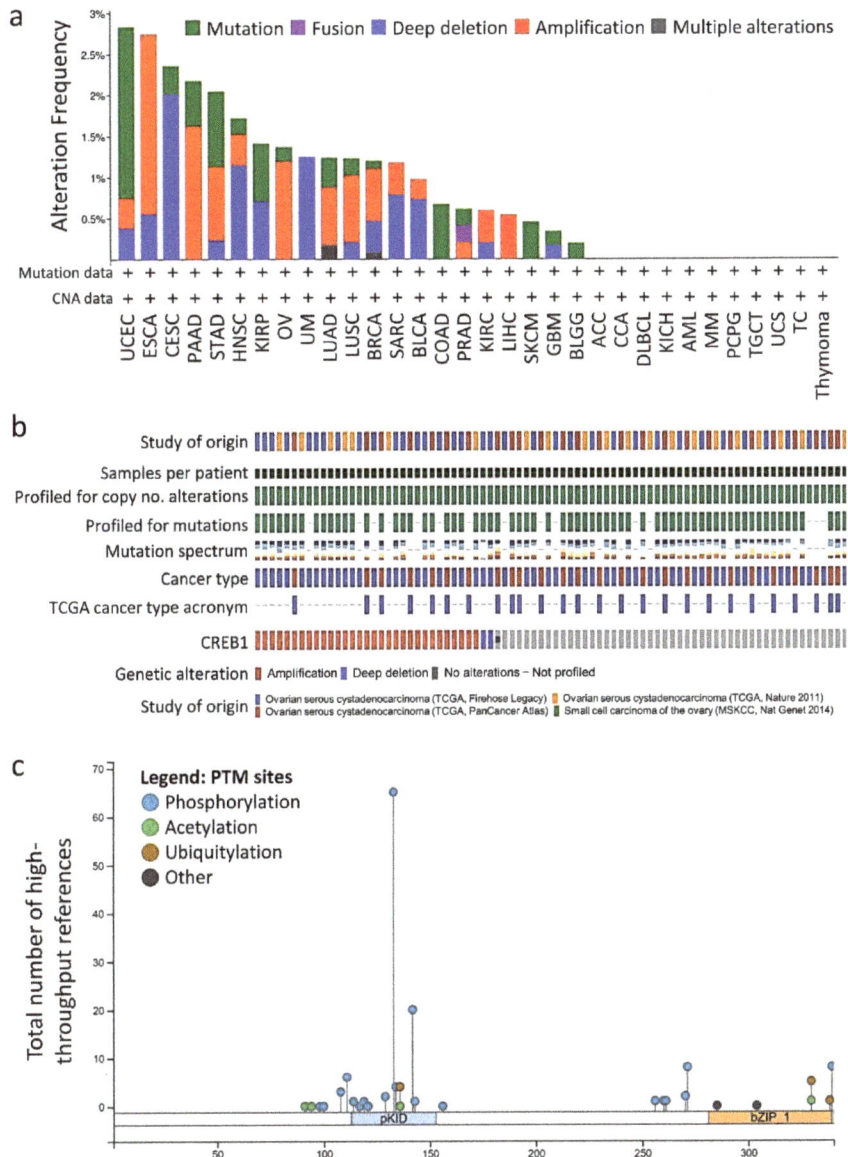

Figure 2. Copy number amplification of CREB1 gene in OV. (**a**) CREB1 gene was analyzed for gene alterations (mutation status and copy number variation) in various cancer types using "TCGA, PanCancer Atlas" data in the cBioPortal cancer genomics database; (**b**) Oncoprint table of significant signature genes. The Oncoprint table summarizes genomic alterations in all queried genes across samples. Red bars indicate gene amplifications, blue bars are homozygous deletions, and green squares are nonsynonymous mutations. (**c**) A cartoon diagram for the gene and protein structures of CREB1. This figure was adapted from the image obtained from the cBioPortal website.

3.3. Distribution and Expression of CREB1 in Cancer Tissues of Patients with OV

To analyze the expression pattern of CREB1 in various cancers, we accessed the TCGA and GEPIA databases. We carried out the comparison of the transcriptional levels of CREB1 in cancers with those in the normal specimens through the use of ONCOMINE databases (Figure 3a). The significant upregulation of the mRNA expression levels of CREB1 was carried out in OV patients. We further acquired the experimental evidence about the sub-localization of CREB1. Meanwhile, sub-localization of CREB1 in human cell line A-431 and U-251 MG demonstrated that CREB1 protein existed at the nucleus of A-431 and U-251 MG cells41 (Figure 3b). In addition, immunohistochemistry of pathological sections from the Human Protein Atlas Database (HPAD) showed that the protein expression of CREB1 was substantially increased in OV tissues of different ages (Figure 3c). Next, we determined the transcriptional expression of the target genes differentially expressed between OV and normal tissues in TCGA. mRNA levels of CREB1 were found to be significantly increased in OV, indicating that these proteins may have potential carcinogenic effects (Figure 3d). Subsequently, OV patients with high levels of CREB1mRNA expression had low overall survival (Figure 3e). GSEA analysis of RNA-seq data was performed to further explore the involved biological pathways and cofactors of CREB1 in OV. High CREB1 expression was defined as TPM in the 1st quartile, and low FBXW4 expression was defined as TPM in the 4th quartile. The results showed that in the patients with high CREB1 expression, the gene sets were significantly enriched in OXPHOS (normalized enrichment score (NES) = 1.862, $p = 0.002$) (Figure 3f).

3.4. Tissue Microarray Analysis of CREB1 Expression

To further confirm the accuracy of the multi-omics analysis, we evaluated CREB1 detected using immunohistochemistry in tumor tissues by using 60 OV commercial tissue microarrays. The results of CREB1 expression in OV tissues in IHC staining are shown in Figure 4a. The expression of CREB1 was significantly higher in early stages than in late stages. At higher IHC scores, CREB1 expression was significantly higher in patients with stage I than in patients with advanced stages ($p < 0.05$; Figure 4b). Similar to the above-mentioned TCGA data, the overall survival rate of OV patients with high CREB1 mRNA expression levels was lower than that of OV patients with low CREB1 mRNA expression levels (Figure 4c). Next, we analyzed the endogenous levels of CREB1 in ovarian cancer cell lines, and the results showed that the mRNA expression levels of CREB1 were higher in early-stage cells than in late-stage cells in different ovarian cancer cell lines (Figure 4d).

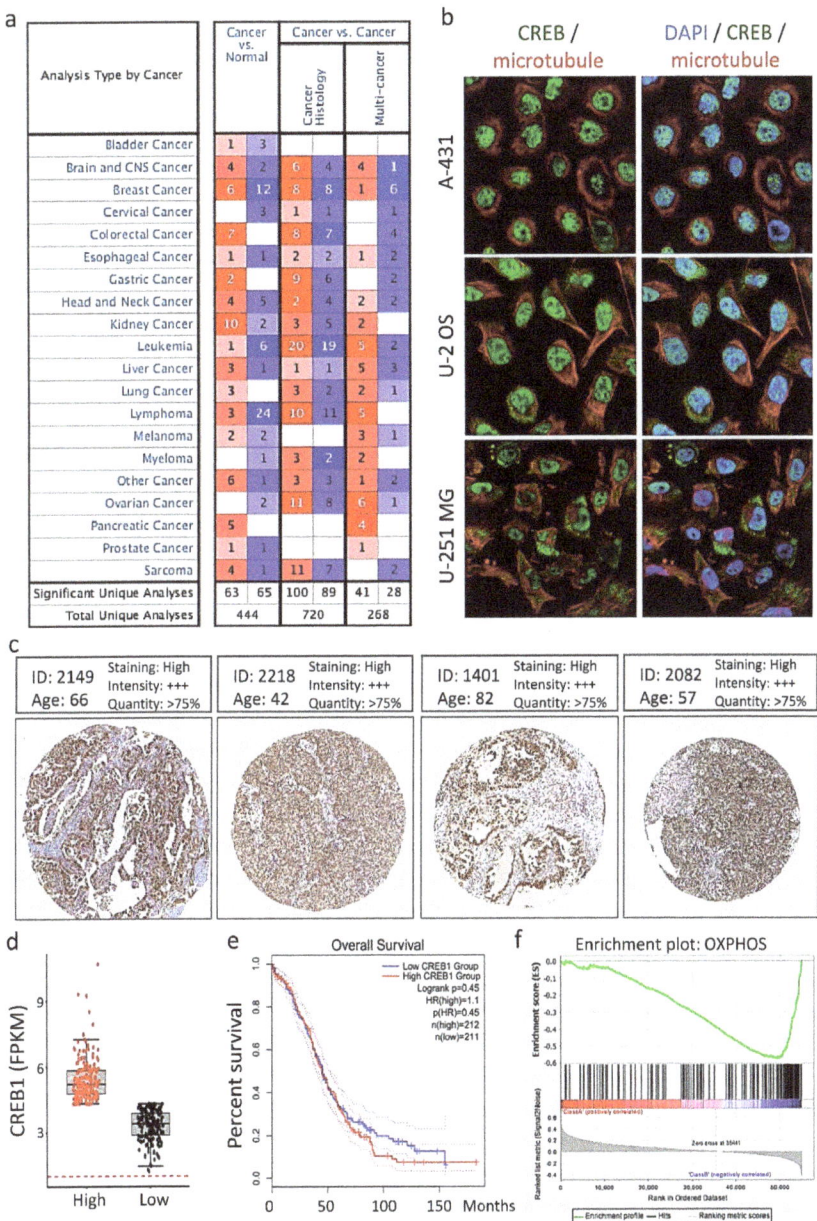

Figure 3. Diagnostic and prognostic value of CREB1 in OV. (**a**) The graphic was generated from the ONCOMINE database, indicating the number of datasets with statistically significant ($p < 0.01$) mRNA overexpression (red) or downregulation (blue) of CREB1. (**b**) The localization of CREB1 protein in human cells. Blue: nucleus; Green: CREB1; Red: microtubules. (**c**) Comparison of immunohistochemistry images of CREB1 in OV tissues with four different patients based on the Human Protein Atlas (+++: strong staining). (**d**) Plots chart showing higher CREB1 expression in OV patients. Data were obtained from TCGA. (**e**) Kaplan–Meier curves was performed to determine differences in OV patients. (**f**) The association between CREB1 gene mutations and OV gene signature. Gene set enrichment analysis (GSEA) was performed to enrich the OV gene signature in the following data sets.

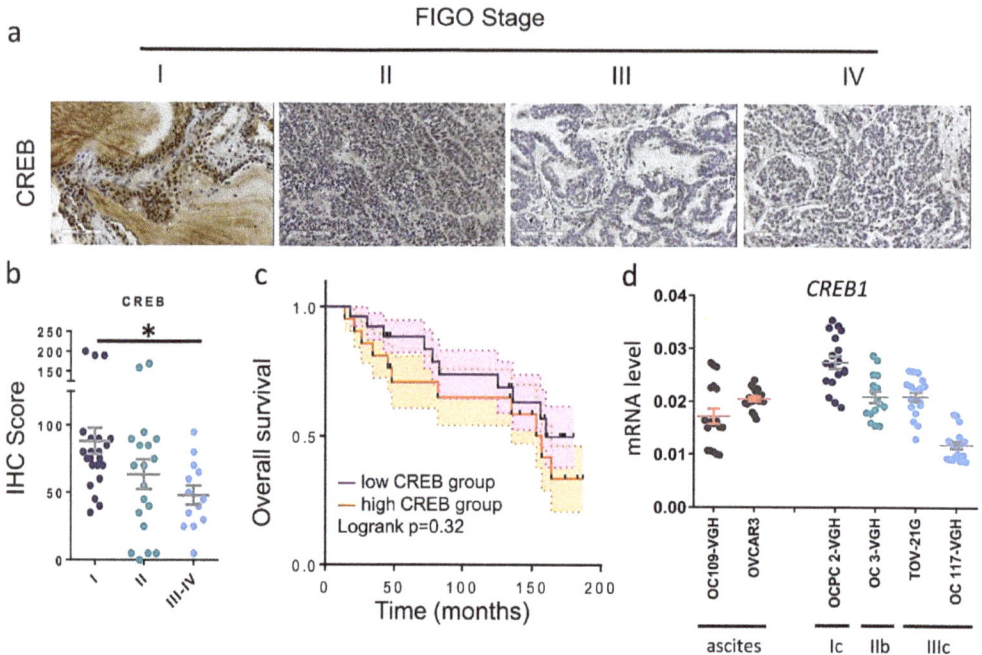

Figure 4. Immunoreactivity of CREB in OV. (**a**) The representative photomicrographs of CREB expression for weak (+), and strong (+++) staining in OV tissues (*n* = 59). (**b**) The IHC score of CREB expression in OV tissue. (**c**) Kaplan–Meier curves were performed to determine differences in OV patients. (**d**) RT-PCR was used to detect the expression levels of different ovarian cancer cells, and U6 small nuclear 1 (RNU6-1) was used as an internal control (*n* = 18). * $p < 0.05$.

3.5. Prediction of Protein–Protein Interaction of CREB1 Mutations and Copy Number Alterations

Next, we conducted Metascape Pathway and process enrichment analysis integrating the gene ontology sources, including GO Biological Process, KEGG pathway, Reactome Gene Sets, and Canonical Pathways. The predicted protein partners of CREB1 were ATF1, ATF2, PRKACB, TSSK4, BARX2, ATF7, NFIL3, NFATC2, NFYA, FAM192A, DYRK1A, ZBTB21, HIST1H2BJ, NIT2, POLR2A, PCK1, CGA, NFATC1, HLF, SOX9, FHL5, JUN, and LAX1. Thus, these predicted interacting protein partners of CREB1 might be involved in the regulation of CREB1-mediated cancer progression and prognosis (Figure 5a). Protein–protein interaction clustering algorithm identified neighborhoods within the networks where the CREB1-regulated genes were densely connected, such as ATF1, ATF2, NFATC1, NFIL3, TP53, JUN, SOX9, etc. (Figure 5b). Top 20 clusters were defined with their representative enriched terms (Figure 5c), including MAPK signaling pathways, calcium signal, and the EGFR pathway. Furthermore, network enrichment captured the interactions between the 20 clusters, as visualized using Cytoscape. These results revealed the novel and essential biological functions of CREB1 in multiple molecular pathways. As shown in Figure 3d, in addition to OV, the high alteration frequency of the CREB1 gene was also found in pancreatic adenocarcinoma (PAAD), esophageal carcinoma (ESCA), lung adenocarcinoma (LUAD), lung squamous cell carcinoma (LUSC), and breast invasive carcinoma (BRCA). The cancer genomics (mutations, copy number variations, and mRNA levels) and patients' survival data in these cancer types were analyzed for the role of CREB1.

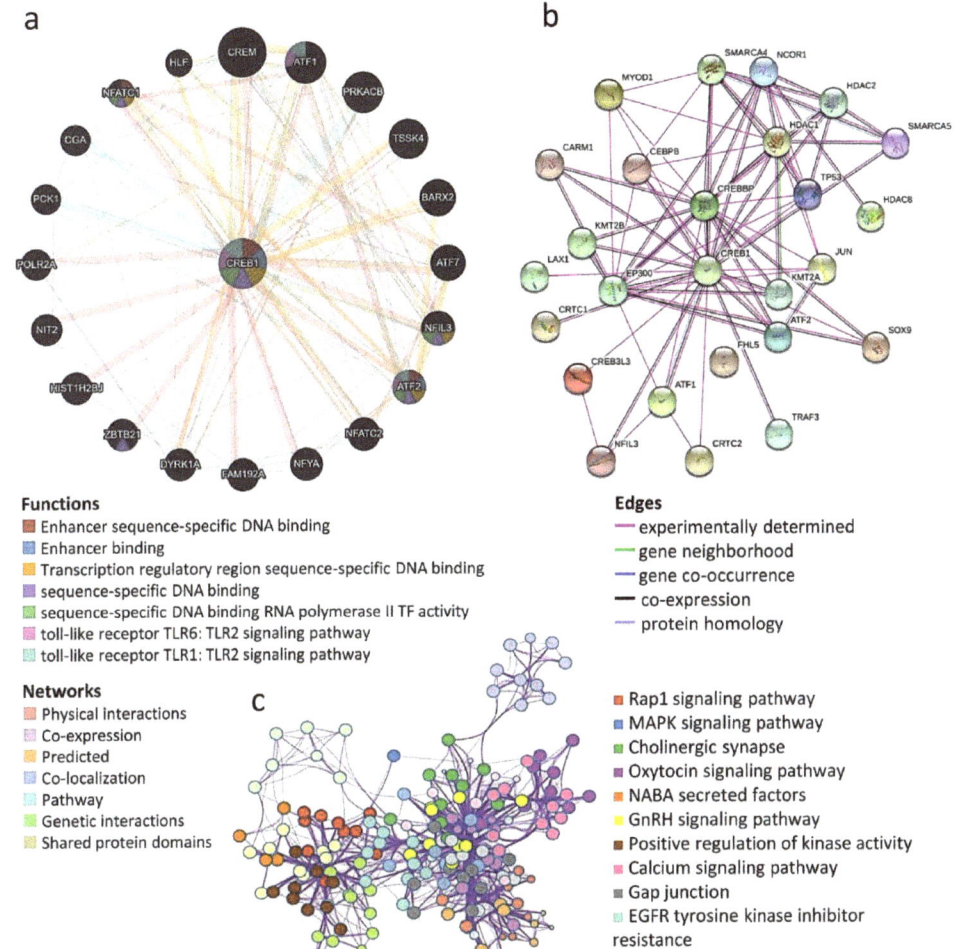

Figure 5. Network data improves the target prediction (**a**) and (**b**) predicted PPI essential for the functions of CREB1 generated from Pathway Commons and String website. (**c**) Metascape functional enrichment analysis and OV clinical relevance of CREB1-regulated DEGs. One term per cluster, colored by p values. Log10 (p) is the p value in log base 10.

3.6. Naloxone Treatment Mimics the Gene Expression Profile of CREB1

To investigate whether naloxone could target CREB1 functional activity, we employed the CMap analysis. The CMap database allows users to query a gene signature and explore the connections between the queried gene signature and drug-driven gene expression [19]. We prepared the differentially expressed genes (DEGs) from CREB1-overexpressing different cancer cells and queried the CMap database. Figure 4a shows the top 10 perturbagens that mimicked the CREB1-driven gene signature, including naloxone with the median connectivity score of 99.9. In contrast, the median connectivity score of fluticasone is −17.98. Therefore, naloxone treatment may mimic the effect of CREB1 overexpression. We further searched the OE and KD gene libraries of 2160 and 3799 oncogenes from the pharmacogenomic database and searched for potential drugs for the treatment of ovarian cancer. As shown in Figure 6b, we found a positive correlation between naloxone and CREB1 OE score of 0.28 ($p < 0.05$); and CREB1 KD score of 0.15. This indicates that CREB1 gene overexpression is positively correlated with naloxone sensitivity. The average transcriptional impact showed a positive correlation between CREB1 mRNA expression and naloxone drug activity. The above result implied that naloxone may reverse the CREB1-associated cancer hallmarks.

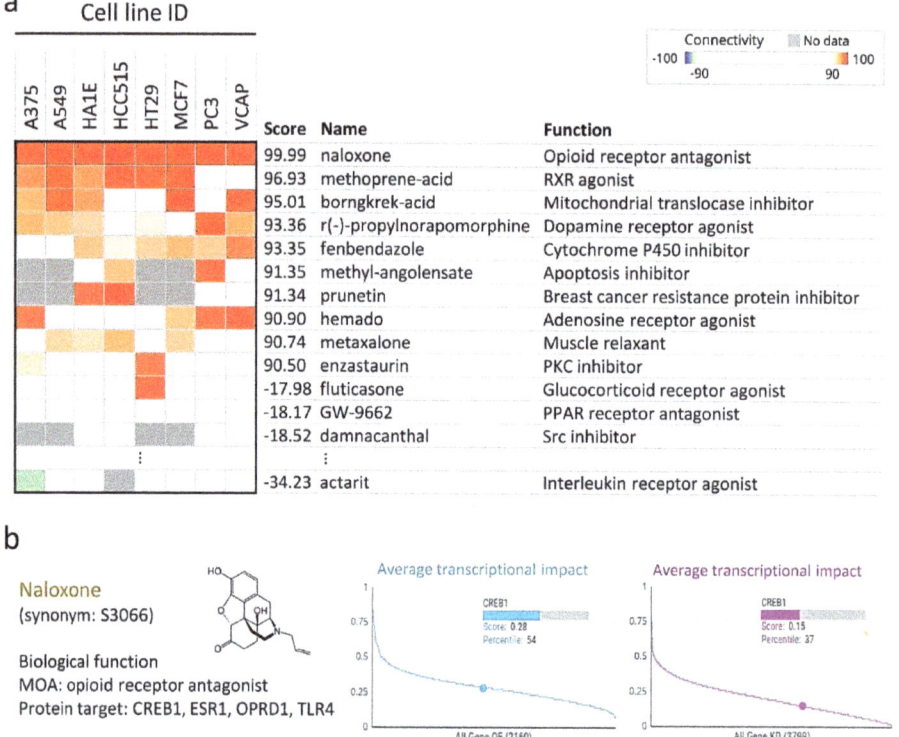

Figure 6. CMap analysis and drug sensitivity profiling in ovarian cancer cells. (a) The CREB1 gene signature was queried using the CMap database to predict potential drugs to reverse this signature. (b) The correlation between CREB1 gene overexpression and knockdown in OV cell lines (The Cancer Therapeutics Response Portal CTRP-OV data from the CTRP database).

4. Discussion

Ovarian cancer is one of the most prevalent malignant tumors in women. With a high degree of malignancy and a short survival period, ovarian cancer is often detected at an advanced stage, and the current treatment is still based on surgery and chemotherapy [20]. Current studies have shown that the abnormal expression of several genes may be closely associated with the survival of ovarian cancer patients through the screening of gene expression profiles, and that these genes are involved in the development and progression of ovarian cancer by promoting or suppressing apoptosis, generating or reversing chemotherapy resistance [21,22]. In this study, we used bioinformatics analysis, ovarian cancer tissue microarray, and multiple types of ovarian cancer cell lines to screen out mRNAs that may be related to the prognosis of ovarian cancer patients and provide a basis for future clinical practice and scientific research.

Previous studies have demonstrated that CREB1 expression levels have an important role in ovarian granulosa cell survival and that follicle-stimulating hormone and luteinizing hormone regulate ovarian function, at least in part, through the cAMP intracellular signaling pathway [23]. Somers et al. found that the mutation of Ser 133 to Ala 133 in the CREB1 sequence, which results in an inability to be activated by phosphorylation, significantly inhibits murine ovarian granulosa cell survival [24]. This also suggests that CREB1 is an important protein that promotes ovarian survival and plays an important role in the development of ovarian cancer [19,25]. Tumor development is not only about the activation of oncogenes and inactivation of oncogenes but also about abnormal expression of apoptosis-related genes and overexpression of growth factors and uncontrolled cell cycle. Most tissue cells undergo malignant transformation with shortened cell cycle and accelerated proliferation.

In addition to reducing metastasis and cell proliferation in bladder cancer cells, it also reduces gastric cancer, esophageal cancer, and glioma by knocking down CREB1 gene levels in tumor cells [26–29]. It is also possible to inhibit CREB1 activity through pharmacological strategies. Previous studies have shown that treatment of cancer cells with 666-15, a CREB inhibitor [30], has potent anticancer activity both in vitro and in vivo. Imperatorin also directly targets CREB1 to inhibit TGFβ2-ERK signaling and inhibit esophageal cancer metastasis [29].

Although CREB1 has been extensively studied in various tumors [31], CREB1 is aberrantly expressed in a variety of human cancers, including solid tumors [9,31–33] and hematologic malignancies [34,35]. In breast cancer studies, CREB1 was found to be highly expressed in metastatic breast cancer cells compared to non-metastatic cells and promoted breast cancer metastasis and subsequent bone destruction [9]. CREB1 was also highly expressed in glioma tissues through the induction of oncogenic microRNA-23a expression and increased the growth survival of glioma cells and colorectal cancer [32,36]. However, there is still evidence that CREB1 inhibits the proliferative effects of the stress-induced acetylcholinesterase variant AChE-R in glioblastoma [33], suggesting a controversial or tissue-specific role for CREB1 in human cancers. In our results, more specifically, there is a significant difference between stage I and stage III/IV, which is directly related to the limited number of patient examinations, in addition to the heterogeneity of both patient and tumor stage that may affect the results. In the future, more ovarian cancer specimens should be collected for confirmation.

In this study, we observed that the expression of CREB1 is dysregulated in pan-cancer, especially in OV. Our study has provided a more detailed picture of the relationship between CREB1 expression and characteristics, prognosis, protein–protein interaction, and hub genes and pathway crosstalk in OV. Our results indicate that CREB1 is highly expressed in OV through tissue microarrays and human ovarian multiple cancer cell lines. Kaplan–Meier analysis indicates that CREB1 may be a potential prognostic factor for OV. Similar to our results, Xu and colleagues reported that CREB1 can be used as a predictor of the prognosis of esophageal cancer [29]. Another study showed that CREB1 plays a vital role in the tumor progression of upper liver cancer [37]. In addition, CREB1 has also been reported to be directly related to the ability of colon, breast, and gastric cancer to metastasize [28,38,39]. These findings showed that CREB1 could serve as a novel biomarker for OV cancer diagnosis and prognosis prediction.

This study has some limitations. First, only in silico and in vitro experimental analyses were performed. Further investigations using animal models specific for OV were required. Second, only 59 cases of OV patients were available for TMA analyses. Third, in this study, a gene expression signature-based approach was used in different cell lines, which should further validate the protein levels as reflecting the results of patient IHC. The differences of genetic mutations, epigenetics, proteomics, and metabolomics should also be considered in future investigations. Finally, although this study was verified by multi-omics, it still requires a large number of clinical specimens and further confirmation through multiple centers.

5. Conclusions

In conclusion, our study shows for the first time that CREB1 is ectopically expressed in OV and is a potential new biomarker for OV survival, providing valuable information to guide research on targeted treatment strategies. Our results warrant further investigation into the mechanisms by which CREB1 promotes tumor progression and metastasis in OV.

Author Contributions: Conceptualization, A.-J.C. and K.-H.T.; methodology, C.-J.L., P.-Y.C. and M.-S.H.; software, C.-J.L.; formal analysis, P.-Y.C. and M.-S.H.; investigation, A.-J.C.; writing—original draft preparation, C.-J.L. and L.-T.L.; writing—review and editing, Z.-H.W. and K.-H.T.; visualization, H.-W.T. and Y.-H.C.; supervision, K.-H.T.; funding acquisition, C.-J.L., L.-T.L. and K.-H.T. All authors have read and agreed to the published version of the manuscript.

Funding: Please add: This research was funded by the Ministry of Science Technology, grant numbers MOST-109-2314-B-075B-014-MY2 and MOST 109-2314-B-075B-002; the Kaohsiung Veterans General Hospital (VGHKS109-103, 109-105, 109-106, 109-D07, 110-088, 110-143, 110-090, and 110-D06).

Institutional Review Board Statement: Not applicable.

Informed Consent Statement: Not applicable.

Data Availability Statement: Not applicable.

Acknowledgments: We acknowledge the laboratory technique service of histology and pathology at Li-Tzung Pathology Laboratory, Kaohsiung, Taiwan.

Conflicts of Interest: The authors declare no conflict of interest.

References

1. Miller, K.D.; Siegel, R.L.; Lin, C.C.; Mariotto, A.B.; Kramer, J.L.; Rowland, J.H.; Stein, K.D.; Alteri, R.; Jemal, A. Cancer treatment and survivorship statistics. *CA Cancer J. Clin.* **2016**, *66*, 271–289. [CrossRef] [PubMed]
2. Yang, Q.; Yang, Y.; Zhou, N.; Tang, K.; Lau, W.B.; Lau, B.; Wang, W.; Xu, L.; Yang, Z.; Huang, S.; et al. Epigenetics in ovarian cancer: Premise, properties, and perspectives. *Mol. Cancer* **2018**, *17*, 109. [CrossRef] [PubMed]
3. Aggarwal, S.; Kim, S.-W.; Ryu, S.-H.; Chung, W.-C.; Koo, J.S. Growth Suppression of Lung Cancer Cells by Targeting Cyclic AMP Response Element-Binding Protein. *Cancer Res.* **2008**, *68*, 981–988. [CrossRef]
4. Montminy, M.R.; Bilezikjian, L.M. Binding of a nuclear protein to the cyclic-AMP response element of the somatostatin gene. *Nat. Cell Biol.* **1987**, *328*, 175–178. [CrossRef] [PubMed]
5. Antony, N.; McDougall, A.R.; Mantamadiotis, T.; Cole, T.J.; Bird, A.D. Creb1 regulates late stage mammalian lung development via respiratory epithelial and mesenchymal-independent mechanisms. *Sci. Rep.* **2016**, *6*, 25569. [CrossRef]
6. Comuzzi, B.; Lambrinidis, L.; Rogatsch, H.; Godoy-Tundidor, S.; Knezevic, N.; Krhen, I.; Mareković, Z.; Bartsch, G.; Klocker, H.; Hobisch, A.; et al. The Transcriptional Co-Activator cAMP Response Element-Binding Protein-Binding Protein Is Expressed in Prostate Cancer and Enhances Androgen- and Anti-Androgen-Induced Androgen Receptor Function. *Am. J. Pathol.* **2003**, *162*, 233–241. [CrossRef]
7. Gubbay, O.; Rae, M.T.; McNeilly, A.S.; Donadeu, F.X.; Zeleznik, A.J.; Hillier, S.G. cAMP response element-binding (CREB) signalling and ovarian surface epithelial cell survival. *J. Endocrinol.* **2006**, *191*, 275–285. [CrossRef]
8. Kim, Y.S.; Lee, H.Y.; Crawley, S.; Hokari, R.; Kwon, S. Kim Bile acid regulates MUC2 transcription in colon cancer cells via positive EGFR/PKC/Ras/ERK/CREB, PI3K/Akt/IκB/NF-κB and p38/MSK1/CREB pathways and negative JNK/c-Jun/AP-1 pathway. *Int. J. Oncol.* **2010**, *36*, 941–953. [CrossRef]
9. Son, J.; Lee, J.; Kim, H.; Ha, H.; Lee, Z. Camp-response-element-binding protein positively regulates breast cancer metastasis and subsequent bone destruction. *Bone* **2011**, *48*, S34. [CrossRef]
10. Lin, Y.; Miao, Z.; Wu, Y.; Ge, F.-F.; Wen, Q.-P. Effect of low dose naloxone on the immune system function of a patient undergoing video-assisted thoracoscopic resection of lung cancer with sufentanil controlled analgesia—A randomized controlled trial. *BMC Anesthesiol.* **2019**, *19*, 236. [CrossRef]
11. Chiang, A.-J.; Li, C.-J.; Tsui, K.-H.; Chang, C.; Chang, Y.-C.I.; Chen, L.-W.; Chang, T.-H.; Sheu, J.J.-C. UBE2C Drives Human Cervical Cancer Progression and Is Positively Modulated by mTOR. *Biomolecules* **2020**, *11*, 37. [CrossRef]
12. Newton, Y.; Novak, A.M.; Swatloski, T.; McColl, D.C.; Chopra, S.; Graim, K.; Weinstein, A.S.; Baertsch, R.; Salama, S.R.; Ellrott, K.; et al. TumorMap: Exploring the Molecular Similarities of Cancer Samples in an Interactive Portal. *Cancer Res.* **2017**, *77*, e111–e114. [CrossRef] [PubMed]
13. Tang, Z.; Li, C.; Kang, B.; Gao, G.; Li, C.; Zhang, Z. GEPIA: A web server for cancer and normal gene expression profiling and interactive analyses. *Nucleic Acids Res.* **2017**, *45*, W98–W102. [CrossRef] [PubMed]
14. Zhou, Y.; Zhou, B.; Pache, L.; Chang, M.; Khodabakhshi, A.H.; Tanaseichuk, O.; Benner, C.; Chanda, S.K. Metascape provides a biologist-oriented resource for the analysis of systems-level datasets. *Nat. Commun.* **2019**, *10*, 1523. [CrossRef]
15. Croft, D.; Mundo, A.F.; Haw, R.; Milacic, M.; Weiser, J.; Wu, G.; Caudy, M.; Garapati, P.; Gillespie, M.; Kamdar, M.R.; et al. The Reactome pathway knowledgebase. *Nucleic Acids Res.* **2014**, *42*, D472–D477. [CrossRef]
16. Fabregat, A.; Jupe, S.; Matthews, L.; Sidiropoulos, K.; Gillespie, M.; Garapati, P.; Haw, R.; Jassal, B.; Korninger, F.; May, B.; et al. The Reactome Pathway Knowledgebase. *Nucleic Acids Res.* **2018**, *46*, D649–D655. [CrossRef]

17. Fabregat, A.; Sidiropoulos, K.; Garapati, P.; Gillespie, M.; Hausmann, K.; Haw, R.; Jassal, B.; Jupe, S.; Korninger, F.; McKay, S.; et al. The Reactome pathway Knowledgebase. *Nucleic Acids Res.* **2016**, *44*, D481–D487. [CrossRef]
18. Subramanian, A.; Narayan, R.; Corsello, S.M.; Peck, D.D.; Natoli, T.E.; Lu, X.; Gould, J.; Davis, J.F.; Tubelli, A.A.; Asiedu, J.K.; et al. A Next Generation Connectivity Map: L1000 Platform and the First 1,000,000 Profiles. *Cell* **2017**, *171*, 1437–1452. [CrossRef]
19. Chen, S.-N.; Chang, R.; Lin, L.-T.; Chern, C.-U.; Tsai, H.-W.; Wen, Z.-H.; Li, Y.-H.; Li, C.-J.; Tsui, K.-H. MicroRNA in Ovarian Cancer: Biology, Pathogenesis, and Therapeutic Opportunities. *Int. J. Environ. Res. Public Health* **2019**, *16*, 1510. [CrossRef]
20. Graffeo, R.; Livraghi, L.; Pagani, O.; Goldhirsch, A.; Partridge, A.H.; Garber, J.E. Time to incorporate germline multigene panel testing into breast and ovarian cancer patient care. *Breast Cancer Res. Treat.* **2016**, *160*, 393–410. [CrossRef]
21. Hsu, H.-C.; Tsai, S.-Y.; Wu, S.-L.; Jeang, S.-R.; Ho, M.-Y.; Liou, W.-S.; Chiang, A.-J.; Chang, T.-H. Longitudinal perceptions of the side effects of chemotherapy in patients with gynecological cancer. *Support. Care Cancer* **2017**, *25*, 3457–3464. [CrossRef] [PubMed]
22. Lin, P.-H.; Lin, L.-T.; Li, C.-J.; Kao, P.-G.; Tsai, H.-W.; Chen, S.-N.; Wen, Z.-H.; Wang, P.-H.; Tsui, K.-H. Combining Bioinformatics and Experiments to Identify CREB1 as a Key Regulator in Senescent Granulosa Cells. *Diagnostics* **2020**, *10*, 295. [CrossRef] [PubMed]
23. Somers, J.P.; DeLoia, J.A.; Zeleznik, A.J. Adenovirus-directed expression of a nonphosphorylatable mutant of CREB (cAMP response element-binding protein) adversely affects the survival, but not the differentiation, of rat granulosa cells. *Mol. Endocrinol.* **1999**, *13*, 1364–1372. [CrossRef] [PubMed]
24. Alper, Ö.; Bergmann-Leitner, E.S.; Abrams, S.; Cho-Chung, Y.S. Apoptosis, growth arrest and suppression of invasiveness by CRE-decoy oligonucleotide in ovarian cancer cells: Protein kinase A downregulation and cytoplasmic export of CRE-binding proteins. *Mol. Cell. Biochem.* **2001**, *218*, 55–63. [CrossRef]
25. Li, J.-Y.; Li, C.-J.; Lin, L.-T.; Tsui, K.-H. Multi-Omics Analysis Identifying Key Biomarkers in Ovarian Cancer. *Cancer Control.* **2020**, *27*, 1073274820976671. [CrossRef] [PubMed]
26. Guo, L.; Yin, M.; Wang, Y. CREB1, a direct target of miR-122, promotes cell proliferation and invasion in bladder cancer. *Oncol. Lett.* **2018**, *16*, 3842–3848. [CrossRef]
27. Mukherjee, S.; Tucker-Burden, C.; Kaissi, E.; Newsam, A.; Duggireddy, H.; Chau, M.; Zhang, C.; Diwedi, B.; Rupji, M.; Seby, S.; et al. CDK5 Inhibition Resolves PKA/cAMP-Independent Activation of CREB1 Signaling in Glioma Stem Cells. *Cell Rep.* **2018**, *23*, 1651–1664. [CrossRef]
28. Rao, M.; Zhu, Y.; Cong, X.; Li, Q. Knockdown of CREB1 inhibits tumor growth of human gastric cancer in vitro and in vivo. *Oncol. Rep.* **2017**, *37*, 3361–3368. [CrossRef]
29. Xu, W.W.; Huang, Z.H.; Liao, L.; Zhang, Q.H.; Li, J.Q.; Zheng, C.C.; He, Y.; Luo, T.T.; Wang, Y.; Hu, H.F.; et al. Direct Targeting of CREB1 with Imperatorin Inhibits TGFbeta2-ERK Signaling to Suppress Esophageal Cancer Metastasis. *Adv. Sci.* **2020**, *7*, 2000925.
30. Li, B.X.; Gardner, R.; Xue, C.; Qian, D.Z.; Xie, F.; Thomas, G.; Kazmierczak, S.C.; Habecker, B.A.; Xiao, X. Systemic Inhibition of CREB is Well-tolerated in vivo. *Sci. Rep.* **2016**, *6*, 34513. [CrossRef] [PubMed]
31. Chhabra, A.; Fernando, H.; Watkins, G.; Mansel, R.E.; Jiang, W.G. Expression of transcription factor CREB1 in human breast cancer and its correlation with prognosis. *Oncol. Rep.* **2007**, *18*, 953–958. [CrossRef] [PubMed]
32. Tan, X.; Wang, S.; Zhu, L.; Wu, C.; Yin, B.; Zhao, J.; Yuan, J.; Qiang, B.; Peng, X. cAMP response element-binding protein promotes gliomagenesis by modulating the expression of oncogenic microRNA-23a. *Proc. Natl. Acad. Sci. USA* **2012**, *109*, 15805–15810. [CrossRef] [PubMed]
33. Perry, C.; Sklan, E.H.; Soreq, H. CREB regulates AChE-R-induced proliferation of human glioblastoma cells. *Neoplasia* **2004**, *6*, 279–286. [CrossRef]
34. Sandoval, S.; Pigazzi, M.; Sakamoto, K.M. CREB: A Key Regulator of Normal and Neoplastic Hematopoiesis. *Adv. Hematol.* **2009**, *2009*, 634292. [CrossRef]
35. Wang, Y.-W.; Chen, X.; Gao, J.-W.; Zhang, H.; Ma, R.-R.; Gao, Z.-H.; Gao, P. High expression of cAMP responsive element binding protein 1 (CREB1) is associated with metastasis, tumor stage and poor outcome in gastric cancer. *Oncotarget* **2015**, *6*, 10646–10657. [CrossRef]
36. Tian, T.; Chen, Z.-H.; Zheng, Z.; Liu, Y.; Zhao, Q.; Liu, Y.; Qiu, H.; Long, Q.; Chen, M.; Li, L.; et al. Investigation of the role and mechanism of ARHGAP5-mediated colorectal cancer metastasis. *Theranostics* **2020**, *10*, 5998–6010. [CrossRef]
37. Li, C.-J.; Lin, H.-Y.; Ko, C.-J.; Lai, J.-C.; Chu, P.-Y. A Novel Biomarker Driving Poor-Prognosis Liver Cancer: Overexpression of the Mitochondrial Calcium Gatekeepers. *Biomedicines* **2020**, *8*, 451. [CrossRef]
38. Tsui, K.-H.; Wu, M.-Y.; Lin, L.-T.; Wen, Z.-H.; Li, Y.-H.; Chu, P.-Y.; Li, C.-J. Disruption of mitochondrial homeostasis with artemisinin unravels anti-angiogenesis effects via auto-paracrine mechanisms. *Theranostics* **2019**, *9*, 6631–6645. [CrossRef] [PubMed]
39. Han, J.; Jiang, Q.; Ma, R.; Zhang, H.; Tong, D.; Tang, K.; Wang, X.; Ni, L.; Miao, J.; Duan, B.; et al. Norepinephrine-CREB1-miR-373 axis promotes progression of colon cancer. *Mol. Oncol.* **2020**, *14*, 1059–1073. [CrossRef] [PubMed]

Article

Low Preoperative Lymphocyte-to-Monocyte Ratio Is Predictive of the 5-Year Recurrence of Bladder Tumor after Transurethral Resection

Kyungmi Kim [†], Jihion Yu [†], Jun-Young Park, Sungwoon Baek, Jai-Hyun Hwang, Woo-Jong Choi * and Young-Kug Kim *

Department of Anesthesiology and Pain Medicine, Asan Medical Center, University of Ulsan College of Medicine, Seoul 05505, Korea; kyungmi_kim@amc.seoul.kr (K.K.); yujihion@gmail.com (J.Y.); parkjy@amc.seoul.kr (J.-Y.P.); baekhans@naver.com (S.B.); jaehyun.hwang.uucm@gmail.com (J.-H.H.)
* Correspondence: woojongchoi@amc.seoul.kr (W.-J.C.); kyk@amc.seoul.kr (Y.-K.K.); Tel.: +82-2-3010-5646 (W.-J.C.); +82-2-3010-5976 (Y.-K.K.)
† The authors contributed equally to this project as co-first authors.

Abstract: Many studies have investigated the prognostic significance of peripheral blood parameters—including lymphocyte-to-monocyte ratio (LMR)—in several cancers in recent decades. We evaluated the prognostic factors for five-year tumor recurrence after the transurethral resection of a bladder tumor (TURBT). In total, 151 patients with non-muscle invasive bladder tumors who underwent TURBT under spinal anesthesia were selected for this retrospective analysis. The time to tumor recurrence was determined by the number of days from surgery until there was a pathological confirmation of tumor recurrence. The preoperative and postoperative laboratory values were defined as results within one month prior to and one month after TURBT. Univariate and multivariate Cox regression analyses were performed. Seventy-one patients (47.0%) developed recurrent bladder tumors within five years after the first TURBT surgery. The multivariate Cox regression analysis revealed that preoperative LMR (hazard ratio, 0.839; 95% confidence interval, 0.739–0.952; $p = 0.006$) and multiple tumor sites (hazard ratio, 2.072; 95% confidence interval, 1.243–3.453; $p = 0.005$) were independent recurrence predictors in patients with recurrent bladder tumors within five years after the TURBT. A low preoperative LMR is an important predictor for the recurrence of a bladder tumor during a five-year follow-up period after surgery.

Keywords: bladder tumor; lymphocyte-to-monocyte ratio; peripheral blood parameters; tumor recurrence; transurethral resection

1. Introduction

A bladder tumor is among the top 10 most common tumors for both genders; in 2018, there were 549,000 newly diagnosed cases and 200,000 deaths worldwide [1]. In particular, a bladder tumor is associated with the ninth highest mortality rate in men. Almost three-quarters of all bladder tumors are non-muscle invasive bladder cancers; this tumor is also well-known for its wide range of tumor biology and heterogeneity. These heterogeneities of non-muscle invasive bladder tumors contribute to a high recurrence rate and expensive economic burden for patients [2]. Thus, it is important to evaluate prognostic factors and prevent bladder tumor recurrence.

Many studies have investigated the prognostic significance of peripheral blood parameters in several tumors in recent decades. They revealed that the inflammatory response is a determining factor of tumor progression and recurrence [2]. While tumors maintain the progress of disease and promote carcinogenesis, a systemic inflammatory response is an essential process and accomplishes the full malignant phenotype, such as tumor tissue remodeling, angiogenesis, metastasis, and the suppression of the innate anticancer immune response [3]. The lymphocyte-to-monocyte ratio (LMR), neutrophil-to-lymphocyte ratio

(NLR), platelet-to-lymphocyte ratio (PLR), and red cell distribution width are valuable prognostic markers for solid tumors [4,5]. In particular, a low LMR is associated with a high tumor mutational burden and an insufficient immune reaction [6,7]. Furthermore, a low LMR is associated with a poor prognosis in several cancers [8,9]. However, no studies to date have reported the association between the LMR and tumor recurrence rate after transurethral resection of bladder tumor (TURBT).

Our previous study found that the five-year recurrence rate was lower in patients who underwent spinal anesthesia for non-muscle invasive bladder tumor resection than in those who underwent general anesthesia [10]. Thus, we designed this study to include only patients who underwent TURBT under spinal anesthesia, and evaluated independent prognostic factors including the peripheral blood parameters for the five-year recurrence of bladder tumor after TURBT.

2. Materials and Methods

2.1. Patient Characteristics

In total, 304 patients who underwent elective TURBT for the first time for non-muscle invasive bladder tumors at Asan Medical Center in Seoul, Korea were selected in January 2000–December 2007. However, 153 patients who underwent general anesthesia, had a tumor stage of T2 or higher, had suffered from other cancer, had combined other urinary cancer, had suffered a urinary tract infection, had taken opioids or analgesics before surgery, or had no medical records within the five-year postoperative period were excluded from the statistical analysis. In total, data from 151 patients who underwent TURBT for the first time were analyzed (Figure 1). The Asan Medical Center Institutional Review Board waived written informed consent and approved this retrospective study (approval number of 2017-1155). This study was performed in accordance with the Strengthening the Reporting of Observational Studies in Epidemiology (STROBE) criteria [11].

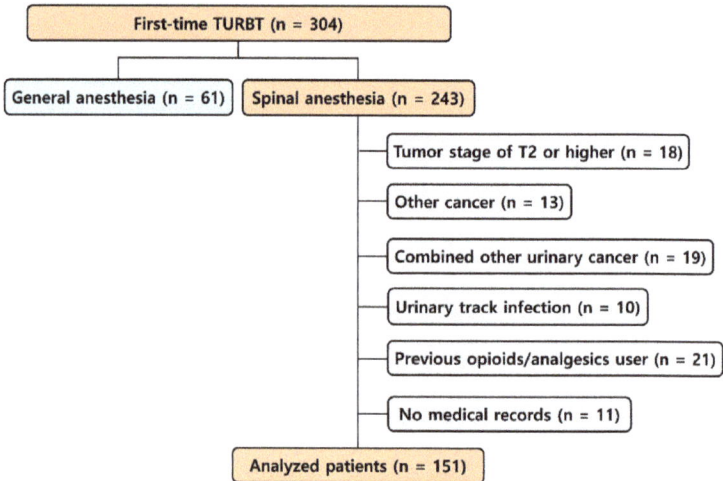

Figure 1. Flowchart of study patients. In January 2000–December 2007, 304 patients who underwent first-time elective TURBT at our institution were assessed. Finally, 151 patients were subjected to the study protocol. TURBT, transurethral resection of bladder tumor.

2.2. Spinal Anesthesia

After essential basic monitoring (electrocardiography, noninvasive blood pressure, and pulse oximetry), all patients underwent spinal anesthesia using 0.5% heavy bupivacaine (8–10 mg). Neuraxial blockade was confirmed by the loss of temperature or pin prick sensation at 10 min after spinal anesthesia. If a patient requested sedation, midazolam (2–5 mg) was administrated intravenously.

2.3. Clinical Data Collection

Tumor recurrence was defined by the pathological confirmation of a newly developed tumor. The time to tumor recurrence was determined by the number of days from surgery until the confirmation of tumor recurrence.

Data regarding demographics, pathologic findings to confirm the tumor grade and histological variant, imaging studies to confirm the tumor stage, intervention methods other than TURBT, and perioperative laboratory values were collected.

The demographic data included age, sex, body mass index, the American Society of Anesthesiologists physical status classification, comorbidities, and smoking status. Data regarding multiple tumor sites, tumor grade, histological variant, tumor stage, chemotherapy, and Bacillus Calmette-Guérin therapy were also collected. The tumor grade was assessed by the 2016 World Health Organization grading system [12]. The tumor stage was distinguished as either multiple bladder tumors (more than two sites) or a pathologic tumor stage. Intervention methods included chemotherapy and a Bacillus Calmette-Guérin intravesical injection.

The perioperative laboratory values included the hemoglobin level, red cell distribution width, platelet count, absolute white blood cell count, differential white blood cell count (neutrophils, lymphocytes, and monocytes), and the calculated NLR, PLR, and LMR. The preoperative laboratory values were defined as the results obtained within one month prior to TURBT. The postoperative values were defined as laboratory results obtained within one month after surgery.

2.4. Statistical Analysis

The continuous values are presented as the mean ± standard deviation, and categorical data are expressed as numbers (percentages). The student's *t*-test or the Mann–Whitney U test was used to analyze continuous variables, and the Chi-square test or Fisher's exact test was conducted to analyze categorical variables. We performed univariate Cox regression analysis and multivariate Cox regression analysis to determine the risk factors for bladder tumor recurrence. The parameters with a *p*-value < 0.05 in the univariate Cox regression analysis were included in the multivariate Cox regression analysis. In other analyses, a *p*-value < 0.05 was considered statistically significant. IBM SPSS Statistics 21.0 software (IBM, Armonk, NY, USA) was used for data management and statistical analysis.

3. Results

In total, 151 patients were enrolled in this study and 71 patients (47.0%) developed recurrent bladder cancer within five years after their first TURBT surgery (Figure 1).

Table 1 describes the patients' demographic characteristics. There were significantly more patients with multiple tumor sites and Bacillus Calmette-Guérin therapy in the recurrent group than in the non-recurrent group.

Table 2 shows the preoperative and postoperative peripheral laboratory values within one month prior to and one month after TURBT. The preoperative LMR and postoperative hemoglobin differed significantly between the recurrent and non-recurrent groups.

Table 1. Demographic data.

	Non-Recurrent Group (n = 80)	Recurrent Group (n = 71)	p-Value
Age (years)	64.1 ± 13.4	63.1 ± 13.0	0.633
Male	71 (88.8)	56 (78.9)	0.120
Body mass index (kg/m^2)	24.0 ± 3.2	23.5 ± 3.0	0.270
ASA physical status			0.191
I or II	77 (96.3)	64 (90.1)	
III	3 (3.8)	7 (9.9)	
Hypertension	21 (26.3)	21 (29.6)	0.717
Diabetes mellitus	8 (10.0)	8 (11.3)	>0.999
Smoking	28 (35.0)	27 (38.0)	0.737
Multiple tumor sites	39 (48.8)	49 (69.0)	0.014
Tumor grade			0.633
I	9 (11.4)	10 (14.1)	
II or III	70 (88.6)	61 (85.9)	
Histological variant			>0.999
Transitional cell carcinoma	79 (98.8)	71 (100.0)	
Adenocarcinoma	1 (1.3)	0 (0.0)	
Tumor stage			0.339
Ta	48 (60.0)	39 (54.9)	
T1	28 (35.0)	31 (43.7)	
Tis	4 (5.0)	1 (1.4)	
Chemotherapy	17 (21.3)	22 (31.0)	0.195
Bacillus Calmette-Guérin therapy	29 (36.3)	38 (53.5)	0.049

Data are presented as the mean ± standard deviation or number (percentage). ASA, American Society of Anesthesiologists; Tis, carcinoma in situ.

Table 2. Preoperative and postoperative peripheral blood parameters.

	Non-Recurrent Group (n = 80)	Recurrent Group (n = 71)	p-Value
Preoperative Values			
Hemoglobin (g/dL)	13.8 ± 2.2	13.2 ± 1.8	0.074
Red cell distribution width (%)	13.2 ± 1.1	13.0 ± 0.9	0.242
Platelet count (10^3/µL)	242.4 ± 87.7	247.1 ± 125.8	0.788
White blood cell count (/mm^3)	6707.5 ± 1961.2	6980.3 ± 2201.3	0.422
Neutrophils (/mm^3)	4061.8 ± 1792.1	4431.1 ± 1888.0	0.220
Lymphocytes (/mm^3)	1964.6 ± 616.8	1834.8 ± 613.0	0.198
Monocytes (/mm^3)	432.1 ± 170.1	476.5 ± 255.6	0.206
NLR	2.3 ± 1.7	2.7 ± 1.8	0.137
PLR	134.5 ± 65.1	141.9 ± 77.2	0.522
LMR	5.1 ± 2.2	4.4 ± 1.8	0.030
Postoperative Values			
Hemoglobin (g/dL)	12.7 ± 2.1	12.1 ± 1.9	0.046
Red cell distribution width (%)	13.1 ± 1.2	13.0 ± 1.0	0.394
Platelet count (10^3/µL)	224.1 ± 69.5	221.5 ± 79.6	0.825
White blood cell count (/mm^3)	7901.3 ± 2438.6	8176.1 ± 3188.9	0.552
Neutrophils (/mm^3)	5149.9 ± 2616.8	5667.4 ± 3134.6	0.271
Lymphocytes (/mm^3)	1920.3 ± 849.4	1706.3 ± 707.7	0.097
Monocytes (/mm^3)	481.4 ± 173.3	538.6 ± 269.0	0.119
NLR	3.5 ± 2.8	4.1 ± 3.3	0.216
PLR	133.5 ± 70.6	150.9 ± 107.8	0.239
LMR	4.4 ± 2.4	3.8 ± 1.9	0.066

Data are presented as mean ± standard deviation. NLR, neutrophil-to-lymphocyte ratio; PLR, platelet-to-lymphocyte ratio; LMR, lymphocyte-to-monocyte ratio.

Table 3 presents the univariate Cox regression analysis. This analysis showed that multiple tumor sites, Bacillus Calmette–Guérin therapy, the preoperative hemoglobin and LMR, and the postoperative hemoglobin and LMR were associated with five-year recurrence after TURBT. The multivariate Cox regression analysis showed that the preoperative LMR (hazard ratio 0.839, 95% confidence interval 0.739–0.952, p = 0.006) and multiple tumor sites (hazard ratio 2.072, 95% confidence interval 1.243–3.453; p = 0.005) were independent predictors of recurrence within five years after TURBT in patients with bladder tumors (Figure 2).

Table 3. Univariate Cox regression analysis for five-year bladder tumor recurrence.

Variables	Univariate Analysis	
	Hazard Ratio (95% CI)	p-Value
Age	1.001 (0.983–1.019)	0.934
Body mass index	0.956 (0.887–1.030)	0.956
ASA physical status		
I or II	1.0	
III	1.888 (0.864–4.126)	0.111
Smoking	1.104 (0.684–1.783)	0.686
Multiple tumor sites	2.200 (1.328–3.645)	0.002
Tumor grade		
I	1.0	
II or III	1.104 (0.566–2.157)	0.771
Tumor stage		
Ta	1.0	
T1	1.488 (0.927–2.387)	0.099
Tis	0.419 (0.058–3.048)	0.390
Chemotherapy	1.382 (0.834–2.287)	0.209
Bacillus Calmette-Guérin therapy	1.726 (1.082–2.753)	0.022
Preoperative Values		
Hemoglobin	0.880 (0.789–0.980)	0.020
While blood cell count	1.027 (0.921–1.145)	0.630
NLR	1.069 (0.967–1.181)	0.191
PLR	1.001 (0.998–1.004)	0.403
LMR	0.847 (0.746–0.961)	0.010
Postoperative Values		
Hemoglobin	0.876 (0.784–0.979)	0.019
While blood cell count	1.037 (0.956–1.124)	0.385
NLR	1.056 (0.987–1.130)	0.117
PLR	1.001 (0.999–1.003)	0.190
LMR	0.869 (0.769–0.983)	0.025

ASA, American Society of Anesthesiologists; Tis, carcinoma in situ; NLR, neutrophil-to-lymphocyte ratio; PLR, platelet-to-lymphocyte ratio; LMR, lymphocyte-to-monocyte ratio; CI, confidence interval.

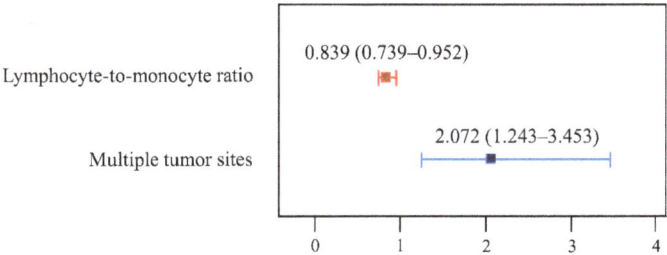

Figure 2. A forest plot of the multivariate Cox regression analysis to evaluate the predictors of the five-year recurrence of non-muscle invasive bladder tumor.

4. Discussion

Our data revealed that preoperative LMR and multiple tumor sites are valuable for predicting five-year tumor recurrence after TURBT. This is the first study to evaluate the association between preoperative LMR and bladder tumor recurrence in patients who underwent TURBT.

Several studies have reported that peripheral inflammatory parameters, particularly the NLR, are associated with a poor prognosis of bladder cancer in patients undergoing radical cystectomy [13–15]. Current theories suggest that the systemic inflammatory response is an important part of carcinogenesis, in which systemic circulating immune molecules and innate immune cells are recruited, leading to the differentiation of leukocytes and causing relative neutrophilia, monocytosis, and lymphocytopenia [16]. This process might be associated with tumor angiogenesis and progression [17]. However, our study showed that the preoperative LMR was more significant for predicting five-year recurrence than the NLR. The lymphocytes in tumor progression play essential roles in anti-tumor reactions by inducing the apoptosis of tumor cells [18,19]. This finding might be due to the differences in the study populations, because our study enrolled only patients who underwent spinal anesthesia.

Hoffmann et al. suggested that lymphocytopenia is related to an inappropriate immunologic response to tumor growth [6]. Moreover, monocytes were identified as regulators in tumor growth and progression. In the early stage of tumor development, monocytes are recruited from the bloodstream to the tumor and differentiate into macrophages. These monocytes enhance tumor proliferation and tumor angiogenesis as immune suppressors and promotors of tumor neovascularization [7]. Therefore, a low LMR represents high tumor mutational burden and an insufficient immune reaction [6,7]. In accordance with our results, Hutterer et al. reported that a low LMR was potentially associated with a poor prognosis in nonmetastatic renal cell carcinoma [8]. Stotz et al. also showed that an elevated LMR was correlated with a longer time to recurrence and a higher overall survival in patients with stage III colon cancer [9].

In addition, the multiplicity of tumors was a significant, poor prognostic factor in our study. Our analysis coincides with previous studies; it showed that the risk of tumor recurrence in patients with multiple tumor sites was 4.8 times higher than those with a single tumor. Numerous studies have reported that tumor multiplicity in a non-muscle invasive bladder tumor was an independent risk factor of tumor recurrence [20–23]. According to the field effect in the tumors, the multiplicity of bladder tumors implied that the overall mucosa had the potential for malignant change [24,25]. Therefore, the multiplicity of tumors was its own factor of pathogenesis in non-muscle invasive bladder tumors and a significant, poor prognostic factor associated with tumor recurrence.

Our study has several limitations. This single-center retrospective study enrolled a small number of patients who underwent only spinal anesthesia. A previous study reported that the type of anesthetic technique affects the tumor recurrence in patients with non-muscle invasive bladder tumors [10]. In particular, spinal anesthesia is associated with a lower recurrence rate of bladder tumors than general anesthesia. Therefore, the recurrence rate among our study patients was limited. However, the results of our study may provide information regarding more influential risk factors, since we validated these risk factors in a low recurrence rate condition.

5. Conclusions

This study found that preoperative LMR is an independent recurrence predictor within five years after TURBT. This result suggests that preoperative LMR provides useful information on bladder tumor recurrence.

Author Contributions: Conceptualization, K.K., J.Y., W.-J.C. and Y.-K.K.; data curation, K.K. and J.Y.; formal analysis, K.K., J.Y. and Y.-K.K.; investigation, K.K., J.Y., J.-Y.P., S.B. and J.-H.H.; methodology, K.K., J.Y., W.-J.C. and Y.-K.K.; project administration, W.-J.C. and Y.-K.K.; supervision, W.-J.C. and Y.-K.K.; writing—original draft, K.K. and J.Y.; writing—review and editing, J.Y., J.-Y.P., S.B., J.-H.H., W.-J.C. and Y.-K.K. All authors have read and agreed to the published version of the manuscript.

Funding: This research received no external funding.

Institutional Review Board Statement: The study was conducted according to the guidelines of the Declaration of Helsinki and approved by the Institutional Review Board of Asan Medical Center (approval number of 2017-1155).

Informed Consent Statement: Patient consent was waived due to this study's retrospective design.

Data Availability Statement: The data used in the present study are available from the corresponding author upon reasonable request.

Conflicts of Interest: The authors declare no conflict of interest.

References

1. Bray, F.; Ferlay, J.; Soerjomataram, I.; Siegel, R.L.; Torre, L.A.; Jemal, A. Global cancer statistics 2018: GLOBOCAN estimates of incidence and mortality worldwide for 36 cancers in 185 countries. *CA Cancer J. Clin.* **2018**, *68*, 394–424. [CrossRef]
2. Vartolomei, M.D.; Porav-Hodade, D.; Ferro, M.; Mathieu, R.; Abufaraj, M.; Foerster, B.; Kimura, S.; Shariat, S.F. Prognostic role of pretreatment neutrophil-to-lymphocyte ratio (NLR) in patients with non-muscle-invasive bladder cancer (NMIBC): A systematic review and meta-analysis. *Urol. Oncol.* **2018**, *36*, 389–399. [CrossRef]
3. Chechlinska, M.; Kowalewska, M.; Nowak, R. Systemic inflammation as a confounding factor in cancer biomarker discovery and validation. *Nat. Rev. Cancer* **2010**, *10*, 2–3. [CrossRef] [PubMed]
4. Kemal, Y.; Demirag, G.; Bas, B.; Onem, S.; Teker, F.; Yucel, I. The value of red blood cell distribution width in endometrial cancer. *Clin. Chem. Lab. Med.* **2015**, *53*, 823–827. [CrossRef]
5. Lee, S.M.; Russell, A.; Hellawell, G. Predictive value of pretreatment inflammation-based prognostic scores (neutrophil-to-lymphocyte ratio, platelet-to-lymphocyte ratio, and lymphocyte-to-monocyte ratio) for invasive bladder carcinoma. *Korean J. Urol.* **2015**, *56*, 749–755. [CrossRef] [PubMed]
6. Hoffmann, T.K.; Dworacki, G.; Tsukihiro, T.; Meidenbauer, N.; Gooding, W.; Johnson, J.T.; Whiteside, T.L. Spontaneous apoptosis of circulating T lymphocytes in patients with head and neck cancer and its clinical importance. *Clin. Cancer Res.* **2002**, *8*, 2553–2562. [PubMed]
7. Lopes-Coelho, F.; Silva, F.; Gouveia-Fernandes, S.; Martins, C.; Lopes, N.; Domingues, G.; Brito, C.; Almeida, A.M.; Pereira, S.A.; Serpa, J. Monocytes as endothelial progenitor cells (EPCs), another brick in the wall to disentangle tumor angiogenesis. *Cells* **2020**, *9*, 107. [CrossRef]
8. Hutterer, G.C.; Stoeckigt, C.; Stojakovic, T.; Jesche, J.; Eberhard, K.; Pummer, K.; Zigeuner, R.; Pichler, M. Low preoperative lymphocyte-monocyte ratio (LMR) represents a potentially poor prognostic factor in nonmetastatic clear cell renal cell carcinoma. *Urol. Oncol.* **2014**, *32*, 1041–1048. [CrossRef] [PubMed]
9. Stotz, M.; Pichler, M.; Absenger, G.; Szkandera, J.; Arminger, F.; Schaberl-Moser, R.; Samonigg, H.; Stojakovic, T.; Gerger, A. The preoperative lymphocyte to monocyte ratio predicts clinical outcome in patients with stage III colon cancer. *Br. J. Cancer* **2014**, *110*, 435–440. [CrossRef]
10. Choi, W.J.; Baek, S.; Joo, E.Y.; Yoon, S.H.; Kim, E.; Hong, B.; Hwang, J.H.; Kim, Y.K. Comparison of the effect of spinal anesthesia and general anesthesia on 5-year tumor recurrence rates after transurethral resection of bladder tumors. *Oncotarget* **2017**, *8*, 87667–87674. [CrossRef]
11. Von Elm, E.; Altman, D.G.; Egger, M.; Pocock, S.J.; Gotzsche, P.C.; Vandenbroucke, J.P.; Initiative, S. The strengthening the reporting of observational studies in epidemiology (STROBE) statement: Guidelines for reporting observational studies. *Int. J. Surg.* **2014**, *12*, 1495–1499. [CrossRef] [PubMed]
12. Comperat, E.M.; Burger, M.; Gontero, P.; Mostafid, A.H.; Palou, J.; Roupret, M.; van Rhijn, B.W.G.; Shariat, S.F.; Sylvester, R.J.; Zigeuner, R.; et al. Grading of Urothelial Carcinoma and The New "World Health Organisation Classification of Tumours of the Urinary System and Male Genital Organs 2016". *Eur. Urol. Focus* **2019**, *5*, 457–466. [CrossRef] [PubMed]
13. Antoni, S.; Ferlay, J.; Soerjomataram, I.; Znaor, A.; Jemal, A.; Bray, F. Bladder cancer incidence and mortality: A global overview and recent trends. *Eur. Urol.* **2017**, *71*, 96–108. [CrossRef] [PubMed]
14. Liu, K.; Zhao, K.; Wang, L.; Sun, E. The prognostic values of tumor-infiltrating neutrophils, lymphocytes and neutrophil/lymphocyte rates in bladder urothelial cancer. *Pathol. Res. Pract.* **2018**, *214*, 1074–1080. [CrossRef] [PubMed]
15. Zhang, J.; Zhou, X.; Ding, H.; Wang, L.; Liu, S.; Liu, Y.; Chen, Z. The prognostic value of routine preoperative blood parameters in muscle-invasive bladder cancer. *BMC Urol.* **2020**, *20*, 31. [CrossRef]
16. Coussens, L.M.; Werb, Z. Inflammation and cancer. *Nature* **2002**, *420*, 860–867. [CrossRef]

17. Liang, W.; Ferrara, N. The complex role of neutrophils in tumor angiogenesis and metastasis. *Cancer Immunol. Res.* **2016**, *4*, 83–91. [CrossRef]
18. Rosenberg, S.A. Progress in human tumour immunology and immunotherapy. *Nature* **2001**, *411*, 380–384. [CrossRef]
19. Zikos, T.A.; Donnenberg, A.D.; Landreneau, R.J.; Luketich, J.D.; Donnenberg, V.S. Lung T-cell subset composition at the time of surgical resection is a prognostic indicator in non-small cell lung cancer. *Cancer Immunol. Immunother.* **2011**, *60*, 819–827. [CrossRef] [PubMed]
20. Youssef, R.F.; Lotan, Y. Predictors of outcome of non-muscle-invasive and muscle-invasive bladder cancer. *Sci. World J.* **2011**, *11*, 369–381. [CrossRef] [PubMed]
21. Chou, R.; Buckley, D.; Fu, R.; Gore, J.L.; Gustafson, K.; Griffin, J.; Grusing, S.; Selph, S. AHRQ comparative effectiveness reviews. In *Emerging Approaches to Diagnosis and Treatment of Non-Muscle-Invasive Bladder Cancer*; Agency for Healthcare Research and Quality (US): Rockville, MD, USA, 2015.
22. Kwon, D.H.; Song, P.H.; Kim, H.T. Multivariate analysis of the prognostic significance of resection weight after transurethral resection of bladder tumor for non-muscle-invasive bladder cancer. *Korean J. Urol.* **2012**, *53*, 457–462. [CrossRef] [PubMed]
23. Nakai, Y.; Nonomura, N.; Kawashima, A.; Mukai, M.; Nagahara, A.; Nakayama, M.; Takayama, H.; Nishimura, K.; Okuyama, A. Tumor multiplicity is an independent prognostic factor of non-muscle-invasive high-grade (T1G3) bladder cancer. *Jpn. J. Clin. Oncol.* **2010**, *40*, 252–257. [CrossRef]
24. Chai, H.; Brown, R.E. Field effect in cancer—An update. *Ann. Clin. Lab. Sci.* **2009**, *39*, 331–337. [PubMed]
25. Vikram, R.; Sandler, C.M.; Ng, C.S. Imaging and staging of transitional cell carcinoma: Part 2, upper urinary tract. *AJR Am. J. Roentgenol.* **2009**, *192*, 1488–1493. [CrossRef] [PubMed]

Review

Tissue- and Liquid-Based Biomarkers in Prostate Cancer Precision Medicine

James Meehan [1,*], Mark Gray [2], Carlos Martínez-Pérez [1,3], Charlene Kay [1,3], Duncan McLaren [4] and Arran K. Turnbull [1,3]

1. Translational Oncology Research Group, Institute of Genetics and Cancer, Western General Hospital, University of Edinburgh, Edinburgh EH4 2XU, UK; carlos.martinez-perez@ed.ac.uk (C.M.-P.); charlene.kay@ed.ac.uk (C.K.); a.turnbull@ed.ac.uk (A.K.T.)
2. The Royal (Dick) School of Veterinary Studies and Roslin Institute, University of Edinburgh, Midlothian EH25 9RG, UK; mark.gray@ed.ac.uk
3. Breast Cancer Now Edinburgh Research Team, Institute of Genetics and Cancer, Western General Hospital, University of Edinburgh, Edinburgh EH4 2XU, UK
4. Edinburgh Cancer Centre, Western General Hospital, NHS Lothian, Edinburgh EH4 2XU, UK; duncan.mclaren@nhslothian.scot.nhs.uk
* Correspondence: james.meehan@ed.ac.uk

Abstract: Worldwide, prostate cancer (PC) is the second-most-frequently diagnosed male cancer and the fifth-most-common cause of all cancer-related deaths. Suspicion of PC in a patient is largely based upon clinical signs and the use of prostate-specific antigen (PSA) levels. Although PSA levels have been criticised for a lack of specificity, leading to PC over-diagnosis, it is still the most commonly used biomarker in PC management. Unfortunately, PC is extremely heterogeneous, and it can be difficult to stratify patients whose tumours are unlikely to progress from those that are aggressive and require treatment intensification. Although PC-specific biomarker research has previously focused on disease diagnosis, there is an unmet clinical need for novel prognostic, predictive and treatment response biomarkers that can be used to provide a precision medicine approach to PC management. In particular, the identification of biomarkers at the time of screening/diagnosis that can provide an indication of disease aggressiveness is perhaps the greatest current unmet clinical need in PC management. Largely through advances in genomic and proteomic techniques, exciting pre-clinical and clinical research is continuing to identify potential tissue, blood and urine-based PC-specific biomarkers that may in the future supplement or replace current standard practices. In this review, we describe how PC-specific biomarker research is progressing, including the evolution of PSA-based tests and those novel assays that have gained clinical approval. We also describe alternative diagnostic biomarkers to PSA, in addition to biomarkers that can predict PC aggressiveness and biomarkers that can predict response to certain therapies. We believe that novel biomarker research has the potential to make significant improvements to the clinical management of this disease in the near future.

Keywords: prostate cancer; precision medicine; tissue-based biomarkers; liquid-based biomarkers

1. Introduction

Prostate cancer (PC) was first reported in 1853 after a histological examination conducted by Dr. J. Adams, a surgeon in The London Hospital [1]. Adams noted in his description that it was 'a very rare disease', a comment that now contrasts greatly to how significant PC has become in the field of oncology. Worldwide, PC is the second-most-frequently diagnosed male cancer and the fifth-most-common cause of cancer-related mortalities. Current estimates indicate that ~1.4 million new cases are diagnosed and 400,000 PC-related deaths occur every year [2]. In the United States of America PC alone accounts for 26% of cancer diagnoses in men [3], while recently in the United Kingdom

PC has overtaken breast cancer to become the most commonly diagnosed cancer [4]. PC is more frequently identified in elderly men, with estimates indicating that ~60% of cases are diagnosed in those older than 65 years of age [5]. Due to the aging nature of the global population, it is thought that the social and economic impact of PC will increase significantly over the coming years.

Prostate Cancer: Risk Classification, Treatment and Challenges

PC is a highly heterogeneous disease with widely varying clinical outcomes. Some PC patients present with slow growing, localised cancers that do not pose an immediate risk to overall health. These patients may never go onto develop clinical symptoms, and in the absence of screening programmes they would never have known that they had PC [6]. Other tumours, however, can grow rapidly, resist treatment, metastasise early and can be fatal. Knowledge of PC aggressiveness is very important in determining the most appropriate treatment for each patient. Current methods for stratifying patient risk involve (i) staging, i.e., determining the extent of the tumour in the body. (ii) Gleason grading, an indication of cancer aggressiveness based on the architecture or pattern of the glands within the prostate. Scores range from 1 to 5, with the most common and second most predominant scores combined to give the final Gleason score (low grade = 6, intermediate grade = 7, high grade $\geq$ 8). (iii) The assessment of prostate specific antigen (PSA) levels. Together, this information is used to determine whether a patient is within a low-, intermediate- or high-risk group [7]. Typically, tumours will be histologically graded using needle core biopsy tissue prior to the patient starting treatment. As Gleason grade continues to be regarded as the greatest predictor of prognosis [8], there is a universal dependence on biopsy samples for risk assessment and treatment selection. However, there are many significant limitations associated with the use of tissue biopsies. PC is different to many other tumour types in that at the time of diagnosis, 60–90% of patients have multiple, separate and potentially diverse cancer foci scattered throughout the prostate. These foci can develop independently and can differ in their aggressiveness [9]. Thus, from a treatment perspective, tumour heterogeneity represents a significant challenge for biopsy-based assays to determine PC aggressiveness, as it can lead to differences in the grade observed between the diagnostic biopsy specimen and the final grade based upon samples acquired following surgery [10]. From the patient's perspective, the acquisition of tissue biopsies is an invasive procedure and can lead to side effects that include rectal bleeding, haematuria, infection and pain [11].

There are a variety of treatment options available to newly diagnosed PC patients [12]. Active surveillance (AS) is one option for low- or favourable-intermediate risk patients; this involves regular testing to assess whether or not their cancer is growing or spreading. If there are signs of disease progression, or if a patient is deemed higher risk, then definitive treatments including radical prostatectomy (RP), radiotherapy (RT) and androgen deprivation therapy (ADT) can be provided. About 87% of patients diagnosed with localised PC are given some form of radical therapy [13]. Unfortunately, treatments such as RT and RP can lead to substantial complications (including urinary, bowel and sexual dysfunction), each of which can significantly affect patient quality of life [14,15].

The focus of PC-specific biomarker research has previously been on disease diagnosis. There is, however, an increasing clinical need for the identification of novel prognostic, predictive and treatment response biomarkers that can be used to provide a precision medicine approach to PC management. Due to the significant complications associated with definitive treatment, the identification of biomarkers at the time of screening/diagnosis that provide an indication of the risk of aggressiveness is perhaps currently the greatest unmet clinical need in PC management. Biomarkers that help fulfil this role would help clinicians determine the most appropriate treatment strategy for newly diagnosed patients (i.e., who should be considered for AS and who should undergo radical treatment).

Largely through advances in genomic and proteomic techniques, exciting pre-clinical and clinical research is continuing to identify potential tissue, blood and urine-based

PC-specific biomarkers that may in the future supplement or replace current standard practices. This review will provide an overview of selected biomarkers that have the potential to increase the likelihood of PC detection, reduce over-diagnosis, predict the risk of progression and recurrence, and also give an indication of treatment response. We will begin by discussing PSA, which is unique as a biomarker as it has can be used for both PC detection, prognosis and to assess the effects of treatment. We will then go onto to discuss other biomarkers that have a role in the pre-diagnostic and post-diagnostic settings. Figure 1 and Table 1 outline where and how each of the biomarkers and their associated tests, discussed in this review, can contribute to PC patient management.

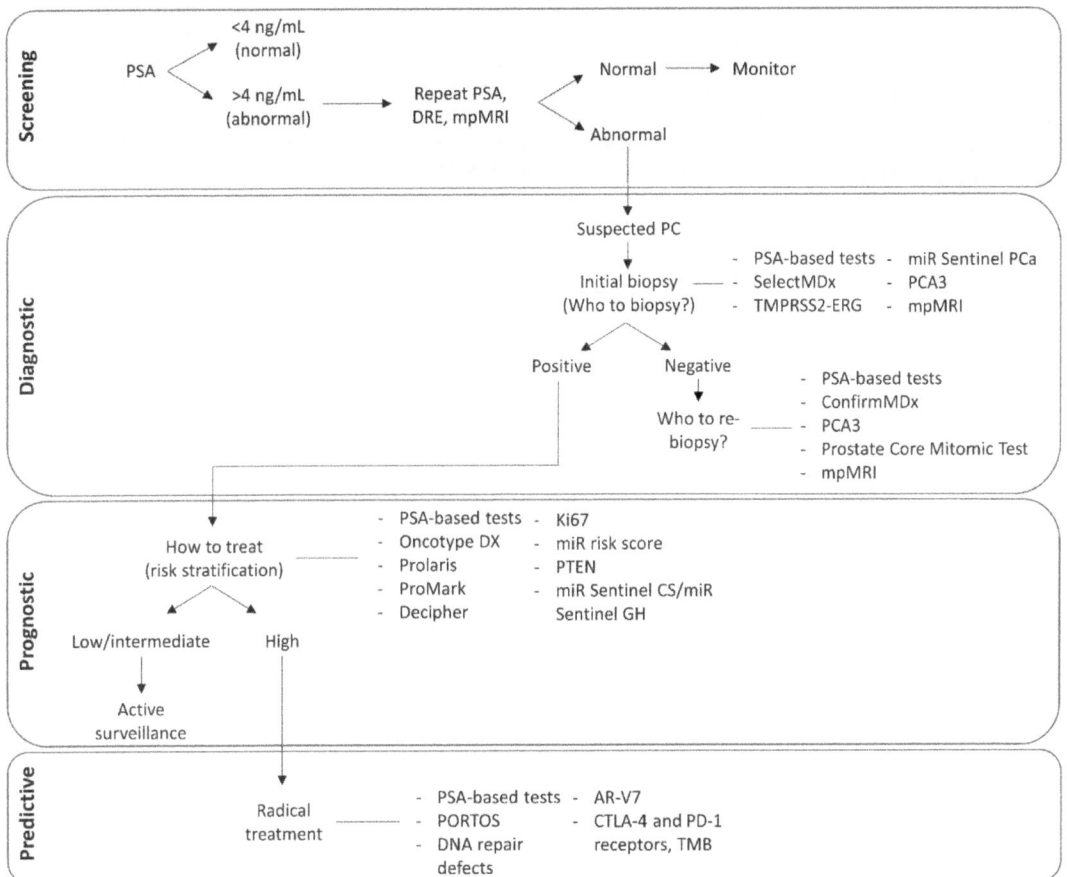

Figure 1. Biomarker assays and their use in PC management. AR-V7, androgen receptor splice variant 7; CTLA-4, cytotoxic T lymphocyte-associated protein-4; DRE, Digital rectal examination; mpMRI, multi-parametric magnetic resonance imaging; PC, Prostate cancer; PD-1, programmed death-1; PORTOS, Post-Operative Radiation Therapy Outcomes Score; PSA, Prostate specific antigen; PTEN, phosphatase and tensin homolog; TMB, Tumour mutational burden.

Table 1. Overview of prostate cancer biomarker assays that are in development or have gained clinical approval. Pre-diagnostic biomarkers (blue), biomarkers used in biopsy-proven prostate cancer cases (yellow), predictive biomarkers (green). 3.4mtΔ, 3.4-kb mitochondrial genome deletion; AR, androgen receptor; AR-V7, androgen receptor splice variant 7; BCR, biochemical recurrence; CTCs, circulating tumour cells; CTLA-4, cytotoxic T lymphocyte-associated protein-4; miRNA, microRNAs; PARP, poly (ADP-ribose) polymerase; PC, prostate cancer; PD-1, programmed death-1; PORTOS, Post-Operative Radiation Therapy Outcomes Score; PTEN, phosphatase and tensin homolog; RP, radical prostatectomy; RT, radiotherapy; sncRNAs, small non-coding RNAs; TMB, Tumour mutational burden.

Test	Analyte	Analyte Source	Outcome Provided by Test
SelectMDX	mRNA	Urine	Probability of detecting PC after prostatic biopsy, tumour grade
TMPRSS2-ERG score	mRNA	Urine	Probability of detecting PC after prostatic biopsy, tumour grade
miR Sentinel PCa	sncRNAs	Urine	Distinguishes patients with PC from subject with no evidence of PC, tumour grade
ConfirmMDx	Methylated DNA	Prostatic biopsy tissue	Separates patients that have PC from those with a true negative biopsy result, tumour grade
PCA3	mRNA	Urine	Probability of detecting PC after prostatic biopsy
Prostate Core Mitomic Test	3.4mtΔ	Prostatic biopsy tissue	Resolves false from true-negative prostatic biopsy results
Oncotype DX Genomic Prostate Score	mRNA	Prostatic biopsy tissue	Tumour grade, BCR, metastasis, recurrence
Prolaris	mRNA	Prostatic biopsy tissue	Tumour aggressiveness, metastasis
ProMark	Protein	Prostatic biopsy tissue	Tumour aggressiveness
Decipher	mRNA	Primary tumour after RP, prostatic biopsy tissue	Tumour aggressiveness, BCR, metastasis
miR Sentinel CS/GH tests	sncRNAs	Urine	Tumour grade
Ki67	Protein	Primary tumour after RP, prostatic biopsy tissue	BCR, metastasis, survival
miR risk score	miRNAs	Serum	Gleason score, BCR
PTEN	Protein	Prostatic biopsy tissue	Gleason score, stage, metastasis, BCR, recurrence
PORTOS	mRNA	Primary tumour after RP	Predict RT response
DNA repair defects	mRNA	Prostatic biopsy tissue	Predict response to PARP inhibitors
AR-V7	Protein, mRNA	Prostatic biopsy tissue, CTCs	Predict resistance to AR signalling inhibitors or sensitivity to taxanes
CTLA-4 and PD-1 receptors, TMB	Protein, mRNA	Prostatic biopsy tissue, CTCs	Response to immunotherapy

2. Prostate Specific Antigen

Prostate specific antigen (PSA) is a blood-based biomarker that can be used in the screening of patients for PC detection, in the surveillance of patients following diagnosis, to assess the risk of PC recurrence, and for monitoring treatment responses. PSA, a kallekrein-like serine protease glycoprotein, is encoded by the prostate-specific gene kallikrein 3 (*KLK3*) [16]. PSA is secreted by prostatic epithelial cells, with low levels of this glycoprotein typically present in blood samples from healthy individuals. Its primary function is to liquefy semen through proteolysis [16]. Although the specific mechanisms are open to debate, raised PSA levels within the blood of men with PC are not due to amplified expression of the protein, but instead result from increased release of PSA into the blood due to the disruption of prostate architecture observed in prostate tumours [17]. While there is no recognised defined cut-off for diagnosing PC, many clinicians consider PSA levels ≤ 4.0 ng/mL as normal, with higher levels indicating a need for further investigation. PC patients with unexpectedly high PSA levels have been encountered, with concentrations as high as 23,126 ng/mL previously reported [18].

2.1. PSA and Screening

In the first large scale investigation of the clinical use of PSA, levels of this protein were found to be associated with the clinical stage of PC, with increased levels correlated with more advanced disease stages [19]. Later studies investigated the use of PSA in terms of its ability to screen the population for disease, with a view to detecting early-stage PC. These reports highlighted that, when used in conjunction with clinical findings, PSA levels of ≥4.0 ng/mL resulted in improved PC detection [20–22]. The assessment of PSA levels was approved by the US Food and Drug Administration (FDA) as a diagnostic tool for PC detection in 1994 [23]. Its use as a screening test among asymptomatic men gained popularity, which in the US alone led to a dramatic increase in PC incidence [24]. The proportion of patients diagnosed at first presentation with metastatic disease also reduced following its use in screening programmes [24]. However, a significant criticism of the widespread use of PSA testing in the population was that it led to a PC diagnosis in men that would never have otherwise been diagnosed with clinically significant PC; the term "over-diagnosis" is often used to describe this situation [24]. Over-diagnosis, in relation to PSA screening programmes, has been reported to range from 20–66% [25,26]. Decreasing the number of PC patients diagnosed with later stage disease, while also increasing the number of patients receiving treatment, led to concerns that PC had become over-treated [27]. As well as having cost implications, over-treating PC can have significant effects on the mental and physical health of patients. As previously mentioned, even diagnostic procedures such as a prostatic biopsy carry risks of complications [11], while the side effects from RP and RT, which can occur in 50% of patients, can be severe. Recognised side effects from these treatments include urinary incontinence, sexual dysfunction and diminished colonic/rectal function [28,29].

Unfortunately, there is still debate on the extent to which PSA screening decreased PC mortality rates observed in the 1990s. The Cluster Randomised Trial of PSA Testing for Prostate Cancer (CAP) [30], the European Randomised Study of Screening for Prostate Cancer (ERSPC) [31,32] and the Prostate, Lung, Colorectal, and Ovarian Cancer Screening Trial (PLCO) [33] were three large randomised prospective trials that assessed the value of PSA screening in asymptomatic men for PC diagnosis. While the ERSPC trial found that screening for PC lowered PC-specific mortality and reduced the risk of developing metastatic disease, the two other trials did not replicate these results. Even though each of these trials assessed asymptomatic men between 50–60 years old, the trials differed greatly in their design, with limitations associated with each of them (for example, a screening/no-screening comparison was not strictly performed in the PLCO trial, as up to 90% of those in the "control group" had at least one PSA test, either before the screening began or over the course of the screening period) [34]. A recent systematic review for the US Preventive Services Task Force (USPSTF) suggested that PSA screening does have the ability to lower PC mortality risk, but it is linked with false-positive results, complications from resulting biopsy procedures and over-diagnosis [26].

In 2018, the USPSTF stated that results from screening trials had failed to show reductions in all-cause mortality and that there was inadequate evidence to suggest a benefit from PSA screening to decrease PC mortality in men over the age of 70. They also concluded that the net benefit of PSA-based screening for PC in men between 55–69 years is small [28]. As a result of the uncertainty over the benefits of PSA in screening, most of the guidelines that have been published are against mass screening, but advocate screening in men over 50 years of age with greater than 10 years life expectancy, only after the potential benefits and harms of screening have been outlined to the patient [28,35–37]. In contrast, the European Association of Urology and the Memorial Sloan Kettering Cancer Center both recommend that PSA screening should begin in well-informed men at 45 years of age, with the interval of testing thereafter dependent on the levels observed in this first test [38,39].

2.2. PSA and Prognosis

As well as having use in patient screening, PSA levels can also be utilised to estimate prognosis in newly diagnosed PC patients. In general, the more elevated the PSA levels are, the poorer the outcome [40–44]. Studies have shown that PSA levels > 20 ng/mL at diagnosis lead to a significant decrease in 5-year survival rates, with PSA concentrations above 98 ng/mL leading to a greater than 50% decrease in survival. The authors concluded that these highly elevated PSA concentrations suggest the presence of more aggressive or occult metastatic disease, thus indicating that these patients might benefit from more aggressive treatments [42]. While this relationship between high PSA levels and poor prognosis is especially relevant in PC patients with low or intermediate grade PC, in patients with high grade disease (Gleason score 8–10), lower PSA levels can actually predict a poorer outcome [45,46]; 10% of PC patients with higher grade disease had PSA readings of ≤2.5 ng/mL [46]. Additionally, reports have indicated that patients who present with PSA levels lower than 4 ng/mL have a greater incidence of distant metastasis than those with PSA concentrations between 4–10 ng/mL, 10–20 ng/mL or >20 ng/mL; Zheng et al. inferred that clinicians should pay particular attention to those patients with lower PSA levels, as their disease may be biologically aggressive [43]. Even though there is a correlation between PSA levels at diagnosis and outcome, PSA has only limited prognostic accuracy when utilised alone. To improve prognostic accuracy within the clinic, tumour histological and clinical factors are assessed alongside PSA levels when predicting outcome [34].

2.3. Use of PSA Following Initial Diagnosis

There are a variety of management options available to patients with newly diagnosed PC. Regardless of the therapy chosen, PSA levels are commonly analysed following the instigation of initial definitive treatment(s). The optimal frequency of PSA testing has yet to be ascertained. After definitive therapy, PSA testing is advised every 6–12 months for the first 5 years, which can then subsequently be reduced to once a year. PSA testing may be carried out more regularly in those patients that are at a higher risk of recurrence (Gleason score 8–10 or PSA > 20 ng/mL) [34]. PSA concentrations observed after therapy differ depending on the treatment given. Within 2 months of RP in patients with localised PC, PSA concentrations generally decrease to undetectable levels (<0.1 ng/mL) [34]. Two successive increasing PSA measurements of >0.2 ng/mL is defined as biochemical recurrence (BCR) after RP [47]. PSA concentrations reduce more slowly after RT or brachytherapy, with concentrations of <0.5 ng/mL generally observed 6 months after treatment. Transient increases in PSA levels may also occur post-RT within 3 years after treatment [48,49]. Increases in PSA levels of 2 ng/mL or more above the PSA nadir (also known as the Phoenix definition of BCR [50]) is regarded as BCR after RT [51].

The clinical management of patients that exhibit BCR after primary treatment is a controversial issue [52]. Even though BCR signifies a higher risk of clinical recurrence, many men remain symptom-free after its manifestation. In one study, only 34% of patients that exhibited BCR later showed signs of clinical recurrence. In those that did suffer recurrence, 8 years was the median duration of time between BCR and metastasis, with an additional median time to death of 5 years [53]. Clinicians therefore face the challenge of preventing or delaying progression in those patients that are deemed to be at risk, while also avoiding the over-treatment of men whose disease might never continue past PSA-only recurrence. There have been attempts to distinguish factors linking BCR to the risk of clinical recurrence; higher Gleason scores and shorter intervals to BCR have been associated with recurrence risk after both RP and RT [54].

Because many patients that exhibit BCR never go on to develop signs of clinical recurrence, there is still debate on whether ADT should be given early, or if clinicians should delay administration until clinical evidence of disease recurrence is present [55,56]. While an initial study comparing immediate ADT (patients treated within 3 months of PSA relapse) to deferred ADT (patients treated when they presented with clinical symptoms) demonstrated that there was no difference in 5 year overall survival rates between the two groups [57], more recent work indicates that prompt treatment with ADT may lead to better outcomes [58]. PSA kinetics and time to PSA nadir are important indicators of response to primary ADT treatment. However, the prognostic significance of PSA kinetics after primary ADT continues to be controversial. Intuitively, many urologists expected that more rapid PSA declines in response to primary ADT would be linked with extended survival. Conversely, reports suggest that these rapid responses to treatment may be indicative of more aggressive disease [59]. Even though ADT is advantageous in most patients exhibiting BCR, there are men whose disease will still progress despite treatment. When this occurs in the absence of any metastatic disease, it is known as non-metastatic castrate-resistant prostate cancer (nmCRPC). Castrate-resistant disease is often distinguished by two successive PSA increases when testosterone levels are <0.5 ng/mL [60].

Whilst the majority of men diagnosed with localised PC may be cured, their risk of treatment failure and death from subsequent metastatic disease increases significantly with their risk grouping at diagnosis (for example, at least 50% of all high-risk patients will not be cured). ADT is the standard initial treatment for those patients that develop distant metastases [61]. Although sequential PSA measurements can be used to assess response to ADT, validated definitions of disease progression or response to treatment with regard to PSA levels have yet to be established for this scenario. However, studies have demonstrated that advanced PC patients with a PSA measurement of <4 ng/mL after around 7 months of ADT have an improved outcome compared to patients with PSA levels > 4 ng/mL [62]. Additional studies have similarly shown that lower PSA measurements after ADT lead to better outcomes [63]. As is the case with localised disease, resistance to ADT also occurs in the metastatic setting, leading to the formation of metastatic castrate-resistant prostate cancer (mCRPC). PSA can be utilised to evaluate the response of mCRPC to treatment [64–67].

3. Techniques to Improve the Diagnostic Accuracy of PC

As previously discussed, over-diagnosis and over-treatment are two well-documented issues of the use of PSA for screening and monitoring programmes. Richard Albin, who is credited with the discovery of PSA [68], published "The Great Prostate Hoax" in 2014, where he discusses how he never intended for his discovery to be used in a PC screening program, highlighting its two major limitations: (i) it is not cancer-specific and (ii) it cannot differentiate between slow growing and aggressive cancers. The low specificity of PSA for detecting disease can lead to a considerable number of men undergoing unnecessary biopsies in order to exclude or verify the presence of malignancy. This situation arises largely because various non-cancerous processes such as trauma, prostatitis and benign prostatic hyperplasia (BPH) can lead to increased serum PSA levels [17]. BPH is a significant confounding factor for PC diagnosis as the occurrence of this condition increases with age, with a prevalence of 8%, 50% and 80% reported in men in their 30s, 50s and 80s, respectively [69]. The number of false positive PSA-based diagnoses will of course depend on the threshold used. In the ERSCP trial a PSA threshold of ≥ 3.0 ng/mL was used to determine if a biopsy was required; approximately 75% of men who presented with PSA levels ≥ 3.0 ng/mL were confirmed as PC negative following a biopsy procedure [31]. PSA screening can also suffer from false negative results; it has been estimated that a cut off of 4.0 ng/mL will miss around 15% of PC cases, of which around 15% will have advanced Gleason scores [70], and that a cut off value of 4.1 ng/mL will only detect ~20% of PC cases [71]. In an attempt to overcome these limitations, studies have investigated the use of various PSA parameters/dynamics, along with the use of additional or adjunct tests, to improve the diagnostic specificity and prognostic potential of PSA (Table 2).

Table 2. Overview of PSA-based diagnostic and prognostic assays. fPSA, free PSA; hK2, human kallikrein 2; iPSA, intact PSA; PHI, Prostate Health Index; PSA, prostate specific antigen; SNPs, single nucleotide polymorphisms.

Test	Fluid	Target
PSA density	Serum	PSA
PSA dynamics	Serum	PSA
%fPSA	Serum	PSA and fPSA
PHI	Serum	PSA, fPSA, [2]proPSA
4Kscore	Serum	PSA, fPSA, iPSA, hK2
STHLM3 model	Serum	PSA, fPSA, iPSA, hK2, beta-microseminoprotein, macrophage inhibitory cytokine 1, 232 SNPs
PSA glycosylation	Serum	α2,3-sialylated PSA
epiCaPture	Urine	PSA and methylated GSTP1, SFRP2, IGFBP3, IGFBP7, APC, PTGS2

3.1. PSA Density

PCs can produce increased levels of PSA per volume of tissue compared to benign prostatic conditions. To take into account prostate volume, PSA density (PSAD) was introduced in the early 1990s by Benson et al. This was done in an attempt to improve the accuracy of serum PSA testing to distinguish between small-volume organ-confined PC and BPH [72]. PSAD is calculated by dividing serum PSA by the volume of the prostate gland, measured by either transrectal ultrasound or magnetic resonance imaging. Studies have shown that PSAD has the potential to influence biopsy decisions by helping to identify men that harbour clinically significant PC [73–76], with PSAD becoming a better marker for predicting clinically significant PC as PSA levels increase [77]. Further work has exhibited the potential of PSAD to determine PC aggressiveness and predict the presence of adverse pathology in patients undergoing RP [78,79]. These results suggest that PSAD may play a role in risk stratification, which could be especially important when deciding which patients may be eligible for AS [79–81]. Overall, PSAD represents a simple, inexpensive tool that, if validated, has the potential to identify patients that may forego unnecessary biopsies.

3.2. PSA Dynamics

Changes in PSA parameters including doubling time (PSADT, time required for PSA levels to double) and velocity (PSAV, the rate of PSA change/year) can provide additional information over the evaluation of total PSA alone. Carter et al. introduced the concept of PSAV in 1992, performing multiple PSA measurements on serum samples obtained from men between 7–25 years prior to histological diagnosis or exclusion of PC; they found that while absolute PSA levels did not significantly differ between men with BPH and PC, the rate of change of PSA was significantly greater in those subjects with PC. They concluded that PSAV may act as an early biomarker for the development of PC [82]. Since this initial study, there has been some debate on the value of PSAV for diagnosing PC or providing a prognosis for PC in patients under AS [83,84]. However, there are studies that indicate that PSAV has potential as a prognostic/predictive biomarker in patients treated with RP [85–88] and RT [89,90]. The evidence thus far indicates that PSAV has better value in the post-treatment setting rather than in the pre-treatment setting. PSADT has shown promise as a predictive biomarker for PC detection on repeat biopsy, thus exhibiting the potential it has in the avoidance of unnecessary biopsies [91]. Studies have also assessed the clinical significance of PSADT before definitive therapy; here patients that exhibit longer pre-operative doubling times have been shown to have a better prognosis following treatment [92]. PSADT can additionally be used to monitor PC recurrence/progression following curative therapy [53,93], with a doubling time of <3 months associated with reduced survival times [93]. More recent work has demonstrated that PSADT can predict the occurrence of metastasis [94–96]. Although measuring PSAV and PSADT can provide additional information over the evaluation of total PSA alone, to date there is a lack of clear evidence to endorse the sole use of PSA dynamics in the clinic. Further prospective studies comparing the analysis of PSAV, PSADT and PSA are required [97].

3.3. Molecular Forms of PSA

PSA can exist in multiple forms within the blood. PSA found in serum can be classified as either free PSA (fPSA) or complexed PSA (cPSA). Whereas fPSA is unbound to carrier molecules/proteins, cPSA is bound to protease inhibitors (α1-antichymotrypsin, α2 macroglobulin or α1-antitrypsin) [98]. Assays that can measure these molecular forms can provide additional information over the assessment of total PSA levels [99–101]. fPSA levels are generally expressed as a percentage of total PSA (%fPSA). In general, men with PC have decreased levels of %fPSA when compared against men without PC [34]. As fPSA levels tend to decrease with PC, it can distinguish PC from BPH [99]. Unfortunately, there are limitations to the assessment of fPSA; this free form is less stable than complexed PSA in the blood, meaning sample processing has to be done soon after collection [102]. Additionally, DRE and biopsy procedures lead to a rise in the amount of fPSA in the blood [103]. Increasing prostate volumes have also been shown to lead to increased %fPSA values; as such, %fPSA is thought to only provide reliable data in patients whose prostate volume is <40 cm^3 [104].

Studies have suggested that the measurement of fPSA can be most beneficial in patients whose PSA levels are between 4–10 ng/mL, with some reporting that the use of fPSA can provide a diagnostic sensitivity of 95% and a specificity of 93%; however, others have reported poorer corresponding values of 75% and 32% [105]. A meta-analysis carried out to assess the accuracy of measuring %fPSA for the diagnosis of PC in men with PSA concentrations ranging from 4–10 ng/mL demonstrated that this test had low sensitivity and specificity. The authors concluded that %fPSA is neither sensitive nor specific enough to be utilised by itself, and the results of these tests need to be combined with additional diagnostic methods in helping to inform whether a prostatic biopsy is required [105]. Oto et al. recently explored the potential of %fPSA when merged with other factors, demonstrating that the combination of %fPSA with total PSA and age in a predictive model increased the diagnostic potential of total PSA [106].

To circumvent some of the issues encountered with %fPSA, studies have investigated the use of molecular forms of PSA in diagnostic assays, including intact PSA (iPSA) and [2]proPSA [107]. The Prostate Health Index (PHI) assay, the 4-kallikrein panel (4Kscore) and the Stockholm-3 (STHLM3) model are each multiplex tests that incorporate various molecular forms of PSA. Each of these assays are detailed in the subsequent sections.

3.4. Prostate Health Index

The PHI assay was developed to aid the detection of clinically significant PC. It is a score derived from total PSA, fPSA and [2]proPSA values using the formula ([2]proPSA/fPSA) $\times$ $\sqrt{\text{total PSA}}$ [108]. [2]proPSA is a peptide precursor to mature PSA that is preferentially produced in malignant cells [109]. The perceived advantage of this test is that it allows clinicians to evaluate individual PSA parameters in combination with the overall score produced. The chief use of PHI within the clinic is to lower the number of unnecessary biopsies acquired from patients with PSA levels that are considered borderline, without losing the detection of aggressive tumours.

The PHI test was approved in 2012 by the FDA for use in patients over 50, with PSA readings between 4–10 ng/mL and a negative DRE. Studies have shown that PHI is superior to %fPSA and total PSA in the detection of PC [110–116]. This greater accuracy in the detection of PC was particularly apparent in patients with PSA levels between 2–10 ng/mL [113]. PHI has also shown increased predictive accuracy for clinically significant/aggressive disease when compared against %fPSA and total PSA [116–120]. Between 15–45% of unnecessary biopsies can be avoided using the PHI test, depending on the cut-off values used [121]. The capacity of PHI-density (determined by dividing the PHI score by the prostate volume) to distinguish clinically significant PC has also been shown [122,123]. The combination of PHI with multi-parametric magnetic resonance imaging (mpMRI) has also been assessed, with PHI helping to determine the need for re-biopsy and improving the detection of clinically significant PC [123,124].

The PHI score has been demonstrated to impact patient management in the clinic, leading to biopsy deferrals when the patient PHI score was low and the decision to carry out a biopsy when the PHI score suggested that there was an intermediate/high probability of PC being present [125]. From a health-economic perspective, the cost-effectiveness of including PHI in the decision-making process for whether a prostatic biopsy is required has also recently been demonstrated [126,127]. As well as lowering the number of unnecessary biopsies, the prediction of BCR following RP is another potential use for the PHI test [128,129]. There are, however, some difficulties associated with the use of this test in the clinic. While it has been shown that PHI is an effective tool for risk stratification in both Asian and European populations, reports indicate that differing PHI reference ranges should be employed for distinct ethnic groups [130]. Like fPSA, studies have demonstrated that [2]proPSA also has some issues with molecular instability [131].

3.5. Four-Kallikrein Panel

Human kallikrein 2 (hK2) is a serine protease that shares 80% sequence homology with that of PSA. Studies have indicated that hK2 may have a role in distinguishing between patients with PC and those without malignant disease, while also having the ability to predict stage, grade and BCR in those patients treated with RP [132]. Using serum samples from the ERSPC trial, a prediction model was produced based on a panel of 4 kallikrein markers: total, free and iPSA in combination with hK2 levels. Commercialised by Opko Diagnostics, the 4-kallikrein panel (4Kscore), in conjunction with patient clinical data (age, DRE and previous biopsy results), generates a risk of the presence of high-grade PC. This model led to a better discrimination of high-grade PC when compared against total PSA and clinical variables alone [133–136].

Like %fPSA and PHI, the primary aim of the 4Kscore is to reduce disease over-detection by helping clinicians decide which patients require a biopsy. Its use is currently recommended in men undergoing either an initial or a repeat biopsy. The results from a large, prospective multi-institutional trial showed that the 4Kscore distinguished patients that had a Gleason score ≥ 7 from those that scored <7. Using a 6% cut-off value, the authors suggested that 30% of biopsies could be avoided whilst delaying a diagnosis of high-grade PC in only 1.3% of patients [137]. Further studies have demonstrated the potential of the 4Kscore to predict the presence of clinically significant PC [138–143]. As with PHI, the 4Kscore test has also been assessed when used in combination with mpMRI, with results showing that the 4Kscore improved the prediction of high-grade PC when utilised alongside mpMRI [144]. The ability of the 4Kscore to identify the presence of aggressive cancers across multi-ethnic populations has also recently been exhibited, thus demonstrating its wide clinical applicability [145].

Studies have established that use of the 4Kscore has the potential to significantly influence clinician and patient decision-making processes, leading to a reduction in the number of biopsies performed, while also increasing the likelihood of identifying aggressive PC [146]. The capacity of the 4Kscore to significantly reduce costs while also enhancing the quality of patient care has also been shown [147,148]. Other studies have investigated the 4Kscore for its ability to predict distant metastasis; 4Kscores from patients assessed at 50 and 60 years of age can stratify men into two cohorts in terms of their risk of developing metastatic disease 20 years following diagnosis [149].

3.6. The STHLM3 Model

Genome-wide association studies have produced convincing evidence for a genetic predisposition for PC in some patients. Single nucleotide polymorphisms (SNPs) have been described which account for around 30% of the hereditary risk for PC, offering novel areas for exploration into the pathogenesis of this disease [150]. The combination of a genetic score centred on these SNPs with PSA to improve the specificity of PSA testing alone has been investigated [151,152]. STHLM3 is a risk-based model for PC screening that combines 232 SNPs, a combination of plasma protein biomarkers (PSA, iPSA, fPSA, hK2, beta-microseminoprotein and macrophage inhibitory cytokine 1) and clinical variables (family history, age, prostate exam and previous biopsies) [153]. Studies have found that this model performed better than PSA alone for the detection of high-risk PC, exhibiting its potential to improve PC diagnosis by significantly reducing the number of unnecessary biopsies taken, while also preserving the same sensitivity to diagnose clinically significant PC [153–157].

3.7. PSA Glycosylation

Glycans are saccharides that can be bound to lipids, proteins and other glycans through glycosylation. Glycosylation is thought to be the most frequent post-translational modification and is essential to nearly all biological processes that occur in the body [158]. Aberrant glycosylation is a widespread characteristic within cancer cells that has been identified in most cancer types, and is often referred to as a "hallmark of cancer" [159]. A SNP that has an effect on PSA glycosylation has recently been linked to PC risk [160]. Developments in mass spectrometry technology have led to further research into glycan structures on tumour-associated proteins; differing studies have assessed whether a glycan signature on PSA may be utilised to improve its clinical efficacy [161]. The extent to which a protein/lipid is glycosylated is dependent on the expression of specific glycosylation enzymes in a cell, as well as the quantity of glycosylation sites present [162]; PSA contains a single N-glycosylation site [161]. Variations in PSA glycosylation states have been shown to occur in both PC cell lines [163] and in blood samples from patients with and without PC [164]. So far, around 50 PSA glycoforms have been defined, with some of these found to be present in aggressive PC. In particular, α2–3-linked sialic acid alterations to PSA in clinical samples have gained the most interest from researchers. The ability of α2,3-sialylated PSA to diagnose PC has been reported [165], with further studies also demonstrating its potential to differentiate high-risk PC from low- and intermediate-risk PC and BPH patients [166,167].

3.8. DNA Methylation

Epigenetic processes can affect the expression of genes, leading to alterations in malignancy-associated phenotypes including angiogenesis, growth, invasion and migration. Numerous alterations in DNA methylation have been distinguished between cancerous and benign prostate tissues [168]. As a result, aberrant DNA methylation is an epigenetic change that has promise as a diagnostic or prognostic PC biomarker [169]. The Epigenetic Cancer of the Prostate Test in Urine (epiCaPture) is a DNA methylation urine test for high-risk PC. It is designed to measure DNA hypermethylation within the regulatory regions of six PC-associated genes (*GSTP1, SFRP2, IGFBP3, IGFBP7, APC* and *PTGS2*) [170]. Increased methylation levels within epiCaPture genes have been shown to be associated with higher PC aggressiveness. The authors concluded that epiCaPture could be used as an adjunct to PSA, aiding in the selection of patients that should undergo a prostatic biopsy [170].

4. Alternative Diagnostic Biomarkers to PSA

There are biomarkers other than PSA that have a role in the pre-diagnostic setting. The ideal biomarker here should have the ability to increase the likelihood of identifying clinically significant PC on biopsy tissues, while also leading to the avoidance of biopsies in men who do not require one due to the absence of clinically significant PC. These types of biomarkers can be categorised into those employed to decide who to biopsy (SelectMDX, TMPRSS2-ERG score and the miR Sentinel test) and those utilised to choose when to re-biopsy (ConfirmMDx, prostate cancer antigen 3 [PCA3] and the Prostate Core Mitomic Test).

4.1. SelectMDx

The SelectMDx assay is a urine-based test designed to give the probability of detecting PC after a biopsy, in addition to the likelihood of low-grade versus high-grade disease. SelectMDx is performed after prostatic massage, with mRNA levels of *DLX1* and *HOXC6* genes (reported to be good predictors for the detection of high-grade PC [171,172]) measured within the urine through qRT-PCR. *DLX1* and *HOXC6* gene expression levels are then combined with clinical parameters (PSA density, age, DRE and family history information). Van Neste et al. postulated that the use of this test could lead to a 42% decrease in the total number of biopsies carried out, with a 53% reduction in the number of unnecessary biopsies [172]. Further studies have shown that this test can help clinicians identify men at risk of clinically significant PC, thus aiding the initial biopsy decisions and helping to reduce the number of unnecessary biopsies [173–175]. Analyses have indicated that the use of SelectMDx before proceeding to biopsy could lead to an increase in quality-adjusted life years (a measure of disease burden that takes into account both the quantity and quality of life lived) while also saving healthcare costs [176–178]. The SelectMDx test was included in the 2020 National Comprehensive Cancer Network (NCCN) guidelines for the early detection of PC. While there have been reports indicating that SelectMDx outperforms other tests such as PHI in screening for the presence of high-grade PC before biopsy [179], more recent papers have led to questions over the worth of the SelectMDx assay [180,181].

4.2. TMPRSS2-ERG Score

Chromosomal translocations are a common occurrence in cancer [182]. Tomlins et al. identified candidate oncogenic genomic rearrangements based on outlier gene expression; through this method, they discovered chromosomal translocations that lead to the fusion of the androgen-regulated gene transmembrane protease serine 2 (*TMPRSS2*) and ETS transcription factors (predominantly ETS-regulated gene [*ERG*]), also known as TMPRSS2-ERG [183]. Experiments indicated that the androgen-responsive promoter of *TMPRSS2* facilitated the overexpression of *ERG* in PC [183]. This chromosomal rearrangement has been identified in pre-cancerous prostatic conditions (e.g., intraepithelial neoplasia) and has been shown to be specific to PC [183–187]. TMPRSS2-ERG gene fusions occur in ~50% of PCs [186,188,189]; in those cases that overexpress ERG, up to 90% will be positive for the gene fusion [183,190–192].

Similar to the SelectMDx test, qRT-PCR can also be used to measure TMPRSS2-ERG mRNA in urine samples following prostatic massage. Simultaneous assessment of PSA mRNA allows a TMPRSS2-ERG score to be generated from the TMPRSS2-ERG mRNA/PSA mRNA ratio. Studies have illustrated that the assessment of TMPRSS2-ERG gene fusions in urine has the potential to predict the diagnosis of PC from subsequent prostatic biopsy samples [188,193]. Others have shown a correlation between TMPRSS2-ERG gene fusion, grade [194,195] and stage [196] at diagnosis, with analysis of the gene fusion also demonstrated to have the ability to predict the risk of clinically relevant PC after a prostatic biopsy [189]. Studies have additionally investigated whether TMPRSS2-ERG gene fusions can be utilised to assess PC aggressiveness in patients undergoing AS, thereby having use as a prognostic biomarker when assessed in prostatic tissues samples [190].

4.3. miR Sentinel Test

Exosomes and prostate-specific exosomes (prostatosomes) are small (30–150 nm) double lipid membrane-bound extracellular vesicles that are generated within cells through internal budding of multi-vesicular body membranes. For endosomal contents to be released from cells, they require endocytosis and fusion of their membranes with the cellular plasma membrane. The contents of prostatosomes can be released into urine, semen and blood, with these prostatosomes containing various molecules including proteins, lipids and nucleic acids [197]. These substances not only play key roles in cellular signalling, but have also been shown to be regulators of tumourigenesis and cancer progression, including immune suppression, angiogenesis, cell migration and invasion [198]. As such, prostatosomes are a rich source of biomarkers for PC diagnosis and prognosis. In comparison to men without disease, PC patients have increased numbers of serum-detected exosomes, with reports indicating that these higher levels may also correlate with higher Gleason scores [199]. Prostatosomal contents including PSA and TMPRSS2-ERG have also been detected within urine-derived exosomes from PC patients [200].

The miR Sentinel test is a recently developed platform that analyses small non-coding RNAs (sncRNAs) acquired from urinary exosomes [201]. This platform consists of three different tests; the Sentinel PCa test (distinguishes patients with PC from those in which there is no evidence of PC), the miR Sentinel CS test (differentiates patients that have PC into those with low-risk disease and those with intermediate/high-risk PC) and the miR Sentinel GH test (classifies patients with PC into those with low- and favourable intermediate-risk disease and those patients with high-risk PC). Each of the tests demonstrated sensitivities and specificities above 90%, highlighting their potential to diagnose and classify PC in a non-invasive manner with great precision [201]. Further validation of these tests is required in other independent patient cohorts and racially diverse patient groups.

4.4. ConfirmMDx

ConfirmMDx (MDxHealth, Inc) is an assay based upon DNA methylation and is designed to separate patients that have PC from those with a true negative biopsy result. The methylation status of Glutathione S-Transferase Pi 1 (*GSTP1*), Ras association (RalGDS/AF-6) domain family member 2 (*RASSF2*) and Adenomatous Polyposis Coli (*APC*) are evaluated using this assay [202]. The assay requires a minimum of eight core biopsy specimens obtained from specific prostatic regions. The advantage of using this assay is that molecular DNA alterations in prostatic cells that are adjacent to PC lesions, which would otherwise be diagnosed as histologically benign, can be identified. This is a result of the "halo effect" that the tumour has on surrounding normal tissues [203]. A positive ConfirmMDx result in biopsy tissue that has been labelled as cancer negative by a pathologist indicates that tumour cells were missed in the biopsy procedure. Thus far, its use has been validated in two different studies [202,204], exhibiting the potential ConfirmMDx has in helping to decrease the number of unnecessary repeat biopsies. In those patients that produce positive results, DNA methylation intensities also aid in the identification of men with high-grade disease [205]. While previous work was predominantly carried out in Caucasian men, recent work has demonstrated that this test is also effective in African American patients [206]. The Prostate Assay Specific Clinical Utility at Launch (PASCUAL) study (NCT02250313) is currently underway, examining the clinical value of the ConfirmMDx test in urologic practices within the US.

4.5. PCA3

The prostate-specific *PCA3* gene encodes a non-coding RNA that exhibits up to a 66-fold upregulation in prostatic tumours, with studies showing it to be present in >90% of PC cases [207–209]. In light of encouragingly high sensitivity and specificity results from tissues, numerous studies investigated the assessment of PCA3 levels non-invasively using urine [209–211]. Through qRT-PCR, PCA3 mRNA can be readily measured in urine samples following prostate massage. A PCA3 score is calculated from the PCA3 mRNA/PSA mRNA ratio, multiplied by 1000. Analysis of PSA mRNA levels, as performed in the TMPRSS2-ERG score assay, is required to control for the quantity of prostate epithelial cells in the urine. A score below the cut-off of 25 is interpreted as a negative result (there is a decreased likelihood of PC being present), with scores ≥25 indicating an increased probability that PC is present. However, there is debate over what PCA3 cut-off score should be used [212,213].

The PCA3 Progensa test was approved by the FDA in 2012 for use in suspect PC cases with equivocal PSA/DRE/biopsy results. Studies have demonstrated that PCA3 has an acceptable diagnostic accuracy and can help guide decisions on whether or not to carry out an initial biopsy, thus reducing the number of unnecessary biopsies [214]. The addition of PCA3 scores to individual risk estimation models, which included clinical factors, age and patient race, has been shown to improve PC stratification [215]. Wei et al. concluded that PCA3 measurement can reduce the under-detection of high-grade disease in initial prostatic biopsies, while also minimising the over-detection of low-grade PC in repeat biopsies [215]. Other studies have also demonstrated that PCA3 can supplement PSA and other clinical information to help give a more accurate prediction of the outcome from repeat biopsies [216,217].

As with the previously discussed biomarkers, the combination of PCA3 score with mpMRI has also been examined. The PCA3 score in men with a suspicious area for PC after mpMRI was higher than that of patients with no suspicious regions post-mpMRI; these results indicated that the PCA3 test could be used to pick those patients that should be referred for an mpMRI scan [218]. The addition of the PCA3 score to mpMRI was also shown to improve the predictive accuracy of mpMRI [219,220]. New methods for PCA3 detection are under development to enable PCA3 tests to be carried out in developing countries and to allow the assay to be used as a point-of-care test [221–224].

Studies have indicated that PCA3 could be employed to influence decisions between AS and more radical treatment options. It has been suggested that a threshold score of 20 could be used to identify men with clinically insignificant PC who would be eligible for AS, while a threshold of 50 could identify men at higher risk of having clinically significant PC who are good candidates for radical therapy [213]. However, the correlation of PCA3 score and PC aggressiveness is under debate, with some studies exhibiting a relationship between PCA3 score and Gleason score [225–229], whilst other do not [230,231]. Additionally, comparative studies indicate that PHI outperforms PCA3; PHI exhibited increased accuracy for PC prediction in initial and repeat biopsies [232], with PHI also superior in the detection of aggressive disease [233]. While it is improbable that PCA3 will replace PSA as the frontline biomarker for PC, the measurement of both PCA3 and PSA could lead to greater specificity for PC diagnosis.

4.6. Combined PCA3 and TMPRSS2-ERG Tests

Considering the significant heterogeneity seen within PCs, and the fact that not all PCs will express PCA3 or possess TMPRSS2-ERG gene fusions, researchers have investigated the use of multiplexed assays using both PCA3 and TMPRSS2-ERG gene fusions to improve PC diagnosis [188,189]. The Mi-Prostate Score and ExoDx Prostate IntelliScore (EPI) test are examples of these assays. The Mi-Prostate Score uses PCA3 and TMPRSS2-ERG urine scores with serum PSA levels; this combination was shown to enhance the ability of serum PSA to predict PC [234,235]. The EPI assay is an exosome-based urine assay which does not require a prostatic massage. It assesses PCA3 and TMPRSS2-ERG mRNA levels, with the SAM pointed domain-containing Ets transcription factor analysed for RNA normalisation. The EPI assay has been suggested for use in men with increased PSA levels in order to give a risk assessment for the presence of clinically significant PC at the initial biopsy [236,237]. The EPI test also has the potential to rule out the presence of high-grade disease using repeat biopsy tissues [238]. Results from the EPI test have been shown to influence biopsy decision making within the clinic [239]. Trials to confirm the performance of the EPI assay in men presenting for initial (NCT04720599) and repeat (NCT04357717) biopsies are currently underway.

4.7. Prostate Core Mitomic Test

Various cumulative genetic and epigenetic alterations within a cell contribute to the process of cell transformation. Although some of these genetic changes lead to cancer formation, early genetic alterations can lead to the growth of pre-neoplastic daughter cells in a particular area of the tumour field. While changes in cellular morphology enable the transformed cells to be diagnosed through histopathology, a population of pre-neoplastic daughter cells may be present that would not be diagnosed using this method, illustrating the concept of field cancerisation [240]. In PC, molecular field characterisation has been described for gene expression profiles and genomic instability. One study demonstrated that a 3.4-kb mitochondrial genome deletion (3.4mtΔ) had potential as a biomarker for PC detection using biopsy samples. As a result of field cancerisation, the levels of 3.4mtΔ in clinical samples from malignant biopsy specimens were similar to the levels that were acquired from samples close to the malignant tissue. The authors concluded that large-scale mitochondrial DNA deletions may have use in the diagnosis of PC through their ability to define benign, malignant and proximal to malignant tissue, thereby helping resolve false from true-negative results [241]. The utility of this 3.4mtΔ in identifying men who do not need a repeat biopsy has been shown [242]. The Mitomic Prostate Core Test was subsequently developed for use in existing negative prostate biopsy tissue to assess if PC was missed in the initial biopsy. Further studies have demonstrated the usefulness of this assay in addressing sampling error issues encountered with prostate needle biopsies, with the test contributing to the earlier detection of PC when clinicians included the test in their re-biopsy decision-making process [243].

5. Biomarkers That Can Predict PC Aggressiveness

Definitive treatment for PC can lead to significant complications. Biomarkers that give an indication of disease aggressiveness in patients who have already been diagnosed would help clinicians decide who should be considered for AS and who should undergo radical treatment. This would assist in the identification of patients who could benefit from treatment, while also reducing the treatment risks and economic costs for those who are unlikely to benefit.

5.1. Oncotype DX Genomic Prostate Score Assay

Predictive gene expression signature assays have been developed to help identify cohorts of patients that gain specific benefits from certain therapies. Signatures of breast cancer RT and chemotherapy response [244,245], and also treatment-predictive signatures for lung cancer [246] are successful, clinically useful examples of these. The Oncotype Dx assay, developed by Genomic Health, is a commercial gene signature assay that has gained significant popularity for identifying cohorts of breast cancer patients that gain benefit from adjuvant chemotherapy [247]. As a result of successful studies in breast, the applicability of an adapted test to PC has been examined. The Oncotype DX Genomic Prostate Score assay is carried out on prostatic biopsy tissue. It was designed to aid treatment selection at the time of diagnosis in patients with low- or intermediate-risk disease, enabling both patients and clinicians to make more informed choices between AS and immediate radical treatment [248]. This test is based on the expression pattern of 12 genes that characterise four separate pathways known to be involved in PC development and progression (proliferation, cellular structure/organisation, stromal interactions and androgen signalling), along with five housekeeper genes. A final Genomic Prostate Score (GPS) ranging from 0–100 is calculated. This GPS can provide predictive information regarding the risk of identifying adverse pathology after RP (higher grade and stage disease) [248–252], aids in determining the risk of PC recurrence after surgery [250], and can also ascertain the risk of BCR and distant metastasis [250,252–254]. The cost-effectiveness of the GPS assay in directing treatment decisions (AS versus immediate treatment) has also been reported [255,256].

However, more recent work has highlighted some limitations of the Oncotype DX GPS. Lin et al. tested the value of the GPS in predicting the presence of higher-grade disease at surgery in low-risk PC patients who were treated with RP after initial surveillance. They found that GPS did not significantly improve the stratification of risk for adverse pathology over the measurement of PSAD and diagnostic Gleason Grade alone [257]. Another study showed that the histopathological features which are present in PC biopsies, but are not usually reported, correlated with the GPS score. The authors suggest that more comprehensive analysis of PC histopathology could be used as a substitute for some of the information obtained from this test [258].

5.2. Prolaris

The Prolaris assay, developed by Myriad Genetics, is a tissue-based test intended for use in patients with newly diagnosed localized low- or intermediate-risk PC. This test is designed to enable clinicians to better define a monitoring/treatment strategy for these patients, identifying those who can be directed safely to AS and those that would benefit from treatment intensification. It is based on the expression patterns of 31 genes involved in cell cycle progression (CCP), in addition to 15 housekeeper genes. Overexpression of the CCP genes suggests that the cancer cells are rapidly dividing, while decreased expression signifies slower growth and a less aggressive cancer [259]. The Prolaris score or CCP score is reported on a scale ranging from 0 to 10, where higher scores are indicative of a more aggressive tumour [260]. The CPP score has been shown to give significant pre-treatment prognostic information that can be used to help determine which patients can be managed conservatively [261,262], with additional studies demonstrating that this assay has the ability to provide prognostic information for men undergoing either RP [259,263–266] or RT [267]. Higher CCP scores have also been shown to be linked with a higher risk of systemic disease [268] and can predict metastasis after either RT or surgery [269]. Results from the Prolaris assay have influenced therapy decisions within the clinic; there has been an increase in the proportion of patients undergoing AS in those that have been classified as low-risk by the Prolaris test, and intensification of treatments in those whose test results indicted the presence of more aggressive cancer [270–272]. While the potential benefits of the Prolaris assay have been exhibited, its value is limited by the retrospective nature of many of the studies performed; largescale, prospective trials are needed [260]. Additionally, the cost-effectiveness of the Prolaris test is still under debate [273,274].

5.3. ProMark

The ProMark quantitative immunofluorescence test was developed in an attempt to give clinicians the ability to predict PC aggressiveness, irrespective of whether biopsy cores came from low- or high-grade tumour regions, therefore accounting for sampling variation and PC heterogeneity. In a study carried out by Shipitsin et al., tissue regions with the lowest and highest grades were isolated in prostatectomy samples from the same patients; a panel of protein biomarkers was identified that predicted PC aggressiveness and outcome from both low- and high-grade areas [275]. This test is based on the expression patterns of eight proteins (DERL1, CUL2, SMAD4, PDSS2, HSPA9, FUS, pS6 and YBOX1) with known functions related to proliferation, tumour-associated signalling pathways and stress response, altogether providing information about tumour aggressiveness from formalin-fixed, paraffin-embedded (FFPE) tissues [276]. The primary function of the ProMark test is to separate candidates for AS from those that require RP, in addition to ascertaining those patients with favourable/non-favourable pathology. Although not yet validated, the test has the potential to accurately stratify low- and high-risk PC patients using biopsy samples.

5.4. Decipher

The Decipher test, developed by GenomeDx, is a genomic signature that was developed to help identify aggressive PC and improve the prediction of early PC metastasis using information from the primary tumour after RP. This test analyses the RNA expression levels of 22 genes (involved in cellular differentiation, proliferation, cell cycle, motility, adhesion, immune modulation and androgen signalling) detected in the primary tumour and was developed by modelling differential RNA expression patterns in early metastatic tissues versus controls [277]. The final Decipher score ranges from 0–1, with higher scores (0.61–1) associated with a higher probability of metastasis. This genomic classifier has gained interest for its use in patients after RP and can predict both the 5- and 10-year metastatic risk [278–280]. A recent meta-analysis conducted by Spratt et al. showed that Decipher can improve the prognostication of patients post-RP; the 10-year cumulative metastatic incidence rates after RP were 5.5%, 15.0% and 26.7% for patients that were deemed low-, intermediate- and high-risk using the Decipher test [281]. These results are supported by another study showing that transcriptional profiles can stratify patients into cohorts, separating those who will develop metastasis after RP from those who will not [282]. A recent study highlighted how Decipher, in combination with standard clinicopathologic variables, can lead to better risk-stratification when combined with current guidelines [283]. While the test was developed from the analysis of primary tissue after RP, the ability of the Decipher test to predict metastasis using biopsy tumour tissue has also been shown [284,285].

The potential of Decipher to predict BCR after surgery has also been established [286]. Patients exhibiting BCR after RP often have varied outcomes and thus present a management dilemma to clinicians; initial studies showed the ability of the Decipher test to predict metastasis in these patients, exhibiting its potential to identify men who require earlier initiation of treatment after BCR [287]. More recent studies have demonstrated that Decipher can be used to predict the absence of adverse pathology in low- and intermediate-risk PC patients, with the authors suggesting that Decipher may have a role in predicting which newly diagnosed patients are good candidates for AS [288]. Furthermore, Decipher scores have been shown to have potential in determining those patients who are most suitable for RT following RP [289,290]. The ability of the Decipher test to alter clinical decisions regarding the use of adjuvant treatments has been reported [291,292]. Altogether, data from several studies has demonstrated the clinical usefulness of the Decipher test, exhibiting its potential to significantly improve the personalisation of PC treatment [293].

5.5. Ki67

Ki67 is a nuclear protein related to ribosomal RNA synthesis. This protein is used as a marker for tumour proliferation, with analysis of Ki67 levels typically carried out through immunohistochemistry on FFPE tissues. Staining is described as the percentage of Ki67-positive cells within the total number of cancer cells present. Ki67 has been shown to be a prognostic and predictive biomarker in breast cancer [294]. Within PC, a higher percentage of Ki67-positive cells seems to have prognostic value for BCR, distant metastasis and survival in patients treated with either surgery or RT [295–298]. A recent meta-analysis incorporating 21 studies, comprised of 5419 patients, demonstrated that after curative-intent treatments, high Ki67 expression was a poor prognostic factor for disease-specific survival, disease-free survival, rate of distant metastases and overall survival. The authors concluded that Ki67 should be integrated into the clinic for use in PC patients [299]. However, despite the fact that Ki67 is one of the best validated prognostic markers that has been in use for over 30 years, some maintain that this protein is not yet ready for use in the clinic. High levels of variability in scores have been observed between different cohorts of PC patients, with scores ranging from 2.1% to 28% [300]. This issue seems to be particularly relevant in high-risk patients, in whom significant inter- and intra-prostatic Ki-67 heterogeneity has been reported [301]. The cut-offs used to distinguish a negative from a positive score also differ greatly between studies; this lack of standardisation across pathology laboratories contributes to the limitations of Ki67 as a PC biomarker [302].

5.6. MicroRNAs

MicroRNAs (miRNAs) are single stranded, small non-coding RNA molecules (~20 nucleotides in length) that function as post-transcriptional gene regulators through their ability to bind to complementary base pairs within specific mRNAs [303]. Alterations in miRNA profiles have been identified in PC. It has been suggested that miRNAs can regulate PC stem cells, cellular proliferation and differentiation, thereby influencing disease development and progression [304,305]. Studies showing that miRNAs are present in human blood in a very stable form [306] led to the development of miRNA signatures from blood samples in an attempt to improve the accuracy of PC diagnosis and prognosis. One such study identified a panel of 14 miRNAs, known as the miR risk score, which was able to discriminate Gleason grade and predict BCR following RP [307]. A further study showed that miR-16, miR-195 and miR-148a expression was correlated with Gleason scores ≥ 8, and that these three miRNAs could stratify patients into intermediate- and high-risk Gleason scores [308]. Several PC studies have also investigated miRNA signatures from urine samples to differentiate healthy patients or those with BPH from those with PC [309,310]

5.7. Phosphatase and Tensin Homolog

Phosphatase and tensin homolog (*PTEN*) is a well characterised tumour suppressor gene involved in the regulation of the phosphatidylinositol 3-kinase (PI3K) pathway. Loss of function of PTEN and the resulting de-regulation of the PI3K pathway is regarded as one of the most common driver events in PC development [311]. Loss of PTEN function has been shown to occur in ~40% of PC cases, especially in those with TMPRSS2-ERG gene fusions [312]. Although immunohistochemistry is typically used to evaluate PTEN loss, fluorescence in situ hybridisation (FISH) can be utilised where ambiguous immunohistochemistry results have been obtained [313]. Several studies have examined the use of PTEN loss as a biomarker in PC; one study suggested that patients exhibiting PTEN loss in Gleason score 6 tumours, identified from biopsy tissue, were at higher risk of having their score upgraded using samples obtained at RP [314]. Other investigations have demonstrated that loss or even just a decrease in PTEN expression is correlated with higher Gleason scores, more advanced disease stage, metastasis, BCR and disease recurrence [315–319]. Furthermore, shorter survival times have been reported in advanced PC with PTEN loss when treated with abiraterone acetate [320]. Apart from the removal of the tumour suppressive function, PTEN loss has also been associated with AR signalling suppression and inhibition of androgenic genes [321]; this may drive PC into an androgen-independent phenotype, ultimately reducing the efficacy of ADT.

6. Predictive Biomarkers

Predictive biomarkers indicate the likelihood of a particular treatment providing a therapeutic benefit. These biomarkers can therefore be used to aid treatment selection, enabling the identification of patients that are most likely to gain benefit from a particular therapy, whilst sparing others from the side effects of ineffectual treatment. Here, we provide an overview of a selection of predictive biomarkers that are currently being researched.

6.1. Post-Operative Radiation Therapy Outcomes Score

Although RT post-RP can significantly improve clinical outcomes, recent work does not support the routine administration of adjuvant RT post-RP [322]. It has been suggested that certain patient cohorts are more likely to gain benefit from its use; identification of these patients will improve their outcome while sparing the risk of developing radiation-induced side effects in those unlikely to gain a clinical benefit. Unfortunately, as of yet no gene signature has been clinically validated to predict RT response in PC patients. To begin to address this clinical issue, one study has developed and initially validated a 24 gene signature to predict RT response. This Post-Operative Radiation Therapy Outcomes Score (PORTOS) was developed using gene expression data from prostatic adenocarcinomas in patients who received a RP with or without adjuvant RT. Results demonstrated that the distant metastatic rate at 10 years for patients with a high PORTOS who received RT was lower than that observed for patients with a high PORTOS who did not receive RT (4% vs. 35%). While the authors suggested that PORTOS could be used to predict outcomes post-RT, thereby identifying which patient cohort should receive RT, they also demonstrated that other prognostic tools such as Decipher and the CCP score did not predict RT response [323].

6.2. DNA Repair Defects

Both pre-clinical and clinical reports indicate that DNA damage response pathways have a significant part to play in the progression of PC [324]. DNA repair defects are thought to be relatively frequent in more advanced PC, with genetic abnormalities that inhibit DNA repair shown to be present in mCRPC tumours [325]. It is thought that the identification of alterations in DNA repair pathways may be predictive of response to certain therapies. Poly (adenosine diphosphate [ADP]–ribose) polymerase (PARP) has a part to play in numerous aspects of DNA repair. PARP inhibitors are a class of anti-cancer agents that work through inducing synthetic lethality; this is a process where the PARP inhibitor, in combination with either an inherent genetic defect or another therapy (such as RT), cause irreparable DNA damage and cell death [326]. PARP inhibitors initially demonstrated their potential as an anti-cancer therapy in patients with BRCA1/2 mutations and they have become a standard treatment for patients suffering from ovarian and breast cancer. Olaparib and rucaparib are PARP inhibitors that have been approved by the FDA for the treatment of mCRPC [326]. The identification of DNA repair defects in mCRPC patients has been shown to predict response to PARP inhibitors; however, not all DNA repair defects have the same impact on the efficacy of treatment [327]. While the majority of data for PARP inhibitors has been generated for mCRPC patients, there will be interest among the scientific and clinical communities on the results of studies concentrating on earlier disease stages.

6.3. Androgen Receptor

The androgen receptor (AR) is a nuclear hormone receptor transcription factor that plays a significant role in the function of prostatic cells through its ability to bind sex steroids and control transcription of androgen-dependent genes [328]. ADT is a common treatment for PC; however, although nearly all PCs respond to this treatment in the beginning, tumour recurrence and progression into castrate-resistant prostate cancer (CRPC) typically occurs [329]. While the progression of androgen-dependent PC to CRPC likely involves various mechanisms, AR and its signalling have been shown to play important roles in disease development, including the acquisition of acquired resistance to various ADT drugs [330]. Within CRPC, AR alterations have been shown to occur through over-expression of wild-type or constitutively active variants (AR-Vs), gene amplification and mutations [331]. AR-Vs, generated from alternative splicing or gene rearrangements, have the ability to regulate transcription. Although these AR-Vs are truncated proteins that lack the AR ligand-binding domain, they still have functional DNA-binding and transcriptional activation domains, resulting in ligand-independent constitutive activation that is not constrained by anti-androgen treatment [331]. The AR-V7 form is frequently detected in mCRPC and has gained clinical interest for its use as a biomarker to help select the most appropriate treatments [332]. A crucial decision in mCRPC management is when to administer an AR signalling inhibitor or a taxane; studies have shown that AR-V7 expression is associated with the resistance of mCRPC to enzalutamide and abiraterone [333–335], while its expression also appears to correlate with increased response to taxane chemotherapies [336]. AR-V7 in CRPC patients can be detected within both prostatic tissue samples and circulating tumours cells (CTCs) [332,337]; however, conflicting findings have been observed between CTC AR-V7 results and AR-V7 protein expression in biopsy samples acquired from the same patient [338]. The OncotypeDX AR-V7 Nucleus Detect (Epic Sciences) and the AdnaTest AR-V7 assay (Qiagen) have been developed for the assessment of the constitutively active AR variant in CTCs.

6.4. Immune Checkpoint Inhibition

Monoclonal antibodies targeting immune checkpoint inhibitors are being considered as a new therapeutic strategy for the treatment of mCRPC. Immune checkpoint inhibitors (cytotoxic T lymphocyte-associated protein-4 (CTLA-4) receptor and programmed death-1 (PD-1) receptor) are present on T lymphocytes; these receptors act as negative regulators of the immune response, setting a balance between an effective immune response (including the response of the immune system to cancer cells) and tolerance to antigens produced by normal cells of the body [339]. The over-expression of ligands for these receptors on cancer cells (leading to the activation of immune checkpoint inhibitors and the inactivation of immune cells) has been observed in PC, contributing to the escape of these cancer cells from the host's immune response [339]. The concept that the CTLA-4 and PD-1 receptors might be utilised by cancer cells to avoid the immune system led to the development of monoclonal antibodies that could inhibit these receptors, with the hope that targeting them would lead to a more effective anti-tumour response from T lymphocytes. Although some studies have demonstrated that Ipilimumab (an anti-CTLA-4 monoclonal antibody) and Nivolumab (an anti-PD-1 monoclonal antibody) are effective treatments for advanced PC [340,341], others have shown mixed results from the use of these agents [339]. It is thought that only certain patients are eligible for immunotherapy: those presenting with either high expression levels of CTLA-4 and PD-1 receptor ligands on cancer/stromal cells, or increased amounts of the immune checkpoint inhibitor receptors on immune cells. As such, it is believed that these proteins may act as biomarkers that could predict/monitor immunotherapy effectiveness [342,343].

Research into the predictive potential of genomic biomarkers for immunotherapy is also ongoing. Tumour mutational burden (TMB) can be used to describe the number of mutations in a tumour cell. Patients suffering from advanced PC have been shown to exhibit higher levels of TMB [344,345]. While the prediction of PC patient reaction to immunotherapy is complex, increased levels of TMB have been linked to better response [346]. It is believed that a higher TMB causes the production of increased levels of neoantigens (mutated antigens that are only expressed by cancer cells), which leads to a higher probability of an effective T-cell-dependent anti-cancer response [347]. Additional genomic predictive biomarkers for response to immunotherapy have also recently been identified; mutations within cyclin-dependent kinase 12 (CDK12), a tumour suppressor protein with roles connected to genomic stability [348], have also been demonstrated to lead to the creation of neoantigens [349]. It is thought that CDK12-altered PCs may respond favourably to immune checkpoint inhibitors [350].

7. Limitations and Future Perspectives of PC Biomarker Assays

The function of the prostate is to perform as a secretory gland, secreting proteins including PSA into seminal fluid. As such, liquid-based biomarkers, such as those acquired from the blood or urine, are well placed to act as PC-specific biomarkers. The identification of biomarkers in liquid biopsies has significant advantages over tissue-based techniques as they can be obtained easily in a less invasive manner. Liquid biopsies can also be routinely taken pre-, post- or on-treatment, meaning continual patient monitoring can be achieved, while tissue biopsies give only a limited snapshot of the tumour. Tumour heterogeneity is a significant problem for tissue-based biopsy tests, as results can only be determined from the area that the tissue samples are acquired from [351,352]. Liquid biopsies, in comparison, have the potential to give a comprehensive view of both primary and metastatic cancers. Urine samples in particular have specific advantages in PC management; as a result of the proximity of the bladder to the prostate, urine can contain biomarkers that reflect PC development and progression.

Of the liquid-based assays, PSA is the best validated and most widely used biomarker employed by clinicians. This is likely to remain the case for the present, despite limitations associated with its use. To overcome some of these issues, studies have examined the use of different PSA parameters/dynamics. The combination of PSA with adjunct tests is also being studied in an attempt to enhance the diagnostic specificity and prognostic potential of PSA. Of the tissue-based biomarker tests discussed, Oncotype DX Prostate, ProMark, Decipher and Prolaris are the best validated thus far. While these tissue-based biomarker assays have the potential to influence the management of PC patients, there are a number of issues that are currently restricting their use: (i) Direct comparison of Oncotype DX, Prolaris and Decipher to one another has shown that prognostic outcomes can differ depending on the test used [353,354]. (ii) Many of these assays were developed and initially validated in cohorts of patients who were mostly white European or white American men, with limited initial research performed into the value of these tests in African American men, who are recognised as having poorer outcomes. While some of the assays have been validated and shown to provide benefit in diverse racial groups [250,355–357], racial differences across the gene expression panels used for PC prognosis have been identified [358]. (iii) Lastly, the clinical usefulness of these multigene signatures has yet to be prospectively validated in a randomised clinical trial. Regardless of these shortcomings, present NCCN recommendations assert that Prolaris, Decipher, ProMark and Oncotype DX Prostate can be used for risk stratification in patients with either low- or favourable intermediate-risk PC [359].

8. Conclusions

Significant advances continue to be made in the field of PC. Although the widespread use of PSA levels for PC diagnosis and management led to criticisms of over-diagnosing and over-treating patients, its use undoubtedly paved the way for investigations into more specific PC biomarkers. The biomarkers discussed in this review have the potential to contribute immensely to PC patient management by (i) cutting down on unnecessary biopsies, (ii) enhancing patient risk assessment and therefore treatment selection and (iii) leading to more selective treatments for PC patients with higher-risk disease.

For any biomarker-based assay to become translated into the clinic and used routinely, studies need to demonstrate specificity, sensitivity and their potential to improve upon current clinical practices. That said, PC biomarker research holds much promise; linking novel PC-specific biomarkers with other techniques, such as clinical data, PSA levels, Gleason grading, disease staging and imaging would undoubtedly help improve the management of PC patients. Ultimately, we need implementation of many of the assays discussed into well designed randomised clinical trials in order to validate them; hopefully it is only a matter of time before this can be achieved.

Author Contributions: J.M., M.G., D.M. and A.K.T. conceptualised the article. J.M. wrote the majority of the manuscript. Figures and tables were composed by J.M. and M.G. Critical revisions were made by J.M., M.G., C.M.-P., C.K., D.M. and A.K.T. All authors have read and agreed to the published version of the manuscript.

Funding: This research was funded by the John Black Foundation, part of The Urology Foundation.

Institutional Review Board Statement: Not applicable.

Informed Consent Statement: Not applicable.

Data Availability Statement: Not applicable.

Conflicts of Interest: The authors declare no conflict of interest.

References

1. Adams, J. The case of scirrhous of the prostate gland with corresponding affliction of the lymphatic glands in the lumbar region and in the pelvis. *Lancet* **1853**, *1*, 393.
2. Sung, H.; Ferlay, J.; Siegel, R.L.; Laversanne, M.; Soerjomataram, I.; Jemal, A.; Bray, F. Global Cancer Statistics 2020: GLOBOCAN Estimates of Incidence and Mortality Worldwide for 36 Cancers in 185 Countries. *CA Cancer J. Clin.* **2021**, *71*, 209–249. [CrossRef]
3. Siegel, R.L.; Miller, K.D.; Fuchs, H.E.; Jemal, A. Cancer Statistics, 2021. *CA Cancer J. Clin.* **2021**, *71*, 7–33. [CrossRef]
4. NICE. 2020. Available online: https://www.nice.org.uk/Media/Default/About/what-we-do/Into-practice/measuring-uptake/prostate-cancer/nice-impact-prostate-cancer.pdf (accessed on 10 June 2021).
5. Rawla, P. Epidemiology of Prostate Cancer. *World J. Oncol.* **2019**, *10*, 63–89. [CrossRef] [PubMed]
6. Jahn, J.L.; Giovannucci, E.L.; Stampfer, M.J. The high prevalence of undiagnosed prostate cancer at autopsy: Implications for epidemiology and treatment of prostate cancer in the Prostate-specific Antigen-era. *Int. J. Cancer* **2015**, *137*, 2795–2802. [CrossRef] [PubMed]
7. Rodrigues, G.; Warde, P.; Pickles, T.; Crook, J.; Brundage, M.; Souhami, L.; Lukka, H. Pre-treatment risk stratification of prostate cancer patients: A critical review. *Can. Urol. Assoc. J.* **2012**, *6*, 121–127. [CrossRef] [PubMed]
8. Chen, N.; Zhou, Q. The evolving Gleason grading system. *Chin. J. Cancer Res.* **2016**, *28*, 58–64. [PubMed]
9. Andreoiu, M.; Cheng, L. Multifocal prostate cancer: Biologic, prognostic, and therapeutic implications. *Hum. Pathol.* **2010**, *41*, 781–793. [CrossRef]
10. Ruijter, E.T.; van de Kaa, C.A.; Schalken, J.A.; Debruyne, F.M.; Ruiter, D.J. Histological grade heterogeneity in multifocal prostate cancer. Biological and clinical implications. *J. Pathol.* **1996**, *180*, 295–299. [CrossRef]

11. Loeb, S.; Vellekoop, A.; Ahmed, H.U.; Catto, J.; Emberton, M.; Nam, R.; Rosario, D.J.; Scattoni, V.; Lotan, Y. Systematic Review of Complications of Prostate Biopsy. *Eur. Urol.* **2013**, *64*, 876–892. [CrossRef]
12. Litwin, M.S.; Tan, H.J. The Diagnosis and Treatment of Prostate Cancer: A Review. *JAMA* **2017**, *317*, 2532–2542. [CrossRef] [PubMed]
13. Mahal, B.A.; Butler, S.; Franco, I.; Spratt, D.E.; Rebbeck, T.R.; D'Amico, A.V.; Nguyen, P.L. Use of Active Surveillance or Watchful Waiting for Low-Risk Prostate Cancer and Management Trends Across Risk Groups in the United States, 2010–2015. *JAMA* **2019**, *321*, 704–706. [CrossRef] [PubMed]
14. Resnick, M.J.; Koyama, T.; Fan, K.H.; Albertsen, P.C.; Goodman, M.; Hamilton, A.S.; Hoffman, R.M.; Potosky, A.L.; Stanford, J.L.; Stroup, A.M.; et al. Long-Term Functional Outcomes after Treatment for Localized Prostate Cancer. *N. Engl. J. Med.* **2013**, *368*, 436–445. [CrossRef] [PubMed]
15. van den Bergh, R.C.; Korfage, I.J.; Roobol, M.J.; Bangma, C.H.; de Koning, H.J.; Steyerberg, E.W.; Essink-Bot, M.L. Sexual function with localized prostate cancer: Active surveillance vs radical therapy. *BJU Int.* **2012**, *110*, 1032–1039. [CrossRef] [PubMed]
16. Balk, S.P.; Ko, Y.-J.; Bubley, G.J. Biology of prostate-specific antigen. *J. Clin. Oncol.* **2003**, *21*, 383–391. [CrossRef] [PubMed]
17. Lilja, H.; Ulmert, D.; Vickers, A.J. Prostate-specific antigen and prostate cancer: Prediction, detection and monitoring. *Nat. Rev. Cancer* **2008**, *8*, 268–278. [CrossRef] [PubMed]
18. Kan, H.-C.; Hou, C.P.; Lin, Y.H.; Tsui, K.H.; Chang, P.L.; Chen, C.L. Prognosis of prostate cancer with initial prostate-specific antigen >1000 ng/mL at diagnosis. *Onco Targets Ther.* **2017**, *10*, 2943–2949. [CrossRef]
19. Stamey, T.A.; Yang, N.; Hay, A.R.; McNeal, J.E.; Freiha, F.S.; Redwine, E. Prostate-specific antigen as a serum marker for adenocarcinoma of the prostate. *N. Engl. J. Med.* **1987**, *317*, 909–916. [CrossRef] [PubMed]
20. Catalona, W.J.; Smith, D.S.; Ratliff, T.L.; Dodds, K.M.; Coplen, D.E.; Yuan, J.J.; Petros, J.A.; Andriole, G.L. Measurement of prostate-specific antigen in serum as a screening test for prostate cancer. *N. Engl. J. Med.* **1991**, *324*, 1156–1161. [CrossRef]
21. Parkes, C.; Wald, N.J.; Murphy, P.; George, L.; Watt, H.C.; Kirby, R.; Knekt, P.; Helzlsouer, K.J.; Tuomilehto, J. Prospective observational study to assess value of prostate specific antigen as screening test for prostate cancer. *BmJ* **1995**, *311*, 1340–1343. [CrossRef] [PubMed]
22. Lopez-Saez, J.-B.; Otero, M.; Senra-Varela, A.; Ojea, A.; Martín, J.S.; Muñoz, B.D.; Fuentes, J.V. Prospective Observational Study to Assess Value of Prostate Cancer Diagnostic Methods. *J. Diagn. Med. Sonogr.* **2004**, *20*, 383–392. [CrossRef]
23. Kouriefs, C.; Sahoyl, M.; Grange, P.; Muir, G. Prostate specific antigen through the years. *Arch. Ital. Urol. Androl.* **2009**, *81*, 195–198.
24. Potosky, A.L.; Feuer, E.J.; Levin, D.L. Impact of screening on incidence and mortality of prostate cancer in the United States. *Epidemiol. Rev.* **2001**, *23*, 181–186. [CrossRef]
25. Draisma, G.; Etzioni, R.; Tsodikov, A.; Mariotto, A.; Wever, E.; Gulati, R.; Feuer, E.; De Koning, H. Lead time and overdiagnosis in prostate-specific antigen screening: Importance of methods and context. *J. Natl. Cancer Inst.* **2009**, *101*, 374–383. [CrossRef] [PubMed]
26. Fenton, J.J.; Weyrich, M.S.; Durbin, S.; Liu, Y.; Bang, H.; Melnikow, J. Prostate-specific antigen–based screening for prostate cancer: Evidence report and systematic review for the US Preventive Services Task Force. *JAMA* **2018**, *319*, 1914–1931. [CrossRef] [PubMed]
27. Klotz, L. Prostate cancer overdiagnosis and overtreatment. *Curr. Opin. Endocrinol. Diabetes Obes.* **2013**, *20*, 204–209. [CrossRef] [PubMed]
28. Grossman, D.C.; Curry, S.J.; Owens, D.K.; Bibbins-Domingo, K.; Caughey, A.B.; Davidson, K.W.; Doubeni, C.A.; Ebell, M.; Epling, J.W.; Kemper, A.R.; et al. Screening for prostate cancer: US Preventive Services Task Force recommendation statement. *JAMA* **2018**, *319*, 1901–1913.
29. Sanda, M.G.; Dunn, R.L.; Michalski, J.; Sandler, H.M.; Northouse, L.; Hembroff, L.; Lin, X.; Greenfield, T.K.; Litwin, M.S.; Saigal, C.S.; et al. Quality of life and satisfaction with outcome among prostate-cancer survivors. *N. Engl. J. Med.* **2008**, *358*, 1250–1261. [CrossRef] [PubMed]
30. Martin, R.M.; Donovan, J.L.; Turner, E.L.; Metcalfe, C.; Young, G.J.; Walsh, E.I.; Lane, J.A.; Noble, S.; Oliver, S.E.; Evans, S.; et al. Effect of a low-intensity PSA-based screening intervention on prostate cancer mortality: The CAP randomized clinical trial. *JAMA* **2018**, *319*, 883–895. [CrossRef] [PubMed]
31. Schröder, F.H.; Hugosson, J.; Roobol, M.J.; Tammela, T.L.; Zappa, M.; Nelen, V.; Kwiatkowski, M.; Lujan, M.; Määttänen, L.; Lilja, H.; et al. Screening and prostate cancer mortality: Results of the European Randomised Study of Screening for Prostate Cancer (ERSPC) at 13 years of follow-up. *Lancet* **2014**, *384*, 2027–2035. [CrossRef]
32. Schröder, F.H.; Hugosson, J.; Carlsson, S.; Tammela, T.; Määttänen, L.; Auvinen, A.; Kwiatkowski, M.; Recker, F.; Roobol, M.J. Screening for prostate cancer decreases the risk of developing metastatic disease: Findings from the European Randomized Study of Screening for Prostate Cancer (ERSPC). *Eur. Urol.* **2012**, *62*, 745–752. [CrossRef] [PubMed]
33. Pinsky, P.F.; Prorok, P.C.; Yu, K.; Kramer, B.S.; Black, A.; Gohagan, J.K.; Crawford, E.D.; Grubb, R.L.; Andriole, G.L. Extended mortality results for prostate cancer screening in the PLCO trial with median follow-up of 15 years. *Cancer* **2017**, *123*, 592–599. [CrossRef]
34. Duffy, M.J. Biomarkers for prostate cancer: Prostate-specific antigen and beyond. *Clin. Chem. Lab. Med.* **2020**, *58*, 326–339. [CrossRef] [PubMed]

35. Sanda, M.G.; Cadeddu, J.A.; Kirkby, E.; Chen, R.C.; Crispino, T.; Fontanarosa, J.; Freedland, S.J.; Greene, K.; Klotz, L.H.; Makarov, D.V.; et al. Clinically Localized Prostate Cancer: AUA/ASTRO/SUO Guideline. Part I: Risk Stratification, Shared Decision Making, and Care Options. *J. Urol.* **2018**, *199*, 683–690. [CrossRef] [PubMed]
36. Smith, R.A.; Andrews, K.S.; Brooks, D.; Fedewa, S.A.; Manassaram-Baptiste, D.; Saslow, D.; Brawley, O.W.; Wender, R.C. Cancer screening in the United States, 2018: A review of current American Cancer Society guidelines and current issues in cancer screening. *CA Cancer J. Clin.* **2018**, *68*, 297–316. [CrossRef]
37. Mottet, N.; Bellmunt, J.; Bolla, M.; Briers, E.; Cumberbatch, M.G.; De Santis, M.; Fossati, N.; Gross, T.; Henry, A.M.; Joniau, S.; et al. EAU-ESTRO-SIOG Guidelines on Prostate Cancer. Part 1: Screening, Diagnosis, and Local Treatment with Curative Intent. *Eur. Urol.* **2017**, *71*, 618–629. [CrossRef]
38. Gandaglia, G.; Albers, P.; Abrahamsson, P.A.; Briganti, A.; Catto, J.W.; Chapple, C.R.; Montorsi, F.; Mottet, N.; Roobol, M.J.; Sønksen, J.; et al. Structured Population-based Prostate-specific Antigen Screening for Prostate Cancer: The European Association of Urology Position in 2019. *Eur. Urol.* **2019**, *76*, 142–150. [CrossRef]
39. Vickers, A.J.; Eastham, J.A.; Scardino, P.T.; Lilja, H. The Memorial Sloan Kettering Cancer Center Recommendations for Prostate Cancer Screening. *Urology* **2016**, *91*, 12–18. [CrossRef]
40. Gasinska, A.; Jaszczynski, J.; Rychlik, U.; Łuczynska, E.; Pogodzinski, M.; Palaczynski, M. Prognostic Significance of Serum PSA Level and Telomerase, VEGF and GLUT-1 Protein Expression for the Biochemical Recurrence in Prostate Cancer Patients after Radical Prostatectomy. *Pathol. Oncol. Res.* **2020**, *26*, 1049–1056. [CrossRef]
41. Liu, D.; Kuai, Y.; Zhu, R.; Zhou, C.; Tao, Y.; Han, W.; Chen, Q. Prognosis of prostate cancer and bone metastasis pattern of patients: A SEER-based study and a local hospital based study from China. *Sci. Rep.* **2020**, *10*, 9104. [CrossRef]
42. Vazquez Martinez, M.A.; Correa, E.; Jeurkar, C.; Shikdar, S.; Jain, M.R.; Topolsky, D.L.; Crilley, P.A.; Ward, K.M.; Styler, M. The prognostic significance of PSA as an indicator of age standardized relative survival: An analysis of the SEER database 2004–2014. *J. Clin. Oncol.* **2018**, *36* (Suppl. 15), e18768. [CrossRef]
43. Zheng, Z.; Zhou, Z.; Yan, W.; Zhou, Y.; Chen, C.; Li, H.; Ji, Z. Tumor characteristics, treatments, and survival outcomes in prostate cancer patients with a PSA level <4 ng/mL: A population-based study. *BMC Cancer* **2020**, *20*, 340. [CrossRef]
44. Ang, M.; Rajcic, B.; Foreman, D.; Moretti, K.; O'Callaghan, M.E. Men presenting with prostate-specific antigen (PSA) values of over 100 ng/mL. *BJU Int.* **2016**, *117*, 68–75. [CrossRef] [PubMed]
45. Kang, Y.; Song, P.; Fang, K.; Yang, B.; Yang, L.; Zhou, J.; Wang, L.; Dong, Q. Survival outcomes of low prostate-specific antigen levels and T stages in patients with high-grade prostate cancer: A population-matched study. *J. Cancer* **2020**, *11*, 6484–6490. [CrossRef]
46. Mahal, B.A.; Yang, D.D.; Wang, N.Q.; Alshalalfa, M.; Davicioni, E.; Choeurng, V.; Schaeffer, E.M.; Ross, A.E.; Spratt, D.E.; Den, R.B.; et al. Clinical and Genomic Characterization of Low-Prostate-specific Antigen, High-grade Prostate Cancer. *Eur. Urol.* **2018**, *74*, 146–154. [CrossRef] [PubMed]
47. McCormick, B.Z.; Mahmoud, A.M.; Williams, S.B.; Davis, J.W. Biochemical recurrence after radical prostatectomy: Current status of its use as a treatment endpoint and early management strategies. *Indian J. Urol.* **2019**, *35*, 6–17.
48. Zietman, A.L.; Christodouleas, J.P.; Shipley, W.U. PSA bounces after neoadjuvant androgen deprivation and external beam radiation: Impact on definitions of failure. *Int. J. Radiat. Oncol. Biol. Phys.* **2005**, *62*, 714–718. [CrossRef] [PubMed]
49. Hanlon, A.L.; Pinover, W.H.; Horwitz, E.M.; Hanks, G.E. Patterns and fate of PSA bouncing following 3D-CRT. *Int. J. Radiat. Oncol. Biol. Phys.* **2001**, *50*, 845–849. [CrossRef]
50. Abramowitz, M.C.; Li, T.; Buyyounouski, M.K.; Ross, E.; Uzzo, R.G.; Pollack, A.; Horwitz, E.M. The Phoenix definition of biochemical failure predicts for overall survival in patients with prostate cancer. *Cancer* **2008**, *112*, 55–60. [CrossRef]
51. Roach, M., 3rd; Hanks, G.; Thames, H., Jr.; Schellhammer, P.; Shipley, W.U.; Sokol, G.H.; Sandler, H. Defining biochemical failure following radiotherapy with or without hormonal therapy in men with clinically localized prostate cancer: Recommendations of the RTOG-ASTRO Phoenix Consensus Conference. *Int. J. Radiat. Oncol. Biol. Phys.* **2006**, *65*, 965–974. [CrossRef]
52. Artibani, W.; Porcaro, A.B.; De Marco, V.; Cerruto, M.A.; Siracusano, S. Management of biochemical recurrence after primary curative treatment for prostate cancer: A review. *Urol. Int.* **2018**, *100*, 251–262. [CrossRef]
53. Pound, C.R.; Partin, A.W.; Eisenberger, M.A.; Chan, D.W.; Pearson, J.D.; Walsh, P.C. Natural history of progression after PSA elevation following radical prostatectomy. *JAMA* **1999**, *281*, 1591–1597. [CrossRef]
54. Van den Broeck, T.; van den Bergh, R.C.; Arfi, N.; Gross, T.; Moris, L.; Briers, E.; Cumberbatch, M.; De Santis, M.; Tilki, D.; Fanti, S.; et al. Prognostic Value of Biochemical Recurrence Following Treatment with Curative Intent for Prostate Cancer: A Systematic Review. *Eur. Urol.* **2019**, *75*, 967–987. [CrossRef]
55. Frydenberg, M.; Woo, H.H. Early Androgen Deprivation Therapy Improves Survival, But How Do We Determine in Whom? *Eur. Urol.* **2018**, *73*, 519–520. [CrossRef] [PubMed]
56. Brand, D.; Parker, C. Management of Men with Prostate-specific Antigen Failure After Prostate Radiotherapy: The Case Against Early Androgen Deprivation. *Eur. Urol.* **2018**, *73*, 521–523. [CrossRef] [PubMed]
57. Garcia-Albeniz, X.; Chan, J.M.; Paciorek, A.; Logan, R.W.; Kenfield, S.A.; Cooperberg, M.R.; Carroll, P.R.; Hernán, M.A. Immediate versus deferred initiation of androgen deprivation therapy in prostate cancer patients with PSA-only relapse. An observational follow-up study. *Eur. J. Cancer* **2015**, *51*, 817–824. [CrossRef]

58. Duchesne, G.M.; Woo, H.H.; Bassett, J.K.; Bowe, S.J.; D'Este, C.; Frydenberg, M.; King, M.; Ledwich, L.; Loblaw, A.; Malone, S.; et al. Timing of androgen-deprivation therapy in patients with prostate cancer with a rising PSA (TROG 03.06 and VCOG PR 01–03 [TOAD]): A randomised, multicentre, non-blinded, phase 3 trial. *Lancet Oncol.* **2016**, *17*, 727–737. [CrossRef]
59. Sasaki, T.; Sugimura, Y. The Importance of Time to Prostate-Specific Antigen (PSA) Nadir after Primary Androgen Deprivation Therapy in Hormone-Naïve Prostate Cancer Patients. *J. Clin. Med.* **2018**, *7*, 565. [CrossRef] [PubMed]
60. Heidenreich, A.; Bastian, P.J.; Bellmunt, J.; Bolla, M.; Joniau, S.; van der Kwast, T.; Mason, M.; Matveev, V.; Wiegel, T.; Zattoni, F.; et al. EAU Guidelines on Prostate Cancer. Part II: Treatment of Advanced, Relapsing, and Castration-Resistant Prostate Cancer. *Eur. Urol.* **2014**, *65*, 467–479. [CrossRef] [PubMed]
61. Morris, M.J.; Rumble, R.B.; Basch, E.; Hotte, S.J.; Loblaw, A.; Rathkopf, D.; Celano, P.; Bangs, R.; Milowsky, M.I. Optimizing Anticancer Therapy in Metastatic Non-Castrate Prostate Cancer: American Society of Clinical Oncology Clinical Practice Guideline. *J. Clin. Oncol.* **2018**, *36*, 1521–1539. [CrossRef]
62. Hussain, M.; Tangen, C.M.; Higano, C.; Schelhammer, P.F.; Faulkner, J.; Crawford, E.D.; Wilding, G.; Akdas, A.; Small, E.J.; Donnelly, B.; et al. Absolute prostate-specific antigen value after androgen deprivation is a strong independent predictor of survival in new metastatic prostate cancer: Data from Southwest Oncology Group Trial 9346 (INT-0162). *J. Clin. Oncol.* **2006**, *24*, 3984–3990. [CrossRef] [PubMed]
63. Harshman, L.C.; Chen, Y.H.; Liu, G.; Carducci, M.A.; Jarrard, D.; Dreicer, R.; Hahn, N.; Garcia, J.A.; Hussain, M.; Shevrin, D.; et al. Seven-Month Prostate-Specific Antigen Is Prognostic in Metastatic Hormone-Sensitive Prostate Cancer Treated With Androgen Deprivation With or Without Docetaxel. *J. Clin. Oncol. Off. J. Am. Soc. Clin. Oncol.* **2018**, *36*, 376–382. [CrossRef] [PubMed]
64. Lorente, D.; Lozano, R.; de Velasco, G.; De Julian, M.; Rodrigo, M.; Sanchez, A.L.; Di Capua, C.; Castro, E.; Sanchez Hernandez, A.; Olmos, D. Prognostic value of PSA progression in metastatic castration-resistant prostate cancer (mCRPC) patients (pts) treated in the COU-AA-302 trial. *J. Clin. Oncol.* **2019**, *37* (Suppl. 15), e16541. [CrossRef]
65. España, S.; de Olza, M.O.; Sala, N.; Piulats, J.M.; Ferrandiz, U.; Etxaniz, O.; Heras, L.; Buisan, O.; Pardo, J.C.; Suarez, J.F.; et al. PSA Kinetics as Prognostic Markers of Overall Survival in Patients with Metastatic Castration-Resistant Prostate Cancer Treated with Abiraterone Acetate. *Cancer Manag. Res.* **2020**, *12*, 10251–10260. [CrossRef]
66. Liu, J.-M.; Lin, C.C.; Liu, K.L.; Lin, C.F.; Chen, B.Y.; Chen, T.H.; Sun, C.C.; Wu, C.T. Second-line Hormonal Therapy for the Management of Metastatic Castration-resistant Prostate Cancer: A Real-World Data Study Using a Claims Database. *Sci. Rep.* **2020**, *10*, 4240. [CrossRef]
67. Scher, H.I.; Morris, M.J.; Stadler, W.M.; Higano, C.; Basch, E.; Fizazi, K.; Antonarakis, E.S.; Beer, T.M.; Carducci, M.A.; Chi, K.N.; et al. Trial Design and Objectives for Castration-Resistant Prostate Cancer: Updated Recommendations From the Prostate Cancer Clinical Trials Working Group 3. *J. Clin. Oncol.* **2016**, *34*, 1402–1418. [CrossRef]
68. Ablin, R.J.; Soanes, W.A.; Bronson, P.; Witebsky, E. Precipitating antigens of the normal human prostate. *J. Reprod. Fertil.* **1970**, *22*, 573–574. [CrossRef]
69. Lim, K.B. Epidemiology of clinical benign prostatic hyperplasia. *Asian J. Urol.* **2017**, *4*, 148–151. [CrossRef]
70. Thompson, I.M.; Pauler, D.K.; Goodman, P.J.; Tangen, C.M.; Lucia, M.S.; Parnes, H.L.; Minasian, L.M.; Ford, L.G.; Lippman, S.M.; Crawford, E.D.; et al. Prevalence of prostate cancer among men with a prostate-specific antigen level ≤ 4.0 ng per milliliter. *N. Engl. J. Med.* **2004**, *350*, 2239–2246. [CrossRef]
71. Thompson, I.M.; Ankerst, D.P.; Chi, C.; Lucia, M.S.; Goodman, P.J.; Crowley, J.J.; Parnes, H.L.; Coltman, C.A. Operating characteristics of prostate-specific antigen in men with an initial PSA level of 3.0 ng/mL or lower. *JAMA* **2005**, *294*, 66–70. [CrossRef] [PubMed]
72. Benson, M.C.; Whang, I.S.; Pantuck, A.; Ring, K.; Kaplan, S.A.; Olsson, C.A.; Cooner, W.H. Prostate specific antigen density: A means of distinguishing benign prostatic hypertrophy and prostate cancer. *J. Urol.* **1992**, *147*, 815–816. [CrossRef]
73. Nordström, T.; Akre, O.; Aly, M.; Grönberg, H.; Eklund, M. Prostate-specific antigen (PSA) density in the diagnostic algorithm of prostate cancer. *Prostate Cancer Prostatic Dis.* **2018**, *21*, 57–63. [CrossRef]
74. Yanai, Y.; Kosaka, T.; Hongo, H.; Matsumoto, K.; Shinojima, T.; Kikuchi, E.; Miyajima, A.; Mizuno, R.; Mikami, S.; Jinzaki, M.; et al. Evaluation of prostate-specific antigen density in the diagnosis of prostate cancer combined with magnetic resonance imaging before biopsy in men aged 70 years and older with elevated PSA. *Mol. Clin. Oncol.* **2018**, *9*, 656–660. [CrossRef]
75. Aminsharifi, A.; Howard, L.; Wu, Y.; De Hoedt, A.; Bailey, C.; Freedland, S.J.; Polascik, T.J. Prostate Specific Antigen Density as a Predictor of Clinically Significant Prostate Cancer When the Prostate Specific Antigen is in the Diagnostic Gray Zone: Defining the Optimum Cutoff Point Stratified by Race and Body Mass Index. *J. Urol.* **2018**, *200*, 758–766. [CrossRef] [PubMed]
76. Yusim, I.; Krenawi, M.; Mazor, E.; Novack, V.; Mabjeesh, N.J. The use of prostate specific antigen density to predict clinically significant prostate cancer. *Sci. Rep.* **2020**, *10*, 20015. [CrossRef] [PubMed]
77. Jue, J.S.; Barboza, M.P.; Prakash, N.S.; Venkatramani, V.; Sinha, V.R.; Pavan, N.; Nahar, B.; Kanabur, P.; Ahdoot, M.; Dong, Y.; et al. Re-examining Prostate-specific Antigen (PSA) Density: Defining the Optimal PSA Range and Patients for Using PSA Density to Predict Prostate Cancer Using Extended Template Biopsy. *Urology* **2017**, *105*, 123–128. [CrossRef] [PubMed]
78. Sfoungaristos, S.; Perimenis, P. PSA density is superior than PSA and Gleason score for adverse pathologic features prediction in patients with clinically localized prostate cancer. *Can. Urol. Assoc. J.* **2012**, *6*, 46–50. [CrossRef]
79. Jin, B.S.; Kang, S.H.; Kim, D.Y.; Oh, H.G.; Kim, C.I.; Moon, G.H.; Kwon, T.G.; Park, J.S. Pathological upgrading in prostate cancer patients eligible for active surveillance: Does prostate-specific antigen density matter? *Korean J. Urol.* **2015**, *56*, 624–629. [CrossRef]

80. Ha, Y.S.; Yu, J.; Salmasi, A.H.; Patel, N.; Parihar, J.; Singer, E.A.; Kim, J.H.; Kwon, T.G.; Kim, W.J.; Kim, I.Y. Prostate-specific antigen density toward a better cutoff to identify better candidates for active surveillance. *Urology* **2014**, *84*, 365–371. [CrossRef] [PubMed]
81. Washington, S.L., 3rd; Baskin, A.S.; Ameli, N.; Nguyen, H.G.; Westphalen, A.C.; Shinohara, K.; Carroll, P.R. MRI-Based Prostate-Specific Antigen Density Predicts Gleason Score Upgrade in an Active Surveillance Cohort. *AJR Am. J. Roentgenol.* **2020**, *214*, 574–578. [CrossRef]
82. Carter, H.B.; Pearson, J.D.; Metter, E.J.; Brant, L.J.; Chan, D.W.; Andres, R.; Fozard, J.L.; Walsh, P.C. Longitudinal evaluation of prostate-specific antigen levels in men with and without prostate disease. *JAMA* **1992**, *267*, 2215–2220. [CrossRef]
83. Vickers, A.J.; Brewster, S.F. PSA Velocity and Doubling Time in Diagnosis and Prognosis of Prostate Cancer. *Br. J. Med. Surg. Urol.* **2012**, *5*, 162–168. [CrossRef] [PubMed]
84. Javaeed, A.; Ghauri, S.K.; Ibrahim, A.; Doheim, M.F. Prostate-specific antigen velocity in diagnosis and prognosis of prostate cancer—A systematic review. *Oncol. Rev.* **2020**, *14*, 449. [CrossRef] [PubMed]
85. D'Amico, A.V.; Chen, M.H.; Roehl, K.A.; Catalona, W.J. Preoperative PSA velocity and the risk of death from prostate cancer after radical prostatectomy. *N. Engl. J. Med.* **2004**, *351*, 125–135. [CrossRef] [PubMed]
86. Sengupta, S.; Myers, R.P.; Slezak, J.M.; Bergstralh, E.J.; Zincke, H.; Blute, M.L. Preoperative prostate specific antigen doubling time and velocity are strong and independent predictors of outcomes following radical prostatectomy. *J. Urol.* **2005**, *174*, 2191–2196. [CrossRef]
87. Patel, D.A.; Presti, J.C., Jr.; McNeal, J.E.; Gill, H.; Brooks, J.D.; King, C.R. Preoperative PSA velocity is an independent prognostic factor for relapse after radical prostatectomy. *J. Clin. Oncol.* **2005**, *23*, 6157–6162. [CrossRef]
88. Berger, A.P.; Deibl, M.; Strasak, A.; Bektic, J.; Pelzer, A.; Steiner, H.; Spranger, R.; Fritsche, G.; Bartsch, G.; Horninger, W. Relapse after radical prostatectomy correlates with preoperative PSA velocity and tumor volume: Results from a screening population. *Urology* **2006**, *68*, 1067–1071. [CrossRef]
89. D'Amico, A.V.; Renshaw, A.A.; Sussman, B.; Chen, M.H. Pretreatment PSA velocity and risk of death from prostate cancer following external beam radiation therapy. *JAMA* **2005**, *294*, 440–447. [CrossRef]
90. Palma, D.; Tyldesley, S.; Blood, P.; Liu, M.; Morris, J.; Pickles, T.; Prostate Cohort Outcomes Initiative. Pretreatment PSA velocity as a predictor of disease outcome following radical radiation therapy. *Int. J. Radiat. Oncol. Biol. Phys.* **2007**, *67*, 1425–1429. [CrossRef]
91. Shimbo, M.; Tomioka, S.; Sasaki, M.; Shima, T.; Suzuki, N.; Murakami, S.; Nakatsu, H.; Shimazaki, J. PSA Doubling Time as a Predictive Factor on Repeat Biopsy for Detection of Prostate Cancer. *Jpn. J. Clin. Oncol.* **2009**, *39*, 727–731. [CrossRef]
92. Takeuchi, H.; Ohori, M.; Tachibana, M. Clinical significance of the prostate-specific antigen doubling time prior to and following radical prostatectomy to predict the outcome of prostate cancer. *Mol. Clin. Oncol.* **2017**, *6*, 249–254. [CrossRef]
93. Makarov, D.V.; Humphreys, E.B.; Mangold, L.A.; Carducci, M.A.; Partin, A.W.; Eisenberger, M.A.; Walsh, P.C.; Trock, B.J. The natural history of men treated with deferred androgen deprivation therapy in whom metastatic prostate cancer developed following radical prostatectomy. *J. Urol.* **2008**, *179*, 156–162. [CrossRef] [PubMed]
94. Markowski, M.C.; Chen, Y.; Feng, Z.; Cullen, J.; Trock, B.J.; Suzman, D.; Antonarakis, E.S.; Paller, C.J.; Rosner, I.; Han, M.; et al. PSA Doubling Time and Absolute PSA Predict Metastasis-free Survival in Men With Biochemically Recurrent Prostate Cancer After Radical Prostatectomy. *Clin. Genitourin. Cancer* **2019**, *17*, 470–475.e1. [CrossRef] [PubMed]
95. Jackson, W.C.; Johnson, S.B.; Li, D.; Foster, C.; Foster, B.; Song, Y.; Schipper, M.; Shilkrut, M.; Sandler, H.M.; Morgan, T.M.; et al. A prostate-specific antigen doubling time of <6 months is prognostic for metastasis and prostate cancer-specific death for patients receiving salvage radiation therapy post radical prostatectomy. *Radiat. Oncol.* **2013**, *8*, 170. [CrossRef]
96. Whitney, C.A.; Howard, L.E.; Freedland, S.J.; DeHoedt, A.M.; Amling, C.L.; Aronson, W.J.; Cooperberg, M.R.; Kane, C.J.; Terris, M.K.; Daskivich, T.J. Thresholds for PSA doubling time in men with non-metastatic castration-resistant prostate cancer. *BJU Int.* **2017**, *120*, E80–E86.
97. McLaren, D.B.; McKenzie, M.; Duncan, G.; Pickles, T. Watchful waiting or watchful progression? *Cancer* **1998**, *82*, 342–348. [CrossRef]
98. Shariat, S.F.; Canto, E.I.; Kattan, M.W.; Slawin, K.M. Beyond prostate-specific antigen: New serologic biomarkers for improved diagnosis and management of prostate cancer. *Rev. Urol.* **2004**, *6*, 58.
99. Catalona, W.J.; Partin, A.W.; Slawin, K.M.; Brawer, M.K.; Flanigan, R.C.; Patel, A.; Richie, J.P.; DeKernion, J.B.; Walsh, P.C.; Scardino, P.T.; et al. Use of the percentage of free prostate-specific antigen to enhance differentiation of prostate cancer from benign prostatic disease: A prospective multicenter clinical trial. *JAMA* **1998**, *279*, 1542–1547. [CrossRef]
100. Partin, A.W.; Brawer, M.K.; Bartsch, G.; Horninger, W.; Taneja, S.S.; Lepor, H.; Babaian, R.; Childs, S.J.; Stamey, T.; Fritsche, H.A.; et al. Complexed prostate specific antigen improves specificity for prostate cancer detection: Results of a prospective multicenter clinical trial. *J. Urol.* **2003**, *170*, 1787–1791. [CrossRef]
101. Prestigiacomo, A.F.; Lilja, H.; Pettersson, K.; Wolfert, R.L.; Stamey, T.A. A comparison of the free fraction of serum prostate specific antigen in men with benign and cancerous prostates: The best case scenario. *J. Urol.* **1996**, *156*, 350–354. [CrossRef]
102. Piironen, T.; Pettersson, K.; Suonpää, M.; Stenman, U.H.; Oesterling, J.E.; Lövgren, T.; Lilja, H. In vitro stability of free prostate-specific antigen (PSA) and prostate-specific antigen (PSA) complexed to α1-antichymotrypsin in blood samples. *Urology* **1996**, *48*, 81–87. [CrossRef]

103. Ornstein, D.K.; Rao, G.S.; Smith, D.S.; Ratliff, T.L.; Basler, J.W.; Catalona, W.J. Effect of digital rectal examination and needle biopsy on serum total and percentage of free prostate specific antigen levels. *J. Urol.* **1997**, *157*, 195–198. [CrossRef]
104. Stephan, C.; Lein, M.; Jung, K.; Schnorr, D.; Loening, S.A. The influence of prostate volume on the ratio of free to total prostate specific antigen in serum of patients with prostate carcinoma and benign prostate hyperplasia. *Cancer* **1997**, *79*, 104–109. [CrossRef]
105. Huang, Y.; Li, Z.Z.; Huang, Y.L.; Song, H.J.; Wang, Y.J. Value of free/total prostate-specific antigen (f/t PSA) ratios for prostate cancer detection in patients with total serum prostate-specific antigen between 4 and 10 ng/mL: A meta-analysis. *Medicine* **2018**, *97*, e0249. [CrossRef]
106. Oto, J.; Fernández-Pardo, Á.; Royo, M.; Hervás, D.; Martos, L.; Vera-Donoso, C.D.; Martínez, M.; Heeb, M.J.; España, F.; Medina, P. A predictive model for prostate cancer incorporating PSA molecular forms and age. *Sci. Rep.* **2020**, *10*, 2463. [CrossRef]
107. Ayyıldız, S.N.; Ayyıldız, A. PSA, PSA derivatives, proPSA and prostate health index in the diagnosis of prostate cancer. *Turk. J. Urol.* **2014**, *40*, 82–88.
108. Ferro, M.; De Cobelli, O.; Lucarelli, G.; Porreca, A.; Busetto, G.M.; Cantiello, F.; Damiano, R.; Autorino, R.; Musi, G.; Vartolomei, M.D.; et al. Beyond PSA: The Role of Prostate Health Index (phi). *Int. J. Mol. Sci.* **2020**, *21*, 1184. [CrossRef]
109. Chan, T.Y.; Mikolajczyk, S.D.; Lecksell, K.; Shue, M.J.; Rittenhouse, H.G.; Partin, A.W.; Epstein, J.I. Immunohistochemical staining of prostate cancer with monoclonal antibodies to the precursor of prostate-specific antigen. *Urology* **2003**, *62*, 177–181. [CrossRef]
110. Jansen, F.H.; van Schaik, R.H.; Kurstjens, J.; Horninger, W.; Klocker, H.; Bektic, J.; Wildhagen, M.F.; Roobol, M.J.; Bangma, C.H.; Bartsch, G. Prostate-specific antigen (PSA) isoform p2PSA in combination with total PSA and free PSA improves diagnostic accuracy in prostate cancer detection. *Eur. Urol.* **2010**, *57*, 921–927. [CrossRef] [PubMed]
111. Lazzeri, M.; Haese, A.; De La Taille, A.; Redorta, J.P.; McNicholas, T.; Lughezzani, G.; Scattoni, V.; Bini, V.; Freschi, M.; Sussman, A.; et al. Serum isoform [-2]proPSA derivatives significantly improve prediction of prostate cancer at initial biopsy in a total PSA range of 2–10 ng/mL: A multicentric European study. *Eur. Urol.* **2013**, *63*, 986–994. [CrossRef] [PubMed]
112. Stephan, C.; Vincendeau, S.; Houlgatte, A.; Cammann, H.; Jung, K.; Semjonow, A. Multicenter evaluation of [-2]proprostate-specific antigen and the prostate health index for detecting prostate cancer. *Clin. Chem.* **2013**, *59*, 306–314. [CrossRef]
113. Filella, X.; Giménez, N. Evaluation of [-2]proPSA and Prostate Health Index (phi) for the detection of prostate cancer: A systematic review and meta-analysis. *Clin. Chem. Lab. Med.* **2013**, *51*, 729–739. [CrossRef] [PubMed]
114. Boegemann, M.; Stephan, C.; Cammann, H.; Vincendeau, S.; Houlgatte, A.; Jung, K.; Blanchet, J.S.; Semjonow, A. The percentage of prostate-specific antigen (PSA) isoform [-2]proPSA and the Prostate Health Index improve the diagnostic accuracy for clinically relevant prostate cancer at initial and repeat biopsy compared with total PSA and percentage free PSA in men aged ≤65 years. *BJU Int.* **2016**, *117*, 72–79. [PubMed]
115. Schulze, A.; Christoph, F.; Sachs, M.; Schroeder, J.; Stephan, C.; Schostak, M.; Koenig, F. Use of the Prostate Health Index and Density in 3 Outpatient Centers to Avoid Unnecessary Prostate Biopsies. *Urol. Int.* **2020**, *104*, 181–186. [CrossRef] [PubMed]
116. Catalona, W.J.; Partin, A.W.; Sanda, M.G.; Wei, J.T.; Klee, G.G.; Bangma, C.H.; Slawin, K.M.; Marks, L.S.; Loeb, S.; Broyles, D.L.; et al. A multicenter study of [-2]pro-prostate specific antigen combined with prostate specific antigen and free prostate specific antigen for prostate cancer detection in the 2.0 to 10.0 ng/mL prostate specific antigen range. *J. Urol.* **2011**, *185*, 1650–1655. [CrossRef]
117. Loeb, S.; Sanda, M.G.; Broyles, D.L.; Shin, S.S.; Bangma, C.H.; Wei, J.T.; Partin, A.W.; Klee, G.G.; Slawin, K.M.; Marks, L.S.; et al. The prostate health index selectively identifies clinically significant prostate cancer. *J. Urol.* **2015**, *193*, 1163–1169. [CrossRef]
118. De La Calle, C.; Patil, D.; Wei, J.T.; Scherr, D.S.; Sokoll, L.; Chan, D.W.; Siddiqui, J.; Mosquera, J.M.; Rubin, M.A.; Sanda, M.G. Multicenter evaluation of the prostate health index to detect aggressive prostate cancer in biopsy naive men. *J. Urol.* **2015**, *194*, 65–72. [CrossRef]
119. Novak, V.; Vesely, S.; Luksanová, H.; Prusa, R.; Capoun, O.; Fiala, V.; Dolejsová, O.; Sedlacková, H.; Kucera, R.; Stejskal, J.; et al. Preoperative prostate health index predicts adverse pathology and Gleason score upgrading after radical prostatectomy for prostate cancer. *BMC Urol.* **2020**, *20*, 144. [CrossRef]
120. Tosoian, J.J.; Druskin, S.C.; Andreas, D.; Mullane, P.; Chappidi, M.; Joo, S.; Ghabili, K.; Agostino, J.; Macura, K.J.; Carter, H.B.; et al. Use of the Prostate Health Index for detection of prostate cancer: Results from a large academic practice. *Prostate Cancer Prostatic Dis.* **2017**, *20*, 228–233. [CrossRef]
121. Olleik, G.; Kassouf, W.; Aprikian, A.; Hu, J.; Vanhuyse, M.; Cury, F.; Peacock, S.; Bonnevier, E.; Palenius, E.; Dragomir, A. Evaluation of New Tests and Interventions for Prostate Cancer Management: A Systematic Review. *J. Natl. Compr. Canc. Netw.* **2018**, *16*, 1340–1351. [CrossRef]
122. Tosoian, J.J.; Druskin, S.C.; Andreas, D.; Mullane, P.; Chappidi, M.; Joo, S.; Ghabili, K.; Mamawala, M.; Agostino, J.; Carter, H.B.; et al. Prostate Health Index density improves detection of clinically significant prostate cancer. *BJU Int.* **2017**, *120*, 793–798. [CrossRef]
123. Druskin, S.C.; Tosoian, J.J.; Young, A.; Collica, S.; Srivastava, A.; Ghabili, K.; Macura, K.J.; Carter, H.B.; Partin, A.W.; Sokoll, L.J.; et al. Combining Prostate Health Index density, magnetic resonance imaging and prior negative biopsy status to improve the detection of clinically significant prostate cancer. *BJU Int.* **2018**, *121*, 619–626. [CrossRef] [PubMed]
124. Gnanapragasam, V.J.; Burling, K.; George, A.; Stearn, S.; Warren, A.; Barrett, T.; Koo, B.; Gallagher, F.A.; Doble, A.; Kastner, C.; et al. The Prostate Health Index adds predictive value to multi-parametric MRI in detecting significant prostate cancers in a repeat biopsy population. *Sci. Rep.* **2016**, *6*, 35364. [CrossRef]

125. White, J.; Shenoy, B.V.; Tutrone, R.F.; Karsh, L.I.; Saltzstein, D.R.; Harmon, W.J.; Broyles, D.L.; Roddy, T.E.; Lofaro, L.R.; Paoli, C.J.; et al. Clinical utility of the Prostate Health Index (phi) for biopsy decision management in a large group urology practice setting. *Prostate Cancer Prostatic Dis.* **2018**, *21*, 78–84. [CrossRef] [PubMed]
126. Huang, D.; Wu, Y.; Lin, X.; Xu, D.; Na, R.; Xu, J. Cost-Effectiveness Analysis of Prostate Health Index in Decision Making for Initial Prostate Biopsy. *Front. Oncol.* **2020**, *10*, 565382. [CrossRef] [PubMed]
127. Teoh, J.Y.-C.; Leung, C.H.; Wang, M.H.; Chiu, P.K.F.; Yee, C.H.; Ng, C.F.; Wong, M.C.S. The cost-effectiveness of prostate health index for prostate cancer detection in Chinese men. *Prostate Cancer Prostatic Dis.* **2020**, *23*, 615–621. [CrossRef]
128. Lughezzani, G.; Lazzeri, M.; Buffi, N.M.; Abrate, A.; Mistretta, F.A.; Hurle, R.; Pasini, L.; Castaldo, L.; De Zorzi, S.Z.; Peschechera, R.; et al. Preoperative prostate health index is an independent predictor of early biochemical recurrence after radical prostatectomy: Results from a prospective single-center study. *Urol. Oncol.* **2015**, *33*, 337.e7–337.e14. [CrossRef] [PubMed]
129. Maxeiner, A.; Kilic, E.; Matalon, J.; Friedersdorff, F.; Miller, K.; Jung, K.; Stephan, C.; Busch, J. The prostate health index PHI predicts oncological outcome and biochemical recurrence after radical prostatectomy—Analysis in 437 patients. *Oncotarget* **2017**, *8*, 79279–79288. [CrossRef]
130. Chiu, P.K.; Ng, C.F.; Semjonow, A.; Zhu, Y.; Vincendeau, S.; Houlgatte, A.; Lazzeri, M.; Guazzoni, G.; Stephan, C.; Haese, A.; et al. A Multicentre Evaluation of the Role of the Prostate Health Index (PHI) in Regions with Differing Prevalence of Prostate Cancer: Adjustment of PHI Reference Ranges is Needed for European and Asian Settings. *Eur. Urol.* **2019**, *75*, 558–561. [CrossRef] [PubMed]
131. Semjonow, A.; Köpke, T.; Eltze, E.; Pepping-Schefers, B.; Bürgel, H.; Darte, C. Pre-analytical in-vitro stability of [-2]proPSA in blood and serum. *Clin. Biochem.* **2010**, *43*, 926–928. [CrossRef]
132. Hong, S.K. Kallikreins as Biomarkers for Prostate Cancer. *BioMed Res. Int.* **2014**, *2014*, 526341. [CrossRef] [PubMed]
133. Vickers, A.J.; Cronin, A.M.; Aus, G.; Pihl, C.G.; Becker, C.; Pettersson, K.; Scardino, P.T.; Hugosson, J.; Lilja, H. A panel of kallikrein markers can reduce unnecessary biopsy for prostate cancer: Data from the European Randomized Study of Prostate Cancer Screening in Göteborg, Sweden. *BMC Med.* **2008**, *6*, 19. [CrossRef]
134. Gupta, A.; Roobol, M.J.; Savage, C.J.; Peltola, M.; Pettersson, K.; Scardino, P.T.; Vickers, A.J.; Schröder, F.H.; Lilja, H. A four-kallikrein panel for the prediction of repeat prostate biopsy: Data from the European Randomized Study of Prostate Cancer screening in Rotterdam, Netherlands. *Br. J. Cancer* **2010**, *103*, 708–714. [CrossRef] [PubMed]
135. Vickers, A.; Cronin, A.; Roobol, M.; Savage, C.; Peltola, M.; Pettersson, K.; Scardino, P.T.; Schröder, F.; Lilja, H. Reducing unnecessary biopsy during prostate cancer screening using a four-kallikrein panel: An independent replication. *J. Clin. Oncol.* **2010**, *28*, 2493–2498. [CrossRef] [PubMed]
136. Vickers, A.J.; Cronin, A.M.; Roobol, M.J.; Savage, C.J.; Peltola, M.; Pettersson, K.; Scardino, P.T.; Schröder, F.H.; Lilja, H. A four-kallikrein panel predicts prostate cancer in men with recent screening: Data from the European Randomized Study of Screening for Prostate Cancer, Rotterdam. *Clin. Cancer Res.* **2010**, *16*, 3232–3239. [CrossRef]
137. Parekh, D.J.; Punnen, S.; Sjoberg, D.D.; Asroff, S.W.; Bailen, J.L.; Cochran, J.S.; Concepcion, R.; David, R.D.; Deck, K.B.; Dumbadze, I.; et al. A multi-institutional prospective trial in the USA confirms that the 4Kscore accurately identifies men with high-grade prostate cancer. *Eur. Urol.* **2015**, *68*, 464–470. [CrossRef]
138. Nordström, T.; Vickers, A.; Assel, M.; Lilja, H.; Grönberg, H.; Eklund, M. Comparison Between the Four-kallikrein Panel and Prostate Health Index for Predicting Prostate Cancer. *Eur. Urol.* **2015**, *68*, 139–146. [CrossRef]
139. Verbeek, J.F.M.; Bangma, C.H.; Kweldam, C.F.; van der Kwast, T.H.; Kümmerlin, I.P.; van Leenders, G.J.; Roobol, M.J. Reducing unnecessary biopsies while detecting clinically significant prostate cancer including cribriform growth with the ERSPC Rotterdam risk calculator and 4Kscore. *Urol. Oncol.* **2019**, *37*, 138–144. [CrossRef] [PubMed]
140. Braun, K.; Sjoberg, D.D.; Vickers, A.J.; Lilja, H.; Bjartell, A.S. A four-kallikrein panel predicts high-grade cancer on biopsy: Independent validation in a community cohort. *Eur. Urol.* **2016**, *69*, 505–511. [CrossRef]
141. Assel, M.; Sjöblom, L.; Murtola, T.J.; Talala, K.; Kujala, P.; Stenman, U.H.; Taari, K.; Auvinen, A.; Vickers, A.; Visakorpi, T.; et al. A Four-kallikrein Panel and β-Microseminoprotein in Predicting High-grade Prostate Cancer on Biopsy: An Independent Replication from the Finnish Section of the European Randomized Study of Screening for Prostate Cancer. *Eur. Urol. Focus* **2019**, *5*, 561–567. [CrossRef]
142. Lin, D.W.; Newcomb, L.F.; Brown, M.D.; Sjoberg, D.D.; Dong, Y.; Brooks, J.D.; Carroll, P.R.; Cooperberg, M.; Dash, A.; Ellis, W.J.; et al. Evaluating the four kallikrein panel of the 4kscore for prediction of high-grade prostate cancer in men in the Canary Prostate Active Surveillance Study. *Eur. Urol.* **2017**, *72*, 448–454. [CrossRef]
143. Zappala, S.M.; Dong, Y.; Linder, V.; Reeve, M.; Sjoberg, D.D.; Mathur, V.; Roberts, R.; Okrongly, D.; Newmark, J.; Sant, G.; et al. The 4Kscore blood test accurately identifies men with aggressive prostate cancer prior to prostate biopsy with or without DRE information. *Int. J. Clin. Pract.* **2017**, *71*, e12943. [CrossRef]
144. Punnen, S.; Nahar, B.; Soodana-Prakash, N.; Koru-Sengul, T.; Stoyanova, R.; Pollack, A.; Kava, B.; Gonzalgo, M.L.; Ritch, C.R.; Parekh, D.J. Optimizing patient's selection for prostate biopsy: A single institution experience with multi-parametric MRI and the 4Kscore test for the detection of aggressive prostate cancer. *PLoS ONE* **2018**, *13*, e0201384. [CrossRef]
145. Darst, B.F.; Chou, A.; Wan, P.; Pooler, L.; Sheng, X.; Vertosick, E.A.; Conti, D.V.; Wilkens, L.R.; Le Marchand, L.; Vickers, A.J.; et al. The Four-Kallikrein Panel Is Effective in Identifying Aggressive Prostate Cancer in a Multiethnic Population. *Cancer Epidemiol. Biomark. Prev.* **2020**, *29*, 1381. [CrossRef] [PubMed]

146. Konety, B.; Zappala, S.M.; Parekh, D.J.; Osterhout, D.; Schock, J.; Chudler, R.M.; Oldford, G.M.; Kernen, K.M.; Hafron, J. The 4Kscore® Test Reduces Prostate Biopsy Rates in Community and Academic Urology Practices. *Rev. Urol.* **2015**, *17*, 231–240. [PubMed]
147. Voigt, J.D.; Dong, Y.; Linder, V.; Zappala, S. Use of the 4Kscore test to predict the risk of aggressive prostate cancer prior to prostate biopsy: Overall cost savings and improved quality of care to the us healthcare system. *Rev. Urol.* **2017**, *19*, 1–10. [PubMed]
148. Voigt, J.D.; Zappala, S.M.; Vaughan, E.D.; Wein, A.J. The Kallikrein Panel for prostate cancer screening: Its economic impact. *Prostate* **2014**, *74*, 250–259. [CrossRef] [PubMed]
149. Stattin, P.; Vickers, A.J.; Sjoberg, D.D.; Johansson, R.; Granfors, T.; Johansson, M.; Pettersson, K.; Scardino, P.T.; Hallmans, G.; Lilja, H. Improving the specificity of screening for lethal prostate cancer using prostate-specific antigen and a panel of kallikrein markers: A nested case–control study. *Eur. Urol.* **2015**, *68*, 207–213. [CrossRef] [PubMed]
150. Al Olama, A.A.; Kote-Jarai, Z.; Berndt, S.I.; Conti, D.V.; Schumacher, F.; Han, Y.; Benlloch, S.; Hazelett, D.J.; Wang, Z.; Saunders, E.; et al. A meta-analysis of 87,040 individuals identifies 23 new susceptibility loci for prostate cancer. *Nat. Genet.* **2014**, *46*, 1103–1109. [CrossRef]
151. Aly, M.; Wiklund, F.; Xu, J.; Isaacs, W.B.; Eklund, M.; D'Amato, M.; Adolfsson, J.; Grönberg, H. Polygenic risk score improves prostate cancer risk prediction: Results from the Stockholm-1 cohort study. *Eur. Urol.* **2011**, *60*, 21–28. [CrossRef]
152. Kader, A.K.; Sun, J.; Reck, B.H.; Newcombe, P.J.; Kim, S.T.; Hsu, F.C.; D'Agostino, R.B., Jr.; Tao, S.; Zhang, Z.; Turner, A.R.; et al. Potential impact of adding genetic markers to clinical parameters in predicting prostate biopsy outcomes in men following an initial negative biopsy: Findings from the REDUCE trial. *Eur. Urol.* **2012**, *62*, 953–961. [CrossRef]
153. Grönberg, H.; Adolfsson, J.; Aly, M.; Nordström, T.; Wiklund, P.; Brandberg, Y.; Thompson, J.; Wiklund, F.; Lindberg, J.; Clements, M.; et al. Prostate cancer screening in men aged 50–69 years (STHLM3): A prospective population-based diagnostic study. *Lancet Oncol.* **2015**, *16*, 1667–1676. [CrossRef]
154. Eklund, M.; Nordström, T.; Aly, M.; Adolfsson, J.; Wiklund, P.; Brandberg, Y.; Thompson, J.; Wiklund, F.; Lindberg, J.; Presti, J.C. The Stockholm-3 (STHLM3) Model can Improve Prostate Cancer Diagnostics in Men Aged 50–69 yr Compared with Current Prostate Cancer Testing. *Eur. Urol. Focus* **2018**, *4*, 707–710. [CrossRef]
155. Ström, P.; Nordström, T.; Aly, M.; Egevad, L.; Grönberg, H.; Eklund, M. The Stockholm-3 Model for Prostate Cancer Detection: Algorithm Update, Biomarker Contribution, and Reflex Test Potential. *Eur. Urol.* **2018**, *74*, 204–210. [CrossRef]
156. Möller, A.; Olsson, H.; Grönberg, H.; Eklund, M.; Aly, M.; Nordström, T. The Stockholm3 blood-test predicts clinically-significant cancer on biopsy: Independent validation in a multi-center community cohort. *Prostate Cancer Prostatic Dis.* **2019**, *22*, 137–142. [CrossRef] [PubMed]
157. Viste, E.; Vinje, C.A.; Lid, T.G.; Skeie, S.; Evjen-Olsen, Ø.; Nordström, T.; Thorsen, O.; Gilje, B.; Janssen, E.A.; Kjosavik, S.R. Effects of replacing PSA with Stockholm3 for diagnosis of clinically significant prostate cancer in a healthcare system—The Stavanger experience. *Scand. J. Prim. Health Care* **2020**, *38*, 315–322. [CrossRef] [PubMed]
158. Varki, A. Biological roles of glycans. *Glycobiology* **2017**, *27*, 3–49. [CrossRef] [PubMed]
159. Munkley, J.; Elliott, D.J. Hallmarks of glycosylation in cancer. *Oncotarget* **2016**, *7*, 35478–35489. [CrossRef]
160. Srinivasan, S.; Stephens, C.; Wilson, E.; Panchadsaram, J.; DeVoss, K.; Koistinen, H.; Stenman, U.H.; Brook, M.N.; Buckle, A.M.; Klein, R.J.; et al. Prostate Cancer Risk-Associated Single-Nucleotide Polymorphism Affects Prostate-Specific Antigen Glycosylation and Its Function. *Clin. Chem.* **2019**, *65*, e1–e9. [CrossRef]
161. Drake, R.R.; Jones, E.E.; Powers, T.W.; Nyalwidhe, J.O. Altered glycosylation in prostate cancer. *Adv. Cancer Res.* **2015**, *126*, 345–382.
162. Pinho, S.S.; Reis, C.A. Glycosylation in cancer: Mechanisms and clinical implications. *Nat. Rev. Cancer.* **2015**, *15*, 540–555. [CrossRef]
163. Peracaula, R.; Tabarés, G.; Royle, L.; Harvey, D.J.; Dwek, R.A.; Rudd, P.M.; de Llorens, R. Altered glycosylation pattern allows the distinction between prostate-specific antigen (PSA) from normal and tumor origins. *Glycobiology* **2003**, *13*, 457–470. [CrossRef] [PubMed]
164. Meany, D.L.; Zhang, Z.; Sokoll, L.J.; Zhang, H.; Chan, D.W. Glycoproteomics for prostate cancer detection: Changes in serum PSA glycosylation patterns. *J. Proteome Res.* **2009**, *8*, 613–619. [CrossRef]
165. Ishikawa, T.; Yoneyama, T.; Tobisawa, Y.; Hatakeyama, S.; Kurosawa, T.; Nakamura, K.; Narita, S.; Mitsuzuka, K.; Duivenvoorden, W.; Pinthus, J.H.; et al. An Automated Micro-Total Immunoassay System for Measuring Cancer-Associated α2,3-linked Sialyl N-Glycan-Carrying Prostate-Specific Antigen May Improve the Accuracy of Prostate Cancer Diagnosis. *Int. J. Mol. Sci.* **2017**, *18*, 470. [CrossRef]
166. Llop, E.; Ferrer-Batallé, M.; Barrabés, S.; Guerrero, P.E.; Ramírez, M.; Saldova, R.; Rudd, P.M.; Aleixandre, R.N.; Comet, J.; de Llorens, R.; et al. Improvement of prostate cancer diagnosis by detecting PSA glycosylation-specific changes. *Theranostics* **2016**, *6*, 1190. [CrossRef] [PubMed]
167. Ferrer-Batallé, M.; Llop, E.; Ramírez, M.; Aleixandre, R.N.; Saez, M.; Comet, J.; De Llorens, R.; Peracaula, R. Comparative Study of Blood-Based Biomarkers, α2,3-Sialic Acid PSA and PHI, for High-Risk Prostate Cancer Detection. *Int. J. Mol. Sci.* **2017**, *18*, 845. [CrossRef]
168. Massie, C.E.; Mills, I.G.; Lynch, A.G. The importance of DNA methylation in prostate cancer development. *J. Steroid Biochem. Mol. Biol.* **2017**, *166*, 1–15. [CrossRef]

169. Kirby, M.K.; Ramaker, R.C.; Roberts, B.S.; Lasseigne, B.N.; Gunther, D.S.; Burwell, T.C.; Davis, N.S.; Gulzar, Z.G.; Absher, D.M.; Cooper, S.J.; et al. Genome-wide DNA methylation measurements in prostate tissues uncovers novel prostate cancer diagnostic biomarkers and transcription factor binding patterns. *BMC Cancer* **2017**, *17*, 273. [CrossRef] [PubMed]
170. O'Reilly, E.; Tuzova, A.V.; Walsh, A.L.; Russell, N.M.; O'Brien, O.; Kelly, S.; Dhomhnallain, O.N.; DeBarra, L.; Dale, C.M.; Brugman, R.; et al. epiCaPture: A Urine DNA Methylation Test for Early Detection of Aggressive Prostate Cancer. *JCO Precis. Oncol.* **2019**. [CrossRef]
171. Leyten, G.H.; Hessels, D.; Smit, F.P.; Jannink, S.A.; de Jong, H.; Melchers, W.J.; Cornel, E.B.; de Reijke, T.M.; Vergunst, H.; Kil, P.; et al. Identification of a Candidate Gene Panel for the Early Diagnosis of Prostate Cancer. *Clin. Cancer Res.* **2015**, *21*, 3061–3070. [CrossRef]
172. Van Neste, L.; Hendriks, R.J.; Dijkstra, S.; Trooskens, G.; Cornel, E.B.; Jannink, S.A.; de Jong, H.; Hessels, D.; Smit, F.P.; Melchers, W.J.; et al. Detection of High-grade Prostate Cancer Using a Urinary Molecular Biomarker-Based Risk Score. *Eur. Urol.* **2016**, *70*, 740–748. [CrossRef]
173. Haese, A.; Trooskens, G.; Steyaert, S.; Hessels, D.; Brawer, M.; Vlaeminck-Guillem, V.; Ruffion, A.; Tilki, D.; Schalken, J.; Groskopf, J.; et al. Multicenter Optimization and Validation of a 2-Gene mRNA Urine Test for Detection of Clinically Significant Prostate Cancer before Initial Prostate Biopsy. *J. Urol.* **2019**, *202*, 256–263. [CrossRef]
174. Hendriks, R.J.; van der Leest, M.M.; Dijkstra, S.; Barentsz, J.O.; Van Criekinge, W.; Hulsbergen-van de Kaa, C.A.; Schalken, J.A.; Mulders, P.F.; van Oort, I.M. A urinary biomarker-based risk score correlates with multiparametric MRI for prostate cancer detection. *Prostate* **2017**, *77*, 1401–1407. [CrossRef]
175. Busetto, G.M.; Del Giudice, F.; Maggi, M.; De Marco, F.; Porreca, A.; Sperduti, I.; Magliocca, F.M.; Salciccia, S.; Chung, B.I.; De Berardinis, E.; et al. Prospective assessment of two-gene urinary test with multiparametric magnetic resonance imaging of the prostate for men undergoing primary prostate biopsy. *World J. Urol.* **2020**, *6*, 1869–1877. [CrossRef] [PubMed]
176. Govers, T.M.; Hessels, D.; Vlaeminck-Guillem, V.; Schmitz-Dräger, B.J.; Stief, C.G.; Martinez-Ballesteros, C.; Ferro, M.; Borque-Fernando, A.; Rubio-Briones, J.; Sedelaar, J.M.; et al. Cost-effectiveness of SelectMDx for prostate cancer in four European countries: A comparative modeling study. *Prostate Cancer Prostatic Dis.* **2019**, *22*, 101–109. [CrossRef] [PubMed]
177. Dijkstra, S.; Govers, T.M.; Hendriks, R.J.; Schalken, J.A.; Van Criekinge, W.; Van Neste, L.; Grutters, J.P.; Sedelaar, J.P.M.; van Oort, I.M. Cost-effectiveness of a new urinary biomarker-based risk score compared to standard of care in prostate cancer diagnostics—A decision analytical model. *BJU Int.* **2017**, *120*, 659–665. [CrossRef] [PubMed]
178. Sathianathen, N.J.; Kuntz, K.M.; Alarid-Escudero, F.; Lawrentschuk, N.L.; Bolton, D.M.; Murphy, D.G.; Weight, C.J.; Konety, B.R. Incorporating Biomarkers into the Primary Prostate Biopsy Setting: A Cost-Effectiveness Analysis. *J. Urol.* **2018**, *200*, 1215–1220. [CrossRef]
179. Hoyer, G.; Crawford, E.D.; Arangua, P.; Stanton, W.; La Rosa, F.G.; Poage, W.; Lucia, M.S.; van Bokhoven, A.; Werahera, P.N. SelectMDx versus Prostate Health Index in the identification of high-grade prostate cancer. *J. Clin. Oncol.* **2019**, *37* (Suppl. 7), 30. [CrossRef]
180. Fasulo, V.; de la Calle, C.M.; Cowan, J.E.; Herlemann, A.; Chu, C.; Gadzinski, A.J.; Au Yeung, R.; Saita, A.; Cooperberg, M.R.; Shinohara, K.; et al. Clinical utility of biomarkers 4K score, SelectMDx and ExoDx with MRI for the detection of high-grade prostate cancer. *J. Clin. Oncol.* **2020**, *38* (Suppl. 6), 307. [CrossRef]
181. Pepe, P.; Dibenedetto, G.; Pepe, L.; Pennisi, M. Multiparametric MRI Versus SelectMDx Accuracy in the Diagnosis of Clinically Significant PCa in Men Enrolled in Active Surveillance. *In Vivo* **2020**, *34*, 393–396. [CrossRef]
182. Zheng, J. Oncogenic chromosomal translocations and human cancer (review). *Oncol. Rep.* **2013**, *30*, 2011–2019. [CrossRef] [PubMed]
183. Tomlins, S.A.; Rhodes, D.R.; Perner, S.; Dhanasekaran, S.M.; Mehra, R.; Sun, X.W.; Varambally, S.; Cao, X.; Tchinda, J.; Kuefer, R.; et al. Recurrent fusion of TMPRSS2 and ETS transcription factor genes in prostate cancer. *Science* **2005**, *310*, 644–648. [CrossRef] [PubMed]
184. Prensner, J.R.; Chinnaiyan, A.M. Oncogenic gene fusions in epithelial carcinomas. *Curr. Opin. Genet. Dev.* **2009**, *19*, 82–91. [CrossRef] [PubMed]
185. Han, B.; Mehra, R.; Lonigro, R.J.; Wang, L.; Suleman, K.; Menon, A.; Palanisamy, N.; Tomlins, S.A.; Chinnaiyan, A.M.; Shah, R.B. Fluorescence in situ hybridization study shows association of PTEN deletion with ERG rearrangement during prostate cancer progression. *Mod. Pathol.* **2009**, *22*, 1083–1093. [CrossRef] [PubMed]
186. Magi-Galluzzi, C.; Tsusuki, T.; Elson, P.; Simmerman, K.; LaFargue, C.; Esgueva, R.; Klein, E.; Rubin, M.A.; Zhou, M. TMPRSS2-ERG gene fusion prevalence and class are significantly different in prostate cancer of Caucasian, African-American and Japanese patients. *Prostate* **2011**, *71*, 489–497. [CrossRef] [PubMed]
187. Mosquera, J.M.; Perner, S.; Genega, E.M.; Sanda, M.; Hofer, M.D.; Mertz, K.D.; Paris, P.L.; Simko, J.; Bismar, T.A.; Ayala, G.; et al. Characterization of TMPRSS2-ERG fusion high-grade prostatic intraepithelial neoplasia and potential clinical implications. *Clin. Cancer Res.* **2008**, *14*, 3380–3385. [CrossRef] [PubMed]
188. Hessels, D.; Smit, F.P.; Verhaegh, G.W.; Witjes, J.A.; Cornel, E.B.; Schalken, J.A. Detection of TMPRSS2-ERG fusion transcripts and prostate cancer antigen 3 in urinary sediments may improve diagnosis of prostate cancer. *Clin. Cancer Res.* **2007**, *13*, 5103–5108. [CrossRef]

189. Tomlins, S.A.; Aubin, S.M.; Siddiqui, J.; Lonigro, R.J.; Sefton-Miller, L.; Miick, S.; Williamsen, S.; Hodge, P.; Meinke, J.; Blase, A.; et al. Urine TMPRSS2:ERG fusion transcript stratifies prostate cancer risk in men with elevated serum PSA. *Sci. Transl. Med.* **2011**, *3*, 94ra72. [CrossRef]

190. Demichelis, F.; Fall, K.; Perner, S.; Andrén, O.; Schmidt, F.; Setlur, S.R.; Hoshida, Y.; Mosquera, J.M.; Pawitan, Y.; Lee, C.; et al. TMPRSS2:ERG gene fusion associated with lethal prostate cancer in a watchful waiting cohort. *Oncogene* **2007**, *26*, 4596–4599. [CrossRef]

191. Lapointe, J.; Kim, Y.H.; Miller, M.A.; Li, C.; Kaygusuz, G.; van de Rijn, M.; Huntsman, D.G.; Brooks, J.D.; Pollack, J.R. A variant TMPRSS2 isoform and ERG fusion product in prostate cancer with implications for molecular diagnosis. *Mod. Pathol.* **2007**, *20*, 467–473. [CrossRef] [PubMed]

192. Winnes, M.; Lissbrant, E.; Damber, J.E.; Stenman, G. Molecular genetic analyses of the TMPRSS2-ERG and TMPRSS2-ETV1 gene fusions in 50 cases of prostate cancer. *Oncol. Rep.* **2007**, *17*, 1033–1036. [CrossRef]

193. Rice, K.R.; Chen, Y.; Ali, A.; Whitman, E.J.; Blase, A.; Ibrahim, M.; Elsamanoudi, S.; Brassell, S.; Furusato, B.; Stingle, N.; et al. Evaluation of the ETS-related gene mRNA in urine for the detection of prostate cancer. *Clin. Cancer Res.* **2010**, *16*, 1572–1576. [CrossRef]

194. Fine, S.W.; Gopalan, A.; Leversha, M.A.; Al-Ahmadie, H.A.; Tickoo, S.K.; Zhou, Q.; Satagopan, J.M.; Scardino, P.T.; Gerald, W.L.; Reuter, V.E. TMPRSS2-ERG gene fusion is associated with low Gleason scores and not with high-grade morphological features. *Mod. Pathol.* **2010**, *23*, 1325–1333. [CrossRef]

195. Gopalan, A.; Leversha, M.A.; Satagopan, J.M.; Zhou, Q.; Al-Ahmadie, H.A.; Fine, S.W.; Eastham, J.A.; Scardino, P.T.; Scher, H.I.; Tickoo, S.K.; et al. TMPRSS2-ERG gene fusion is not associated with outcome in patients treated by prostatectomy. *Cancer Res.* **2009**, *69*, 1400–1406. [CrossRef]

196. Pettersson, A.; Graff, R.E.; Bauer, S.R.; Pitt, M.J.; Lis, R.T.; Stack, E.C.; Martin, N.E.; Kunz, L.; Penney, K.L.; Ligon, A.H.; et al. The TMPRSS2:ERG rearrangement, ERG expression, and prostate cancer outcomes: A cohort study and meta-analysis. *Cancer Epidemiol. Biomark. Prev.* **2012**, *21*, 1497–1509. [CrossRef]

197. Vader, P.; Breakefield, X.O.; Wood, M.J. Extracellular vesicles: Emerging targets for cancer therapy. *Trends Mol. Med.* **2014**, *20*, 385–393. [CrossRef]

198. Kahlert, C.; Kalluri, R. Exosomes in tumor microenvironment influence cancer progression and metastasis. *J. Mol. Med.* **2013**, *91*, 431–437. [CrossRef]

199. Tavoosidana, G.; Ronquist, G.; Darmanis, S.; Yan, J.; Carlsson, L.; Wu, D.; Conze, T.; Ek, P.; Semjonow, A.; Eltze, E.; et al. Multiple recognition assay reveals prostasomes as promising plasma biomarkers for prostate cancer. *Proc. Natl. Acad. Sci. USA* **2011**, *108*, 8809–8814. [CrossRef]

200. Nilsson, J.; Skog, J.; Nordstrand, A.; Baranov, V.; Mincheva-Nilsson, L.; Breakefield, X.O.; Widmark, A. Prostate cancer-derived urine exosomes: A novel approach to biomarkers for prostate cancer. *Br. J. Cancer* **2009**, *100*, 1603–1607. [CrossRef]

201. Wang Wei-Lin, W.; Sorokin, I.; Aleksic, I.; Fisher, H.; Kaufman, R.P., Jr.; Winer, A.; McNeill, B.; Gupta, R.; Tilki, D.; Fleshner, N.; et al. Expression of Small Noncoding RNAs in Urinary Exosomes Classifies Prostate Cancer into Indolent and Aggressive Disease. *J. Urol.* **2020**, *204*, 466–475. [CrossRef]

202. Stewart, G.D.; Van Neste, L.; Delvenne, P.; Delrée, P.; Delga, A.; McNeill, S.A.; O'Donnell, M.; Clark, J.; Van Criekinge, W.; Bigley, J.; et al. Clinical utility of an epigenetic assay to detect occult prostate cancer in histopathologically negative biopsies: Results of the MATLOC study. *J. Urol.* **2013**, *189*, 1110–1116. [CrossRef]

203. Yang, B.; Bhusari, S.; Kueck, J.; Weeratunga, P.; Wagner, J.; Leverson, G.; Huang, W.; Jarrard, D.F. Methylation profiling defines an extensive field defect in histologically normal prostate tissues associated with prostate cancer. *Neoplasia* **2013**, *15*, 399–408. [CrossRef]

204. Partin, A.W.; Van Neste, L.; Klein, E.A.; Marks, L.S.; Gee, J.R.; Troyer, D.A.; Rieger-Christ, K.; Jones, J.S.; Magi-Galluzzi, C.; Mangold, L.A.; et al. Clinical Validation of an Epigenetic Assay to Predict Negative Histopathological Results in Repeat Prostate Biopsies. *J. Urol.* **2014**, *192*, 1081–1087. [CrossRef]

205. Van Neste, L.; Partin, A.W.; Stewart, G.D.; Epstein, J.I.; Harrison, D.J.; Van Criekinge, W. Risk score predicts high-grade prostate cancer in DNA-methylation positive, histopathologically negative biopsies. *Prostate* **2016**, *76*, 1078–1087. [CrossRef]

206. Waterhouse, R.L.; Van Neste, L.; Moses, K.A.; Barnswell, C.; Silberstein, J.L.; Jalkut, M.; Tutrone, R.; Sylora, J.; Anglade, R.; Murdock, M.; et al. Evaluation of an Epigenetic Assay for Predicting Repeat Prostate Biopsy Outcome in African American Men. *Urology* **2019**, *128*, 62–65. [CrossRef]

207. Bussemakers, M.J.; Van Bokhoven, A.; Verhaegh, G.W.; Smit, F.P.; Karthaus, H.F.; Schalken, J.A.; Debruyne, F.M.; Ru, N.; Isaacs, W.B. Dd3:: A new prostate-specific gene, highly overexpressed in prostate cancer. *Cancer Res.* **1999**, *59*, 5975–5979.

208. De Kok, J.B.; Verhaegh, G.W.; Roelofs, R.W.; Hessels, D.; Kiemeney, L.A.; Aalders, T.W.; Swinkels, D.W.; Schalken, J.A. DD3PCA3, a very sensitive and specific marker to detect prostate tumors. *Cancer Res.* **2002**, *62*, 2695–2698.

209. Hessels, D.; Gunnewiek, J.M.K.; van Oort, I.; Karthaus, H.F.; van Leenders, G.J.; van Balken, B.; Kiemeney, L.A.; Witjes, J.A.; Schalken, J.A. DD3(PCA3)-based molecular urine analysis for the diagnosis of prostate cancer. *Eur. Urol.* **2003**, *44*, 8–15; discussion 15–16. [CrossRef]

210. Tinzl, M.; Marberger, M.; Horvath, S.; Chypre, C. DD3PCA3 RNA analysis in urine—A new perspective for detecting prostate cancer. *Eur. Urol.* **2004**, *46*, 182–186; discussion 187. [CrossRef]

211. Groskopf, J.; Aubin, S.M.; Deras, I.L.; Blase, A.; Bodrug, S.; Clark, C.; Brentano, S.; Mathis, J.; Pham, J.; Meyer, T.; et al. APTIMA PCA3 molecular urine test: Development of a method to aid in the diagnosis of prostate cancer. *Clin. Chem.* **2006**, *52*, 1089–1095. [CrossRef]
212. Auprich, M.; Bjartell, A.; Chun, F.K.H.; de la Taille, A.; Freedland, S.J.; Haese, A.; Schalken, J.; Stenzl, A.; Tombal, B.; van der Poel, H. Contemporary role of prostate cancer antigen 3 in the management of prostate cancer. *Eur. Urol.* **2011**, *60*, 1045–1054. [CrossRef] [PubMed]
213. van Poppel, H.; Haese, A.; Graefen, M.; de la Taille, A.; Irani, J.; de Reijke, T.; Remzi, M.; Marberger, M. The relationship between Prostate CAncer gene 3 (PCA3) and prostate cancer significance. *BJU Int.* **2012**, *109*, 360–366. [CrossRef]
214. Rodríguez, S.V.M.; García-Perdomo, H.A. Diagnostic accuracy of prostate cancer antigen 3 (PCA3) prior to first prostate biopsy: A systematic review and meta-analysis. *Can. Urol. Assoc. J.* **2020**, *14*, e214–e219. [CrossRef]
215. Wei, J.T.; Feng, Z.; Partin, A.W.; Brown, E.; Thompson, I.; Sokoll, L.; Chan, D.W.; Lotan, Y.; Kibel, A.S.; Busby, J.E.; et al. Can urinary PCA3 supplement PSA in the early detection of prostate cancer? *J. Clin. Oncol.* **2014**, *32*, 4066–4072. [CrossRef] [PubMed]
216. Gittelman, M.C.; Hertzman, B.; Bailen, J.; Williams, T.; Koziol, I.; Henderson, R.J.; Efros, M.; Bidair, M.; Ward, J.F. PCA3 molecular urine test as a predictor of repeat prostate biopsy outcome in men with previous negative biopsies: A prospective multicenter clinical study. *J. Urol.* **2013**, *190*, 64–69. [CrossRef]
217. Pepe, P.; Aragona, F. PCA3 Score vs PSA Free/Total Accuracy in Prostate Cancer Diagnosis at Repeat Saturation Biopsy. *Anticancer Res.* **2011**, *31*, 4445.
218. Leyten, G.H.; Wierenga, E.A.; Sedelaar, J.P.; Van Oort, I.M.; Futterer, J.J.; Barentsz, J.O.; Schalken, J.A.; Mulders, P.F. Value of PCA3 to predict biopsy outcome and its potential role in selecting patients for multiparametric MRI. *Int. J. Mol. Sci.* **2013**, *14*, 11347–11355. [CrossRef]
219. Kaufmann, S.; Bedke, J.; Gatidis, S.; Hennenlotter, J.; Kramer, U.; Notohamiprodjo, M.; Nikolaou, K.; Stenzl, A.; Kruck, S. Prostate cancer gene 3 (PCA3) is of additional predictive value in patients with PI-RADS grade III (intermediate) lesions in the MR-guided re-biopsy setting for prostate cancer. *World J. Urol.* **2016**, *34*, 509–515. [CrossRef]
220. De Luca, S.; Passera, R.; Cattaneo, G.; Manfredi, M.; Mele, F.; Fiori, C.; Bollito, E.; Cirillo, S.; Porpiglia, F. High prostate cancer gene 3 (PCA3) scores are associated with elevated Prostate Imaging Reporting and Data System (PI-RADS) grade and biopsy Gleason score, at magnetic resonance imaging/ultrasonography fusion software-based targeted prostate biopsy after a previous negative standard biopsy. *BJU Int.* **2016**, *118*, 723–730.
221. Yamkamon, V.; Htoo, K.P.P.; Yainoy, S.; Suksrichavalit, T.; Tangchaikeeree, T.; Eiamphungporn, W. Urinary PCA3 detection in prostate cancer by magnetic nanoparticles coupled with colorimetric enzyme-linked oligonucleotide assay. *Excli. J.* **2020**, *19*, 501–513.
222. Soares, J.C.; Soares, A.C.; Rodrigues, V.C.; Melendez, M.E.; Santos, A.C.; Faria, E.F.; Reis, R.M.; Carvalho, A.L.; Oliveira, O.N., Jr. Detection of the Prostate Cancer Biomarker PCA3 with Electrochemical and Impedance-Based Biosensors. *ACS Appl. Mater. Interfaces* **2019**, *11*, 46645–46650. [CrossRef]
223. Fu, X.; Wen, J.; Li, J.; Lin, H.; Liu, Y.; Zhuang, X.; Tian, C.; Chen, L. Highly sensitive detection of prostate cancer specific PCA3 mimic DNA using SERS-based competitive lateral flow assay. *Nanoscale* **2019**, *11*, 15530–15536. [CrossRef]
224. Htoo, K.P.P.; Yamkamon, V.; Yainoy, S.; Suksrichavalit, T.; Viseshsindh, W.; Eiamphungporn, W. Colorimetric detection of PCA3 in urine for prostate cancer diagnosis using thiol-labeled PCR primer and unmodified gold nanoparticles. *Clin. Chim. Acta* **2019**, *488*, 40–49. [CrossRef]
225. Merola, R.; Tomao, L.; Antenucci, A.; Sperduti, I.; Sentinelli, S.; Masi, S.; Mandoj, C.; Orlandi, G.; Papalia, R.; Guaglianone, S.; et al. PCA3 in prostate cancer and tumor aggressiveness detection on 407 high-risk patients: A National Cancer Institute experience. *J. Exp. Clin. Cancer Res.* **2015**, *34*, 15. [CrossRef]
226. Chevli, K.K.; Duff, M.; Walter, P.; Yu, C.; Capuder, B.; Elshafei, A.; Malczewski, S.; Kattan, M.W.; Jones, J.S. Urinary PCA3 as a predictor of prostate cancer in a cohort of 3073 men undergoing initial prostate biopsy. *J. Urol.* **2014**, *191*, 1743–1748. [CrossRef]
227. Haese, A.; de la Taille, A.; Van Poppel, H.; Marberger, M.; Stenzl, A.; Mulders, P.F.; Huland, H.; Abbou, C.C.; Remzi, M.; Tinzl, M.; et al. Clinical utility of the PCA3 urine assay in European men scheduled for repeat biopsy. *Eur. Urol.* **2008**, *54*, 1081–1088. [CrossRef]
228. Tosoian, J.J.; Patel, H.D.; Mamawala, M.; Landis, P.; Wolf, S.; Elliott, D.J.; Epstein, J.I.; Carter, H.B.; Ross, A.E.; Sokoll, L.J.; et al. Longitudinal assessment of urinary PCA3 for predicting prostate cancer grade reclassification in favorable-risk men during active surveillance. *Prostate Cancer Prostatic Dis.* **2017**, *20*, 339–342. [CrossRef] [PubMed]
229. Chunhua, L.; Zhao, H.; Zhao, H.; Lu, Y.; Wu, J.; Gao, Z.; Li, G.; Zhang, Y.; Wang, K. Clinical Significance of Peripheral Blood PCA3 Gene Expression in Early Diagnosis of Prostate Cancer. *Transl. Oncol.* **2018**, *11*, 628–632. [CrossRef] [PubMed]
230. Hessels, D.; van Gils, M.P.; van Hooij, O.; Jannink, S.A.; Witjes, J.A.; Verhaegh, G.W.; Schalken, J.A. Predictive value of PCA3 in urinary sediments in determining clinico-pathological characteristics of prostate cancer. *Prostate* **2010**, *70*, 10–16. [CrossRef] [PubMed]
231. Foj, L.; Milà, M.; Mengual, L.; Luque, P.; Alcaraz, A.; Jiménez, W.; Filella, X. Real-time PCR PCA3 assay is a useful test measured in urine to improve prostate cancer detection. *Clin. Chim. Acta* **2014**, *435*, 53–58. [CrossRef]
232. Scattoni, V.; Lazzeri, M.; Lughezzani, G.; De Luca, S.; Passera, R.; Bollito, E.; Randone, D.; Abdollah, F.; Capitanio, U.; Larcher, A.; et al. Head-to-head comparison of prostate health index and urinary PCA3 for predicting cancer at initial or repeat biopsy. *J. Urol.* **2013**, *190*, 496–501. [CrossRef] [PubMed]

233. Cantiello, F.; Russo, G.I.; Ferro, M.; Cicione, A.; Cimino, S.; Favilla, V.; Perdonà, S.; Bottero, D.; Terracciano, D.; De Cobelli, O.; et al. Prognostic accuracy of Prostate Health Index and urinary Prostate Cancer Antigen 3 in predicting pathologic features after radical prostatectomy. *Urol. Oncol.* **2015**, *33*, 163.e15–163.e23. [CrossRef] [PubMed]
234. Salami, S.S.; Schmidt, F.; Laxman, B.; Regan, M.M.; Rickman, D.S.; Scherr, D.; Bueti, G.; Siddiqui, J.; Tomlins, S.A.; Wei, J.T.; et al. Combining urinary detection of TMPRSS2: ERG and PCA3 with serum PSA to predict diagnosis of prostate cancer. In *Urologic Oncology: Seminars and Original Investigations*; Elsevier: Amsterdam, The Netherlands, 2013.
235. Tomlins, S.A.; Day, J.R.; Lonigro, R.J.; Hovelson, D.H.; Siddiqui, J.; Kunju, L.P.; Dunn, R.L.; Meyer, S.; Hodge, P.; Groskopf, J.; et al. Urine TMPRSS2:ERG Plus PCA3 for Individualized Prostate Cancer Risk Assessment. *Eur. Urol.* **2016**, *70*, 45–53. [CrossRef]
236. McKiernan, J.; Donovan, M.J.; O'Neill, V.; Bentink, S.; Noerholm, M.; Belzer, S.; Skog, J.; Kattan, M.W.; Partin, A.; Andriole, G.; et al. A novel urine exosome gene expression assay to predict high-grade prostate cancer at initial biopsy. *JAMA Oncol.* **2016**, *2*, 882–889. [CrossRef]
237. McKiernan, J.; Donovan, M.J.; Margolis, E.; Partin, A.; Carter, B.; Brown, G.; Torkler, P.; Noerholm, M.; Skog, J.; Shore, N.; et al. A Prospective Adaptive Utility Trial to Validate Performance of a Novel Urine Exosome Gene Expression Assay to Predict High-grade Prostate Cancer in Patients with Prostate-specific Antigen 2–10 ng/mL at Initial Biopsy. *Eur. Urol.* **2018**, *74*, 731–738. [CrossRef]
238. McKiernan, J.; Noerholm, M.; Tadigotla, V.; Kumar, S.; Torkler, P.; Sant, G.; Alter, J.; Donovan, M.J.; Skog, J. A urine-based Exosomal gene expression test stratifies risk of high-grade prostate Cancer in men with prior negative prostate biopsy undergoing repeat biopsy. *BMC Urol.* **2020**, *20*, 138. [CrossRef]
239. Tutrone, R.; Donovan, M.J.; Torkler, P.; Tadigotla, V.; McLain, T.; Noerholm, M.; Skog, J.; McKiernan, J. Clinical utility of the exosome based ExoDx Prostate(IntelliScore) EPI test in men presenting for initial Biopsy with a PSA 2–10 ng/mL. *Prostate Cancer Prostatic Dis.* **2020**, *23*, 607–614. [CrossRef]
240. Dakubo, G.D.; Jakupciak, J.P.; Birch-Machin, M.A.; Parr, R.L. Clinical implications and utility of field cancerization. *Cancer Cell Int.* **2007**, *7*, 2. [CrossRef]
241. Maki, J.; Robinson, K.; Reguly, B.; Alexander, J.; Wittock, R.; Aguirre, A.; Diamandis, E.P.; Escott, N.; Skehan, A.; Prowse, O.; et al. Mitochondrial genome deletion aids in the identification of false- and true-negative prostate needle core biopsy specimens. *Am. J. Clin. Pathol.* **2008**, *129*, 57–66. [CrossRef]
242. Robinson, K.; Creed, J.; Reguly, B.; Powell, C.; Wittock, R.; Klein, D.; Maggrah, A.; Klotz, L.; Parr, R.L.; Dakubo, G.D. Accurate prediction of repeat prostate biopsy outcomes by a mitochondrial DNA deletion assay. *Prostate Cancer Prostatic Dis.* **2010**, *13*, 126–131. [CrossRef]
243. Legisi, L.; DeSa, E.; Qureshi, M.N. Use of the Prostate Core Mitomic Test in Repeated Biopsy Decision-Making: Real-World Assessment of Clinical Utility in a Multicenter Patient Population. *Am. Health Drug Benefits* **2016**, *9*, 497–502. [PubMed]
244. Eschrich, S.A.; Fulp, W.J.; Pawitan, Y.; Foekens, J.A.; Smid, M.; Martens, J.W.; Echevarria, M.; Kamath, V.; Lee, J.H.; Harris, E.E.; et al. Validation of a Radiosensitivity Molecular Signature in Breast Cancer. *Clin. Cancer Res.* **2012**, *18*, 5134–5143. [CrossRef] [PubMed]
245. Weichselbaum, R.R.; Ishwaran, H.; Yoon, T.; Nuyten, D.S.; Baker, S.W.; Khodarev, N.; Su, A.W.; Shaikh, A.Y.; Roach, P.; Kreike, B.; et al. An interferon-related gene signature for DNA damage resistance is a predictive marker for chemotherapy and radiation for breast cancer. *Proc. Natl. Acad. Sci. USA* **2008**, *105*, 18490–18495. [CrossRef]
246. Pitroda, S.P.; Pashtan, I.M.; Logan, H.L.; Budke, B.; Darga, T.E.; Weichselbaum, R.R.; Connell, P.P. DNA repair pathway gene expression score correlates with repair proficiency and tumor sensitivity to chemotherapy. *Sci. Transl. Med.* **2014**, *6*, 229ra42. [CrossRef] [PubMed]
247. Paik, S.; Tang, G.; Shak, S.; Kim, C.; Baker, J.; Kim, W.; Cronin, M.; Baehner, F.L.; Watson, D.; Bryant, J.; et al. Gene expression and benefit of chemotherapy in women with node-negative, estrogen receptor-positive breast cancer. *J. Clin. Oncol.* **2006**, *24*, 3726–3734. [CrossRef] [PubMed]
248. Klein, E.A.; Cooperberg, M.R.; Magi-Galluzzi, C.; Simko, J.P.; Falzarano, S.M.; Maddala, T.; Chan, J.M.; Li, J.; Cowan, J.E.; Tsiatis, A.C.; et al. A 17-gene assay to predict prostate cancer aggressiveness in the context of Gleason grade heterogeneity, tumor multifocality, and biopsy undersampling. *Eur. Urol.* **2014**, *66*, 550–560. [CrossRef]
249. Eggener, S.; Karsh, L.I.; Richardson, T.; Shindel, A.W.; Lu, R.; Rosenberg, S.; Goldfischer, E.; Korman, H.; Bennett, J.; Newmark, J.; et al. A 17-gene Panel for Prediction of Adverse Prostate Cancer Pathologic Features: Prospective Clinical Validation and Utility. *Urology* **2019**, *126*, 76–82. [CrossRef]
250. Cullen, J.; Rosner, I.L.; Brand, T.C.; Zhang, N.; Tsiatis, A.C.; Moncur, J.; Ali, A.; Chen, Y.; Knezevic, D.; Maddala, T.; et al. A biopsy-based 17-gene genomic prostate score predicts recurrence after radical prostatectomy and adverse surgical pathology in a racially diverse population of men with clinically low-and intermediate-risk prostate cancer. *Eur. Urol.* **2015**, *68*, 123–131. [CrossRef]
251. CMoschovas, M.C.; Chew, C.; Bhat, S.; Sandri, M.; Rogers, T.; Dell'Oglio, P.; Roof, S.; Reddy, S.; Sighinolfi, M.C.; Rocco, B.; et al. Association Between Oncotype DX Genomic Prostate Score and Adverse Tumor Pathology After Radical Prostatectomy. *Eur. Urol. Focus* **2021**. [CrossRef]
252. Kornberg, Z.; Cooperberg, M.R.; Cowan, J.E.; Chan, J.; Shinohara, K.; Simko, J.P.; Tenggara, I.; Carroll, P.R. A 17-Gene Genomic Prostate Score as a Predictor of Adverse Pathology in Men on Active Surveillance. *J. Urol.* **2019**, *202*, 702–709. [CrossRef]

253. Eeden, S.K.V.D.; Lu, R.; Zhang, N.; Quesenberry, C.P.; Shan, J.; Han, J.S.; Tsiatis, A.C.; Leimpeter, A.D.; Lawrence, H.J.; Febbo, P.G.; et al. A Biopsy-based 17-gene Genomic Prostate Score as a Predictor of Metastases and Prostate Cancer Death in Surgically Treated Men with Clinically Localized Disease. *Eur. Urol.* **2018**, *73*, 129–138. [CrossRef]
254. Cullen, J.; Kuo, H.-C.; Shan, J.; Lu, R.; Aboushwareb, T.; Eeden, S.K.V.D. The 17-Gene Genomic Prostate Score Test as a Predictor of Outcomes in Men with Unfavorable Intermediate Risk Prostate Cancer. *Urology* **2020**, *143*, 103–111. [CrossRef] [PubMed]
255. Chang, E.M.; Punglia, R.S.; Steinberg, M.L.; Raldow, A.C. Cost Effectiveness of the Oncotype DX Genomic Prostate Score for Guiding Treatment Decisions in Patients With Early Stage Prostate Cancer. *Urology* **2019**, *126*, 89–95. [CrossRef]
256. Albala, D.; Kemeter, M.J.; Febbo, P.G.; Lu, R.; John, V.; Stoy, D.; Denes, B.; McCall, M.; Shindel, A.W.; Dubeck, F. Health Economic Impact and Prospective Clinical Utility of Oncotype DX® Genomic Prostate Score. *Rev. Urol.* **2016**, *18*, 123–132. [PubMed]
257. Lin, D.W.; Zheng, Y.; McKenney, J.K.; Brown, M.D.; Lu, R.; Crager, M.; Boyer, H.; Tretiakova, M.; Brooks, J.D.; Dash, A.; et al. 17-Gene Genomic Prostate Score Test Results in the Canary Prostate Active Surveillance Study (PASS) Cohort. *J. Clin. Oncol.* **2020**, *38*, 1549–1557. [CrossRef]
258. Greenland Nancy, Y.; Zhang, L.; Cowan, J.E.; Carroll, P.R.; Stohr, B.A.; Simko, J.P. Correlation of a Commercial Genomic Risk Classifier with Histological Patterns in Prostate Cancer. *J. Urol.* **2019**, *202*, 90–95. [CrossRef]
259. Cuzick, J.; Swanson, G.; Fisher, G.; Brothman, A.R.; Berney, D.; E Reid, J.; Mesher, D.; Speights, V.; Stankiewicz, E.; Foster, C.S.; et al. Prognostic value of an RNA expression signature derived from cell cycle proliferation genes in patients with prostate cancer: A retrospective study. *Lancet Oncol.* **2011**, *12*, 245–255. [CrossRef]
260. NICE, NICE Advice—Prolaris gene expression assay for assessing long-term risk of prostate cancer progression: © NICE (2016) Prolaris gene expression assay for assessing long-term risk of prostate cancer progression. *BJU Int.* **2018**, *122*, 173–180. [CrossRef]
261. Cuzick, J.; On behalf of the Transatlantic Prostate Group; Stone, S.B.; Fisher, G.; Yang, Z.H.; North, B.V.; Berney, D.; Beltran, L.E.; Greenberg, D.S.; Moller, H.; et al. Validation of an RNA cell cycle progression score for predicting death from prostate cancer in a conservatively managed needle biopsy cohort. *Br. J. Cancer* **2015**, *113*, 382–389. [CrossRef] [PubMed]
262. Lin, D.W.; Crawford, E.D.; Keane, T.; Evans, B.; Reid, J.; Rajamani, S.; Brown, K.; Gutin, A.; Tward, J.; Scardino, P.; et al. Identification of men with low-risk biopsy-confirmed prostate cancer as candidates for active surveillance. *Urol. Oncol.* **2018**, *36*, 310.e7–310.e13. [CrossRef]
263. Bishoff, J.T.; Freedland, S.J.; Gerber, L.; Tennstedt, P.; Reid, J.; Welbourn, W.; Graefen, M.; Sangale, Z.; Tikishvili, E.; Park, J.; et al. Prognostic utility of the cell cycle progression score generated from biopsy in men treated with prostatectomy. *J. Urol.* **2014**, *192*, 409–414. [CrossRef] [PubMed]
264. Cooperberg, M.R.; Simko, J.P.; Cowan, J.E.; Reid, J.E.; Djalilvand, A.; Bhatnagar, S.; Gutin, A.; Lanchbury, J.S.; Swanson, G.; Stone, S.; et al. Validation of a cell-cycle progression gene panel to improve risk stratification in a contemporary prostatectomy cohort. *J. Clin. Oncol.* **2013**, *31*, 1428–1434. [CrossRef]
265. Tosoian, J.J.; Chappidi, M.R.; Bishoff, J.T.; Freedland, S.J.; Reid, J.; Brawer, M.; Stone, S.; Schlomm, T.; Ross, A.E. Prognostic utility of biopsy-derived cell cycle progression score in patients with National Comprehensive Cancer Network low-risk prostate cancer undergoing radical prostatectomy: Implications for treatment guidance. *BJU Int.* **2017**, *120*, 808–814. [CrossRef]
266. Léon, P.; Cancel-Tassin, G.; Drouin, S.; Audouin, M.; Varinot, J.; Comperat, E.; Cathelineau, X.; Rozet, F.; Vaessens, C.; Stone, S.; et al. Comparison of cell cycle progression score with two immunohistochemical markers (PTEN and Ki-67) for predicting outcome in prostate cancer after radical prostatectomy. *World J. Urol.* **2018**, *36*, 1495–1500. [CrossRef]
267. Freedland, S.J.; Gerber, L.; Reid, J.; Welbourn, W.; Tikishvili, E.; Park, J.; Younus, A.; Gutin, A.; Sangale, Z.; Lanchbury, J.S.; et al. Prognostic utility of cell cycle progression score in men with prostate cancer after primary external beam radiation therapy. *Int. J. Radiat. Oncol. Biol. Phys.* **2013**, *86*, 848–853. [CrossRef]
268. Koch, M.O.; Cho, J.S.; Kaimakliotis, H.Z.; Cheng, L.; Sangale, Z.; Brawer, M.; Welbourn, W.; Reid, J.; Stone, S. Use of the cell cycle progression (CCP) score for predicting systemic disease and response to radiation of biochemical recurrence. *Cancer Biomark* **2016**, *17*, 83–88. [CrossRef]
269. Canter, D.J.; Freedland, S.; Rajamani, S.; Latsis, M.; Variano, M.; Halat, S.; Tward, J.; Cohen, T.; Stone, S.; Schlomm, T.; et al. Analysis of the prognostic utility of the cell cycle progression (CCP) score generated from needle biopsy in men treated with definitive therapy. *Prostate Cancer Prostatic Dis.* **2020**, *23*, 102–107. [CrossRef] [PubMed]
270. Shore, N.D.; Kella, N.; Moran, B.; Boczko, J.; Bianco, F.J.; Crawford, E.D.; Davis, T.; Roundy, K.M.; Rushton, K.; Grier, C.; et al. Impact of the Cell Cycle Progression Test on Physician and Patient Treatment Selection for Localized Prostate Cancer. *J. Urol.* **2016**, *195*, 612–618. [CrossRef] [PubMed]
271. Kaul, S.; Wojno, K.J.; Stone, S.; Evans, B.; Bernhisel, R.; Meek, S.; D'Anna, R.E.; Ferguson, J.; Glaser, J.; Morgan, T.M.; et al. Clinical outcomes in men with prostate cancer who selected active surveillance using a clinical cell cycle risk score. *Per. Med.* **2019**, *16*, 491–499. [CrossRef] [PubMed]
272. Crawford, E.D.; Scholz, M.C.; Kar, A.J.; Fegan, J.E.; Haregewoin, A.; Kaldate, R.R.; Brawer, M.K. Cell cycle progression score and treatment decisions in prostate cancer: Results from an ongoing registry. *Curr. Med. Res. Opin.* **2014**, *30*, 1025–1031. [CrossRef]
273. Gustavsen, G.; Taylor, K.; Cole, D.; Gullet, L.; Lewine, N. Health economic impact of a biopsy-based cell cycle gene expression assay in localized prostate cancer. *Future Oncol.* **2020**, *16*, 3061–3074. [CrossRef]
274. Health Quality Ontario. Prolaris Cell Cycle Progression Test for Localized Prostate Cancer: A Health Technology Assessment. *Ont. Health Technol. Assess. Ser.* **2017**, *17*, 1–75.

275. Shipitsin, M.; E Small, C.; Choudhury, S.; Giladi, E.; Friedlander, S.F.; Nardone, J.; Hussain, S.; Hurley, A.D.; Ernst, C.; E Huang, Y.; et al. Identification of proteomic biomarkers predicting prostate cancer aggressiveness and lethality despite biopsy-sampling error. *Br. J. Cancer* **2014**, *111*, 1201–1212. [CrossRef] [PubMed]
276. Blume-Jensen, P.; Berman, D.; Rimm, D.L.; Shipitsin, M.; Putzi, M.; Nifong, T.P.; Small, C.; Choudhury, S.; Capela, T.; Coupal, L.; et al. Development and clinical validation of an in situ biopsy-based multimarker assay for risk stratification in prostate cancer. *Clin. Cancer Res.* **2015**, *21*, 2591–2600. [CrossRef] [PubMed]
277. Erho, N.; Crisan, A.; Vergara, I.A.; Mitra, A.P.; Ghadessi, M.; Buerki, C.; Bergstralh, E.J.; Kollmeyer, T.; Fink, S.; Haddad, Z.; et al. Discovery and validation of a prostate cancer genomic classifier that predicts early metastasis following radical prostatectomy. *PLoS ONE* **2013**, *8*, e66855. [CrossRef]
278. Klein, E.A.; Yousefi, K.; Haddad, Z.; Choeurng, V.; Buerki, C.; Stephenson, A.J.; Li, J.; Kattan, M.; Magi-Galluzzi, C.; Davicioni, E. A genomic classifier improves prediction of metastatic disease within 5 years after surgery in node-negative high-risk prostate cancer patients managed by radical prostatectomy without adjuvant therapy. *Eur. Urol.* **2015**, *67*, 778–786. [CrossRef]
279. Broeck, T.V.D.; Moris, L.; Gevaert, T.; Tosco, L.; Smeets, E.; Fishbane, N.; Liu, Y.; Helsen, C.; Margrave, J.; Buerki, C.; et al. Validation of the Decipher Test for Predicting Distant Metastatic Recurrence in Men with High-risk Nonmetastatic Prostate Cancer 10 Years After Surgery. *Eur. Urol. Oncol.* **2019**, *2*, 589–596. [CrossRef]
280. Karnes, R.J.; Bergstralh, E.J.; Davicioni, E.; Ghadessi, M.; Buerki, C.; Mitra, A.P.; Crisan, A.; Erho, N.; Vergara, I.; Lam, L.L.; et al. Validation of a genomic classifier that predicts metastasis following radical prostatectomy in an at risk patient population. *J. Urol.* **2013**, *190*, 2047–2053. [CrossRef]
281. Spratt, D.E.; Yousefi, K.; Deheshi, S.; Ross, A.E.; Den, R.; Schaeffer, E.M.; Trock, B.J.; Zhang, J.; Glass, A.G.; Dicker, A.P.; et al. Individual Patient-Level Meta-Analysis of the Performance of the Decipher Genomic Classifier in High-Risk Men After Prostatectomy to Predict Development of Metastatic Disease. *J. Clin. Oncol.* **2017**, *35*, 1991–1998. [CrossRef]
282. Alshalalfa, M.; Crisan, A.; Vergara, I.; Ghadessi, M.; Buerki, C.; Erho, N.; Yousefi, K.; Sierocinski, T.; Haddad, Z.; Black, P.C.; et al. Clinical and genomic analysis of metastatic prostate cancer progression with a background of postoperative biochemical recurrence. *BJU Int.* **2015**, *116*, 556–567. [CrossRef]
283. Spratt, D.E.; Zhang, J.; Santiago-Jiménez, M.; Dess, R.T.; Davis, J.W.; Den, R.B.; Dicker, A.P.; Kane, C.J.; Pollack, A.; Stoyanova, R.; et al. Development and Validation of a Novel Integrated Clinical-Genomic Risk Group Classification for Localized Prostate Cancer. *J. Clin. Oncol.* **2018**, *36*, 581–590. [CrossRef]
284. Klein, E.A.; Haddad, Z.; Yousefi, K.; Lam, L.L.; Wang, Q.; Choeurng, V.; Palmer-Aronsten, B.; Buerki, C.; Davicioni, E.; Li, J.; et al. Decipher genomic classifier measured on prostate biopsy predicts metastasis risk. *Urology* **2016**, *90*, 148–152. [CrossRef] [PubMed]
285. Nguyen, P.L.; Haddad, Z.; Ross, A.E.; Martin, N.E.; Deheshi, S.; Lam, L.L.; Chelliserry, J.; Tosoian, J.J.; Lotan, T.L.; Spratt, D.E.; et al. Ability of a Genomic Classifier to Predict Metastasis and Prostate Cancer-specific Mortality after Radiation or Surgery based on Needle Biopsy Specimens. *Eur. Urol.* **2017**, *72*, 845–852. [CrossRef] [PubMed]
286. Jambor, I.; Falagario, U.; Ratnani, P.; Msc, I.M.P.; Demir, K.; Merisaari, H.; Sobotka, S.; Haines, G.K.; Martini, A.; Beksac, A.T.; et al. Prediction of biochemical recurrence in prostate cancer patients who underwent prostatectomy using routine clinical prostate multiparametric MRI and decipher genomic score. *J. Magn. Reson. Imaging* **2020**, *51*, 1075–1085. [CrossRef]
287. Ross, A.E.; Feng, F.Y.; Ghadessi, M.; Erho, N.; Crisan, A.; Buerki, C.; Sundi, D.; Mitra, A.P.; Vergara, I.; Thompson, D.J.S.; et al. A genomic classifier predicting metastatic disease progression in men with biochemical recurrence after prostatectomy. *Prostate Cancer Prostatic Dis.* **2014**, *17*, 64–69. [CrossRef] [PubMed]
288. Kim, H.L.; Li, P.; Huang, H.-C.; Deheshi, S.; Marti, T.; Knudsen, B.; Abou-Ouf, H.; Alam, R.; Lotan, T.; Lam, L.L.C.; et al. Validation of the Decipher Test for predicting adverse pathology in candidates for prostate cancer active surveillance. *Prostate Cancer Prostatic Dis.* **2019**, *22*, 399–405. [CrossRef]
289. Den, R.B.; Yousefi, K.; Trabulsi, E.J.; Abdollah, F.; Choeurng, V.; Feng, F.Y.; Dicker, A.P.; Lallas, C.D.; Gomella, L.G.; Davicioni, E.; et al. Genomic classifier identifies men with adverse pathology after radical prostatectomy who benefit from adjuvant radiation therapy. *J. Clin. Oncol.* **2015**, *33*, 944. [CrossRef] [PubMed]
290. Michalopoulos, S.N.; Michalopoulos, S.N.; Kella, N.; Payne, R.; Yohannes, P.; Singh, A.; Hettinger, C.; Yousefi, K.; Hornberger, J.; On behalf of the PRO-ACT Study Group. Influence of a genomic classifier on post-operative treatment decisions in high-risk prostate cancer patients: Results from the PRO-ACT study. *Curr. Med. Res. Opin.* **2014**, *30*, 1547–1556. [CrossRef]
291. Badani, K.K.; Thompson, D.J.; Brown, G.; Holmes, D.; Kella, N.; Albala, D.; Singh, A.; Buerki, C.; Davicioni, E.; Hornberger, J. Effect of a genomic classifier test on clinical practice decisions for patients with high-risk prostate cancer after surgery. *BJU Int.* **2015**, *115*, 419–429. [CrossRef]
292. Badani, K.; Thompson, D.J.S.; Buerki, C.; Davicioni, E.; Garrison, J.; Ghadessi, M.; Mitra, A.P.; Wood, P.J.; Hornberger, J. Impact of a genomic classifier of metastatic risk on postoperative treatment recommendations for prostate cancer patients: A report from the DECIDE study group. *Oncotarget* **2013**, *4*, 600. [CrossRef]
293. Jairath, N.K.; Pra, A.D.; Vince, R.; Dess, R.T.; Jackson, W.C.; Tosoian, J.J.; McBride, S.M.; Zhao, S.G.; Berlin, A.; Mahal, B.A.; et al. A Systematic Review of the Evidence for the Decipher Genomic Classifier in Prostate Cancer. *Eur. Urol.* **2021**, *79*, 374–383. [CrossRef]
294. Yerushalmi, R.; Woods, R.; Ravdin, P.M.; Hayes, M.M.; Gelmon, K.A. Ki67 in breast cancer: Prognostic and predictive potential. *Lancet Oncol.* **2010**, *11*, 174–183. [CrossRef]

295. Li, R.; Heydon, K.; Hammond, M.E.; Grignon, D.J.; Roach, M.; Wolkov, H.B.; Sandler, H.M.; Shipley, W.U.; Pollack, A. Ki-67 staining index predicts distant metastasis and survival in locally advanced prostate cancer treated with radiotherapy: An analysis of patients in radiation therapy oncology group protocol 86–10. *Clin. Cancer Res.* **2004**, *10*, 4118–4124. [CrossRef]
296. Pollack, A.; DeSilvio, M.; Khor, L.-Y.; Li, R.; Al-Saleem, T.; Hammond, M.; Venkatesan, V.; Lawton, C.; Roach, M.; Shipley, W.; et al. Ki-67 staining is a strong predictor of distant metastasis and mortality for men with prostate cancer treated with radiotherapy plus androgen deprivation: Radiation Therapy Oncology Group Trial 92–02. *J. Clin. Oncol.* **2004**, *22*, 2133–2140. [CrossRef] [PubMed]
297. Verhoven, B.; Yan, Y.; Ritter, M.; Khor, L.-Y.; Hammond, E.; Jones, C.; Amin, M.; Bahary, J.-P.; Zeitzer, K.; Pollack, A. Ki-67 is an independent predictor of metastasis and cause-specific mortality for prostate cancer patients treated on Radiation Therapy Oncology Group (RTOG) 94–08. *Int. J. Radiat. Oncol. Biol. Phys.* **2013**, *86*, 317–323. [CrossRef] [PubMed]
298. Mathieu, R.; Shariat, S.F.; Seitz, C.; Karakiewicz, P.I.; Fajkovic, H.; Sun, M.; Lotan, Y.; Scherr, D.; Tewari, A.; Montorsi, F.; et al. Multi-institutional validation of the prognostic value of Ki-67 labeling index in patients treated with radical prostatectomy. *World J. Urol.* **2015**, *33*, 1165–1171. [CrossRef] [PubMed]
299. Berlin, A.; Berlin, A.; Castro-Mesta, J.F.; Rodriguez-Romo, L.; Hernandez-Barajas, D.; González-Guerrero, J.F.; Rodríguez-Fernández, I.A.; González-Conchas, G.; Verdines-Perez, A.; Vera-Badillo, F.E. Prognostic role of Ki-67 score in localized prostate cancer: A systematic review and meta-analysis. *Urol. Oncol.* **2017**, *35*, 499–506. [CrossRef]
300. Epstein, J.I.; Amin, M.B.; Fine, S.W.; Algaba, F.; Aron, M.; Baydar, D.E.; Beltran, A.L.; Brimo, F.; Cheville, J.C.; Colecchia, M.; et al. The 2019 Genitourinary Pathology Society (GUPS) White Paper on Contemporary Grading of Prostate Cancer. *Arch. Pathol. Lab. Med.* **2021**, *145*, 461–493. [CrossRef] [PubMed]
301. Mesko, S.; Kupelian, P.; Demanes, D.J.; Huang, J.; Wang, P.C.; Kamrava, M. Quantifying the ki-67 heterogeneity profile in prostate cancer. *Prostate Cancer* **2013**, *2013*, 717080. [CrossRef]
302. Van der Kwast, T.H. Prognostic prostate tissue biomarkers of potential clinical use. *Virchows Arch.* **2014**, *464*, 293–300. [CrossRef]
303. Esteller, M. Non-coding RNAs in human disease. *Nat. Rev. Genet.* **2011**, *12*, 861–874. [CrossRef]
304. Liu, C.; Kelnar, K.; Liu, B.; Chen, X.; Calhoun-Davis, T.; Li, H.; Patrawala, L.; Yan, H.; Jeter, C.; Honorio, S.; et al. The microRNA miR-34a inhibits prostate cancer stem cells and metastasis by directly repressing CD44. *Nat. Med.* **2011**, *17*, 211–215. [CrossRef] [PubMed]
305. Kong, D.; Heath, E.; Chen, W.; Cher, M.L.; Powell, I.; Heilbrun, L.; Li, Y.; Ali, S.; Sethi, S.; Hassan, O.; et al. Loss of let-7 up-regulates EZH2 in prostate cancer consistent with the acquisition of cancer stem cell signatures that are attenuated by BR-DIM. *PLoS ONE* **2012**, *7*, e33729. [CrossRef]
306. Mitchell, P.S.; Parkin, R.K.; Kroh, E.M.; Fritz, B.R.; Wyman, S.K.; Pogosova-Agadjanyan, E.L.; Peterson, A.; Noteboom, J.; O'Briant, K.C.; Allen, A.; et al. Circulating microRNAs as stable blood-based markers for cancer detection. *Proc. Natl. Acad. Sci. USA* **2008**, *105*, 10513. [CrossRef]
307. Mihelich, B.L.; Maranville, J.C.; Nolley, R.; Peehl, D.M.; Nonn, L. Elevated serum microRNA levels associate with absence of high-grade prostate cancer in a retrospective cohort. *PLoS ONE* **2015**, *10*, e0124245. [CrossRef]
308. Al-Qatati, A.; Akrong, C.; Stevic, I.; Pantel, K.; Awe, J.; Saranchuk, J.; Drachenberg, D.; Mai, S.; Schwarzenbach, H. Plasma micro RNA signature is associated with risk stratification in prostate cancer patients. *Int. J. Cancer* **2017**, *141*, 1231–1239. [CrossRef] [PubMed]
309. Salido-Guadarrama, A.I.; Morales-Montor, J.G.; Rangel-Escareño, C.; Langley, E.; Peralta-Zaragoza, O.; Cruz Colin, J.L.; Rodriguez-Dorantes, M. Urinary microRNA-based signature improves accuracy of detection of clinically relevant prostate cancer within the prostate-specific antigen grey zone. *Mol. Med. Rep.* **2016**, *13*, 4549–4560. [CrossRef] [PubMed]
310. Foj, L.; Ferrer, F.; Serra, M.; Arévalo, A.; Gavagnach, M.; Gimenez, N.; Filella, X. Exosomal and non-exosomal urinary miRNAs in prostate cancer detection and prognosis. *Prostate* **2017**, *77*, 573–583. [CrossRef]
311. Wise, H.M.; Hermida, M.A.; Leslie, N.R. Prostate cancer, PI3K, PTEN and prognosis. *Clin. Sci.* **2017**, *131*, 197–210. [CrossRef]
312. Taylor, B.S.; Schultz, N.; Hieronymus, H.; Gopalan, A.; Xiao, Y.; Carver, B.S.; Arora, V.K.; Kaushik, P.; Cerami, E.; Reva, B.; et al. Integrative genomic profiling of human prostate cancer. *Cancer Cell* **2010**, *18*, 11–22. [CrossRef]
313. Lotan, T.L.; Wei, W.; Ludkovski, O.; Morais, C.L.; Guedes, L.B.; Jamaspishvili, T.; Lopez, K.; Hawley, S.T.; Feng, Z.; Fazli, L.; et al. Analytic validation of a clinical-grade PTEN immunohistochemistry assay in prostate cancer by comparison with PTEN FISH. *Mod. Pathol.* **2016**, *29*, 904–914. [CrossRef]
314. Lotan, T.L.; Carvalho, F.L.; Peskoe, S.B.; Hicks, J.L.; Good, J.; Fedor, H.L.; Humphreys, E.B.; Han, M.; Platz, E.A.; Squire, J.; et al. PTEN loss is associated with upgrading of prostate cancer from biopsy to radical prostatectomy. *Mod. Pathol.* **2015**, *28*, 128–137. [CrossRef]
315. Yoshimoto, M.; Cunha, I.W.; A Coudry, R.; Fonseca, F.P.; Torres, C.H.; Soares, F.A.; A Squire, J. FISH analysis of 107 prostate cancers shows that PTEN genomic deletion is associated with poor clinical outcome. *Br. J. Cancer* **2007**, *97*, 678–685. [CrossRef]
316. Köksal, I.T.; Dirice, E.; Yasar, D.; Sanlioglu, A.D.; Ciftcioglu, A.; Gulkesen, K.H.; O Ozes, N.; Baykara, M.; Luleci, G.; Sanlioglu, S. The assessment of PTEN tumor suppressor gene in combination with Gleason scoring and serum PSA to evaluate progression of prostate carcinoma. In *Urologic Oncology: Seminars and Original Investigations*; Elsevier: Amsterdam, The Netherlands, 2004.
317. Lotan, T.; Gurel, B.; Sutcliffe, S.; Esopi, D.; Liu, W.; Xu, J.; Hicks, J.L.; Park, B.H.; Humphreys, E.; Partin, A.W.; et al. PTEN protein loss by immunostaining: Analytic validation and prognostic indicator for a high risk surgical cohort of prostate cancer patients. *Clin. Cancer Res.* **2011**, *17*, 6563–6573. [CrossRef] [PubMed]

318. Chaux, A.; Peskoe, S.B.; Gonzalez-Roibon, N.; Schultz, L.; Albadine, R.; Hicks, J.; De Marzo, A.M.; A Platz, E.; Netto, G.J. Loss of PTEN expression is associated with increased risk of recurrence after prostatectomy for clinically localized prostate cancer. *Mod. Pathol.* **2012**, *25*, 1543–1549. [CrossRef]
319. Xie, H.; Xie, B.; Liu, C.; Wang, J.; Xu, Y. Association of PTEN expression with biochemical recurrence in prostate cancer: Results based on previous reports. *Onco Targets Ther.* **2017**, *10*, 5089. [CrossRef] [PubMed]
320. Ferraldeschi, R.; Rodrigues, D.N.; Riisnaes, R.; Miranda, S.; Figueiredo, I.; Rescigno, P.; Ravi, P.; Pezaro, C.; Omlin, A.; Lorente, D.; et al. PTEN protein loss and clinical outcome from castration-resistant prostate cancer treated with abiraterone acetate. *Eur. Urol.* **2015**, *67*, 795–802. [CrossRef] [PubMed]
321. Mulholland, D.J.; Tran, L.M.; Li, Y.; Cai, H.; Morim, A.; Wang, S.; Plaisier, S.; Garraway, I.P.; Huang, J.; Graeber, T.; et al. Cell autonomous role of PTEN in regulating castration-resistant prostate cancer growth. *Cancer Cell* **2011**, *19*, 792–804. [CrossRef]
322. Parker, C.C.; Clarke, N.W.; Cook, A.D.; Kynaston, H.G.; Petersen, P.M.; Catton, C.; Cross, W.; Logue, J.; Parulekar, W.; Payne, H.; et al. Timing of radiotherapy after radical prostatectomy (RADICALS-RT): A randomised, controlled phase 3 trial. *Lancet* **2020**, *396*, 1413–1421. [CrossRef]
323. Zhao, S.G.; Chang, S.L.; E Spratt, D.; Erho, N.; Yu, M.; Ashab, H.A.-D.; Alshalalfa, M.; Speers, C.; A Tomlins, S.; Davicioni, E.; et al. Development and validation of a 24-gene predictor of response to postoperative radiotherapy in prostate cancer: A matched, retrospective analysis. *Lancet Oncol.* **2016**, *17*, 1612–1620. [CrossRef]
324. Zhang, W.; van Gent, D.C.; Incrocci, L.; van Weerden, W.M.; Nonnekens, J. Role of the DNA damage response in prostate cancer formation, progression and treatment. *Prostate Cancer Prostatic Dis.* **2020**, *23*, 24–37. [CrossRef] [PubMed]
325. Cheng, H.H. The resounding effect of DNA repair deficiency in prostate cancer. *Urol. Oncol.* **2018**, *36*, 385–388. [CrossRef] [PubMed]
326. Nizialek, E.; Antonarakis, E.S. PARP Inhibitors in Metastatic Prostate Cancer: Evidence to Date. *Cancer Manag. Res.* **2020**, *12*, 8105–8114. [CrossRef]
327. Teyssonneau, D.; Margot, H.; Cabart, M.; Anonnay, M.; Sargos, P.; Vuong, N.-S.; Soubeyran, I.; Sevenet, N.; Roubaud, G. Prostate cancer and PARP inhibitors: Progress and challenges. *J. Hematol. Oncol.* **2021**, *14*, 51. [CrossRef] [PubMed]
328. Fujita, K.; Nonomura, N. Role of Androgen Receptor in Prostate Cancer: A Review. *World J. Mens Health* **2019**, *37*, 288–295. [CrossRef]
329. Crawford, E.D.; Heidenreich, A.; Lawrentschuk, N.; Tombal, B.; Pompeo, A.C.L.; Mendoza-Valdes, A.; Miller, K.; Debruyne, F.M.J.; Klotz, L. Androgen-targeted therapy in men with prostate cancer: Evolving practice and future considerations. *Prostate Cancer Prostatic Dis.* **2019**, *22*, 24–38. [CrossRef] [PubMed]
330. Coutinho, I.; Day, T.K.; Tilley, W.; Selth, L.A. Androgen receptor signaling in castration-resistant prostate cancer: A lesson in persistence. *Endocr. Relat. Cancer* **2016**, *23*, T179–T197. [CrossRef]
331. Karantanos, T.; Evans, C.P.; Tombal, B.; Thompson, T.C.; Montironi, R.; Isaacs, W.B. Understanding the mechanisms of androgen deprivation resistance in prostate cancer at the molecular level. *Eur. Urol.* **2015**, *67*, 470–479. [CrossRef]
332. Luo, J. Development of AR-V7 as a putative treatment selection marker for metastatic castration-resistant prostate cancer. *Asian J. Androl.* **2016**, *18*, 580. [CrossRef]
333. Antonarakis, E.S.; Lu, C.; Wang, H.; Luber, B.; Nakazawa, M.; Roeser, J.C.; Chen, Y.; Mohammad, T.A.; Chen, Y.; Fedor, H.L.; et al. AR-V7 and resistance to enzalutamide and abiraterone in prostate cancer. *N. Engl. J. Med.* **2014**, *371*, 1028–1038. [CrossRef]
334. Antonarakis, E.S.; Lu, C.; Luber, B.; Wang, H.; Chen, Y.; Zhu, Y.; Silberstein, J.L.; Taylor, M.N.; Maughan, B.L.; Denmeade, S.R.; et al. Clinical Significance of Androgen Receptor Splice Variant-7 mRNA Detection in Circulating Tumor Cells of Men With Metastatic Castration-Resistant Prostate Cancer Treated With First- and Second-Line Abiraterone and Enzalutamide. *J. Clin. Oncol.* **2017**, *35*, 2149–2156. [CrossRef]
335. Armstrong, A.J.; Halabi, S.; Luo, J.; Nanus, D.M.; Giannakakou, P.; Szmulewitz, R.Z.; Danila, D.C.; Healy, P.; Anand, M.; Rothwell, C.J.; et al. Prospective Multicenter Validation of Androgen Receptor Splice Variant 7 and Hormone Therapy Resistance in High-Risk Castration-Resistant Prostate Cancer: The PROPHECY Study. *J. Clin. Oncol.* **2019**, *37*, 1120–1129. [CrossRef] [PubMed]
336. Armstrong, A.J.; Luo, J.; Anand, M.; Antonarakis, E.S.; Nanus, D.M.; Giannakakou, P.; Szmulewitz, R.Z.; Danila, D.C.; Healy, P.; Berry, W.R.; et al. AR-V7 and prediction of benefit with taxane therapy: Final analysis of PROPHECY. *J. Clin. Oncol.* **2020**, *38* (Suppl. 6), 184. [CrossRef]
337. Scher, H.I.; Lu, D.; Schreiber, N.A.; Louw, J.; Graf, R.P.; Vargas, H.A.; Johnson, A.; Jendrisak, A.; Bambury, R.; Danila, D.; et al. Association of AR-V7 on Circulating Tumor Cells as a Treatment-Specific Biomarker With Outcomes and Survival in Castration-Resistant Prostate Cancer. *JAMA Oncol.* **2016**, *2*, 1441–1449. [CrossRef] [PubMed]
338. Sharp, A.; Welti, J.C.; Lambros, M.B.; Dolling, D.; Rodrigues, D.N.; Pope, L.; Aversa, C.; Figueiredo, I.; Fraser, J.; Ahmad, Z.; et al. Clinical Utility of Circulating Tumour Cell Androgen Receptor Splice Variant-7 Status in Metastatic Castration-resistant Prostate Cancer. *Eur. Urol.* **2019**, *76*, 676–685. [CrossRef]
339. Surdacki, G.; Szudy-Szczyrek, A.; Goracy, A.; Chyl-Surdacka, K.; Hus, M. The role of immune checkpoint inhibitors in prostate cancer. *Ann. Agric. Environ. Med.* **2019**, *26*, 120–124. [CrossRef]
340. Sharma, P.; Pachynski, R.K.; Narayan, V.; Fléchon, A.; Gravis, G.; Galsky, M.D.; Mahammedi, H.; Patnaik, A.; Subudhi, S.K.; Ciprotti, M.; et al. Nivolumab Plus Ipilimumab for Metastatic Castration-Resistant Prostate Cancer: Preliminary Analysis of Patients in the CheckMate 650 Trial. *Cancer Cell* **2020**, *38*, 489–499.e3. [CrossRef]

341. Fizazi, K.; Drake, C.G.; Beer, T.M.; Kwon, E.D.; Scher, H.I.; Gerritsen, W.R.; Bossi, A.; Eertwegh, A.J.V.D.; Krainer, M.; Houede, N.; et al. Final Analysis of the Ipilimumab Versus Placebo Following Radiotherapy Phase III Trial in Postdocetaxel Metastatic Castration-resistant Prostate Cancer Identifies an Excess of Long-term Survivors. *Eur. Urol.* **2020**, *78*, 822–830. [CrossRef]
342. Patel, S.P.; Kurzrock, R. PD-L1 Expression as a Predictive Biomarker in Cancer Immunotherapy. *Mol. Cancer Ther.* **2015**, *14*, 847–856. [CrossRef]
343. Zhang, T.; Agarwal, A.; Almquist, R.G.; Runyambo, D.; Park, S.; Bronson, E.; Boominathan, R.; Rao, C.; Anand, M.; Oyekunle, T.; et al. Expression of immune checkpoints on circulating tumor cells in men with metastatic prostate cancer. *Biomark. Res.* **2021**, *9*, 14. [CrossRef]
344. Wang, L.; Pan, S.; Zhu, B.; Yu, Z.; Wang, W. Comprehensive analysis of tumour mutational burden and its clinical significance in prostate cancer. *BMC Urol.* **2021**, *21*, 29. [CrossRef]
345. Ryan, M.J.; Bose, R. Genomic Alteration Burden in Advanced Prostate Cancer and Therapeutic Implications. *Front. Oncol.* **2019**, *9*, 1287. [CrossRef]
346. Goodman, A.M.; Kato, S.; Bazhenova, L.; Patel, S.P.; Frampton, G.M.; Miller, V.; Stephens, P.J.; Daniels, G.A.; Kurzrock, R. Tumor Mutational Burden as an Independent Predictor of Response to Immunotherapy in Diverse Cancers. *Mol. Cancer Ther.* **2017**, *16*, 2598–2608. [CrossRef] [PubMed]
347. Hellmann, M.D.; Nathanson, T.; Rizvi, H.; Creelan, B.C.; Sanchez-Vega, F.; Ahuja, A.; Ni, A.; Novik, J.B.; Mangarin, L.M.; Abu-Akeel, M.; et al. Genomic Features of Response to Combination Immunotherapy in Patients with Advanced Non-Small-Cell Lung Cancer. *Cancer Cell* **2018**, *33*, 843–852.e4. [CrossRef] [PubMed]
348. Blazek, D.; Kohoutek, J.; Bartholomeeusen, K.; Johansen, E.; Hulinkova, P.; Luo, Z.; Cimermancic, P.; Ule, J.; Peterlin, B.M. The Cyclin K/Cdk12 complex maintains genomic stability via regulation of expression of DNA damage response genes. *Genes Dev.* **2011**, *25*, 2158–2172. [CrossRef] [PubMed]
349. Wu, Y.-M.; Cieślik, M.; Lonigro, R.J.; Vats, P.; Reimers, M.A.; Cao, X.; Ning, Y.; Wang, L.; Kunju, L.P.; de Sarkar, N.; et al. Inactivation of CDK12 Delineates a Distinct Immunogenic Class of Advanced Prostate Cancer. *Cell* **2018**, *173*, 1770–1782.e14. [CrossRef] [PubMed]
350. Antonarakis, E.S.; Velho, P.I.; Fu, W.; Wang, H.; Agarwal, N.; Santos, V.S.; Maughan, B.L.; Pili, R.; Adra, N.; Sternberg, C.N.; et al. CDK12-Altered Prostate Cancer: Clinical Features and Therapeutic Outcomes to Standard Systemic Therapies, Poly (ADP-Ribose) Polymerase Inhibitors, and PD-1 Inhibitors. *JCO Precis. Oncol.* **2020**, *4*, 370–381. [CrossRef]
351. Salami, S.S.; Hovelson, D.H.; Kaplan, J.B.; Mathieu, R.; Udager, A.M.; Curci, N.E.; Lee, M.; Plouffe, K.R.; De La Vega, L.L.; Susani, M.; et al. Transcriptomic heterogeneity in multifocal prostate cancer. *JCI Insight* **2018**, *3*, 3. [CrossRef]
352. Wei, L.; Wang, J.; Lampert, E.; Schlanger, S.; DePriest, A.D.; Hu, Q.; Gomez, E.C.; Murakam, M.; Glenn, S.T.; Conroy, J.; et al. Intratumoral and Intertumoral Genomic Heterogeneity of Multifocal Localized Prostate Cancer Impacts Molecular Classifications and Genomic Prognosticators. *Eur. Urol.* **2017**, *71*, 183–192. [CrossRef]
353. Alam, S.; Tortora, J.; Staff, I.; McLaughlin, T.; Wagner, J. Prostate cancer genomics: Comparing results from three molecular assays. *Can. J. Urol.* **2019**, *26*, 9758–9762.
354. Lehto, T.K.; Stürenberg, C.; Malén, A.; Erickson, A.M.; Koistinen, H.; Mills, I.G.; Rannikko, A.; Mirtti, T. Transcript analysis of commercial prostate cancer risk stratification panels in hard-to-predict grade group 2–4 prostate cancers. *Prostate* **2021**, *81*, 368–376. [CrossRef] [PubMed]
355. Murphy, A.B.; Carbunaru, S.; Nettey, O.S.; Gornbein, C.; Dixon, M.A.; Macias, V.; Sharifi, R.; Kittles, R.A.; Yang, X.; Kajdacsy-Balla, A.; et al. A 17-Gene Panel Genomic Prostate Score Has Similar Predictive Accuracy for Adverse Pathology at Radical Prostatectomy in African American and European American Men. *Urology* **2020**, *142*, 166–173. [CrossRef] [PubMed]
356. Rayford, W.; Greenberger, M.; Bradley, R.V. Bradley, Improving risk stratification in a community-based African American population using cell cycle progression score. *Transl. Androl. Urol.* **2018**, *7* (Suppl. 4), S384–S391. [CrossRef] [PubMed]
357. Canter, D.J.; Reid, J.; Latsis, M.; Variano, M.; Halat, S.; Rajamani, S.; Gurtner, K.E.; Sangale, Z.; Brawer, M.; Stone, S.; et al. Comparison of the Prognostic Utility of the Cell Cycle Progression Score for Predicting Clinical Outcomes in African American and Non-African American Men with Localized Prostate Cancer. *Eur. Urol.* **2019**, *75*, 515–522. [CrossRef] [PubMed]
358. Creed, J.H.; Berglund, A.E.; Rounbehler, R.J.; Awasthi, S.; Cleveland, J.L.; Park, J.Y.; Yamoah, K.; Gerke, T.A. Commercial Gene Expression Tests for Prostate Cancer Prognosis Provide Paradoxical Estimates of Race-Specific Risk. *Cancer Epidemiol. Biomark. Prev.* **2020**, *29*, 246–253. [CrossRef] [PubMed]
359. Mohler, J.L.; Antonarakis, E.S.; Armstrong, A.J.; D'Amico, A.V.; Davis, B.J.; Dorff, T.; Eastham, J.A.; Enke, C.A.; Farrington, T.A.; Higano, C.S.; et al. Prostate cancer, version 2.2019, NCCN clinical practice guidelines in oncology. *J. Natl. Compr. Cancer Netw.* **2019**, *17*, 479–505. [CrossRef] [PubMed]

Review

Prognostic Genomic Tissue-Based Biomarkers in the Treatment of Localized Prostate Cancer

Gianluca Ingrosso [1,*], Emanuele Alì [1], Simona Marani [1], Simonetta Saldi [2], Rita Bellavita [2] and Cynthia Aristei [1]

1. Radiation Oncology Section, Department of Medicine and Surgery, Perugia General Hospital, University of Perugia, 06156 Perugia, Italy; emanuele.ali.06@gmail.com (E.A.); maranisimona@libero.it (S.M.); cynthia.aristei@unipg.it (C.A.)
2. Radiation Oncology Section, Perugia General Hospital, 06156 Perugia, Italy; saldisimonetta@gmail.com (S.S.); rita.bellavita@ospedale.perugia.it (R.B.)
* Correspondence: ingrosso.gianluca@gmail.com or gianluca.ingrosso@unipg.it; Tel.: +39-0755-783259

Simple Summary: In clinically localized prostate cancer, risk stratification (low-, intermediate- and high-risk) is crucial for the management of such a heterogenic disease, and it is based only on clinicopathologic features (i.e., baseline prostate-specific antigen (PSA), Gleason score and clinical stage of the tumor). New prognostic tools have been developed, mainly based on genomic tissue analysis. The aim of the present overview report is to focus on commercially available tissue-based biomarkers and more specifically on mRNA-based gene expression classifiers: Decipher (GenomeDX Biosciences), Prolaris (Myriad Genetics), and Oncotype Dx (Genomic Health). These new prognostic tests are going to be incorporated in clinicopathologic nomograms to better design the individualized treatment strategy for the cure of localized prostate cancer.

Abstract: In localized prostate cancer clinicopathologic variables have been used to develop prognostic nomograms quantifying the probability of locally advanced disease, of pelvic lymph node and distant metastasis at diagnosis or the probability of recurrence after radical treatment of the primary tumor. These tools although essential in daily clinical practice for the management of such a heterogeneous disease, which can be cured with a wide spectrum of treatment strategies (i.e., active surveillance, RP and radiation therapy), do not allow the precise distinction of an indolent instead of an aggressive disease. In recent years, several prognostic biomarkers have been tested, combined with the currently available clinicopathologic prognostic tools, in order to improve the decision-making process. In the following article, we reviewed the literature of the last 10 years and gave an overview report on commercially available tissue-based biomarkers and more specifically on mRNA-based gene expression classifiers. To date, these genomic tests have been widely investigated, demonstrating rigorous quality criteria including reproducibility, linearity, analytical accuracy, precision, and a positive impact in the clinical decision-making process. Albeit data published in literature, the systematic use of these tests in prostate cancer is currently not recommended due to insufficient evidence.

Keywords: localized prostate cancer; prognostic factors; tissue-based biomarkers

1. Introduction

In clinically localized prostate cancer (PCa), risk stratification (low-, intermediate- and high-risk) is based on baseline prostate-specific antigen (PSA), Gleason score and clinical stage of the tumor [1]. These clinicopathologic variables have been used to develop nomograms (e.g., Partin tables, Briganti nomogram) quantifying the probability of locally advanced disease (i.e., extracapsular extension, seminal vesicles involvement), and of pelvic lymph node and distant metastasis at diagnosis of localized PCa [2–4]. Some other calculators such as the Cancer of the Prostate Risk Assessment (CAPRA) or the Stephenson nomogram are used in the post-operative setting to predict the probability of recurrence after radical prostatectomy [5,6]. In the localized setting, these risk assessment tools are of pivotal importance for the management of PCa patients who can be cured with a wide spectrum of treatment strategies including active surveillance (AS), radical prostatectomy (RP) and radiation therapy (RT) [1,7]. Although essential in the daily clinical practice, nomograms based on clinical parameters do not allow the precise distinction of an indolent instead of an aggressive disease [8]. Prognostic biomarkers estimating the likelihood of an adverse outcome, combined with the currently available prognostic tools, might help in the decision-making process providing more tailored treatment for an individual patient. In recent years, several urine, blood, and tissue-based biomarkers have been introduced. In the following overview report, we focus on commercially available tissue-based biomarkers and more specifically on mRNA-based gene expression classifiers: Decipher (GenomeDX Biosciences, San Francisco, CA, USA), Prolaris (Myriad Genetics, Salt Lake City, UT, USA), and Oncotype Dx (Genomic Health, Redwood City, CA, USA).

2. Materials and Methods

We reviewed the current literature and gave an overview report on commercially available genomic tissue-based biomarkers in patients affected by localized PCa. We limited the scope of our search to Decipher, Prolaris, and Oncotype Dx because of the current availability of data in the literature about these genomic tests demonstrating rigorous quality criteria, including reproducibility, linearity, analytical accuracy, and precision [9,10]. We performed a PubMed literature search according to the preferred reporting items and meta-analysis (PRISMA) guidelines [11] of the available data for each selected biomarker. Keywords used were: "prostate cancer" or "prostatic cancer" or "prostatic carcinoma" or "prostate carcinoma" and "tissue-based biomarker" or "genetic tissue-based biomarker" or "genomic tissue-based biomarker" or "tissue-based markers". Our inclusion criteria were as follows: full articles in the English language published within the last 10 years up to 31 May 2021. Titles and abstracts were used to screen for initial study inclusion. Clinical studies published in English language journals were identified and screened for duplicates. Reference lists of the retrieved reports were also manually searched and cross-referenced to ensure completeness. Once a comprehensive list of abstracts has been retrieved and reviewed, any studies meeting inclusion criteria were obtained and reviewed in full. Reviews, commentaries, letters, and conference abstracts were excluded. Two authors (E.A. and S.M.) independently performed the study selection. Disagreements were resolved by consensus with two authors (G.I. and C.A.). We reviewed the full version of each article. The flowchart of the systematic review is reported in Figure 1. Data extraction was completed independently by two reviewers (E.A. and S.M.) to establish inter-rater reliability using a standardized form to obtain: (1) general information, author name, year and type of publication, literature source; (2) clinical data, including number of patients, patients' subset, analyzed tissue-type, and follow-up; (3) study endpoints and statistical methods. Disagreements were resolved by discussion and re-review of the literature. Data were summarized in evidence tables and described in the text.

For risk of bias assessment, we used the star-based Newcastle Ottawa Scale (NOS) (Table S1). A maximum of one star can be given for each item, except for comparability, for which one or two stars can be given. The risk of bias was considered as low, intermediate, or high for the scores $\geq 7–9$, 4–6, and <4, respectively (Table S1, Supplemental material).

Data about the predictive power of each single tissue-based biomarker were extracted and reported in tables. These included concordance index (c-index), which is a measure of goodness of fit for binary outcomes in a logistic regression model, corresponding to the area under the receiver operating characteristic (ROC) curve (ranges from 0.5 to 1). Hazard ratios (HRs) were also used in several studies to define the prediction of biochemical recurrence or prostate cancer specific survival. Eventually, we performed the weighted average of c-indices and of HRs to summarize results.

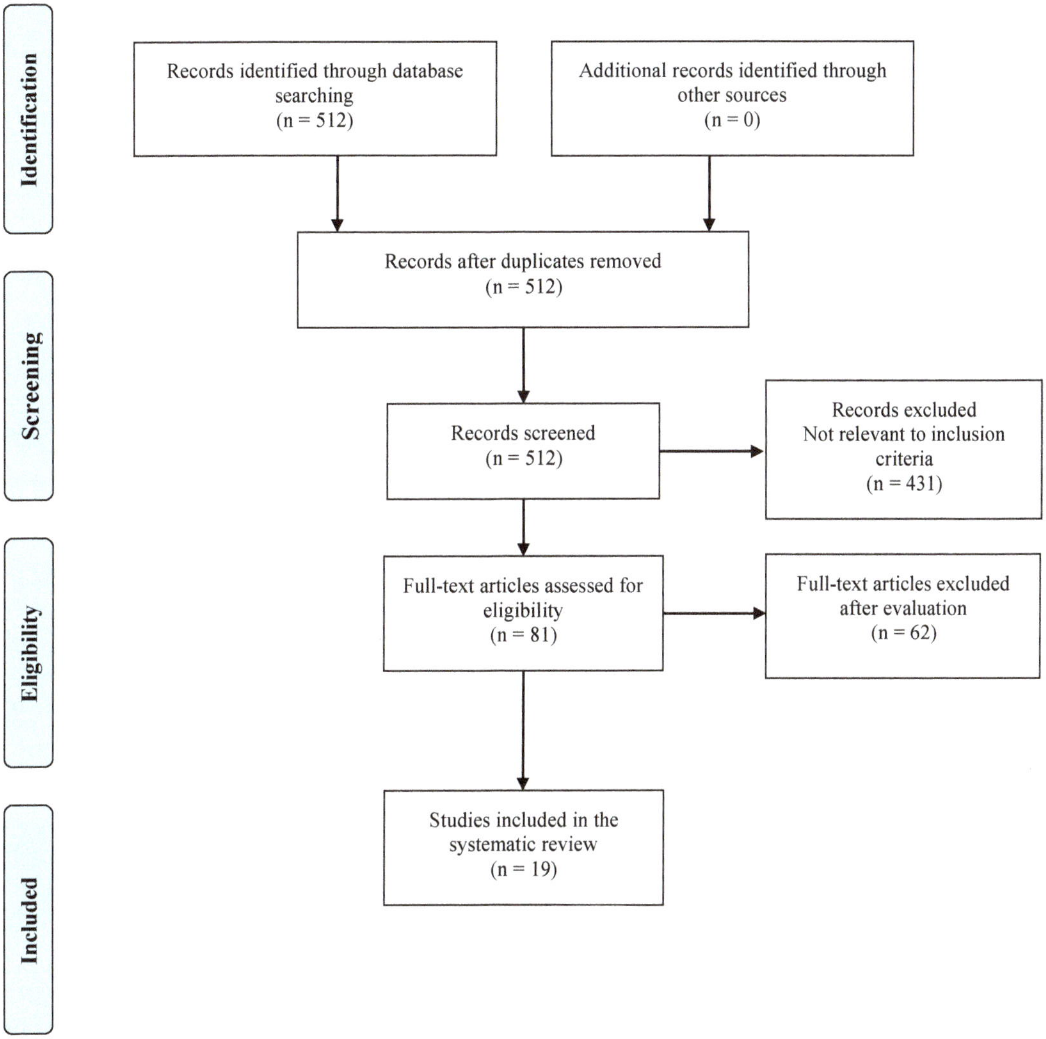

Figure 1. Flowchart of inclusion of studies in the systematic review.

3. Data Synthesis

3.1. Decipher

Decipher (Decipher Biosciences, San Diego, CA, USA) is a genomic classifier (GC) of a 22-gene panel predicting the probability of metastatic progression after primary treatment for localized PCa (Table 1). It is a tissue-based assay obtained from formalin-fixed paraffin embedded (FFPE) primary prostate cancer [12], which was developed using a high-density transcriptome-wide microarray analysis.

More specifically, PCa cancer tissue specimen from 545 patients undergone RP at Mayo Clinic between 1987 and 2001 were analyzed profiling the expression of about 1.4 million RNA features. After training and validation sets, selected features were assembled into a classifier using a random forest algorithm. The final GC was based on the expression of 22 RNA biomarkers involved in cell proliferation and differentiation, cell cycle progression, androgen receptor signaling, cell structure and adhesion, immune response (LASP1,

IQGAP3, NFIB, S1PR4, THBS2, ANO7, PCDH7, MYBPC1, EPPK1, TSBP, PBX1, NUSAP1, ZWILCH, UBE2C, CAMK2N1, RABGAP1, PCAT-32, GLYATL1P4, PCAT-80, TNFRSF19). The analysis performed after a median follow-up of 16.9 years demonstrated that the GC could better predict metastasis onset than clinical variables alone, and that it could be used as a prognostic tool assessing the probability of systemic progression after primary treatment with a score range between 0 and 1, patients with a score > 0.6 having a high-risk of developing metastatic disease.

Table 1. Tissue-based mRNA genomic classifiers.

Tissue Biomarker	Laboratory	Tested Genes	Score Report	Clinical Use
Decipher 22 genes	GenomeDx (San Diego, CA, USA)	LASP1, IQGAP3, NFIB, S1PR4, THBS2, ANO7, PCDH7, MYBPC1, EPPK1, TSBP, PBX1, NUSAP1, ZWILCH, UBE2C, CAMK2N1, RABGAP1, PCAT-32, GLYATL1P4, PCAT-80, TNFRSF19	GC score: 0–1	Post-RP: to predict the probability of disease recurrence after primary treatment. At localized PCa diagnosis: to categorize patients into risk groups and better define for AS vs. treatment and treatment intensification.
Prolaris 31 genes	Myriad Gentics (Salt Lake City, UT, USA)	FOXM1, CDC20, CDKN3, CDC2, KIF11, KIAA0101, NUSAP1, CENPF, ASPM, BUB1B, RRM2, DLGAP5, BIRC5, KIF20A, PLK1, TOP2A, TK1, PBK, ASF1B, C18orf24, RAD54L, PTTG1, CDCA3, MCM10, PRC1, DTL, CEP55, RAD51, CENPM, CDCA8, ORC6L	CCP score: 0–6	To predict the risk of metastasis and CSM. To better define for treatment after primary therapy.
Oncotype Dx 17 genes	Genomic Health, Redwood City, CA, USA	ARF1, ATP5E, CLTC, GPS1, PGK1, AZGP1, KLK2, SRD5A2, FAM13C, FLNC, GSN, TPM2, GSTM2, TPX2, BGN, COL1A1, SFRP4	GPS score: 0–100	To predict the risk of adverse pathological features (EPE, SVI) after RP.

RP: radical prostatectomy; PCa: prostate cancer; AS: active surveillance; CSM: cancer-specific survival; EPE: extra-prostatic extension; SVI: seminal vesicles involvement.

In the post-RP setting, Decipher has been tested by several researchers to predict the development of distant metastases (Table 2) compared with clinical variables and with clinical-derived nomograms such as the Stephenson and the CAPRA. Many of the studies are retrospective and monocentric focusing on the subset of high-risk disease, like the one by Karnes et al. [13] evaluating the efficacy of Decipher in the 5-year metastasis prediction after RP in 219 high-risk patients, compared with clinical variables. Using survival receiver operating characteristic (ROC) curves assessing classifier discrimination, the c-index for Decipher was 0.79 (95%CI, 0.68–0.87) outperforming clinical variables.

In the study by Karnes [13] (Table 2) as well as in the others reported in the literature [12–19], Decipher was the predominant predictor of metastasis at multivariate analysis. In the series by Thomas Jefferson University [14], the GC has been retrospectively tested in 139 patients affected by adverse risk factors (pT3 stage or positive margins) after RP and treated with post-operative radiotherapy. The c-index for distant metastasis endpoint was 0.78 (95%CI, 0.64–0.91) for Decipher compared with 0.70 (95%CI, 0.49–0.90) for the post-RP Stephenson model and 0.65 (95%CI, 0.44–0.86) for CAPRA-S. Stratifying patients by the three GC risk-groups (low-risk: <0.4; intermediate-risk: 0.4–0.6; high-risk: >0.6), the 8-years cumulative incidence of distant metastases was 0%, 12% and 17%, respectively ($p = 0.032$). Eventually, high-risk (score > 0.6) patients with undetectable PSA (≤ 0.2 ng/mL) before post-operative RT had a distant metastasis cumulative incidence of 3% compared with a rate of 23% for those with detectable PSA ($p = 0.03$).

Taking into account all the c-indices of retrospective studies on the post-operative setting, the weighted average of c-indices was 0.77.

Table 2. Decipher studies.

	Study Type	No of Pts	Setting	Tissue Type	Disease State	Median Fu	Endpoint	Decipher c-Index (95%CI)
Erho 2013 [12]	Retrospective, nested-case control (Mayo Clinic)	545	Post-RP	RP	All risk classes	16.9 year	Metastasis prediction	0.75 (0.67–0.83)
Karnes 2013 [13]	Retrospective (Mayo Clinic)	219	Post-RP	RP	High-risk	6.7 year	5-year metastasis prediction compared with clin. variables	0.79 (0.68–0.87)
Cooperberg 2014 [14]	Retrospective (Mayo Clinic)	185	Post-RP	RP	High-risk	6.4 year	PCSM prediction compared with CAPRA-S	0.78 (0.68–0.87)
Ross 2014 [15]	Retrospective (Mayo Clinic)	85	BCR after RP	RP	High risk with BCR	NA	Metastasis prediction compared with clin. variables, CAPRA-S and Stephenson	0.82 (0.77–0.86)
Den 2014 [16]	Retrospective (Thomas Jefferson University)	139	Post-RP + PORT	RP	adverse risk factors after RP	NA	Metastasis and BCR prediction compared with CAPRA-S and Stephenson	0.78 (0.64–0.91)
Klein 2015 [17]	Retrospective (Cleveland Clinic)	169	Post-RP	RP	High-risk	NA	5-year metastasis prediction compared with CAPRA-S and Stephenson	0.77 (0.66–0.87)
Ross 2016 [18]	Retrospective (John Hopkins)	260	Post-RP	RP	Intermediate and high-risk	9 year	Metastasis prediction	0.76 (0.66–0.84)
Den 2015 [19]	Retrospective (Bi-institutional)	188	Post-RP + PORT	RP	adverse risk factors after RP	8 year	Metastasis prediction compared with CAPRA-S	0.83 (0.27–0.89)

RP: radical prostatectomy; PORT: post-operative radiotherapy; PCSM: prostate cancer-specific survival; BCR: biochemical recurrence; NA: not available.

3.2. Decipher Role in Clinical Practice

After RP, the use of effective prognostic tools in clinical practice might have an important impact on decision-making for therapy intensification based on the estimated risk of disease recurrence. For instance, Gore et al. [20] prospectively evaluated the effect of Decipher on treatment recommendation in the adjuvant (ART) and salvage (SRT) settings, revealing that high Decipher score was associated with treatment intensification. Eventually, the GC score was an independent predictor for change in management for ART and SRT, at multivariate analysis [20].

Decipher has also been tested in localized PCa to improve prognostication for primary treatment decision-making.

Recently, the applicability of this genomic test in biopsy-derived tissue has been demonstrated, with a high correlation between information derived from RP and biopsy specimens [21]. Klein et al. [17] evaluated at eight years of follow-up the ability of the GC in predicting metastasis from needle biopsy-derived tumor tissue of 57 patients affected by localized PCa (Table 2). The combination of Decipher and National Comprehensive Cancer Network (NCCN) predictive models had an improved c-index of 0.88 (95%CI, 0.77–0.96) compared to NCCN alone (C-index 0.75, 95%CI 0.64–0.87). On multivariate analysis, the GC was the only significant predictor of metastasis when adjusting for age, preoperative PSA and biopsy Gleason score [17].

In 2018, Spratt et al. [22] proposed a three-tier clinical-genomic risk grouping system of distant metastasis and PCa-specific mortality (PCSM) based on genomic and clinicopathological features, demonstrating that Decipher consistently improves prognostic performance over clinicopathological (NCCN classification, and CAPRA-score) variables alone. On a total cohort of 6928 patients studied for development and validation of the prognostic scoring system, c-indices for the three-tier (low-, intermediate-, and high-risk) clinical-genomic risk grouping system significantly outperformed those of NCCN and CAPRA, with 30% of patients being reclassified.

Testing Decipher on biopsy cores, a multi-institutional study on 855 patients affected by localized PCa showed that a high-risk GC score was independently associated with shorter time to treatment in those undergone AS, and with a worse time to failure in those undergone radical therapy [22].

3.3. Prolaris

The Prolaris (Myriad Genetics, Salt Lake City, UT, USA) is a multigene test commercially available in the USA and in Europe which uses prostate tissue samples from biopsy or prostatectomy to give prognostic information about PCa (Table 1). It measures the expression of 31 cell cycle progression (CCP) genes with a score range from 0 to 10, a high score correlating with tumor aggressiveness and with the risk of progression. Cuzick et al. first tried to build a CCP score by a gene signature in order to improve PCa patients' stratification risk [23]. The rationale for the development of such a CCP score is based on the assumption that the measurement of actively growing cells (showing high CCP score) within a tumor gives information about disease aggressiveness and prognosis [10,23].

In the study by Cuzick et al. [23], 126 CCP genes from the Gene Expression Omnibus database were tested on 96 commercially available FFPE PCa sections, creating a gene signature with 31 selected cell cycle genes (FOXM1, CDC20, CDKN3, CDC2, KIF11, KIAA0101, NUSAP1, CENPF, ASPM, BUB1B, RRM2, DLGAP5, BIRC5, KIF20A, PLK1, TOP2A, TK1, PBK, ASF1B, C18orf24, RAD54L, PTTG1, CDCA3, MCM10, PRC1, DTL, CEP55, RAD51, CENPM, CDCA8, and ORC6L), and a predefined score of disease outcome prediction. The genetic signature was then assessed retrospectively in two localized PCa patients' cohorts (366 patients undergone RP, and 337 patients with PCa diagnosis performed by transurethral resection of the prostate (TURP) undergone watchful waiting).

In the post-prostatectomy setting, the increase in hazard ratio (HR) for a 1-unit change in CCP score proved to be predictive of biochemical recurrence in both univariate (HR 1.89 95%CI 1.54–2.31, $p = 5.6 \times 10^{-9}$) and multivariate analysis (HR 1.77 95%CI 1.40–2.22, $p = 4.3 \times 10^{-6}$). Similarly, in the TURP setting the CCP score was strictly related to cancer specific survival (HR 2.57 95%CI 1.93 to 3.43, $p = 8.2 \times 10^{-11}$) [23]. Up to date, several research works have focused on the use of CCP as a prognostic tool for PCa management and it has been tested on tissue samples deriving not only from RP or TURP but also from biopsies. Bishoff et al. [20–24] retrospectively tested the CCP score on biopsy specimens from three cohorts (283 patients from Martini Clinic, 176 from Durham Veterans Affairs Medical and 123 from Intermountain Healthcare), with a total of 582 localized PCa patients treated with RP (Table 3). For each cohort, at multivariate analysis the CCP score proved to be a strong predictor of biochemical recurrence and metastatic disease (Table 3). One of the limitations of this study was the use, in one of the three cohorts, of simulated biopsies resulting from post-operative tissue blocks. However, the combined analysis carried out excluding this cohort confirmed the CCP score as a strong predictor of biochemical recurrence (BCR), both at univariate (HR 1.45, 95%CI 1.18–1.79, $p = 5.7 \times 10^{-4}$) and multivariate analyses (HR 1.40, 95%CI 1.1–1.74, $p = 0.0032$). A strong association was also found for metastatic disease but only on univariate analysis (HR 4.69, 95%CI 2.28–9.64, $p = 1.6 \times 10^{-5}$) [24].

Similar results in terms of BCR prediction were reported by Freedland et al. evaluating the Prolaris test on biopsy samples from 141 localized PCa patients treated with external beam radiotherapy as primary curative therapy (Table 3). The authors obtained a strong correlation between high CCP score and biochemical recurrence (HR for BCR of 2.55 for 1-unit increase in CCP score), which was confirmed at multivariate analysis after adjustments for pretreatment PSA level, Gleason, percent positive cores, and concurrent androgen deprivation therapy (Table 3) [25].

As the stratification risk of localized PCa patients is mainly based on clinical parameters such as preoperative PSA, pathologic Gleason score and pathologic parameters, several authors tried to find an association between the Prolaris test and clinical nomograms such as CAPRA score in order to improve the therapeutic decision-making process. Cooperberg et al. showed in 413 patients undergone RP the usefulness of CCP score to stratify patients with low clinical risk defined by CAPRA score ≤ 2 (HR 2.3, 95%CI 1.4–3.7), moreover they validated a combined CAPRA + CCP score that proved to be more predictive than the CAPRA score alone ($p < 0.001$) [26]. Another validation study conducted on biopsy samples from 585 patients affected by localized PCa reported at multivariate analysis adjusted for clinical parameters a strong correlation of CCP score with cancer-specific survival evaluated as primary endpoint (HR 2.17, 95%CI 1.83–2.57, $p < 0.0001$) (Table 3). The authors also validated the clinical-cell-cycle-risk (CCR) score, which resulted from a combination of CAPRA and CCP score showing a strict relation with death from prostate cancer (HR for a 1-unit change in CCR score 2.17, 95%CI 1.83–2.57; $p = 4.1 \times 10^{-21}$) [27]. Recently, Canter et al. evaluated CCP and CCR scores as predictive factors for clinical outcomes after prostate cancer treatment. They analyzed 4 cohorts with 1062 patients undergone RP, using biopsy or simulated biopsy samples for Prolaris testing. The authors showed that CCP and CCR score were strictly related to progression to metastatic disease at univariate and at multivariate analysis adjusted for all significant variables (HR for a 1-unit change in CCP score 2.21, 95%CI 1.64–2.98, $p = 1.9 \times 10^{-6}$; HR for a 1-unit change in CCR score 3.63, 95%CI 2.60–5.05, $p = 2.1 \times 10^{-16}$) [28]. Regarding BCR, the weighted average of HRs of available studies was 1.68.

Table 3. Prolaris studies.

	Study Type	No of Pts	Setting	Tissue Type	Median Fu	Endpoint	CCP Results
Cuzick 2011 [23]	Retrospective monocentric	366 337	Post-RP Post-TURP	RP TURP	NA	BCR CSS	MVA: HR for a 1-unit change in CCP score 1.77 95%CI 1.40–2.22, $p = 4.3 \times 10^{-6}$ MVA: HR for a 1-unit change in CCP score 2.56 95%CI 1.85–3.53, $p = 1.3 \times 10^{-8}$
Bishoff 2014 [24]	Retrospective multicentric	582	Clinically localized	Biopsy	61-88 mo	BCR DMS	MVA: HR 1.47 95%CI 1.23–1.76, $p < 10^{-4}$ MVA: HR 4.19 95%CI 2.08–8.45, $p < 10^{-5}$
Freedland 2013 [25]	Retrospective monocentric	141	Clinically localized	Biopsy		BCR	MVA: HR for a 1-unit change in CCP score 2.11, 95%CI 1.04–4.25, $p < 0.034$
Cuzick 2015 [27]	Retrospective multicentric	585	Clinically localized	Biopsy	9.52 mo	PCSM	MVA adjusted for CAPRA score: HR for a 1-unit change in CCR score 2.17, 95%CI 1.83–2.57; $p = 4.1 \times 10^{-21}$
Cooperberg 2013 [26]	Retrospective multicentric	413	Post-RP	RP	85 mo	Biochemical/clinical recurrence	MVA adjusted for CAPRA score: HR for a 1-unit change in CCP score 1.7, 95%CI 1.3–2.3; $p < 0.001$
Canter 2020 [28]	Retrospective multicentric	1062	Post-RP	Biopsy or simulated biopsy		Progression to metastatic disease	MVA adjusted for CAPRA score: HR for a 1-unit change in CCP score 2.21, 95%CI 1.64–2.98; $p = 1.9 \times 10^{-6}$

RP: radical prostatectomy; CSS: cancer-specific survival; BCR: biochemical recurrence; PCSM: prostate cancer-specific survival; MVA: multivariate analysis; NA: not available.

3.4. Prolaris Role in Clinical Practice

Many studies support the hypothesis that Prolaris test could be used to stratify PCa patients and to guide the therapeutic approach. Several outcomes have been tested, such as BCR and distant metastasis survival (DMS) and a strict association with CCP score has been reported. Prolaris was evaluated on different specimen types (biopsy or RP samples) with no reported differences in terms of predictive utility of the test, although it seems of pivotal importance to adequately perform prostate biopsy in order to reduce the risk of under-sampling errors, which might affect the GC validity [29,30].

In localized PCa, Prolaris might be used as a treatment decision-making tool for primary therapy. For instance, it might help in the better definition of low-risk patients otherwise defined as intermediate or high-risk according to purely clinical variables such as Gleason score, PSA, Ki67 or CAPRA score and this is crucial for the therapeutic decision-making process because an active treatment could be avoided, limiting adverse events and improving patients' quality of life without neglecting the curative intent.

3.5. Oncotype

Oncotype DX (Genomic Health, Redwood City, CA, USA) platform is made up of multi-gene real-time polymerase chain reaction (RT-PCR) assays (Oncotype DX® Assays) used in the treatment-decision process for patients affected by breast or colon cancer. It was first evaluated on a retrospective series of hormone-responsive breast cancer patients with negative lymph nodes, randomly assigned to placebo vs. tamoxifen or tamoxifen vs. cyclophosphamide, methotrexate, fluorouracil, and tamoxifen (CMFT) [31,32]. The risk (i.e., low-, intermediate- and high-risk) of relapse in these 2 cohorts was evaluated based on the expression level of 21 genes on reverse transcriptase-polymerase chain reaction (RT-PCR) from tissue blocks of the primary tumor [32].

In prostate cancer, Oncotype DX (Table 1) integrates with traditional clinical and pathological diagnostic features (PSA, Gleason score, cTNM) in order to better discriminate between indolent and aggressive disease. Compared with the Oncotype DX platform used for breast cancer, the multiplex preamp step has been introduced to create more copies of the initial RNA prior to quantitative evaluation of gene expression allowing the processing of very small samples (5 μm sections). The test analyzes the expression of 17 genes (five housekeeping genes and 12 genes related to prostate cancer) through RT-PCR on fixed, FFPE prostate needle biopsy tissue. The five housekeeping genes (ARF1, ATP5E, CLTC, GPS1, and PGK1) were selected for their low inter-patient variability, lack of relationship to clinical outcome, and robust analytical performance. The 12 cancer-related genes belong to four distinct biological pathways with a role in prostate tumorigenesis: the androgen pathway (AZGP1, KLK2, SRD5A2, and FAM13C), cell organization (FLNC, GSN, TPM2, and GSTM2), proliferation (TPX2) and stromal response (BGN, COL1A1 and SFRP4) (Table 1). The combination of the expression of these genes is used to calculate the Genomic Prostate Score (GPS), which ranges from 0 to 100.

Knezevic et al. demonstrated that Oncotype DX is able to reliably and accurately measure gene expression over a wide range of PCa populations and using very small amounts of RNA, based on the average amplification efficiency of the 17 gene tests (93%), a high analytical sensitivity, and a wide linear range and low bias (less than 9.7%) [33].

In localized PCa, Oncotype DX has been clinically validated to predict the risk of disease recurrence and of prostate cancer death [34]. In the clinical validation study, Oncotype DX was tested using three cohorts of patients: a prostatectomy ($n = 441$), a biopsy ($n = 167$), and a prospectively designed, independent clinical validation cohort ($n = 395$). GPS predicted high-grade (odds ratio [OR] per 20 GPS units: 2.3; 95% confidence interval [CI], 1.5–3.7; $p < 0.001$) and high-stage (OR per 20 GPS units: 1.9; 95%CI, 1.3–3.0; $p = 0.003$) at RP pathology [34]. Cullen et al. demonstrated that Oncotype DX predicts time to biochemical recurrence at univariate analysis (hazard ratio per 20 GPS units [HR/20 units]: 2.9; $p < 0.001$) and time to metastases (HR/20 units: 3.8; $p = 0.032$) after primary treatment [35]. In a RP retrospective cohort of 279 localized PCa patients, Van Den Eeden et al. assessed the association between Oncotype DX and time to metastases and PCSM (Table 4). Analyzing a total of 259 GPS valid results, they demonstrated a strong correlation of GPS score with the two endpoints. Moreover, at the time of analysis (median follow-up 9.8 years) no patient with low- or intermediate risk and GPS score < 20 developed metastases or died of PCa. At ROC analysis, the combination of Oncotype DX with CAPRA score significantly improved the c-statistic of CAPRA alone from 0.65 to 0.73 for metastasis prediction and from 0.78 to 0.84 for CSM [36].

Table 4. Oncotype DX studies.

	Study Type	No of Pts	Setting	Tissue Type	Disease State	Median Fu	Endpoint	Oncotype DX AUC at ROC Curve
Klein 2014 [34]	Retrospective	441	Post-RP	Biopsy	All risk classes	NA	Clinical recurrence, adverse pathology, PCSM	NA
Van Den Eeden 2018 [36]	Retrospective	279	Post-RP	Biopsy	All risk classes	9.8 year	Metastasis and PCSM prediction compared with clinical variables only	Metastasis: 0.73 PCSM: 0.84
Brooks 2021 [37]	Retrospective	428	Post-RP	RP index lesion	All risk classes	15.5 year	Metastasis and PCSM prediction compared with clinical variables only	Metastasis: 0.82 PCSM: 0.82
Covas Moschovas 2021 [38]	Retrospective	749	Post-RP	biopsy	All risk classes	Median time between GPS test and RP: 176 days	Prediction of adverse pathology features (EPE, PSM, SVI) compared with clinical variables only	EPE: 0.70 SVI: 0.78 PCSM: not improved
Cullen 2021 [35]	Retrospective	431	Post-RP	Biopsy	Low-, intermediate-risk	5.2 year	BCR, adverse pathology	Adverse pathology: 0.72 BCR: 0.68

RP: radical prostatectomy; BCR: biochemical recurrence; PCSM: prostate cancer-specific survival; NA: not available.

In a retrospective analysis testing GPS on 428 patients undergone RP between 1987 and 2004, GPS score was significantly associated with the risk of distant metastasis and CSM at 20 years of follow-up. Eventually, GPS score < 20 indicated a low risk of both outcomes, whereas a score > 40 indicated a high-risk of developing distant metastases and of dying of PCa [37]. For PCSM, the weighted average of AUC at ROC curves was 0.81 and for adverse pathology after RP was 0.76.

3.6. Oncotype DX Role in Clinical Practice

Oncotype DX has been used to better discriminate between indolent and aggressive disease in the primary setting as well as in the post-operative setting. For instance, Moschovas et al. investigated the capability of Oncotype DX in predicting adverse pathological features (ie extraprostatic extension (EPE), positive surgical margin (PSM) and seminal vesicle invasion (SVI)) in patients treated with RP for localized PCa (Table 4). Multivariate analysis assessing the odds ratio per 20-points change in Oncotype DX genomic score showed that GPS is an independent predictor of adverse pathological features after RP, and specifically for EPE and SVI. At ROC analysis, GPS score did not increase the area under the curve (AUC) of PCSM [38].

Recently, Oncotype DX has been tested as a predictor of outcome for patients in active surveillance. In the PASS trial, GPS scores available in 432 patients were evaluated for the association with adverse pathological features at RP after a period of active surveillance. At the time of analysis, on total of 101 RP with central pathology review Oncotype DX was not significantly associated with adverse pathological features neither with upgrading in surveillance biopsy [39].

4. Conclusions

A prognostic biomarker must estimate the likelihood of a disease characteristic being present or absent more accurately determining the prognosis. Molecular information, providing specific insights into the underlying tumor biology, combined with clinicopathologic features might improve the decision-making process and clinical outcomes. In recent years, tissue-based mRNA-GC have been widely investigated as new tools for localized PCa prognosis. More specifically, they have been tested in the context of newly diagnosed prostate cancer and of surgically treated patients, in order to better define risk stratification and to guide clinical management especially in borderline scenarios such as AS for specific subsets of localized PCa patients or treatment intensification after RP.

In our systematic review, Decipher, Prolaris, and Oncotype Dx, which are commercially available tissue-based biomarkers demonstrating rigorous quality criteria, seem to be reliable prognostic tools for the prediction of biochemical recurrence or prostate cancer specific survival. Despite advances in tissue-based mRNA-GC validation and data published in literature, the systematic use of these tests in prostate cancer is currently not recommended due to insufficient evidence. About validation, many of the results are based on White Caucasian cohorts. About data published in the literature, evidence of efficacy derives from retrospective monocentric studies with short median follow-up and low number of events. Prospective randomized trials are needed for the safe and effective use of these tools in clinical practice [10]. Moreover, tissue-based biomarkers results have an important intrinsic limitation coming from their dependence on the sampled tumor, showing PCa multifocality and high intratumoral heterogeneity [30]. To date, no comparison studies between tissue-based GCs have been published and there is currently uncertainty regarding the potential specific role of each available biomarker. The American Society of Clinical Oncology (ASCO) guidelines on molecular biomarkers in localized PCa, which have been recently published, recommend the use of tissue-based biomarkers "only in situations in which a specific assay result, when considered in combination with routine clinical factors, will clearly affect the management decision" [40].

In the future, it is likely these prognostic biomarkers will be incorporated in clinicopathologic nomograms to better design the personalized diagnostic and treatment strategy for each single patient.

Supplementary Materials: The following are available online at https://www.mdpi.com/article/10.3390/jpm12010065/s1, Table S1: Newcastle Ottawa Scale for risk of bias assessment of the included studies (scores ≥ 7–9, 4–6, <4 are considered as low, intermediate, and high risk, respectively).

Author Contributions: Conceptualization, C.A. and G.I.; methodology, G.I. and E.A.; resources, S.S., S.M. and R.B.; draft preparation, G.I., E.A. and S.M.; review and editing, G.I. and C.A.; visualization, R.B. and S.S.; supervision, C.A. and G.I. All authors have read and agreed to the published version of the manuscript.

Funding: This research received no external funding.

Data Availability Statement: Not applicable.

Conflicts of Interest: The authors declare no conflict of interest.

References

1. Mottet, N.; van den Bergh, R.C.N.; Briers, E.; Van den Broeck, T.; Cumberbatch, M.G.; De Santis, M.; Fanti, S.; Fossati, N.; Gandaglia, G.; Gillessen, S.; et al. Guidelines on Prostate Cancer-2020 Update. Part 1: Screening, Diagnosis, and Local Treatment with Curative Intent. *Eur. Urol.* **2021**, *79*, 243–262. [CrossRef]
2. Partin, A.W.; Yoo, J.; Carter, H.B.; Pearson, J.D.; Chan, D.W.; Epstein, J.I.; Walsh, P.C. Clinical stage and Gleason score to predict pathological stage in men with localized prostate cancer. *J. Urol.* **1993**, *150*, 110–114. [CrossRef]
3. Tosoian, J.J.; Chappidi, M.; Feng, Z.; Humphreys, E.B.; Han, M.; Pavlovich, C.P.; Epstein, J.I.; Partin, A.W.; Trock, B.J. Prediction of pathological stage based on clinical stage, serum prostate-specific antigen, and biopsy Gleason score: Partin Tables in the contemporary era. *BJU Int.* **2017**, *119*, 676–683. [CrossRef]

4. Briganti, A.; Larcher, A.; Abdollah, F.; Capitanio, U.; Gallina, A.; Suardi, N.; Bianchi, M.; Sun, M.; Freschi, M.; Salonia, A.; et al. Updated nomogram predicting lymph node invasion in patients with prostate cancer undergoing extended pelvic lymph node dissection: The essential importance of percentage of positive cores. *Eur. Urol.* **2012**, *61*, 480–487. [CrossRef] [PubMed]
5. University of California San Francisco. Prostate Cancer Risk Assessment and the UCSF-CAPRA Score. Available online: https://urology.ucsf.edu/research/cancer/prostate-cancer-risk-assessment-and-the-ucsf-capra-score (accessed on 10 September 2021).
6. Stephenson, A.J.; Scardino, P.T.; Eastham, J.A.; Bianco, F.J., Jr.; Dotan, Z.A.; DiBlasio, C.J.; Reuther, A.; Klein, E.A.; Kattan, M.W. Postoperative Nomogram Predicting the 10-Year Probability of Prostate Cancer Recurrence After RP. *J. Clin. Oncol.* **2005**, *23*, 7005–7012. [CrossRef] [PubMed]
7. National Comprehensive Cancer Network 2021. Available online: https://www.nccn.org/professionals/physician_gls/pdf/prostate.pdf (accessed on 10 September 2021).
8. Wang, S.Y.; Cowan, J.E.; Cary, K.C.; Chan, J.M.; Carroll, P.R.; Cooperberg, M.R. Limited ability of existing nomograms to predict outcomes in men undergoing active surveillance for prostate cancer. *BJU Int.* **2014**, *114*, E18–E24. [CrossRef] [PubMed]
9. Canfield, S.E.; Kibel, A.S.; Kemeter, M.J.; Febbo, P.G.; Lawrence, H.J.; Moul, J.W. A guide for clinicians in the evaluation of emerging molecular diagnostics for newly diagnosed prostate cancer. *Rev. Urol.* **2014**, *16*, 172–180.
10. Lamy, P.J.; Allory, Y.; Gauchez, A.S.; Asselain, B.; Beuzeboc, P.; de Cremoux, P.; Fontugne, J.; Georges, A.; Hennequin, C.; Lehmann-Che, J.; et al. Prognostic Biomarkers Used for Localised Prostate Cancer Management: A Systematic Review. *Eur. Urol. Focus* **2018**, *4*, 790–803. [CrossRef]
11. Moher, D.; Liberati, A.; Tetzlaff, J.; Altman, D.G. Preferred reporting items for systematic reviews and meta-analyses: The PRISMA statement. *Ann. Intern. Med.* **2009**, *151*, 264–269. [CrossRef]
12. Erho, N.; Crisan, A.; Vergara, I.A.; Mitra, A.P.; Ghadessi, M.; Buerki, C.; Bergstralh, E.J.; Kollmeyer, T.; Fink, S.; Haddad, Z.; et al. Discovery and validation of a prostate cancer genomic classifier that predicts early metastasis following RP. *PLoS ONE* **2013**, *8*, e66855. [CrossRef]
13. Karnes, R.J.; Bergstralh, E.J.; Davicioni, E.; Ghadessi, M.; Buerki, C.; Mitra, A.P.; Crisan, A.; Erho, N.; Vergara, I.A.; Lam, L.L.; et al. Validation of a genomic classifier that predicts metastasis following RP in an at risk patient population. *J. Urol.* **2013**, *190*, 2047–2053. [CrossRef]
14. Cooperberg, M.R.; Davicioni, E.; Crisan, A.; Jenkins, R.B.; Ghadessi, M.; Karnes, R.J. Combined value of validated clinical and genomic risk stratification tools for predicting prostate cancer mortality in a high-risk prostatectomy cohort. *Eur. Urol.* **2015**, *67*, 326–333. [CrossRef]
15. Ross, A.E.; Feng, F.Y.; Ghadessi, M.; Erho, N.; Crisan, A.; Buerki, C.; Sundi, D.; Mitra, A.P.; Vergara, I.A.; Thompson, D.J.S. A genomic classifier predicting metastatic disease progression in men with biochemical recurrence after prostatectomy. *Prostate Cancer Prostatic Dis.* **2014**, *17*, 64–69. [CrossRef]
16. Den, R.B.; Feng, F.Y.; Showalter, T.N.; Mishra, M.V.; Trabulsi, E.J.; Lallas, C.D.; Gomella, L.G.; Kelly, W.K.; Birbe, R.C.; McCue, P.A.; et al. Genomic prostate cancer classifier predicts biochemical failure and metastases in patients after postoperative radiation therapy. *Int. J. Radiat. Oncol. Biol. Phys.* **2014**, *89*, 1038–1046. [CrossRef]
17. Klein, E.A.; Haddad, Z.; Yousefi, K.; Lam, L.L.; Wang, Q.; Choeurng, V.; Palmer-Aronsten, B.; Buerki, C.; Davicioni, E.; Li, J.; et al. Decipher Genomic Classifier Measured on Prostate Biopsy Predicts Metastasis Risk. *Urology* **2016**, *90*, 148–152. [CrossRef]
18. Ross, A.E.; Johnson, M.H.; Yousefi, K.; Davicioni, E.; Netto, G.J.; Marchionni, L.; Fedor, H.L.; Glavaris, S.; Choeurng, V.; Buerki, C.; et al. Tissue-based Genomics Augments Post-prostatectomy Risk Stratification in a Natural History Cohort of Intermediate- and High-Risk Men. *Eur. Urol.* **2016**, *69*, 157–165. [CrossRef] [PubMed]
19. Den, R.B.; Yousefi, K.; Trabulsi, E.J.; Abdollah, F.; Choeurng, V.; Feng, F.Y.; Dicker, A.P.; Lallas, C.D.; Gomella, L.G.; Davicioni, E.; et al. Genomic classifier identifies men with adverse pathology after RP who benefit from adjuvant radiation therapy. *J. Clin. Oncol.* **2015**, *33*, 944–951. [CrossRef]
20. Gore, J.L.; du Plessis, M.; Santiago-Jiménez, M.; Yousefi, K.; Thompson, D.J.S.; Karsh, L.; Lane, B.R.; Franks, M.; Chen, D.Y.T.; Bandyk, M.; et al. Decipher test impacts decision making among patients considering adjuvant and salvage treatment after RP: Interim results from the Multicenter Prospective PRO-IMPACT study. *Cancer* **2017**, *123*, 2850–2859. [CrossRef] [PubMed]
21. Knudsen, B.S.; Kim, H.L.; Erho, N.; Shin, H.; Alshalalfa, M.; Lam, L.L.C.; Tenggara, I.; Chadwick, K.; Van Der Kwast, T.; Fleshner, N.; et al. Application of a Clinical Whole-Transcriptome Assay for Staging and Prognosis of Prostate Cancer Diagnosed in Needle Core Biopsy Specimens. *J. Mol. Diagn* **2016**, *18*, 395–406. [CrossRef] [PubMed]
22. Spratt, D.E.; Zhang, J.; Santiago-Jiménez, M.; Dess, R.T.; Davis, J.W.; Den, R.B.; Dicker, A.P.; Kane, C.J.; Pollack, A.; Stoyanova, R.; et al. Development and Validation of a Novel Integrated Clinical-Genomic Risk Group Classification for Localized Prostate Cancer. *J. Clin. Oncol.* **2018**, *36*, 581–590. [CrossRef]
23. Cuzick, J.; Swanson, G.P.; Fisher, G.; Brothman, A.R.; Berney, D.M.; Reid, J.E.; Mesher, D.; Speights, V.O.; Stankiewicz, E.; Foster, C.S.; et al. Prognostic value of an RNA expression signature derived from cell cycle proliferation genes in patients with prostate cancer: A retrospective study. *Lancet Oncol.* **2011**, *12*, 245–255. [CrossRef]
24. Bishoff, J.T.; Freedland, S.J.; Gerber, L.; Tennstedt, P.; Reid, J.; Welbourn, W.; Graefen, M.; Sangale, Z.; Tikishvili, E.; Park, J.; et al. Prognostic utility of the cell cycle progression score generated from biopsy in men treated with prostatectomy. *J. Urol.* **2014**, *192*, 409–414. [CrossRef]

25. Freedland, S.J.; Gerber, L.; Reid, J.; Welbourn, W.; Tikishvili, E.; Park, J.; Younus, A.; Gutin, A.; Sangale, Z.; Lanchbury, J.S.; et al. Prognostic utility of cell cycle progression score in men with prostate cancer after primary external beam radiation therapy. *Int. J. Radiat. Oncol. Biol. Phys.* **2013**, *86*, 848–853. [CrossRef]
26. Cooperberg, M.R.; Simko, J.P.; Cowan, J.E.; Reid, J.E.; Djalilvand, A.; Bhatnagar, S.; Gutin, A.; Lanchbury, J.S.; Swanson, G.P.; Stone, S.; et al. Validation of a cell-cycle progression gene panel to improve risk stratification in a contemporary prostatectomy cohort. *J. Clin. Oncol.* **2013**, *31*, 1428–1434. [CrossRef]
27. Cuzick, J.; Stone, S.; Fisher, G.; Yang, Z.H.; North, B.V.; Berney, D.M.; Beltran, L.; Greenberg, D.; Møller, H.; Reid, J.E.; et al. Validation of an RNA cell cycle progression score for predicting death from prostate cancer in a conservatively managed needle biopsy cohort. *Br. J. Cancer* **2015**, *113*, 382–389. [CrossRef]
28. Canter, D.J.; Freedland, S.; Rajamani, S.; Latsis, M.; Variano, M.; Halat, S.; Tward, J.; Cohen, T.; Stone, S.; Schlomm, T.; et al. Analysis of the prognostic utility of the cell cycle progression (CCP) score generated from needle biopsy in men treated with definitive therapy. *Prostate Cancer Prostatic Dis.* **2020**, *23*, 102–107. [CrossRef] [PubMed]
29. Sommariva, S.; Tarricone, R.; Lazzeri, M.; Ricciardi, W.; Montorsi, F. Prognostic Value of the Cell Cycle Progression Score in Patients with Prostate Cancer: A Systematic Review and Meta-analysis. *Eur. Urol.* **2016**, *69*, 107–115. [CrossRef]
30. Wei, L.; Wang, J.; Lampert, E.; Schlanger, S.; DePriest, A.D.; Hu, Q.; Gomez, E.C.; Murakam, M.; Glenn, S.T.; Conroy, J.; et al. Intratumoral and Intertumoral Genomic Heterogeneity of Multifocal Localized Prostate Cancer Impacts Molecular Classifications and Genomic Prognosticators. *Eur. Urol.* **2017**, *71*, 183–192. [CrossRef] [PubMed]
31. Fisher, B.; Jeong, J.H.; Bryant, J.; Anderson, S.; Dignam, J.; Fisher, E.R.; Wolmark, N. National Surgical Adjuvant Breast and Bowel Project randomised clinical trials. Treatment of lymph-node-negative, oestrogen-receptor-positive breast cancer: Long-term findings from National Surgical Adjuvant Breast and Bowel Project randomised clinical trials. *Lancet* **2004**, *364*, 858–868. [CrossRef]
32. Ross, D.T.; Kim, C.Y.; Tang, G.; Bohn, O.L.; Beck, R.A.; Ring, B.Z.; Seitz, R.S.; Paik, S.; Costantino, J.P.; Wolmark, N. Chemosensitivity and stratification by a five monoclonal antibody immunohistochemistry test in the NSABP B14 and B20 trials. *Clin. Cancer Res.* **2008**, *14*, 6602–6609. [CrossRef]
33. Knezevic, D.; Goddard, A.D.; Natraj, N.; Cherbavaz, D.B.; Clark-Langone, K.M.; Snable, J.; Watson, D.; Falzarano, S.M.; Magi-Galluzzi, C.; Klein, E.A.; et al. Analytical validation of the Oncotype DX prostate cancer assay—A clinical RT-PCR assay optimized for prostate needle biopsies. *BMC Genom.* **2013**, *14*, 690. [CrossRef]
34. Klein, E.A.; Cooperberg, M.R.; Magi-Galluzzi, C.; Simko, J.P.; Falzarano, S.M.; Maddala, T.; Chan, J.M.; Li, J.; Cowan, J.E.; Tsiatis, A.C.; et al. A 17-gene assay to predict prostate cancer aggressiveness in the context of Gleason grade heterogeneity, tumor multifocality, and biopsy undersampling. *Eur. Urol.* **2014**, *66*, 550–560. [CrossRef]
35. Cullen, J.; Rosner, I.L.; Brand, T.C.; Zhang, N.; Tsiatis, A.C.; Moncur, J.; Ali, A.; Chen, Y.; Knezevic, D.; Maddala, T.; et al. A Biopsy-based 17-gene Genomic Prostate Score Predicts Recurrence After RP and Adverse Surgical Pathology in a Racially Diverse Population of Men with Clinically Low- and Intermediate-risk Prostate Cancer. *Eur. Urol.* **2015**, *68*, 123–131. [CrossRef] [PubMed]
36. Van Den Eeden, S.K.; Lu, R.; Zhang, N.; Quesenberry, C.P., Jr.; Shan, J.; Han, J.S.; Tsiatis, A.C.; Leimpeter, A.D.; Lawrence, H.J.; Febbo, P.G.; et al. A Biopsy-based 17-gene Genomic Prostate Score as a Predictor of Metastases and Prostate Cancer Death in Surgically Treated Men with Clinically Localized Disease. *Eur. Urol.* **2018**, *73*, 129–138. [CrossRef]
37. Brooks, M.A.; Thomas, L.; Magi-Galluzzi, C.; Li, J.; Crager, M.R.; Lu, R.; Abran, J.; Aboushwareb, T.; Klein, E.A. GPS Assay Association with Long-Term Cancer Outcomes: Twenty-Year Risk of Distant Metastasis and Prostate Cancer-Specific Mortality. *JCO Precis. Oncol.* **2021**, *5*, 442–449. [CrossRef]
38. Moschovas, C.M.; Chew, C.; Bhat, S.; Sandri, M.; Rogers, T.; Dell'Oglio, P.; Roof, S.; Reddy, S.; Sighinolfi, M.C.; Rocco, B.; et al. Association Between Oncotype DX Genomic Prostate Score and Adverse Tumor Pathology After RP. *Eur. Urol. Focus* **2021**, *21*, S2405–S4569. [CrossRef]
39. Lin, D.W.; Zheng, Y.; McKenney, J.K.; Brown, M.D.; Lu, R.; Crager, M.; Boyer, H.; Tretiakova, M.; Brooks, J.D.; Dash, A.; et al. 17-Gene Genomic Prostate Score Test Results in the Canary Prostate Active Surveillance Study (PASS) Cohort. *J. Clin. Oncol.* **2020**, *38*, 1549–1557. [CrossRef] [PubMed]
40. Eggener, S.E.; Rumble, R.B.; Armstrong, A.J.; Morgan, T.M.; Crispino, T.; Cornford, P.; van der Kwast, T.; Grignon, D.J.; Rai, A.J.; Agarwal, N.; et al. Molecular Biomarkers in Localized Prostate Cancer: ASCO Guideline. *J. Clin. Oncol.* **2020**, *38*, 1474–1494. [CrossRef] [PubMed]

Article

Circulating Tumor Cell Persistence Associates with Long-Term Clinical Outcome to a Therapeutic Cancer Vaccine in Prostate Cancer

Ingrid Jenny Guldvik [1,2,*], Lina Ekseth [3], Amar U. Kishan [4], Andreas Stensvold [5], Else Marit Inderberg [6] and Wolfgang Lilleby [7]

1. Department of Tumor Biology, Institute of Cancer Research, Oslo University Hospital, 0379 Oslo, Norway
2. Institute of Clinical Medicine, University of Oslo, 0318 Oslo, Norway
3. Faculty of Clinical Medicine, University of Stettin, 70-111 Szczecin, Poland; lina.ekseth@gmail.com
4. Department of Radiotherapy, University of California, Los Angeles, CA 90095, USA; AUKishan@mednet.ucla.edu
5. Department of Oncology, Østfold Hospital Trust, 1714 Kalnes, Norway; andreas.stensvold@so-hf.no
6. Translational Research Unit, Department of Cellular Therapy, Oslo University Hospital, 0379 Oslo, Norway; Suso.Else.Marit.Inderberg@rr-research.no
7. Department of Oncology, Oslo University Hospital, 0379 Oslo, Norway; WLL@ous-hf.no
* Correspondence: ingrid.jenny.guldvik@rr-research.no

Abstract: De novo metastatic or recurrence of prostate cancer (PC) remains life-threatening. Circulating tumor cells (CTCs) are noninvasive biomarkers and provide unique information that could enable tailored treatment. This study evaluated the impact of CTCs in PC patients eligible for peptide vaccine therapy. Twenty-seven patients were tested for CTCs with the CellCollector® device (Detector CANCER01(DC01)) during short-term androgen deprivation therapy (ADT) before cancer vaccine treatment (cohort 1) or salvage radiation (cohort 2). CTC counts were compared to clinicopathological parameters. In cohort 1, CTCs were correlated to immune responses, serum protein profiles, and clinical outcomes. In cohort 2, captured CTCs were further profiled for expression of PSMA, PAP, and PD-L1. Nine out of 22 patients (40.9%) in cohort 1 were CTC positive. These patients demonstrated vaccine-specific immune response ($p = 0.009$) and long-term prostate cancer-specific survival (log-rank, $p = 0.008$). All five patients in cohort 2 had CTCs at recurrence (count range 18–31), and 4/5 had CTCs positive for PSMA, PAP, and PD-L1. The DC01 CTC detection provides information beyond current clinical practice. Despite the small size of cohort 1, a correlation between CTC detection and outcome was shown.

Keywords: circulating tumor cells; prostate cancer; cancer vaccine; immune response; biomarker

1. Introduction

Prostate cancer is among the most common occurring malignancies globally, and despite the high effectiveness of definitive treatments in the primary setting, the disease will recur in 20–30% of patients [1]. Moreover, owing to the lack of screening programs for early detection of PC, emerging worrying statistics demonstrate that a larger proportion of patients present with more advanced PC and metastatic PC [2]. In Norway, PC is the second leading cause of cancer-related deaths after lung cancer. One man out of seven will develop PC during his lifetime, and more than 100,000 men die of prostate cancer in Europe each year. The probability of developing PC sharply increases in the sixth decade of life and further increases after age 70 [3]. The aging of the current population means that the disease will become an even more significant public health issue in the future.

Additional predictive biomarkers are urgently needed to improve the standard clinical decision model used in the routine staging of this disease (T stage, Gleason score, serum prostate-specific antigen (PSA), and bone scan) [4].

Several studies have confirmed the predictive and prognostic value of circulating tumor cell (CTC) detection as a monitoring method for treatment response in castration-resistant PC patients [5,6].

Recent reports have shown the efficacy of an in vivo capture device (CellCollector®, Detector CANCER01, DC01, Gilupi GmbH, Potsdam, Germany) in men with high-risk non-metastatic PC treated with definitive therapy [7]. This novel antibody-coated medical assay captures epithelial cell adhesion molecule (EpCAM) positive circulating cells that allow enumeration and further characterization of these cells.

We have applied the DC01 to detect CTCs in patients enrolled in two different studies with de novo metastatic PC (mPC, cohort 1) receiving ADT and a synthetic long peptide vaccine that targets telomerase (UV1®), and with biochemical relapse after radical prostatectomy (bPC, cohort 2) (Figure 1). Here, we report the prevalence of CTCs and evaluate the associations to immune responses, serum protein, and long-term clinical outcome (cohort 1) and explore molecular features of CTCs present in biochemically relapsed PC before salvage radiation (cohort 2).

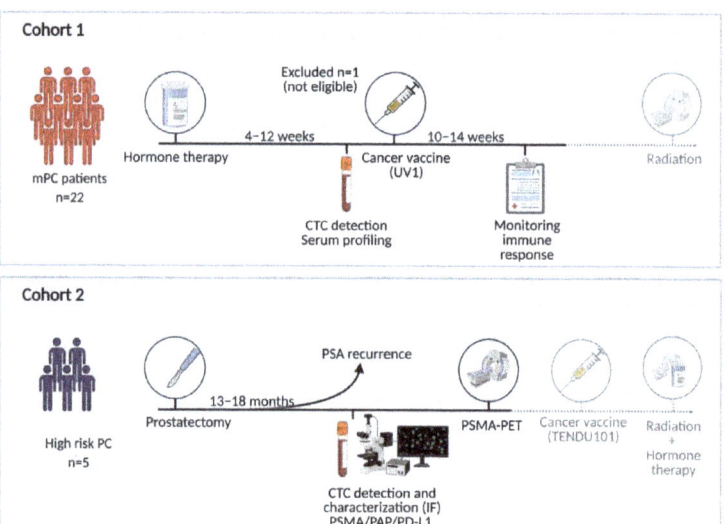

Figure 1. Overview of study. Two independent patient cohorts has been investigated for the presence of CTC. Figure created with Biorender.com.

2. Materials and Method

2.1. Patients

2.1.1. Cohort 1 (mPC)

In the mPC cohort, twenty-two patients participated in a phase I study with a therapeutic cancer vaccine (UV1®, Ultimovacs, Oslo, Norway), a second-generation hTERT peptide-based cancer vaccine [8]. The primary objective of this study was to determine the maximum-tolerated dose and safety. CTC capture and enumeration were performed at enrollment and blood samples were biobanked for biomarker discovery. The study was approved by the institutional protocol board, the Ethical Committee (EudraCT 2012-002411-26), and the National Medical Agents Authority (NoMA), and the study was registered at clinicaltrials.gov (NCT01784913) [9]. The study was approved by the Ethics Committee of Health Region South-East (protocol code A 2013/112 of date 17.03.2013). Written consent was obtained from all participants.

2.1.2. Cohort 2 (bPC)

This pilot study consisted of five men (bPC) referred to salvage radiotherapy after prostatectomy with high-risk features defined by the EAU guidelines [4]. The study was initiated to test the possible utility of the DC01 to detect CTCs in a planned first-in-man phase I study with a new therapeutic peptide vaccine (TENDU101®, EudraCT 2020-000918-15, NCT04701021). The study was approved by the Ethics Committee of Health Region South-East (protocol code D 2020/143561 of date 9 September 2020). Written consent was obtained from all participants.

2.2. Laboratory Analyses

2.2.1. Capture of CTCs

The detection of CTCs was performed utilizing the novel in vivo device CellCollector® CANCER01 (DC01) (Gilupi GmbH, Potsdam, Germany) that captures and enables enumeration of EpCAM positive tumor cells in the circulation [7]. An intravenous catheter (20-gauge, BD-Venflon™, Stockholm, Sweden) was placed into a cubital vein, and the DC01 was inserted into the catheter, dwelling for 30 min in the bloodstream. After being applied in patients, the tip of the CellCollector® (DC01) was washed three times in phosphate-buffered saline (PBS) (Gibco) including 1.6 mg/mL (final) ethylenediaminetetraacetic acid (EDTA) (Roth). Bound cells were fixed with Acetone (VWR) for 10 min at room temperature, dried, stored at -20 °C, and transferred on dry ice to Gilupi GmbH, Potsdam, Germany for further processing.

2.2.2. Immunocytochemical Analysis and CTC Enumeration in Cohort 1

All procedures were performed by an experienced operator who was blinded to patient characteristics. In brief, the cells were blocked and permeabilized with 3% bovine serum albumin (BSA) (Roth) and 0.1% Triton X-100 (Fluka) in PBS for 30 min. Primary antibodies, including anti-pan-cytokeratin-Alexa 488 (CK4, 5, 6, 8, 10, 13, 18) (C-11) (Exbio), anti-CK19-Alexa 488 (A53-B/A2) (Exbio), anti-CK7- fluorescein isothiocyanate (LP5K) (FITC) (Millipore), anti-EpCAM- fluorescein isothiocyanate (FITC) (HEA125) (Acris), and anti-CD45- Alexa 647 (MEM-28, Exbio), were added for 30 min. The DC01 was then rinsed three times with 3 mL of PBS, and the nuclei were counterstained with Hoechst 33342 (Invitrogen). CTCs were identified using a Zeiss Axio imager fluorescent microscope with a 20× objective. Fluorescent images were recorded with a Zeiss MRm camera. A cell was considered to be a CTC if it was positively stained for cytokeratin and/or EpCAM, it was negative for CD45, and certain morphological criteria for tumor cells were met: the presence of a nucleus with a round or ellipsoid shape and a cell size ranging from 4 to 50 µm. Leukocytes were defined as nucleated (Hoechst-positive), CD45-positive, and cytokeratin and/or EpCAM-negative cells, and were not counted.

2.2.3. Immunofluorescence Staining for PSMA, PD-L1, and PAP in Cohort 2

Immunocytology was combined with immunofluorescence (IF) staining for PSMA, PAP, and PD-L1. The following criteria defined tumor cells: intact morphology, diverse cells (large cell bodies, irregular cell shapes, several cells next to each other/cluster), cell diameter ≥ 4 µm, distinct and positive nuclei staining by Hoechst, and at least one positive marker (PAP or PD-L1 or PSMA). Nucleated cells with tumor cell-like morphology, but lacking IF staining were reported, but counted as negative.

Cells attached to the DC01 were permeabilized in 1× PBS/0.1% GIBCO™ for 10 min, washed three times in 1× PBS, and blocked with 1× PBS/3% BSA (Roth) for 30 min. Cells were further blocked with PBS/3% normal goat serum (Invitrogen) for 30 min. Immunolabeling was performed for 30 min at room temperature in 45 µL of PBS/3% BSA (Roth) containing primary antibodies (mouse IgG1-anti PSAP/PAP (clone PASE/4LJ, unconjugated, Invitrogen, dilution 1:25) and rabbit-anti-PD-L1 XP® (clone E1L3N®, unconjugated, Cell Signaling Technology, dilution 1:300)). Afterwards, samples were washed with PBS (Life Technologies: Carlsbad, CA, USA) twice for 10 min at room temperature with agitation and subsequently incubated for 30 min at room temperature with secondary antibodies protected from light. Goat-anti-mouse IgG1, Alexa Flour®647 (Invitrogen, diluted 1:300), and goat anti-rabbit IgG (H + L), Alexa Flour®555 (Life Technologies, dilution 1:400), were also prepared in 45 µL of PBS/3% BSA. Following two wash steps in PBS for 10 min at room temperature with agitation, the samples were incubated for 30 min at room temperature protected from light with the conjugated antibody solution PSMA-Alexa Flour®488 (clone k1h7, Novus Biological, diluted 1:100) in PBS/3% BSA. After washing with PBS for 1 min at room temperature, cells were counterstained with Hoechst 33342 (Sigma Aldrich, final concentration: 1 µg/mL in PBS/3% BSA) for 5 min at room temperature, washed with PBS for 1 min at room temperature, and air-dried for 5 min each (all steps protected from light). Images of stained cells were acquired using a fluorescent microscope (Axio Imager Carl Zeiss AG, Jena, Germany) combined with a monochrome camera. Filter set (Carl Zeiss AG) numbers used for microscopic evaluation were 49 (Blue), 52 (Green), 43 (Orange), and 50 (Dark.RED).

2.2.4. Detection of UV1® Vaccine-Specific T-Cell Response in Cohort 1

Peripheral blood in acid citrate dextrose (ACD) tubes was taken from patients before UV1®-vaccination, two weeks after vaccination, and then monthly until week 26, then every three months. A detailed description of the procedures mentioned herein can be reviewed in Lilleby et al., 2017 [9]. Peripheral blood mononuclear cells (PBMCs) were isolated, frozen, and stored before further analysis. PBMCs were then thawed and pre-stimulated with the three UV1® vaccine peptides, and the UV1®-specific T cell proliferative response was tested. Briefly, PBMCs were pre-stimulated with UV1® peptide at 10 µM for 10–12 days, and cytokines (IL-2, IL-7) were added on day 3. On day 10–12, the T cells were then re-stimulated with irradiated, peptide-loaded autologous antigen-presenting cells (APCs), and T-cell proliferation was determined in 3H-Thymidine incorporation assays. The stimulation index (SI) was calculated by dividing the counts of wells with either the mix of UV1®-peptides or the three single peptides comprising the vaccine by the mean count of wells containing no peptide. An SI ≥ 3 was considered a positive, peptide-specific response.

2.2.5. Targeted Serum Profiling in Cohort 1

Relative quantification of serum proteins known to be implicated in the interplay between the immune system and tumorigenic processes was performed by proximity extension assay (PEA) technology (Olink Bioscience Service Center Uppsala, Uppsala, Sweden) [10]. Briefly, one microliter serum drawn at study inclusion was profiled by the Immuno-Oncology panel (v.1). All sample handling and laboratory analyses were performed blinded. Data were normalized to minimize both intra- and inter-assay variation and presented as normalized protein expression values (NPX), an arbitrary unit on a log2 scale. NPX values of the different proteins within each patient were then analyzed to associate CTC findings and immune response.

2.3. Statistical Analysis

All statistical analyses were performed using SPSS (version 20) or R (version 3.3.1). All tests were two-sided, and a p-value < 0.05 was considered to be statistically significant. CTC counts were stratified as negative or positive, with positive meaning at least one cell to meet the criteria as CTC: In cohort 1, a CTC was defined as EpCAM+/panCK+/CD45-, whereas a CTC was defined as PSMA+/PD-L1+/PAP+ in cohort 2. In addition, all cells had to show normal morphology by Hoechst. Mann–Whitney U test (MWU) was used to assess differences in continuous variables between CTC positive and CTC negative patients. Chi-square test and Fisher's exact test were used to evaluate associations between categorical coded variables. Spearman rank correlations were used to determine correlations between the number of CTCs detected, serum levels of proteins, and the number of peptides involved in immune reactivity towards the cancer vaccine. Kaplan–Meier survival analysis with prostate cancer specific survival (PCSS) and overall survival (OS) as endpoints was used to evaluate surviving proportions of patients stratified by CTC status. Log-rank test was used to test for statistical differences in surviving proportions. Univariate Cox proportional hazards (Cox PH) modelling was used to calculate crude hazard ratios (HRs) and evaluate the individual association of CTC with PCSS and OS.

3. Results

Two cohorts of patients were investigated (Figure 1). Patients' characteristics are summarized in Table 1. The median age of cohort 1 was 67 years and 66 years in cohort 2.

Table 1. Clinical characteristics of patients with metastatic prostate cancer (mPC) or bPC stratified by circulating tumor cell (CTC) status.

	Cohort 1 (mPC)			Cohort 2 (bPC)	
	CTC Status			CTC Status	
	Positive	Negative	p=	Positive	Negative
n (%)	9 (40.9)	13 (59.1)	<0.001	5 (100)	0 (0.00)
Time FU (month, median [IQR])	77.6 [52.6, 82.5]	46.6 [30.2, 76.9]			
Age (yr, median [IQR])	66.9 [59.2, 75.4]	66.8 [63.9, 73.9]	0.85	65.8 [57.4, 74.7]	
T stage (%)			0.72 *		
cT2	1 (11.1)	1 (7.7)			
cT3	6 (66.7)	7 (53.8)			
cT4	2 (22.2)	5 (38.5)			
pT2				2 (40.0)	
pT3				3 (60.0)	
Gleason grade group (%)			0.61 **		
2 + 3	0 (0.00)	0 (0.00)		3 (60.0)	
4	3 (33.3)	2 (14.3)		1 (20.0)	
5	6 (66.7)	11 (84.6)		1 (20.0)	
PSA values (ng/mL, median [IQR])					
PSA at diagnosis	26.0 [12.0, 72.0]	33.0 [12.0, 58.0]	0.95		
PSA after ADT	1.10 [0.40, 7.60]	3.00 [0.60, 9.20]	0.66		
PSA at relapse				0.26 [0.20, 0.54]	
Time ADT/DC01 (mo., median [IQR])	3.42 [1.74, 3.72]	2.30 [1.64, 4.77]	0.92		
Time RP/DC01 (mo., median [IQR])				13.2 [7.02, 14.5]	
No. reactive peptides in IR (%)			0.009 *		
0	0 (0.0)	3 (23.1)			
1	1 (11.1)	6 (46.2)			
2	5 (55.6)	3 (23.1)			
3	3 (33.3)	1 (7.7)			

ADT: Androgen-deprivation therapy; IQR: interquartile range; FU: follow-up; IR: immune reaction; mo.: months; PSA: prostate-specific antigen; RP: radical prostatectomy; yr: years. * Chi-square test, ** Fisher's exact test.

3.1. CTC Presence Predicts Long-Term Survival Benefit of a Therapeutic Cancer Vaccine

3.1.1. CTC Detection Predicts Broad Immune Response to Therapeutic Cancer Vaccine

In cohort 1, the median PSA was 3 ng/mL after starting with ADT (median duration on ADT 3.2 months, IQR 1.7–3.72) and 9 out of 22 patients (40.9%) had detectable CTCs. CTC positivity was associated with immunity towards two out of three UV1® vaccine peptides ($p = 0.009$, Figure 2A). There was a direct correlation between the number of CTCs detected and the number of peptides involved in the immunity (Spearman rho 0.59, $p = 0.004$, Figure 2B).

In order to assess the circulatory immune microenvironment for CTCs, a targeted serum profiling by Olink technology was performed on samples collected in parallel of

CTC capture. Both serum levels of CXCL5 and CD70 were significantly elevated in patients with CTC detected, whereas IL-18, ADGRG1, and HO1 were all downregulated (Table S1). Intriguingly, when assessing the relationship between a broad immune response (defined as reactivity to two or more peptides in the UV1®vaccine) and serum protein levels (Table S2), CXCL5 was also elevated in patients with a broad immune response. CXCL5 levels correlated positively both to the number of CTCs detected and to the number of peptides involved in the immune response (Figure 2C,D, respectively).

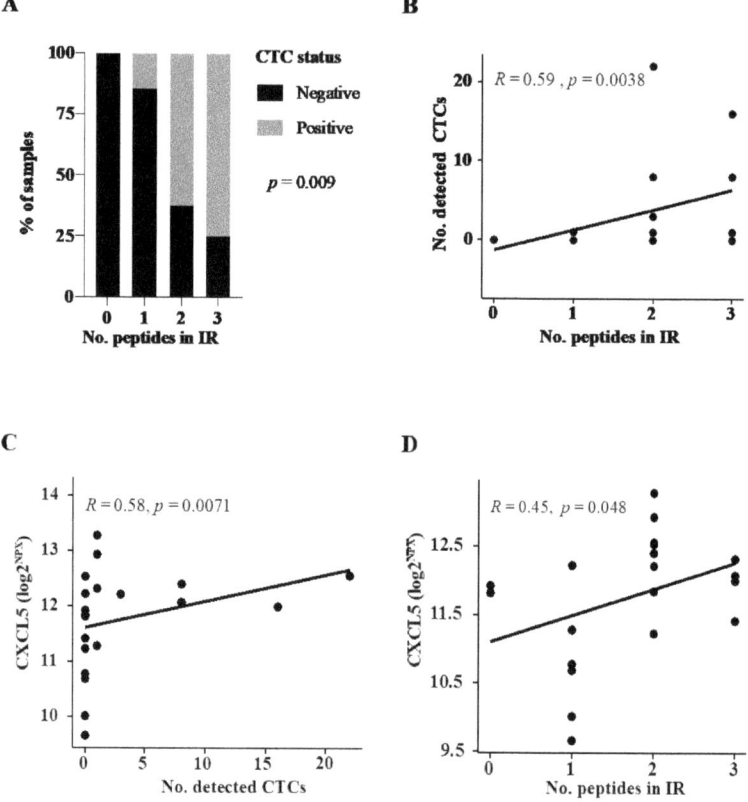

Figure 2. Assessment of association between circulating tumor cell (CTC) status, immune response, and serum proteins in cohort 1. (**A**) Percentage of CTC-positive patients according to the number of reactive UV1®vaccine peptides in immune response (IR). *p*-value reported on Chi-square test. Dot plot to assess the distribution of (**B**) the number detected CTCs according to the number of reactive peptides in immune response (IR) in vaccinated patients, (**C**) serum levels of CXCL5 across the number of CTCs detected, and (**D**) CXCL5 according to the number of reactive peptides in IR. Correlation coefficient and *p*-values are reported on the Spearman rank test (**B–D**).

3.1.2. Patients with CTC Show Survival Benefit of Therapeutic Cancer Vaccination

Next, patients were categorized based on their CTC status and the long-term survival was assessed. Despite a small cohort, patients with CTC positivity illustrated long-term PCSS, with only 1 out of 9 succumbing to the disease at 80 months follow-up, whereas 10 out of 13 patients negative for CTC had PC-specific death ($p = 0.008$, Figure 3A). Overall survival was also improved for CTC-positive patients, with 3 out of 9 dead within the patient group positive for CTC and 10 out of 13 among the CTC negative patient group ($p = 0.058$, Figure 3B).

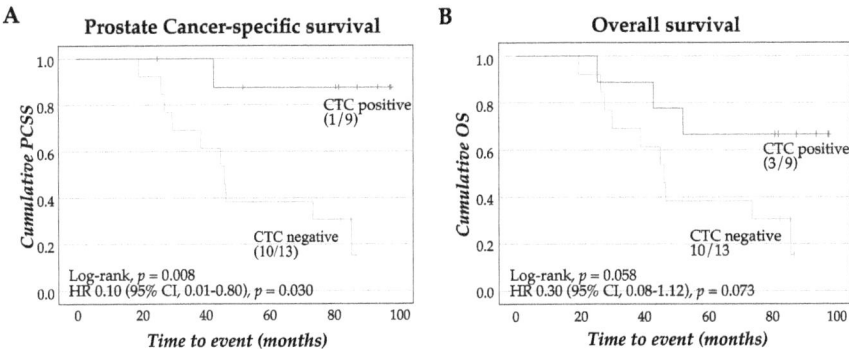

Figure 3. Survival estimates of patients stratified by CTC status. Kaplan–Meier plot and crude hazard ratio (HR) with confidence intervals (CIs) for PCSS (**A**) and OS (**B**) in cohort 1 (mPC) grouped according to CTC detection. PCSS: prostate cancer-specific survival, OS: overall survival.

3.2. CTCs are Present at Biochemical Relapse and Express PSMA, PD-L1, and PAP

In a pilot study, CTC presence was assessed in five men referred to salvage radiotherapy after prostatectomy. All five patients presented with CTCs at biochemical relapse (range 18–31). Two out of the five patients had a negative PSMA-PET scan. Membrane staining of PSMA, PAP, and PD-L1 was assessed on the captured CTCs, as well as clusters (Table 2 and Figure 4). Four CTC samples were positive for all three markers, and patients with high Gleason grade group (4 and 5) had fewer CTCs with clusters (range 0–1), whereas patients with grade group 2 and 3 had CTCs with more clusters (range 3–4) ($p < 0.0001$). Further, the proportion of CTCs stained with all three markers increased with the Gleason grade ($p = 0.005$). In all samples, nucleated cells negative for PSMA, PD-L1, and PAP were found (Figure 4A).

Table 2. Summary of immunofluorescence stained cells in individual patients.

Patient ID	1	3	5	2	4	$p=$ *
Gleason Grade Group	2	3	3	4	5	
PSMA-PET status	+	−	+	−	+	
Total no. CTC	31	19	18	27	29	
PSMA, n (%)	0 (0.0)	0 (0.0)	5 (27.8)	0 (0.0)	0 (0.0)	0.96
PD-L1, n (%)	20 (64.5)	3 (15.8)	3 (16.7)	0 (0.0)	0 (0.0)	<0.001
PAP, n (%)	0 (0.0)	6 (31.6)	4 (22.2)	3 (11.1)	2 (6.9)	0.92
PSMA/PD-L1, n (%)	0 (0.0)	0 (0.0)	0 (0.0)	0 (0.0)	0 (0.0)	-
PSMA/PAP, n (%)	6 (19.4)	3 (15.8)	6 (33.3)	0 (0.0)	0 (0.0)	0.005
PD-L1/PAP, n (%)	0 (0.0)	0 (0.0)	0 (0.0)	0 (0.0)	0 (0.0)	-
PSMA/PD-L1/PAP, n (%)	5 (16.1)	7 (36.8)	0 (0.0)	24 (88.9)	27 (93.1)	0.005
Clusters (no.)	4 (12.3)	3 (15.8)	4 (22.2)	1 (3.7)	0 (0.0)	<0.0001
Nuclei+ cells (no. PSMA/PD-L1/PAP) †	4	5	8	11	0	-

* Chi-square test for trend; † not counted as positive CTC.

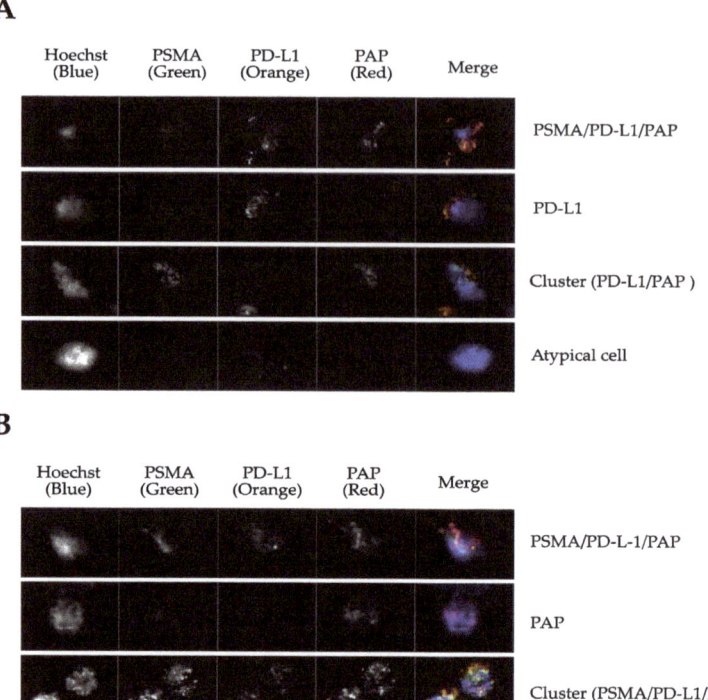

Figure 4. Immunofluorescence staining for PSMA, PD-L1, and PAP on captured CTCs in men with biochemical failure post-prostatectomy (cohort 2). (**A**) Examples of stained CTCs captured in patients with Gleason grade group 2; (**B**) examples of stained CTCs captured in Gleason grade group 5.

4. Discussion

In the present study, we assessed the prevalence of CTCs in two clinically relevant settings: (a) patients with de novo metastatic PC and (b) patients with biochemical relapse PC referred to postoperative radiation. In both groups, we found considerable context-dependent evidence of CTCs even after ADT had been commenced (cohort 1) or when low serum PSA levels signaled tumor recurrence (cohort 2). Notably, targeting EpCAM is currently recognized as the only FDA-approved marker for detecting CTCs [11] and is recommended by the prostate cancer working group (PCWG3) guidelines [12]. Cancer cells of epithelial origin can retain stem cell-like features and constitute to further insight into the development of phenotypes and therapy failure [13,14]. The PCWG emphasizes the importance of clinical trials with a biomarker context. In the present study, CTC enumeration, immune response, and serum proteins were embedded in the disease state model.

Here, we could show that CTCs were detectable in 40.9% of patients with onset metastatic disease treated short-term with ADT and a detection rate of 100% in patients with biochemical relapse after prostatectomy. In both scenarios, patients had low serum PSA owing to ADT or resection of the prostate. The results give new insights into the biological behavior of PC, in patients with low serum PSA both due to ADT or prostatectomy.

It has been established that CTC is a prognostic marker in metastatic castration-resistant prostate cancer [6,9]. Surprisingly, patients positively stained for CTCs after the onset of ADT and before initiating the UV1® vaccine had a survival benefit at median six-year FU post-vaccination. This is contrary to recent findings where a high count of CTCs signaled poor outcomes in those treated with life-long ADT [15]. We speculate that

CTCs with epithelial features could be a source to antigens and act synergistically with the therapeutic peptide vaccine in stimulating cancer-specific immune cells, improving the outcome in some patients.

In support of this unexpected observation, we found that the number of CTCs detected in cohort 1 patients correlated to the broadness of the vaccine IR, demonstrated by the number of vaccine peptides involved [8]. Goldkorn et al. found that telomerase activity independently predicted overall survival in men with detectable CTCs [16]. The CTC presence could thus potentially boost the induction of an anti-telomerase IR by the UV1® vaccine. This opens up for further investigation on whether CTCs can be used to predict patients that will have a favorable response to arising immune therapies in prostate cancer [17].

The high prevalence of CTCs in cohort 1 during ADT raises some intriguing questions. It is well established that immunosenescence in older men leads to thymic involution and is related to the predominantly significant clinical detection of cancer [18]. However, ADT can reverse thymic involution, thereby recruiting naïve T-cells capable of forming lymphocyte infiltrates in the primary tumor [19,20]. Of note, the typical immune-pathological cell picture is governed by suppressed immunity in prostate cancer patients when treated with ADT [21]. However, ADT can lead to androgen receptor amplification and programmed cell death. Increased antigen presentation can assist in the often-seen sustained response in biochemical responding patients [22].

On the other hand, in those with a negative CTC finding (59.1% in cohort 1), ADT could reset cancer cells to senescence, shedding typical epithelial surface markers contributing to an immune mimicry. It has been recently described that secretory stimuli in the microenvironment of minimal residual disease can induce senescence [23]. Moreover, depending on the driver-mutation, the senescence-associated secretory phenotype of senescent tumor cells can have pro- and antitumorigenic effects [24]. Besides, cancer-associated fibroblasts producing CXCL5 are involved in promoting PD-L1 upregulation in tumor cells [25]. CXCL5 is often elevated in metastatic PC patients, increases with tumor apoptosis, and is thus considered as a relevant therapeutic target [26]. CXCL5 is involved in recruiting immune cells to the tumor, including myeloid-derived suppressor cells, contributing to tumor immunoresistance (reviewed in [27]). Our study found the chemokine CXCL5 to be associated with the number of CTCs and immune response. In line with previous reports, this could be the response of the cancer and its microenvironment to an immune attack, suggesting that CXCL5 is a potential Achilles' heel. Combination treatment could be required to overcome such resistance mechanisms and to have a sustained and broad immune response.

In cohort 2, the number of CTCs detected was independent of PSA, and could also be detected at low PSA levels. CTCs survive only for a short time in the blood circulation [28]. Chen et al. showed that the finding of CTCs by the DC01 was reproducible at different timepoints [29]. The in vivo DC01 device, previously tested in men after surgery, had a detection rate of 34% obtained three months after prostatectomy. In our study, a significant number of CTCs were detected more than one year after surgery in a prognostic high-risk group. Therefore, they likely originate from clinically significant minimal residual disease after primary radical resection of the prostate. In line with this, the CTC positivity in our study was supported by observing that CTC count was independent of PSA. Many recent basic science findings point toward the possible early genesis of a so-called immune tolerance [30]. This is in line with Benko et al., who found higher expression of EpCAM positivity in patients with high-grade Gleason score and T stage, and that EpCAM expression was a significant predictor of shorter biochemical recurrence-free survival [31]. Despite the normally long tumorigenesis of primary PC, CTCs may have accelerated clonal evolution, enabling them to spread.

Interestingly, CTCs from cohort 2 stained positively for PSMA with PD-L1 and PAP. The presence of CTCs was independent of PSA values or PSMA-PET findings. This is in line with the findings of Cieslikowski et al., who found presence of CTCs in patients with no evidence of metastasis by imaging [32]. PSMA is a transmembrane glycoprotein with catalytic properties, named glutamate carboxypeptidase II. It is not specific to prostate cancer, but has proven useful as it is highly overexpressed in prostate cancer cells in about 95% of the patients [33,34]. PSMA-PET has a sensitivity level that depends on the tumor volume. Sensitivity ranged from 42 to 98% and specificity from 71 to 99%. Thus, in patients with early biochemical relapse, not all will have sufficient minimal residual disease to be detected by PSMA-PET.

The expression of PD-L1 on CTCs has been linked to tumor immune evasion [10]. In the present study, Gleason grade and IF markers were correlated. The finding of a distinct phenotype in CTCs could provide a protective mechanism of CTC survival outside the tumor microenvironment.

Our study has limitations. Apart from the small sample size and the lack of baseline CTC measurement before initiation of ADT (cohort 1), CTCs are heterogeneous and might not at all express the epithelial marker. Using EpCAM as a positive selection marker may introduce a bias, but the prolonged in vivo detection time can lead to favorable enrichment of CTCs. Thus, by a pre-defined set of criteria, the probability of a false positive CTC decreased. Longitudinal observation of CTCs using the DC01 will be performed in the ongoing phase I TENDU101 study (NCT04701021).

In cohort 2, some nucleated cells captured by the DC01 were suspicious, but did not stain for either PSMA, PD-L1, or PAP and were disregarded as CTC. As these cells were not counterstained with CD45 or pan-cytokeratin, leukocyte origin cannot be excluded.

Although the survival data reported herein show great potential in the small study of the UV1®vaccine (cohort 1), we cannot exclude that immune responses triggered by other prostate antigens not covered by our assay contribute to improved clinical outcome in this cohort.

5. Conclusions

Our results indicate that implementation of the CTC detection could improve the shared decision-making process addressing targeted therapy for men with de novo and relapsed PC after prostatectomy. Notably, presence of CTCs during the onset of ADT and before the start of peptide vaccine was correlated to outcome. We found a substantial number of CTCs with the DC01 device, which could be a valuable clinical tool for assessing relapse, contextual treatment efficacy, and tailored therapy in men with PC.

Supplementary Materials: The following are available online at https://www.mdpi.com/article/10.3390/jpm11070605/s1, Table S1: Targeted serum profiling for differences between CTC status; Table S2: Targeted serum profiling for differences in immune response.

Author Contributions: Conceptualization, I.J.G., L.E., A.U.K., A.S., E.M.I. and W.L.; methodology, I.J.G., L.E., A.U.K., A.S., E.M.I. and W.L.; software, I.J.G., E.M.I. and W.L.; validation, I.J.G., L.E., E.M.I. and W.L.; formal analysis, I.J.G., L.E., E.M.I. and W.L.; investigation, L.E., A.S. and W.L.; resources, I.J.G., A.S., E.M.I. and W.L.; data curation, I.J.G., L.E., E.M.I. and W.L.; writing—original draft preparation, I.J.G., L.E., A.U.K., E.M.I. and W.L.; writing—review and editing, I.J.G., L.E., A.U.K., A.S., E.M.I. and W.L.; visualization, I.J.G., E.M.I. and W.L.; supervision, I.J.G., E.M.I. and W.L.; project administration, I.J.G., E.M.I. and W.L.; funding acquisition, A.S. and W.L. All authors have read and agreed to the published version of the manuscript.

Funding: The generous support of the Magne and Bodil Foundation is highly appreciated.

Institutional Review Board Statement: The studies reported herein were conducted according to the guidelines of the Declaration of Helsinki, and approved by the Ethics Committee of Health Region South-East (protocol code A 2013/112 of date 17.03.2013 and D 2020/143561 of date 09.09.2020).

Informed Consent Statement: Informed consent was obtained from all subjects involved in the study.

Acknowledgments: We are grateful to Hedvig Vidarsdotter Juul and Grete Berntsen for performing screening of UV1®-specific immune responses.

Conflicts of Interest: E.M.I. is inventor of a UV1 vaccine patent. All remaining authors have declared no conflict of interest.

References

1. Broeck, T.V.D.; Bergh, R.C.V.D.; Briers, E.; Cornford, P.; Cumberbatch, M.; Tilki, D.; De Santis, M.; Fanti, S.; Fossati, N.; Gillessen, S.; et al. Biochemical Recurrence in Prostate Cancer: The European Association of Urology Prostate Cancer Guidelines Panel Recommendations. *Eur. Urol. Focus* **2020**, *6*, 231–234. [CrossRef] [PubMed]
2. Culp, M.B.; Soerjomataram, I.; Efstathiou, J.A.; Bray, F.; Jemal, A. Recent Global Patterns in Prostate Cancer Incidence and Mortality Rates. *Eur. Urol.* **2020**, *77*, 38–52. [CrossRef] [PubMed]
3. Cancer Registry of Norway. *Cancer in Norway 2019—Cancer Incidence, Mortality, Survival and Prevalence in Norway*; Cancer Registry of Norway: Oslo, Norway, 2020.
4. Mottet, N.; Bellmunt, J.; Bolla, M.; Briers, E.; Cumberbatch, M.G.; De Santis, M.; Fossati, N.; Gross, T.; Henry, A.; Joniau, S.; et al. EAU-ESTRO-SIOG Guidelines on Prostate Cancer. Part 1: Screening, Diagnosis, and Local Treatment with Curative Intent. *Eur. Urol.* **2017**, *71*, 618–629. [CrossRef] [PubMed]
5. De Bono, J.S.; Scher, H.I.; Montgomery, R.B.; Parker, C.; Miller, M.C.; Tissing, H.; Doyle, G.V.; Terstappen, L.W.; Pienta, K.; Raghavan, D. Circulating Tumor Cells Predict Survival Benefit from Treatment in Metastatic Castration-Resistant Prostate Cancer. *Clin. Cancer Res.* **2008**, *14*, 6302–6309. [CrossRef] [PubMed]
6. Danila, D.C.; Heller, G.; Gignac, G.A.; Gonzalez-Espinoza, R.; Anand, A.; Tanaka, E.; Lilja, H.; Schwartz, L.; Larson, S.; Fleisher, M.; et al. Circulating Tumor Cell Number and Prognosis in Progressive Castration-Resistant Prostate Cancer. *Clin. Cancer Res.* **2007**, *13*, 7053–7058. [CrossRef] [PubMed]
7. Kuske, A.; Gorges, T.M.; Tennstedt, P.; Tiebel, A.-K.; Pompe, R.S.; Preißer, F.; Prues, S.; Mazel, M.; Markou, A.; Lianidou, E.; et al. Improved detection of circulating tumor cells in non-metastatic high-risk prostate cancer patients. *Sci. Rep.* **2016**, *6*, 39736. [CrossRef]
8. Lilleby, W.; Gaudernack, G.; Brunsvig, P.F.; Vlatkovic, L.; Schulz, M.; Mills, K.; Hole, K.H.; Inderberg, E.M. Phase I/IIa clinical trial of a novel hTERT peptide vaccine in men with metastatic hormone-naive prostate cancer. *Cancer Immunol. Immunother.* **2017**, *66*, 891–901. [CrossRef]
9. Scher, H.I.; Lu, D.; Schreiber, N.A.; Louw, J.; Graf, R.P.; Vargas, H.A.; Johnson, A.; Jendrisak, A.; Bambury, R.; Danila, D.; et al. Association of AR-V7 on Circulating Tumor Cells as a Treatment-Specific Biomarker With Outcomes and Survival in Castration-Resistant Prostate Cancer. *JAMA Oncol.* **2016**, *2*, 1441–1449. [CrossRef]
10. Zhang, T.; Agarwal, A.; Almquist, R.G.; Runyambo, D.; Park, S.; Bronson, E.; Boominathan, R.; Rao, C.; Anand, M.; Oyekunle, T.; et al. Expression of immune checkpoints on circulating tumor cells in men with metastatic prostate cancer. *Biomark. Res.* **2021**, *9*, 1–12. [CrossRef]
11. Millner, L.M.; Linder, M.W.; Valdes, R. Circulating Tumor Cells: A Review of Present Methods and the Need to Identify Heterogeneous Phenotypes. *Ann. Clin. Lab. Sci.* **2013**, *43*, 295–304.
12. Scher, H.I.; Morris, M.J.; Stadler, W.M.; Higano, C.; Basch, E.; Fizazi, K.; Antonarakis, E.S.; Beer, T.M.; Carducci, M.A.; Chi, K.N.; et al. Trial Design and Objectives for Castration-Resistant Prostate Cancer: Updated Recommendations from the Prostate Cancer Clinical Trials Working Group 3. *J. Clin. Oncol.* **2016**, *34*, 1402–1418. [CrossRef]
13. Imrich, S.; Hachmeister, M.; Gires, O. EpCAM and its potential role in tumor-initiating cells. *Cell Adhes. Migr.* **2012**, *6*, 30–38. [CrossRef]
14. Munz, M.; Baeuerle, P.A.; Gires, O. The Emerging Role of EpCAM in Cancer and Stem Cell Signaling. *Cancer Res.* **2009**, *69*, 5627–5629. [CrossRef]
15. Scher, H.I.; Heller, G.; Molina, A.; Attard, G.; Danila, D.C.; Jia, X.; Peng, W.; Sandhu, S.; Olmos, D.; Riisnaes, R.; et al. Circulating Tumor Cell Biomarker Panel as an Individual-Level Surrogate for Survival in Metastatic Castration-Resistant Prostate Cancer. *J. Clin. Oncol.* **2015**, *33*, 1348–1355. [CrossRef]
16. Goldkorn, A.; Ely, B.; Tangen, C.M.; Tai, Y.-C.; Xu, T.; Li, H.; Twardowski, P.; Van Veldhuizen, P.J.; Agarwal, N.; Carducci, M.A.; et al. Circulating tumor cell telomerase activity as a prognostic marker for overall survival in SWOG 0421: A phase III metastatic castration resistant prostate cancer trial. *Int. J. Cancer* **2014**, *136*, 1856–1862. [CrossRef] [PubMed]
17. Li, J.; Gregory, S.G.; Garcia-Blanco, M.A.; Armstrong, A.J. Using circulating tumor cells to inform on prostate cancer biology and clinical utility. *Crit. Rev. Clin. Lab. Sci.* **2015**, *52*, 191–210. [CrossRef] [PubMed]
18. Fulop, T.; Kotb, R.; Fortin, C.F.; Pawelec, G.; De Angelis, F.; Larbi, A. Potential role of immunosenescence in cancer development. *Ann. N. Y. Acad. Sci.* **2010**, *1197*, 158–165. [CrossRef]
19. Drake, C.G.; Doody, A.D.; Mihalyo, M.A.; Huang, C.-T.; Kelleher, E.; Ravi, S.; Hipkiss, E.L.; Flies, D.B.; Kennedy, E.P.; Long, M.; et al. Androgen ablation mitigates tolerance to a prostate/prostate cancer-restricted antigen. *Cancer Cell* **2005**, *7*, 239–249. [CrossRef]
20. Mercader, M.; Bodner, B.K.; Moser, M.T.; Kwon, P.S.; Park, E.S.Y.; Manecke, R.G.; Ellis, T.M.; Wojcik, E.M.; Yang, D.; Flanigan, R.C.; et al. T cell infiltration of the prostate induced by androgen withdrawal in patients with prostate cancer. *Proc. Natl. Acad. Sci. USA* **2001**, *98*, 14565–14570. [CrossRef] [PubMed]
21. Ebelt, K.; Babaryka, G.; Frankenberger, B.; Stief, C.G.; Eisenmenger, W.; Kirchner, T.; Schendel, D.J.; Noessner, E. Prostate cancer lesions are surrounded by FOXP3+, PD-1+ and B7-H1+ lymphocyte clusters. *Eur. J. Cancer* **2009**, *45*, 1664–1672. [CrossRef]

22. Olson, B.M.; Gamat-Huber, M.; Seliski, J.; Sawicki, T.; Jeffery, J.; Ellis, L.; Drake, C.G.; Weichert, J.; McNeel, D.G. Prostate Cancer Cells Express More Androgen Receptor (AR) Following Androgen Deprivation, Improving Recognition by AR-Specific T Cells. *Cancer Immunol. Res.* **2017**, *5*, 1074–1085. [CrossRef] [PubMed]
23. Nardella, C.; Clohessy, J.G.; Alimonti, A.; Pandolfi, P.P. Pro-senescence therapy for cancer treatment. *Nat. Rev. Cancer* **2011**, *11*, 503–511. [CrossRef] [PubMed]
24. Toso, A.; DI Mitri, D.; Alimonti, A. Enhancing chemotherapy efficacy by reprogramming the senescence-associated secretory phenotype of prostate tumors: A way to reactivate the antitumor immunity. *Oncoimmunology* **2015**, *4*, e994380. [CrossRef] [PubMed]
25. Li, Z.; Zhou, J.; Zhang, J.; Li, S.; Wang, H.; Du, J. Cancer-associated fibroblasts promote PD-L1 expression in mice cancer cells via secreting CXCL5. *Int. J. Cancer* **2019**, *145*, 1946–1957. [CrossRef] [PubMed]
26. Roca, H.; Jones, J.D.; Purica, M.C.; Weidner, S.; Koh, A.J.; Kuo, R.; Wilkinson, J.E.; Wang, Y.; Daignault-Newton, S.; Pienta, K.J.; et al. Apoptosis-induced CXCL5 accelerates inflammation and growth of prostate tumor metastases in bone. *J. Clin. Investig.* **2017**, *128*, 248–266. [CrossRef] [PubMed]
27. Wang, G.; Zhao, D.; Spring, D.J.; Depinho, R.A. Genetics and biology of prostate cancer. *Genes Dev.* **2018**, *32*, 1105–1140. [CrossRef] [PubMed]
28. Rejniak, K.A. Circulating Tumor Cells: When a Solid Tumor Meets a Fluid Microenvironment. *Adv. Exp. Med. Biol.* **2016**, *936*, 93–106. [CrossRef]
29. Chen, S.; Tauber, G.; Langsenlehner, T.; Schmölzer, L.M.; Pötscher, M.; Riethdorf, S.; Kuske, A.; Leitinger, G.; Kashofer, K.; Czyż, Z.T.; et al. In Vivo Detection of Circulating Tumor Cells in High-Risk Non-Metastatic Prostate Cancer Patients Undergoing Radiotherapy. *Cancers* **2019**, *11*, 933. [CrossRef]
30. Klein, C.A. Parallel progression of primary tumours and metastases. *Nat. Rev. Cancer* **2009**, *9*, 302–312. [CrossRef]
31. Benko, G.; Spajić, B.; Krušlin, B.; Tomas, D. Impact of the EpCAM expression on biochemical recurrence-free survival in clinically localized prostate cancer. *Urol. Oncol. Semin. Orig. Investig.* **2013**, *31*, 468–474. [CrossRef]
32. Cieślikowski, W.A.; Budna-Tukan, J.; Świerczewska, M.; Ida, A.; Hrab, M.; Jankowiak, A.; Mazel, M.; Nowicki, M.; Milecki, P.; Pantel, K.; et al. Circulating Tumor Cells as a Marker of Disseminated Disease in Patients with Newly Diagnosed High-Risk Prostate Cancer. *Cancers* **2020**, *12*, 160. [CrossRef] [PubMed]
33. Sheikhbahaei, S.; Afshar-Oromieh, A.; Eiber, M.; Solnes, L.B.; Javadi, M.S.; Ross, A.E.; Pienta, K.J.; Allaf, M.E.; Haberkorn, U.; Pomper, M.G.; et al. Pearls and pitfalls in clinical interpretation of prostate-specific membrane antigen (PSMA)-targeted PET imaging. *Eur. J. Nucl. Med. Mol. Imaging* **2017**, *44*, 2117–2136. [CrossRef] [PubMed]
34. Sheikhbahaei, S.; Werner, R.A.; Solnes, L.B.; Pienta, K.J.; Pomper, M.G.; Gorin, M.A.; Rowe, S.P. Prostate-Specific Membrane Antigen (PSMA)-Targeted PET Imaging of Prostate Cancer: An Update on Important Pitfalls. *Semin. Nucl. Med.* **2019**, *49*, 255–270. [CrossRef] [PubMed]

Article

Preoperative Predicting the WHO/ISUP Nuclear Grade of Clear Cell Renal Cell Carcinoma by Computed Tomography-Based Radiomics Features

Claudia-Gabriela Moldovanu [1,2], Bianca Boca [1,2,*], Andrei Lebovici [2,3,*], Attila Tamas-Szora [4], Diana Sorina Feier [2,3], Nicolae Crisan [5], Iulia Andras [5] and Mircea Marian Buruian [1,6]

1. Department of Radiology and Medical Imaging, Faculty of Medicine, George Emil Palade University of Medicine, Pharmacy, Science and Technology of Târgu Mureș, 540139 Târgu Mureș, Romania; moldovanu_claudia@yahoo.com (C.-G.M.); mircea.buruian@umfst.ro (M.M.B.)
2. Department of Radiology, Emergency Clinical County Hospital of Cluj-Napoca, 400006 Cluj-Napoca, Romania; diana.feier@umfcluj.ro
3. Department of Radiology, Faculty of Medicine, Iuliu Hațieganu University of Medicine and Pharmacy, 400012 Cluj-Napoca, Romania
4. Department of Radiology, Clinical Municipal Hospital, 400139 Cluj-Napoca, Romania; attitamas@yahoo.com
5. Department of Urology, Faculty of Medicine, Iuliu Hațieganu University of Medicine and Pharmacy, 400012 Cluj-Napoca, Romania; drnicolaecrisan@gmail.com (N.C.); dr.iuliaandras@gmail.com (I.A.)
6. Department of Radiology, Emergency Clinical County Hospital Târgu Mureș, 540136 Târgu Mureș, Romania
* Correspondence: bianca.petresc@gmail.com (B.B.); andrei1079@yahoo.com (A.L.)

Citation: Moldovanu, C.-G.; Boca, B.; Lebovici, A.; Tamas-Szora, A.; Feier, D.S.; Crisan, N.; Andras, I.; Buruian, M.M. Preoperative Predicting the WHO/ISUP Nuclear Grade of Clear Cell Renal Cell Carcinoma by Computed Tomography-Based Radiomics Features. *J. Pers. Med.* **2021**, *11*, 8. https://dx.doi.org/10.3390/jpm11010008

Received: 22 November 2020
Accepted: 21 December 2020
Published: 23 December 2020

Publisher's Note: MDPI stays neutral with regard to jurisdictional claims in published maps and institutional affiliations.

Copyright: © 2020 by the authors. Licensee MDPI, Basel, Switzerland. This article is an open access article distributed under the terms and conditions of the Creative Commons Attribution (CC BY) license (https://creativecommons.org/licenses/by/4.0/).

Abstract: Nuclear grade is important for treatment selection and prognosis in patients with clear cell renal cell carcinoma (ccRCC). This study aimed to determine the ability of preoperative four-phase multiphasic multidetector computed tomography (MDCT)-based radiomics features to predict the WHO/ISUP nuclear grade. In all 102 patients with histologically confirmed ccRCC, the training set ($n = 62$) and validation set ($n = 40$) were randomly assigned. In both datasets, patients were categorized according to the WHO/ISUP grading system into low-grade ccRCC (grades 1 and 2) and high-grade ccRCC (grades 3 and 4). The feature selection process consisted of three steps, including least absolute shrinkage and selection operator (LASSO) regression analysis, and the radiomics scores were developed using 48 radiomics features (10 in the unenhanced phase, 17 in the corticomedullary (CM) phase, 14 in the nephrographic (NP) phase, and 7 in the excretory phase). The radiomics score (Rad-Score) derived from the CM phase achieved the best predictive ability, with a sensitivity, specificity, and an area under the curve (AUC) of 90.91%, 95.00%, and 0.97 in the training set. In the validation set, the Rad-Score derived from the NP phase achieved the best predictive ability, with a sensitivity, specificity, and an AUC of 72.73%, 85.30%, and 0.84. We constructed a complex model, adding the radiomics score for each of the phases to the clinicoradiological characteristics, and found significantly better performance in the discrimination of the nuclear grades of ccRCCs in all MDCT phases. The highest AUC of 0.99 (95% CI, 0.92–1.00, $p < 0.0001$) was demonstrated for the CM phase. Our results showed that the MDCT radiomics features may play a role as potential imaging biomarkers to preoperatively predict the WHO/ISUP grade of ccRCCs.

Keywords: clear cell renal cell carcinoma; radiomics; WHO/ISUP nuclear grade; multiphasic multidetector computed tomography

1. Introduction

Clear cell renal cell carcinoma (ccRCC) encompasses around 70% of all renal cell carcinomas, making it the most common pathological subtype [1,2]. It has the worst prognosis of all types of RCC, and its biological aggressiveness significantly changes the prognosis [3].

Tumor grading is among the most important prognostic factors as an independent predictor of cancer-specific survival for ccRCC stages [4]. The World Health Organization/International Society of Urological Pathology (WHO/ISUP) grading system for ccRCC [5] has improved interobserver reproducibility compared to the former Fuhrman grading system, being easier to apply and more clinically relevant. This four-grade system is based primarily on nucleolar prominence assessed to determine grades 1–3. Grade 4 is defined by the presence of highly atypical "pleomorphic" cells and/or sarcomatoid or rhabdoid morphology. Grades 1–2 are classified as low grades, and grades 3–4 are classified as high grades. Patients with low-grade ccRCC may be candidates for less invasive procedures, such as nephron-saving surgery, radiofrequency ablation, or active surveillance, whereas radical interventions are acceptable in patients with high-grade ccRCC [6].

Percutaneous renal mass biopsy is an accurate procedure that can identify the histology of the lesions [7]. Due to the heterogeneity of ccRCCs, the accuracy of tumor grading through biopsy is controversial, as the biopsy shows some discrepancies of the resection sample for grading systems. Some studies focusing on renal tumor biopsies and tumor grading [8–11] have reported that biopsies usually underestimate the final grade and less often overestimate the final grade. The percentage of accurate biopsy grading was reported between 43% and 75%, and the percentage of differentiation between low and high grade was reported between 64% and 87%. Moreover, different parts of a tumor have distinct molecular characteristics and such differences change over time. Thus, optimal characterization of tumor grading by percutaneous biopsy cannot be obtained properly because it is not possible to biopsy each part of a tumor at different times [12,13].

The field of medical and biological image analysis has recently grown exponentially, and a new method called radiomics has been developed [14–16]. Radiomics is a promising technique that extracts and analyzes large numbers of imaging features to provide more information than only human imaging evaluation can offer. This method uses high-throughput extraction of large numbers of quantitative radiomics features obtained from medical images using advanced mathematical algorithms to determine tumor phenotypes [17–19]. Thus, the heterogeneity of the entire tumor volume is assessed compared to biopsies that assess the heterogeneity in a small portion of the tumor and at a single anatomical site [20–24]. Several previous studies [25–32] have shown that radiomics features based on multiphasic multidetector computed tomography (MDCT) images perform efficiently in differentiating between high-/low-grade ccRCC tumors. Given these promising results, we assume that MDCT-based radiomics features may play a feasible role in predicting high-/low-grade ccRCCs. This study aims to evaluate if radiomics features extracted from a four-phase MDCT study may be helpful to preoperatively differentiate the WHO/ISUP nuclear grades of ccRCCs.

2. Materials and Methods

The ethical approval for this retrospective study was obtained from the Institutional Review Board of Clinical Municipal Hospital of Cluj-Napoca (Approval Code: No. 15/2020; Approval Date: 11 June 2020). No formal written consent was required for this study.

2.1. Study Population

We performed a retrospective analysis of the medical database for patients with pathologically proven ccRCC from January 2018 to February 2020. The inclusion criteria were as follows: patients with four-phase MDCT scan before surgery; WHO/ISUP nuclear grades, which were available from the pathology reports. The exclusion criteria were: significant artifacts on images (motion or metal artifacts), previous tumor treatment, and patients with cystic lesions. Our study comprised 102 patients (mean age: 61.92 ± 13.03), which were divided into two groups: the training set (62 patients) and the validation set (40 patients).

2.2. Image Acquisition

MDCT scans were performed with a 64-row scanner (Somatom Sensation, Siemens, Erlangen, Germany) using: a 120 kV variable tube current (variable setting from 200 to 400 mAs based on patient size); section collimation, 0.6 mm; table feed, 5 mm/s; slice thickness, 3.0 mm; and a pitch of 1. Nonionic contrast material was injected via an antecubital vein at a rate of 3.0 mL/s using a CT-compatible power injector with a total volume of 80–150 mL. A region of interest (ROI) in the thoracoabdominal aorta junction was placed, with a trigger set to begin at 150 HU. The renal mass protocol consisted of a four-phase study: an unenhanced (UN) scan followed by contrast-enhanced acquisitions during the corticomedullary (CM, 30 s delay), nephrographic (NP, 90 s delay), and excretory (EX, 8 min delay) phases.

2.3. Histopathological Assessment of Nuclear Grade

WHO/ISUP nuclear grades were obtained from the pathology reports of the histopathological examination. The samples were obtained from the partial nephrectomy of 22 patients, total nephrectomy of 13 patients, and radical nephrectomy of 67 patients. All tumors were divided into low-grade ccRCC (WHO/ISUP grades 1 and 2) and high-grade ccRCC (WHO/ISUP grades 3 and 4).

2.4. Tumor Segmentation, Preprocessing, and Radiomics Feature Extraction

From the pictured archiving and communication system (PACS, Carestream, Canada), all MDCT acquisitions were exported and transferred to a workstation to be segmented using the open-source 3D Slicer software, version 4.10.2 (www.slicer.org). For each renal mass, the volume of interest (VOI) segmentation was manually and slightly delineated slice by slice by a radiology resident (Claudia-Gabriela Moldovanu), in accordance with a senior radiologist with 9 years of experience in urogenital imaging (Attila Tamas-Szora) to ensure the accuracy of the tumor boundaries. The two radiologists were blinded to the pathological results. To minimize the partial volume effect from surrounding structures, the segmentations were carefully delineated, reducing the size of the tumors by 1 mm from the current visible edge. The nephrographic phase was used for segmentation because it provides an adequate delimitation between the tumor and uninvolved adjacent parenchyma (Figure 1).

Prior to radiomics features extraction, the images of each patient were preprocessed: first, the images and VOIs were resampled to an isotropic voxel size of $1 \times 1 \times 1$ mm^3 using B-Spline interpolation; then, normalization of the images was performed by centering in the chosen place by division through standard deviation; finally, image discretization was performed of the gray level by a fixed bin width of 25 in the histogram.

A total of 4184 radiomics features of the four-phase MDCT study per patient (1046 features per phase) were extracted from the VOIs and divided into four groups: (1) image intensity (first-order statistics features); (2) shape and size-based features; (3) second-order statistics features (textural features); and (4) higher-order statistical features (obtained after applying filters and mathematical transforms to the images). We used Laplacian transforms of Gaussian-filter- and wavelet-transformed images. The Laplacian of Gaussian (LoG) filter was used with values of 2 mm, 4 mm, and 6 mm, representing fine, medium, and coarse patterns, respectively. Wavelet-based texture features were generated using eight different frequency band combinations, applying either a high- or low-pass filter in each of the three dimensions including high–high–high, high–high–low, high–low–low, high–low–high, low–high–low, low–high–high, low–low–high, and low–low–low. Radiomics features were extracted from images with and without preprocessing filters from all four MDCT phases separately. PyRadiomics version 2.1.2. was used for both preprocessing and feature extraction.

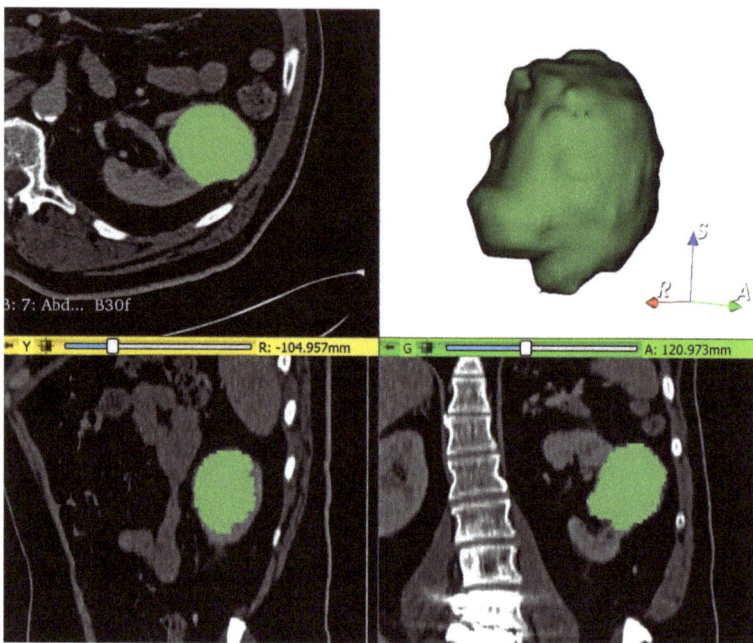

Figure 1. Example of volume of interest (VOI) segmentation in the nephrographic (NP) phase of a pathologically proven clear cell renal cell carcinoma (ccRCC).

2.5. Reliability Validation of Texture Features

According to previously published guidelines [33–35], the reproducibility of texture features was calculated using the interclass correlation coefficient (ICC) of the radiomics features. Another radiologist (Andrei Lebovici, with 8 years of experience in urogenital imaging) independently resegmented all renal masses and extracted radiomics features, also blinded by the pathological results. Thus, for each extracted texture features, the ICC was calculated. For the feature selection process, the features with an ICC value of ≥ 0.75 were included, indicating excellent reproducibility, resulting in a total of 3429 features (826 in the UN phase, 861 in the CM phase, 864 in the NP phase, and 878 in the EX phase).

2.6. Statistical Analysis

Statistical analysis was performed using SPSS Statistics software for Windows, version 18.0 (SPSS Inc., Chicago, IL, USA) and R software version 3.6.3 using the "glmnet" package. The Mann–Whitney U-test was used for univariate analysis to identify the features with a significant difference between low/high-grade ccRCC groups. The Benjamini–Hochberg (BH) correction method was applied to control the false discovery rate in multiple hypothesis testing. BH-adjusted p-values < 0.05 were considered significant. Spearman's correlation coefficient was used to assess the correlation between all radiomics features. This was performed between any two features, and when the Spearman coefficient was $> 0.9/< -0.9$, the feature with the higher p-value in the univariate analysis was eliminated. For standard comparison and mitigating the effects of the data splitting, the radiomics scores were built using the least absolute shrinkage and selection operator (LASSO) performed by 10-fold cross-validation. The radiomics score (Rad-Score) was computed for each MDCT phase of each patient through a linear combination of features weighted by their LASSO coefficients. To evaluate the predictive performance of the radiomics score for the differentiating ability of low/high-grade ccRCC in the training and validation sets, the area under the receiver operating characteristic (ROC) curve (AUC) was used, and $p < 0.05$ was considered statistically significant. Multivariate analysis using binary logistic regression (enter method) was conducted to detect independent predictors of the WHO/ISUP nuclear grade of ccRCCs, including the clinicoradiological characteristics and radiomics score as independent variables.

3. Results

3.1. Patients Characteristics

A total of 102 patients (mean age: 61.92 ± 13.03) were included in this study, divided into training sets and validation sets based on the random split method. Thus, 62 patients constituted the training set (40 men, 22 women; mean age: 61.09 ± 12.64), whereas 40 patients constituted the validation set (27 men, 13 women; mean age: 63.2 ± 13.66). In the training set, 40 patients were classified according to the WHO/ISUP grading system as low-grade ccRCC, and the remaining 22 patients were classified as high-grade ccRCC. The validation set comprised 29 patients with low-grade ccRCC and 11 patients with high-grade ccRCC. The baseline characteristics of training and validation sets are provided in Table 1.

Table 1. Demographic and clinicoradiological characteristics of the study population. * $p < 0.05$ was considered statistically significant.

	Training Set			Validation Set		
Characteristic	Low Grade	High Grade	p-Value	Low Grade	High Grade	p-Value
Number	40	22		29	11	
Age (years)	58.2 ± 12.92	66.36 ± 10.45	0.009 *	61.89 ± 13.30	66.63 ± 14.66	0.36
Gender			0.91			1.00
Male	26 (65%)	14 (35%)		20 (74.1%)	7 (25.9%)	
Female	14 (63.6%)	8 (36.4%)		9 (69.2%)	4 (30.8%)	
Tumor size (mm)	46.65 ± 28.53	73.22 ± 26.25	0.001 *	53.17 ± 22.68	79.45 ± 25.15	0.008 *
Tumor stage (n)			0.001 *			0.01 *
1	30 (85.7%)	5 (14.3%)		18 (94.7%)	1 (5.3%)	
2	3 (50%)	3 (50%)		6 (100%)	-	
3	7 (35%)	13 (65%)		5 (35.7)	9 (64.3%)	
4	-	1 (100%)		-	1 (100%)	
Vein thrombosis			0.02 *			0.009 *
No	33 (73.3%)	12 (26.6%)		22 (88%)	3 (12%)	
Yes	7 (41.1%)	10 (58.8%)		7 (46.6%)	8 (53.3%)	
Tumor necrosis			0.47			1.00
No	7 (77.7%)	2 (22.2%)		2 (66.6%)	1 (33.3%)	
Yes	33 (60%)	22 (40%)		27 (72.9%)	10 (27.0%)	
Perinephritic invasion			0.009 *			0.49
No	34 (73.9%)	12 (26.0%)		12 (80%)	3 (20%)	
Yes	6 (37.5%)	10 (62.5%)		17 (68%)	8 (32%)	
Intratumoral neovascularity			0.003 *			0.29
No	30 (78.9%)	8 (21.0%)		11 (85.6%)	2 (15.3%)	
Yes	10 (41.6%)	14 (58.3%)		18 (66.6%)	9 (33.3%)	
Hemorrhage			0.01 *			0.48
No	32 (74.4%)	11 (25.5%)		16 (80%)	4 (20%)	
Yes	8 (42.1%)	11 (57.8%)		13 (65%)	7 (35%)	
Lymphadenopathy			0.05			0.12
No	35 (71.4%)	14 (28.5%)		27 (77.1%)	8 (22.8%)	
Yes	5 (38.4%)	8 (61.5%)		2 (40%)	3 (60%)	

No significant difference in the gender of the patients, N stage, and intratumoral necrosis between low- and high-grade ccRCC in both the training and validation sets was observed. However, significant differences were observed in the ages of the patients, tumor size, tumor stage, vein thrombosis, perinephric fat invasion, intratumoral neovascularity, and intratumoral hemorrhage in the training set. These results are partially confirmed in the validation set, where tumor size, tumor stage, and vein thrombosis were the only significantly different characteristics.

3.2. Feature Selection and Radiomics Score Building: Training Set

Feature selection and radiomics score building were separately performed on each MDCT phase of each patient. According to the standard of ICC $\geq$ 0.75 in the inter-reader agreement evaluation, 826 radiomics features from the UN phase, 861 features from the CM phase, 864 features from the NP phase, and 878 features from the EX phase were highly reproducible and selected for further analysis.

To develop the radiomics signature, univariate analysis was performed to assess the potential of the radiomics features to differentiate between the low- and high-grade ccRCC groups. Excluding those with an adjusted p-value > 0.05, the number of features was further decreased to 1241 features (228 in the UN phase, 387 in the CM phase, 340 in the NP phase, and 286 in the EX phase). These features were included in the further selection process.

After applying the Spearman correlation analysis, these features were secondly reduced to 302 potential predictors (46 in the UN phase, 110 in the CM phase, 85 in the NP phase, and 61 in the EX phase). Furthermore, the LASSO binary logistic regression algorithm was used to reduce the dimensionality of the above high-dimensional features; thus, the best features were selected based on the optimal λ parameters. Forty-eight radiomics features with nonzero coefficients were then selected to construct the radiomics scores across all MDCT phases (10 in the UN phase, 17 in the CM phase, 14 in the NP phase, and 7 in the EX phase). Most of the features included in the radiomics scores were obtained from filtered images using LoG and wavelet-transformed filters, being mainly texture and first-order features (Table 2).

Table 2. List of selected radiomics features and their coefficients for calculating the radiomics score.

Radiomic Group	Radiomic Feature	Associated Filter	Coefficient
UN phase			
	Intercept		−0.872
Texture feature	JointAverage	LoG filter (2 mm)	0.409
Texture feature	SizeZoneNonUniformity	LoG filter (2 mm)	0.010
Texture feature	DependenceVariance	LoG filter (4 mm)	0.362
First-order	Minimum	LoG filter (4 mm)	−0.296
Texture feature	LongRunEmphasis	LoG filter (4 mm)	0.477
Texture feature	SmallAreaHighGrayLevelEmphasis	LoG filter (4 mm)	0.091
Texture feature	LargeAreaLowGrayLevelEmphasis	Wavelet-LHL	0.039
Texture feature	LongRunLowGrayLevelEmphasis	Wavelet-LLH	−0.431
Texture feature	SmallAreaLowGrayLevelEmphasis	Wavelet-LLH	−0.349
Texture feature	LongRunLowGrayLevelEmphasis	Wavelet-HHL	0.343
CM phase			
	Intercept		−1.184
Texture feature	SmallAreaLowGrayLevelEmphasis	Original	−0.387
First-order	Skewness	LoG filter (2 mm)	−0.311
First-order	Minimum	LoG filter (2 mm)	−0.052
First-order	10Percentile	LoG filter (2 mm)	0.303
Texture feature	LowGrayLevelEmphasis	LoG filter (4 mm)	−0.373
Texture feature	LongRunHighGrayLevelEmphasis	Wavelet-HLL	0.306
Texture feature	LowGrayLevelZoneEmphasis	Wavelet-HLL	−0.076
Texture feature	Imc2	Wavelet-LHL	0.797
First-order	Mean	Wavelet-LHL	0.516

Table 2. Cont.

Radiomic Group	Radiomic Feature	Associated Filter	Coefficient
Texture feature	GrayLevelNonUniformity	Wavelet-LHL	−0.153
Texture feature	SmallAreaEmphasis	Wavelet-LHL	0.823
Texture feature	LongRunLowGrayLevelEmphasis	Wavelet-LLH	−0.429
First-order	Maximum	Wavelet-HLH	0.583
Texture feature	GrayLevelVariance	Wavelet-HHL	0.084
First-order	Entropy	Wavelet-HHL	0.049
Texture feature	RunVariance	Wavelet-HHL	0.064
Texture feature	ShortRunLowGrayLevelEmphasis	Wavelet-LLL	−0.379
NP phase			
	Intercept		−0.765
Texture feature	HighGrayLevelRunEmphasis	Original	0.325
Texture feature	ShortRunHighGrayLevelEmphasis	LoG filter (6 mm)	−0.087
Texture feature	Imc2	Wavelet-HLL	0.191
Texture feature	ShortRunHighGrayLevelEmphasis	Wavelet-HLL	0.225
Texture feature	Contrast	Wavelet-LHL	0.192
Texture feature	SmallAreaHighGrayLevelEmphasis	Wavelet-LHL	0.353
Texture feature	ZoneEntropy	Wavelet-LHL	0.049
First-order	Entropy	Wavelet-LHH	0.070
Texture feature	DependenceNonUniformityNormalized	Wavelet-LLH	0.013
Texture feature	SumEntropy	Wavelet-HLH	−0.190
Texture feature	Imc2	Wavelet-HLH	0.331
Texture feature	GrayLevelVariance	Wavelet-HHL	0.019
Texture feature	Idn	Wavelet-LLL	0.223
Texture feature	SmallAreaLowGrayLevelEmphasis	Wavelet-LLL	−0.189
EX phase			
	Intercept		−0.653
Texture feature	DependenceVariance	LoG filter (4 mm)	−0.234
First-order	Kurtosis	LoG filter (4 mm)	0.139
Texture feature	RunVariance	LoG filter (4 mm)	0.163
Texture feature	SizeZoneNonUniformity	LoG filter (4 mm)	0.032
Texture feature	DependenceNonUniformityNormalized	Wavelet-HLL	0.165
Texture feature	SmallDependenceLowGrayLevelEmphasis	Wavelet-HLL	−0.046
Texture feature	SmallAreaHighGrayLevelEmphasis	Wavelet-LHL	0.028

A significant difference in the radiomics scores between low- and high-grade ccRCCs in all MDCT phases, with patients from the second group having higher values (Table 3), was observed. Rad-Score was calculated according to the following formula:

$$\text{Rad} - \text{Score} = \sum_{c=0}^{a} Cc * Xc + b$$

where a is the number of radiomics features with nonzero coefficients for each MDCT phase (10 for the UN phase, 17 for the CM phase, 14 for the NP phase, and 7 for the EX phase), Cc is the coefficient of the cth feature, Xc the cth feature, and b the intercept.

3.3. Performance of the Radiomics Scores: Training Set

To compare the detection performance, the Rad-Scores were validated in terms of ROC curve and AUC in the training set (Figure 2). Sensitivity, specificity, positive predictive value (PPV), and negative predictive value (NPV) were also calculated. The radiomics scores showed a favorable predictive efficacy for differentiating low- from high-grade ccRCC based on each phase of the MDCT protocol. The results are summarized in Table 4. In the training set, the Rad-Scores derived from the UN and CM phases achieved the best predictive ability, with a sensitivity, specificity, and an AUC of 81.82%, 92.50%, and 0.89 in the UN phase and 90.91%, 95.00%, and 0.97 in the CM phase.

Table 3. Difference of the radiomics score (Rad-Score) between low- and high-grade ccRCC in the training and validation sets.

WHO/ISUP Nuclear Grades	Radiomic Score Mean ± SD	p-Value	MDCT Phase
Training set			
Low grade (n = 40)	−1.68 ± 1.16	$p < 0.001$	UN
	−2.50 ± 1.95	$p < 0.001$	CM
	−1.26 ± 0.68	$p < 0.001$	NP
	−0.92 ± 0.53	$p < 0.001$	EX
High grade (n = 22)	0.60 ± 1.34	$p < 0.001$	UN
	1.21 ± 1.29	$p < 0.001$	CM
	0.15 ± 1.18	$p < 0.001$	NP
	−0.16 ± 0.51	$p < 0.001$	EX
Validation set			
Low grade (n = 29)	−1.18 ± 1.70	$p = 0.051$	UN
	−2.21 ± 2.42	$p < 0.001$	CM
	−1.12 ± 0.72	$p = 0.001$	NP
	−0.80 ± 0.62	$p = 0.009$	EX
High grade (n = 11)	−0.03 ± 1.32	$p = 0.051$	UN
	1.53 ± 3.43	$p < 0.001$	CM
	0.19 ± 1.56	$p = 0.001$	NP
	−0.24 ± 0.45	$p = 0.009$	EX

Table 4. Radiomic score performance in the training and validation sets in all MDCT phases.

Variable	AUC (95% CI)	Se (95% CI)	Sp (95% CI)	PPV (95% CI)	NPV (95% CI)	Cut-Off Value	p-Value
Training set							
Radiomic score: UN phase	0.89 (0.796–0.961)	81.82 (59.7–94.8)	92.50 (79.6–98.4)	85.7 (63.7–97.0)	90.2 (76.9–97.3)	−0.34	<0.001
Radiomic score: CM phase	0.97 (0.89–0.99)	90.91 (70.8–98.9)	95.00 (83.1–99.4)	90.9 (70.8–98.9)	95.0 (83.1–99.4)	−0.25	<0.001
Radiomic score: NP phase	0.87 (0.76–0.94)	81.82 (59.7–94.8)	92.50 (79.6–98.4)	85.7 (63.7–97.0)	90.2 (76.9–97.3)	−0.55	<0.001
Radiomic score: EX phase	0.85 (0.73–0.92)	72.73 (49.8–89.3)	87.50 (73.2–95.8)	76.2 (52.8–91.8)	85.4 (70.8–94.4)	−0.34	<0.001
Validation set							
Radiomic score: UN phase	0.72 (0.56–0.85)	72.73 (39.0–94.0)	72.41 (52.8–87.3)	50.0 (24.7–75.3)	87.5 (67.6–97.3)	−0.43	0.0157
Radiomic score: CM phase	0.81 (0.66–0.92)	72.73 (39.0–94.0)	75.90 (56.5–89.7)	53.3 (26.6–78.7)	88.0 (68.8–97.5)	−0.85	<0.001
Radiomic score: NP phase	0.84 (0.69–0.93)	72.73 (39.0–94.0)	85.30 (68.3–96.1)	66.7 (34.9–90.1)	89.3 (71.8–97.7)	−0.58	<0.001
Radiomic score: EX phase	0.77 (0.61–0.89)	100.0 (71.5–100.0)	51.72 (32.5–70.6)	44.0 (24.4–65.1)	100.0 (78.2–100.0)	−0.88	<0.001

Using the variables with a significant difference among low- and high-grade ccRCCs in the training set (including age, tumor size, vein thrombosis, perinephric invasion, tumor stage (2–4), intratumoral neovascularity, and hemorrhage), we conducted a multivariate logistic regression analysis to develop a clinicoradiological model for the preoperative prediction of the WHO/ISUP nuclear grade of ccRCCs (Table 5). Further, we constructed four complex models, adding the radiomics score of each phase to the clinicoradiological model (Table 6).

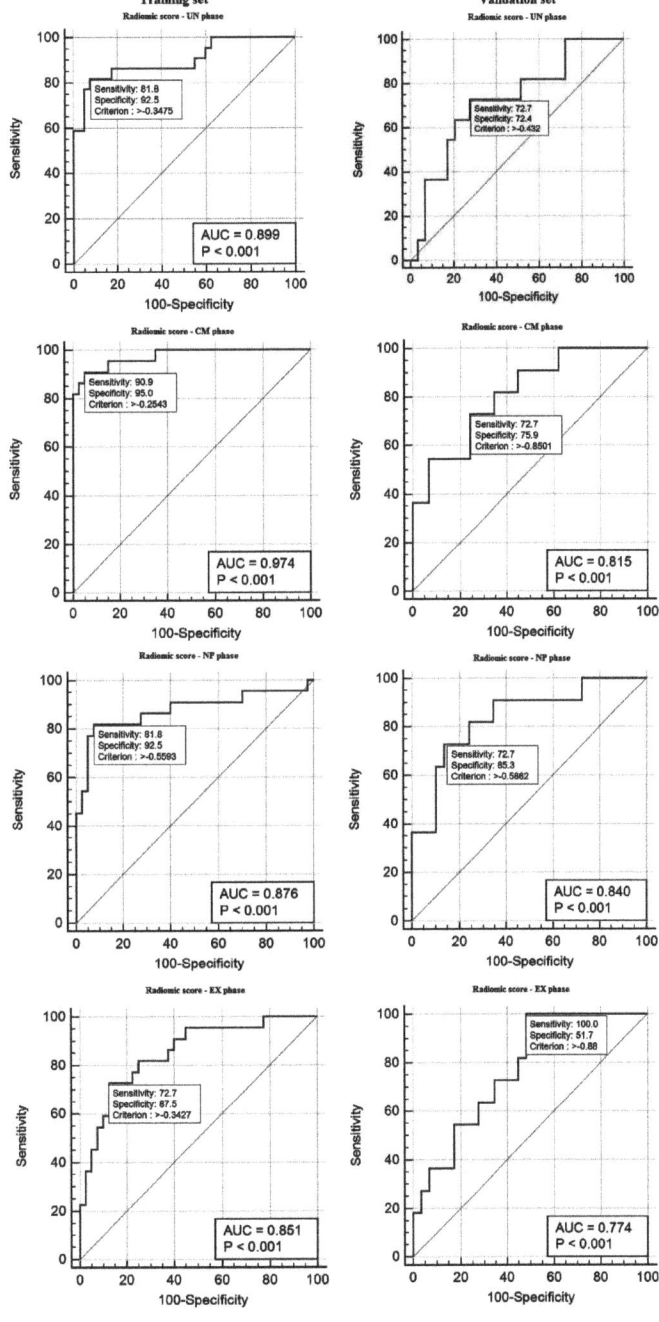

Figure 2. ROC curves of radiomics scores of all MDCT phases in the training and validation sets. ROC, receiver operating characteristic, AUC area under the curve.

Table 5. Multivariate logistic regression analysis for the preoperatively prediction of the WHO/ISUP nuclear grade of ccRCCs: clinicoradiological model.

Variable	Coefficient	Std. Error	p-Value	Odds Ratio (OR)
Age (years)	0.05	0.03	0.10	1.05
Tumor size (mm)	0.006	0.01	0.74	1.00
Vein thrombosis: positive	−1.32	1.22	0.27	0.26
Perinephric invasion: positive	1.60	1.39	0.25	4.98
Tumor stage (2, 3, or 4)	2.09	1.07	0.05	8.13
Intratumoral neovascularity: positive	1.04	0.99	0.29	2.85
Hemorrhage: positive	−1.81	1.55	0.24	0.16
Constant	−5.55			

Table 6. Multivariate logistic regression analysis for the preoperative prediction of the WHO/ISUP nuclear grade of ccRCCs: complex model.

Variable	Coefficient	Std. Error	p-Value	Odds Ratio (OR)
UN phase				
Age (years)	0.03	0.05	0.4997	1.03
Tumor size (mm)	−0.02	0.02	0.4156	0.97
Vein thrombosis: positive	1.64	1.81	0.3653	5.16
Perinephric invasion: positive	−0.17	1.73	0.9176	0.83
Tumor stage (2, 3, or 4)	0.99	1.42	0.4860	2.70
Intratumoral neovascularity: positive	0.07	1.17	0.9485	1.07
Hemorrhage: positive	−1.75	1.87	0.34	0.17
Radiomic score: UN phase	1.83	0.59	0.0021	6.27
Constant	−0.90			
CM phase				
Age (years)	0.12	0.09	0.1772	1.13
Tumor size (mm)	−0.08	0.07	0.2576	0.91
Vein thrombosis: positive	1.13	16.78	0.9460	3.11
Perinephric invasion: positive	2.76	17.36	0.8735	15.86
Tumor stage (2, 3, or 4)	3.62	3.38	0.2837	37.59
Intratumoral neovascularity: positive	−2.36	2.85	0.4075	3.11
Hemorrhage: positive	−4.07	17.52	0.8161	0.01
Radiomic score: CM phase	4.92	2.37	0.0384	137.75
Constant	−1.50			
NP phase				
Age (years)	0.11	0.05	0.03	1.12
Tumor size (mm)	−0.03	0.02	0.2121	0.96
Vein thrombosis: positive	−3.15	1.68	0.0610	0.04
Perinephric invasion: positive	3.88	2.23	0.0827	48.76
Tumor stage (2, 3, or 4)	2.19	1.50	0.1449	8.98
Intratumoral neovascularity: positive	0.34	1.24	0.7830	1.41
Hemorrhage: positive				
Radiomic score: NP phase	2.78	0.91	0.0023	16.17
Constant	−4.64			
EX phase				
Age (years)	0.05	0.04	0.2408	1.05
Tumor size (mm)	−0.05	0.03	0.0879	0.94
Vein thrombosis: positive	−0.24	1.42	0.8639	0.78
Perinephric invasion: positive	1.01	1.57	0.5184	2.77
Tumor stage (2, 3, or 4)	1.87	1.19	0.1170	6.49
Intratumoral neovascularity: positive	0.09	1.07	0.9263	1.10
Hemorrhage: positive	−1.20	1.71	0.4832	0.29
Radiomic score: EX phase	4.64	1.75	0.0081	103.88
Constant	1.40			

The ability of the clincoradiological model and the complex model to categorize nuclear grades was evaluated by the AUC of the ROC curves (Figure 3). The clincoradiological model showed a high performance in the discrimination of low- and high-grade ccRCCs with an AUC of 0.83 (95% CI, 0.71–0.91, $p < 0.0001$). We found that the addition of radiomics score to the clincoradiological characteristics improved their performance in all MDCT phases: AUC = 0.93 (95% CI, 0.84–0.98, $p < 0.0001$) vs. AUC = 0.89 (95% CI, 0.79–0.96, $p < 0.0001$) in the UN phase, AUC = 0.99 (95% CI, 0.92–1.00, $p < 0.0001$) vs. AUC = 0.97 (95% CI, 0.89–0.99, $p < 0.0001$) in the CM phase, AUC = 0.91 (95% CI, 0.81–0.97, $p < 0.0001$) vs. AUC = 0.87 (95% CI, 0.76–0.94, $p < 0.0001$) in the NP phase, AUC = 0.87 (95% CI, 0.77–0.94, $p < 0.0001$) vs. AUC = 0.85 (95% CI, 0.73–0.92, $p < 0.0001$) in the EX phase.

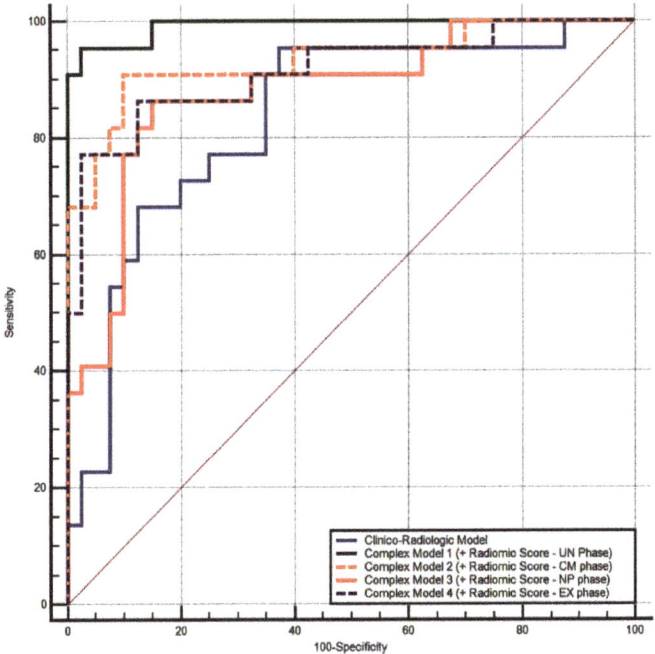

Figure 3. ROC curves of the clincoradiological model and complex models to categorize the nuclear grades of ccRCC.

3.4. Validation of the Radiomics Score

The performance of the Rad-Scores for the discrimination of low- and high-grade ccRCCs was confirmed in the validation set in each MDCT phase of each patient (Table 3). ROC curve analysis was conducted, and the AUC, sensitivity, specificity, PPV, and NPV for the determined cut-off values were calculated. The results are presented in Figure 2 and Table 4. Compared with the training set, in the validation set, the Rad-Scores derived from the CM and NP phases achieved the best predictive ability, with a sensitivity, specificity, and an AUC of 72.73%, 75.90%, and 0.81 in the CM phase and 72.73%, 85.30%, and 0.84 in the NP phase.

4. Discussion

In this study, we evaluated if radiomics features extracted from a four-phase MDCT study may be helpful to preoperatively differentiate the WHO/ISUP nuclear grades of ccRCC. In the era of personalized medicine, radiomics features, along with metabolic, histopathologic, and genetic datasets, may be useful to improve patient management, a biomarker that could be useful in tumor characterization, treatment selection, and prognosis [36–39]. Many radiomics features have proven to be useful in differentiating between early- and advanced-stage diseases of various types of cancers [40–43]. In recent years, concerning renal imaging, little research [25–32] has investigated the radiomics potential based on MDCT to predict the ccRCC nuclear grade. Regarding the histologic tumor grading system, the majority of studies used the Fuhrman classification system as a pathological reference. Although the Fuhrman and WHO/ISUP grading systems are linboth used in current medical practice for ccRCC grading, some studies [44–47] have reported that the Fuhrman grading system has poor interobserver reproducibility compared to the new WHO/ISUP grading system.

In recent years, the WHO/ISUP grading system has been accepted in current medical practice, replacing the former Fuhrman grading system. To the best of our knowledge, there are only a few published papers that have studied radiomics features based on MDCT for predicting the ccRCC WHO/ISUP nuclear grade [29,48–51]. However, no previous work used parameters extracted from a four-phase MDCT study to develop the prediction model, as our study does.

Our results show that our constructed MDCT-based radiomics scores using a four-phase protocol achieved a considerably promising performance in differentiating high- from low-grade ccRCCs. The Rad-Scores derived from the UN and CM phases achieved the best predictive ability in the training set. However, in the validation set, the Rad-Scores from the CM and NP phases achieved the best predictive ability. We found that the best predictive ability with an AUC of 0.94 was for the CM phase in the training dataset and 0.84 was for the NP phase in the validation datasets. This diversity illuminates that the CM and NP phases are valuable and necessary for ccRCC grading. Our results are in concordance with the results of previous studies on ccRCC grading using texture analysis or machine learning (ML), which reported an accuracy between 0.78 and 0.82 and an AUC between 0.71 and 0.98 [29,48,49].

Our feature selection results showed that the first-order features and second-order statistics features were significantly associated with the WHO/ISUP grade. In building our radiomics scores, most of the features included were obtained from filtered images, especially from wavelet-transformed filters. Shu et al. [50] used two predictive models constructed by radiomics features extracted from the nephrographic and medullary phases and reported no significant difference in the AUC between them to differentiate low- from high-grade ccRCC. Conversely, they showed that the combined model of radiomics features from two certain phases had the highest differential diagnostic efficiency (AUC: 0.82 (95% CI: 0.76–0.86). A recent study [51] showed that the value of the NP phase is limited in predicting the ISUP grade. This may be due to two reasons: firstly, regarding tumor delineation, Sun et al. used a single-slice approach (largest cross-section diameter of the tumor) and did not perform data analysis of the entire tumor VOI. Although VOIs segmentations are time-consuming processes, we believe that the single-slice approach does not fully reflect the heterogeneity of the tumor, and the information obtained from the VOI might be more reliable for the characterization of the tumor. Secondly, their features extraction algorithm is different; they extracted the radiomics features from original and wavelet-filtered images, without the use of LoG filters. It is known that filtered-based images can limit the impact of technical noise [52]. More and more studies are using them, but a current technical standardization regarding their use has not yet been established [53].

MRI-derived ADC values are useful in characterizing tumor activity [54]. Some studies [55,56] that evaluated the utility of ADC to differentiate low- from high-grade ccRCC reported that MRI has a favorable predictive accuracy in detecting high-grade ccRCC (AUC = 0.80). With all its advantages, MRI is not as widely used as MDCT for the analysis of renal masses, being used only in selected cases. Cui et al. [48] used MRI- and CT-based radiomics models to differentiate low- from high-grade ccRCCs, and then the authors compared their performance. They reported that radiomics models based on a three-phase

MDCT performed better than the radiomics model based on a single-phase MDCT, with an ACC ranging from 77 to 79% in internal validation and 61 to 69% in external validation.

Similarly, the radiomics model based on all-sequence MRI was also superior to the radiomics model based on single-sequence MRI, with an ACC ranging from 71 to 73% in internal validation and 64 to 74% in external validation. When comparing the performance between MDCT and MRI, they found that the MRI-based radiomics model had performed better than the MDCT-based radiomics model for diagnosing low-grade ccRCC and showed a similar ability for diagnosing high-grade ccRCC.

Regarding the statistical approach, our study included one classification method: the binary logistic regression method. This algorithm is used to predict the probability of the class of a categorical dependent variable [57]. Several studies [58,59] have assessed the performance of quantitative MDCT texture analysis combined with different machine-learning-based classifiers to discriminate low- from high-grade ccRCC. It was observed that the highest predictive performance was obtained by the support vector machine classifier. However, these results were obtained for the Fuhrman grading system of ccRCC. Despite differences in the procedure followed, we believe that all studies support each other with the same conclusion that MDCT-based radiomics features may be a promising noninvasive method in predicting preoperative ccRCC grades.

In this study, the radiomics scores combined with the clinicoradiological characteristics showed a high performance in the discrimination of ccRCC grades. Two characteristics (age and tumor stage) were consistent with previous studies [29,56]. Li et al. proved a correlation between radiological characteristics and the ccRCC nuclear grade [28]. Shape, margin, and necrosis may be independent predictors of high-grade ccRCC, whereas a regular shape can often be seen in low-grade ccRCC lesions [28]. Another paper [60] demonstrated statistically significant differences in WHO/ISUP grading and pT staging between ccRCCs. In addition, they found that coagulative necrosis often occurs in high-grade and high-stage tumors.

This study may have important practical implications. The new WHO/ISUP grading system is a prognostic factor for ccRCCs. ccRCC grades were strongly related to patient outcomes and tumor biological behavior [61,62]. If low-grade tumors can be identified preoperatively, the treatment would consist of less invasive procedures. Moreover, partial nephrectomy can preserve partial renal function, thus reducing overall mortality and the incidence of cardiovascular disease [63]. Therefore, medical images can become a valuable source of information, and radiomics features may be used as a noninvasive method for characterizing and classifying lesions. However, further larger prospective studies to validate the performance of our proposed radiomics model in differentiating high from low-grade ccRCC are necessary for the future.

The present study has some limitations. (1) It was a single-center retrospective study with a small sample size of patients. (2) The statistical approach included one classification method, the binary logistic regression method, and advanced classifiers may offer better prediction performance. (3) External validation in more centers with more samples size is needed to overcome these limitations and validate the results in order to improve generalization and evaluate the potential for clinical translation of our radiomics models. (4) Volume effect interference cannot be completely avoided due to the fact that the tumor boundary was manually drawn. (5) The four-phase MDCT renal mass protocol involves a high dose of radiation to the patient and should be performed where it is necessary to discriminate the lesions before treatment selection.

5. Conclusions

Although there are limitations with regard to sample size, we have shown that radiomics features extracted from the four-phase MDCT study may play a role as a potential imaging biomarker to predict preoperatively the WHO/ISUP grade of ccRCCs, helping urologists to better stratify patients and choose the best treatment.

Author Contributions: Conceptualization, C.-G.M., A.L., A.T.-S., D.S.F., and M.M.B.; methodology, C.-G.M., A.L., A.T.-S., N.C., D.S.F., and I.A.; software, C.-G.M., A.T.-S., and A.L.; formal analysis, B.B. and D.S.F.; investigation, C.-G.M., A.L., and A.T.-S.; resources, C.-G.M.; data curation, C.-G.M. and B.B.; writing—original draft preparation, C.-G.M.; writing—review and editing, A.L., A.T.-S., N.C., D.S.F., B.B., I.A., and M.M.B.; visualization, C.-G.M. and M.M.B.; supervision, M.M.B.; project administration, M.M.B. All authors have read and agreed to the published version of the manuscript.

Funding: This research received no external funding.

Conflicts of Interest: The authors declare no conflict of interest.

References

1. Medina-Rico, M.; López-Ramos, H.; Lobo, M.; Romo, J.; Prada, J.G. Epidemiology of renal cancer in developing countries: Review of the literature. *Can. Urol. Assoc. J.* **2018**, *12*, 154–162. [CrossRef] [PubMed]
2. Znaor, A.; Lortet-Tieulent, J.; Laversanne, M.; Jemal, A.; Bray, F. International Variations and Trends in Renal Cell Carcinoma Incidence and Mortality. *Eur. Urol.* **2015**, *67*, 519–530. [CrossRef] [PubMed]
3. Muglia, V.F.; Prando, A. Renal cell carcinoma: Histological classification and correlation with imaging findings. *Radiol. Bras.* **2015**, *48*, 166–174. [CrossRef] [PubMed]
4. Zhang, G.; Wu, Y.; Zhang, J.; Fang, Z.; Liu, Z.; Xu, Z.; Fan, Y. Nomograms for predicting long-term overall survival and disease-specific survival of patients with clear cell renal cell carcinoma. *Onco Targets Ther.* **2018**, *11*, 5535–5544. [CrossRef]
5. Perrino, C.M.; Cramer, H.M.; Chen, S.; Idrees, M.T.; Wu, H.H. World Health Organization (WHO)/International Society of Urological Pathology (ISUP) grading in fine-needle aspiration biopsies of renal masses. *Diagn. Cytopathol.* **2018**, *46*, 895–900. [CrossRef]
6. Bhatt, J.R.; Finelli, A. Landmarks in the diagnosis and treatment of renal cell carcinoma. *Nat. Rev. Urol.* **2014**, *11*, 517–525. [CrossRef]
7. Neuzillet, Y.; Lechevallier, E.; Andre, M.; Daniel, L.; Coulange, C. Accuracy and Clinical Role of Fine Needle Percutaneous Biopsy with Computerized Tomography Guidance of Small (Less Than 4.0 Cm) Renal Masses. *J. Urol.* **2004**, *171*, 1802–1805. [CrossRef]
8. Lebret, T.; Poulain, J.E.; Molinié, V.; Herve, J.M.; Denoux, Y.; Guth, A.; Scherrer, A.; Botto, H. Percutaneous Core Biopsy for Renal Masses: Indications, Accuracy and Results. *J. Urol.* **2007**, *178*, 1184–1188. [CrossRef]
9. Blumenfeld, A.J.; Guru, K.; Fuchs, G.J.; Kim, H.L. Percutaneous Biopsy of Renal Cell Carcinoma Underestimates Nuclear Grade. *Urology* **2010**, *76*, 610–613. [CrossRef]
10. Ficarra, V.; Brunelli, M.; Novara, G.; D'Elia, C.; Segala, D.; Gardiman, M.; Artibani, W.; Martignoni, G. Accuracy of on-bench biopsies in the evaluation of the histological subtype, grade, and necrosis of renal tumours. *Pathology* **2011**, *43*, 149–155. [CrossRef]
11. Jeldres, C.; Sun, M.; Liberman, D.; Leghezzani, G.; de la Taille, A.; Tostain, J.; Valeri, A.; Cindolo, L.; Ficarra, V.; Artibani, W.; et al. Can renal mass biopsy assessment of tumor grade be safely substituted for by a predictive model? *J. Urol.* **2009**, *182*, 2585–2589. [CrossRef] [PubMed]
12. Millet, I.; Curros, F.; Serre, I.; Taourel, P.; Thuret, R. Can Renal Biopsy Accurately Predict Histological Subtype and Fuhrman Grade of Renal Cell Carcinoma? *J. Urol.* **2012**, *188*, 1690–1694. [CrossRef] [PubMed]
13. Kutikov, A.; Smaldone, M.C.; Uzzo, R.G.; Haifler, M.; Bratslavsky, G.; Leibovich, B.C. Renal Mass Biopsy: Always, Sometimes, or Never? *Eur. Urol.* **2016**, *70*, 403–406. [CrossRef] [PubMed]
14. Rizzo, S.; Botta, F.; Raimondi, S.; Origgi, D.; Fanciullo, C.; Morganti, A.G.; Bellomi, M. Radiomics: The facts and the challenges of image analysis. *Eur. Radiol. Exp.* **2018**, *2*, 1–8. [CrossRef] [PubMed]
15. Keek, S.A.; Leijenaar, R.T.; Jochems, A.; Woodruff, H.C. A review on radiomics and the future of theranostics for patient selection in precision medicine. *Brit. J. Radiol.* **2018**, *91*, 20170926. [CrossRef]
16. Gillies, R.J.; Kinahan, P.E.; Hricak, H. Radiomics: Images Are More than Pictures, They Are Data. *Radiology* **2016**, *278*, 563–577. [CrossRef]
17. Mayerhoefer, M.E.; Materka, A.; Langs, G.; Häggström, I.; Szczypiński, P.; Gibbs, P.; Cook, G. Introduction to Radiomics. *J. Nucl. Med.* **2020**, *61*, 488–495. [CrossRef]
18. Lubner, M.G.; Smith, A.D.; Sandrasegaran, K.; Sahani, D.V.; Pickhardt, P.J. CT Texture Analysis: Definitions, Applications, Biologic Correlates, and Challenges. *Radiographics* **2017**, *37*, 1483–1503. [CrossRef]
19. Capobianco, E.; Dominietto, M. From Medical Imaging to Radiomics: Role of Data Science for Advancing Precision Health. *J. Pers. Med.* **2020**, *10*, 15. [CrossRef]
20. Van Timmeren, J.E.; Cester, D.; Tanadini-Lang, S.; Alkadhi, H.; Baessler, B. Radiomics in medical imaging—"How-to" guide and critical reflection. *Insights Imaging* **2020**, *11*, 1–16. [CrossRef]
21. Liu, Z.; Wang, S.; Dong, D.; Wei, J.; Fang, C.; Zhou, X.; Sun, K.; Li, L.; Li, B.; Wang, M.; et al. The Applications of Radiomics in Precision Diagnosis and Treatment of Oncology: Opportunities and Challenges. *Theranostics* **2019**, *9*, 1303–1322. [CrossRef] [PubMed]
22. Bera, K.; Velcheti, V.; Madabhushi, A. Novel Quantitative Imaging for Predicting Response to Therapy: Techniques and Clinical Applications. *Am. Soc. Clin. Oncol. Educ. Book* **2018**, *38*, 1008–1018. [CrossRef] [PubMed]

23. Aerts, H.J.; Velazquez, E.R.; Leijenaar, R.T.H.; Parmar, C.; Grossmann, P.; Carvalho, S.; Bussink, J.; Monshouwer, R.; Haibe-Kains, B.; Rietveld, D.; et al. Decoding tumour phenotype by noninvasive imaging using a quantitative radiomics approach. *Nat. Commun.* **2014**, *5*, 4006. [CrossRef] [PubMed]
24. Forghani, R.; Savadjiev, P.; Chatterjee, A.; Muthukrishnan, N.; Reinhold, C.; Forghani, B. Radiomics and Artificial Intelligence for Biomarker and Prediction Model Development in Oncology. *Comput. Struct. Biotechnol. J.* **2019**, *17*, 995–1008. [CrossRef] [PubMed]
25. Lin, F.; Cui, E.-M.; Lei, Y.; Luo, L. CT-based machine learning model to predict the Fuhrman nuclear grade of clear cell renal cell carcinoma. *Abdom. Radiol.* **2019**, *44*, 2528–2534. [CrossRef] [PubMed]
26. Ding, J.; Xing, Z.; Jiang, Z.; Chen, J.; Pan, L.; Qiu, J.; Xing, W. CT-based radiomic model predicts high grade of clear cell renal cell carcinoma. *Eur. J. Radiol.* **2018**, *103*, 51–56. [CrossRef] [PubMed]
27. Shu, J.; Tang, Y.; Cui, J.; Yang, R.; Meng, X.; Cai, Z.; Zhang, J.; Xu, W.; Wen, D.; Yin, H. Clear cell renal cell carcinoma: CT-based radiomics features for the prediction of Fuhrman grade. *Eur. J. Radiol.* **2018**, *109*, 8–12. [CrossRef]
28. Li, Q.; Liu, Y.; Dong, D.; Bai, X.; Huang, Q.; Guo, A.; Ye, H.; Tian, J.; Wang, H.-Y. Multiparametric MRI Radiomic Model for Preoperative Predicting WHO / ISUP Nuclear Grade of Clear Cell Renal Cell Carcinoma. *J. Magn. Reson. Imaging* **2020**, *52*, 1557–1566. [CrossRef]
29. Zhou, H.; Mao, H.; Dong, D.; Fang, M.; Gu, D.; Liu, X.; Xu, M.; Yang, S.; Zou, J.; Yin, R.; et al. Development and External Validation of Radiomics Approach for Nuclear Grading in Clear Cell Renal Cell Carcinoma. *Ann. Surg. Oncol.* **2020**, *27*, 4057–4065. [CrossRef]
30. Feng, Z.; Shen, Q.; Li, Y.; Hu, Z. CT texture analysis: A potential tool for predicting the Fuhrman grade of clear-cell renal carcinoma. *Cancer Imaging* **2019**, *19*, 1–7. [CrossRef]
31. Han, D.; Yu, Y.; Yu, N.; Dang, S.; Wu, H.; Jialiang, R.; He, T. Prediction models for clear cell renal cell carcinoma ISUP/WHO grade: Comparison between CT radiomics and conventional contrast-enhanced CT. *Br. J. Radiol.* **2020**, *93*, 20200131. [CrossRef] [PubMed]
32. Huhdanpaa, H.; Hwang, D.; Cen, S.Y.; Quinn, B.; Nayyar, M.; Zhang, X.; Chen, F.; Desai, B.; Liang, G.; Gill, I.S.; et al. CT prediction of the Fuhrman grade of clear cell renal cell carcinoma (RCC): Towards the development of computer-assisted diagnostic method. *Abdom. Imaging* **2015**, *40*, 3168–3174. [CrossRef] [PubMed]
33. Koo, T.K.; Li, M.Y. A Guideline of Selecting and Reporting Intraclass Correlation Coefficients for Reliability Research. *J. Chiropr. Med.* **2016**, *15*, 155–163. [CrossRef] [PubMed]
34. Zou, G. Sample size formulas for estimating intraclass correlation coefficients with precision and assurance. *Stat. Med.* **2012**, *31*, 3972–3981. [CrossRef] [PubMed]
35. Yen, M.; Lo, L.-H. Examining Test-Retest Reliability. *Nurs. Res.* **2002**, *51*, 59–62. [CrossRef] [PubMed]
36. Arimura, H.; Soufi, M.; Ninomiya, K.; Kamezawa, H.; Yamada, M. Potentials of radiomics for cancer diagnosis and treatment in comparison with computer-aided diagnosis. *Radiol. Phys. Technol.* **2018**, *11*, 365–374. [CrossRef] [PubMed]
37. Kumar, V.; Gu, Y.; Basu, S.; Berglund, A.; Eschrich, S.A.; Schabath, M.B.; Forster, K.; Aerts, H.J.; Dekker, A.; Fenstermacher, D.; et al. Radiomics: The process and the challenges. *Magn. Reson. Imaging* **2012**, *30*, 1234–1248. [CrossRef]
38. Lambin, P.; Leijenaar, R.T.; Deist, T.M.; Peerlings, J.; De Jong, E.E.; Van Timmeren, J.; Sanduleanu, S.; LaRue, R.T.H.M.; Even, A.J.; Jochems, A.; et al. Radiomics: The bridge between medical imaging and personalized medicine. *Nat. Rev. Clin. Oncol.* **2017**, *14*, 749–762. [CrossRef]
39. Shinagare, A.B.; Krajewski, K.M.; Braschi-Amirfarzan, M.; Ramaiya, N.H. Advanced Renal Cell Carcinoma: Role of the Radiologist in the Era of Precision Medicine. *Radiology* **2017**, *284*, 333–351. [CrossRef]
40. Dong, X.; Xing, L.; Wu, P.; Fu, Z.; Wan, H.; Li, D.; Yin, Y.; Sun, X.; Yu, J. Three-dimensional positron emission tomography image texture analysis of esophageal squamous cell carcinoma. *Nucl. Med. Commun.* **2013**, *34*, 40–46. [CrossRef]
41. Mu, W.; Chen, Z.; Liang, Y.; Shen, W.; Yang, F.; Dai, R.; Wu, N.; Tian, J. Staging of cervical cancer based on tumor heterogeneity characterized by texture features on18F-FDG PET images. *Phys. Med. Biol.* **2015**, *60*, 5123–5139. [CrossRef] [PubMed]
42. Petresc, B.; Lebovici, A.; Caraiani, C.; Feier, D.S.; Graur, F.; Buruian, M.M. Pre-Treatment T2-WI Based Radiomics Features for Prediction of Locally Advanced Rectal Cancer Non-Response to Neoadjuvant Chemoradiotherapy: A Preliminary Study. *Cancers* **2020**, *12*, 1894. [CrossRef] [PubMed]
43. Ganeshan, B.; Abaleke, S.; Young, R.C.; Chatwin, C.R.; Miles, K.A. Texture analysis of non-small cell lung cancer on unenhanced computed tomography: Initial evidence for a relationship with tumour glucose metabolism and stage. *Cancer Imaging* **2010**, *10*, 137–143. [CrossRef] [PubMed]
44. Lang, H.; Lindner, V.; De Fromont, M.; Molinié, V.; Letourneux, H.; Meyer, N.; Martin, M.; Jacqmin, D. Multicenter determination of optimal interobserver agreement using the Fuhrman grading system for renal cell carcinoma. *Cancer* **2005**, *103*, 625–629. [CrossRef] [PubMed]
45. Letourneux, H.; Lindner, V.; Lang, H.; Massfelder, T.; Meyer, N.; Saussine, C.; Jacqmin, D. Reproductibilité du grade nucléaire de Fuhrman. Avantages d'un regroupement en deux grades (Reproducibility of Fuhrman nuclear grade: Advantages of a two-grade system). *Prog. Urol.* **2006**, *16*, 281–285. [PubMed]
46. Al-Aynati, M.; Chen, V.; Salama, S.; Shuhaibar, H.; Treleaven, D.; Vincic, L. Interobserver and intraobserver variability using the Fuhrman grading system for renal cell carcinoma. *Arch. Pathol. Lab. Med.* **2003**, *127*, 593–596. [PubMed]

47. Bektaş, S.; Bahadir, B.; Kandemir, N.O.; Barut, F.; Gul, A.E.; Ozdamar, S.O. Intraobserver and Interobserver Variability of Fuhrman and Modified Fuhrman Grading Systems for Conventional Renal Cell Carcinoma. *Kaohsiung J. Med Sci.* **2009**, *25*, 596–600. [CrossRef]
48. Cui, E.; Li, Z.; Ma, C.; Li, Q.; Lei, Y.; Lan, Y.; Yu, J.; Zhou, Z.; Li, R.; Long, W.; et al. Predicting the ISUP grade of clear cell renal cell carcinoma with multiparametric MR and multiphase CT radiomics. *Eur. Radiol.* **2020**, *30*, 2912–2921. [CrossRef]
49. He, X.; Zhang, H.; Zhang, T.; Han, F.; Song, B. Predictive models composed by radiomic features extracted from multi-detector computed tomography images for predicting low-and high-grade clear cell renal cell carcinoma: A STARD-compliant article. *Medicine* **2019**. [CrossRef]
50. Shu, J.; Wen, D.; Xi, Y.; Xia, Y.; Cai, Z.; Xu, W.; Meng, X.; Liu, B.; Yin, H. Clear cell renal cell carcinoma: Machine learning-based computed tomography radiomics analysis for the prediction of WHO/ISUP grade. *Eur. J. Radiol.* **2019**, *121*, 108738. [CrossRef]
51. Sun, X.; Liu, L.; Xu, K.; Li, W.; Huo, Z.; Liu, H.; Shen, T.; Pan, F.; Jiang, Y.; Zhang, M. Prediction of ISUP grading of clear cell renal cell carcinoma using support vector machine model based on CT images. *Medicine* **2019**, *98*, e15022. [CrossRef] [PubMed]
52. Yip, S.S.F.; Aerts, H.J.W.L. Applications and limitations of radiomics. *Phys. Med. Biol.* **2016**, *61*, R150–R166. [CrossRef] [PubMed]
53. Lubner, M.G. Radiomics and Artificial Intelligence for Renal Mass Characterization. *Radiol. Clin. North Am.* **2020**, *58*, 995–1008. [CrossRef] [PubMed]
54. Yoshida, R.; Yoshizako, T.; Hisatoshi, A.; Mori, H.; Tamaki, Y.; Ishikawa, N.; Kitagaki, H. The additional utility of apparent diffusion coefficient values of clear-cell renal cell carcinoma for predicting metastasis during clinical staging. *Acta Radiol. Open* **2017**, *6*. [CrossRef] [PubMed]
55. Rosenkrantz, A.B.; Niver, B.E.; Fitzgerald, E.F.; Babb, J.S.; Chandarana, H.; Melamed, J. Utility of the Apparent Diffusion Coefficient for Distinguishing Clear Cell Renal Cell Carcinoma of Low and High Nuclear Grade. *Am. J. Roentgenol.* **2010**, *195*, W344–W351. [CrossRef]
56. Maruyama, M.; Yoshizako, T.; Uchida, K.; Araki, H.; Tamaki, Y.; Ishikawa, N.; Shiina, H.; Kitagaki, H. Comparison of utility of tumor size and apparent diffusion coefficient for differentiation of low- and high-grade clear-cell renal cell carcinoma. *Acta Radiol.* **2015**, *56*, 250–256. [CrossRef] [PubMed]
57. Larsen, K.; Petersen, J.H.; Budtz-Jørgensen, E.; Endahl, L. Interpreting Parameters in the Logistic Regression Model with Random Effects. *Biometrics* **2000**, *56*, 909–914. [CrossRef]
58. Bektas, C.T.; Kocak, B.; Yardimci, A.H.; Turkcanoglu, M.H.; Yucetas, U.; Koca, S.B.; Erdim, C.; Kilickesmez, O. Clear Cell Renal Cell Carcinoma: Machine Learning-Based Quantitative Computed Tomography Texture Analysis for Prediction of Fuhrman Nuclear Grade. *Eur. Radiol.* **2019**, *29*, 1153–1163. [CrossRef]
59. Nazari, M.; Shiri, I.; Hajianfar, G.; Oveisi, N.; Abdollahi, H.; Deevband, M.R.; Oveisi, M.; Zaidi, H. Noninvasive Fuhrman grading of clear cell renal cell carcinoma using computed tomography radiomic features and machine learning. *Radiol. Med.* **2020**, *125*, 754–762. [CrossRef]
60. Xu, K.; Liu, L.; Li, W.; Sun, X.; Shen, T.; Pan, F.; Jiang, Y.; Guo, Y.; Ding, L.; Zhang, M. CT-Based Radiomics Signature for Preoperative Prediction of Coagulative Necrosis in Clear Cell Renal Cell Carcinoma. *Korean J. Radiol.* **2020**, *21*, 670–683. [CrossRef]
61. Frank, I.; Blute, M.L.; Cheville, J.C.; Lohse, C.M.; Weaver, A.L.; Zincke, H. An outcome prediction model for patients with clear cell renal cell carcinoma treated with radical nephrectomy based on tumor stage, size, grade and necrosis: The SSIGN score. *J. Urol.* **2002**, *168*, 2395–2400. [CrossRef]
62. Klatte, T.; Patard, J.-J.; De Martino, M.; Bensalah, K.; Verhoest, G.; De La Taille, A.; Abbou, C.-C.; Allhoff, E.P.; Carrieri, G.; Riggs, S.B.; et al. Tumor Size Does Not Predict Risk of Metastatic Disease or Prognosis of Small Renal Cell Carcinomas. *J. Urol.* **2008**, *179*, 1719–1726. [CrossRef] [PubMed]
63. Motzer, R.J.; Jonasch, E.; Agarwal, N.; Bhayani, S.; Bro, W.P.; Chang, S.S.; Choueiri, T.K.; Costello, B.A.; Derweesh, I.H.; Fishman, M.; et al. Kidney Cancer, Version 2.2017, NCCN Clinical Practice Guidelines in Oncology. *J. Natl. Compr. Cancer Netw.* **2017**, *15*, 804–834. [CrossRef] [PubMed]

Article

IL13Rα2 Is Involved in the Progress of Renal Cell Carcinoma through the JAK2/FOXO3 Pathway

Mi-Ae Kang [1,†], Jongsung Lee [2,†], Chang Min Lee [3,†], Ho Sung Park [4,5,6], Kyu Yun Jang [4,5,6,*] and See-Hyoung Park [3,*]

1. Department of Biological Science, Gachon University, Seongnam 13120, Korea; makang53@hanmail.net
2. Department of Integrative Biotechnology, Sungkyunkwan University, Suwon 16419, Korea; bioneer@skku.edu
3. Department of Bio and Chemical Engineering, Hongik University, Sejong 30016, Korea; yycc456@naver.com
4. Department of Pathology, Jeonbuk National University Medical School, Jeonju 54896, Korea; hspark@jbnu.ac.kr
5. Research Institute of Clinical Medicine of Jeonbuk National University, Jeonju 54896, Korea
6. Biomedical Research Institute of Jeonbuk National University Hospital, Jeonju 54896, Korea
* Correspondence: kyjang@jbnu.ac.kr (K.Y.J.); shpark74@hongik.ac.kr (S.-H.P.); Tel.: +82-63-270-3136 (K.Y.J.); +82-44-860-2126 (S.-H.P.)
† These authors contributed equally to this work.

Abstract: Previously, we reported a close relationship between type II IL4Rα and IL13Rα1 complex and poor outcomes in renal cell carcinoma (RCC). In this study, we investigated the clinicopathologically significant oncogenic role of IL13Rα2, a kind of the independent receptor for IL13, in 229 RCC patients. The high expression of IL13Rα2 was closely related to relapse-free survival in specific cancers in univariate and multivariate analysis. Then, the oncogenic role of IL13Rα2 was evaluated by performing in vitro assays for cell proliferation, cell cycle arrest, and apoptosis in A498, ACHN, Caki1, and Caki2, four kinds of RCC cells after transfection of siRNA against IL13Rα2. Cell proliferation was suppressed, and apoptosis was induced in A498, ACHN, Caki1, and Caki2 cells by knockdown of IL13Rα2. Interestingly, the knockdown of IL13Rα2 decreased the phosphorylation of JAK2 and increased the expression of FOXO3. Furthermore, the knockdown of IL13Rα2 reduced the protein interaction among IL13Rα2, phosphorylated JAK2, and FOXO3. Since phosphorylation of JAK2 was regulated by IL13Rα2, we tried to screen a novel JAK2 inhibitor from the FDA-approved drug library and selected telmisartan, a clinically used medicine against hypertension, as one of the strongest candidates. Telmisartan treatment decreased the cell proliferation rate and increased apoptosis in A498, ACHN, Caki1, and Caki2 cells. Mechanistically, telmisartan treatment decreased the phosphorylation of JAK2 and increased the expression of FOXO3. Taken together, these results suggest that IL13Rα2 regulates the progression of RCC via the JAK2/FOXO3-signaling path pathway, which might be targeted as the novel therapeutic option for RCC patients.

Keywords: IL13Rα2; renal cell carcinoma; JAK2; FOXO3; telmisartan

1. Introduction

Every year, there are more than 300,000 new renal cell carcinoma cases (RCC) diagnosis globally [1]. Among them, about 30% of patients were diagnosed with metastatic RCC [2]. Moreover, the 5-year survival rate of patients with metastatic RCC is lower than 10% [3]. The prognosis of RCC patients is divided into several categories, such as favorable, intermediate, and poor-risk disease according to well-characterized clinical and laboratory risk factors [4]. Approximately 75% of patients with RCC have a poor-risk disease, and their prognosis is worse than that with a favorable-risk disease [5,6]. Over the past decade, there have been marked advances in the treatment of metastatic RCC. Sorafenib, sunitinib, bevacizumab, and axitinib are effective inhibitors of vascular endothelial growth factor (VEGF) and its receptor (VEGFR) [7]. Everolimus and temsirolimus inhibited the mechanistic target of rapamycin complex 1 (mTORC1) [8]. However, the mortality rate of metastatic RCC is still high because of resistance to conventional chemotherapy and the side effect of radiation therapy [9]. Therefore, we still need to consider the efficient treatment option for RCC.

IL13Rα2 is a membrane-bound protein encoded by the IL13Rα2 gene [10]. IL13Rα2 is closely associated with IL13Rα1, a subunit of type II IL4Rα and IL13Rα1 complex [10]. IL13 binds IL13Rα2 with high-affinity [10]. Recently, IL13Rα2 has been considered an important target for cancer treatment in various clinical studies [11]. A recent study indicated that IL13Rα2 is a potential marker and therapeutic target for human melanoma treatment [12]. It was reported that IL13Rα2 was overexpressed in metastatic colorectal cancer and inhibition of IL13 binding to IL13Rα2 showed the therapeutic activity in colorectal cancer by reducing metastatic spread [13]. Furthermore, it was demonstrated that targeting IL13Rα2 depletion suppressed breast tumor growth and IL13Rα2 activated IL13-mediated STAT6-signaling pathway, and knockdown of IL13Rα2 suppressed breast cancer metastasis into the lung [14]. Therefore, IL13Rα2 could be a potential biomarker to diagnose various cancers. However, there is not enough study for the clinical analysis, biological function, and molecular mechanisms of IL13Rα2 in RCC development.

Drug repositioning means the application of the drugs that have been clinically used to other diseases by elucidating the novel activities and target proteins [15]. Since the clinically used drugs were approved by the US Food and Drug Administration (FDA), drug repositioning has many advantages, such as no need to test toxicity and to evaluate pharmacokinetics. Furthermore, there are many previous reports and patents for studying the metabolism and interactions of the old drug, which could help researchers to examine the possible working mechanism of the old drug in the new application. Thus, drug repositioning could considerably save the cost and time for researchers to develop efficient drugs leading to improve the success rate [16]. For the proof-of-concept trial, we successfully selected telmisartan, a clinically used medicine against hypertension, as the strongest JAK2 inhibitor. Telmisartan is known as an agonist of angiotensin II receptor, but not reported on the possible involvement of-signaling pathway, including the regulation of JAK2 [17].

In this study, we investigated the clinical implication and oncogenic role of IL13Rα2 in RCC progression. Interestingly, IL13Rα2 seemed to increase the phosphorylation of JAK2 and decrease the expression of FOXO3. These results suggest that IL13Rα2 regulates RCC progression through JAK2/FOXO3-signaling pathway. Since JAK2 was regulated by IL13Rα2 and type II IL4Rα and IL13Rα1 complex, we tried to screen an FDA-approved drug library with a JAK2 kinase assay kit to identify the novel candidates that were possibly inhibiting JAK2 in RCC cells. Here, we show that telmisartan has the potential for antiproliferative activity in RCC cells, which could broaden the therapeutic options for RCC patients.

2. Materials and Methods

2.1. RCC Patients and Tissue Samples

RCC patients who operated between July 1998 and August 2011 at Jeonbuk National University Hospital were analyzed in this study. Medical records, histologic and tissue samples were available in 229 cases and included in this study. The clinicopathologic information for patients with RCC was obtained by analyzing medical records and original histologic slides. Tumor stage and histopathologic factors were re-evaluated according to the World Health Organization classification of the renal tumor [18] and the 8th edition of the staging system of the American Joint Committee on Cancer [19]. Histological subtypes of RCCs included in this study were 201 cases of clear cell RCC (CCRCC), 16 cases of chromophobe RCC, and twelve cases of papillary RCC. This study obtained institutional review board approval from Jeonbuk National University Hospital (IRB No., CUH 2019-11-039) and was performed according to the Declaration of Helsinki. The approval contained a waiver for written informed consent based on the retrospective and anonymous character of this study.

2.2. Immunohistochemical Staining and Scoring

Immunohistochemical staining in RCC tissue was performed using tissue microarray sections. One 3.0 mm core per case was arrayed in tissue microarray. The tissue microarray core was obtained from the area of the original paraffin-embedded tissue block, mainly composed of tumor cells with the highest histologic grade. The histologic sections were deparaffinized and boiled with the microwave oven for 20 min in pH 6.0 antigen retrieval solution (DAKO, Glostrup, Denmark) to induce antigen retrieval. Thereafter, the tissue sections are incubated with anti-IL13Rα2 primary antibody (1:100 dilution, Santa Cruz Biotechnology, Santa Cruz, CA, USA) and visualized using the enzyme-substrate 3-amino-9-ethylcarbazole. Immunohistochemical staining scoring was performed by two pathologists (HSP and KYJ) with consensus by observing in a multi-viewing microscope. The scoring was performed without clinicopathologic information. The score obtained by adding staining intensity point (point 0; no staining, point 1; weak, point 2; intermediate, point 3; strong) and staining area point (point 0; no staining, point 1; 1%, point 2; 2–10%, point 3:11–33%, point 4; 34–66%, point 5; 67–100%) [20–22]. Therefore, the immunohistochemical staining score ranged from zero to eight.

2.3. Chemical Reagents, Antibodies, and Plasmid DNAs

The FDA-approved drug library (SCREEN-WELL FDA-approved drug library V2, 821 drugs) was purchased from Enzo Life Sciences (Farmingdale, NY, USA). Mouse anti-β-actin antibody, mouse anti-Myc antibody, mouse anti-HA antibody, protease inhibitors, phosphatase inhibitors, AZD1480, telmisartan, the following chemicals, and solvents (non-fat dry milk powder, dimethyl sulfoxide (DMSO), ethylenediaminetetraacetic acid (EDTA), glycerol, glycine, sodium chloride, Trizma base, Triton X-100, sodium dodecyl sulfate (SDS), crystal violet, 4% paraformaldehyde solution, 4′,6-diamidino-2-phenylindole (DAPI), propidium iodide (PI), and Tween-20) were from Sigma (St. Louis, MO, USA). Control siRNA, siRNA against IL13Rα2, protein A or G-agarose beads, rabbit anti-IL13Rα2, and rabbit anti-FOXO3 antibodies were purchased from Santa Cruz Biotechnology (Santa Cruz, CA, USA). Rabbit anti-JAK2, rabbit anti-phospho-JAK2 (pJAK2), rabbit anti-cleaved PARP1, rabbit anti-cleaved caspase3, and rabbit anti-p27 antibodies were purchased from Cell Signaling Technology (Danvers, MA, USA). Goat anti-rabbit and goat anti-mouse horseradish peroxidase (HRP)-conjugated IgG (heavy/light or light chain-specific) were from Jackson ImmunoResearch (West Grove, PA, USA). Enhanced chemiluminescence (ECL) reagent was from GE Healthcare (Little Chalfont, United Kingdom). pCMV3-C-HA and pCMV3-JAK2-C-HA plasmid DNA were from Sino Biological (Wayne, PA, USA). pCMV6-C-Myc-Flag and pCMV6-IL13Rα2-C-Myc-Flag plasmid DNA were from OriGene (Rockville, MD, USA).

2.4. Cell Culture

A498, ACHN, Caki1, Caki2, and 293T cells were purchased from ATCC (Manassas, VA, USA) and were grown in Dulbecco's modified Eagle's media (DMEM, Invitrogen, Carlsbad, CA, USA) media containing 10% fetal bovine serum (FBS, Invitrogen) and 1% streptomycin/penicillin. The cells were cultured in a humidified incubator (5% CO_2, 37 °C). We performed all experiments with early passages cells (passages 4–10).

2.5. Transfection of siRNA and Plasmid DNA

Cells were plated (5.0×10^5 cells/well) in 60 mm cell culture dishes and incubated for 18 h in an incubator. After 18 h of incubation, cells were transfected with siRNAs (siRNA against IL13Rα2: sc-63339, control siRNA: sc-37007 from Santa Cruz, 1 μL) or plasmid DNAs (pCMV3-C-HA empty/HA-JAK2 plasmid DNA, 1 μg). siRNAs or plasmid DNAs were mixed with 3 μL of lipofectamine 2000 (Invitrogen), respectively, in 600 μL of serum-free media for 20 min. After PBS washing twice, the cells were incubated with the media containing siRNAs or DNAs for 6 h in a humidified incubator. After 6 h, cell culture media was removed, and fresh media containing 10% FBS was added. After then, the cells were incubated for 18 h.

2.6. WST-1 Assay

Cells were plated (1×10^3 cells/well) in 96-well plates and incubated for 18 h in a humidified incubator. After incubation, cells were transfected with control/IL13Rα2 siRNA or treated with DMSO (0.1%) control/the indicated treatment for 24, 48, or 72 h. After incubation, 20 µL of EZ-Cytox (DoGenBio, Republic of Korea) was added to the medium. After 4 h, absorbance was measured at 460 nm wavelength by a microplate reader (Bio-Rad Laboratories, Hercules, CA, USA).

2.7. Cell Counting Assay

Cells were plated (2×10^4 cells/well) in 60 mm culture dishes and incubated for 18 h in an incubator. After incubation, cells were transfected with control/IL13Rα2 siRNA or treated with DMSO (0.1%) control/the indicated treatment for 14 days. The number of cells was counted by a hemocytometer.

2.8. Colony Formation Assay

Cells were plated (5×10^2 cells/well) in 60 mm culture dishes and incubated for 18 h in a humidified incubator. After incubation, cells were transfected with control/IL13Rα2 siRNA or treated with DMSO (0.1%) control/the indicated treatment for 2 weeks. Cells were transfected with IL13Rα2 or control siRNA every other day, changing cell culture media. Similarly, Cells were treated with telmisartan or the same volume of DMSO vehicle every other day, changing cell culture media. The cells were fixed with 4% formaldehyde (Sigma) and stained using 1% crystal violet (Sigma). The number of colonies was counted.

2.9. Cell Cycle Analysis

Cells were plated (5×10^5 cells/well) in 60 mm cell culture dishes and incubated for 18 h in a humidified incubator. After incubation, cells were transfected with control/IL13Rα2 siRNA or treated with DMSO (0.1%) control/the indicated treatment for 48 h. Then, the cells were trypsinized and fixed in 70% ice-cold absolute ethanol overnight at -20 °C. After then, centrifugation was carried out (1000 rpm, 5 min), and the cells were suspended with propidium iodide (PI) solution for 30 min at 37 °C. After staining, cell cycle distribution was analyzed by a FACSCalibur (BD Biosciences, San Jose, CA, USA), and the data were analyzed using the FlowJo program (De Novo Software, Glendale, CA, USA).

2.10. TUNEL Assay

Cells were plated (5×10^5 cells/well) in 60 mm cell culture dishes and incubated for 18 h in a humidified incubator. After incubation, cells were transfected with control/IL13Rα2 siRNA or treated with DMSO (0.1%) control/the indicated treatment for 48 h. After transfection, the cells were fixed in 4% formaldehyde solution at 4 °C for 20 min. After fixation, the cells were permeabilized with 0.2% Triton X100 (Sigma). DNA strand breaks labeling was performed using a TUNEL assay kit (Promega, Madison, WI, USA). Nuclei were dyed with DAPI.

2.11. Annexin V Staining Analysis

Cells were plated (5×10^5 cells/well) in 60 mm culture dishes and incubated for 18 h in a humidified incubator. After incubation, cells were transfected with control/IL13Rα2 siRNA or treated with DMSO (0.1%) control/the indicated treatment for 48 h. The cells were trypsinized and resuspended in annexin V-binding buffer. The percentage of apoptotic cells was evaluated by a FITC annexin V apoptosis detection kit I (BD Biosciences) with PI according to the manufacturer's protocol. 1×10^4 events were collected for each run. Cells were analyzed by a FACSCalibur (BD Biosciences), and FlowJo software (De Novo Software) was used to analyze the data.

2.12. Western Blotting Analysis

Cells were lysed in lysis buffer (RIPA buffer, Cell Signaling Technology, USA) containing protease and phosphatase inhibitors. Centrifugation (10,000× g, 4 °C, 10 min) was carried out, and protein lysates were separated on 10% NuPAGE pre-casting gels (Invitrogen) and transferred to nitrocellulose membranes (Bio-Rad Laboratories). The membranes were blocked with 3% defatted dry milk powder at room temperature for 1 h, and immunoblotting was performed with specific primary antibodies (overnight, 4 °C). Membranes were incubated with HRP-conjugated anti-mouse or anti-rabbit IgG in 3% defatted dry milk powder at room temperature for 1 h. Finally, the bands were detected using ECL solution (GE Healthcare, Chicago, IL, USA) and ChemiDoc system (Bio-Rad Laboratories).

2.13. Immunoprecipitation Analysis

Cells were lysed in lysis buffer (RIPA buffer, Cell Signaling Technology) containing protease and phosphatase inhibitors. Centrifugation (10,000× g, 4 °C, 10 min) was carried out, and protein lysates were separated. The protein lysates were incubated with a specific primary antibody by rotating at 4 °C overnight. After then, 20 µL of 50% protein A or G-agarose slurry (Santa Cruz) was added to the lysates and rotating for 2 h at 4 °C. Protein A or G-agaroses containing antigen–antibody complexes were collected and rinsed with PBS. Immunoprecipitants were analyzed by Western blotting.

2.14. JAK2 Kinase Inhibition Assay

Inhibitory activity of AZD1480 and telmisartan against JAK2 was evaluated by JAK2 kinase assay kit (BPS Bioscience, San Diego, CA, USA) and Glo-Max kinase assay kit (Promega). Briefly, according to the manufacturer's instructions, recombinant JAK2 protein was incubated with the indicated concentration of AZD1480 or telmisartan, peptide substrate, and ATP for 30 min at 37 °C. After incubation, the reaction mixture was incubated with Glo-Max solution for 30 min at room temperature to stop the reaction. Then, the remaining ATP level in each reaction was measured by a microplate reader for luminescence (Bio-Rad Laboratories).

2.15. Statistical Analysis

The immunohistochemical staining score for IL13Rα2 in the RCC tissue sample was grouped into negative and positive cases with receiver operating characteristic curve analysis [22–24]. The cutoff point for IL13Rα2 immunostaining score to discriminate negative or positive cases was determined at the point that significantly estimates patients' death from RCC. The cutoff point has the highest area under the curve in the receiver operating characteristic curve analysis. The survival analysis was conducted for cancer-specific survival (CSS) and relapse-free survival (RFS) through December 2013. The duration for CSS was calculated from the date of diagnosis to the date of the patient's last contact or death. The event in CSS analysis was the death of patients from RCC. The death of patients from other causes or alive of patients finally contact was censored in CSS analysis. The duration for RFS was calculated from the date of diagnosis to the date of the last contact without relapse, the date of the first relapse, or patients' death. The event in RFS analysis was a relapse of RCC or death of patients from RCC. Patients' death from other causes or alive of patients finally contact without relapse were censored in RFS analysis. The survival analysis was performed with univariate and multivariate Cox proportional hazards regression analyses and Kaplan–Meier survival analysis using SPSS software (version 20.0, IBM, CA, USA). The association between clinicopathological factors was analyzed by Pearson's chi-squared test using SPSS software, and all statistical tests were two-sided. The values of P lower than 0.05 were considered statistically significant.

3. Results

3.1. Immunohistochemical Expression of IL13Rα2 Is Associated with Poor Prognosis of RCC Patients

The immunohistochemical staining for IL13Rα2 was seen in tumor cells of all histologic subtypes of RCC (Figure 1A). The cut-off point for IL13Rα2 immunostaining was seven in receiver operating characteristic curve analysis (Figure 1B). The cases have immunohistochemical staining scores equal to, or greater than, seven were grouped as positive for IL13Rα2 staining. In this cut-off value, IL13Rα2-positivity was significantly associated with tumor size ($P = 0.004$), tumor stage ($P = 0.002$), histologic nuclear grade of tumor cells ($P < 0.001$), and histologic subtype of RCC ($P = 0.005$) in 229 cases of RCCs (Table 1). CCRCC is the major histologic subtype of RCC, and there were 201 cases of CCRCC in this study. Therefore, we also evaluated in CCRCC subgroup of RCCs. In CCRCC subgroup, IL13Rα2-positivity was significantly associated with tumor size ($P = 0.005$), tumor stage ($P = 0.003$), and histologic nuclear grade of tumor cells ($P < 0.001$) (Table 1). In 229 overall RCCs, the factors significantly associated with CCS or RFS in univariate analysis were age (CSS, $P < 0.001$; RFS, $P = 0.005$), tumor size (CSS, $P < 0.001$; RFS, $P < 0.001$), tumor stage (CSS, $P < 0.001$; RFS, $P < 0.001$), lymph node metastasis (CSS, $P = 0.615$; RFS, $P < 0.001$), histologic nuclear grade (CSS, overall $P = 0.032$; RFS, overall $P = 0.008$), tumor necrosis (CSS, $P < 0.001$; RFS, $P = 0.004$), and IL13Rα2-positivity (CSS, $P = 0.002$; RFS, $P < 0.001$) (Table 2). The IL13Rα2-positivity showed a 3.726-fold (95% confidence interval [95% CI]; 1.636–8.489, $P = 0.002$) greater risk of death and a 3.625-fold (95% CI; 1.806–7.278, $P < 0.001$) greater risk of relapse or death of RCC patients (Table 2). The Kaplan–Meier survival curves for CSS and RFS according to IL13Rα2-positivity in overall RCC are presented in Figure 1C. In 201 CCRCCs, the factors significantly associated with CCS or RFS in univariate analysis were age (CSS, $P = 0.004$; RFS, $P = 0.012$), tumor size (CSS, $P < 0.001$; RFS, $P < 0.001$), tumor stage (CSS, $P < 0.001$; RFS, $P < 0.001$), lymph node metastasis (CSS, $P = 0.721$; RFS, $P = 0.011$), histologic nuclear grade (CSS, overall $P = 0.170$; RFS, overall $P = 0.028$), tumor necrosis (CSS, $P = 0.005$; RFS, $P = 0.063$), and IL13Rα2-positivity (CSS, $P = 0.003$; RFS, $P < 0.001$) (Table 2). The IL13Rα2-positivity had a 3.591-fold (95% CI; 1.546–8.342, $P = 0.003$) greater risk of death from CCRCC and a 3.518-fold (95% CI; 1.724–7.181, $P < 0.001$) greater risk of relapse or death from CCRCC (Table 2). The Kaplan–Meier survival analysis also showed significant prognostic significance of IL13Rα2 expression for CSS and RFS in CCRCC subgroups (Figure 2A). However, in chromophobe RCC and papillary RCC, despite relatively shorter survival of IL13Rα2-positive subgroups compared with IL13Rα2-negative subgroups, there was no significant difference in survival of patients (Figure 2B,C). Multivariate analysis was performed with the factors significantly associated with CSS or RFS in univariate analysis. The factors included in multivariate analysis were age, tumor size, tumor stage, lymph node metastasis, histologic nuclear grade, tumor necrosis, and immunohistochemical expression of IL13Rα2. In 272 overall RCCs, age (CSS, $P = 0.018$), tumor stage (CSS, $P = 0.005$; RFS, $P < 0.001$), tumor necrosis (CSS, $P = 0.005$; RFS, $P = 0.015$), and IL13Rα2 expression (CSS, $P = 0.025$; RFS, $P = 0.004$) were significantly associated with CSS or RFS (Table 3). The IL13Rα2-positivity had a 2.627-fold (95% CI; 1.132–6.097) greater risk of death and a 2.801-fold (95% CI; 1.3795.688) greater risk of relapse or death of RCC patients (Table 3). In 201 CCRCCs, age (CSS, $P = 0.042$), tumor stage (CSS, $P = 0.010$; RFS, $P < 0.001$), tumor necrosis (CSS, $P = 0.006$; RFS, $P = 0.054$), and IL13Rα2 expression (CSS, $P = 0.019$; RFS, $P = 0.005$) were significantly associated with CSS or RFS (Table 3). The IL13Rα2-positivity showed a 2.792-fold (95% CI; 1.182–6.595) greater risk of death and a 2.838-fold (95% CI; 1.372–5.870, $P < 0.001$) greater risk of relapse or death of CCRCC patients (Table 3). Taken together, we investigated the clinicopathologically significant oncogenic role of IL13Rα2 in 229 RCC patients and the high expression of IL13Rα2 was significantly associated with cancer-specific survival and relapse-free survival in univariate and multivariate analysis.

Table 1. Clinicopathologic variables and the expression status of IL13Rα2 in renal cell carcinomas.

Characteristics		Overall Renal Cell Carcinoma (n = 229)			Clear Cell Renal Cell Carcinoma (n = 201)		
		No.	IL13Rα2 Positive	P	No.	IL13Rα2 Positive	P
Sex	Male	156	86 (55%)	0.411	140	71 (51%)	0.530
	Female	73	36 (49%)		61	28 (46%)	
Age, y	≤55	95	46 (48%)	0.215	82	35 (43%)	0.122
	>55	134	76 (57%)		119	64 (54%)	
Tumor size, cm	≤7	193	95 (49%)	0.004	169	76 (45%)	0.005
	>7	36	27 (75%)		32	23 (72%)	
TNM stage	I	183	88 (48%)	0.002	163	72 (44%)	0.003
	II-IV	46	34 (74%)		38	27 (71%)	
LN metastasis	Absence	226	119 (53%)	0.103	199	97 (49%)	0.149
	Presence	3	3 (100%)		2	2 (100%)	
Nuclear grade	1	45	17 (38%)	<0.001	36	10 (28%)	<0.001
	2	134	67 (50%)		123	59 (48%)	
	3 and 4	50	38 (76%)		42	30 (71%)	
Necrosis	Absence	196	102 (52%)	0.362	174	85 (49%)	0.772
	Presence	33	20 (61%)		27	14 (52%)	
Histologic type	Clear cell	201	99 (49%)	0.005			
	Chromophobe	16	13 (81%)				
	Papillary	12	10 (83%)				

Table 2. Univariate Cox regression analysis of cancer-specific survival and relapse-free survival in renal cell carcinoma patients.

Characteristics.	No.	CSS		RFS	
		HR (95% CI)	P	HR (95% CI)	P
Overall RCC (n = 229)					
Sex, male (vs. female)	156/229	0.564 (0.258–1.234)	0.152	0.513 (0.255–1.030)	0.060
Age, y, >55 (vs. ≤55)	134/229	4.386 (1.828–10.524)	<0.001	2.537 (1.319–4.880)	0.005
Tumor size, >7 cm (vs. ≤7 cm)	36/229	3.415 (1.736–6.715)	<0.001	3.984 (2.218–7.155)	<0.001
TNM stage, I (vs. II-IV)	46/229	4.231 (2.219–8.068)	<0.001	5.166 (2.930–9.018)	<0.001
LN metastasis, presence (vs. absence)	3/229	1.670 (0.226–12.308)	0.615	17.410 (3.874–78.249)	<0.001
Nuclear grade, 1	45/229	1	0.032	1	0.008
2	134/229	0.943 (0.347–2.564)	0.909	1.172 (0.476–2.883)	0.730
3 and 4	50/229	2.327 (0.836–6.476)	0.106	2.846 (1.128–7.179)	0.027
Necrosis, presence (vs. absence)	33/229	3.620 (1.842–7.114)	<0.001	2.542 (1.345–4.807)	0.004
Histologic type, clear cell	201/229	1	0.654	1	0.328
chromophobe	16/229	0.808 (0.193–3.382)	0.771	0.585 (0.141–2.421)	0.460
papillary	12/229	1.570 (0.553–4.462)	0.397	1.802 (0.711–4.565)	0.214
IL13Rα2, positive (vs. negative)	122/229	3.726 (1.636–8.489)	0.002	3.625 (1.806–7.278)	<0.001
Clear cell RCC (n = 201)					
Sex, male (vs. female)	140/201	0.541 (0.222–1.319)	0.177	0.523 (0.241–1.132)	0.100
Age, y, >55 (vs. ≤55)	119/201	4.152 (1.593–10.822)	0.004	2.491 (1.220–5.084)	0.012
Tumor size, >7 cm (vs. ≤7 cm)	32/201	3.977 (1.928–8.204)	<0.001	4.773 (2.560–8.900)	<0.001
TNM stage, I (vs. II-IV)	38/201	3.964 (1.953–8.049)	<0.001	5.199 (2.814–9.604)	<0.001
LN metastasis, presence (vs. absence)	2/201	$0.049 (0.000–7.516 \times 10^5)$	0.721	14.681 (1.841–117.039)	0.011
Nuclear grade, 1	36/201	1	0.170	1	0.028
2	123/201	1.028 (0.344–3.075)	0.961	1.122 (0.423–2.978)	0.817
3 and 4	42/201	2.111 (0.661–6.739)	0.207	2.655 (0.955–7.380)	0.061
Necrosis, presence (vs. absence)	27/201	3.044 (1.401–6.617)	0.005	2.016 (0.962–4.225)	0.063
IL13Rα2, positive (vs. negative)	99/201	3.591 (1.546–8.342)	0.003	3.518 (1.724–7.181)	<0.001

Abbreviations: CSS, cancer-specific survival; RFS, relapse-free survival; HR, hazard ratio; 95% CI, 95% confidence interval; RCC, renal cell carcinoma; LN, lymph node.

Table 3. Multivariate Cox regression analysis of cancer-specific survival and relapse-free survival in renal cell carcinoma patients.

Characteristics	CSS		RFS	
	HR (95% CI)	P	HR (95% CI)	P
Overall RCC (n = 229) *				
Age, y, >55 (vs. ≤55)	2.941 (1.200–7.209)	0.018		
TNM stage, I (vs. II-IV)	2.600 (1.331–5.077)	0.005	4.036 (2.260–7.209)	<0.001
Necrosis, presence (vs. absence)	2.686 (1.350–5.345)	0.005	2.240 (1.172–4.278)	0.015
IL13Rα2, positive (vs. negative)	2.627 (1.132–6.097)	0.025	2.801 (1.379–5.688)	0.004
Clear cell RCC (n = 201) **				
Age, y, >55 (vs. ≤55)	2.779 (1.036–7.453)	0.042		
TNM stage, I (vs. II-IV)	2.616 (1.255–5.451)	0.010	4.214 (2.257–7.867)	<0.001
Necrosis, presence (vs. absence)	3.002 (1.361–6.618)	0.006	2.088 (0.988–4.414)	0.054
IL13Rα2, positive (vs. negative)	2.792 (1.182–6.595)	0.019	2.838 (1.372–5.870)	0.005

Abbreviations: CSS, cancer-specific survival; RFS, relapse-free survival; HR, hazard ratio; 95% CI, 95% confidence interval; RCC, renal cell carcinoma. * The variables included in the multivariate analysis were age, tumor size, tumor stage, histologic nuclear grade, tumor necrosis, and the expression of IL13Rα2. ** The variables included in the multivariate analysis were age, tumor size, tumor stage, histologic nuclear grade, tumor necrosis, and the expression of IL13Rα2.

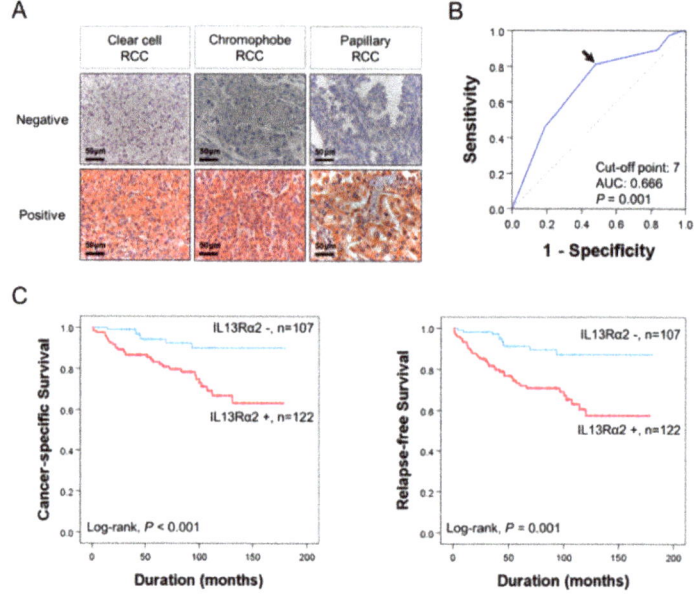

Figure 1. Immunohistochemical expression and survival analysis for the expression of IL13Rα2 in renal cell carcinomas. (**A**) Immunohistochemical expression of IL13Rα2 in clear cell renal cell carcinoma, chromophobe renal cell carcinoma, and papillary renal cell carcinoma tissue. Original magnification, ×400. (**B**) Receiver operator characteristic curve analysis to determine the cutoff point of IL13Rα2 immunostaining. The cutoff point is determined to predict cancer-specific survival of renal cell carcinoma patients. The cutoff point has the highest area under the curve (AUC). Arrow indicates a cutoff point for the IL13Rα2 immunostaining. (**C**) Kaplan–Meier survival analysis for cancer-specific survival and relapse-free survival according to the immunohistochemical positivity for IL13Rα2 in 229 cell renal cell carcinomas.

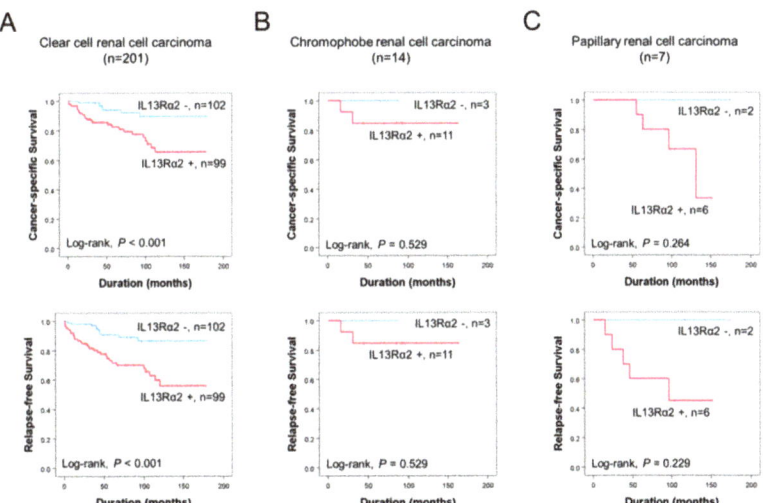

Figure 2. Kaplan–Meier survival analysis in histologic subtypes of renal cell carcinomas. Kaplan–Meier survival curves for cancer-specific survival (CSS) and relapse-free survival (RFS) according to the expression of IL13Rα2 in clear cell renal cell carcinoma (**A**), chromophobe renal cell carcinoma (**B**), and papillary renal cell carcinoma (**C**).

3.2. Knockdown of IL13Rα2 Displays the AntiProliferative Activity in A498, ACHN, Caki1, and Caki2 Cells

In 229 cases of human RCC, a significant association between the expression IL13Rα2 and poor prognosis was observed by tissue microarray. Hence, as the next step, we tried to investigate the possible oncogenic role of IL13Rα2 by performing in vitro assays for cell proliferation, cell cycle arrest, and apoptosis in RCC cells after transfection of siRNA against IL13Rα2. WST-1 and cell counting assay were conducted to evaluate the antiproliferative activity of the knockdown of IL13Rα2. Cells were transfected with control or siRNA against IL13Rα2 and incubated for the indicated time. As shown in Figure 3A,B, compared to the control, cells transfected with siRNA against IL13Rα2 showed a decreased proliferation rate, which was confirmed by performing colony formation assay (Figure 3C). Cell cycle analysis showed that knockdown of IL13Rα2 with siRNA increased G2/M population in A498, ACHN, Caki1, and Caki2 cells compared to control siRNA (Figure 3D). TUNEL and annexin V staining assay results showed that knockdown of IL13Rα2 with siRNA increased the apoptosis in A498, ACHN Caki1, and Caki2 cells compared to control siRNA (Figure 3E,F). Western blotting analysis indicated that knockdown of IL13Rα2 with siRNA increased the expression of cleaved PARP1, cleaved caspase3, FOXO3, and p27 (Figure 3G). Overall, these results indicate that knockdown of IL13Rα2 with siRNA transfection could regulate proliferation, cell cycle arrest, and apoptosis in A498, ACHN, Cak1, and Caki2 RCC cells.

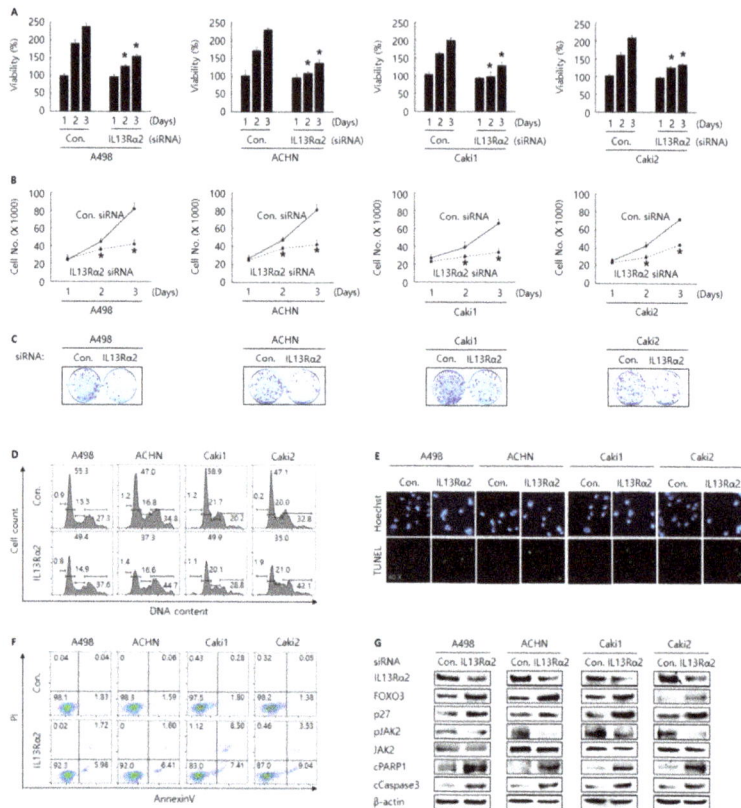

Figure 3. Antiproliferative effect by transfection of siRNA against IL13Rα2 in A498, ACHN, Caki1, and Caki2 cells. Cell viability and proliferation rate were determined by WST-1 (**A**), cell counting assay (**B**) for 24, 48, and 72 h, and Colony formation assay for 14 days (**C**). This result is representative data of at least three independent experiments, and the error bar indicates mean ± standard error (STE). * stands for the P-value < 0.05. Cell cycle arrest for 48 h after transfection was determined by cell cycle analysis (**D**). Apoptosis for 48 h after transfection was determined by Annexin V staining analysis (**E**) and Terminal deoxynucleotidyl transferase dUTP nick end labeling (TUNEL) assay (**F**). This result represents at least three independent experiments (**G**) Western blotting analysis of proteins related to cell cycle arrest and apoptosis for 48 h after transfection. β-actin was used for a gel-loading control.

3.3. Knockdown of IL13Rα2 Attenuates the Protein Interaction Among IL13Rα2, pJAK2, and FOXO3 in A498, ACHN, Caki1, and Caki2 Cells

In the previous report, we found that pJAK2 interacts with FOXO3, which was regulated by type II IL4R and IL13Rα1 heterodimeric receptor complex [21]. Since IL13Rα2 can accept IL13 as the same ligand with type II IL4R and IL13Rα1 complex, we examined whether the phosphorylation level of JAK2 was regulated by knockdown of siRNA against IL13Rα2 in RCC cells. When A498, ACHN, Caki1, Caki2, and 293T cell lysates were analyzed by Western blotting for IL13Rα2, pJAK2, JAK2, and FOXO3, there seemed the correlation pattern between IL13Rα2 and pJAK2 except for Caki2 cell lysates (Supplementary Figure S1A). In contrast, the expression of pJAK2 and FOXO3 was reversely correlated. In addition, as shown in Figure 3G, the expression of pJAK2 was significantly downregulated by transfection of IL13Rα2 with siRNA in A498, ACHN Caki1, and Caki2 cells compared to control siRNA. Then, to investigate the protein interaction among IL13Rα2, pJAK2, and FOXO3, we performed co-immunoprecipitation experiments with an antibody against IL13Rα2, JAK2, and FOXO3 followed by immunoblot analysis with an antibody against IL13Rα2, pJAK2, JAK2, and FOXO3 in A498, ACHN Caki1, and Caki2 cells transfected with siRNA against IL13Rα2. As shown in Figure 4A–C, the protein interaction among IL13Rα2, pJAK2 and FOXO3 was weakened in RCC cells transfected with siRNA against IL13Rα2 compared to the control siRNA. Furthermore, we could observe that the level of protein interaction between IL13Rα2 and JAK2 was increased in 293T cells co-transfected with overexpression plasmid DNA for IL13Rα2 or JAK2 (Figure 3D). Collectively, these results implicate that IL13Rα2 interacts with JAK2, which may regulate the protein expression level of FOXO3.

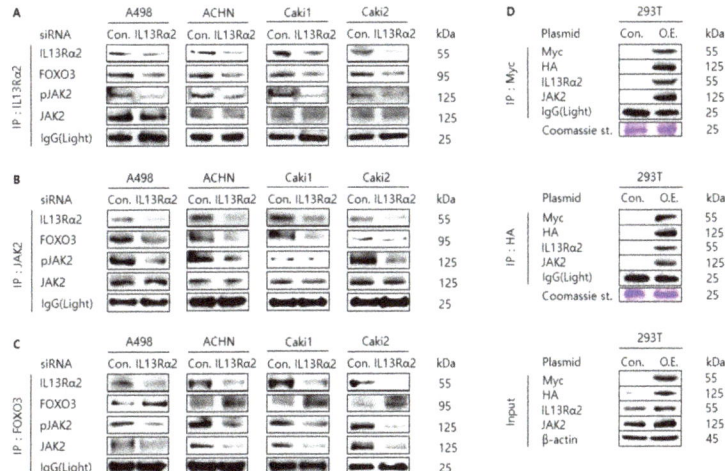

Figure 4. Protein interaction between IL13Rα2 and JAK2. Knock-down of IL4Rα2 in A498, ACHN, Caki1, and Caki2 cells reduced the interaction between IL13Rα2 and JAK2. Cells were transfected with siRNA against IL4Rα2 or control siRNA. Then cell lysates were immunoprecipitated with antibodies against IL4Rα2 (**A**), JAK2 (**B**), or FOXO3 (**C**). The immunoprecipitated proteins were immunoblotted by IL4Rα2, pJAK2, JAK2, and FOXO3 antibodies. Light chain of IgG was used for the loading control. (**D**) 293T cells were co-transfected with Myc-IL4Rα2 and HA-JAK2 (O.E.) or a control plasmid DNA (pCMV6-C-Myc-Flag and pCMV3-C-HA, Con.) as indicated. Then cell lysates were immunoprecipitated with antibodies against Myc or HA. The immunoprecipitated proteins were immunoblotted by Myc, HA, IL4Rα2, JAK2 antibodies. Light chain of IgG and Coomassie Blue staining of SDS–PAGE were used for the loading control.

3.4. Telmisartan Suppresses Cell Proliferation and Induces Apoptosis and Cell Cycle Arrest in A498, ACHN, Caki1, and Caki2 Cells Via Inhibition of JAK2

Previously, we reported that type II IL4Rα and IL13Rα1 complex are involved in RCC progress through regulation JAK2/FOXO3 pathway [21]. In addition, in this study, we showed that JAK2 was regulated by IL13Rα2. Thus, we thought that JAK2 was the common downstream-signaling kinase under the type II IL4Rα and IL13Rα1 complex and IL13Rα2. Hence, we tried to find the novel chemical inhibitor against JAK2 as the therapeutic way to treat RCC by screening an FDA-approved drug library (821 drugs) with a JAK2 kinase assay kit. After narrowing down the possible candidates, telmisartan, a clinically used medicine against hypertension, could be selected as one of the strongest JAK2 inhibitors from 821 drugs. As shown in Supplementary Figure S1B, telmisartan reduced ATP consumption in a dose-dependent manner in vitro. In fact, telmisartan treatment decreased the phosphorylation level of JAK2 in A498, ACHN, and 293T transfected with JAK2 overexpression plasmid DNA (Supplementary Figure S1C). To determine the anti-carcinogenic effect of telmisartan, we conducted in vitro assays for cell proliferation, cell cycle arrest, and apoptosis in RCC cells after telmisartan treatment. Cells were treated with 0, 20, and 40 μM of telmisartan and incubated for the indicated time. As shown in Figure 5A–C, telmisartan treatment decreased cell proliferation rate in a dose and time-dependent manner. We found that telmisartan treatment increased the G2/M population in A498, ACHN, Caki1, and Caki2 cells compared to DMSO control (Figure 5D). TUNEL and annexin V staining assay results showed that telmisartan treatment increased the apoptosis in A498, ACHN Caki1, and Caki2 cells compared to DMSO control (Figure 5E,F). Western blotting analysis indicated that telmisartan increased the expression of cleaved PARP1, cleaved caspase3, FOXO3, and p27, whereas decreased the expression of IL13Rα2 and pJAK2 (Figure 5G). Overall, these results indicate that telmisartan treatment could regulate proliferation, cell cycle arrest, and apoptosis in A498, ACHN, Cak1, and Caki2 RCC cells via inhibition of JAK2.

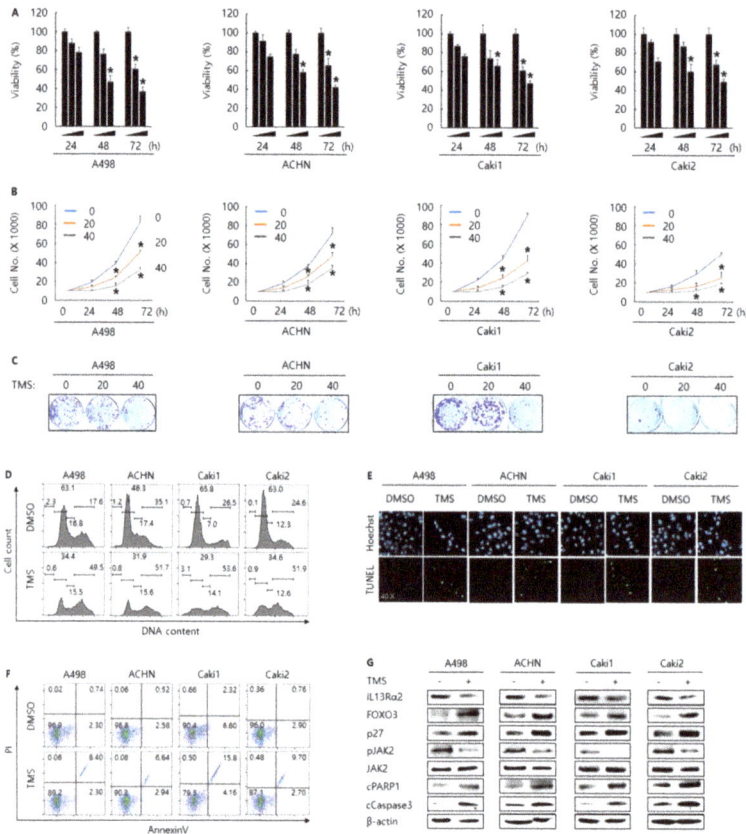

Figure 5. Antiproliferative effect by telmisartan treatment in A498, ACHN, Caki1, and Caki2 cells. Cell viability and proliferation rate were determined by WST-1 (A), cell counting assay (B) for 24, 48, and 72 h, and Colony formation assay for 14 days (C) after treatment of telmisartan (0, 20, and 40 μM). This result is representative data of at least three independent experiments, and the error bar indicates mean ± standard error (STE). * stands for the P-value < 0.05. Cell cycle arrest for 48 h after treatment of telmisartan (40 μM) was determined by cell cycle analysis (D). Apoptosis for 48 h after treatment of telmisartan (40 μM) was determined by Annexin V staining analysis (E) and Terminal deoxynucleotidyl transferase dUTP nick end labeling (TUNEL) assay (F). This result is representative data of at least three independent experiments. (G) Western blotting analysis of proteins related to cell cycle arrest and apoptosis for 48 h after treatment of telmisartan (40 μM). β-actin was used for a gel-loading control.

4. Discussion

It has been reported that IL13Rα2 was overexpressed in various cancers, such as glioblastoma, metastatic colorectal cancer, and ovarian cancer, which suggests that IL13Rα2 can play crucial roles in the development of various cancer types [25,26]. A recent study revealed that IL13Rα2 might be an important therapeutic target in a perineural invasion, the invasion of cancer to nerves [27]. In another study, IL13Rα2 was closely related to cancer cell migration, which indicated that IL13Rα2 might be a key factor in metastasis in cancers [28]. It was reported that IL13Rα2 was a functional receptor-mediating-signaling pathway in human pancreatic cancer cell lines [29]. They showed that IL13 induced the activation of transforming growth factor-β (TGFβ) through the AP-1 pathway, which can promote tumorigenesis caused by immunosuppression. In another study using the mouse model, it was demonstrated that two kinds of humanized scFv based chimeric antigen receptor (CAR) T cells targeting IL13Rα2 inhibited tumor growth in vitro and in vivo [30]. Sunitinib is an agent for treating metastatic or unresectable clear cell RCC, and IL13Rα2 can be a potential target to overcome sunitinib resistance [31]. However, the exact mechanism related to IL13Rα2 has not been investigated in RCC development. As shown in Figures 1 and 2, immunohistochemical expression of IL13Rα2 was highly associated with cancer-specific survival and relapse-free survival by univariate and multivariate analysis in 229 RCC patients. In addition, the oncogenic role of IL13Rα2 was confirmed by the in vitro cell assay. Knock-down of IL13Rα2 showed the antiproliferative activity in A498, ACHN, Caki1, and Caki2 cells (Figure 3). As shown in Figure 3G, the expression of pJAK2 was significantly downregulated by transfection of IL13Rα2 with siRNA in RCC cells. Mechanistically, IL13Rα2 seemed to interact with JAK2 in RCC cells to activate the phosphorylation of JAK2, which may downregulate FOXO3, a representative tumor-suppressive transcriptional factor. To the best of our knowledge, this is the first research to demonstrate the IL13Rα2/JAK2/FOXO3-signaling pathway in cancer development.

Atopic dermatitis (AD) has been the most common type of chronic inflammatory skin disease [32]. JAK2 inhibitors have been identified as effective reagents for the treatment of atopic dermatitis [33]. A recent study showed that JTE-052, which is a novel JAK inhibitor suppressed skin inflammation and had therapeutic effects on chronic dermatitis in rodent models [34]. Interestingly, the recent clinical report has shown that cream containing ruxolitinib that is JAK1/JAK2 inhibitor alleviated AD symptoms and itch effectively in AD patients [35]. These studies suggested that JAK2 inhibitor could be a promising reagent for developing effective drugs for AD treatment. Furthermore, JAK2 inhibitor has been considered a promising therapeutic reagent for arthritis treatment [36]. A recent study has reported that ferulic acid showed anti-arthritic activity in rats induced arthritis through inhibition of the JAK/STAT pathway [37]. It was also reported that the Ershiwuwei Lvxue pill (ELP) that is Tibetan traditional medicine, reduced collagen-induced arthritis through JAK2/STAT3-signaling pathway inhibition [38]. These studies indicated that JAK2 inhibitor also could be considered an effective reagent for arthritis treatment.

IL-13 has been known as a crucial cytokine in chronic airway inflammation, and it plays an important role in AD pathogenesis [39,40]. Because IL-13 is a pivotal cytokine involved in allergic responses, it is important to find an effective way to alleviate immune responses by inhibiting IL-13 [41]. A recent study demonstrated that inhibition of IL-13 for AD is a new pathway, which suggested that IL-13 inhibitors could be an effective reagent for AD treatment [42]. It was reported that lebrikizumab is an IL-13 inhibitor that has the potential to treat moderate-to-severe AD with fewer side effects [43]. A clinical report showed that tralokinumab is the other IL-13 inhibitor that shows promising results of alleviating moderate-to-severe AD in adult patients. In short, these results supported that IL-13 inhibitor appears to have the potential to be a promising reagent for the development of new drugs for AD treatment.

Janus kinases, often referred as JAK, have been known as cytoplasmic tyrosine kinase combined with intracellular domains of various cytokine receptors [44]. JAK family member is divided into JAK1, JAK2, JAK3, and TYK2 [45]. According to recent studies,

JAK2/STAT3 signaling pathway played critical roles in metastasis and progression of cancers, which implied that JAK2 might be a crucial therapeutic target for treatment of cancer [46–49]. Recent study showed that salidroside had anti-cancer effects and suppressed RCC proliferation through inhibition of JAK2/STAT3 signaling pathway [50]. The data presented in this study indicated that salidroside decreased the levels of phosphorylated STAT3 and JAK2 in A498 and 786-0 RCC cells. It was also reported that thymoquinone, a natural compound extracted from black seed oil, possessed anti-cancer effects in RCC cells [51]. According to them, inhibition of JAK2/STAT3 signaling pathway was observed after treatment of thymoquinone in Caki2 cells. Furthermore, recent studies have reported that the synthetic JAK2 inhibitor was considered as the therapeutic agent for other cancer types [52–54]. It was reported that treatment of JAK inhibitors CEP-33779 and NVP-BSK805 helped vincristine work effectively by sensitizing drug-resistant KBV20C oral cancer cells [55]. AG490, JAK2 inhibitor, also inhibited the proliferation and invasion of gallbladder cancer cells through inhibition of JAK2/STAT3 signaling pathway [54]. Thus, our current study supported that JAK2 has a potential to be an important target for various cancer treatment.

Telmisartan is angiotensin II receptor blocker and selectively inhibits the binding of angiotensin II into AT1 receptor [56]. Telmisartan was approved by FDA in 1998 and it has been used to treat high blood pressure and heart failure [57–59]. It also has been reported that telmisartan has anti-cancer effect against several cancer cell lines [60–62]. Recent study showed that telmisartan has cytotoxic effect through generation of reactive oxygen species (ROS) and upregulation of death receptor 5 (DR5) in human lung cancer A549 cells [63]. It was reported that telmisartan inhibited cancer cell growth and induced DNA damage in HHUA human endometrial cancer cells [64]. In another study, telmisartan downregulated Bcl-2 and induced apoptosis in 786-0 RCC cells [65]. Also, recent study has shown that telmisartan exhibited anti-cancer effect in MKN74 gastric cancer cells in vitro and in vivo [66]. Interestingly, this study showed that telmisartan inhibited tumor growth through cell cycle arrest in a mouse xenograft model of gastric cancer. Furthermore, growth inhibitory effect of telmisartan was observed in esophageal squamous cell carcinoma xenograft mouse model [67]. Similar with the previous studies, we observed that telmisartan treatment suppressed cell proliferation and induced cell cycle arrest and apoptosis via inhibition of JAK2 in human RCC cells. However, we still need to perform in vivo experiments using RCC mouse model to prove the anti-cancer activity of telmisartan. We selected telmisartan as one of the strongest JAK2 inhibitors from 821 FDA approved drugs. Since we adopted the screening way based on the assay to measure ATP consumption by JAK2, we thought that telmisartan might compete with ATP to bind the ATP binding site in JAK2. We are planning to conduct the competitive enzyme assay and simulate in silico docking model to prove this hypothesis. As shown in Figure 5G, interestingly, telmisartan treatment caused the downregulation of IL13Rα2. It seems that inhibition of JAK2 by telmisartan might induce the transcriptional downregulation of IL13Rα2 through inhibition of the phosphorylation of STAT3 transcriptional factor. So, we plan to perform other experiments demonstrating that STAT3 bind to the promoter region of IL13Rα2 and whether the binding affinity of STAT3 on the promoter region was weakened by JAK2 inhibition or not.

Since telmisartan has been used to treat heart disease for 22 years, there are lots of previous reports for researcher to examine the possible working mechanism of telmisartan in terms of anti-cancer activity. The relationship between JAK2 and angiotensin II signaling pathway has been investigated in various studies [68–71]. It has been reported that angiotensin II activates STAT3 through the IL6/gp130/JAK2 signaling pathway in cardiomyocytes [72]. AG490, well-known JAK2 inhibitor, inhibited angiotensin II-induced differentiation of bone marrow-derived mesenchymal stem cells (BM-MSCs) into keratinocytes, which suggested that JAK2 is associated with angiotensin II signaling pathway [73]. Recent study has shown that angiotensin II upregulated nitroxidative stress via JAK2/STAT3 signaling pathway leading to the hyperproliferation of vascular smooth muscle cells (VSMCs) [74]. In another study, it is demonstrated that inhibition of angiotensin II through JAK2/STAT3 signaling pathway suppressed tubular epithelial myofibroblast trans-differentiation mediated by hepatocyte growth factor (HGF) [75]. Thus, we thought that blocking of angiotensin II binding into AT1 receptor by telmisartan might cause the inhibition of JAK2 through direct or indirect signaling pathway in RCC cells. We might need to investigate the change of the phosphorylation status of JAK2 under knock-down of AT1 receptor in RCC cells. Peroxisome proliferator-activated receptor γ (PPARγ) is also well-known agonistic target of telmisartan. PPARγ is a member of nuclear receptor family and it plays an important role in regulating lipid metabolism [76]. According to a previous research, activation of JAK2/STAT3 signaling pathway was associated with downregulation of PPARγ, which promoted fibrosis in rats [77]. Furthermore, PPARγ decreased the protein expression of suppressor of cytokine signaling 3 (SOCS3) through inhibition of JAK2/STAT3 signaling pathway leading to alleviation of hepatocyte steatosis [78]. Additionally, it was reported that pioglitazone, one of PPARγ agonists, inhibited breast cancer growth by regulating JAK2/STAT signaling pathway in vitro and in vivo [79]. For the further study, we are trying to examine that rosiglitazone, FDA approved hypoglycemic agent as PPARγ agonists, has anti-cancer activity against RCC through inhibition of JAK2 phosphorylation.

In this study, we demonstrated the clinicopathologically significance of IL13Rα2, a kind of the independent receptor for IL13, in RCC progression. Mechanistically, downregulation of IL13Rα2 in RCC cells seemed to decrease the phosphorylation of JAK2 and increase expression of FOXO3, suggesting that IL13Rα2 probably is involved in the progression of RCC through JAK2/FOXO3 pathway (Figure 6). In addition, we screened an FDA approved drug library to identify the novel candidates inhibiting JAK2 in RCC cells and selected telmisartan as the one of strongest JAK2 inhibitors. Telmisartan displayed the anti-proliferative activity in RCC cells, which could be one of the therapeutic options for RCC patients.

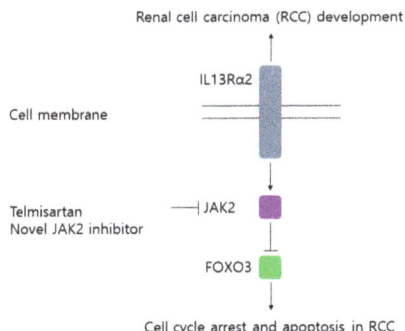

Figure 6. A diagram for the possible oncogenic role of IL13Rα2 in renal cell carcinoma (RCC) by activation of JAK2 and inhibition of FOXO3.

Supplementary Materials: The following are available online at https://www.mdpi.com/article/10.3390/jpm11040284/s1, Figure S1: (A). The correlation pattern between expression of IL13Rα2, pJAK2, JAK2, and FOXO3 in A498, ACHN, Caki1, and Caki2 cells. Western blotting analysis of IL13Rα2, pJAK2, JAK2, and FOXO3 in each cell line. β-actin was used for a gel-loading control; (B) Reduction of ATP consumption by telmisartan in a dose-dependent manner in vitro. JAK2 protein was incubated with the indicated concentration of AZD1480 or telmisartan, peptide substrate, and ATP for 30 min at 37 °C. After incubation, the reaction mixture was incubated with Glo-Max solution for 30 min at room temperature to stop the reaction. Then, the remaining ATP level in each reaction was measured by a microplate reader for luminescence; (C). Reduction of phosphorylation of JAK2 by telmisartan. Western blotting analysis of pJAK2 and JAK2 after telmisartan treatment (0, 10, 20, and 40 μM). β-actin was used for a gel-loading control.

Author Contributions: Conceptualization, M.-A.K., J.L., C.M.L., H.S.P., K.Y.J., and S.-H.P.; methodology, M.-A.K., J.L., C.M.L., H.S.P., K.Y.J., and S.-H.P.; software, M.-A.K., J.L., C.M.L., H.S.P., K.Y.J., and S.-H.P.; validation, M.-A.K., J.L., C.M.L., H.S.P., K.Y.J., and S.-H.P.; formal analysis, M.-A.K., J.L., C.M.L., H.S.P., K.Y.J., and S.-H.P.; investigation, M.-A.K., J.L., C.M.L., H.S.P., K.Y.J., and S.-H.P.; resources, M.-A.K., J.L., C.M.L., H.S.P., K.Y.J., and S.-H.P.; data curation, M.-A.K., J.L., C.M.L., H.S.P., K.Y.J., and S.-H.P.; writing—original draft preparation, M.-A.K., J.L., C.M.L., H.S.P., K.Y.J., and S.-H.P.; writing—review and editing, M.-A.K., J.L., C.M.L., H.S.P., K.Y.J., and S.-H.P.; visualization, M.-A.K., J.L., C.M.L., H.S.P., K.Y.J., and S.-H.P.; supervision, K.Y.J. and S.-H.P.; project administration, K.Y.J. and S.-H.P.; funding acquisition, K.Y.J. and S.-H.P. All authors have read and agreed to the published version of the manuscript.

Funding: This research was supported by the Basic Science Research Program (NRF-2014R1A6A3A04054307) through the National Research Foundation of Korea (NRF), funded by the Ministry of Science and ICT (MSIP). This research was supported by the Medical Research Center Program (NRF-2017R1A5A2015061) through the National Research Foundation of Korea (NRF), funded by the Ministry of Science and ICT (MSIP).

Institutional Review Board Statement: The study was conducted according to the guidelines of the Declaration of Helsinki, and approved by the Institutional Review Board of Jeonbuk National University Hospital (IRB number: CUH 2019-11-039, date of approval: 10 November 2019).

Informed Consent Statement: The IRB approval (CUH 2019-11-039) contained a waiver for written informed consent based on the retrospective and anonymous character of the study.

Data Availability Statement: The datasets used in the current study are available from the corresponding author upon reasonable request.

Conflicts of Interest: The authors declare no conflict of interest.

References

1. Ferlay, J.; Soerjomataram, I.; Dikshit, R.; Eser, S.; Mathers, C.; Rebelo, M.; Parkin, D.M.; Forman, D.; Bray, F. Cancer incidence and mortality worldwide: Sources, methods and major patterns in GLOBOCAN 2012. *Int. J. Cancer* **2014**, *136*, E359–E386. [CrossRef] [PubMed]
2. Fisher, R.; Gore, M.; Larkin, J. Current and future systemic treatments for renal cell carcinoma. *Semin. Cancer Biol.* **2013**, *23*, 38–45. [CrossRef] [PubMed]
3. Escudier, B.; Eisen, T.; Stadler, W.M.; Szczylik, C.; Oudard, S.; Siebels, M.; Negrier, S.; Chevreau, C.; Solska, E.; Desai, A.A.; et al. Sorafenib in Advanced Clear-Cell Renal-Cell Carcinoma. *N. Engl. J. Med.* **2007**, *356*, 125–134. [CrossRef] [PubMed]
4. Choueiri, T.K.; Motzer, R.J. Systemic Therapy for Metastatic Renal-Cell Carcinoma. *N. Engl. J. Med.* **2017**, *376*, 354–366. [CrossRef]
5. Heng, D.Y.; Xie, W.; Regan, M.M.; Warren, M.A.; Golshayan, A.R.; Sahi, C.; Eigl, B.J.; Ruether, J.D.; Cheng, T.; North, S.; et al. Prognostic Factors for Overall Survival in Patients With Metastatic Renal Cell Carcinoma Treated With Vascular Endothelial Growth Factor–Targeted Agents: Results From a Large, Multicenter Study. *J. Clin. Oncol.* **2009**, *27*, 5794–5799. [CrossRef]
6. Heng, D.Y.; Xie, W.; Regan, M.M.; Harshman, L.C.; Bjarnason, G.; Vaishampayan, U.N.; Mackenzie, M.; Wood, L.; Donskov, F.; Tan, M.-H.; et al. External validation and comparison with other models of the International Metastatic Renal-Cell Carcinoma Database Consortium prognostic model: A population-based study. *Lancet Oncol.* **2013**, *14*, 141–148. [CrossRef]
7. Hsieh, J.J.; Purdue, M.P.; Signoretti, S.; Swanton, C.; Albiges, L.; Schmidinger, M.; Heng, D.Y.; Larkin, J.; Ficarra, V. Renal cell carcinoma. *Nat. Rev. Dis. Prim.* **2017**, *3*, 1–19. [CrossRef]
8. Qi, W.-X.; Huang, Y.-J.; Yao, Y.; Shen, Z.; Min, D.-L. Incidence and Risk of Treatment-Related Mortality with mTOR Inhibitors Everolimus and Temsirolimus in Cancer Patients: A Meta-Analysis. *PLoS ONE* **2013**, *8*, e65166. [CrossRef]
9. Escudier, B.; Szczylik, C.; Porta, C.; Gore, M. Treatment selection in metastatic renal cell carcinoma: Expert consensus. *Nat. Rev. Clin. Oncol.* **2012**, *9*, 327–337. [CrossRef]
10. Hershey, G.K. IL-13 receptors and signaling pathways: An evolving web. *J. Allergy Clin. Immunol.* **2003**, *111*, 677–690. [CrossRef]
11. Bhardwaj, R.; Suzuki, A.; Leland, P.; Joshi, B.H.; Puri, R.K. Identification of a novel role of IL-13Rα2 in human Glioblastoma multiforme: Interleukin-13 mediates signal transduction through AP-1 pathway. *J. Transl. Med.* **2018**, *16*, 1–13. [CrossRef]
12. Okamoto, H.; Yoshimatsu, Y.; Tomizawa, T.; Kunita, A.; Takayama, R.; Morikawa, T.; Komura, D.; Takahashi, K.; Oshima, T.; Sato, M.; et al. Interleukin-13 receptor α2 is a novel marker and potential therapeutic target for human melanoma. *Sci. Rep.* **2019**, *9*, 1–13. [CrossRef]
13. Bartolomé, R.A.; Jaén, M.; Casal, J.I. An IL13Rα2 peptide exhibits therapeutic activity against metastatic colorectal cancer. *Br. J. Cancer* **2018**, *119*, 940–949. [CrossRef]
14. Papageorgis, P.; Ozturk, S.; Lambert, A.W.; Neophytou, C.M.; Tzatsos, A.; Wong, C.K.; Thiagalingam, S.; Constantinou, A.I. Targeting IL13Ralpha2 activates STAT6-TP63 pathway to suppress breast cancer lung metastasis. *Breast Cancer Res.* **2015**, *17*, 98. [CrossRef]

15. Park, S.-H.; Chung, Y.M.; Ma, J.; Yang, Q.; Berek, J.S.; Hu, M.C.-T. Pharmacological activation of FOXO3 suppresses triple-negative breast cancer in vitro and in vivo. *Oncotarget* **2016**, *7*, 42110–42125. [CrossRef]
16. Tang, M.; Hu, X.; Wang, Y.; Yao, X.; Zhang, W.; Yu, C.; Cheng, F.; Li, J.; Fang, Q. Ivermectin, a potential anticancer drug derived from an antiparasitic drug. *Pharmacol. Res.* **2021**, *163*, 105207. [CrossRef]
17. Borém, L.M.A.; Neto, J.F.R.; Brandi, I.V.; Lelis, D.F.; Santos, S.H.S. The role of the angiotensin II type I receptor blocker telmisartan in the treatment of non-alcoholic fatty liver disease: A brief review. *Hypertens. Res.* **2018**, *41*, 394–405. [CrossRef]
18. Moch, H.; Humphrey, P.A.; Ulbright, T.M. *WHO Classification of Tumours of the Urinary System and Male Genital Organs*, 4th ed.; World Health Organization: Geneva, Switzerland, 2016; Volume 70, pp. 93–105.
19. Amin, M.B.; American Joint Committee on Cancer; American Cancer Society. *AJCC Cancer Staging Manual*; Springer: Chicago, IL, USA, 2017; 1024p, p. xvii.
20. Allred, D.C.; Harvey, J.M.; Berardo, M.; Clark, G.M. Prognostic and predictive factors in breast cancer by immunohistochemical analysis. *Mod. Pathol.* **1998**, *11*, 155–168.
21. Kang, M.-A.; Lee, J.; Ha, S.H.; Lee, C.M.; Kim, K.M.; Jang, K.Y.; Park, S.-H. Interleukin4Rα (IL4Rα) and IL13Rα1 Are Associated with the Progress of Renal Cell Carcinoma through Janus Kinase 2 (JAK2)/Forkhead Box O3 (FOXO3) Pathways. *Cancers* **2019**, *11*, 1394. [CrossRef]
22. Kim, K.M.; Hussein, U.K.; Park, S.-H.; Kang, M.A.; Moon, Y.J.; Zhang, Z.; Song, Y.; Park, H.S.; Bae, J.S.; Park, B.-H.; et al. FAM83H is involved in stabilization of β-catenin and progression of osteosarcomas. *J. Exp. Clin. Cancer Res.* **2019**, *38*, 1–15. [CrossRef]
23. Park, H.J.; Bae, J.S.; Kim, K.M.; Moon, Y.J.; Park, S.-H.; Ha, S.H.; Hussein, U.K.; Zhang, Z.; Park, H.S.; Park, B.-H.; et al. The PARP inhibitor olaparib potentiates the effect of the DNA damaging agent doxorubicin in osteosarcoma. *J. Exp. Clin. Cancer Res.* **2018**, *37*, 107. [CrossRef] [PubMed]
24. Kim, K.M.; Hussein, U.K.; Bae, J.S.; Park, S.-H.; Kwon, K.S.; Ha, S.H.; Park, H.S.; Lee, H.; Chung, M.J.; Moon, W.S.; et al. The Expression Patterns of FAM83H and PANX2 Are Associated with Shorter Survival of Clear Cell Renal Cell Carcinoma Patients. *Front. Oncol.* **2019**, *9*, 14. [CrossRef] [PubMed]
25. Sai, K.K.S.; Sattiraju, A.; Almaguel, F.G.; Xuan, A.; Rideout, S.; Krishnaswamy, R.S.; Zhang, J.; Herpai, D.M.; Debinski, W.; Mintz, A. Peptide-based PET imaging of the tumor restricted IL13RA2 biomarker. *Oncotarget* **2017**, *8*, 50997–51007. [CrossRef] [PubMed]
26. Bartolomé, R.A.; Martín-Regalado, Á.; Jaén, M.; Zannikou, M.; Zhang, P.; Ríos, V.D.L.; Balyasnikova, I.V.; Casal, J.I. Protein Tyrosine Phosphatase-1B Inhibition Disrupts IL13Rα2-Promoted Invasion and Metastasis in Cancer Cells. *Cancers* **2020**, *12*, 500. [CrossRef]
27. Fujisawa, T.; Shimamura, T.; Goto, K.; Nakagawa, R.; Muroyama, R.; Ino, Y.; Horiuchi, H.; Endo, I.; Maeda, S.; Harihara, Y.; et al. A Novel Role of Interleukin 13 Receptor alpha2 in Perineural Invasion and its Association with Poor Prognosis of Patients with Pancreatic Ductal Adenocarcinoma. *Cancers* **2020**, *12*, 1294. [CrossRef]
28. Chong, S.T.; Tan, K.M.; Kok, C.Y.L.; Guan, S.P.; Lai, S.H.; Lim, C.; Hu, J.; Sturgis, C.; Eng, C.; Lam, P.Y.P.; et al. IL13RA2 Is Differentially Regulated in Papillary Thyroid Carcinoma vs Follicular Thyroid Carcinoma. *J. Clin. Endocrinol. Metab.* **2019**, *104*, 5573–5584. [CrossRef]
29. Shimamura, T.; Fujisawa, T.; Husain, S.R.; Joshi, B.H.; Puri, R.K. Interleukin 13 Mediates Signal Transduction through Interleukin 13 Receptor α2 in Pancreatic Ductal Adenocarcinoma: Role of IL-13 Pseudomonas Exotoxin in Pancreatic Cancer Therapy. *Clin. Cancer Res.* **2010**, *16*, 577–586. [CrossRef]
30. Yin, Y.; Boesteanu, A.C.; Binder, Z.A.; Xu, C.; Reid, R.A.; Rodriguez, J.L.; Cook, D.R.; Thokala, R.; Blouch, K.; McGettigan-Croce, B.; et al. Checkpoint Blockade Reverses Anergy in IL-13Rα2 Humanized scFv-Based CAR T Cells to Treat Murine and Canine Gliomas. *Mol. Ther. Oncolytics* **2018**, *11*, 20–38. [CrossRef]
31. Shibasaki, N.; Yamasaki, T.; Kanno, T.; Arakaki, R.; Sakamoto, H.; Utsunomiya, N.; Inoue, T.; Tsuruyama, T.; Nakamura, E.; Ogawa, O.; et al. Role of IL13RA2 in Sunitinib Resistance in Clear Cell Renal Cell Carcinoma. *PLoS ONE* **2015**, *10*, e0130980. [CrossRef]
32. Rodrigues, M.A.; Torres, T. JAK/STAT inhibitors for the treatment of atopic dermatitis. *J. Dermatol. Treat.* **2019**, *31*, 33–40. [CrossRef]
33. Cotter, D.G.; Schairer, D.; Eichenfield, L. Emerging therapies for atopic dermatitis: JAK inhibitors. *J. Am. Acad. Dermatol.* **2018**, *78*, S53–S62. [CrossRef]
34. Tanimoto, A.; Shinozaki, Y.; Yamamoto, Y.; Katsuda, Y.; Taniai-Riya, E.; Toyoda, K.; Kakimoto, K.; Kimoto, Y.; Amano, W.; Konishi, N.; et al. A novel JAK inhibitor JTE-052 reduces skin inflammation and ameliorates chronic dermatitis in rodent models: Comparison with conventional therapeutic agents. *Exp. Dermatol.* **2017**, *27*, 22–29. [CrossRef]
35. Kim, B.S.; Howell, M.D.; Sun, K.; Papp, K.; Nasir, A.; Kuligowski, M.E. Treatment of atopic dermatitis with ruxolitinib cream (JAK1/JAK2 inhibitor) or triamcinolone cream. *J. Allergy Clin. Immunol.* **2020**, *145*, 572–582. [CrossRef]
36. Malemud, C.J. The role of the JAK/STAT signal pathway in rheumatoid arthritis. *Ther. Adv. Musculoskelet. Dis.* **2018**, *10*, 117–127. [CrossRef]
37. Zhu, L.; Zhang, Z.; Xia, N.; Zhang, W.; Wei, Y.; Huang, J.; Ren, Z.; Meng, F.; Yang, L. Anti-arthritic activity of ferulic acid in complete Freund's adjuvant (CFA)-induced arthritis in rats: JAK2 inhibition. *Inflammopharmacology* **2019**, *28*, 463–473. [CrossRef]
38. Liu, C.; Zhao, Q.; Zhong, L.; Li, Q.; Li, R.; Li, S.; Li, Y.; Li, N.; Su, J.; Dhondrup, W.; et al. Tibetan medicine Ershiwuwei Lvxue Pill attenuates collagen-induced arthritis via inhibition of JAK2/STAT3 signaling pathway. *J. Ethnopharmacol.* **2021**, *270*, 113820. [CrossRef]

39. Wollenberg, A.; Howell, M.D.; Guttman-Yassky, E.; Silverberg, J.I.; Kell, C.; Ranade, K.; Moate, R.; van der Merwe, R. Treatment of atopic dermatitis with tralokinumab, an anti–IL-13 mAb. *J. Allergy Clin. Immunol.* **2019**, *143*, 135–141. [CrossRef]
40. Marone, G.; Granata, F.; Pucino, V.; Pecoraro, A.; Heffler, E.; Loffredo, S.; Scadding, G.W.; Varricchi, G. The Intriguing Role of Interleukin 13 in the Pathophysiology of Asthma. *Front. Pharmacol.* **2019**, *10*, 1387. [CrossRef]
41. Bieber, T. Interleukin-13: Targeting an underestimated cytokine in atopic dermatitis. *Allergy* **2020**, *75*, 54–62. [CrossRef]
42. Ratnarajah, K.; Le, M.; Muntyanu, A.; Mathieu, S.; Nigen, S.; Litvinov, I.V.; Jack, C.S.; Netchiporouk, E. Inhibition of IL-13: A New Pathway for Atopic Dermatitis. *J. Cutan. Med. Surg.* **2020**, *10*, 1203475420982553. [CrossRef]
43. Loh, T.Y.; Hsiao, J.L.; Shi, V.Y. Therapeutic Potential of Lebrikizumab in the Treatment of Atopic Dermatitis. *J. Asthma Allergy* **2020**, *13*, 109–114. [CrossRef]
44. Clark, J.D.; Flanagan, M.E.; Telliez, J.-B. Discovery and Development of Janus Kinase (JAK) Inhibitors for Inflammatory Diseases. *J. Med. Chem.* **2014**, *57*, 5023–5038. [CrossRef]
45. Roskoski, R. Janus kinase (JAK) inhibitors in the treatment of inflammatory and neoplastic diseases. *Pharmacol. Res.* **2016**, *111*, 784–803. [CrossRef]
46. Yuan, K.; Ye, J.; Liu, Z.; Ren, Y.; He, W.; Xu, J.; He, Y.; Yuan, Y. Complement C3 overexpression activates JAK2/STAT3 pathway and correlates with gastric cancer progression. *J. Exp. Clin. Cancer Res.* **2020**, *39*, 1–15. [CrossRef]
47. Zeng, Y.-T.; Liu, X.-F.; Yang, W.-T.; Zheng, P.-S. REX1 promotes EMT-induced cell metastasis by activating the JAK2/STAT3-signaling pathway by targeting SOCS1 in cervical cancer. *Oncogene* **2019**, *38*, 6940–6957. [CrossRef]
48. Liu, K.; Gao, H.; Wang, Q.; Wang, L.; Zhang, B.; Han, Z.; Chen, X.; Han, M.; Gao, M. Retracted: Hispidulin suppresses cell growth and metastasis by targeting PIM 1 through JAK 2/ STAT 3 signaling in colorectal cancer. *Cancer Sci.* **2018**, *109*, 1369–1381. [CrossRef]
49. Zhou, X.; Yan, T.; Huang, C.; Xu, Z.; Wang, L.; Jiang, E.; Wang, H.; Chen, Y.; Liu, K.; Shao, Z.; et al. Melanoma cell-secreted exosomal miR-155-5p induce proangiogenic switch of cancer-associated fibroblasts via SOCS1/JAK2/STAT3 signaling pathway. *J. Exp. Clin. Cancer Res.* **2018**, *37*, 1–15. [CrossRef]
50. Lv, C.; Huang, Y.; Liu, Z.-X.; Yu, D.; Bai, Z.-M. Salidroside reduces renal cell carcinoma proliferation by inhibiting JAK2/STAT3 signaling. *Cancer Biomark.* **2016**, *17*, 41–47. [CrossRef]
51. Chae, I.G.; Chun, K.-S. Abstract 4844: Thymoquinone induces apoptosis through inhibition of JAK2/STAT3 signaling via production of ROS in human renal cancer Caki cells. *Cancer Chem.* **2016**, *76*, 4844. [CrossRef]
52. Morgan, E.L.; Macdonald, A. JAK2 Inhibition Impairs Proliferation and Sensitises Cervical Cancer Cells to Cisplatin-Induced Cell Death. *Cancers* **2019**, *11*, 1934. [CrossRef]
53. Kim, J.W.; Gautam, J.; Kim, J.; Kang, K.W. Inhibition of tumor growth and angiogenesis of tamoxifen-resistant breast cancer cells by ruxolitinib, a selective JAK2 inhibitor. *Oncol. Lett.* **2019**, *17*, 3981–3989. [CrossRef] [PubMed]
54. Fu, L.X.; Lian, Q.W.; Pan, J.D.; Xu, Z.L.; Zhou, T.M.; Ye, B. JAK2 tyrosine kinase inhibitor AG490 suppresses cell growth and invasion of gallbladder cancer cells via inhibition of JAK2/STAT3 signaling. *J. Boil. Regul. Homeost. Agents* **2017**, *31*, 51–58.
55. Cheon, J.H.; Kim, K.S.; Yadav, D.K.; Kim, M.; Kim, H.S.; Yoon, S. The JAK2 inhibitors CEP-33779 and NVP-BSK805 have high P-gp inhibitory activity and sensitize drug-resistant cancer cells to vincristine. *Biochem. Biophys. Res. Commun.* **2017**, *490*, 1176–1182. [CrossRef] [PubMed]
56. McClellan, K.J.; Markham, A. Telmisartan. *Drugs* **1998**, *56*, 1039–1044. [CrossRef]
57. Sharpe, M.; Jarvis, B.; Goa, K.L. Telmisartan. *Drugs* **2001**, *61*, 1501–1529. [CrossRef]
58. Sukumaran, V.; Veeraveedu, P.T.; Gurusamy, N.; Yamaguchi, K.; Lakshmanan, A.P.; Ma, M.; Suzuki, K.; Kodama, M.; Watanabe, K. Cardioprotective Effects of Telmisartan against Heart Failure in Rats Induced By Experimental Autoimmune Myocarditis through the Modulation of Angiotensin-Converting Enzyme-2/Angiotensin 1-7/Mas Receptor Axis. *Int. J. Biol. Sci.* **2011**, *7*, 1077–1092. [CrossRef]
59. McFarlane, S. Telmisartan and cardioprotection. *Vasc. Heal. Risk Manag.* **2011**, *7*, 677–683. [CrossRef]
60. Lee, L.D.; Mafura, B.; Lauscher, J.C.; Seeliger, H.; Kreis, M.E.; Gröne, J. Antiproliferative and apoptotic effects of telmisartan in human colon cancer cells. *Oncol. Lett.* **2014**, *8*, 2681–2686. [CrossRef]
61. Matsuyama, M.; Funao, K.; Kuratsukuri, K.; Tanaka, T.; Kawahito, Y.; Sano, H.; Chargui, J.; Touraine, J.-L.; Yoshimura, N.; Yoshimura, R. Telmisartan inhibits human urological cancer cell growth through early apoptosis. *Exp. Ther. Med.* **2010**, *1*, 301–306. [CrossRef]
62. Wu, T.T.-L.; Niu, H.-S.; Chen, L.-J.; Cheng, J.-T.; Tong, Y.-C. Increase of human prostate cancer cell (DU145) apoptosis by telmisartan through PPAR-delta pathway. *Eur. J. Pharmacol.* **2016**, *775*, 35–42. [CrossRef]
63. Rasheduzzaman, M.; Moon, J.-H.; Lee, J.-H.; Nazim, U.M.; Park, S.-Y. Telmisartan generates ROS-dependent upregulation of death receptor 5 to sensitize TRAIL in lung cancer via inhibition of autophagy flux. *Int. J. Biochem. Cell Biol.* **2018**, *102*, 20–30. [CrossRef]
64. Koyama, N.; Nishida, Y.; Ishii, T.; Yoshida, T.; Furukawa, Y.; Narahara, H. Telmisartan Induces Growth Inhibition, DNA Double-Strand Breaks and Apoptosis in Human Endometrial Cancer Cells. *PLoS ONE* **2014**, *9*, e93050. [CrossRef] [PubMed]
65. Júnior, R.F.D.A.; Oliveira, A.L.C.L.; Silveira, R.F.D.M.; Rocha, H.A.D.O.; Cavalcanti, P.D.F.; De Araújo, A.A. Telmisartan induces apoptosis and regulates Bcl-2 in human renal cancer cells. *Exp. Biol. Med.* **2014**, *240*, 34–44. [CrossRef]

66. Fujita, N.; Fujita, K.; Iwama, H.; Kobara, H.; Fujihara, S.; Chiyo, T.; Namima, D.; Yamana, H.; Kono, T.; Takuma, K.; et al. Antihypertensive drug telmisartan suppresses the proliferation of gastric cancer cells in vitro and in vivo. *Oncol. Rep.* **2020**, *44*, 339–348. [CrossRef]
67. Matsui, T.; Chiyo, T.; Kobara, H.; Fujihara, S.; Fujita, K.; Namima, D.; Nakahara, M.; Kobayashi, N.; Nishiyama, N.; Yachida, T.; et al. Telmisartan Inhibits Cell Proliferation and Tumor Growth of Esophageal Squamous Cell Carcinoma by Inducing S-Phase Arrest In Vitro and In Vivo. *Int. J. Mol. Sci.* **2019**, *20*, 3197. [CrossRef]
68. Marrero, M.B.; Venema, V.J.; Ju, H.; Eaton, D.C.; Venema, R.C. Regulation of angiotensin II-induced JAK2 tyrosine phosphorylation: Roles of SHP-1 and SHP-2. *Am. J. Physiol. Content* **1998**, *275*, C1216–C1223. [CrossRef]
69. McWhinney, C.D.; Dostal, D.; Baker, K. Angiotensin II Activates Stat5 Through Jak2 Kinase in Cardiac Myocytes. *J. Mol. Cell. Cardiol.* **1998**, *30*, 751–761. [CrossRef]
70. Shaw, S.S.; Schmidt, A.M.; Banes, A.K.; Wang, X.; Stern, D.M.; Marrero, M.B. S100B-RAGE-Mediated Augmentation of Angiotensin II-Induced Activation of JAK2 in Vascular Smooth Muscle Cells Is Dependent on PLD2. *Diabetes* **2003**, *52*, 2381–2388. [CrossRef]
71. Banes-Berceli, A.K.L.; Ketsawatsomkron, P.; Ogbi, S.; Patel, B.; Pollock, D.M.; Marrero, M.B. Angiotensin II and endothelin-1 augment the vascular complications of diabetes via JAK2 activation. *Am. J. Physiol. Circ. Physiol.* **2007**, *293*, H1291–H1299. [CrossRef]
72. Han, J.; Ye, S.; Zou, C.; Chen, T.; Wang, J.; Li, J.; Jiang, L.; Xu, J.; Huang, W.; Wang, Y.; et al. Angiotensin II Causes Biphasic STAT3 Activation Through TLR4 to Initiate Cardiac Remodeling. *Hypertension* **2018**, *72*, 1301–1311. [CrossRef]
73. Jiang, X.; Wu, F.; Xu, Y.; Yan, J.-X.; Wu, Y.-D.; Li, S.-H.; Liao, X.; Liang, J.-X.; Li, Z.-H.; Liu, H.-W. A novel role of angiotensin II in epidermal cell lineage determination: Angiotensin II promotes the differentiation of mesenchymal stem cells into keratinocytes through the p38 MAPK, JNK and JAK2 signalling pathways. *Exp. Dermatol.* **2019**, *28*, 59–65. [CrossRef] [PubMed]
74. Hossain, E.; Li, Y.; Anand-Srivastava, M.B. Role of JAK2/STAT3 pathway in angiotensin II-induced enhanced expression of Giα proteins and hyperproliferation of aortic vascular smooth muscle cells. *Can. J. Physiol. Pharmacol.* **2021**, *99*, 237–246. [CrossRef] [PubMed]
75. Wang, H.-Y.; Zhang, C.; Xiao, Q.-F.; Dou, H.; Chen, Y.; Gu, C.-M.; Cui, M.-J. Hepatocyte growth factor inhibits tubular epithelial-myofibroblast transdifferentiation by suppression of angiotensin II via the JAK2/STAT3 signaling pathway. *Mol. Med. Rep.* **2017**, *15*, 2737–2743. [CrossRef] [PubMed]
76. Harris, S.G.; Phipps, R.P. The nuclear receptor PPAR gamma is expressed by mouse T lymphocytes and PPAR gamma agonists induce apoptosis. *Eur. J. Immunol.* **2001**, *31*, 1098–1105. [CrossRef]
77. Mahmoud, A.M.; Desouky, E.M.; Hozayen, W.G.; Bin-Jumah, M.; El-Nahass, E.-S.; Soliman, H.A.; Farghali, A.A. Mesoporous Silica Nanoparticles Trigger Liver and Kidney Injury and Fibrosis Via Altering TLR4/NF-κB, JAK2/STAT3 and Nrf2/HO-1 Signaling in Rats. *Biomolecules* **2019**, *9*, 528. [CrossRef]
78. Bi, J.; Sun, K.; Wu, H.; Chen, X.; Tang, H.; Mao, J. PPARγ alleviated hepatocyte steatosis through reducing SOCS3 by inhibiting JAK2/STAT3 pathway. *Biochem. Biophys. Res. Commun.* **2018**, *498*, 1037–1044. [CrossRef]
79. Jiao, X.X.; Lin, S.Y.; Lian, S.X.; Qiu, Y.R.; Li, Z.H.; Chen, Z.H.; Lu, W.Q.; Zhang, Y.; Deng, L.; Jiang, Y.; et al. Inhibition of the breast cancer by PPARγ agonist pioglitazone through JAK2/STAT3 pathway. *Neoplasma* **2020**, *67*, 834–842. [CrossRef]

Review

Predictive and Diagnostic Biomarkers of Anastomotic Leakage: A Precision Medicine Approach for Colorectal Cancer Patients

Mark Gray [1,*], Jamie R. K. Marland [2], Alan F. Murray [3], David J. Argyle [1] and Mark A. Potter [4]

1. The Royal (Dick) School of Veterinary Studies and Roslin Institute, University of Edinburgh, Easter Bush, Roslin, Midlothian, Edinburgh EH25 9RG, UK; david.argyle@roslin.ed.ac.uk
2. School of Engineering, Institute for Integrated Micro and Nano Systems, University of Edinburgh, Scottish Microelectronics Centre, King's Buildings, Edinburgh EH9 3FF, UK; jamie.marland@ed.ac.uk
3. School of Engineering, Institute for Bioengineering, University of Edinburgh, Faraday Building, The King's Buildings, Edinburgh EH9 3DW, UK; alan.murray@ed.ac.uk
4. Department of Surgery, Western General Hospital, Crewe Road, Edinburgh EH4 2XU, UK; mark.potter@ed.ac.uk
* Correspondence: mark.gray@ed.ac.uk

Abstract: Development of an anastomotic leak (AL) following intestinal surgery for the treatment of colorectal cancers is a life-threatening complication. Failure of the anastomosis to heal correctly can lead to contamination of the abdomen with intestinal contents and the development of peritonitis. The additional care that these patients require is associated with longer hospitalisation stays and increased economic costs. Patients also have higher morbidity and mortality rates and poorer oncological prognosis. Unfortunately, current practices for AL diagnosis are non-specific, which may delay diagnosis and have a negative impact on patient outcome. To overcome these issues, research is continuing to identify AL diagnostic or predictive biomarkers. In this review, we highlight promising candidate biomarkers including ischaemic metabolites, inflammatory markers and bacteria. Although research has focused on the use of blood or peritoneal fluid samples, we describe the use of implantable medical devices that have been designed to measure biomarkers in peri-anastomotic tissue. Biomarkers that can be used in conjunction with clinical status, routine haematological and biochemical analysis and imaging have the potential to help to deliver a precision medicine package that could significantly enhance a patient's post-operative care and improve outcomes. Although no AL biomarker has yet been validated in large-scale clinical trials, there is confidence that personalised medicine, through biomarker analysis, could be realised for colorectal cancer intestinal resection and anastomosis patients in the years to come.

Keywords: colorectal cancer; intestinal anastomosis; anastomotic leak; biomarkers; precision medicine

1. Introduction

Colorectal cancer is the fourth most commonly diagnosed cancer in the world, with ~1.8 million new cases and ~0.7 million cancer-related deaths occurring per year. The disease accounts for 10% of all newly diagnosed cancers, meaning it is a significant social and economic burden for many countries throughout the world [1]. In this review, we briefly discuss disease staging, colorectal cancer treatments, pathophysiology of normal intestinal healing and the consequences of abnormal intestinal healing. We then go on to describe in depth how this knowledge has led to the identification of diagnostic and predictive biomarkers of anastomotic leakage, which could be used to provide a precision medicine approach for managing colorectal cancer patients.

2. Colorectal Cancer Staging and Treatment

Before instigating treatment, patients undergo investigations to define the stage of the cancer. This is typically done using the tumour, node, metastasis (TNM) classification system (developed by the Union for Interventional Cancer Control) whereby data are collected from physical examinations, imaging and endoscopy. Pathological classification will be based on histopathology from biopsy samples typically obtained during endoscopy. Depending on disease stage, various treatment options are available; however, for curative intent strategies, surgery will be the treatment of choice. UK estimates indicate that 66% of colon cancer and 63% of rectal cancer patients will receive surgery as part of their primary care [2]. Surgery encompasses the excision of diseased intestinal segments containing the tumour (resection) with the subsequent re-joining of the disease-free intestinal ends (anastomosis). This intestinal resection and anastomosis procedure can be performed either with hand-placed sutures, automatic stapling devices or through robotically assisted techniques. Regardless of which technique is used, the procedure aims to re-establish luminal and mural intestinal continuity. Records from the Association of Coloproctology of Great Britain and Ireland (ACPGBI) have shown that, within Ireland and the UK, ~20,000 patients undergo a large bowel resection and anastomosis every year. The majority of these procedures are performed to treat colorectal cancers. Colorectal cancer patient outcomes have improved over the years through advances in peri-operative management, the use of neoadjuvant and adjuvant radiotherapy and chemotherapy and through modifications of the surgical procedure. These advancements have undoubtably contributed to the improved 5-year survival rate, which is now almost 60% [2]. Unfortunately, no matter how safe the surgical procedure is regarded to be, complications can still occur. One such life-threatening complication that typically occurs following failure of the anastomotic site to heal correctly is termed an anastomotic leak (AL).

3. Anastomotic Leaks

The exact definition of what an AL is continues to be debated in the literature. The UK Surgical Infection Study Group defined an AL as 'a leak of luminal contents from a surgical join between 2 hollow viscera' [3]. However, a subsequent review of 97 papers highlighted a lack of definition consistency between studies, with 56 different terms being identified [4]. The lack of standardised terminology creates problems when comparing results generated between different studies. A more recent attempt by the International Multispecialty Anastomotic Leak Global Improvement Exchange Group has re-defined an AL as 'a defect of continuity localised at the surgical site of the anastomosis, which creates a communication between intra-luminal and extra-luminal compartments.' Using this classification method, three grades of AL, increasing in severity from A to C, have been described. Whereas grade A can be left untreated, grade B requires medical management and grade C requires revision surgery [5,6].

Whatever the definition used, an AL is typically diagnosed 5–8 days post-surgery, although some case reports have demonstrated that a delayed presentation beyond 30 days is possible [7]. While AL can occur in up to 24% of patients undergoing distal rectal surgery, combined rates for surgery performed at any level of the intestinal tract are accepted to be ~6–7% [8,9]. The development of an AL not only results in increased morbidity [10–12] and 30-day mortality rates [13], but in cancer patients, it has also been associated with higher local recurrence rates and decreased long-term survival, but not with distant recurrence [8,14–17]. One large study involving 1984 colorectal cancer patients showed that 5-year cancer-specific survival was 57.4% in those that developed an AL compared with 72% that recovered uneventfully. The 5-year local recurrence rates were also increased from 1.9% to 4.7% in those that developed an AL [18]. Several explanations for these poorer survival times and increased local recurrence rates have been proposed. As viable cancer cells have been identified within the intestinal lumen and on staple/suture lines, it is possible that, following an AL, these cells could exfoliate to extra-luminal tissues. Implantation of these cells in the serosal surface of the intestine, peritoneum or pelvis could lead to the development of local recurrence [19–25]. The inflammatory response related to an AL has been proposed to stimulate tumour proliferation and evolution to distant metastasis [26–30], with elevated levels of inflammatory markers such as C-reactive protein

(CRP) associated with higher recurrence rates and impaired disease-free survival [31,32]. Intra-abdominal bacterial infections have also been suggested to stimulate neoangiogenesis, which may increase the risk of disease recurrence [33].

Revision surgery will be required in ~85–95% of AL patients, with 50% of symptomatic AL cases requiring permanent stoma formation [34]. Complications such as multi-organ failure, pneumonia, renal and cardiac issues, localised/generalised sepsis, wound infections and surgical site dehiscence are also commonly encountered secondary to an AL [11]. The intensive care and revision surgery needed to manage these conditions, as well as the AL itself, increases hospitalisation periods [35,36] and total treatment costs [36–41]. If a patient develops an AL, then early diagnosis is essential to decrease mortality rates and achieve a positive outcome [7,42–45]. One study suggested that a 2.5-day delay in instigating AL-specific treatments increased mortality rates from 24 to 39% [13], while a further study identified that a 7.6% decrease in survival was associated with every hour of delay from septic shock onset to when antibiotics were administered [46].

4. Intestinal Healing

Research into anastomotic healing and AL development has been acknowledged as a priority by numerous healthcare providers, including the National Health Service, National Institute for Health Research, the ACPGBI and the Colorectal Therapies Healthcare Technology Co-operative. Following a resection and anastomosis, intestinal healing has been described to occur in four stages [47–49].

➢ Stage 1. Haemostasis. Occurring immediately after intestinal injury, this stage involves platelet and coagulation cascade activation.
➢ Stage 2. Inflammation. Occurring within 10 days after intestinal injury, this stage involves surgical site recruitment of lymphocytes, neutrophils and macrophages.
➢ Stage 3. Proliferation. Occurring from 5 to 21 days after intestinal injury, this stage involves intestinal re-epithelisation through fibroblast recruitment and endothelial cell proliferation.
➢ Stage 4. Remodelling. The final stage of intestinal healing occurs from 21 days after intestinal injury and continues for up to 1 year. Here, collagen deposition and tissue remodelling can restore intestinal integrity.

5. Anastomotic Leak Pathophysiology and Risk Factors

In contrast to the well-documented and characterised stages of uneventful intestinal healing, relatively little is known about AL pathophysiology. Studies, however, have identified several surgical and patient-related risk factors that can influence AL development (Table 1) [8,9].

Table 1. Risk factors associated with the development of an anastomotic leak.

Patient Factors		Surgical Factors	
Age	Cardiovascular disease	Poor anastomotic blood supply	Intra-operative sepsis
Malnutrition	Gender	Concurrent surgical procedures	Peritonitis
Steroid use	Alcohol use	Poor colonic preparation	Operative time >3 h
Diabetes	ASA fitness score	Peri-operative blood transfusion	Pre-operative radiotherapy
Hypertension	Diverticulitis	Anastomotic ischaemia or tension	Anastomotic location
Tobacco use	Leukocytosis	Emergency resection	Bowel obstruction

5.1. Patient-Related Factors

Patient age as well as gender have been identified as AL risk factors, with men and patients of either sex >60 years old being at increased risk of AL. Although the exact reason for this is unknown, it is thought that the narrower male pelvis and androgenic hormonal effects on the intestinal microvascular blood supply may play roles in AL development in male patients [50–55]. Multiple studies have also shown that the American Society of Anaesthesiologists (ASA) fitness score is also an independent AL risk factor. Patient scores

≥III are associated with a 2.5-fold increased AL risk [55,56]. ASA scores are generated using multiple patient-specific factors including nutritional status and medical history, which have themselves been identified as independent AL risk factors.

Adequate nutrition is an important factor for intestinal healing as it contributes to collagen synthesis and immune responses. Various studies have shown that patients who are malnourished (including obese patients), have pre-operative weight loss [57–59], anaemia or low albumin levels are at increased risk of AL [60,61]. Neo-adjuvant, pre-operative chemo-radiotherapy has also been shown to be an independent risk factor for AL. Radiotherapy causes poor intestinal healing and increased fibrosis by damaging the local intestinal vascular system and impairing fibroblast function [51,62–64]. Pre-operative blood transfusions, advanced tumour stage and tumours >5 cm have also been identified as AL risk factors [56,65]. Currently, there is no consensus as to whether metabolic diseases, such as diabetes mellitus, increase AL risk through impaired wound healing [66,67]; however, patients with pre-existing renal disease, or those that smoke or drink alcohol excessively, have been identified as high-risk for AL development [67–70].

Although the intestinal microbiome plays an important role in the health, physiology and healing of the intestine [71], specific bacterial infections have been demonstrated to increase AL risk. One early study exhibited that rats inoculated with *Pseudomonas aeruginosa*, 24 h following gastrectomy and oesophagoduodenostomy, demonstrated higher AL rates compared with rats that were also inoculated with *Pseudomonas aeruginosa* but received peri-operative antibiotics (95% vs. 6%) [72]. A subsequent human clinical trial supported these pre-clinical results by showing that reduced AL incidence (10.6% vs. 2.9%) and mortality rates (10.6% vs. 4.9%) were achieved in gastrectomy and oesophagojejunostomy patients treated with peri-operative antibiotics [73]. The authors suggested that antibiotics may play a protective role against AL development. Although the mechanisms by which bacterial infections contribute to AL development are not fully understood, matrix metalloprotease (MMP) activation and collagenolytic substances produced by anastomotic site bacteria may play a role [74]. Using a pre-clinical rat model, one study demonstrated that antibiotics, with efficacy against *Enterococcus faecalis* (a bacterial strain with high collagen-degrading activity), placed topically at the colorectal anastomotic site, reduced AL incidence, whereas intravenous antibiotics failed to eliminate anastomotic site *Enterococcus faecalis* and reduce AL rates [75]. Following these results, MMP inhibitors have undergone investigations for their ability to prevent AL. One meta-analysis concluded that although anastomotic strength in animal models can be improved through MMP inhibitors, human clinical trials have yet to demonstrate a role in decreasing AL rates [76].

5.2. Surgery-Related Factors

A significant AL risk factor is the anatomical location of where the anastomosis is performed in the gastrointestinal tract [77]. One systematic review identified that the highest rate of AL occurred in coloanal and colorectal anastomoses (5–19%). This rate was significantly greater than that seen in enteroentero (1–2%), ileorectal (3–7%), ileocolic (1–4%) and colocolic (2–3%) anastomoses [78]. Multiple studies have also shown that anastomotic position in relation to the anal verge is important; cancer resections performed in the mid/low rectum [79] or <6 cm from the anal verge [80] have been associated with significantly higher AL rates. Patients that require an emergency resection and anastomosis at any level of the gastrointestinal tract are also at higher risk [55].

When considering the surgical procedure itself, studies have failed to show AL rate differences between hand-sewn or stapled anastomoses [81,82] or between open abdominal procedures or laparoscopic surgery [83–85]. Studies investigating the advantages of robotically performed colorectal anastomoses have failed to show AL rate differences compared with laparoscopic resections [86–88]. Conflicting results have been reported as to what extent surgical experience can influence AL rates. Whilst one study demonstrated that surgery performed by high-volume colorectal surgeons may reduce AL, another failed to demonstrate AL rate differences when surgeon experience was taken into account [89,90].

Multiple firings of the stapling device and surgical times >3 h have also been identified as AL risk factors [56,65].

Poor intestinal tissue oxygenation (partial pressure of O_2 in tissue; ptO_2) has also been suggested to contribute to AL development. Iatrogenic surgical disruption of the peri-anastomotic microvascular blood supply or tension at the anastomotic site can compromise intestinal tissue perfusion. If local blood supply is unable to meet intestinal O_2 requirements, this situation can lead to peri-anastomotic ischaemia and necrosis [48,49,91,92]. Adequate ptO_2 is also required for collagen production, with O_2 levels <15–20 mmHg associated with compromised synthesis. As submucosal collagen is the predominant tissue layer for anchoring sutures/staples in the early stages of anastomotic healing, inadequate collagen production could contribute to AL incidence [93].

6. Diagnosis

As already mentioned, early AL diagnosis and subsequent management is essential to reduce patient morbidity and mortality. Unfortunately, early diagnosis can be extremely difficult as there are no pathognomonic signs which can be specifically attributed to an AL. Patients can initially be asymptomatic while non-specific clinical signs can range from abdominal pain, ileus, pyrexia and cardiorespiratory issues to acute organ failure and sepsis. These wide-ranging clinical symptoms can be difficult to distinguish from those caused by normal post-operative inflammatory and physiological responses [94]. Based on clinical assessments, one study demonstrated that 69% of AL patients had a delayed diagnosis, of which the majority of patients presented with only cardiovascular symptoms [95]. Clinical assessment, regardless of experience and training, is therefore regarded as an inadequate technique for identifying high-risk AL patients or for its early diagnosis [96].

As clinical signs cannot be relied upon for AL diagnosis, clinicians use a variety of blood tests assessing inflammatory markers such as CRP and leukocytes. Unfortunately, these markers are again non-specific, with raised levels commonly occurring secondary to various post-operative complications, including chest, urinary and surgical site infections. Rather than using individual markers, one study assessed leukocyte number, creatinine levels, CRP levels, core temperature, urine production and systemic inflammatory response syndrome components. This combined approach was able to reduce the delay in AL diagnosis from 4 to 1.5 days [13]. Scoring systems have also been designed to predict AL risk. One study generated a scoring system based upon data from 1060 patients who underwent an anterior resection. Using known AL risk factors (intra-operative haemorrhage, gender and level of anastomosis), this study classified patients into low- (0–1), intermediate- (2–3) and high-risk (4–5 score) cohorts with AL rates of 1.9%, 8% and 16.1%, respectively [97]. The Colon Leak Score, which incorporates surgical and patient-specific risk factors, has also been developed to predict AL risk [98]. As well as these predictive scoring systems, others have looked to diagnose AL. The Dutch Leakage Score and the Modified Dutch Leakage Score have, unlike the previously mentioned predictive scoring systems, undergone clinical validation. Using clinical and physiological data with laboratory results, the derived Dutch Leakage Score has been shown to have a sensitivity of 97%, specificity of 53.5%, positive predictive value (PPV) of 16.1% and negative predictive value (NPV) of 99.5% for AL diagnosis (depending on the score cut-off values used). Meanwhile, the much simpler Modified Dutch Leakage Score, again depending on the score cut-off, could still produce a sensitivity of 97%, specificity of 56.8%, PPV of 17.2% and NPP of 99.5% for AL diagnosis [13,99,100].

Current clinical practices for AL diagnosis rely on abdominal imaging (plain radiographs, computed tomography (CT) scans or water-soluble contrast enemas (WSCE)), in conjunction with clinical and biochemical evaluation. Although CT is perhaps the most commonly used imaging modality for AL diagnosis, studies have shown it to have variable sensitivity and specificity. One retrospective study reported that only 47% of CT scans performed within 72 h of a patient requiring repeat surgery were diagnostic for an AL [101]. CT and rectal contrast radiography have been shown to have comparable sensitivity (57–60%) and specificity (100%) rates for AL diagnosis, greater than those of using clinical assessments alone (50% sensitivity and 89% specificity). The authors of this study suggested that whilst these imaging techniques gave false negative results, both were equally good for AL diagnosis [102]. Another large study in the 1970s analysed data from almost 2000 anterior resection patients. From the results, the authors suggested that although contrast studies could provide an indication of leak severity, they offered no diagnostic advantage over sigmoidoscopy and/or digital rectal examination, especially in patients that received a low anastomosis [103]. WSCE have also been shown to have higher false positive rates compared with digital rectal examination (6.4% vs. 3.5%) [104]. As plain abdominal X-rays can identify disrupted staple lines, a further study suggested that WSCE may only be required when intact staple lines, identified on radiographs, occur in conjunction with unrelenting clinical signs [105]. Another study demonstrated that WSCE used in colorectal or left-sided colonic anastomoses had sensitivity and specificity values of 52.2% and 86.7%, respectively, leading the authors to suggest that the test had little impact on improving early patient morbidity [106]. A retrospective study using data from colorectal patients demonstrated that WSCE detected ~83% of leaks, whereas only ~15% were identified using CT. This difference was most apparent for distal ALs, leading the authors to conclude that WSCE may be more beneficial in evaluating low anastomoses [107]. Several studies have highlighted that CT-based AL diagnosis is challenging [107,108]. These studies have indicated that the only reliable CT sign of an AL was the presence of peri-anastomotic liquid and air; extravasated contrast material from the intestinal luminal into the abdomen was not always present. Similarities in CT data between patients with and without AL were also observed. These results indicated that CT interpretation requires an experienced radiologist, and that radiological interpretation should be performed with knowledge of clinical data.

As a result of these conflicting findings, there is still no definitive consensus on which imaging modality should be used for AL diagnosis. Furthermore, a reluctance by clinicians to perform multiple scans due to cost, logistics and patient radiation exposure, combined with inherent delays incurred from the time of imaging to the interpretation of results, can significantly hinder prompt AL diagnosis. As a result of these imaging-based limitations, researchers and clinicians are looking at novel AL predictive and diagnostic methods that could lead to a more refined, precision medicine approach to patient management (Table 2).

Table 2. Peri-operative techniques for AL risk prediction and diagnosis.

Pre-Operative	Intra-Operative	Post-Operative
Surgical factors	Tissue appearance	Scoring systems
Patient factors	Air leak test	Clinical assessment
Predictive scoring systems	Endoscopy	Routine bloodwork
Blood samples	Intestinal tissue perfusion	Imaging
Urine samples	Intestinal tissue oxygenation	Biomarkers: ischaemic, inflammatory, bacterial

7. Precision Medicine, Prognostic, Predictive and Pharmacodynamic Biomarkers

Current post-operative management of intestinal resection and anastomotic patients is principally focused on improving global and local tissue perfusion [109]. Post-operative preservation of normovolaemia [110], normothermia [111], delivering supplemental O_2 [112] and goal-directed intravenous fluid therapy [113,114] can all reduce morbidity and improve outcomes. A pig model has also shown that intravenous colloids can increase perfusion and ptO_2 in normal and peri-anastomotic colonic tissue [115]. Although these generic peri-operative patient management strategies may help to promote anastomotic healing and decrease AL rates, patient outcomes are likely to be improved by clinicians adopting a precision or personalised peri-operative treatment strategy.

Precision and personalised medicine encompass the idea that optimal patient management requires patient- and/or disease-specific factors to be considered. Although similar, specific personalised and precision medicine definitions highlight key differences in their concepts; whereas personalised medicine considers individual patient genetics, patient beliefs, social background, preferences and attitudes, precision medicine emphasises the importance of data collection and analysis [116]. Although both concepts have subtle differences, the term 'precision medicine' has become more extensively used. This has principally been due to unease amongst clinicians thinking that patients might misinterpret 'personalised medicine' as a technique by which drugs are developed for specific individuals [117,118]. Perhaps the clearest definition of what precision medicine is came from The National Research Council in America, who described it as 'the tailoring of medical treatment to the individual characteristics of each patient ... to classify individuals into subpopulations that differ in their susceptibility to a particular disease or their response to a specific treatment. Preventative or therapeutic interventions can then be concentrated on those who will benefit, sparing expense and side effects for those who will not' [118]. To achieve the aims of this precision medicine concept, research has largely focused on the use of patient- or disease-specific biomarkers.

A biomarker can be any measurable tissue or bodily fluid biological substance that represents normal or abnormal physiological processes or pathological conditions [119]. There are four classical biomarker categories [120–122]:

1. Diagnostic. Identifies the presence of disease;
2. Predictive. Indicates the likely benefit of a specific treatment;
3. Prognostic. Indicates patient outcome, irrespective of treatment;
4. Pharmacodynamic. Allows monitoring treatment effectiveness.

A clinically useful biomarker is one that is obtained non-invasively, is easily assayed and provides results that have high sensitivity and specificity. Broadly speaking, for AL, these can be biomarkers of ischaemia, inflammation, tissue repair and the presence of bacterial contamination (Figure 1). All these potential biomarkers can be assessed through either blood or peritoneal fluid samples. These types of biomarkers have the potential to be assessed either intra-operatively, to predict which patients are at high risk of complications, or post-operatively, to identify which patients may require additional management to prevent an AL from developing or allow for its early diagnosis. In conjunction with clinical status, physiological parameters and imaging results, these types of biomarkers could be used to achieve a precision medicine approach for patients undergoing a resection and anastomosis.

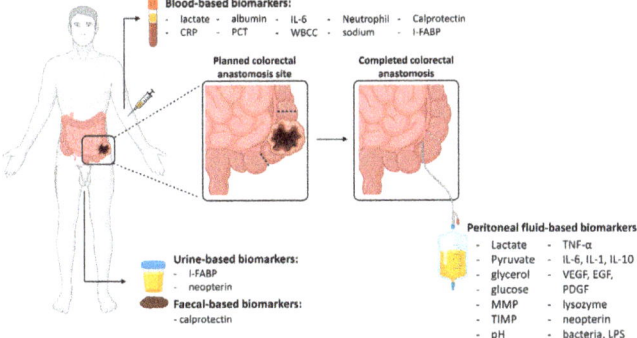

Figure 1. Patient samples used for biomarker assessment following a colorectal anastomosis in the treatment of colon cancer. (IL; interleukin, CRP; C-reactive protein, PCT; procalcitonin, WBCC; white blood cell count, I-FABP; intestinal fatty acid binding protein, MMP; matrix metalloproteinases, TIMP; tissue inhibitor of metalloproteinases, VEGF; vascular endothelial growth factor, EGF; epidermal growth factor, PDGF; platelet-derived growth factor, LPS; lipopolysaccharide). Figure created in Biorender.

8. Intra-Operative Techniques

During surgery, immediately following anastomosis, surgeons evaluate intestinal integrity through assessment of anastomotic doughnut completeness, air leak testing and/or endoscopic visualisation. Although air leak testing alone can reduce post-operative AL rates from 14% to 4% [123,124], intra-operative endoscopy can be performed with an air leak test and allows the surgeon to assess for vascular insufficiency, staple line bleeding, adequate tumour margins, iatrogenic intestinal injury and missed pathology [125–127]. One study demonstrated that routine intra-operative endoscopy identified anastomotic issues that required correcting in 10% of patients [125].

Further intra-operative assessment techniques have predominantly focused on intestinal ptO_2 levels as a way of predicting which patients are at high risk of AL. Although surgeons evaluate macroscopic tissue appearance (colour, intestinal bleeding and palpable mesenteric pulses) as a surrogate for intestinal perfusion, these subjective techniques are unable to predict AL risk. To overcome this issue, various techniques have been developed to objectively measure intestinal tissue oxygen saturation (StO_2) (visible light and near infrared spectroscopy) [128,129], tissue perfusion (laser fluorescence angiography, laser Doppler flowmetry) [130,131] and arterial haemoglobin O_2 saturation (wireless handheld pulse oximeters) [132].

Visible light spectroscopy used in colorectal anastomoses has demonstrated that reduced tissue O_2 saturation immediately after resection can predict AL. Interestingly, this study also showed that patients who recovered uneventfully demonstrated a significant intra-operative rise in StO_2 in the proximal part of the anastomosis, which was not seen in those who developed an AL [128]. Animal studies have supported these results through comparing intestinal tissue oxygenation with staple size and by using wireless pulse oximeters [132,133]. In a recent human study, intra-operative colonic O_2 saturation was measured with a pulse oximetry device placed on the colonic wall and evaluated for its ability to assess tissue viability and predict AL in colorectal anastomotic patients. The results showed that the risk of developing an AL was 4.2 times higher when post-anastomotic colonic StO_2 was ≤90% of the pre-resection values. The authors suggested that low intra-operative colonic StO_2 values were associated with AL occurrence [134]. Laser fluorescence angiography has also been shown, in a retrospective clinical trial of 402 patients, to reduce the number of patients that developed an AL. Out of the 22 patients that developed an AL, only seven (3.5%) were in the imaging group, compared with 15 (7.5%) in the control group [130]. Near infrared (NIR) fluorescent imaging has also been investigated for its intra-operative use as the energy range it uses is capable of penetrating deep into the intestinal walls and mesenteric tissues without causing thermal damage [135]. Coupled with indocyanine green, veins, arteries and capillaries can be identified, with vascular streams being used as an approximation of tissue perfusion. In animal models, this technique has been shown to predict the viability of ischaemic intestine [136] and an ongoing human clinical trial is assessing the use of NIR laparoscopy–indocyanine green to minimise leak occurrence compared with conventional white-light laparoscopy [137].

Clark O_2 electrodes have also been investigated for their ability to measure intra-operative intestinal ptO_2. In pre-clinical animal models, gradual intestinal perfusion reduction through sequential accurate intestinal artery ligation demonstrated that intestinal ptO_2 measured before performing an anastomosis could predict AL occurrence [93,138,139]. In humans, Clark O_2 electrodes have been used to provide intra-operative ptO_2 reference values for the majority of the gastrointestinal tract [140]. Clinical trials have also shown that colonic ptO_2 of less than either 20 mmHg, 50% of pre-resection values, 15% of arterial oxygen partial pressure (PaO_2) or 40% of ptO_2 at a control site were all associated with AL development. This study provided evidence that Clark O_2 electrodes could be used to measure peri-anastomotic colonic ptO_2 before, during and immediately after performing a resection and anastomosis and that, using defined cut-off values, the risk of developing an AL could be predicted [93].

Although these intra-operative techniques are well established, there are no standard guidelines as to which should be used. As a result of this, there is considerable variation between surgeons and hospitals [141]. To begin to address this, the European Society of Coloproctology Safe-anastomosis Programme in Colorectal Surgery (EAGLE) has been set up. Launched in 2019, this is an international, multicentre, cluster randomised controlled sequence study. EAGLE aims to recruit at least 2000 surgeons from 300 hospitals in order to collect data from >4500 patients who have undergone a right colectomy and ileocecal resection. The study results will be used as a quality improvement programme aimed at pre-operative risk stratification and standardising surgical techniques used for these patients [142].

9. Post-Operative Techniques

Many pre-clinical and clinical research studies have provided clear evidence that intra-operative assessment of anastomotic integrity and peri-anastomotic tissue perfusion can predict AL risk. Unfortunately, these techniques ultimately fail to encompass a precision medicine approach to patient care as they cannot be used to assess intestinal healing in the post-operative period. To overcome this, researchers are now looking at biomarkers which can be used post-operatively that allow continual patient monitoring and assessment of intestinal healing. These techniques have the goal of identifying high-risk patients and provide a means of early AL diagnosis.

10. Biomarkers of Ischaemia

Under aerobic conditions, adenosine triphosphate (ATP) is efficiently generated during the conversion of glucose to pyruvate through glycolysis and the Krebs cycle. However, when O_2 and glucose supply are limited and unable to meet the metabolic demands of a tissue, cells have to rely on anaerobic metabolism. Here, ATP is generated less efficiently from pyruvate being converted into lactate, with CO_2 being released in the process. Ischaemic tissue microenvironments are therefore typically regarded as having low glucose and pyruvate levels in the presence of high lactate concentrations. Accumulated amounts of CO_2 will also cause a reduction in tissue pH. If ischaemic conditions are prolonged, cells become damaged and, with the breakdown of their cell membranes, phospholipids are released, generating glycerol and free fatty acids. Although these individual ischaemic biomarkers can be measured, calculating lactate/pyruvate ratios (LP ratio) can characterise the aerobic/anaerobic metabolic balance, with higher values signifying ischaemia. Unfortunately, assessment of ischaemic biomarkers in blood has been shown to lack specificity for AL diagnosis [143]. To address this, ischaemic biomarker levels in peritoneal fluid have been investigated. Animal studies using microdialysis catheter fluid (small probes with dialysis membranes inserted into/onto the intestine) have shown that lactate, glucose and glycerol levels change with the metabolic alterations that occur in hypoxic [144] and ischaemia [145] conditions. Most human clinical studies have now focused on assaying ischaemic biomarkers in microdialysis catheter fluid or peritoneal fluid from abdominal drains. In a pilot study of eight patients undergoing right hemicolectomy, metabolic changes consistent with visceral ischaemia were identified in microdialysis catheter fluid several hours before clinical signs of AL became apparent [146]. A subsequent study characterised microdialysis catheter fluid reference ranges for the first 45 h following surgery in patients that recovered uneventfully from a variety of elective gastrointestinal operations [147].

10.1. Lactate/Pyruvate Ratio

High peritoneal LP ratios have been associated with AL in multiple clinical studies. In one study, patients undergoing anterior rectal resections had their LP ratio and glucose levels assessed for the first 6 days following surgery. Results indicated that the LP ratio in patients that developed an AL was significantly higher on days 5 and 6 following surgery. Unfortunately, due to low patient numbers, LP ratio cut-off values for predicting an AL could not be determined [148]. A further study using microdialysis catheters in 45 low anterior resection patients obtained fluid samples from the anastomotic site every 4 h. Lactate and LP ratios were found to be significantly raised in the four patients that developed an AL. Interestingly, in three patients who developed an AL >10 days following surgery, raised lactate and LP ratios were detected several days before clinical symptoms developed. Lactate levels in the remaining AL patient increased 18 h before clinical signs, with LP ratios only becoming elevated once clinical symptoms became evident [149]. A similar study again using microdialysis catheters obtained peritoneal fluid samples from patients every 2 h for the first 2 days, then every 4 h until 7 days following colorectal surgery. Higher peritoneal lactate and LP ratios and lower glycerol levels were seen immediately following surgery in patients that went on to develop an AL. These levels became significantly raised by the 4th day following surgery [150]. In another study that contained colorectal anastomoses, abdominal aortic aneurysm repairs, gastric procedures and cholecystectomy, results showed that increased peritoneal LP ratios and decreased glycerol levels were associated with 'major intra-abdominal complications' [151]. A further study involving 88 patients who underwent various abdominal procedures, including an intestinal resection and anastomosis, had their post-operative peritoneal and serum lactate levels assessed. Patients that had peritoneal/serum lactate level >4.5 or peritoneal lactate level >9.1 mM in the presence of pyrexia (>38.3 °C), raised white cell count (>12 × 10^9/L), delayed passage of flatus (>72 h) and abdominal pain by the 4th day post-surgery were significantly associated with post-operative complications that required revision surgery (AL were included in this group) [152]. In slight contrast to these results, a further study which measured lactate, pyruvate, glycerol and glucose levels every 4 h for 5 days after patients underwent a left-sided colorectal anastomosis demonstrated that, in the three AL patients, lactate levels but not LP ratios were significantly elevated. Interestingly, in all the patients which developed an AL, the raised lactate levels occurred in the first 3 days following surgery [153].

A recent prospective study has also compared peritoneal lactate, pyruvate, glucose and glycerol assessment with daily clinical scoring (leak scores and the Dutch Leakage Score system). This study showed that, in cases of AL, peritoneal lactate concentration increases over time and its assessment can have greater sensitivity, specificity and better PPV and NPV than clinical scoring systems. The median day for an AL diagnosis with a change in lactate ≥6.3 mM was 1.6, whereas for leak scores and for the Dutch Leakage score system, it was 3.3 and 7 days, respectively [154].

10.2. pH

To investigate pH as an ischaemic AL biomarker, one study has measured intestinal mucosal pH with tonometry. pH measurements were taken using a catheter placed at a colorectal anastomosis site through the anus. Imaging performed on the 6th day following surgery was used for symptomatic and asymptomatic AL diagnosis. Results indicated that in the first 24 h, mucosal pH values were significantly reduced in patients who subsequently developed an AL. Using a pH cut-off value of <7.28 in the first 24 h was associated with a 22-times greater risk of AL, with a sensitivity of 28% and specificity of 98% for AL prediction [155]. A further study measured peritoneal drain fluid pH in the first 12 days following colorectal surgery. Similar to the previous study, results indicated that pH values were significantly lower in patients which developed an AL that needed revision surgery. Using a cut-off pH value of <6.978 on the 3rd day following surgery had a sensitivity of 98.7% and specificity of 94.7% for predicting an AL. Interestingly, a decline in pH was seen in all patients preceding their AL diagnosis [156].

10.3. Tissue Oxygenation

The intra-operative use of Clark O_2 electrodes, as previously described, has been investigated for their ability to predict AL. However, these studies overlooked their applications for post-operative use. The concept of using miniaturised Clark O_2 sensors to provide post-operative intestinal ptO$_2$ measurements has begun to be investigated by the Implantable Microsystems for Personalised Anti-Cancer Therapy (IMPACT) programme. Our group has developed novel implantable miniaturised Clark-type electrochemical O$_2$ sensors [157] and methylene blue-based electrochemical and ion sensitive field-effect transistor (ISFET) pH sensors [158]. The idea that these sensors could be placed intra-operatively around the anastomotic site and be left in situ would allow clinicians to continuously monitor peri-anastomotic intestinal ptO$_2$ and pH throughout the post-operative recovery period. This type of continuous monitoring system would help to identify patients at risk of developing an AL due to poor or deteriorating peri-anastomotic intestinal ptO$_2$. It would also allow clinicians the ability to assess the efficacy of interventions designed to improve intestinal ptO$_2$ and prevent a leak from occurring. The electrochemical O$_2$ sensor has undergone initial *in vivo* validation in a rat model [159]. The results from this study showed that sensors, placed on intestinal serosal surfaces, were able to provide continuous, real-time ptO$_2$ readings. These sensors also recognised and reported dynamic intestinal ptO$_2$ changes that occurred with hypoxaemic and ischaemic challenges. The authors suggested that although further research is required, this pre-clinical study demonstrated the potential use of miniaturised implantable medical devices for intestinal surgery.

11. Biomarkers of Inflammation

A wide range of inflammatory mediators, such as acute-phase proteins, cytokines and growth factors, are released into the peritoneal cavity and bloodstream following abdominal surgery [160]. If these substances are to be used as part of a precision medicine approach in intestinal surgery, then studies have to show their ability to differentiate the normal physiological responses to surgery from clinically important complications such as AL.

11.1. C-Reactive Protein, Albumin and Procalcitonin

CRP, a hepatic acute-phase reactant with a half-life of 19 h, is typically found at low levels (0.8 mg/L) in the blood of healthy individuals. CRP levels can rise dramatically in response to inflammatory cytokines such as interleukin (IL)-6 (IL-6), tumour necrosis factor-α (TNF-α) and IL-1β. This can occur as part of an acute-phase inflammatory response due to infection, tissue damage and neoplasia [161]. Post-operative serum CRP levels are routinely assessed as part of standard care practices to provide information on clinically significant inflammation and post-operative complications. Unfortunately, its use in resection and anastomotic patients is still contentious, with studies demonstrating poor CRP specificity for AL diagnosis, with levels only becoming significantly raised when clinical symptoms become apparent [162–164].

In contrast to these results, recent research has shown that serum CRP levels can become elevated several days before clinical AL diagnosis and are significantly raised in comparison to patients who have an uneventful post-operative recovery [165–177]. Currently, the main issue with using serum CRP levels for AL prediction or diagnosis is the lack of definitive cut-off values. Cut-off values in these studies alone ranged from 123 to 245 mg/L, which were measured between 3 and 5 days following surgery. In a meta-analysis of seven clinical trials which included 2483 patients, results suggested that serum CRP cut-off values of 172 mg/L, 124 mg/L and 144 mg/L on the 3rd, 4th and 5th days following surgery possessed an NPV of 97% for excluding an AL [178]. Furthermore, in a recent prospective international study of 933 colorectal resection and anastomosis patients, of which 41 developed an AL, serum CRP levels were assessed pre-operatively and continued for 5 days post-surgery. Results indicated that a change in CRP levels >50 mg/L over any 2 post-operative days had a sensitivity of 85% for diagnosing an AL and an NPV of 99% for ruling it out. A change in CRP >50 mg/L between days 3 and 4 or 4 and 5 had an even higher specificity of 97%. The authors highlighted the value of CRP trajectory assessment for its ability to rule out an AL [179].

Albumin, also an acute phase protein, has been proposed as an indicator of surgical stress and can be used to predict the development of post-operative complications [180]. Hypoalbuminemia has also been suggested to be an AL risk factor for colorectal resections as part of treatment for ovarian cancer [181]. A novel indicator, the C-reactive protein:albumin ratio (CAR), has been used to identify patients at risk of post-operative complications after colorectal surgery [182]. This study showed that CAR measurement provided greater diagnostic accuracy than assessing CRP or albumin levels alone. In one recent retrospective study of 1068 elderly colorectal anastomotic patients, the AL predictive value of CAR was investigated. Using a pre-operative CAR cut-off value of 2.44, the assay had a sensitivity of 61% and specificity of 80% for predicting an AL. Surgical time and pre-operative CAR were also both identified as independent AL risk factors [183].

Procalcitonin (PCT), the prohormone of calcitonin, is produced by thyroid parafollicular C-cells. PCT is typically found in low levels (<0.05 ng/mL) in the blood of healthy individuals. Bacterial infections have been shown to induce PCT release from all differentiated cell types, which can occur within 2–3 h following infection and is related to the presence of bacterial endotoxins and inflammatory cytokines such as TNF and IL-6. Patients with serum PCT levels >2 ng/mL have been associated with bacterial infections, but levels >700 ng/mL can be seen in cases of severe sepsis [184]. Serum PCT levels, in contrast to CRP, do not become raised secondary to inflammation of a non-infectious origin and its use for early AL diagnosis has been investigated.

One study involving 157 colorectal resection and anastomotic patients demonstrated that serum PCT levels in the range of 1.4–4.62 ng/mL measured on the 1st day following surgery predicted those that subsequently developed an AL. These values were significantly higher than that seen in patients who recovered uneventfully (0.09–0.44 ng/mL). Using a PCT cut-off value of 1.09 ng/mL on the 1st day following surgery gave sensitivity and specificity values of 87% for AL prediction. The authors suggested that PCT could be used at this early post-operative time point to identify high-risk patients [185]. A similar study, again involving colorectal cancer resection and anastomosis patients, demonstrated that PCT measured on the 3rd day following surgery could identify patients at low risk of AL development. Using 3.83 ng/mL as a PCT cut-off value gave a sensitivity of 75% and specificity of 100% for AL prediction [168]. These results are supported by another study that also concluded that PCT levels measured over the first 5 days following surgery are a reliable predictor of AL after colorectal surgery. Using a PCT cut-off value of 0.31 ng/mL on the 5th day following surgery was shown to have 100% sensitivity, 72% specificity, 100% NPV and 17% PPV for AL. The authors suggested that patients with elevated serum PCT levels on post-operative days 3–5 warranted further assessment before discharge [186]. These results have been supported in other studies [166,170].

Many of these studies assessed both CRP and PCT levels simultaneously and it has been proposed that measuring both can improve AL diagnosis. In the recent PREDICS study involving 504 colorectal resection and anastomosis patients, the study demonstrated that a PCT cut-off value of 2.7 ng/mL had an NPV of 96.9% and specificity of 91.7% on the 3rd day following surgery, whereas a cut-off value of 2.3 ng/mL on day 5 had an NPV of 98.3% and specificity of 93% for AL diagnosis. CRP also exhibited good NPV 96.4% on the 3rd day (cut-off value 169 mg/L) and 98.4% on the 5th day (cut-off value 125 mg/L). Combined CRP and PCT assessment further improved AL diagnosis [170]. These results have been supported by a more recent study that suggested that CRP and PCT levels were higher on post-operative days 1–3 in patients who subsequently developed an AL. The authors suggested that these markers could be used to allow early patient discharge in those with low risk of developing an AL [187].

11.2. Cytokines, Tumour Necrosis Factor-α and Growth Factors

Cytokines such as IL-1, IL-6, IL-10 and TNF-α are polypeptides with known roles in immune responses. In response to surgical trauma, they regulate physiological responses and induce the production of hepatic acute-phase proteins, whilst, in response to sepsis, they can mediate systemic inflammatory responses [188–190]. Raised IL-1b, IL-6 and TNF-α levels have also been associated with surgical stress, including lengthier operating times, haemorrhage and high peritoneal bacterial counts [191–193]. Within the first few hours following abdominal surgery, these substances are released from the surgical site. During this period, studies have shown that peritoneal cytokine levels are raised to a greater degree than serum levels. This provides evidence, similar to the ischaemic biomarkers, that peritoneal rather than serum biomarker levels are more representative of localised tissue changes [194,195]. In patients that have uncomplicated post-operative recoveries, peritoneal cytokine levels typically decrease within 24 h following surgery [195]. However, cytokine dynamics that occur with an AL follow a significantly different course.

Raised peritoneal levels of IL-6 and TNF-α have been shown in numerous studies to occur as early as the 1st day following surgery in patients who go on to develop an AL [148,196–199]. Further studies, however, have demonstrated that their levels only become elevated from the 3rd post-operative day [200,201]. An important observation from all these studies was that, in AL patients, IL-6 and TNF-α levels for the first 5 post-operative days remained elevated, whereas, in patients that recovered uneventfully, their levels remained low or even decreased. Although a further study observed no differences between a control group and patients who developed an AL in their IL-6 and TNF-α levels over the first 7 days following surgery, the results demonstrated that TNF-α levels rapidly rose 24 h before a surgical diagnosis of AL was made [202]. A more recent case–control study investigated serum and peritoneal IL-6 levels on the 2nd and 4th days following colorectal surgery. In total, 30 AL and intra-abdominal abscesses (infection group) were compared with 30 uneventful recovery patients (control group). These results demonstrated that higher peritoneal levels in the infection group were seen on the 2nd and 4th days, whereas serum levels only became significantly elevated on the 4th day [203]. A further study identified that serum IL-6 levels on the 3rd post-operative day were significantly elevated in AL patients with similar sensitivity to that of CRP. Interestingly, the relative change in pre-operative to post-operative IL-6 levels was significantly higher in AL patients, with granulocyte-colony stimulating factor also showing similar changes [204]. Increased peritoneal levels of IL-1, IL-10, vascular endothelial growth factor, epidermal growth factor and platelet-derived growth factor have also been suggested to occur in patients who develop an AL and sepsis following colorectal surgery [148,193,196–198,201,203].

11.3. Leukocytes, Neutrophils and Intestinal Damage Markers

White blood cells (WBC) play a crucial role in wound healing through microorganism elimination. It had been proposed that the WBC count (WBCC) can reflect the extent of inflammation within the body or surgical site. As an AL creates significant inflammatory responses, several studies have investigated whether assessing leukocyte and/or neutrophil numbers in blood can aid AL diagnosis.

In one retrospective study of 1187 colorectal cancer patients, CRP levels and WBCC were assessed for the first 5 days following surgery. CRP levels, in line with other studies, measured on the 4th day provided the highest diagnostic accuracy for identifying post-operative complications, whereas WBCC contributed little information [174]. These results were supported by other retrospective [176] and prospective studies [170]. A further study demonstrated that in patients who developed an AL, increased WBCC only occurred after 6 days following surgery [165]. In a smaller study of 129 laparoscopic colorectal surgery patients, systemic CRP levels and WBCC were assessed. Using a CRP cut-off value of >200 mg/L on the 3rd day following surgery had a sensitivity of 68% and a specificity of 74% for predicting septic complications, whilst using a WBCC cut-off value of >12 × 10^9 on the 2nd day had a sensitivity of 90% and a specificity of 62%. The authors concluded that systemic CRP levels and WBCC were poor early diagnostic markers for predicting post-operative septic complications (including AL) [164]. Assessing neutrophil:lymphocyte ratios (NLR) has also been described as a method for AL prediction. One study demonstrated that NLR on the 4th day following surgery had prognostic value, with higher NLR identified in AL patients. An NLR cut-off value of 6.5 had a sensitivity of 69%, specificity of 78%, PPV of 49% and NPV of 88% for AL diagnosis. NLR were also significantly higher at this time point in patients who subsequently died in the post-operative period [171].

Calprotectin makes up ~60% of the cytosolic proteins found within neutrophils and is a recognised marker of neutrophil activation [205]. Studies have begun to investigate calprotectin as an inflammatory biomarker for early AL diagnosis. In one retrospective study of 84 colorectal cancer patients, serum CRP and calprotectin levels were assessed for 5 days following surgery. In patients that developed an AL, calprotectin levels became significantly elevated on the 2nd day following surgery (588 ng/mL) compared to those that went on to recover uneventfully (366 ng/mL). Calprotectin levels in AL patients also remained elevated throughout the 5 days. Although the authors suggested that calprotectin levels could be used to diagnose an AL, improved diagnostic accuracy was obtained when combined calprotectin and CRP assessment was performed on the 3rd day following surgery. This assay provided a sensitivity of 100%, specificity of 89%, positive likelihood ratio of 9.09 and negative likelihood ratio of 0.00 [167]. Similar results have been seen in a further study which identified raised pre-operative and post-operative calprotectin levels on the 1st, 3rd and 5th days following surgery in patients which developed an AL, whereas CRP levels only became elevated on the 3rd and 5th days, with no WBCC changes being observed. The authors again suggested that combined calprotectin and CRP assessment might aid early AL diagnosis [206].

Faecal calprotectin has also been used for assessing inflammation secondary to colorectal cancer and inflammatory bowel disease and its role in predicting AL has been investigated. In one study of 100 colorectal anastomotic patients, in which 11 developed an AL, faecal calprotectin levels were assessed 4 days after surgery. Results indicated that faecal calprotectin was significantly higher (>300 μg/g) in patients who developed an AL compared with those that recovered uneventfully (<90 μg/g). Faecal calprotectin levels assessed in combination with CRP using a cut-off value of 120 mg/L provided a sensitivity of 85%, specificity of 95% and an NPV of 95% for AL diagnosis [207].

Plasma markers of intestinal damage, such as liver, ileal bile acid and intestinal fatty acid-binding proteins, have also been investigated as predictive AL biomarkers. Using a pre-clinical intestinal resection and anastomosis rat model, post-operative serum intestinal fatty acid-binding protein level was shown to be raised as early as the 3rd day following surgery in those that developed an AL [208]. One human study demonstrated that pre-operative intestinal fatty acid-binding protein levels >882 pg/mL had a sensitivity of 50% and specificity of 100% for predicting AL [167]. A further study demonstrated that urinary intestinal fatty acid-binding protein and the urinary intestinal fatty acid-binding protein:creatinine ratio on the 3rd day following colorectal surgery were significantly elevated in patients with an AL. The authors suggested that this urine-based assay could be used as a non-invasive assay for AL diagnosis [209].

11.4. Macrophage Biomarkers

Produced by macrophages, lysozyme is a substance which disrupts the cell wall of Gram-negative bacteria. As lysozyme plays an important role in the inflammatory response to sepsis and trauma, studies have begun investigating whether it could be used as an AL biomarker. One study demonstrated that peritoneal lysozyme levels in patients on the 1st day following a low anterior resection who had an uneventful post-operative recovery were 5.5 mg/L. Significantly higher levels were seen in patients with clinical (180 mg/L) and radiological (153 mg/L) evidence of AL [210]. Although the authors suggested that lysozyme could be used for early AL diagnosis, the electrophoretic technique used had significant practical restraints in terms of its usefulness as a rapid AL diagnostic test as the gel required overnight soaking.

Neopterin, also produced by macrophages, is recognised as a biomarker of T helper cell activation and plays a significant role in mediating inflammatory responses. Neopterin production is stimulated by interferon-γ and can be detected in urine, cerebrospinal fluid and blood. Increased neopterin levels have been associated with viral, bacterial and parasitic infections, autoimmune diseases, cancer [211], sepsis [212] and multiple organ dysfunction syndrome [213]. In terms of AL, one study has investigated pre- and post-operative blood, urine and peritoneal fluid neopterin levels in colorectal resection and anastomosis patients [214]. Results demonstrated that the pre-operative urinary neopterin:creatinine ratio was significantly higher in patients that went on to develop an AL compared to those that recovered uneventfully (139.5 µmol/mol vs. 114.8 µmol/mol). Patients that developed complications also had higher urinary and peritoneal neopterin levels following surgery. The authors suggested that high pre-operative levels of urinary neopterin could identify AL high-risk patients and that monitoring post-operative urinary and peritoneal fluid neopterin levels could be useful for early AL diagnosis.

11.5. Hyponatraemia

Hyponatremia, although a commonly diagnosed electrolyte disorder, has been proposed as an inflammatory biomarker and a potential indicator of peritonitis [215]. Sodium levels are predominantly maintained via osmotic vasopressin release mediated by intravascular volume. However, research has shown potential immune-neuroendocrine pathways involving IL-6 which may have a role in non-osmotic driven vasopressin release in response to inflammation [216,217]. Hyponatremia (<136 mmol/l) and leukocytosis (>10 × 10^9/l) have subsequently been investigated as predictive AL biomarkers following colorectal surgery [218]. Results from this study of 1025 patients identified that 23% ($n = 19$) of AL patients and 15% ($n = 69$) of patients who recovered uneventfully had hyponatremia. Leukocytosis was identified in 12 of the 19 patients with hyponatremia and an AL. Hyponatraemia on the 5th day following surgery had a sensitivity of 23%, specificity of 93%, NPV 97% and PPV of 5% for AL diagnosis. The combined presence of hyponatremia and leukocytosis had a sensitivity of 68%, specificity of 75%, PPV of 18% and NPV of 97%. The authors suggested that, due to low sensitivity (23%), hyponatremia absence cannot exclude the presence of an AL. However, as the specificity of hyponatremia for an AL was high, especially when it occurred in the presence of leucocytosis, this result should raise suspicion of an AL being present. Further prospective trials are needed to confirm these results.

12. Biomarkers of Tissue Repair

MMPs are a group of zinc-dependent endopeptidases that are involved with extracellular matrix (ECM) remodelling. Secreted as an inactive pro-enzyme, they become active following proteolytic cleavage [219]. Physiological and pathological processes involving tissue repair and regeneration depend on the balance between MMP proteolysis and its prevention by tissue inhibitors of metalloproteinase (TIMP) [220]. MMPs have been suggested to play a role in AL development through inhibition of collagen synthesis. Although collagen type I and III genes are normally overexpressed at anastomotic sites, in a rat model, maximal gene expression was not reached until 7 days following surgery [221]. Further animal models have demonstrated that colonic peri-anastomotic healing (as shown by higher bursting pressures, improved structural layers and increased collagen production) was improved through MMP inhibition [222–225]. Furthermore, human colonic tissue from patients with poor anastomotic healing has demonstrated higher mucosal MMP-1 and MMP-2 expression and higher submucosal MMP-2 and MMP-9 expression. Interestingly, colonic samples from AL sites demonstrated a significantly lower collagen type I:III ratio compared to uncomplicated anastomotic sites [226].

In a study of 58 colorectal anastomotic patients, peritoneal levels of MMP-1, 2, 3, 8 and 9 and TIMP-1 and 2 were assessed for 8 days following surgery. Differential levels of MMP and TIMP were assessed on each day along with total MMP activity. Their levels were shown to vary depending on the operation type and duration, amount of haemorrhage and with the occurrence of post-operative complications. Only MMP-2 and MMP-9 levels positively correlated with the development of post-operative complications, whereas

TIMP-1 and TIMP-2 levels demonstrated a negative correlation. The authors suggested that peritoneal MMP and TIMP may act as biomarkers of intestinal wound healing and surgical outcome. However, as the patient cohort within the study was heterogeneous and because the types of post-operative complications were not specified, further studies are required [227]. In contrast to these results, a pilot study of 29 low anterior resection patients had their peritoneal fluid levels of MMP-1, 2, 3, 7, 8, 9 and 13 assessed every 4 h following surgery. Only MMP-8 and 9 were significantly increased in the 10 patients who developed an AL compared with the 19 patients who had an uneventful post-operative recovery [228].

In a recent systematic review, which included animal and human studies, the role of tissue, blood and peritoneal MMP levels in the development of AL was investigated. The results from 12 studies suggested that elevated MMP-9 levels were the most consistent finding in patients that developed an AL [229]. The authors claimed that although these studies suggested that tissue or peritoneal fluid levels of MMP and/or TIMP could act as biomarkers for AL, the number of studies and number of patients used were small. In addition, the inconsistent results for specific MMPs suggests that further investigations are required.

13. The Intestinal Microbiome and Bacterial Contamination

The intestinal microbiome has been shown to play a role in the development of AL and can be influenced by multiple peri-operative factors [71]. During intestinal surgery, inadvertent spillage of intestinal contents can cause bacterial contamination of the abdominal/pelvic cavities. In the majority of patients, immune responses deal with this contamination and their post-operative recovery is not compromised. However, in patients that develop anastomotic dehiscence, irrespective of its cause, significant and ongoing bacterial contamination can overwhelm the patient's immune system. A 5-year prospective trial of patients diagnosed with abdominal sepsis syndrome (inflammatory response with organ failure secondary to digestive tract perforation) identified multiple micro-organisms to be present within their abdominal fluid. The peritoneal microbial flora composition of these critically ill patients also varied depending on site of the intestinal perforation. Following a colorectal perforation ~70% of intra-operative fluid samples contained aerobic Gram-negative bacteria (*Escherichia coli*, *Klebsiella* and *Pseudomonas* species) whilst the predominant anaerobic species was *Bacteroides*. Gram-positive bacteria (*Enterococci* and *Staphylococci*) were found in ~50% of colorectal perforation cases. Antibiotic treatment was also shown to change the microbial flora, causing Gram-negative bacterial counts to drop whilst Gram-positive bacterial counts increased [230]. These results are supported by another study which identified similar peritoneal microbial flora constituents following intestinal perforation [231].

Bacterial Load

Assessment of peritoneal bacterial contamination has been investigated as an early AL diagnostic biomarker. One study obtained post-operative peritoneal fluid samples for microbial culture from 56 low anterior resection patients. In eight patients that had an AL confirmed by imaging, *Escherichia coli*, *Bacteroides*, *Klebsiella* and *Pseudomonas* species were identified on the 1st, 3rd and 5th days following surgery. Unfortunately, the specificity of using culture results as an AL diagnostic test was low as several false positive cases occurred in which all four bacterial species were identified in a patient without an AL [200]. The clinical applicability and usefulness of peritoneal microbial cultures for rapid AL diagnosis is also severely limited by the time required to grow laboratory cultures. To overcome this issue, studies have investigated other techniques in which bacteria or bacterial components can be identified.

In a pilot study of 17 colorectal anastomotic patients, a reverse transcription-polymerase chain reaction (RT-PCR) assay designed to identify *Escherichia coli* and *Enterococcus faecalis* was performed on 10 culture-positive and 7 culture-negative peritoneal fluid samples.

While the RT-PCR results agreed with microbiological culture results, the assay suffered from a lack of specificity, with four false positive results being identified. Although these false positives all resulted from samples originating from a single patient with a surgical site infection, the authors suggested that RT-PCR may be too sensitive for AL diagnosis, leading to over-diagnosis and over-treatment [232]. To further investigate this, the authors used the same *Escherichia coli* and *Enterococcus faecalis* RT-PCR assay in a multicentre study involving 243 left-sided colonic anastomotic patients. In the 19 patients that developed a symptomatic AL, *Escherichia coli* concentration was significantly increased on the 4th and 5th days following surgery, whereas *Enterococcus faecalis* was significantly increased on days 2, 3 and 4. The authors suggested that *Enterococcus faecalis* on the 3rd day had the highest diagnostic accuracy, with a sensitivity of 92.9% and NPV of 98.7% of AL diagnosis. Although a number of false positives were still identified, the authors further suggested that the absence of *Enterococcus faecalis* on day 3 could potentially exclude the presence of an AL [233].

The use of online infrared absorption to spectroscopically detect bacteria in peritoneal fluid samples has also been investigated as a means of identifying bacterial contamination. To provide proof-of-principle, one study demonstrated that this technique could differentiate between peritoneal fluid samples obtained from a patient who recovered uneventfully from those of a patient who developed post-operative complications with highly contaminated peritoneal fluid. A significant increase in infrared absorption occurred as contamination levels increased. The authors suggested that although the technique cannot provide information on the source of the contamination, it has the potential to be used as a bedside AL early-warning system [234]. Further studies are required to assess the use of optical systems as AL diagnostic tools.

Peritoneal levels of endotoxins and lipopolysaccharide (LPS), which forms part of the cell wall of Gram-negative bacteria (including intestinal commensals), have been suggested as diagnostic biomarkers of peritonitis and AL [235]. One study measuring peritoneal LPS levels from 22 colonic resection patients demonstrated significantly raised LPS levels on the 1st and 3rd days following surgery in three patients that developed an AL. Although LPS differences between patients that recovered uneventfully and those that developed an AL were great, standard deviations between patient groups were large. Two patients also had surgery for perforated sigmoid diverticulitis, so elevated LPS levels may have been due to pre-existing bacterial contamination [236]. Currently, LPS is not routinely measured in clinical laboratories and further studies are required before its usability as an AL biomarker can be determined.

14. Limitations of Biomarkers of Anastomotic Leakage and Future Perspectives

Biomarkers that can monitor intestinal healing, identify patients at high risk of developing an AL or provide early AL diagnosis, have the potential to significantly change how we manage resection and anastomosis patients. Although pre-clinical and clinical research continues to identify novel biomarkers for these purposes, none have made it into clinical use. Stumbling blocks for the translation of study results into practice changing policies is complicated but can be related to study design and the usability of the assay itself.

Direct comparison between biomarker studies is difficult, not only because many use heterogeneous patient populations, but also because of a lack of a single, clearly defined AL definition. Some studies use asymptomatic patients with diagnosis based on imaging, whilst others only use patients exhibiting clinical signs. These clinical signs can also be wide-ranging, from non-specific to the presence of a faecal fistula or multi-organ failure. A significant number of studies are also retrospective in nature. Although this means large sample sizes can be obtained, studies can run for several years, over which time the surgical team, surgical techniques and post-operative management practices can change significantly. Studies also differ in the timings of blood tests and/or peritoneal fluid analysis, with biomarker levels rarely evaluated specifically in terms of anastomotic position (colonic and rectal resections) or the underlying disease process. They also fail to account for medications or treatments that may alter inflammatory responses, such as statins, steroids, chemotherapy or radiotherapy. All these considerations are especially important when AL cut-off values are determined from study results; these differences will undoubtedly have contributed to the significant variations in AL cut-off values reported across these studies. Standardised, multicentre prospective studies are needed to overcome these issues.

A large number of studies suggest that peritoneal fluid samples obtained from abdominal drains provide a better indication of the peri-anastomotic tissue environment than blood samples. Although this may be true, studies typically fail to document drain location and type, which makes comparing results from different studies challenging. It has also been shown that drain location can influence drain fluid composition [237]. As gross body movements, including coughing, can affect drain location, this means that changes in drain fluid biomarker levels may be solely due to changes in drain location rather than fluctuations in patient status or intestinal healing. The clinical value of using peritoneal drains after a resection and anastomosis also remains a contentious issue. Several studies and meta-analyses have not shown any benefit of peritoneal drainage in decreasing AL incidence [238–240]. If surgeons are unwilling to routinely place them at surgery, then basing a biomarker assay on drain fluid will ultimately fail to reach clinical use. Strong clinical evidence proving that peritoneal drain fluid analysis is useful for the management of these patients is needed to allow peritoneal drainage to become routine and no longer controversial.

In terms of developing a clinically usable assay, certain biomarkers have inherent problems. Biomarkers such as cytokines and MMP are labile, meaning that peritoneal fluid analysis has to be performed immediately. Expensive and labour-intensive assays such as enzyme-linked immunosorbent assays (ELISA) or PCR technologies also have clinical limitations as laboratory processing, even if available in the hospital, incurs inherent time delays in reporting results. Studies that have investigated bacterial contamination have also shown a lack of clinical usability, either through time delays associated with growing cultures or through high false positive rates with RT-PCR assays.

Researchers are continuing to investigate methods to overcome these issues. Multidisciplinary projects involving engineers, chemists and clinicians are looking at ways in which implantable medical devices and sensor technologies could be utilised for such purposes. Studies such as the IMPACT project have already provided initial results regarding the development of miniaturised O_2 and pH sensors. Further research will undoubtedly lead into the creation of sensors for the detection of the most promising AL biomarkers, such as CRP, lactate and pyruvate (Figure 2). Wireless technology also creates the possibility of producing a biodegradable implant, which could be fixed around the anastomosis to remotely provide information about the tissue environment. Although this research is still in its infancy, technological advancements may ultimately deliver a simple, acceptable and low-cost method of measuring known AL biomarkers from peritoneal fluid directly surrounding an anastomosis or from peri-anastomotic serosal surfaces on which the sensors are placed.

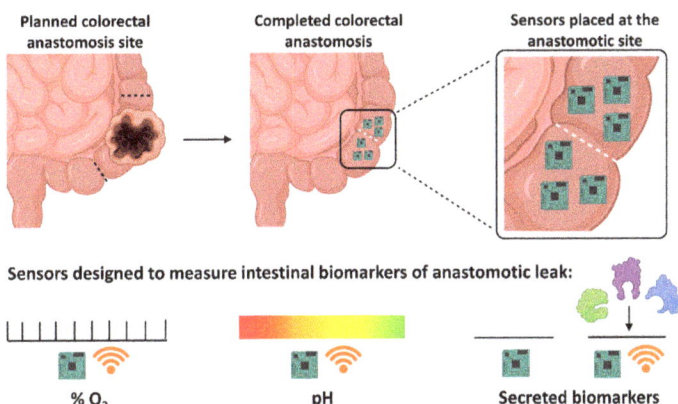

Figure 2. Future applications of advanced technologies for measuring anastomotic leak biomarkers. Implantable sensors placed intra-operatively around the anastomotic site could be left in situ throughout the post-operative recovery period. This concept would allow clinicians to continuously monitor peri-anastomotic biomarkers such as O_2, pH, C-reactive protein, lactate and pyruvate levels. This type of continuous monitoring system would help to identify patients at risk of developing an AL due to poor or deteriorating peri-anastomotic intestinal ptO_2. It would also allow clinicians the ability to assess the efficacy of interventions designed to improve intestinal ptO_2 and prevent a leak from occurring. Figure created in Biorender.

15. Conclusions

In the field of colorectal cancer research, significant advancements have been made in the identification of diagnostic and predictive biomarkers of AL. This research is driven by the clinical need to identify patients at high risk of developing an AL and to diagnose AL earlier than current protocols allow. The ideal biomarker would allow for rapid, cost-effective and reliable prediction or detection of an AL in a time frame that allows clinicians to instigate interventions that minimise patient morbidity and mortality. Here, we have highlighted the current, most promising potential candidate biomarkers, including ischaemic metabolites, inflammatory markers and bacterial components. Although none of these biomarkers have yet been validated in large-scale clinical trials, with none in routine clinical use, ongoing biomarker research in the field of intestinal surgery holds much promise. The incorporation of such biomarkers outlined in our review with other techniques, such as clinical status, routine haematological and biochemical analysis and imaging, has the potential to deliver an overall precision medicine package that could significantly enhance the effectiveness of a patient's post-operative care. There is a need, now more than ever, to utilise our knowledge of these biomarkers in carefully designed prospective, multicentre studies. These trials should be designed to investigate whether proactive post-operative patient management based on predictive biomarker levels can be used to reduce AL rates. There is confidence within the scientific community that precision medicine, through the incorporation of biomarker analysis, will finally be realised for intestinal resection and anastomosis patients in the decades to come.

Author Contributions: M.A.P., D.J.A. and A.F.M. secured funding for this research. M.G. wrote the majority of the manuscript and composed the figures, with significant contributions from J.R.K.M. Critical revisions were made by M.G., J.R.K.M., A.F.M., D.J.A. and M.A.P. All authors have read and agreed to the published version of the manuscript.

Funding: This research was funded by the UK Engineering and Physical Sciences Research Council, through the IMPACT programme grant (EP/K-34510/1), and a project grant from Bowel and Cancer Research.

Institutional Review Board Statement: Not applicable.

Informed Consent Statement: Not applicable.

Data Availability Statement: Not applicable.

Conflicts of Interest: The authors declare no conflict of interest.

References

1. Bray, F.; Ferlay, J.; Soerjomataram, I.; Siegel, R.L.; Torre, L.A.; Jemal, A. Global cancer statistics 2018: GLOBOCAN estimates of incidence and mortality worldwide for 36 cancers in 185 countries. *CA Cancer J. Clin.* **2018**, *68*, 394–424. [CrossRef]
2. Cancer Research UK. Cancer Statistics for the UK. Available online: www.cancerresearchuk.org (accessed on 12 March 2021).
3. Peel, A.L.; Taylor, E.W. Proposed definitions for the audit of postoperative infection: A discussion paper. Surgical Infection Study Group. *Ann. R. Coll. Surg. Engl.* **1991**, *73*, 385–388.
4. Bruce, J.; Krukowski, Z.H.; Al-Khairy, G.; Russell, E.M.; Park, K.G.M. Systematic review of the definition and measurement of anastomotic leak after gastrointestinal surgery. *BJS* **2002**, *88*, 1157–1168. [CrossRef] [PubMed]
5. Rahbari, N.N.; Weitz, J.; Hohenberger, W.; Heald, R.J.; Moran, B.; Ulrich, A.; Holm, T.; Wong, W.D.; Tiret, E.; Moriya, Y.; et al. Definition and grading of anastomotic leakage following anterior resection of the rectum: A proposal by the International Study Group of Rectal Cancer. *Surgery* **2010**, *147*, 339–351. [CrossRef] [PubMed]
6. Kulu, Y.; Ulrich, A.; Bruckner, T.; Contin, P.; Welsch, T.; Rahbari, N.N.; Büchler, M.W.; Weitz, J. Validation of the International Study Group of Rectal Cancer definition and severity grading of anastomotic leakage. *Surgery* **2013**, *153*, 753–761. [CrossRef]
7. Hyman, N.; Manchester, T.L.; Osler, T.; Burns, B.; Cataldo, P.A. Anastomotic Leaks after Intestinal Anastomosis: It's Later than You Think. *Ann. Surg.* **2007**, *245*, 254–258. [CrossRef]
8. Boccola, M.A.; Buettner, P.G.; Rozen, W.M.; Siu, S.K.; Stevenson, A.R.L.; Stitz, R.; Ho, Y.-H. Risk Factors and Outcomes for Anastomotic Leakage in Colorectal Surgery: A Single-Institution Analysis of 1576 Patients. *World J. Surg.* **2011**, *35*, 186–195. [CrossRef] [PubMed]
9. Kingham, T.P.; Pachter, H.L. Colonic Anastomotic Leak: Risk Factors, Diagnosis, and Treatment. *J. Am. Coll. Surg.* **2009**, *208*, 269–278. [CrossRef]
10. Hallböök, O.; Sjödahl, R. Anastomotic leakage and functional outcome after anterior resection of the rectum. *BJS* **1996**, *83*, 60–62. [CrossRef]
11. Kube, R.; Mroczkowski, P.; Granowski, D.; Benedix, F.; Sahm, M.; Schmidt, U.; Gastinger, I.; Lippert, H. Anastomotic leakage after colon cancer surgery: A predictor of significant morbidity and hospital mortality, and diminished tumour-free survival. *Eur. J. Surg. Oncol.* **2010**, *36*, 120–124. [CrossRef]
12. Marra, F.; Steffen, T.; Kalak, N.; Warschkow, R.; Tarantino, I.; Lange, J.; Zünd, M. Anastomotic leakage as a risk factor for the long-term outcome after curative resection of colon cancer. *Eur. J. Surg. Oncol.* **2009**, *35*, 1060–1064. [CrossRef] [PubMed]
13. Dulk, M.D.; Noter, S.; Hendriks, E.; Brouwers, M.; van der Vlies, C.; Oostenbroek, R.; Menon, A.; Steup, W.; van de Velde, C. Improved diagnosis and treatment of anastomotic leakage after colorectal surgery. *Eur. J. Surg. Oncol.* **2009**, *35*, 420–426. [CrossRef] [PubMed]
14. Bell, S.W.; Walker, K.G.; Rickard, M.J.F.X.; Sinclair, G.; Dent, O.F.; Chapuis, P.H.; Bokey, E.L. Anastomotic leakage after curative anterior resection results in a higher prevalence of local recurrence. *BJS* **2003**, *90*, 1261–1266. [CrossRef]
15. Petersen, S.; Freitag, M.; Hellmich, G.; Ludwig, K. Anastomotic leakage: Impact on local recurrence and survival in surgery of colorectal cancer. *Int. J. Colorectal Dis.* **1998**, *13*, 160–163. [CrossRef] [PubMed]
16. Branagan, G.; Finnis, D. Prognosis after Anastomotic Leakage in Colorectal Surgery. *Dis. Colon Rectum* **2005**, *48*, 1021–1026. [CrossRef]
17. Wang, S.; Liu, J.; Wang, S.; Zhao, H.; Ge, S.; Wang, W. Adverse Effects of Anastomotic Leakage on Local Recurrence and Survival After Curative Anterior Resection for Rectal Cancer: A Systematic Review and Meta-analysis. *World J. Surg.* **2016**, *41*, 277–284. [CrossRef]
18. Ramphal, W.; Boeding, J.R.; Gobardhan, P.D.; Rutten, H.J.; de Winter, L.J.B.; Crolla, R.M.; Schreinemakers, J.M. Oncologic outcome and recurrence rate following anastomotic leakage after curative resection for colorectal cancer. *Surg. Oncol.* **2018**, *27*, 730–736. [CrossRef]
19. Umpleby, H.C.; Fermor, B.; Symes, M.O.; Williamson, R.C.N. Viability of exfoliated colorectal carcinoma cells. *BJS* **2005**, *71*, 659–663. [CrossRef]
20. Symes, M.O.; Fermor, B.; Umpleby, H.C.; Tribe, C.R.; Williamson, R.C. Cells exfoliated from colorectal cancers can proliferate in immune deprived mice. *Br. J. Cancer* **1984**, *50*, 423–425. [CrossRef]
21. Fermor, B.; Umpleby, H.C.; Lever, J.V.; Symes, M.O.; Williamson, R.C.N. Proliferative and Metastatic Potential of Exfoliated Colorectal Cancer Cells. *J. Natl. Cancer Inst.* **1986**, *76*, 347–349. [CrossRef] [PubMed]
22. Gertsch, P.; Baer, H.U.; Kraft, R.; Maddern, G.J.; Altermatt, H.J. Malignant cells are collected on circular staplers. *Dis. Colon Rectum* **1992**, *35*, 238–241. [CrossRef] [PubMed]
23. Tol, P.M.V.D.; Van Rossen, E.E.M.; Van Eijck, C.H.J.; Bonthuis, F.; Marquet, R.L.; Jeekel, H. Reduction of Peritoneal Trauma By Using Nonsurgical Gauze Leads to Less Implantation Metastasis of Spilled Tumor Cells. *Ann. Surg.* **1998**, *227*, 242–248. [CrossRef] [PubMed]

24. Skipper, D.; Cooper, A.J.; Marston, J.E.; Taylor, I. Exfoliated cells and in vitro growth in colorectal cancer. *BJS* **1987**, *74*, 1049–1052. [CrossRef]
25. Baskaranathan, S.; Philips, J.; McCredden, P.; Solomon, M. Free Colorectal Cancer Cells on the Peritoneal Surface: Correlation with Pathologic Variables and Survival. *Dis. Colon Rectum* **2004**, *47*, 2076–2079. [CrossRef] [PubMed]
26. Aggarwal, B.B.; Vijayalekshmi, R.V.; Sung, B. Targeting Inflammatory Pathways for Prevention and Therapy of Cancer: Short-Term Friend, Long-Term Foe. *Clin. Cancer Res.* **2009**, *15*, 425–430. [CrossRef]
27. Wu, Y.; Zhou, B.P. Inflammation: A driving force speeds cancer metastasis. *Cell Cycle* **2009**, *8*, 3267–3273. [CrossRef]
28. Mantovani, A.; Allavena, P.; Sica, A.; Balkwill, F. Cancer-related inflammation. *Nature* **2008**, *454*, 436–444. [CrossRef] [PubMed]
29. Karin, M. Nuclear factor-κB in cancer development and progression. *Nature* **2006**, *441*, 431–436. [CrossRef] [PubMed]
30. Coussens, L.M.; Werb, Z. Inflammation and cancer. *Nature* **2002**, *420*, 860–867. [CrossRef] [PubMed]
31. McMillan, D.C.; Canna, K.; McArdle, C.S. Systemic inflammatory response predicts survival following curative resection of colorectal cancer. *BJS* **2003**, *90*, 215–219. [CrossRef] [PubMed]
32. Canna, K.; McMillan, D.C.; McKee, R.F.; McNicol, A.-M.; Horgan, P.G.; McArdle, C.S. Evaluation of a cumulative prognostic score based on the systemic inflammatory response in patients undergoing potentially curative surgery for colorectal cancer. *Br. J. Cancer* **2004**, *90*, 1707–1709. [CrossRef] [PubMed]
33. Bohle, B.; Pera, M.; Pascual, M.; Alonso, S.; Mayol, X.; Salvado, M.; Schmidt, J.; Grande, L. Postoperative intra-abdominal infection increases angiogenesis and tumor recurrence after surgical excision of colon cancer in mice. *Surgery* **2010**, *147*, 120–126. [CrossRef] [PubMed]
34. Lindgren, R.; Hallböök, O.; Rutegård, J.; Sjödahl, R.; Matthiessen, P. What Is the Risk for a Permanent Stoma After Low Anterior Resection of the Rectum for Cancer? A Six-Year Follow-Up of a Multicenter Trial. *Dis. Colon Rectum* **2011**, *54*, 41–47. [CrossRef] [PubMed]
35. Choi, H.-K.; Law, W.-L.; Ho, J.W.C. Leakage after Resection and Intraperitoneal Anastomosis for Colorectal Malignancy: Analysis of Risk Factors. *Dis. Colon Rectum* **2006**, *49*, 1719–1725. [CrossRef]
36. La Regina, D.; Di Giuseppe, M.; Lucchelli, M.; Saporito, A.; Boni, L.; Efthymiou, C.; Cafarotti, S.; Marengo, M.; Mongelli, F. Financial Impact of Anastomotic Leakage in Colorectal Surgery. *J. Gastroint. Surg.* **2018**, *23*, 580–586. [CrossRef]
37. Brisinda, G.; Vanella, S.; Cadeddu, F.; Civello, I.M.; Brandara, F.; Nigro, C.; Mazzeo, P.; Marniga, G.; Maria, G. End-to-end versus end-to-side stapled anastomoses after anterior resection for rectal cancer. *J. Surg. Oncol.* **2009**, *99*, 75–79. [CrossRef]
38. Koperna, T. Cost-effectiveness of defunctioning stomas in low anterior resections for rectal cancer: A call for benchmarking. *Arch. Surg.* **2003**, *138*, 1334–1338. [CrossRef]
39. Ashraf, S.; Burns, E.; Jani, A.; Altman, S.; Young, J.; Cunningham, C.; Faiz, O.; Mortensen, N. The economic impact of anastomotic leakage after anterior resections in English NHS hospitals: Are we adequately remunerating them? *Colorectal Dis.* **2013**, *15*, e190–e198. [CrossRef]
40. Nesbakken, A.; Nygaard, K.; Lunde, O.C. Outcome and late functional results after anastomotic leakage following mesorectal excision for rectal cancer. *BJS* **2002**, *88*, 400–404. [CrossRef]
41. Hammond, J.; Lim, S.; Wan, Y.; Gao, X.; Patkar, A. The Burden of Gastrointestinal Anastomotic Leaks: An Evaluation of Clinical and Economic Outcomes. *J. Gastroint. Surg.* **2014**, *18*, 1176–1185. [CrossRef]
42. Murrell, Z.A.; Stamos, M.J. Reoperation for Anastomotic Failure. *Clin. Colon Rectal Surg.* **2006**, *19*, 213–216. [CrossRef] [PubMed]
43. Macarthur, D.C.; Nixon, S.J.; Aitken, R.J. Avoidable deaths still occur after large bowel surgery. *BJS* **2003**, *85*, 80–83. [CrossRef] [PubMed]
44. Alves, A.; Panis, Y.; Pocard, M.; Regimbeau, J.-M.; Valleur, P. Management of anastomotic leakage after nondiverted large bowel resection. *J. Am. Coll. Surg.* **1999**, *189*, 554–559. [CrossRef]
45. Alves, Y.P.A.; Panis, Y.; Trancart, D.; Regimbeau, J.-M.; Pocard, M.; Valleur, P. Factors Associated with Clinically Significant Anastomotic Leakage after Large Bowel Resection: Multivariate Analysis of 707 Patients. *World J. Surg.* **2002**, *26*, 499–502. [CrossRef]
46. Dellinger, R.P.; The Surviving Sepsis Campaign Guidelines Committee including The Pediatric Subgroup; Levy, M.M.; Rhodes, A.; Annane, D.; Gerlach, H.; Opal, S.M.; Sevransky, J.E.; Sprung, C.L.; Douglas, I.S.; et al. Surviving Sepsis Campaign: International Guidelines for Management of Severe Sepsis and Septic Shock, 2012. *Intensive Care Med.* **2013**, *39*, 165–228. [CrossRef]
47. Lundy, J.B. A Primer on Wound Healing in Colorectal Surgery in the Age of Bioprosthetic Materials. *Clin. Colon Rectal Surg.* **2014**, *27*, 125–133. [CrossRef]
48. Thompson, S.K.; Chang, E.Y.; Jobe, B.A. Clinical review: Healing in gastrointestinal anastomoses, Part I. *Microsurgery* **2006**, *26*, 131–136. [CrossRef]
49. Enestvedt, C.K.; Thompson, S.K.; Chang, E.Y.; Jobe, B.A. Clinical review: Healing in gastrointestinal anastomoses, Part II. *Microsurgery* **2006**, *26*, 137–143. [CrossRef]
50. Trencheva, K.; Morrissey, K.P.; Wells, M.; Mancuso, C.A.; Lee, S.W.; Sonoda, T.; Michelassi, F.; Charlson, M.E.; Milsom, J.W. Identifying important predictors for anastomotic leak after colon and rectal resection: Prospective study on 616 patients. *Ann. Surg.* **2013**, *257*, 108–113. [CrossRef]
51. Park, J.S.; Choi, G.-S.; Kim, S.H.; Kim, H.R.; Kim, N.K.; Lee, K.Y.; Kang, S.B.; Kim, J.Y.; Lee, K.Y.; Kim, B.C. Multicenter analysis of risk factors for anastomotic leakage after laparoscopic rectal cancer excision: The Korean laparoscopic colorectal surgery study group. *Ann. Surg.* **2013**, *257*, 665–671. [CrossRef]

52. Jung, S.H.; Yu, C.S.; Choi, P.W.; Kim, D.D.; Park, I.J.; Kim, H.C.; Kim, J.C. Risk Factors and Oncologic Impact of Anastomotic Leakage after Rectal Cancer Surgery. *Dis. Colon Rectum* **2008**, *51*, 902–908. [CrossRef]
53. Law, W.-L.; Chu, K.-W.; Ho, J.W.; Chan, C.-W. Risk factors for anastomotic leakage after low anterior resection with total mesorectal excision. *Am. J. Surg.* **2000**, *179*, 92–96. [CrossRef]
54. Ba, Z.F.; Yokoyama, Y.; Toth, B.; Rue, L.W.; Bland, K.I.; Chaudry, I.H. Gender differences in small intestinal endothelial function: Inhibitory role of androgens. *Am. J. Physiol. Liver Physiol.* **2004**, *286*, G452–G457. [CrossRef] [PubMed]
55. Bakker, I.S.; Grossmann, I.; Henneman, D.; Havenga, K.; Wiggers, T. Risk factors for anastomotic leakage and leak-related mortality after colonic cancer surgery in a nationwide audit. *BJS* **2014**, *101*, 424–432. [CrossRef] [PubMed]
56. Buchs, N.C.; Gervaz, P.; Secic, M.; Bucher, P.; Mugnier-Konrad, B.; Morel, P. Incidence, consequences, and risk factors for anastomotic dehiscence after colorectal surgery: A prospective monocentric study. *Int. J. Colorectal Dis.* **2008**, *23*, 265–270. [CrossRef]
57. Kwag, S.-J.; Kim, J.-G.; Kang, W.-K.; Lee, J.-K.; Oh, S.-T. The nutritional risk is a independent factor for postoperative morbidity in surgery for colorectal cancer. *Ann. Surg. Treat. Res.* **2014**, *86*, 206–211. [CrossRef] [PubMed]
58. Veyrie, N.; Ata, T.; Muscari, F.; Couchard, A.-C.; Msika, S.; Hay, J.-M.; Fingerhut, A.; Dziri, C. Anastomotic Leakage after Elective Right Versus Left Colectomy for Cancer: Prevalence and Independent Risk Factors. *J. Am. Coll. Surg.* **2007**, *205*, 785–793. [CrossRef]
59. Kang, C.Y.; Halabi, W.J.; Chaudhry, O.O.; Nguyen, V.; Pigazzi, A.; Carmichael, J.; Mills, S.; Stamos, M.J. Risk Factors for Anastomotic Leakage After Anterior Resection for Rectal Cancer. *JAMA Surg.* **2013**, *148*, 65–71. [CrossRef]
60. Hennessey, D.B.; Burke, J.P.; Ni-Dhonochu, T.; Shields, C.; Winter, D.C.; Mealy, K. Preoperative hypoalbuminemia is an independent risk factor for the development of surgical site infection following gastrointestinal surgery: A multi-institutional study. *Ann. Surg.* **2010**, *252*, 325–329. [CrossRef]
61. Hayden, D.M.; Pinzon, M.C.M.; Francescatti, A.B.; Saclarides, T.J. Patient factors may predict anastomotic complications after rectal cancer surgery: Anastomotic complications in rectal cancer. *Ann. Med. Surg.* **2015**, *4*, 11–16. [CrossRef]
62. Tibbs, M.K. Wound healing following radiation therapy: A review. *Radiother. Oncol.* **1997**, *42*, 99–106. [CrossRef]
63. Hu, M.-H.; Huang, R.-K.; Zhao, R.-S.; Yang, K.-L.; Wang, H. Does neoadjuvant therapy increase the incidence of anastomotic leakage after anterior resection for mid and low rectal cancer? A systematic review and meta-analysis. *Colorectal Dis.* **2017**, *19*, 16–26. [CrossRef] [PubMed]
64. Pommergaard, H.-C.; Gessler, B.; Burcharth, J.; Angenete, E.; Haglind, E.; Rosenberg, J. Preoperative risk factors for anastomotic leakage after resection for colorectal cancer: A systematic review and meta-analysis. *Colorectal Dis.* **2014**, *16*, 662–671. [CrossRef]
65. Sciuto, A.; Merola, G.; De Palma, G.D.; Sodo, M.; Pirozzi, F.; Bracale, U.M. Predictive factors for anastomotic leakage after laparoscopic colorectal surgery. *World J. Gastroenterol.* **2018**, *24*, 2247–2260. [CrossRef] [PubMed]
66. Volk, A.; Kersting, S.; Held, H.C.; Saeger, H.D. Risk factors for morbidity and mortality after single-layer continuous suture for ileocolonic anastomosis. *Int. J. Colorectal Dis.* **2010**, *26*, 321–327. [CrossRef] [PubMed]
67. Ziegler, M.A.; Catto, J.A.; Riggs, T.W.; Gates, E.R.; Grodsky, M.B.; Wasvary, H.J. Risk factors for anastomotic leak and mortality in diabetic patients undergoing colectomy: Analysis from a statewide surgical quality collaborative. *Arch. Surg.* **2012**, *147*, 600–605. [CrossRef]
68. Richards, C.H.; Campbell, V.; Ho, C.; Hayes, J.; Elliott, T.; Thompson-Fawcett, M. Smoking is a major risk factor for anastomotic leak in patients undergoing low anterior resection. *Colorectal Dis.* **2011**, *14*, 628–633. [CrossRef]
69. Biondo, S.; Parés, D.; Kreisler, E.; Ragué, J.M.; Fraccalvieri, D.; Ruiz, A.G.; Jaurrieta, E. Anastomotic Dehiscence After Resection and Primary Anastomosis in Left-Sided Colonic Emergencies. *Dis. Colon Rectum* **2005**, *48*, 2272–2280. [CrossRef]
70. Sørensen, L.T.; Jørgensen, T.; Kirkeby, L.T.; Skovdal, J.; Vennits, B.; Wille-Jørgensen, P. Smoking and alcohol abuse are major risk factors for anastomotic leakage in colorectal surgery. *BJS* **2002**, *86*, 927–931. [CrossRef]
71. Gershuni, V.M.; Friedman, E.S. The Microbiome-Host Interaction as a Potential Driver of Anastomotic Leak. *Curr. Gastroenterol. Rep.* **2019**, *21*, 4. [CrossRef]
72. Schardey, H.M.; Kamps, T.; Rau, H.G.; Gatermann, S.; Baretton, G.; Schildberg, F.W. Bacteria: A major pathogenic factor for anastomotic insufficiency. *Antimicrob. Agents Chemother.* **1994**, *38*, 2564–2567. [CrossRef]
73. Schardey, H.M.; Joosten, U.; Finke, U.; Staubach, K.H.; Schauer, R.; Heiss, A.; Kooistra, A.; Rau, H.G.; Nibler, R.; Lüdeling, S.; et al. The Prevention of Anastomotic Leakage after Total Gastrectomy with Local Decontamination. A prospective, randomized, double-blind, placebo-controlled multicenter trial. *Ann. Surg.* **1997**, *225*, 172–180. [CrossRef]
74. Shogan, B.D.; Carlisle, E.M.; Alverdy, J.C.; Umanskiy, K. Do We Really Know Why Colorectal Anastomoses Leak? *J. Gastrointest. Surg.* **2013**, *17*, 1698–1707. [CrossRef]
75. Shogan, B.D.; Belogortseva, N.; Luong, P.M.; Zaborin, A.; Lax, S.; Bethel, C.; Ward, M.; Muldoon, J.P.; Singer, M.; An, G. Collagen degradation and MMP9 activation by Enterococcus faecalis contribute to intestinal anastomotic leak. *Sci. Transl. Med.* **2015**, *7*, 286ra68. [CrossRef]
76. Øines, M.N.; Krarup, P.-M.; Jorgensen, L.N.; Ågren, M.S. Pharmacological interventions for improved colonic anastomotic healing: A meta-analysis. *World J. Gastroenterol.* **2014**, *20*, 12637–12648. [CrossRef]
77. Damen, N.; Spilsbury, K.; Levitt, M.; Makin, G.; Salama, P.; Tan, P.; Penter, C.; Platell, C. Anastomotic leaks in colorectal surgery. *ANZ J. Surg.* **2014**, *84*, 763–768. [CrossRef]

78. McDermott, F.D.; Heeney, A.; Kelly, M.E.; Steele, R.J.; Carlson, G.L.; Winter, D.C. Systematic review of preoperative, intraoperative and postoperative risk factors for colorectal anastomotic leaks. *BJS* **2015**, *102*, 462–479. [CrossRef]
79. Akiyoshi, T.; Ueno, M.; Fukunaga, Y.; Nagayama, S.; Fujimoto, Y.; Konishi, T.; Kuroyanagi, H.; Yamaguchi, T. Incidence of and risk factors for anastomotic leakage after laparoscopic anterior resection with intracorporeal rectal transection and double-stapling technique anastomosis for rectal cancer. *Am. J. Surg.* **2011**, *202*, 259–264. [CrossRef]
80. Tortorelli, A.P.; Alfieri, S.; Sanchez, A.M.; Rosa, F.; Papa, V.; Di Miceli, D.; Bellantone, C.; Doglietto, G.B. Anastomotic Leakage after Anterior Resection for Rectal Cancer with Mesorectal Excision: Incidence, Risk Factors, and Management. *Am. Surg.* **2015**, *81*, 41–47. [CrossRef]
81. Naumann, D.N.; Bhangu, A.; Kelly, M.; Bowley, D.M. Stapled versus handsewn intestinal anastomosis in emergency laparotomy: A systemic review and meta-analysis. *Surgery* **2015**, *157*, 609–618. [CrossRef]
82. Slesser, A.A.P.; Pellino, G.; Shariq, O.; Cocker, D.; Kontovounisios, C.; Rasheed, S.; Tekkis, P.P. Compression versus hand-sewn and stapled anastomosis in colorectal surgery: A systematic review and meta-analysis of randomized controlled trials. *Tech. Coloproctol.* **2016**, *20*, 667–676. [CrossRef]
83. Jayne, D.G.; Thorpe, H.C.; Copeland, J.; Quirke, P.; Brown, J.M.; Guillou, P.J. Five-year follow-up of the Medical Research Council CLASICC trial of laparoscopically assisted versus open surgery for colorectal cancer. *BJS* **2010**, *97*, 1638–1645. [CrossRef] [PubMed]
84. Vasiliu, E.C.Z.; Zarnescu, N.O.; Costea, R.; Neagu, S. Review of Risk Factors for Anastomotic Leakage in Colorectal Surgery. *Chir.* **2015**, *110*, 319.
85. Vennix, S.; Pelzers, L.; Bouvy, N.; Beets, G.L.; Pierie, J.-P.; Wiggers, T.; Breukink, S. Laparoscopic versus open total mesorectal excision for rectal cancer. *Cochrane Database Syst. Rev.* **2014**, *4*, CD005200. [CrossRef] [PubMed]
86. Kim, N.-K.; Kang, J. Optimal total mesorectal excision for rectal cancer: The role of robotic surgery from an expert's view. *J. Korean Soc. Coloproctol* **2010**, *26*, 377. [CrossRef] [PubMed]
87. Cho, M.S.; Baek, S.J.; Hur, H.; Min, B.S.; Baik, S.H.; Lee, K.Y.; Kim, N.K. Short and long-term outcomes of robotic versus laparoscopic total mesorectal excision for rectal cancer: A case-matched retrospective study. *Medicine* **2015**, *94*, e522. [CrossRef]
88. Lim, D.R.; Bae, S.U.; Hur, H.; Min, B.S.; Baik, S.H.; Lee, K.Y.; Kim, N.K. Long-term oncological outcomes of robotic versus laparoscopic total mesorectal excision of mid–low rectal cancer following neoadjuvant chemoradiation therapy. *Surg. Endosc.* **2016**, *31*, 1728–1737. [CrossRef]
89. Smith, J.A.E.; King, P.M.; Lane, R.H.S.; Thompson, M.R. Evidence of the effect of 'specialization' on the management, surgical outcome and survival from colorectal cancer in Wessex. *BJS* **2003**, *90*, 583–592. [CrossRef]
90. Singh, K.K.; Aitken, R.J. Outcome in patients with colorectal cancer managed by surgical trainees. *BJS* **2002**, *86*, 1332–1336. [CrossRef]
91. Fujiwata, H.; Kuga, T.; Esato, K. High submucosal blood flow and low anastomotic tension prevent anastomotic leakage in rabbits. *Surg. Today* **1997**, *27*, 924–929. [CrossRef] [PubMed]
92. Attard, J.-A.P.; Raval, M.J.; Martin, G.R.; Kolb, J.; Afrouzian, M.; Buie, D.W.; Sigalet, D.L. The Effects of Systemic Hypoxia on Colon Anastomotic Healing: An Animal Model. *Dis. Colon Rectum* **2005**, *48*, 1460–1470. [CrossRef] [PubMed]
93. Sheridan, W.G.; Lowndes, R.H.; Young, H.L. Tissue oxygen tension as a predictor of colonic anastomotic healing. *Dis. Colon Rectum* **1987**, *30*, 867–871. [CrossRef] [PubMed]
94. Khan, A.A.; Wheeler, J.M.D.; Cunningham, C.; George, B.; Kettlewell, M.; Mortensen, N.J.M. The management and outcome of anastomotic leaks in colorectal surgery. *Colorectal Dis.* **2008**, *10*, 587–592. [CrossRef] [PubMed]
95. Sutton, C.D.; Marshall, L.J.; Williams, N.; Berry, D.P.; Thomas, W.M.; Kelly, M.J. Colo-rectal anastomotic leakage often masquerades as a cardiac complication. *Colorectal Dis.* **2003**, *6*, 21–22. [CrossRef]
96. Karliczek, A.; Harlaar, N.J.; Zeebregts, C.J.; Wiggers, T.; Baas, P.C.; Van Dam, G.M. Surgeons lack predictive accuracy for anastomotic leakage in gastrointestinal surgery. *Int. J. Colorectal Dis.* **2009**, *24*, 569–576. [CrossRef]
97. Liu, Y.; Wan, X.; Wang, G.; Ren, Y.; Cheng, Y.; Zhao, Y.; Han, G. A scoring system to predict the risk of anastomotic leakage after anterior resection for rectal cancer. *J. Surg. Oncol.* **2014**, *109*, 122–125. [CrossRef]
98. Dekker, J.W.T.; Liefers, G.J.; Otterloo, J.C.D.M.V.; Putter, H.; Tollenaar, R.A. Predicting the Risk of Anastomotic Leakage in Left-sided Colorectal Surgery Using a Colon Leakage Score. *J. Surg. Res.* **2011**, *166*, e27–e34. [CrossRef]
99. Martin, G.; Dupré, A.; Mulliez, A.; Prunel, F.; Slim, K.; Pezet, D. Validation of a score for the early diagnosis of anastomotic leakage following elective colorectal surgery. *J. Visc. Surg.* **2015**, *152*, 5–10. [CrossRef]
100. Dulk, M.D.; Witvliet, M.J.; Kortram, K.; Neijenhuis, P.A.; De Hingh, I.H.; Engel, A.F.; Van De Velde, C.J.; De Brauw, L.M.; Putter, H.; Brouwers, M.A.; et al. The DULK (Dutch leakage) and modified DULK score compared: Actively seek the leak. *Colorectal Dis.* **2013**, *15*, e528–e533. [CrossRef]
101. Khoury, W.; Ben-Yehuda, A.; Ben-Haim, M.; Klausner, J.M.; Szold, O. Abdominal Computed Tomography for Diagnosing Postoperative Lower Gastrointestinal Tract Leaks. *J. Gastrointest. Surg.* **2009**, *13*, 1454–1458. [CrossRef]
102. Nesbakken, A.; Nygaard, K.; Lunde, O.C.; Blucher, J.; Gjertsen, O.; Dullerud, R. Anastomotic leak following mesorectal excision for rectal cancer: True incidence and diagnostic challenges. *Colorectal Dis.* **2005**, *7*, 576–581. [CrossRef]
103. Gollgher, J.; Graham, N.; de Dombal, F. Anastomotic dehiscence after anterior resection of rectum and sigmold. *Br. J. Surg.* **1970**, *57*, 109–118. [CrossRef]

104. Tang, C.-L.; Seow-Choen, F. Digital rectal examination compares favourably with conventional water-soluble contrast enema in the assessment of anastomotic healing after low rectal excision: A cohort study. *Int. J. Colorectal Dis.* **2004**, *20*, 262–266. [CrossRef]
105. Williams, C.E.; Makin, C.A.; Reeve, R.G.; Ellenbogen, S.B. Over-utilisation of radiography in the assessment of stapled colonic anastomoses. *Eur. J. Radiol.* **1991**, *12*, 35–37. [CrossRef]
106. Akyol, A.M.; McGregor, J.R.; Galloway, D.J.; George, W.D. Early postoperative contrast radiology in the assessment of colorectal anastomotic integrity. *Int. J. Colorectal Dis.* **1992**, *7*, 141–143. [CrossRef]
107. Nicksa, G.A.; Dring, R.V.; Johnson, K.H.; Sardella, W.V.; Vignati, P.V.; Cohen, J.L. Anastomotic Leaks: What is the Best Diagnostic Imaging Study? *Dis. Colon Rectum* **2007**, *50*, 197–203. [CrossRef]
108. Power, N.; Atri, M.; Ryan, S.; Haddad, R.; Smith, A. CT assessment of anastomotic bowel leak. *Clin. Radiol.* **2007**, *62*, 37–42. [CrossRef]
109. Foster, M.E.; Laycock, J.R.D.; Silver, I.A.; Leaper, D.J. Hypovolaemia and healing in colonic anastomoses. *BJS* **1985**, *72*, 831–834. [CrossRef]
110. Brandstrup, B.; Tønnesen, H.; Beier-Holgersen, R.; Hjortsø, E.; Ørding, H.; Lindorff-Larsen, K.; Rasmussen, M.S.; Langn, C.; Wallin, L.; Iversen, L.H. Effects of intravenous fluid restriction on postoperative complications: Comparison of two perioperative fluid regimens: A randomized assessor-blinded multicenter trial. *Ann. Surg.* **2003**, *238*, 641–648. [CrossRef]
111. Kurz, A.; Sessler, D.I.; Lenhardt, R. Perioperative Normothermia to Reduce the Incidence of Surgical-Wound Infection and Shorten Hospitalization. *N. Engl. J. Med.* **1996**, *334*, 1209–1216. [CrossRef] [PubMed]
112. Belda, F.J.; Aguilera, L.; De La Asunción, J.G.; Alberti, J.; Vicente, R.; Ferrándiz, L.; Rodríguez, R.; Sessler, D.I.; Aguilar, G.; Botello, S.G. Supplemental perioperative oxygen and the risk of surgical wound infection: A randomized controlled trial. *JAMA* **2005**, *294*, 2035–2042. [CrossRef]
113. Gan, T.J.; Soppitt, A.; Maroof, M.; El-Moalem, H.; Robertson, K.M.; Moretti, E.; Dwane, P.; Glass, P.S.A. Goal-directed Intraoperative Fluid Administration Reduces Length of Hospital Stay after Major Surgery. *J. Am. Soc. Anesthesiol.* **2002**, *97*, 820–826. [CrossRef]
114. Donati, A.; Loggi, S.; Preiser, J.-C.; Orsetti, G.; Munch, C.; Gabbanelli, V.; Pelaia, P.; Pietropaoli, P. Goal-Directed Intraoperative Therapy Reduces Morbidity and Length of Hospital Stay in High-Risk Surgical Patients. *Chest* **2007**, *132*, 1817–1824. [CrossRef]
115. Kimberger, M.D.O.; Arnberger, M.D.M.; Brandt, M.D.S.; Plock, M.D.J.; Gisli, M.D.P.D.; Sigurdsson, H.; Kurz, M.D.A.; Hiltebrand, M.D.L. Goal-directed Colloid Administration Improves the Microcirculation of Healthy and Perianastomotic Colon. *Anesthesiology* **2009**, *110*, 496–504. [CrossRef]
116. Ghasemi, M.; Nabipour, I.; Omrani, A.; Alipour, Z.; Assadi, M. Precision medicine and molecular imaging: New targeted approaches toward cancer therapeutic and diagnosis. *Am. J. Nucl. Med. Mol. Imaging* **2016**, *6*, 310–327.
117. Katsnelson, A. Momentum grows to make 'personalized' medicine more 'precise'. *Nat. Med.* **2013**, *19*, 249. [CrossRef]
118. National Research Council. *The National Academies Collection: Reports Funded by National Institutes of Health, Toward Precision Medicine: Building a Knowledge Network for Biomedical Research and a New Taxonomy of Disease*; National Academy of Sciences: Washington, DC, USA, 2011.
119. Atkinson, A.J., Jr.; Colburn, W.A.; DeGruttola, V.G.; DeMets, D.L.; Downing, G.J.; Hoth, D.F.; Oates, J.A.; Peck, C.C.; Schooley, R.T. Biomarkers and surrogate endpoints: Preferred definitions and conceptual framework. *Clin. Pharmacol. Ther.* **2001**, *69*, 89–95.
120. Mandrekar, S.J.; Sargent, D.J. Clinical Trial Designs for Predictive Biomarker Validation: Theoretical Considerations and Practical Challenges. *J. Clin. Oncol.* **2009**, *27*, 4027–4034. [CrossRef]
121. Polley, M.-Y.C.; Freidlin, B.; Korn, E.L.; Conley, B.A.; Abrams, J.S.; McShane, L.M. Statistical and Practical Considerations for Clinical Evaluation of Predictive Biomarkers. *J. Natl. Cancer Inst.* **2013**, *105*, 1677–1683. [CrossRef]
122. Ludwig, J.A.; Weinstein, J.N. Biomarkers in Cancer Staging, Prognosis and Treatment Selection. *Nat. Rev. Cancer* **2005**, *5*, 845–856. [CrossRef]
123. Beard, J.D.; Nicholson, M.L.; Sayers, R.D.; Lloyd, D.; Everson, N.W. Intraoperative air testing of colorectal anastomoses: A prospective, randomized trial. *BJS* **1990**, *77*, 1095–1097. [CrossRef] [PubMed]
124. Ivanov, D.; Cvijanovic, R.; Gvozdenovic, L. Intraoperative air testing of colorectal anastomoses. *Srp. Arh. Za Celok. Lek.* **2011**, *139*, 333–338. [CrossRef] [PubMed]
125. Li, V.K.M.; Wexner, S.D.; Pulido, N.; Wang, H.; Jin, H.Y.; Weiss, E.G.; Nogueras, J.J.; Sands, D.R. Use of routine intraoperative endoscopy in elective laparoscopic colorectal surgery: Can it further avoid anastomotic failure? *Surg. Endosc.* **2009**, *23*, 2459–2465. [CrossRef] [PubMed]
126. Shamiyeh, A.; Szabo, K.; Wayand, W.U.; Zehetner, J. Intraoperative Endoscopy for the Assessment of Circular-stapled Anastomosis in Laparoscopic Colon Surgery. *Surg. Laparosc. Endosc. Percutaneous Tech.* **2012**, *22*, 65–67. [CrossRef] [PubMed]
127. Brugiotti, C.; Corazza, V.; Peruzzi, A.; Lozano, C.; Casadevall, E.; Ques, F.; Ferrer, S.; Canaves, J.; Amurrio, R.; Rodriguez, J. The efficacy of intraoperative endoscopic control of the colorectal stapled anastomosis after anterior resection of the rectum for rectal cancer: P31. *Colorectal Dis.* **2011**, *13*, 31–32.
128. Karliczek, A.; Benaron, D.A.; Baas, P.C.; Zeebregts, C.J.; Wiggers, T.; Van Dam, G.M. Intraoperative assessment of microperfusion with visible light spectroscopy for prediction of anastomotic leakage in colorectal anastomoses. *Colorectal Dis.* **2009**, *12*, 1018–1025. [CrossRef]

129. Hirano, Y.; Omura, K.; Tatsuzawa, Y.; Shimizu, J.; Kawaura, Y.; Watanabe, G. Tissue Oxygen Saturation during Colorectal Surgery Measured by Near-infrared Spectroscopy: Pilot Study to Predict Anastomotic Complications. *World J. Surg.* **2006**, *30*, 457–461. [CrossRef]
130. Kudszus, S.; Roesel, C.; Schachtrupp, A.; Höer, J.J. Intraoperative laser fluorescence angiography in colorectal surgery: A noninvasive analysis to reduce the rate of anastomotic leakage. *Langenbeck's Arch. Surg.* **2010**, *395*, 1025–1030. [CrossRef]
131. Boyle, N.; Manifold, D.; Jordan, M.; Mason, R. Intraoperative assessment of colonic perfusion using scanning laser doppler flowmetry during colonic resection11No competing interests declared. *J. Am. Coll. Surg.* **2000**, *191*, 504–510. [CrossRef]
132. Servais, E.L.; Rizk, N.P.; Oliveira, L.; Rusch, V.W.; Bikson, M.; Adusumilli, P.S. Real-time intraoperative detection of tissue hypoxia in gastrointestinal surgery by wireless pulse oximetry. *Surg. Endosc.* **2011**, *25*, 1383–1389. [CrossRef]
133. Myers, C.; Mutafyan, G.; Petersen, R.; Pryor, A.; Reynolds, J.; DeMaria, E. Real-time probe measurement of tissue oxygenation during gastrointestinal stapling: Mucosal ischemia occurs and is not influenced by staple height. *Surg. Endosc.* **2009**, *23*, 2345–2350. [CrossRef] [PubMed]
134. Salusjärvi, J.M.; Carpelan-Holmström, M.A.; Louhimo, J.M.; Kruuna, O.; Scheinin, T.M. Intraoperative colonic pulse oximetry in left-sided colorectal surgery: Can it predict anastomotic leak? *Int. J. Colorectal Dis.* **2018**, *33*, 333–336. [CrossRef]
135. Cahill, A.R.; Mortensen, N.J. Intraoperative augmented reality for laparoscopic colorectal surgery by intraoperative near-infrared fluorescence imaging and optical coherence tomography. *Minerva Chir.* **2010**, *65*, 451–462.
136. Matsui, A.; Winer, J.H.; Laurence, R.G.; Frangioni, J.V. Predicting the survival of experimental ischaemic small bowel using intraoperative near-infrared fluorescence angiography. *BJS* **2011**, *98*, 1725–1734. [CrossRef]
137. Armstrong, G.; Croft, J.; Corrigan, N.; Brown, J.M.; Goh, V.; Quirke, P.; Hulme, C.; Tolan, D.; Kirby, A.; Cahill, R.; et al. IntAct: Intra-operative fluorescence angiography to prevent anastomotic leak in rectal cancer surgery: A randomized controlled trial. *Colorectal Dis.* **2018**, *20*, O226–O234. [CrossRef]
138. Shandall, A.; Lowndes, R.; Young, H.L. Colonic anastomotic healing and oxygen tension. *BJS* **2005**, *72*, 606–609. [CrossRef] [PubMed]
139. Locke, R.; Hauser, C.J.; Shoemaker, W.C. The Use of Surface Oximetry to Assess Bowel Viability. *Arch. Surg.* **1984**, *119*, 1252. [CrossRef] [PubMed]
140. Sheridan, W.G.; Lowndes, R.H.; Young, H.L. Intraoperative tissue oximetry in the human gastrointestinal tract. *Am. J. Surg.* **1990**, *159*, 314–319. [CrossRef]
141. Clifford, R.E.; Fowler, H.; Manu, N.; Sutton, P.; Vimalachandran, D. Intra-operative assessment of left-sided colorectal anastomotic integrity: A systematic review of available techniques. *Colorectal Dis.* **2020**, *23*, 582–591. [CrossRef] [PubMed]
142. 2019 Safe-Anastomosis Programme in Colorectal Surgery (EAGLE). The European Society of Coloproctology. Available online: https://www.escp.eu.com/research/cohort-studies/2019-escp-safe-anastomosis-programme-in-colorectal-surgery (accessed on 8 April 2021).
143. Corke, C.; Glenister, K. Monitoring intestinal ischaemia. *Crit. Care Resusc.* **2001**, *3*, 176–180.
144. Klaus, S.; Heringlake, M.; Gliemroth, J.; Bruch, H.-P.; Bahlmann, L. Intraperitoneal microdialysis for detection of splanchnic metabolic disorders. *Langenbeck's Arch. Surg.* **2002**, *387*, 276–280. [CrossRef]
145. Sommer, T.; Larsen, J.F. Intraperitoneal and intraluminal microdialysis in the detection of experimental regional intestinal ischaemia. *BJS* **2004**, *91*, 855–861. [CrossRef]
146. Jansson, K.; Ungerstedt, J.; Jonsson, T.; Redler, B.; Andersson, M.; Ungerstedt, U.; Norgren, L. Human intraperitoneal microdialysis: Increased lactate/pyruvate ratio suggests early visceral ischaemia. *Scand. J. Gastroenterol.* **2003**, *38*, 1007–1011.
147. Jansson, K.; Jansson, M.; Andersson, M.; Magnuson, A.; Ungerstedt, U.; Norgren, L. Normal values and differences between intraperitoneal and subcutaneous microdialysis in patients after non-complicated gastrointestinal surgery. *Scand. J. Clin. Lab. Investig.* **2005**, *65*, 273–282. [CrossRef]
148. Matthiessen, P.; Strand, I.; Jansson, K.; Törnquist, C.; Andersson, M.; Rutegård, J.; Norgren, L. Is Early Detection of Anastomotic Leakage Possible by Intraperitoneal Microdialysis and Intraperitoneal Cytokines After Anterior Resection of the Rectum for Cancer? *Dis. Colon Rectum* **2007**, *50*, 1918–1927. [CrossRef] [PubMed]
149. Pedersen, M.E.; Qvist, N.; Bisgaard, C.; Kelly, U.; Bernhard, A.; Pedersen, S.M. Peritoneal Microdialysis. Early diagnosis of Anastomotic Leakage after Low Anterior Resection for Rectosigmoid Cancer. *Scand. J. Surg.* **2009**, *98*, 148–154. [CrossRef] [PubMed]
150. Oikonomakis, I.; Jansson, D.; Hörer, T.M.; Skoog, P.; Nilsson, K.; Jansson, K. Results of postoperative microdialysis intraperitoneal and at the anastomosis in patients developing anastomotic leakage after rectal cancer surgery. *Scand. J. Gastroenterol.* **2019**, *54*, 1261–1268. [CrossRef]
151. Hörer, T.M.; Norgren, L.; Jansson, K. Intraperitoneal glycerol levels and lactate/pyruvate ratio: Early markers of postoperative complications. *Scand. J. Gastroenterol.* **2011**, *46*, 913–919. [CrossRef]
152. Bini, R.; Ferrari, G.; Aprà, F.; Viora, T.; Leli, R.; Cotogni, P. Peritoneal lactate as a potential biomarker for predicting the need for reintervention after abdominal surgery. *J. Trauma Acute Care Surg.* **2014**, *77*, 376–380. [CrossRef]
153. Daams, F.; Wu, Z.; Cakir, H.; Karsten, T.M.; Lange, J.F. Identification of anastomotic leakage after colorectal surgery using microdialysis of the peritoneal cavity. *Tech. Coloproctol.* **2013**, *18*, 65–71. [CrossRef]
154. Ellebæk, M.B.; Rahr, H.B.; Boye, S.; Fristrup, C.; Qvist, N. Detection of early anastomotic leakage by intraperitoneal microdialysis after low anterior resection for rectal cancer: A prospective cohort study. *Colorectal Dis.* **2019**, *21*, 1387–1396. [CrossRef] [PubMed]

155. Millan, M.; García-Granero, E.; Flor, B.; García-Botello, S.; Lledo, S. Early prediction of anastomotic leak in colorectal cancer surgery by intramucosal pH. *Dis. Colon Rectum* **2006**, *49*, 595–601. [CrossRef] [PubMed]
156. Yang, L.; Huang, X.-E.; Xu, L.; Zhou, X.; Zhou, J.-N.; Yu, D.-S.; Li, D.-Z.; Guan, X. Acidic Pelvic Drainage as a Predictive Factor For Anastomotic Leakage after Surgery for Patients with Rectal Cancer. *Asian Pac. J. Cancer Prev.* **2013**, *14*, 5441–5447. [CrossRef] [PubMed]
157. Marland, J.R.K.; Gray, M.E.; Dunare, C.; Blair, E.O.; Tsiamis, A.; Sullivan, P.; González-Fernández, E.; Greenhalgh, S.N.; Gregson, R.; Clutton, R.E.; et al. Real-time measurement of tumour hypoxia using an implantable microfabricated oxygen sensor. *Sens. Bio-Sens. Res.* **2020**, *30*, 1–12.
158. Marland, J.R.K.; Blair, E.O.; Flynn, B.W.; González-Fernández, E.; Huang, L.; Kunkler, I.H.; Smith, S.; Staderini, M.; Tsiamis, A.; Ward, C.; et al. Implantable Microsystems for Personalised Anticancer Therapy. In *CMOS Circuits for Biological Sensing and Processing*; Mitra, S., Cumming, D.R.S., Eds.; Springer International Publishing: Cham, Switzerland, 2018; pp. 259–286.
159. Gray, M.E.; Marland, J.R.K.; Dunare, C.; Blair, E.O.; Meehan, J.; Tsiamis, A.; Kunkler, I.H.; Murray, A.F.; Argyle, D.; Dyson, A.; et al. In vivo validation of a miniaturized electrochemical oxygen sensor for measuring intestinal oxygen tension. *Am. J. Physiol. Liver Physiol.* **2019**, *317*, G242–G252. [CrossRef]
160. Badia, J.M.; Whawell, S.A.; Scott-Coombes, D.M.; Abel, P.D.; Williamson, R.C.N.; Thompson, J.N. Peritoneal and systemic cytokine response to laparotomy. *BJS* **1996**, *83*, 347–348. [CrossRef]
161. Welsch, T.; Müller, S.A.; Ulrich, A.; Kischlat, A.; Hinz, U.; Kienle, P.; Büchler, M.W.; Schmidt, J.; Schmied, B.M. C-reactive protein as early predictor for infectious postoperative complications in rectal surgery. *Int. J. Colorectal Dis.* **2007**, *22*, 1499–1507. [CrossRef]
162. Matthiessen, P.; Henriksson, M.; Hallböök, O.; Grunditz, E.; Norén, B.; Arbman, G. Increase of serum C-reactive protein is an early indicator of subsequent symptomatic anastomotic leakage after anterior resection. *Colorectal Dis.* **2007**, *10*, 75–80. [CrossRef]
163. Woeste, G.; Müller, C.; Bechstein, W.O.; Wullstein, C. Increased Serum Levels of C-Reactive Protein Precede Anastomotic Leakage in Colorectal Surgery. *World J. Surg.* **2009**, *34*, 140–146. [CrossRef]
164. Pedersen, T.; Roikjær, O.; Jess, P. Increased levels of C-reactive protein and leukocyte count are poor predictors of anastomotic leakage following laparoscopic colorectal resection. *Dan. Med. J.* **2012**, *59*, 1–4.
165. Almeida, A.; Faria, G.; Moreira, H.; Pinto-De-Sousa, J.; Correia-Da-Silva, P.; Maia, J.C. Elevated serum C-reactive protein as a predictive factor for anastomotic leakage in colorectal surgery. *Int. J. Surg.* **2012**, *10*, 87–91. [CrossRef]
166. Lagoutte, N.; Facy, O.; Ravoire, A.; Chalumeau, C.; Jonval, L.; Rat, P.; Ortega-Deballon, P. C-reactive protein and procalcitonin for the early detection of anastomotic leakage after elective colorectal surgery: Pilot study in 100 patients. *J. Visc. Surg.* **2012**, *149*, e345–e349. [CrossRef]
167. Reisinger, K.W.; Poeze, M.; Hulsewé, K.W.; van Acker, B.A.; van Bijnen, A.A.; Hoofwijk, A.G.; Stoot, J.; Derikx, J.P. Accurate Prediction of Anastomotic Leakage after Colorectal Surgery Using Plasma Markers for Intestinal Damage and Inflammation. *J. Am. Coll. Surg.* **2014**, *219*, 744–751. [CrossRef]
168. Zawadzki, M.; Czarnecki, R.; Rzaca, M.; Obuszko, Z.; Velchuru, V.R.; Witkiewicz, W. C-reactive protein and procalcitonin predict anastomotic leaks following colorectal cancer resections–a prospective study. *Videosurg. Other Miniinvasive Tech.* **2015**, *10*, 567. [CrossRef]
169. Waterland, P.; Ng, J.; Jones, A.; Broadley, G.; Nicol, D.; Patel, H.; Pandey, S. Using CRP to predict anastomotic leakage after open and laparoscopic colorectal surgery: Is there a difference? *Int. J. Colorectal Dis.* **2016**, *31*, 861–868. [CrossRef]
170. Giaccaglia, V.; Salvi, P.F.; Antonelli, M.S.; Nigri, G.R.; Corcione, F.; Pirozzi, F.; De Manzini, N.; Casagranda, B.; Balducci, G.; Ziparo, V. Procalcitonin reveals early dehiscence in colorectal surgery: The PREDICS study. *J. Am. Coll. Surg.* **2014**, *219*, e8. [CrossRef]
171. Mik, M.; Dziki, L.; Berut, M.; Trzcinski, R.; Dziki, A. Neutrophil to Lymphocyte Ratio and C-Reactive Protein as Two Predictive Tools of Anastomotic Leak in Colorectal Cancer Open Surgery. *Dig. Surg.* **2017**, *35*, 77–84. [CrossRef]
172. Smith, S.R.; Pockney, P.; Holmes, R.; Doig, F.; Attia, J.; Holliday, E.; Carroll, R.; Draganic, B. Biomarkers and anastomotic leakage in colorectal surgery: C-reactive protein trajectory is the gold standard. *ANZ J. Surg.* **2018**, *88*, 440–444. [CrossRef]
173. Reynolds, I.S.; Boland, M.R.; Reilly, F.; Deasy, A.; Majeed, M.H.; Deasy, J.; Burke, J.P.; McNamara, D.A. C-reactive protein as a predictor of anastomotic leak in the first week after anterior resection for rectal cancer. *Colorectal Dis.* **2017**, *19*, 812–818. [CrossRef]
174. Warschkow, R.; Tarantino, I.; Torzewski, M.; Näf, F.; Lange, J.; Steffen, T. Diagnostic accuracy of C-reactive protein and white blood cell counts in the early detection of inflammatory complications after open resection of colorectal cancer: A retrospective study of 1187 patients. *Int. J. Colorectal Dis.* **2011**, *26*, 1405–1413. [CrossRef]
175. Kørner, H.; Nielsen, H.J.; Søreide, J.A.; Nedrebø, B.S.; Søreide, K.; Knapp, J.C. Diagnostic Accuracy of C-reactive Protein for Intraabdominal Infections After Colorectal Resections. *J. Gastrointest. Surg.* **2009**, *13*, 1599–1606. [CrossRef] [PubMed]
176. Ho, Y.M.; Laycock, J.; Kirubakaran, A.; Hussain, L.; Clark, J. Systematic use of the serum C-reactive protein concentration and computed tomography for the detection of intestinal anastomotic leaks. *ANZ J. Surg.* **2019**, *90*, 109–112. [CrossRef] [PubMed]
177. Messias, B.A.; Botelho, R.V.; Saad, S.S.; Mocchetti, E.R.; Turke, K.C.; Waisberg, J. Serum C-reactive protein is a useful marker to exclude anastomotic leakage after colorectal surgery. *Sci. Rep.* **2020**, *10*, 1–8. [CrossRef]
178. Singh, P.P.; Zeng, I.S.; Srinivasa, S.; Lemanu, D.P.; Connolly, A.B.; Hill, A.G. Systematic review and meta-analysis of use of serum C-reactive protein levels to predict anastomotic leak after colorectal surgery. *BJS* **2014**, *101*, 339–346. [CrossRef]
179. Stephensen, B.; Reid, F.; Shaikh, S.; Carroll, R.; Smith, S.; Pockney, P. C-reactive protein trajectory to predict colorectal anastomotic leak. *Br. J. Surg.* **2020**, *107*, 1832–1837. [CrossRef] [PubMed]

180. Hübner, M.; Mantziari, S.; Demartines, N.; Pralong, F.; Coti-Bertrand, P.; Schäfer, M. Postoperative Albumin Drop Is a Marker for Surgical Stress and a Predictor for Clinical Outcome: A Pilot Study. *Gastroenterol. Res. Pract.* **2016**, *2016*, 1–8. [CrossRef]
181. Lago, V.; Fotopoulou, C.; Chiantera, V.; Minig, L.; Gil-Moreno, A.; Cascales-Campos, P.; Jurado, M.; Tejerizo, A.; Padilla-Iserte, P.; Malune, M.; et al. Risk factors for anastomotic leakage after colorectal resection in ovarian cancer surgery: A multi-centre study. *Gynecol. Oncol.* **2019**, *153*, 549–554. [CrossRef]
182. Ge, X.; Cao, Y.; Wang, H.; Ding, C.; Tian, H.; Zhang, X.; Gong, J.; Zhu, W.; Li, N. Diagnostic accuracy of the postoperative ratio of C-reactive protein to albumin for complications after colorectal surgery. *World J. Surg. Oncol.* **2017**, *15*, 1–7. [CrossRef]
183. Yu, Y.; Wu, Z.; Shen, Z.; Cao, Y. Preoperative C-reactive protein-to-albumin ratio predicts anastomotic leak in elderly patients after curative colorectal surgery. *Cancer Biomark.* **2020**, *27*, 295–302. [CrossRef]
184. Maruna, P.; Nedelníková, K.; Gürlich, R. Physiology and genetics of procalcitonin. *Physiol. Res.* **2000**, *49*, S57–S62. [PubMed]
185. Zielińska-Borkowska, U.; Dib, N.; Tarnowski, W.; Skirecki, T. Monitoring of procalcitonin but not interleukin-6 is useful for the early prediction of anastomotic leakage after colorectal surgery. *Clin. Chem. Lab. Med.* **2017**, *55*, 1053–1059. [CrossRef] [PubMed]
186. Garcia-Granero, A.; Frasson, M.; Flor-Lorente, B.; Blanco, F.; Puga, R.; Carratalá, A.; Garcia-Granero, E. Procalcitonin and C-reactive protein as early predictors of anastomotic leak in colorectal surgery: A prospective observational study. *Dis. Colon Rectum* **2013**, *56*, 475–483. [CrossRef] [PubMed]
187. Muñoz, J.L.; Alvarez, M.O.; Cuquerella, V.; Miranda, E.; Picó, C.; Flores, R.; Resalt-Pereira, M.; Moya, P.; Pérez, A.; Arroyo, A. Procalcitonin and C-reactive protein as early markers of anastomotic leak after laparoscopic colorectal surgery within an enhanced recovery after surgery (ERAS) program. *Surg. Endosc.* **2018**, *32*, 4003–4010. [CrossRef] [PubMed]
188. Dinarello, A.C. Role of pro- and anti-inflammatory cytokines during inflammation: Experimental and clinical findings. *J. Boil. Regul. Homeost. Agents* **1997**, *11*, 91–103.
189. Lin, E.; Calvano, S.E.; Lowry, S.F. Inflammatory cytokines and cell response in surgery. *Surgery* **2000**, *127*, 117–126. [CrossRef] [PubMed]
190. Hack, C.E.; Aarden, L.A.; Thus, L.G. Role of Cytokines in Sepsis. *Dev. Funct. Myeloid Subsets* **1997**, *66*, 101–195. [CrossRef]
191. Tsukada, K.; Katoh, H.; Shiojima, M.; Suzuki, T.; Takenoshita, S.; Nagamachi, Y. Concentrations of cytokines in peritoneal fluid after abdominal surgery. *Eur. J. Surg.* **1993**, *159*, 475–479. [PubMed]
192. Tsukada, K.; Takenoshita, S.-I.; Nagamachi, Y. Peritoneal interleukin-6, interleukin-8 and granulocyte elastase activity after elective abdominal surgery. *APMIS* **1994**, *102*, 837–840. [CrossRef] [PubMed]
193. Baker, E.A.; Gaddal, S.E.-; Aitken, D.G.; Leaper, D.J. Growth factor profiles in intraperitoneal drainage fluid following colorectal surgery: Relationship to wound healing and surgery. *Wound Repair Regen.* **2003**, *11*, 261–267. [CrossRef]
194. Wiik, H.; Karttunen, R.; Haukipuro, K.; Syrjälä, H. Maximal local and minimal systemic cytokine response to colorectal surgery: The influence of perioperative filgrastim. *Cytokine* **2001**, *14*, 188–192. [CrossRef]
195. Jansson, K.; Redler, B.; Truedsson, L.; Magnuson, A.; Matthiessen, P.; Andersson, M.; Norgren, L. Intraperitoneal cytokine response after major surgery: Higher postoperative intraperitoneal versus systemic cytokine levels suggest the gastrointestinal tract as the major source of the postoperative inflammatory reaction. *Am. J. Surg.* **2004**, *187*, 372–377. [CrossRef] [PubMed]
196. Herwig, R.; Glodny, B.; Kühle, C.; Schlüter, B.; Brinkmann, O.A.; Strasser, H.; Senninger, N.; Winde, G. Early Identification of Peritonitis by Peritoneal Cytokine Measurement. *Dis. Colon Rectum* **2002**, *45*, 514–521. [CrossRef] [PubMed]
197. Uğraş, B.; Giriş, M.; Erbil, Y.; Gökpınar, M.; Çıtlak, G.; İşsever, H.; Bozbora, A.; Öztezcan, S. Early prediction of anastomotic leakage after colorectal surgery by measuring peritoneal cytokines: Prospective study. *Int. J. Surg.* **2008**, *6*, 28–35. [CrossRef] [PubMed]
198. Sammour, T.; Singh, P.P.; Zargar-Shoshtari, K.; Su'A, B.; Hill, A.G. Peritoneal Cytokine Levels Can Predict Anastomotic Leak on the First Postoperative Day. *Dis. Colon Rectum* **2016**, *59*, 551–556. [CrossRef]
199. Rettig, T.C.D.; Verwijmeren, L.; Dijkstra, I.M.; Boerma, D.; van de Garde, E.M.W.; Noordzij, P.G. Postoperative Interleukin-6 Level and Early Detection of Complications After Elective Major Abdominal Surgery. *Ann. Surg.* **2016**, *263*, 1207–1212. [CrossRef] [PubMed]
200. Fouda, E.; El Nakeeb, A.; Magdy, A.; Hammad, E.A.; Othman, G.; Farid, M. Early Detection of Anastomotic Leakage after Elective Low Anterior Resection. *J. Gastrointest. Surg.* **2010**, *15*, 137–144. [CrossRef] [PubMed]
201. Yamamoto, T.; Umegae, S.; Matsumoto, K.; Saniabadi, A.R. Peritoneal cytokines as early markers of peritonitis following surgery for colorectal carcinoma: A prospective study. *Cytokine* **2011**, *53*, 239–242. [CrossRef]
202. Bertram, P.; Junge, K.; Schachtrupp, A.; Götze, C.; Kunz, D.; Schumpelick, V. Peritoneal release of TNFα and IL-6 after elective colorectal surgery and anastomotic leakage. *J. Investig. Surg.* **2003**, *16*, 65–69. [CrossRef]
203. Alonso, S.; Pascual, M.; Salvans, S.; Mayol, X.; Mojal, S.; Gil, M.; Grande, L.; Pera, M. Postoperative intra-abdominal infection and colorectal cancer recurrence: A prospective matched cohort study of inflammatory and angiogenic responses as mechanisms involved in this association. *Eur. J. Surg. Oncol.* **2015**, *41*, 208–214. [CrossRef]
204. Zawadzki, M.; Krzystek-Korpacka, M.; Gamian, A.; Witkiewicz, W. Serum cytokines in early prediction of anastomotic leakage following low anterior resection. *Videosurg. Other Miniinvasive Tech.* **2018**, *13*, 33–43. [CrossRef]
205. Fagerhol, M.K. Calprotectin, a faecal marker of organic gastrointestinal abnormality. *Lancet* **2000**, *356*, 1783–1784. [CrossRef]
206. Cikot, M.; Kones, O.; Gedikbası, A.; Kocatas, A.; Karabulut, M.; Temizgonul, K.B.; Alis, H.; Information, P.E.K.F.C. The marker C-reactive protein is helpful in monitoring the integrity of anastomosis: Plasma calprotectin. *Am. J. Surg.* **2016**, *212*, 53–61. [CrossRef] [PubMed]

207. Morandi, E.; Monteleone, M.; Merlini, D.; Vignati, G.; D'Aponte, T.; Castoldi, M. P0071 Faecal calprotectin as an early biomarker of colorectal anastomotic leak. *Eur. J. Cancer* **2015**, *51*, e15. [CrossRef]
208. Popescu, G.A.; Jung, I.; Cordoş, B.A.; Azamfirei, L.; Huţanu, A.; Gurzu, S. Intestinal fatty acid-binding protein, as a marker of anastomotic leakage after colonic resection in rats. *Rom. J. Morphol. Embryol.* **2018**, *59*, 1075–1081. [PubMed]
209. Plat, V.D.; Derikx, J.P.M.; Jongen, A.C.; Nielsen, K.; Sonneveld, D.J.A.; Tersteeg, J.J.C.; Crolla, R.M.P.H.; van Dam, D.A.; Cense, H.A.; de Meij, T.G.J.; et al. Diagnostic accuracy of urinary intestinal fatty acid binding protein in detecting colorectal anastomotic leakage. *Tech. Coloproctol.* **2020**, *24*, 449–454. [CrossRef]
210. Miller, K.; Arrer, E.; Leitner, C. Early detection of anastomotic leaks after low anterior resection of the rectum. *Dis. Colon Rectum* **1996**, *39*, 1081–1085. [CrossRef] [PubMed]
211. Murr, C.; Widner, B.; Wirleitner, B.; Fuchs, D. Neopterin as a Marker for Immune System Activation. *Curr. Drug Metab.* **2002**, *3*, 175–187. [CrossRef]
212. Baydar, T.; Yuksel, O.; Sahin, T.T.; Dikmen, K.; Girgin, G.; Sipahi, H.; Kurukahvecioglu, O.; Bostanci, H.; Şare, M. Neopterin as a prognostic biomarker in intensive care unit patients. *J. Crit. Care* **2009**, *24*, 318–321. [CrossRef]
213. Girgin, G.; Sahin, T.T.; Fuchs, D.; Yuksel, O.; Kurukahvecioglu, O.; Şare, M.; Baydar, T. Tryptophan degradation and serum neopterin concentrations in intensive care unit patients. *Toxicol. Mech. Methods* **2010**, *21*, 231–235. [CrossRef]
214. Dusek, T.; Örhalmi, J.; Sotona, O.; Krčmová, L.K.; Javorska, L.; Dolejs, J.; Páral, J. Neopterin, kynurenine and tryptophan as new biomarkers for early detection of rectal anastomotic leakage. *Videosurg. Other Miniinvasive Tech.* **2018**, *13*, 44–52. [CrossRef]
215. Käser, S.A.; Furler, R.; Evequoz, D.C.; Maurer, C.A. Hyponatremia Is a Specific Marker of Perforation in Sigmoid Diverticulitis or Appendicitis in Patients Older than 50 Years. *Gastroenterol. Res. Pract.* **2013**, *2013*, 1–4. [CrossRef]
216. Swart, R.M.; Hoorn, E.J.; Betjes, M.G.; Zietse, R. Hyponatremia and Inflammation: The Emerging Role of Interleukin-6 in Osmoregulation. *Nephron* **2011**, *118*, p45–p51. [CrossRef]
217. Sharshar, T.; Blanchard, A.; Paillard, M.; Raphael, J.C.; Gajdos, P.; Annane, D. Circulating vasopressin levels in septic shock. *Crit. Care Med.* **2003**, *31*, 1752–1758. [CrossRef]
218. Käser, S.A.; Nitsche, U.; Maak, M.; Michalski, C.W.; Späth, C.; Müller, T.C.; Maurer, C.A.; Janssen, K.P.; Kleeff, J.; Friess, H.; et al. Could hyponatremia be a marker of anastomotic leakage after colorectal surgery? A single center analysis of 1,106 patients over 5 years. *Langenbeck's Arch. Surg.* **2014**, *399*, 783–788. [CrossRef] [PubMed]
219. Clark, I.M.; Swingler, T.E.; Sampieri, C.L.; Edwards, D. The regulation of matrix metalloproteinases and their inhibitors. *Int. J. Biochem. Cell Biol.* **2008**, *40*, 1362–1378. [CrossRef]
220. Chakraborti, S.; Mandal, M.; Das, S.; Mandal, A.; Chakraborti, T. Regulation of matrix metalloproteinases: An overview. *Mol. Cell. Biochem.* **2003**, *253*, 269–285. [CrossRef] [PubMed]
221. Braskén, P.; Renvall, S.; Sandberg, M. Fibronectin and collagen gene expression in healing experimental colonic anastomoses. *BJS* **1991**, *78*, 1048–1052. [CrossRef] [PubMed]
222. Kiyama, T.; Onda, M.; Tokunaga, A.; Efron, D.T.; Barbul, A. Effect of matrix metalloproteinase inhibition on colonic anastomotic healing in rats. *J. Gastrointest. Surg.* **2001**, *5*, 303–311. [CrossRef]
223. De Hingh, I.H.J.T.; Siemonsma, M.A.; De Man, B.M.; Lomme, R.M.L.M.; Hendriks, T. The matrix metalloproteinase inhibitor BB-94 improves the strength of intestinal anastomoses in the rat. *Int. J. Colorectal Dis.* **2002**, *17*, 348–354. [CrossRef]
224. Krarup, P.-M.; Eld, M.; Jorgensen, L.N.; Hansen, M.B.; Ågren, M.S. Selective matrix metalloproteinase inhibition increases breaking strength and reduces anastomotic leakage in experimentally obstructed colon. *Int. J. Colorectal Dis.* **2017**, *32*, 1277–1284. [CrossRef]
225. Krarup, P.-M.; Eld, M.; Heinemeier, K.M.; Jorgensen, L.N.; Hansen, M.B.; Ågren, M.S. Expression and inhibition of matrix metalloproteinase (MMP)-8, MMP-9 and MMP-12 in early colonic anastomotic repair. *Int. J. Colorectal Dis.* **2013**, *28*, 1151–1159. [CrossRef] [PubMed]
226. Stumpf, M.; Klinge, U.; Wilms, A.; Zabrocki, R.; Rosch, R.; Junge, K.; Krones, C.; Schumpelick, V. Changes of the extracellular matrix as a risk factor for anastomotic leakage after large bowel surgery. *Surgery* **2005**, *137*, 229–234. [CrossRef]
227. Baker, E.A.; Leaper, D.J. Profiles of matrix metalloproteinases and their tissue inhibitors in intraperitoneal drainage fluid: Relationship to wound healing. *Wound Repair Regen.* **2003**, *11*, 268–274. [CrossRef] [PubMed]
228. Pasternak, B.; Matthiessen, P.; Jansson, K.; Andersson, M.; Aspenberg, P. Elevated intraperitoneal matrix metalloproteinases-8 and -9 in patients who develop anastomotic leakage after rectal cancer surgery: A pilot study. *Colorectal Dis.* **2010**, *12*, 93–98. [CrossRef] [PubMed]
229. Edomskis, P.; Goudberg, M.R.; Sparreboom, C.L.; Menon, A.G.; Wolthuis, A.M.; D'Hoore, A.; Lange, J.F. Matrix metalloproteinase-9 in relation to patients with complications after colorectal surgery: A systematic review. *Int. J. Colorectal Dis.* **2021**, *36*, 1–10. [CrossRef] [PubMed]
230. De Ruiter, J.; Weel, J.; Manusama, E.; Kingma, W.P.; Van Der Voort, P.H.J. The Epidemiology of Intra-Abdominal Flora in Critically Ill Patients with Secondary and Tertiary Abdominal Sepsis. *Infection* **2009**, *37*, 522–527. [CrossRef]
231. Brook, I.; Frazier, E.H. Aerobic and anaerobic microbiology in intra-abdominal infections associated with diverticulitis. *J. Med. Microbiol.* **2000**, *49*, 827–830. [CrossRef] [PubMed]
232. Komen, N.; Morsink, M.; Beiboer, S.; Miggelbrink, A.; Willemsen, P.; Van Der Harst, E.; Lange, J.; Van Leeuwen, W. Detection of colon flora in peritoneal drain fluid after colorectal surgery: Can RT-PCR play a role in diagnosing anastomotic leakage? *J. Microbiol. Methods* **2009**, *79*, 67–70. [CrossRef]

233. Komen, N.; Slieker, J.; Willemsen, P.; Mannaerts, G.; Pattyn, P.; Karsten, T.; de Wilt, H.; van der Harst, E.; van Leeuwen, W.; Decaestecker, C.; et al. Polymerase chain reaction for Enterococcus faecalis in drain fluid: The first screening test for symptomatic colorectal anastomotic leakage. The Appeal-study: Analysis of Parameters Predictive for Evident Anastomotic Leakage. *Int. J. Colorectal Dis.* **2014**, *29*, 15–21. [CrossRef]
234. Pakula, M.; Tanase, D.; Kraal, K.; De Graaf, G.; Lange, J.; French, P. Optical Measurements on Drain Fluid for the Detection of Anastomotic Leakage. In Proceedings of the 2005 3rd IEEE/EMBS Special Topic Conference on Microtechnology in Medicine and Biology, Oahu, HI, USA, 12–15 May 2005; pp. 72–75.
235. Muhammed, K.O.; Özener, Ç.; Akoglu, E. Diagnostic Value of Effluent Endotoxin Level in Gram-Negative Peritonitis in Capd Patients. *Perit. Dial. Int.* **2001**, *21*, 154–158. [CrossRef]
236. Junger, W.G.; Miller, K.; Bahrami, S.; Redl, H.; Schlag, G.; Moritz, E. Early detection of anastomotic leaks after colorectal surgery by measuring endotoxin in the drainage fluid. *Hepatogastroenterology* **1996**, *43*, 1523–1529. [PubMed]
237. Jansson, K.; Strand, I.; Redler, B.; Magnuson, A.; Ungerstedt, U.; Norgren, L. Results of intraperitoneal microdialysis depend on the location of the catheter. *Scand. J. Clin. Lab. Investig.* **2004**, *64*, 63–70. [CrossRef] [PubMed]
238. Merad, F.; Yahchouchi, E.; Hay, J.-M.; Fingerhut, A.; Laborde, Y.; Langlois-Zantain, O. Prophylactic Abdominal Drainage After Elective Colonic Resection and Supraporomontory AnastomosisA Multicenter Study Controlled by Randomization. *Arch. Surg.* **1998**, *133*, 309–314. [CrossRef] [PubMed]
239. Petrowsky, H.; Demartines, N.; Rousson, V.; Clavien, P.-A. Evidence-based value of prophylactic drainage in gastrointestinal surgery: A systematic review and meta-analyses. *Ann. Surg.* **2004**, *240*, 1074. [CrossRef] [PubMed]
240. Jesus, E.; Karliczek, A.; Matos, D.; Castro, A.; Atallah, A. Prophylactic anastomotic drainage for colorectal surgery. *Cochrane Database Syst. Rev.* **2004**, CD002100. [CrossRef]

Article

Identification of Novel Mutations in Colorectal Cancer Patients Using AmpliSeq Comprehensive Cancer Panel

Bader Almuzzaini [1,*,†], Jahad Alghamdi [2,†], Alhanouf Alomani [3], Saleh AlGhamdi [4], Abdullah A. Alsharm [5], Saeed Alshieban [6], Ahood Sayed [2], Abdulmohsen G. Alhejaily [7], Feda S. Aljaser [8], Manal Abudawood [8], Faisal Almajed [9], Abdulhadi Samman [10], Mohammed A. Al Balwi [1] and Mohammad Azhar Aziz [11,*]

1. King Abdullah International Medical Research Center, Medical Genomics Research Department, Ministry of National Guard Health Affairs, King Saud Bin Abdulaziz University for Health Sciences, Riyadh 11481, Saudi Arabia; BalwiM@NGHA.MED.SA
2. King Abdullah International Medical Research Center, Saudi Biobank, King Saud Bin Abdulaziz University for Health Sciences, Ministry of National Guard Health Affairs, Riyadh 11481, Saudi Arabia; alghamdija@ngha.med.sa (J.A.); SayedA@NGHA.MED.SA (A.S.)
3. Department of Pathology, College of Medicine, Imam Mohammad Ibn Saud Islamic University (IMSIU), Riyadh 13318, Saudi Arabia; Alhanoufomani@hotmail.com
4. Clinical Research Department, Research Center, King Fahad Medical City, Riyadh 11564, Saudi Arabia; saghamdi@kfmc.med.sa
5. Comprehensive Cancer Center, King Fahad Medical City, Riyadh 11564, Saudi Arabia; aalsharm@kfmc.med.sa
6. King Abdul Aziz Medical City-National Guard Health Affairs (NGHA), King Abdullah International Medical Research Center, King Saud Bin Abdul Aziz University for Health Sciences (KSAU-HS), Riyadh 14611, Saudi Arabia; shieban@hotmail.com
7. Faculty of Medicine, King Fahad Medical City, Riyadh 11564, Saudi Arabia; aalhejaily@kfmc.med.sa
8. Department of Clinical Laboratory Sciences, Chair of Medical and Molecular Genetics Research, College of Applied Medical Sciences, King Saud University Riyadh, Riyadh 11564, Saudi Arabia; faljaser@ksu.edu.sa (F.S.A.); mabudawood@ksu.edu.sa (M.A.)
9. Department of Clinical Laboratory Sciences, College of Applied Medical Sciences, King Saud Bin Abdulaziz University for Health Sciences, Ministry of National Guard Health Affairs, Riyadh 11481, Saudi Arabia; Faisal.almajed@gmail.com
10. Department of Pathology, Faculty of Medicine, University of Jeddah, Jeddah 23218, Saudi Arabia; dr.a.samman@hotmail.com
11. King Abdullah International Medical Research Center, Colorectal Cancer Research Program, Department of Cellular Therapy and Cancer Research, Ministry of National Guard Health Affairs, King Saud Bin Abdulaziz University for Health Sciences, Riyadh 11481, Saudi Arabia
* Correspondence: MuzainiB@NGHA.MED.SA (B.A.); azizmo@ngha.med.sa (M.A.A.); Tel.: +966-11-429-4533 (B.A.); +966-11-429-4582 (M.A.A.)
† Contributed equally.

Citation: Almuzzaini, B.; Alghamdi, J.; Alomani, A.; AlGhamdi, S.; Alsharm, A.A.; Alshieban, S.; Sayed, A.; Alhejaily, A.G.; Aljaser, F.S.; Abudawood, M.; et al. Identification of Novel Mutations in Colorectal Cancer Patients Using AmpliSeq Comprehensive Cancer Panel. *J. Pers. Med.* **2021**, *11*, 535. https://doi.org/10.3390/jpm11060535

Academic Editor: James Meehan

Received: 14 April 2021
Accepted: 30 May 2021
Published: 9 June 2021

Publisher's Note: MDPI stays neutral with regard to jurisdictional claims in published maps and institutional affiliations.

Copyright: © 2021 by the authors. Licensee MDPI, Basel, Switzerland. This article is an open access article distributed under the terms and conditions of the Creative Commons Attribution (CC BY) license (https://creativecommons.org/licenses/by/4.0/).

Abstract: Biomarker discovery would be an important tool in advancing and utilizing the concept of precision and personalized medicine in the clinic. Discovery of novel variants in local population provides confident targets for developing biomarkers for personalized medicine. We identified the need to generate high-quality sequencing data from local colorectal cancer patients and understand the pattern of occurrence of variants. In this report, we used archived samples from Saudi Arabia and used the AmpliSeq comprehensive cancer panel to identify novel somatic variants. We report a comprehensive analysis of next-generation sequencing results with a coverage of >300X. We identified 466 novel variants which were previously unreported in COSMIC and ICGC databases. We analyzed the genes associated with these variants in terms of their frequency of occurrence, probable pathogenicity, and clinicopathological features. Among pathogenic somatic variants, 174 were identified for the first time in the large intestine. APC, RET, and EGFR genes were most frequently mutated. A higher number of variants were identified in the left colon. Occurrence of variants in ERBB2 was significantly correlated with those of EGFR and ATR genes. Network analyses of the identified genes provide functional perspective of the identified genes and suggest affected pathways and probable biomarker candidates. This report lays the ground work for biomarker discovery and identification of driver gene mutations in local population.

Keywords: colorectal cancer; personalized medicine; biomarker; AmpliSeq

1. Introduction

Colorectal cancer (CRC) is a heterogeneous disease. Inter-patient heterogeneity has been one of the major obstacles towards developing therapeutic strategies. Different populations have been found to show varied response towards standard of care regimens [1]. This variation has largely been attributed to the difference in underlying gene mutations and genetic changes which determines the progression of CRC. CRC progresses with continuing accumulation of genomic and epi-genomic alterations, which eventually induce oncogenic transformation of the normal colon cell into tumor cells followed by metastasis. Pathways responsible to initiate CRC are well known, based on the evidence of mutations and chromosomal changes observed in patients. The mechanistic role of signaling pathways in causing CRC has constantly been enriched with better understanding of the underlying gene mutations. These gene mutations have been used as biomarkers to predict disease progression and outcome of therapeutic regimens.

KRAS mutation status is routinely used for administering antibodies to inhibit epidermal growth factor receptor (EGFR). Successful use of these antibodies (cetuximab and panitumumab), only in KRAS wild-type patients, had set the stage of precision and personalized medicine. However, not all patients with the wild-type KRAS gene respond to anti-EGFR therapy. Therefore, there is a pertinent need to identify biomarkers that can capture the population heterogeneity and facilitate the practice of precision and personalized medicine. Earlier studies have taken up population-based mutational profiling of CRC to develop the concept of precision medicine [2,3]. Population-specific mutational analysis of colorectal cancer is scarce in Saudi Arabia but highly pertinent to develop the precision and personalized medicine paradigm [4–6]. With the technological advancement in detecting mutations at an unprecedented scale, the possibility of practicing precision medicine through biomarkers has further increased. Precision medicine in colorectal cancer is more relevant than in other cancers, owing to its heterogeneity in development as well as response to therapy. Colorectal cancer metastasis has been largely unimproved with new developments in therapeutics, and precision medicine holds optimism especially for these patients [7,8]. There is better precision and accuracy in detecting mutations in patients that can be used as predictive and prognostic biomarkers. Use of non-invasive biomarkers for colorectal cancers is one of the most promising strategies in treating CRC [9].

Next-generation sequencing (NGS) technology can be used with DNA-enrichment methods to generate deep sequencing of target genes or genomic regions of interest, such as the exome or identified cancer hotspots. For the targeted detection of mutations in known cancer genes, a comprehensive cancer panel (IonAmpliSeq) is available. Gene panels allow simultaneous detection of relevant mutations with unprecedented accuracy and sensitivity. This comprehensive cancer panel (CCP) is designed to target coding DNA sequences (CDS) and splice variants from 409 tumor suppressor and onco-genes that are frequently mutated. The requirement of small amount of input DNA (only 40 ng) per reaction enables challenging analysis of formalin-fixed, paraffin-embedded (FFPE) tissues. The use of the IonAmpliSeq™ Cancer Panel and NGS using the IonTorrent platform provides a fast, easy, and cost-effective sequencing workflow for detecting genomic hotspot regions that are frequently mutated in human cancer. A previous study from Jeddah, Saudi Arabia used the Ion AmpliSeq™ Cancer Hotspot Panel v2 which spans only 50 frequently mutated genes [10].

In this study, we used IonAmpliSeq™ CCP to sequence samples from 99 archived patient samples from two hospitals in Riyadh, Saudi Arabia, over a duration of two years. The confirmation of well-known mutations point towards the chromosomal instability pathway as the predominant mechanism of the development of CRC in this cohort. We provide comprehensive analyses of novel variants that can be useful for biomarker discovery and identification of driver genes. Discovery of biomarkers and identification of driver genes from local population is critical in developing precision and personalized medicine approach towards addressing colorectal cancer.

2. Materials and Methods

2.1. Patient Description and Sample Collection

A total of 100 patient tumor samples were retrospectively recruited in this study, and, after exclusion of 1 sample due to low DNA quality, we sequenced 99 samples, and clinicopathological characteristics were available from 95 patients. Sequencing data from 90 of these samples qualified for coverage requirement and was used for further analyses. All samples were collected in the period between 2016 and 2018 at King Abdulaziz Medical City (KAMC) and King Fahad Medical City (KFMC), Riyadh, Saudi Arabia. All samples were diagnosed as primary colorectal adenocarcinoma at histopathology level. Patients were excluded if: (i) they had been treated with chemotherapy or radiotherapy prior to tumor resection, (ii) they had familial adenomatous polyposis (FAP) or hereditary non-polyposis colorectal cancer (HNPCC), or (iii) the formalin-fixed, paraffin-embedded tissue (FFPE) samples, patients' clinical and pathological data, or written informed consent form signed by patient to access the archival samples were not available.

The formalin-fixed, paraffin-embedded tissues (FFPE) blocks from patients with colorectal adenocarcinomas were retrieved from the archives of the Department of Anatomical Pathology Laboratory in KAMC and KFMC. All slides were revised and marked by a histopathologist before DNA extraction. We selected only marked tissue with tumor percentage more than 40% and used 1–2 slides for extraction based on tissue size. Chart reviews were done after obtaining the ethical approval to collect the demographic and clinicopathological features from the hospital information system BESTCare 2.0 A at KAMC, including age at diagnosis, gender, tumor stage, site, and metastasis grade.

2.2. Ethical Approval

Full Institutional Review Board (IRB) approval was given by King Abdullah International Medical Research center (KAIMRC), Ministry of National Guard, Health Affairs (IRB protocol #RC13/249/R). All patients' data were secured and accessed only by research investigators.

2.3. DNA Extraction

Genomic DNA was extracted from FFPE samples that were assessed by a pathologist to select the appropriate block to assure presence of colorectal cancer cells and excluded the insufficient necrotic tissue for NGS. DNA was extracted either from slide sample using Ion AmpliSeq ™ Direct FFPE DNA Kit (Thermo Fisher Scientific Inc, Toronto, ON, Canada) according to the manufacturer's instructions. In case of FFPE block samples, DNA was extracted from FFPE blocks using 8μm of tissue ribbon using QIAamp DNA FFPE Tissue Kit (QIAGEN) following manufacturer's instruction. Measurement of the DNA quality and concentration was done by using Qubit® 3.0 Fluorometer (Life Technologies).

2.4. Comprehensive Cancer Panel (CCP) and Data Availability

The pre-designed comprehensive cancer panel (CCP) from Ion AmpliSeq™ (Life Technologies) was used. This panel comprises 16,000 primer pairs in four primer pools for 409 genes, which covers approximately 15,749 somatic mutations reported in The Catalogue of Somatic Mutations in Cancer (COSMIC). For the complete list of 409 genes, see Supplementary Table S1. All sequencing data generated from 90 patients is deposited in SRA database (reference PRJNA685957, https://www.ncbi.nlm.nih.gov/sra/PRJNA685957 Last accessed on 1 June 2021)

2.5. Library Preparation and NGS Data Analysis

The library was constructed using Ion AmpliSeq™ (CCP) Library Kit 2.0 (Life Technologies) and Ion Xpress™ Barcode Adapter 1–16 Kit (Life Technologies) according to manufacturer's instructions. Library quantification was done using the Ion Library TaqMan Quantitation Kit (Life Technologies) following standard procedure available. The qualified library was sequenced by the use of Ion S5XL Semiconductor Sequencer following the manufacturer's user guide.

2.6. Variant Calling and Annotation

Variants were called by Torrent Suite Variant (TSV) (version 5.8) [11]. Variants with a coverage of more than 300X and read quality more than 50 were included in this study to enhance the quality of identified somatic variants. Variants that passed this quality metrics were annotated by using Ensemble Variant Effect Predictor (VEP) tool (version 102). This tool uses gnomAD (version r2.1) and the Catalogue Of Somatic Mutations (COSMIC) databases (version 90) [12]. We excluded common variants previously reported in Ensemble (v102) and only included variants classified as confirmed somatic or pathogenic by COSMIC database. This classification is based on functional analysis through hidden Markov models (FATHMM). Further, variants were classified into colorectal cancer associated or other organ sites. The potential damaging effect at protein level of the variants were assessed using prediction software using Sorting Intolerant From Tolerant (SIFT; v5.2.2) and Polymorphism Phenotyping v2 (PolyPhen2; v2.2.2) scores [13,14]. These scores predict the impact of detected missense variants on the human protein structure. All variants which showed as deleterious on SIFT and/or damaging on PolyPhen2 were included for downstream analysis.

2.7. Molecular Profiling and Statistical Analysis

Descriptive statistics were applied to summarize patient characteristics based on clinicopathological features. Summary statistics of the identified genetic variants were carried out in PLINK [15] to calculate the minor allele frequencies (MAF) and Hardy–Weinberg equilibrium p-value [3]. Associations between mutations and CRC or histological features were determined using Fisher's exact test. Due to the limited sample size, tumor stages were grouped into early (stage I–II) and late (stage III–IV). Site of tumor was classified as left, right, and others. The involvement of lymph nodes and secondary metastases were analyzed as a dichotomous trait. All analyses were conducted using JMP Prostatistical software (JMP®, Version 13. SAS Institute Inc., Cary, NC, 1989–2019). Sequence Kernel Association—Optimal unified test (SKAT-O) was used to perform gene-based association analysis [16]. The association of rare variants with tumor stage (defined as late versus early), gender (female versus male), age group (young < 50 years versus old), and tumor location (left versus right) was analyzed. The variants were weighted based on their allele frequency, where rare variants were assigned higher weight than common variants. To account for multiple testing, an adjusted p-value of 0.0001 was considered as a significant threshold, reflecting the Bonferroni correction of 409 genes.

2.8. Ingenuity Pathway Analysis

The networks for mutated genes were generated through the use of IPA (QIAGEN Inc., https://www.qiagenbio-informatics.com/products/ingenuity-pathway-analysis. Last accessed on 1 June 2021) [17]. Networks were created using following filter: Species = Human AND Disease = Cancer AND mutation = hemizygous OR in-frame OR gain-of-function OR frameshift OR missense OR homozygous OR null mutation OR silent OR heterozygous OR loss of function OR knockout OR nonsense. Two networks were generated—one with the 27 most frequently mutated genes and another with 75 genes harboring pathogenic mutations reported in large intestine. The connect function was used to investigate the known interactions among these genes. The overlay function was used to find the association of these genes with canonical pathways and finding candidate biomarkers.

3. Results

3.1. Cohort Characteristics

The baseline characteristics of the analyzed samples are shown in Table 1. The median age of patients was 62 years, with 58 of them being male (61%). According to TNM staging system, 65% of the patients were classified as T3, with 59% showing no spread to regional lymph node (T0), and 96% were without distant metastases. The highest proportion of patients were diagnosed as stage III (39%), and more tumors were located in left colon (52%).

Table 1. Clinicopathological features of CRC patients.

Age, Years (SD)	62 (14)
Male, n (%)	58 (61%)
Stage	
I, n (%)	17 (18%)
II, n (%)	32 (34%)
III, n (%)	37 (39%)
IV, n (%)	8 (9%)
Primary Tumor	
T1, n (%)	2 (2%)
T2, n (%)	18 (19%)
T3, n (%)	61 (65%)
T4, n (%)	13 (14%)
Lymph Node	
N0, n (%)	55 (59%)
N1, n (%)	33 (35%)
N2, n (%)	6 (6%)
Distant Metastasis	
M0, n (%)	90 (96%)
M1, n (%)	4 (4%)
Site	
Left colon, n (%)	47 (52%)
Right colon, n (%)	30 (33%)
Rectum, n (%)	13 (14%)

T, tumor; N, node (0, no nodes; 1, 1 node; 2, 2 nodes); M, metastasis (0, no metastasis; 1, metastasis); SD, standard deviation.

3.2. Novel Variants Identified in Colorectal Cancer Patient Cohort

From a panel of 409 genes, we identified 4256 variants. Among these, 483 variants were classified as novel, as they were not found in the COSMIC database. However, 17 of these variants were reported in the international cancer genome consortium (ICGC) database. All novel variants are provided in Supplementary Table S2. We checked for the probability of these variants to be germline by analyzing their variant allele frequency (VAF). A total of 69 variants presented in at least one patient with a VAF between 49–51 or 99–100, indicating that they could be germline mutations, which is also supported by the MAF (>1%) among 45 of these variants (Supplementary Table S3).

Among the 4256 identified variants, 299 variants were classified as pathogenic. A total of 174 variants from 299 pathogenic variants were found to be identified for the first time in the large intestine, representing novel variants in colorectal cancer (Figure 1 and Supplementary Table S4). We employed two different methods (SIFT and PolyPhen) for classifying 561 somatic variants. Both methods suggest the detected variants to be either synonymous (n = 240) or missense (n = 247) (Figure 2A). According to the PolyPhen scoring method, 143 mutations were predicted to be benign, and the rest could be pathogenic (Figure 2B). The SIFT prediction method also provided similar categorization, with 111 variants listed as tolerated, and 130 variants were classified as deleterious (Figure 2C).

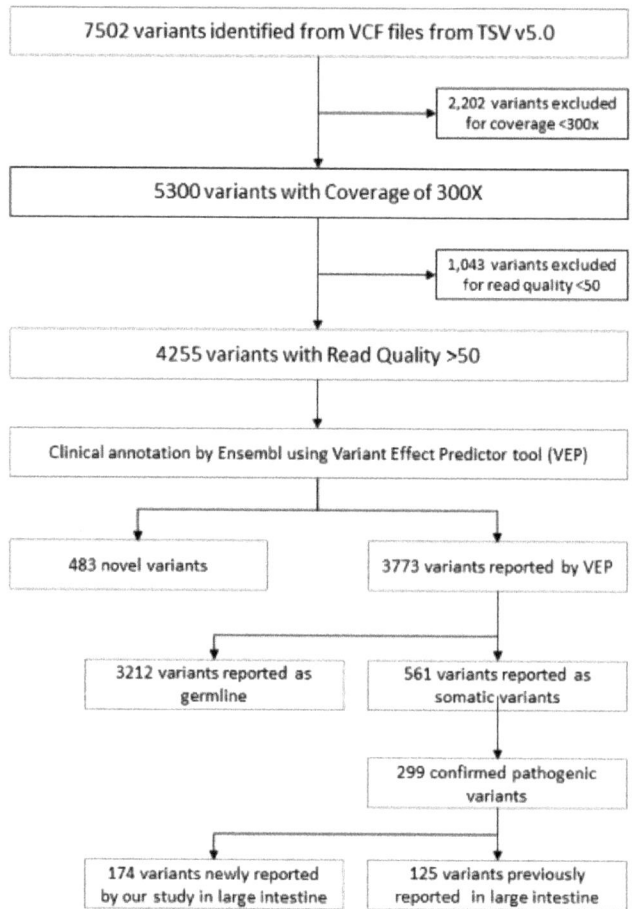

Figure 1. Variant filtration analysis workflow. Schematic illustration of variants identified in this study. A total of 483 novel variants were identified, and 561 somatic variants were observed. This study focused on pathogenic variants that were identified as novel in the large intestine.

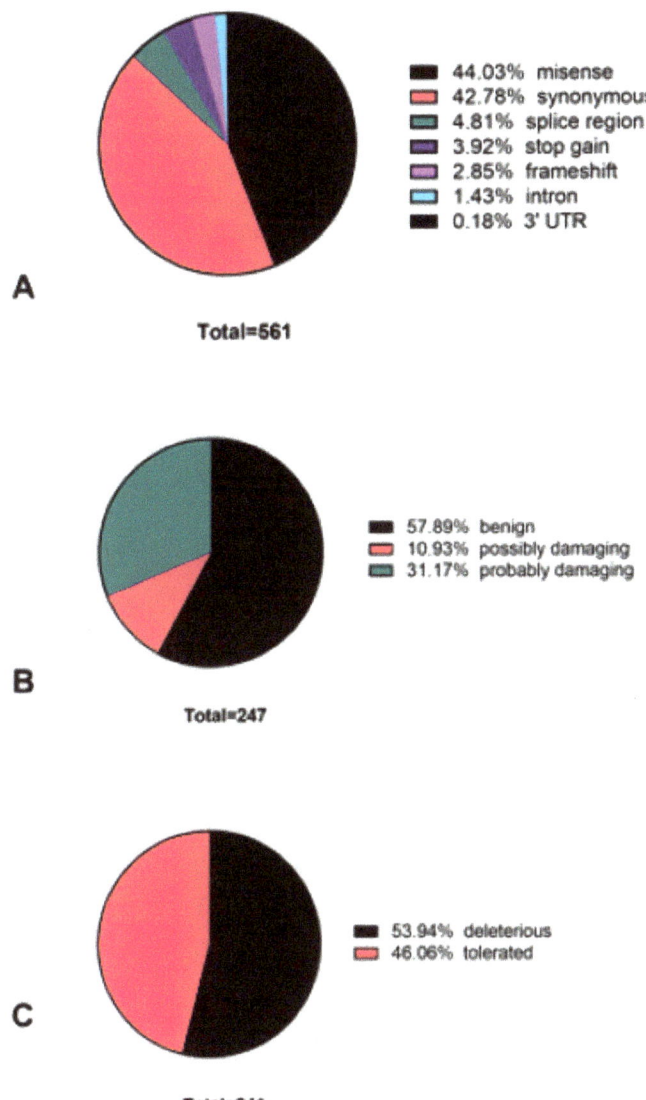

Figure 2. Classification of somatic variants. A total of 561 variants were classified by consequence (**A**), PolyPhen score (**B**), and SIFT score (**C**).

3.3. Novel Variants Identified in Most Commonly Mutated APC, RET, and EGFR Genes

The highest mutated genes (n = 20) among the patients were identified based on the presence of at least one confirmed pathogenic variant and arranged according to decreasing trend of frequency in the patient cohort (Table 2). A total of 96% of the patient samples had at least one confirmed pathogenic variant within the APC gene. We identified 5 novel (defined as previously unreported in COSMIC database) variants out of total 38 variants detected in the APC gene. These novel variants include c.1696G>A (p.V566I) missense mutation at exon 14, c.1697delT (p.V566X) frame shift mutation at exon 14, c.2680_2681delGTinsTA (p.Val894Ter) stop gain mutation at exon 16, c.3917delA (p.E1306X), frame shift mutation at exon 16, and c.4320-4341del ACCACCTCCTCAAACAGCTCAA

(p. PPPPQTAQ1440-1447X). A total of 23 of 38 variants were confirmed as somatic variants in COSMIC. Of the 23 variants, 15 were confirmed as pathogenic, and 12 were confirmed as tissue-specific pathogenic variants for the large intestine. RET gene mutations were found in 53% of the patient samples. Out of 13 detected variants in RET gene, 3 were somatic and one of was pathogenic (p.L769 = 0.53% patient samples harbored EGFR gene mutation. A total of 17 variants were detected, of which 6 were somatic, and 2 variants were specific for the large intestine. One of these variants was a high-impact nonsense mutation (p.R1068*). This comprehensive analysis and finding of novel variants within known genes could open up avenues to develop biomarkers that will be relevant for local population.

Table 2. List of twenty genes with variants in order of frequency in the sample cohort.

Gene	Description of Variants — Variants and Individuals for the Top 20 Genes									
	Variants								Individual [a]	
	Total	Novel	Pathogenic		Somatic	PolyPhen Damaging	SIFT Deleterious	Non-Synonymous	Pathogenic %	Somatic %
			All Tissue	Tissue-Specific						
APC	38	5	15	12	23	1	2	30	0.96	0.99
RET	13	0	2	0	3	3	3	6	0.53	0.83
EGFR	17	0	4	2	6	0	0	7	0.53	0.69
LRP1B	70	12	10	1	22	3	1	48	0.52	0.86
ERBB2	14	3	3	3	4	1	2	10	0.51	0.52
ATR	31	6	4	2	11	1	0	19	0.46	0.68
CSMD3	42	5	4	0	8	5	3	34	0.44	0.52
RALGDS	13	1	2	1	3	1	2	6	0.36	0.42
HIF1A	8	0	3	0	3	0	0	6	0.36	0.36
FGFR3	18	2	2	2	3	1	0	11	0.33	0.34
KRAS	7	1	1	1	2	0	1	5	0.28	0.28
PIK3CG	14	1	4	1	5	1	4	7	0.22	0.23
TP53	24	1	16	16	18	1	9	23	0.21	0.31
HNF1A	13	0	5	3	4	0	0	9	0.20	0.27
PIK3R1	10	2	2	0	3	0	0	6	0.19	0.30
KDM6A	11	1	3	2	3	1	0	8	0.19	0.19
ATM	31	3	6	3	3	0	4	26	0.17	0.16
MLH1	9	1	2	1	3	0	1	8	0.16	0.17
PRDM1	5	0	2	0	2	0	1	3	0.16	0.16
JAK1	11	0	3	0	7	0	0	3	0.14	0.44

[a] Percentage of samples with at least one pathogenic or somatic variant within the gene.

3.4. Colorectal-Cancer-Specific Variants Mapped to Twenty-Seven Genes

The distribution of pathogenic mutations found in the large intestine across gender, age, tumor stage, site, lymph node, and metastasis is described in Figure 3. A total of 73 variants specific for colon and rectum were identified within 27 genes. Tissue-specific pathogenic variants in the studied population show that the APC gene was the highest mutated, with variants detected in 66% of the samples, followed by ERBB2 (51%), ATR (45%), EGFR (40%), and FGFR3 (30%) genes. It is known that APC gene mutation is the initial event in CRC progression and is well depicted in our results. We observed variants in APC, ATR, KRAS, ATM, and KIT genes in the left colon of young female patients (<50 years age) in stage 1. However, no mutation was observed in young male patients in the left colon in early stage (I and II), but mutations were observed in these patients in the right colon and rectum. This detailed catalogue of variants analyzed according to clinicopathological features could be further used for molecular classification of patients.

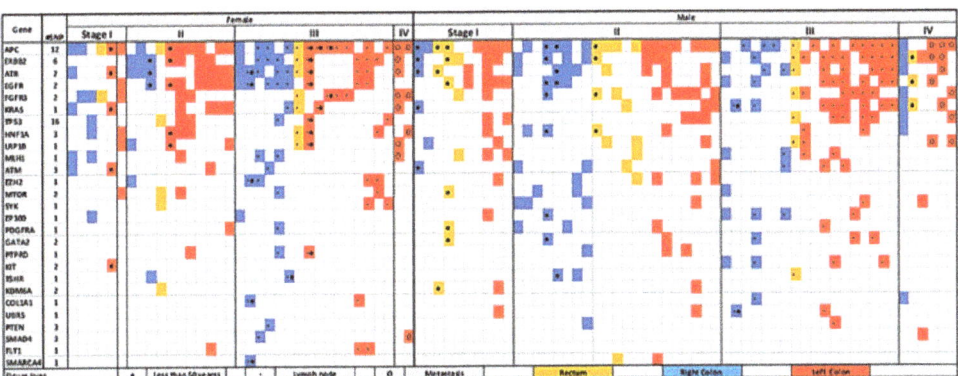

Figure 3. Mutation distribution based on gender, age, stage, site, lymph node, and metastasis. This figure shows only variants reported as pathogenic and located in the large intestine in COSMIC database and found in three individuals or more.

3.5. Left Colon Exhibits Higher Mutation Load

We identified 27 genes with at least one confirmed pathogenic variant presented in at least three patients. We found that patients with left side of the colon had higher prevalence of mutated genes, with the exception of ATR, MLH1, ATM, MTOR, PDGFRA, EP300, COL1A1, PTEN, and TSHR genes (Figure 4A). A significantly higher number of mutations were observed in FGFR3 gene in the left side and EP300 and TSHR genes on the right side of the colon. While comparing the early- and late-stage tumors, the prevalence of mutated genes were almost similar except for significantly higher COL1A1 gene mutations among patients in early stage when compared to late stage (Figure 4B).

3.6. Pathogenic Variants in ERBB2 Were Significantly Correlated with Mutations in EGFR and ATR

Gene correlation analysis showed that occurrence of pathogenic gene mutations was correlated (Figure 5). Presence of pathogenic variants at ERBB2 was significantly correlated with mutations in EGFR and ATR (r^2 = 0.39 and 0.26; *p*-values = 0.0001 and 0.01, respectively). High correlation was found between KDM6A and UBR5 gene mutations (r^2 = 0.47, *p*-value = 2.3×10^{-6}). FGFR3 gene was the most correlated. It was found positively correlated with HNF1A and TP53, whereas EGFR and ATR were negatively correlated.

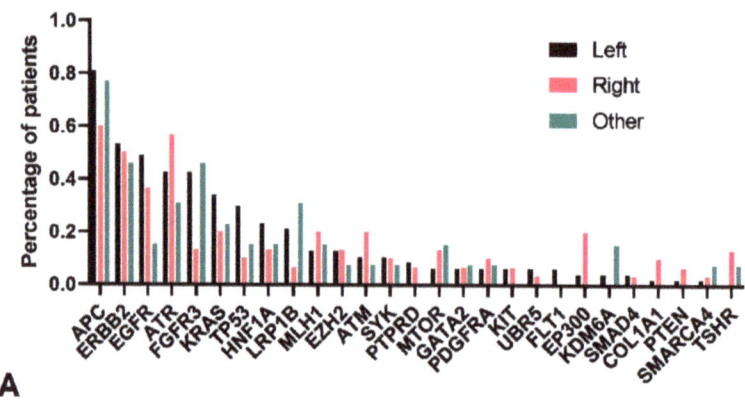

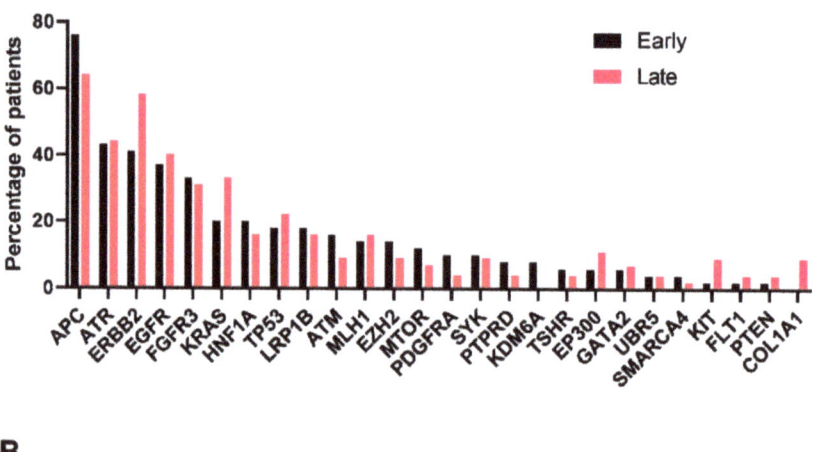

Figure 4. Frequency of variants in 27 genes among samples with at least one confirmed pathogenic variant for the large intestine. Frequency of variants based on tumor location (**A**) and stage (**B**). Y-axis denotes the number of samples with at least one confirmed pathogenic variant for the large intestine for that particular gene. For location, each bar is divided into left, right, and other categories, whereas for stage, they were grouped into early (stage I and II) and late (stage III and IV).

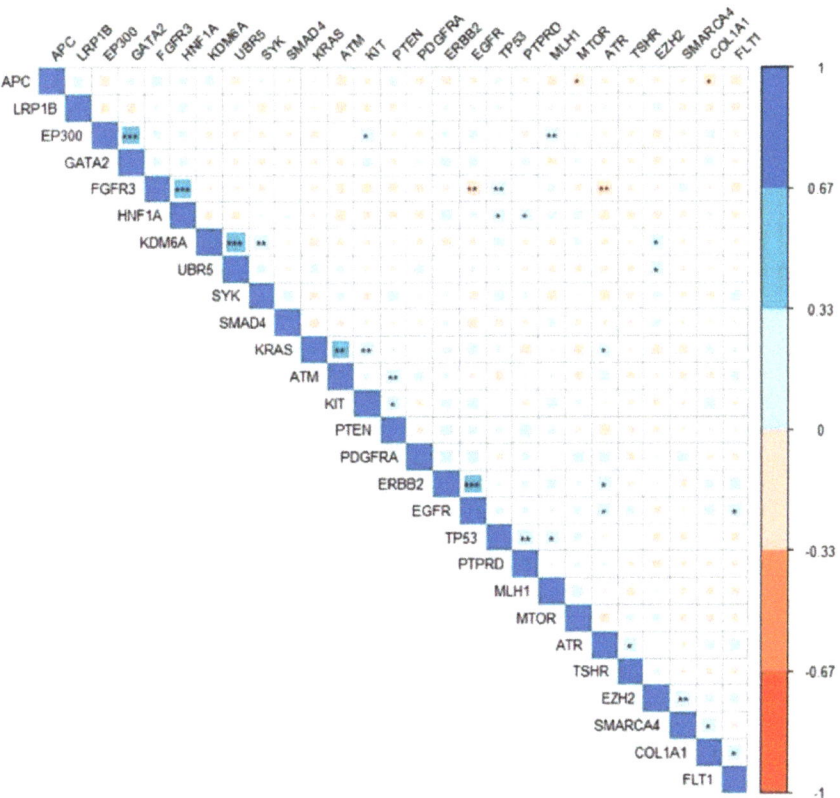

Figure 5. Correlation between mutated genes. The Pearson correlation between presence of a tissue-specific pathogenic variant between genes. Significant correlations are noted with * for p-values < 0.05, ** for p-values < 0.01, and *** for p-values less than 0.001. Color and size of the square denotes the value of correlation as indicated in the bar legend.

We tested the association of mutated genes with clinicopathological variables and found a significant association between ERBB2 mutation and tumor late stage (Fisher's exact t-test; p value = 0.04). Mutations in EP300 and TSHR mutations were found to be significantly associated with right colon tumor (chi-square; p = 0.02 and 0.01; respectively). Increased FGFR3 mutations was observed in the left colon (chi-square; p = 0.01)

For the gene-based rare variants analysis (SKAT-O), no gene was associated with clinicopathological variable at the significant threshold. However, suggestive significance was found between PIK3CB and colorectal cancer on the left side (p = 0.0007), androgen receptor (AR) and female gender (p = 0.0002), TGM7 and young patient (p = 0.002), and EXT1 and late stage (Table 3).

Table 3. Association of mutated genes with clinicopathological.

	SetID	p.Value	N.Marker.All	N.Marker.Test	MAC	m	Method.bin	MAP
Top genes associated with female versus male (232).	AR	0.0002	5	5	37	19	QA	−1
	BTK	0.003	1	1	5	5	ER	0.003523
	SAMD9	0.003	14	14	66	30	QA	−1
	PAX7	0.006605	6	6	41	26	QA	−1
	KDM6A	0.010983	11	11	58	27	QA	−1
Top genes associated with young group (<50 years old) versus old (285).	5TGM7	0.002186	4	4	85	43	QA	−1
	MRE11A	0.006959	8	8	177	62	QA	−1
	NBN	0.011436	11	11	171	47	QA	−1
	VHL	0.015733	2	2	2	2	ER	0.015733
	IDH1	0.019368	7	7	16	12	ER	1.32×10^{-10}
Top genes associated with left colorectal cancer versus right (204).	PIK3CB	0.0007	9	9	9	8	ER	8.76×10^{-5}
	PIK3CA	0.001	12	12	86	39	QA	−1
	RNF213	0.001	61	56	709	78	MA	−1
	AURKB	0.003	5	5	101	39	QA	−1
	ERBB4	0.005	11	11	41	27	QA	−1
Top genes associated with late stage versus early stage (180).	EXT1	0.003027	7	7	73	51	MA	−1
	RNASEL	0.013275	4	4	34	29	ER.A	−1
	CDH5	0.018248	8	8	135	62	MA	−1
	BUB1B	0.021125	18	18	178	48	MA	−1
	MUTYH	0.027108	11	11	108	51	MA	−1

N.Marker.All, number of all variants within that gene; N.Marker.Test, number of variants entered the analysis (in our case, we did not exclude common variants, but we assigned them lower weight, so it will be similar as N.Marker.All); MAC, total minor allele count (MAC); m, the number of individuals with minor alleles; method.bin, a type of method to be used to compute the *p*-value; MAP, minimum possible *p*-values. The number in the bracket shows the number of effective tests (we chose to select a *p* value that is equal to 0.05/409).

3.7. Network Analysis of Mutated Genes

Using ingenuity pathway analysis, we created an information-based network of 27 highly mutated genes and found TP53 was the most connected node (Figure 6A). This network identified 16 druggable target genes. A network of 75 genes with pathogenic mutations in the large intestine also exhibited TP53 as a highly connected node. Thirty-three of these network genes were identified as target molecules (Figure 6B). Both networks identified the TSHR gene as a potential druggable target (Supplementary Table S5).

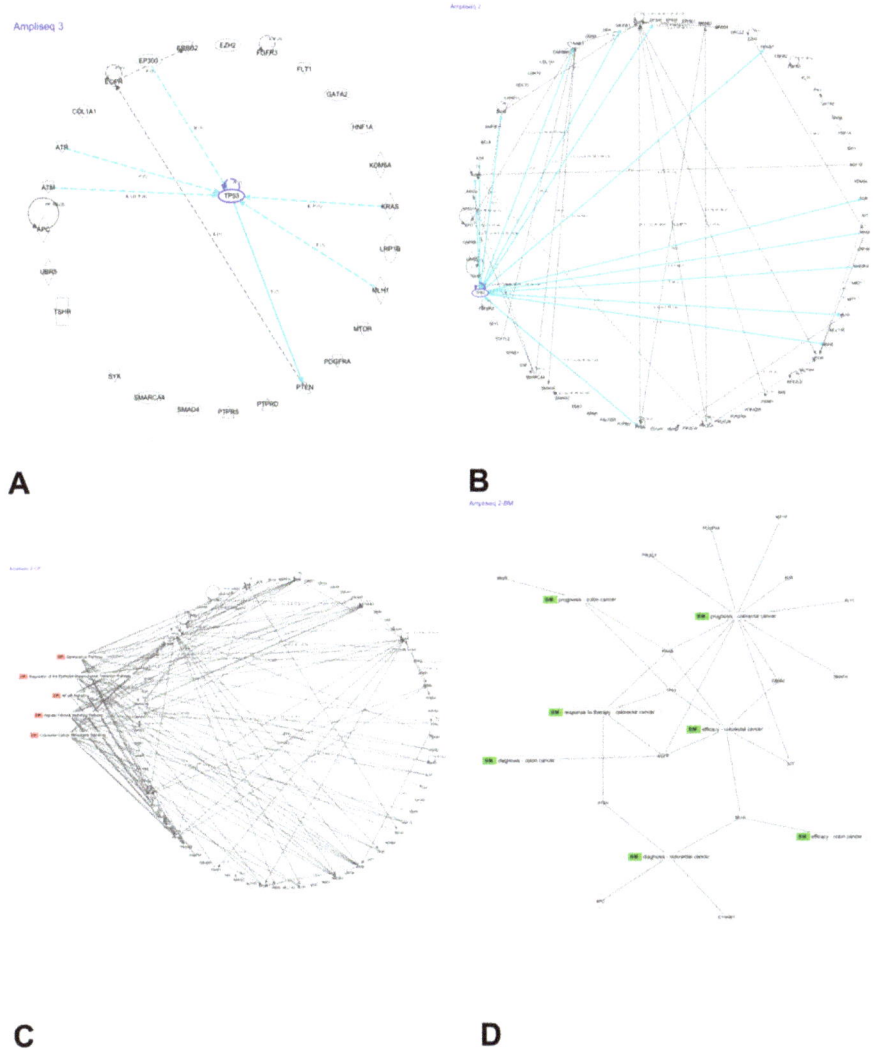

Figure 6. Network analyses of genes with reported variants. Network of 27 most frequently mutated genes (**A**). Network of 75 genes harboring pathogenic mutations reported in the large intestine (**B**). Association of 75 genes with canonical signaling pathways (**C**). Possible biomarker candidates for diagnosis, prognosis, efficacy, and response to drugs for colon and colorectal cancer (**D**).

Genes with pathogenic mutations in the large intestine were found to be associated with hepatic fibrosis signaling, CRC metastasis, senescence, NF-kB, and regulation of epithelial-to-mesenchymal transition pathways. Genes associated with these pathways are shown in Figure 6C. Biomarker analysis of these 75 genes revealed 16 candidate molecules, some of which are already in clinical use (Figure 6D and Table 4). These biomarkers have potential use in determining diagnosis, prognosis, efficacy, and response to drugs.

Table 4. Candidate biomarkers from the list of 75 genes with pathogenic mutations in the large intestine.

Symbol	Entrez Gene Name	Location	Family	Entrez Gene ID for Human
APC	APC regulator of WNT signaling pathway	Nucleus	Enzyme	324
BRAF	B-Raf proto-oncogene, serine/threonine kinase	Cytoplasm	Kinase	673
CTNNB1	Catenin beta 1	Nucleus	Transcription regulator	1499
EGFR	epidermal growth factor receptor	Plasma membrane	Kinase	1956
ERBB2	Erb-b2 receptor tyrosine kinase 2	Plasma membrane	Kinase	2064
FLT1	Fms-related receptor tyrosine kinase 1	Plasma membrane	Kinase	2321
IGF1R	Insulin-like growth factor 1 receptor	Plasma membrane	Transmembrane receptor	3480
KDR	Kinase insert domain receptor	Plasma membrane	Kinase	3791
KIT	KIT proto-oncogene, receptor tyrosine kinase	Plasma membrane	Transmembrane receptor	3815
KRAS	KRAS proto-oncogene, GTPase	Cytoplasm	Enzyme	3845
MLH1	MutL homolog 1	Nucleus	Enzyme	4292
PDGFRA	Platelet-derived growth factor receptor alpha	Plasma membrane	Kinase	5156
PIK3CA	Phosphatidylinositol-4,5-bisphosphate 3-kinase catalytic subunit alpha	Cytoplasm	Kinase	5290
PTEN	Phosphatase and tensin homolog	Cytoplasm	Phosphatase	5728
SMAD4	SMAD family member 4	Nucleus	Transcription regulator	4089
TP53	Tumor protein p53	Nucleus	Transcription regulator	7157

4. Discussion

Tumorigenesis and progression of cancer is suggested to be driven and supported by gene mutations [18–23]. Somatic mutations that are observed in cancer cells help to understand the cause and severity of the disease. Colorectal cancer is well known to have specific gene mutations associated with particular stages of the disease. In the present study, we aimed to provide comprehensive analysis of gene variants as studied in a cohort of patients in Riyadh, Saudi Arabia. We employed NGS on the eAmpliSeq comprehensive cancer panel to unravel the information locked in FFPE samples. This study provides successful evidence to support the use of archived samples and sequencing technology to generate information that is relevant for the local population. While we aimed to understand the mutational profile in the local population, we found results that confirmed existing evidence supporting the initiation and progression of CRC. We also report novel variants in our population, which is suggestive of a unique genomic landscape of patients and supports the idea of precision and personalized medicine [5,6,24].

As determined by two separate prediction methods (PolyPhen [14] and SIFT [25]), most of the detected mutations were missense and synonymous. This is in conformity with a recent pan-cancer analysis [26] and opens up avenues to further study the effect of point mutations in CRC. These point mutations could be responsible for changes in gene expression and mRNA secondary structures. Similar studies from other populations have also reported the predominance of synonymous and missense mutations [2]. However, the challenge to separately identify driver mutations from passenger mutations with precision and accuracy is still an ongoing area of intense research [27–29].

We identified the APC gene as highly mutated in our cohort, with less-common mutation frequency for RET, EGFR, LRP1B, and ERBB2 genes. The APC gene mutation is well known to be one of the very early events in initiation of CRC. RET gene fusions have been associated with a subtype of CRC on the right side of the colon. EGFR was earlier identified as one of the highest mutated genes in a cohort of patients from Jeddah, Saudi Arabia, and confirms our results [10]. This study adds to the evidence of detected variants in a previous similar study from a different geographical location. More studies from different regions of the country are needed, as there is an observed disparity in incidence and mortality of CRC in other regions within Saudi Arabia [4]. Identification of novel mutations in APC, RET, and EGFR mutations may lead us to develop predictive and/or prognostic biomarkers for CRC. Mutations in these genes were earlier studied in detail for their use as biomarkers [30–33]. Most of the previous studies are associated with the common mutation of the APC gene except two studies on an Arab cohort that showed APC mutation frequency was the second highest (34%) after P53 gene. Another study from gulf region patients showed 27.3% mutation frequency and higher mutation of TP53 (52.5%).

We found more CRC cases are localized at the left site compared to the right side or rectum site. This is suggestive of the involvement of the CIN pathway and is evident from our results, which show APC, KRAS, and P53 as highly mutated genes. However, PIK3CA do not appear among the highly mutated genes, but three pathogenic mutations were identified among PIK3CA (p.R88Q, p.I102F, and p.PI04L). All these mutations are reported in the large intestine except (p.I102F mutation). This could be due to the population-specific nature of the mutations and suggest further study to understand the mechanism of CRC progression in these patients. Left-sided colorectal cancers have better prognosis and response to 5-Fluorouracil-based and targeted therapies [34]. Our results are therefore very significant in understanding and predicting the prognosis of local patients who primarily exhibited mutations suggesting left-sided CRC. KRAS mutations, EGFR/HER2 amplifications, and a high level of amphiregulin and epiregulin expression has been observed in left-sided CRC [35,36]. Treatment strategies widely differ according to the location of the tumor, and hence understanding the molecular differences in local population is pertinent.

Increasing incidence of CRC cases in early ages has caused the guidelines for screening to be revised [37]. Our observation regarding young patients suggests presence of mutations in the left colon of female patients in stage I, whereas young male patients did not show any mutations in left colon in stage I and II. This can be an important finding that can be studied further in larger cohorts to develop early diagnostic tests. Our catalog of reported variants have enriched the database for CRC and would be useful in building up larger studies for finding actionable targets and biomarkers. Information regarding these variants will need to be complemented with further levels of evidence to prove their role in CRC or identify them as drug targets. Multiomics approach is therefore recommended to be carried out on the same samples for further proof of evidence [38,39].

The gene correlations observed in our cohort and network analysis could provide clues for the possible mechanism of CRC development. These networks and correlation analyses should be done at gene-expression level to further understand the mechanistic details and effect of variants [40]. Network analyses confirms the probable effect of the detected variants through well-known pathways. We report mutations that can be associated with senescent pathway and point towards development of therapeutic strategies. Targeting senescent pathways has been suggested as an anticancer therapy and point towards their role in senescence and metastasis. Biomarker candidate molecules need to be further validated and tested for advancing into a clinical setting.

Though our study is limited with a smaller number of patient samples, it does exhibit the heterogeneous nature of CRC [41]. Another major limitation of our study is lack of matched normal samples to account for possible germline mutations. This is one of the bargains for utilizing the treasure of formalin-fixed samples. Using a matched normal sample is a requirement for accurately classifying somatic mutations and ruling out germline mutations. However, the availability of matched normal tissue is a limitation when using archived and fresh samples [42]. Computational methods have been developed that are arguably better than matched normal tissue [43]. Most of the studies have relied on increasingly rich databases to identify novel mutations in absence of matched normal samples [44]. In order to address this issue, we used public databases and also employed an earlier reported method where the VAF corresponding to 50 or 100% may indicate their probability to be germline mutation [45].

This study provides evidence that can be useful for developing biomarker-based precision medicine as well as allowing us to appreciate the heterogeneity in CRC and hence develop strategies accordingly.

Supplementary Materials: The following are available online at https://www.mdpi.com/article/10.3390/jpm11060535/s1. Table S1: Ion AmpliSeq comprehensive cancer panel genes list (409) used on this study. Table S2: Details of reported novel variants. Table S3: VAF of reported novel variants Table S4: novel variants in colorectal cancer. Table S5: Target molecules in reported pathogenic mutations in the large intestine.

Author Contributions: B.A.: funding acquisition and project management. B.A., J.A., S.A. (Saleh AlGhamdi), A.G.A., F.S.A., M.A.A., F.A., A.A.A., S.A. (Saeed Alshieban), A.S. (Ahood Sayed), M.A. and A.A.: contributed to study concept and study design and patient data collection. B.A., A.S. (Abdulhadi Samman), M.A.A.B. and A.A.: design experiment, sample collection, sample processing, library preparation, and sequencing. J.A.: statistical analysis. B.A., J.A., and M.A.A.: data analysis and original manuscript draft preparation, figures and tables preparation. All authors read and approved the final manuscript. All authors have read and agreed to the published version of the manuscript.

Funding: This study was supported by King Abdullah International Medical Research Center grant (RC 13/249/R) awarded to B.M.

Institutional Review Board Statement: Full Institutional Review Board (IRB) approval was given by King Abdullah International Medical Research center (KAIMRC), Ministry of National Guard, Health Affairs (IRB protocol #RC13/249/R). All patients' data were secured and accessed only by research investigators.

Informed Consent Statement: Informed consent was obtained from all subjects involved in the study.

Data Availability Statement: All data is available in the repository as mentioned in methods.

Acknowledgments: We would like to acknowledge the help of Mamoon Rashid in submitting the data to SRA repository.

Conflicts of Interest: The authors declare no conflict of interest.

References

1. Rashid, M.; Vishwakarma, R.K.; Deeb, A.M.; Hussein, M.A.; Aziz, M.A. Molecular classification of colorectal cancer using the gene expression profile of tumor samples. *Exp. Biol. Med.* **2019**, *244*, 1005–1016. [CrossRef]
2. Zhunussova, G.; Afonin, G.; Abdikerim, S.; Jumanov, A.; Perfilyeva, A.; Kaidarova, D.; Djansugurova, L. Mutation Spectrum of Cancer-Associated Genes in Patients With Early Onset of Colorectal Cancer. *Front. Oncol.* **2019**, *9*, 673. [CrossRef]
3. Dos Santos, W.; Sobanski, T.; de Carvalho, A.C.; Evangelista, A.F.; Matsushita, M.; Berardinelli, G.N.; de Oliveira, M.A.; Reis, R.M.; Guimaraes, D.P. Mutation profiling of cancer drivers in Brazilian colorectal cancer. *Sci. Rep.* **2019**, *9*, 13687. [CrossRef] [PubMed]
4. Alyabsi, M.; Alhumaid, A.; Allah-Bakhsh, H.; Alkelya, M.; Aziz, M.A. Colorectal cancer in Saudi Arabia as the proof-of-principle model for implementing strategies of predictive, preventive, and personalized medicine in healthcare. *EPMA J.* **2020**, *11*, 119–131. [CrossRef]
5. Aziz, M.A.; Allah-Bakhsh, H. Colorectal cancer: A looming threat, opportunities, and challenges for the Saudi population and its healthcare system. *Saudi J. Gastroenterol.* **2018**, *24*, 196–197. [CrossRef]
6. Aziz, M.A. Precision medicine in colorectal cancer. *Saudi J. Gastroenterol.* **2019**, *25*, 139–140. [CrossRef]
7. Di Nicolantonio, F.; Vitiello, P.P.; Marsoni, S.; Siena, S.; Tabernero, J.; Trusolino, L.; Bernards, R.; Bardelli, A. Precision oncology in metastatic colorectal cancer—From biology to medicine. *Nat. Rev. Clin. Oncol.* **2021**. [CrossRef]
8. Sammarco, G.; Gallo, G.; Vescio, G.; Picciariello, A.; De Paola, G.; Trompetto, M.; Curro, G.; Ammendola, M. Mast Cells, microRNAs and Others: The Role of Translational Research on Colorectal Cancer in the Forthcoming Era of Precision Medicine. *J. Clin. Med.* **2020**, *9*, 2852. [CrossRef] [PubMed]

9. Pellino, G.; Gallo, G.; Pallante, P.; Capasso, R.; De Stefano, A.; Maretto, I.; Malapelle, U.; Qiu, S.; Nikolaou, S.; Barina, A.; et al. Noninvasive Biomarkers of Colorectal Cancer: Role in Diagnosis and Personalised Treatment Perspectives. *Gastroenterol. Res. Pract.* **2018**, *2018*, 2397863. [CrossRef] [PubMed]
10. Dallol, A.; Buhmeida, A.; Al-Ahwal, M.S.; Al-Maghrabi, J.; Bajouh, O.; Al-Khayyat, S.; Alam, R.; Abusanad, A.; Turki, R.; Elaimi, A.; et al. Clinical significance of frequent somatic mutations detected by high-throughput targeted sequencing in archived colorectal cancer samples. *J. Transl. Med.* **2016**, *14*, 118. [CrossRef]
11. *Torrent Suite Version 5.8*; Thermo Fisher Scientific: South San Francisco, CA, USA, 2018.
12. McLaren, W.; Gil, L.; Hunt, S.E.; Riat, H.S.; Ritchie, G.R.; Thormann, A.; Flicek, P.; Cunningham, F. The Ensembl Variant Effect Predictor. *Genome Biol.* **2016**, *17*, 122. [CrossRef] [PubMed]
13. Ng, P.C.; Henikoff, S. Predicting deleterious amino acid substitutions. *Genome Res.* **2001**, *11*, 863–874. [CrossRef] [PubMed]
14. Adzhubei, I.; Jordan, D.M.; Sunyaev, S.R. Predicting functional effect of human missense mutations using PolyPhen-2. *Curr. Protoc. Hum. Genet.* **2013**, *76*, 7–20. [CrossRef]
15. Purcell, S.; Neale, B.; Todd-Brown, K.; Thomas, L.; Ferreira, M.A.; Bender, D.; Maller, J.; Sklar, P.; de Bakker, P.I.; Daly, M.J.; et al. PLINK: A tool set for whole-genome association and population-based linkage analyses. *Am. J. Hum. Genet.* **2007**, *81*, 559–575. [CrossRef]
16. Lee, S.; Emond, M.J.; Bamshad, M.J.; Barnes, K.C.; Rieder, M.J.; Nickerson, D.A.; Christiani, D.C.; Wurfel, M.M.; Lin, X. Optimal Unified Approach for Rare-Variant Association Testing with Application to Small-Sample Case-Control Whole-Exome Sequencing Studies. *Am. J. Hum. Genet.* **2012**, *91*, 224–237. [CrossRef]
17. Kramer, A.; Green, J.; Pollard, J., Jr.; Tugendreich, S. Causal analysis approaches in Ingenuity Pathway Analysis. *Bioinformatics* **2014**, *30*, 523–530. [CrossRef]
18. Baker, S.J.; Preisinger, A.C.; Jessup, J.M.; Paraskeva, C.; Markowitz, S.; Willson, J.K.; Hamilton, S.; Vogelstein, B. p53 gene mutations occur in combination with 17p allelic deletions as late events in colorectal tumorigenesis. *Cancer Res.* **1990**, *50*, 7717–7722.
19. Fearon, E.R.; Vogelstein, B. A genetic model for colorectal tumorigenesis. *Cell* **1990**, *61*, 759–767. [CrossRef]
20. Huang, J.; Papadopoulos, N.; McKinley, A.J.; Farrington, S.M.; Curtis, L.J.; Wyllie, A.H.; Zheng, S.; Willson, J.K.; Markowitz, S.D.; Morin, P.; et al. APC mutations in colorectal tumors with mismatch repair deficiency. *Proc. Natl. Acad. Sci. USA* **1996**, *93*, 9049–9054. [CrossRef]
21. Jen, J.; Kim, H.; Piantadosi, S.; Liu, Z.F.; Levitt, R.C.; Sistonen, P.; Kinzler, K.W.; Vogelstein, B.; Hamilton, S.R. Allelic loss of chromosome 18q and prognosis in colorectal cancer. *N. Engl. J. Med.* **1994**, *331*, 213–221. [CrossRef] [PubMed]
22. Morin, P.J.; Vogelstein, B.; Kinzler, K.W. Apoptosis and APC in colorectal tumorigenesis. *Proc. Natl. Acad. Sci. USA* **1996**, *93*, 7950–7954. [CrossRef]
23. Vogelstein, B.; Papadopoulos, N.; Velculescu, V.E.; Zhou, S.; Diaz, L.A., Jr.; Kinzler, K.W. Cancer genome landscapes. *Science* **2013**, *339*, 1546–1558. [CrossRef] [PubMed]
24. Aziz, M.A.; Yousef, Z.; Saleh, A.M.; Mohammad, S.; Al Knawy, B. Towards personalized medicine of colorectal cancer. *Crit. Rev. Oncol. Hematol.* **2017**, *118*, 70–78. [CrossRef] [PubMed]
25. Kumar, P.; Henikoff, S.; Ng, P.C. Predicting the effects of coding non-synonymous variants on protein function using the SIFT algorithm. *Nat. Protoc.* **2009**, *4*, 1073–1081. [CrossRef]
26. Sharma, Y.; Miladi, M.; Dukare, S.; Boulay, K.; Caudron-Herger, M.; Gross, M.; Backofen, R.; Diederichs, S. A pan-cancer analysis of synonymous mutations. *Nat. Commun.* **2019**, *10*, 2569. [CrossRef]
27. Foo, J.; Liu, L.L.; Leder, K.; Riester, M.; Iwasa, Y.; Lengauer, C.; Michor, F. An Evolutionary Approach for Identifying Driver Mutations in Colorectal Cancer. *PLoS Comput. Biol.* **2015**, *11*, e1004350. [CrossRef]
28. Huang, D.; Sun, W.; Zhou, Y.; Li, P.; Chen, F.; Chen, H.; Xia, D.; Xu, E.; Lai, M.; Wu, Y.; et al. Mutations of key driver genes in colorectal cancer progression and metastasis. *Cancer Metastasis Rev.* **2018**, *37*, 173–187. [CrossRef]
29. Carethers, J.M.; Jung, B.H. Genetics and Genetic Biomarkers in Sporadic Colorectal Cancer. *Gastroenterology* **2015**, *149*, 1177–1190.e1173. [CrossRef]
30. Aghagolzadeh, P.; Radpour, R. New trends in molecular and cellular biomarker discovery for colorectal cancer. *World J. Gastroenterol.* **2016**, *22*, 5678–5693. [CrossRef] [PubMed]
31. Jauhri, M.; Bhatnagar, A.; Gupta, S.; Shokeen, Y.; Minhas, S.; Aggarwal, S. Targeted molecular profiling of rare genetic alterations in colorectal cancer using next-generation sequencing. *Med. Oncol.* **2016**, *33*, 106. [CrossRef]
32. Song, H.N.; Lee, C.; Kim, S.T.; Kim, S.Y.; Kim, N.K.; Jang, J.; Kang, M.; Jang, H.; Ahn, S.; Kim, S.H.; et al. Molecular characterization of colorectal cancer patients and concomitant patient-derived tumor cell establishment. *Oncotarget* **2016**, *7*, 19610–19619. [CrossRef]
33. Song, E.K.; Tai, W.M.; Messersmith, W.A.; Bagby, S.; Purkey, A.; Quackenbush, K.S.; Pitts, T.M.; Wang, G.; Blatchford, P.; Yahn, R.; et al. Potent antitumor activity of cabozantinib, a c-MET and VEGFR2 inhibitor, in a colorectal cancer patient-derived tumor explant model. *Int. J. Cancer* **2015**, *136*, 1967–1975. [CrossRef]
34. Baran, B.; Mert Ozupek, N.; Yerli Tetik, N.; Acar, E.; Bekcioglu, O.; Baskin, Y. Difference Between Left-Sided and Right-Sided Colorectal Cancer: A Focused Review of Literature. *Gastroenterol. Res.* **2018**, *11*, 264–273. [CrossRef]
35. Missiaglia, E.; Jacobs, B.; D'Ario, G.; Di Narzo, A.F.; Soneson, C.; Budinska, E.; Popovici, V.; Vecchione, L.; Gerster, S.; Yan, P.; et al. Distal and proximal colon cancers differ in terms of molecular, pathological, and clinical features. *Ann. Oncol.* **2014**, *25*, 1995–2001. [CrossRef]

36. LaPointe, L.C.; Dunne, R.; Brown, G.S.; Worthley, D.L.; Molloy, P.L.; Wattchow, D.; Young, G.P. Map of differential transcript expression in the normal human large intestine. *Physiol. Genomics* **2008**, *33*, 50–64. [CrossRef] [PubMed]
37. Young, G.P.; Rabeneck, L.; Winawer, S.J. The Global Paradigm Shift in Screening for Colorectal Cancer. *Gastroenterology* **2019**, *156*, 843–851.e842. [CrossRef] [PubMed]
38. Bian, S.; Hou, Y.; Zhou, X.; Li, X.; Yong, J.; Wang, Y.; Wang, W.; Yan, J.; Hu, B.; Guo, H.; et al. Single-cell multiomics sequencing and analyses of human colorectal cancer. *Science* **2018**, *362*, 1060–1063. [CrossRef]
39. Hu, W.; Yang, Y.; Li, X.; Huang, M.; Xu, F.; Ge, W.; Zhang, S.; Zheng, S. Multi-omics Approach Reveals Distinct Differences in Left- and Right-Sided Colon Cancer. *Mol. Cancer Res.* **2018**, *16*, 476–485. [CrossRef]
40. Jia, P.; Zhao, Z. Impacts of somatic mutations on gene expression: An association perspective. *Brief Bioinform.* **2017**, *18*, 413–425. [CrossRef] [PubMed]
41. Molinari, C.; Marisi, G.; Passardi, A.; Matteucci, L.; De Maio, G.; Ulivi, P. Heterogeneity in Colorectal Cancer: A Challenge for Personalized Medicine? *Int. J. Mol. Sci.* **2018**, *19*, 3733. [CrossRef]
42. Grizzle, W.E.; Bell, W.C.; Sexton, K.C. Issues in collecting, processing and storing human tissues and associated information to support biomedical research. *Cancer Biomark* **2010**, *9*, 531–549. [CrossRef] [PubMed]
43. Hiltemann, S.; Jenster, G.; Trapman, J.; van der Spek, P.; Stubbs, A. Discriminating somatic and germline mutations in tumor DNA samples without matching normals. *Genome Res.* **2015**, *25*, 1382–1390. [CrossRef]
44. Kumar, A.; White, T.A.; MacKenzie, A.P.; Clegg, N.; Lee, C.; Dumpit, R.F.; Coleman, I.; Ng, S.B.; Salipante, S.J.; Rieder, M.J.; et al. Exome sequencing identifies a spectrum of mutation frequencies in advanced and lethal prostate cancers. *Proc. Natl. Acad. Sci. USA* **2011**, *108*, 17087–17092. [CrossRef] [PubMed]
45. He, M.M.; Li, Q.; Yan, M.; Cao, H.; Hu, Y.; He, K.Y.; Cao, K.; Li, M.M.; Wang, K. Variant Interpretation for Cancer (VIC): A computational tool for assessing clinical impacts of somatic variants. *Genome Med.* **2019**, *11*, 53. [CrossRef] [PubMed]

Article

Lincp21-RNA as Predictive Response Marker for Preoperative Chemoradiotherapy in Rectal Cancer

Jose Carlos Benitez [1,2,†], Marc Campayo [2,*,†], Tania Díaz [3], Carme Ferrer [4], Melissa Acosta-Plasencia [3], Mariano Monzo [3], Luis Cirera [2], Benjamin Besse [1,5] and Alfons Navarro [3,*]

1. Department of Cancer Medicine, Gustave Roussy Cancer Center, 94805 Villejuif, France; josecarlos.benitez-montanez@gustaveroussy.fr (J.C.B.); benjamin.besse@gustaveroussy.fr (B.B.)
2. Department of Medical Oncology, Mutua Terrassa University Hospital, University of Barcelona, 08221 Terrassa, Spain; lcirera@mutuaterrassa.es
3. Molecular Oncology and Embryology Laboratory, Human Anatomy Unit, Faculty of Medicine and Health Sciences, University of Barcelona, IDIBAPS, 08036 Barcelona, Spain; tdiaz@ub.edu (T.D.); melissaacostaplasencia@gmail.com (M.A.-P.); mmonzo@ub.edu (M.M.)
4. Department of Pathology, Mutua Terrassa University Hospital, University of Barcelona, 08221 Terrassa, Barcelona, Spain; carmeferrer@mutuaterrassa.es
5. Faculty of Science, Orsay Campus, Paris-Saclay University, 91400 Orsay, France
* Correspondence: mcampayo@mutuaterrassa.es (M.C.); anavarroponz@ub.edu (A.N.)
† These two authors contributed equally to this work.

Abstract: Preoperative chemoradiotherapy (CRT) is a standard treatment for locally advanced rectal cancer (RC) patients, but its use in non-responders can be associated with increased toxicities and resection delay. LincRNA-p21 is a long non-coding RNA involved in the p53 pathway and angiogenesis regulation. We aimed to study whether lincRNA-p21 expression levels can act as a predictive biomarker for neoadjuvant CRT response. We analyzed RNAs from pretreatment biopsies from 70 RC patients treated with preoperative CRT. Pathological response was classified according to the tumor regression grade (TRG) Dworak classification. LincRNA-p21 expression was determined by RTqPCR. The results showed that lincRNA-p21 was upregulated in stage III tumors ($p = 0.007$) and in tumors with the worst response regarding TRG ($p = 0.027$) and downstaging ($p = 0.016$). ROC curve analysis showed that lincRNA-p21 expression had the capacity to distinguish a complete response from others (AUC:0.696; $p = 0.014$). LincRNA-p21 was shown as an independent marker of preoperative CRT response ($p = 0.047$) and for time to relapse (TTR) ($p = 0.048$). In conclusion, lincRNA-p21 is a marker of advanced disease, worse response to neoadjuvant CRT, and shorter TTR in locally advanced RC patients. The study of lincRNA-p21 may be of value in the individualization of pre-operative CRT in RC.

Keywords: lincRNA-p21; rectal cancer; chemoradiotherapy; colorectal cancer; long non-coding RNA; p53; predictive biomarker

1. Introduction

Rectal cancer (RC) accounts for approximately one-third of all colorectal tumors (CRC) and remains the third most common cancer worldwide and the second leading cause of cancer-related death in the world [1]. RC differs in etiologies and risk factors due to odd environmental exposures [2,3] and may have unique genetics and epigenetics factors [4]. However, during the past decade, reduction in mortality for RC has slowed [1] owing to a high rate of distant metastasis (29–39%) [5,6]. Long-term analysis has shown that preoperative chemoradiotherapy (CRT) followed by surgery of primary tumor results in persistent local control [5] and has become the standard of care for locally advanced tumors (T3-T4 or N+) [7]. The most frequently used chemotherapy agent is 5-fluorouracil in combination with concurrent fractionation radiotherapy [7]. Preoperative CRT achieves a higher radiosensitivity of tissues before surgery, a lower rate of toxicities, and a higher probability of sphincter preservation due to tumor downstaging [8]. Of note, the rate of pathological response after neoadjuvant treatment has been associated with prognosis [8,9]. Pathological complete response (pCR; ypT0N0), which occurs in 15–25% of patients, has been linked with lower rates of local recurrences [9,10]. Indeed, to achieve a complete response after preoperative CRT has been associated with better disease-free survival (DFS) and overall survival (OS) rates [9]. Nonetheless, survival outcomes of patients with an assessed pCR compared to those without have not been properly compared; therefore, selection of patients to avoid unnecessary toxicities and to perform suitable management remains uncertain. Furthermore, despite the adoption of adjuvant postoperative chemotherapy, patients are more than twice as likely to present with a distant recurrence rather than tumor regrowth at the primary site [5,6]. This situation emphasizes the urgency of devising upfront treatment strategies aimed at controlling obscure micro-metastases. Identifying patients who will not respond to treatment is crucial to avoid unnecessary treatment, potential toxicities, and a delay of surgery. Biomarkers to identify patients at high risk of relapse or lack of response are needed to guide treatment options and improve survival rates [11], and non-coding RNAs are promising candidates [12,13].

Non-coding RNAs comprise 97% of the transcriptome, while protein-coding messenger RNAs (mRNA) account for only 3% [14]. Long non-coding RNAs (lncRNAs) have been related to the main hallmarks of cancer [15] and have been described as key in the tumorigenesis of different solid tumors, including RC [16,17]. Indeed, lncRNAs have been shown to be highly tissue specific [18,19], being able to discriminate between tumor and normal cells [20]. The long intergenic non-coding RNA p21 (lincRNA-p21) acts as a regulator for p53-mediated apoptosis [21], angiogenesis [22], and HIF1A-mediated response to hypoxia in cancer cells [23]. However, the role of lincRNA-p21 in RC remains poorly understood and explored only in vitro or using small cohorts of patients [24,25]. In this setting, lncRNAs, and especially lincRNA-p21, could serve as predictive biomarkers to select the most optimal treatment in each case in order to individualize therapy. We aimed to evaluate whether lincRNA-p21 can act as a predictive biomarker for CRT response in a 70-patient cohort of RC treated before resection.

2. Materials and Methods

2.1. Study Population

Seventy patients diagnosed from December 2006 to October 2016, with RC and available baseline endoscopy biopsy from Mutua Terrassa University Hospital, were included in the present study. All selected patients suffered with rectal adenocarcinoma in a clinical stage II or III (uT3-T4 and/or uN+) and were consecutively treated at Mutua Terrassa University Hospital. Although the study population was collected in Barcelona (Europe), ethnical information was not considered for patient inclusion within the study. All samples were stored as paraffin-embedded blocks until use. All patients had received neoadjuvant chemotherapy with 5-fluorouracil 225 mg/m^2/day × 7 days in continuous infusion and in combination with pelvic locoregional radiotherapy (45–50 Gy). Six to eight weeks after completion, all patients underwent surgery. All surgical specimens were evaluated and classified according to TNM 7th edition, and the pathological response was graded according to the tumor regression grade (TRG) Dworak classification [26]. Approval for the study was obtained from the Institutional Review Board of the Mutua Terrassa University Hospital, Barcelona, Spain.

2.2. RNA Extraction and lincRNA-p21 Quantification

Total RNA was extracted from formalin-fixed, paraffin-embedded, tumor tissues from pretreatment endoscopy biopsies using a RecoverAll Total Nucleic Acid Isolation Kit (Ambion, ThermoFisher Scientific, Waltham, MA, USA) as previously reported [27] and quantified using a NanoDrop ND-1000 Spectrophotometer (NanoDrop Technologies, Wilmington, DE). Total cDNA was obtained from 250 ng of RNA using the High-Capacity cDNA Reverse Transcription Kit (Applied Biosystems, Foster City, CA, USA). LincRNAp21 expression was determined as previously described [22]. LincRNA-p21 expression was calculated using $2^{-\Delta\Delta Ct}$ using B2M (beta-2-microglobulin) (Hs99999907_m1) (Applied Biosystems) as endogenous control.

2.3. Statistical Methods

Assumptions of distributional normality were tested using the Shapiro–Wilk test and quantile–quantile plot. Continuous data were tested with the T-test (two groups) or ANOVA (more than two groups) when normally distributed and the Mann–Whitney U-test or Kruskal–Wallis test when not normally distributed. ROC curves were calculated using R package pROC [28]. The multivariate analysis for treatment response was performed by using binary logistic regression. Time to relapse (TTR) was defined as the time between resection and recurrence or last follow-up. Overall survival (OS) was calculated from the time of resection to the date of death or last follow-up. Optimal cutoffs of lincRNA-p21 expression data for TTR and OS were obtained using X-Tile software [29]. Kaplan–Meier curves for TTR and OS were plotted and compared with log-rank test. The multivariate analysis was performed using the stepwise proportional hazard Cox regression model to determine hazard ratios (HR) with their 95% confidence intervals (CI). Statistical significance was set at $p \leq 0.05$. All statistical analyses were performed using IBM SPSS Statistics 26 (SPSS Inc., IBM, Chicago, IL, USA), R 4.0.2 and GraphPad Prism v9.1.0.

3. Results

3.1. Patient Characteristics

Samples from 70 patients were analyzed, most of whom were males ($n = 49$, 70%). Median age at diagnosis was 66 (range: 38 to 82) years. Sixty-one (87.1%) patients reported stage III and 9 (12.9%) stage II; 52 (74.3%) patients were assessed for TRG 0–3, and 18 (25.7%) reported pathological complete response (TRG 4, ypT0N0); 64.3% of downstaging was shown. Finally, 43 (66.2%) patients received adjuvant chemotherapy after primary tumor resection. Table 1 shows further main characteristics of the 70 patients included in the study. Median follow-up time was 105.40 months (IQR: 78.63–127.33).

Table 1. Main clinical characteristics of the 70 patients included in the study with their associated time to relapse (TTR) and overall survival (OS) according to the univariate analyses (log rank). Significant p-values are shown in bold. RC: rectal cancer; CRT: Chemo-radiotherapy.

Characteristic		Number of Patients (%)	TTR p-Value	OS p-Value
Sex	Male	49 (70)	0.203	0.269
	Female	21 (30)		
Median age (range)	66 (38–82)			
	<60	19 (27.1)	0.679	0.815
	>60	51 (72.9)		
Clinical stage pre-CRT	II	9 (12.9)	0.585	0.497
	III	61 (87.1)		
Adjuvant therapy	No	27 (33.8)	0.776	0.130
	5-FLU	7 (8.7)		
	FOLFOX	40 (50)		
	Other	6 (7.5)		

Table 1. Cont.

Characteristic		Number of Patients (%)	TTR p-Value	OS p-Value
ypT	ypT0	18 (25.7)	0.015	0.051
	ypT1–2	18 (25.7)		
	ypT3–4	34 (48.6)		
ypN	ypN0	48 (68.6)	0.003	0.044
	ypN1–2	22 (31.4)		
Pathological stage after neoadjuvant CRT	ypT0N0	17 (24.2)	0.024	0.133
	I	16 (22.9)		
	II	14 (20)		
	III	23 (32.9)		
Downstaging	No	25 (35.7)	0.001	0.010
	Yes	45 (64.3)		
Tumor regression grade (TRG)	0–3	52 (74.3)	0.324	0.161
	4	18 (25.7)		

3.2. LincRNA-p21 Expression Levels

The correlation of lincRNA-p21 levels in tumor tissue with the main clinicopathological characteristics showed a significant association with disease stage, ypT, ypN, pathological stage (ypTNM), downstaging, and pathological response. LincRNA-p21 was upregulated in stage III compared to stage II tumors ($p = 0.007$) (Figure 1A). Significant differences in lincRNA-p21 levels were observed according to ypT, where the ypT0 group had the lowest levels ($p = 0.0493$, Figure 1B). Patients with ypN1–2 showed higher levels of lincRNA-p21 ($p = 0.02$). Furthermore, patients with pathological stage III had higher lincRNA-p21 levels ($p = 0.0171$). Tumors with the worst response to CRT regarding negative downstaging and TRG 0–3 showed higher levels of lincRNA-p21 than tumors with positive downstaging ($p = 0.0165$; Figure 1E) and TRG4 (TRG0–3, $n = 52$ vs. TRG4, $n = 18$, $p = 0.027$; Figure 1F).

3.3. Predictive Ability of lincRNA-p21 for Response to CRT

Receiver operating characteristic (ROC) curves were generated to investigate the potential of lincRNA-p21 as a marker for neoadjuvant treatment response. The area under the curve (AUC) value showed that lincRNA-p21 expression had capacity to distinguish patients with complete response (TRG4) from others (AUC: 0.696; 95% confidence interval (CI) = 0.558–0.833; $p = 0.014$). In the optimum truncation point (-0.1), the sensitivity and specificity were 83.3% and 57.7%, respectively (Figure 2A). Using the best threshold identified by the ROC curve analysis, we divided the patients into two groups, observing that there were differences in TRG proportions allocation between low or high lincRNA-p21 levels ($p = 0.026$, Figure 2B). Among patients with low levels of lincRNA-p21, 39.5% had a TRG 4 vs. only 9.4% in the group with a high lincRNA-p21 expression value.

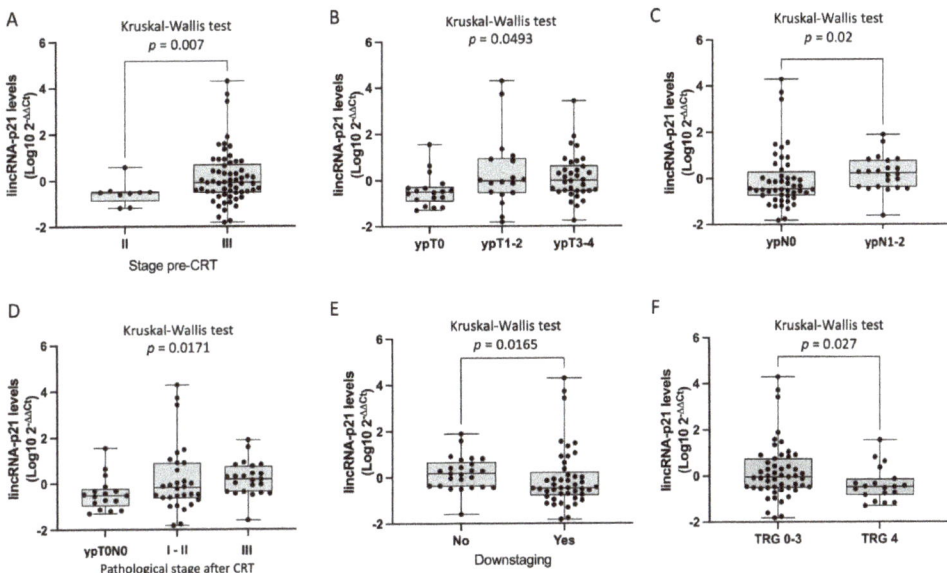

Figure 1. LincRNA-p21 levels and clinicopathological characteristics. (**A**) LincRNA-p21 expression in (**A**) stage III vs. stage II; (**B**) ypT0 vs. ypT-1–2 vs. ypT3–4; (**C**) ypN0 vs. ypN1–2; (**D**) ypT0N0 vs. I-II vs. III; (**E**) downstaging no vs. yes; (**F**) TRG 0–3 vs. TRG 4.

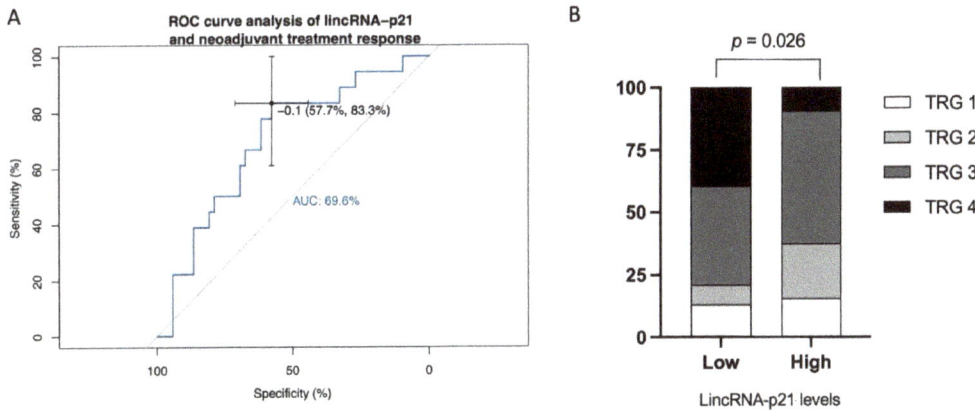

Figure 2. Predictive analyses for response to neoadjuvant treatment. (**A**) ROC curve analyses to evaluate the potential utility of lincRNA-p21 to distinguish patients with maximum response to neoadjuvant treatment (TRG4) from others (TRG 0–3). (**B**) Percentage of patients with each TRG according to low vs. high lincRNA-p21 expression, dichotomized using optimum truncation point obtained in the ROC curve analysis (−0.1). AUC, area under the curve. TRG, tumor regression grade.

Finally, we performed a multivariate analysis of response to neoadjuvant treatment including sex, age, pre-CRT stage, CEA levels pre-CRT, and lincRNA-p21 levels (Table 2). Only lincRNA-p21 levels emerged as an independent marker of neoadjuvant treatment response (odds ratio (OR): 0.485; 95% CI: 0.237–0.992; $p = 0.047$).

Table 2. Results obtained in the multivariate logistic analysis for complete response to neoadjuvant treatment (TRG4 vs. others).

Factors	OR (95% CI)	*p*-Value
Stage II at diagnosis	1.703 (0.363–8.003)	0.500
Age	0.980 (0.918–1.046)	0.549
Gender male	2.756 (0.682–11.137)	0.155
CEA at baseline	0.930 (0.809–1.068)	0.301
LincRNA-p21 levels	0.485 (0.237–0.992)	0.047
Constant	0.307	<0.001

3.4. LincRNA-p21 Expression and Survival

In our cohort, overall, median TTR and median OS were not reached (NR). Overall, mean TTR was 136.5 months (95% CI: 127.8–145.2) and mean OS was 124.3 months (95% CI: 114–134.6).

Using the optimal cutoff values identified by X-Tile, the patients were classified in two groups as having high or low lincRNA-p21 levels. Among the 70 RC patients, 26 were classified as low, and 44 as high. Patients with high lincRNA-p21 levels had significantly shorter TTR ($p = 0.014$). TTR for patients with high levels was 104.4 months (95% CI 86.4–122.5), while it was 126.2 months (95% CI 115.7–136.6) for those with low levels (Figure 3A). No significant differences were observed for OS ($p = 0.284$), but patients with high lincRNA-p21 levels had shorter OS (116.9 vs. 129.5 months; Figure 3B).

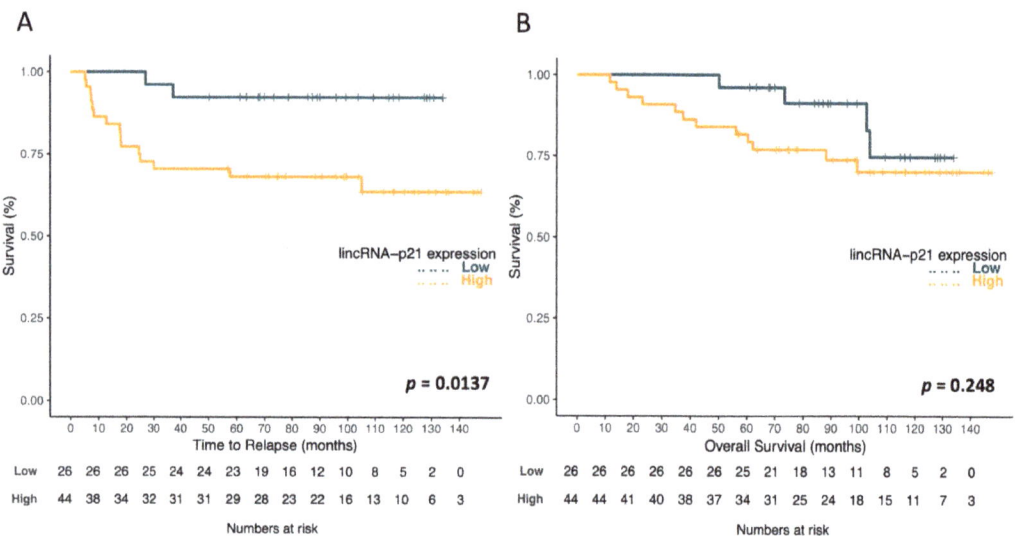

Figure 3. Kaplan–Meier curves for time to relapse (TTR) (**A**) and overall survival (OS) (**B**) according to lincRNA-p21 expression levels in 70 rectal cancer patients. The log-rank test was used to calculate whether significant differences in survival times between high or low lincRNA-p21 levels were achieved.

3.5. Multivariate Analysis of TTR and OS

In the univariate analysis, there were statistically significant differences in TTR and OS related to tumor pathological stage (ypT), lymph node pathological stage after CRT (ypN), pathological stage after CRT (ypTNM), and downstaging. The *p*-values are summarized in Table 1. Since ypT and ypN are included in the calculation of pathological stage, ypTNM, we decided to include only the pathological stage, downstaging, and the lincRNA-p21 expression in the Cox multivariate analysis (Table 3). The multivariate analysis showed that lincRNA-p21 levels (HR, 4.458; 95% CI, 1.014–19.603; p = 0.048) and stage (HR, 4.430; 95% CI: 1.266–15.497; p = 0.020) were independent prognostic factors for TTR, while downstaging (HR, 3.512; 95% CI: 1.275–9.673; p = 0.015) was the unique independent prognostic factor for OS.

Table 3. Multivariate analysis for TTR and OS.

Time to Relapse	HR (95% CI)	p-Value
Pathological stage > I	4.430 (1.266–15.497)	0.020
No downstaging	1.737 (0.350–8.621)	0.499
High lincRNA-p21	4.458 (1.014–19.603)	0.048
Overall Survival	**HR (95% CI)**	***p*-Value**
Pathological stage > I	2.020 (0.362–11.273)	0.423
No downstaging	3.512 (1.275–9.673)	0.015
High lincRNA-p21	1.387 (0.411–4.679)	0.598

4. Discussion

We showed the potential use of lincRNA-p21 expression levels in tumor tissue from baseline biopsies of RC patients as a predictive marker of CRT response and as a prognostic biomarker for TTR. Firstly, we observed that higher lincRNA-p21 levels were found in patients with stage III pre-CRT, and, interestingly, after CRT treatment, the highest lincRNA-p21 levels were reported for patients presenting pathological stage III, and the lowest levels were found in patients with ypT0N0. Indeed, higher lincRNA-p21 levels were observed in patients with ypT3–4 and in ypN1–2 patients. These results are in line with previous reports in CRC [17,25]. In a cohort of 66 patients with CRC, including 39% (26/66) of RC [25], higher lincRNA-p21 levels were associated with poor prognostic factors, such as a poorer stage (stage III vs. I), tumor size (pT), and vascular invasion [25]. In another study, Li et al. analyzed 177 CRC tumors samples from surgical resection, of which 81 (45.7%) were RC; lincRNA-p21 was found as a marker of advanced disease, as higher lincRNA-p21 levels were observed in stage III patients and in N+ patients, and worse survival [17]. However, although these reports are in line with our results, we must take into account that we studied a different RC population, namely, patients receiving neoadjuvant treatment before surgery. Of note, this group of patients was excluded from both previous reports.

Secondly, we observed that lower lincRNA-p21 levels were found in patients who underwent tumor downstaging and complete pathological response after CRT treatment. Locally advanced rectal cancer patients are commonly explored with a rectal endoscopy, which provides sufficient tissue samples for diagnosis and biomarker analyses. Currently, there are no clinically validated biomarkers to correctly identify those patients that will not respond. LincRNA-p21 emerged as a predictive biomarker for CRT response, and when it was compared to other predictive factors at diagnosis such as baseline stage or CEA levels, it was shown as an independent predictor factor. The neoadjuvant CRT treatment in our cohort was based on 5-fluorouracil combined with locoregional radiotherapy. Wang et al. carried out an in vitro study aiming to evaluate the role of lincRNA-p21 in radiotherapy response [24]; in contrast to our results, they described that lincRNA-p21 expression level may affect the sensitivity to radiotherapy. In this study, the authors observed that after X-ray treatment, the levels of lincRNA-p21 became upregulated in two colorectal cancer cell lines, SW1116 and LOVO. When researchers overexpressed lincRNA-p21 in the SW1116 cell line and treated the cells with X-rays, they noted a higher apoptosis rate than in control cells; nonetheless, this result was not validated by the authors when they silenced lincRNA-p21 before X-ray treatment on the same cell line (no differences in apoptosis rate were observed between the silenced and control group). Our group has reported results in this line; however, we used a different cohort of patients (resected CRC patients not receiving neoadjuvant treatment) [17]. We observed that patients with tumors with high expression of lincRNA-p21 demonstrated an increased benefit of CRT as an adjuvant therapy (longer OS compared to those patients not receiving CRT after surgical resection [17]). Nonetheless, these results are not comparable with the present work since our correlation was obtained in tumor tissue isolated prior to adjuvant CRT treatment administration. Moreover, the two previous references focused on the potential role of lincRNA-p21 and radiotherapy response in relation to its role in the p53 pathway [21,30–32]; nonetheless, we cannot ignore that rectal cancer patients included in the cohort also received 5-fluorouracil. Lee and colleagues analyzed the pattern of lncRNAs in 5-fluorouracil-resistant colon cancer cell lines and observed that lincRNA-p21 was significantly upregulated in SNU-C5 5-FU-resistant cells compared to its parental cell line [33]. This provides an important insight into the involvement of lincRNA-p21 within 5-FU resistance of colon cancer cells and allows us to speculate the following: the better response rates observed in patients with low levels of lincRNA-p21 could be associated with, at least partially, an enhanced sensitivity to 5-fluorouracil. The role of lincRNA-p21 in 5-fluorouracil resistance and its effect when 5-FU is combined with radiotherapy deserves further study, but this is out of the scope of the present paper.

Finally, we found a correlation between high expression of lincRNA-p21 levels and shorter TTR. In this regard, high lincRNA-p21 levels have been previously related to a worst outcome in CRC [17] and also in other solid tumors such as non-small-cell lung cancer [22], bladder carcinoma [34], or hepatocellular carcinoma [35]. In CRC, Li et al. observed that lincRNA-p21 was found as a marker of advanced disease and worse survival outcomes, especially for RC where high lincRNA-p21 levels were linked to shorter DFS and shorter OS [17].

We are conscious that the present study has several limitations, including the small number of samples analyzed (n = 70), which can affect the robustness of the multivariate analysis. The results obtained in the multivariate analysis, despite being informative, need to be validated in a larger cohort. Moreover, an additional limitation is that lincRNA-p21 was analyzed in a retrospective cohort of paraffin-embedded samples. Nonetheless, no related studies have been published for RC patient cohorts in neoadjuvant settings, and our study may provide new evidence of epigenetic pathways behind the tumor response to CRT. LincRNA-p21 may be a promising predictive biomarker of CRT benefit, avoiding delay of resection and unnecessary comorbidities for those patients with tumors and reporting high expression levels of lincRNA-p21 at baseline.

5. Conclusions

LincRNA-p21 is a marker of advanced disease, worse response to neoadjuvant CRT, and shorter TTR in locally advanced rectal cancer patients. The study of lincRNA-p21 in endoscopy samples obtained prior to treatment decision may be of value in the individualization of pre-operative CRT in rectal cancer.

Author Contributions: Conceptualization, J.C.B., M.C. and A.N.; data curation, J.C.B., T.D., C.F. and A.N.; formal analysis, T.D. and A.N.; funding acquisition, M.C. and A.N.; investigation, T.D., M.A.-P. and A.N.; methodology, J.C.B., M.C. and A.N.; project administration, J.C.B., M.C. and A.N.; resources, J.C.B., M.M. and A.N.; supervision, J.C.B., M.C. and A.N.; validation, M.C. and A.N.; writing—original draft, J.C.B.; writing—review and editing, J.C.B., M.C., M.M., L.C., B.B. and A.N. All authors have read and agreed to the published version of the manuscript.

Funding: This work was supported by grants from the Ministry of Economy, Industry, and Competition, Agencia Estatal de Investigación co-financed with the European Union FEDER funds SAF2017-88606-P (AEI/FEDER, UE). None of the funding bodies had a role in the design of the study; in the collection, analysis, and interpretation of data; or in writing the manuscript.

Institutional Review Board Statement: The study was approved by the Institutional Review Board of the University Hospital of Terrassa.

Informed Consent Statement: Considering the retrospective nature of this research protocol with no impact on patients' treatment and the use of anonymized data only, written consent was not required. However, the CEIC from Hospital Mutua de Terrasa approved the study. Moreover, before the study started, all cases were de-identified and coded by an oncologist staff member (first author), and all data were accessed anonymously.

Data Availability Statement: Not applicable.

Conflicts of Interest: All of the authors have reviewed and approved this version of the manuscript and agree with the decision to submit. None of the authors declare any conflicts of interest.

References

1. Siegel, R.L.; Miller, K.D.; Jemal, A. Cancer statistics, 2020. *CA Cancer J. Clin.* **2020**, *70*, 7–30. [CrossRef] [PubMed]
2. Wei, E.K.; Giovannucci, E.; Wu, K.; Rosner, B.; Fuchs, C.S.; Willett, W.C.; Colditz, G.A. Comparison of risk factors for colon and rectal cancer. *Int. J. Cancer* **2003**, *108*, 433–442. [CrossRef] [PubMed]
3. Kirkegaard, H.; Johnsen, N.F.; Christensen, J.; Frederiksen, K.; Overvad, K.; Tjønneland, A. Association of adherence to lifestyle recommendations and risk of colorectal cancer: A prospective Danish cohort study. *BMJ* **2010**, *341*, c5504. [CrossRef]
4. Guinney, J.; Dienstmann, R.; Wang, X.; De Reyniès, A.; Schlicker, A.; Soneson, C.; Marisa, L.; Roepman, P.; Nyamundanda, G.; Angelino, P.; et al. The consensus molecular subtypes of colorectal cancer. *Nat. Med.* **2015**, *21*, 1350–1356. [CrossRef]
5. Bosset, J.-F.; Calais, G.; Mineur, L.; Maingon, P.; Stojanovic-Rundic, S.; Bensadoun, R.-J.; Bardet, E.; Beny, A.; Ollier, J.-C.; Bolla, M.; et al. Fluorouracil-based adjuvant chemotherapy after preoperative chemoradiotherapy in rectal cancer: Long-term results of the EORTC 22921 randomised study. *Lancet Oncol.* **2014**, *15*, 184–190. [CrossRef]
6. Sainato, A.; Nunzia, V.C.L.; Valentini, V.; De Paoli, A.; Maurizi, E.R.; Lupattelli, M.; Aristei, C.; Vidali, C.; Conti, M.; Galardi, A.; et al. No benefit of adjuvant Fluorouracil Leucovorin chemotherapy after neoadjuvant chemoradiotherapy in locally advanced cancer of the rectum (LARC): Long term results of a randomized trial (I-CNR-RT). *Radiother. Oncol.* **2014**, *113*, 223–229. [CrossRef]
7. Glynne-Jones, R.; Wyrwicz, L.; Tiret, E.; Brown, G.; Rödel, C.; Cervantes, A.; Arnold, D. Rectal cancer: ESMO Clinical Practice Guidelines for diagnosis, treatment and follow-up. *Ann. Oncol.* **2017**, *28*, iv22–iv40. [CrossRef]
8. Rödel, C.; Martus, P.; Papadoupolos, T.; Füzesi, L.; Klimpfinger, M.; Fietkau, R.; Liersch, T.; Hohenberger, W.; Raab, R.; Sauer, R.; et al. Prognostic Significance of Tumor Regression After Preoperative Chemoradiotherapy for Rectal Cancer. *J. Clin. Oncol.* **2005**, *23*, 8688–8696. [CrossRef] [PubMed]
9. Sell, N.M.; Qwaider, Y.Z.; Goldstone, R.N.; Cauley, C.E.; Cusack, J.C.; Ricciardi, R.; Bordeianou, L.G.; Berger, D.L.; Kunitake, H. Ten-year survival after pathologic complete response in rectal adenocarcinoma. *J. Surg. Oncol.* **2021**, *123*, 293–298. [CrossRef]
10. Kasi, A.; Abbasi, S.; Handa, S.; Al-Rajabi, R.; Saeed, A.; Baranda, J.; Sun, W. Total Neoadjuvant Therapy vs Standard Therapy in Locally Advanced Rectal Cancer: A Systematic Review and Meta-analysis. *JAMA Netw. Open* **2020**, *3*, e2030097. [CrossRef]
11. Arnold, M.; Sierra, M.S.; Laversanne, M.; Soerjomataram, I.; Jemal, A.; Bray, F. Global patterns and trends in colorectal cancer incidence and mortality. *Gut* **2017**, *66*, 683–691. [CrossRef]
12. Deng, S.; Calin, G.A.; Croce, C.M.; Coukos, G.; Zhang, L. Mechanisms of microRNA deregulation in human cancer. *Cell Cycle* **2008**, *7*, 2643–2646. [CrossRef] [PubMed]
13. Bartel, D.P. MicroRNAs: Target Recognition and Regulatory Functions. *Cell* **2009**, *136*, 215–233. [CrossRef]
14. Djebali, S.; Davis, C.A.; Merkel, A.; Dobin, A.; Lassmann, T.; Mortazavi, A.; Tanzer, A.; Lagarde, J.; Lin, W.; Schlesinger, F.; et al. Landscape of transcription in human cells. *Nature* **2012**, *489*, 101–108. [CrossRef]
15. Gutschner, T.; Diederichs, S. The hallmarks of cancer: A long non-coding RNA point of view. *RNA Biol.* **2012**, *9*, 703–719. [CrossRef]
16. Galamb, O.; Barták, B.K.; Kalmár, A.; Nagy, Z.B.; Szigeti, K.A.; Tulassay, Z.; Igaz, P.; Molnár, B. Diagnostic and prognostic potential of tissue and circulating long non-coding RNAs in colorectal tumors. *World J. Gastroenterol.* **2019**, *25*, 5026–5048. [CrossRef] [PubMed]
17. Li, Y.; Castellano, J.J.; Moreno, I.; Martínez-Rodenas, F.; Hernandez, R.; Canals, J.; Diaz, T.; Han, B.; Muñoz, C.; Biete, A.; et al. LincRNA-p21 Levels Relates to Survival and Post-Operative Radiotherapy Benefit in Rectal Cancer Patients. *Life* **2020**, *10*, 172. [CrossRef]
18. Necsulea, A.; Soumillon, M.; Warnefors, M.; Liechti, A.; Daish, T.; Zeller, U.; Baker, J.C.; Grützner, F.; Kaessmann, H. The evolution of lncRNA repertoires and expression patterns in tetrapods. *Nat. Cell Biol.* **2014**, *505*, 635–640. [CrossRef] [PubMed]
19. Washietl, S.; Kellis, M.; Garber, M. Evolutionary dynamics and tissue specificity of human long noncoding RNAs in six mammals. *Genome Res.* **2014**, *24*, 616–628. [CrossRef]
20. Yan, X.; Hu, Z.; Feng, Y.; Hu, X.; Yuan, J.; Zhao, S.D.; Zhang, Y.; Yang, L.; Shan, W.; He, Q.; et al. Comprehensive Genomic Characterization of Long Non-coding RNAs across Human Cancers. *Cancer Cell* **2015**, *28*, 529–540. [CrossRef]
21. Huarte, M.; Guttman, M.; Feldser, D.; Garber, M.; Koziol, M.J.; Kenzelmann-Broz, D.; Khalil, A.M.; Zuk, O.; Amit, I.; Rabani, M.; et al. A Large Intergenic Noncoding RNA Induced by p53 Mediates Global Gene Repression in the p53 Response. *Cell* **2010**, *142*, 409–419. [CrossRef]

22. Castellano, J.J.; Navarro, A.; Viñolas, N.; Marrades, R.M.; Moises, J.; Santanach, A.C.; Saco, A.; Muñoz, C.; Fuster, D.; Molins, L.; et al. LincRNA-p21 Impacts Prognosis in Resected Non-Small Cell Lung Cancer Patients through Angiogenesis Regulation. *J. Thorac. Oncol.* **2016**, *11*, 2173–2182. [CrossRef]
23. Yang, F.; Zhang, H.; Mei, Y.; Wu, M. Reciprocal Regulation of HIF-1α and LincRNA-p21 Modulates the Warburg Effect. *Mol. Cell* **2014**, *53*, 88–100. [CrossRef] [PubMed]
24. Wang, G.; Li, Z.; Zhao, Q.; Zhu, Y.; Zhao, C.; Li, X.; Ma, Z.; Li, X.; Zhang, Y. LincRNA-p21 enhances the sensitivity of radiotherapy for human colorectal cancer by targeting the Wnt/β-catenin signaling pathway. *Oncol. Rep.* **2014**, *31*, 1839–1845. [CrossRef] [PubMed]
25. Zhai, H.; Fesler, A.; Schee, K.; Fodstad, Ø.; Flatmark, K.; Ju, J. Clinical Significance of Long Intergenic Noncoding RNA-p21 in Colorectal Cancer. *Clin. Color. Cancer* **2013**, *12*, 261–266.
26. Dworak, O.; Keilholz, L.; Hoffmann, A. Pathological features of rectal cancer after preoperative radiochemotherapy. *Int. J. Color. Dis.* **1997**, *12*, 19–23. [CrossRef]
27. Campayo, M.; Navarro, A.; Benítez, J.C.; Santasusagna, S.; Ferrer, C.; Monzó, M.; Cirera, L. miR-21, miR-99b and miR-375 combination as predictive response signature for preoperative chemoradiotherapy in rectal cancer. *PLoS ONE* **2018**, *13*, e0206542. [CrossRef]
28. Robin, X.A.; Turck, N.; Hainard, A.; Tiberti, N.; Lisacek, F.; Sanchez, J.-C.; Muller, M.J. pROC: An open-source package for R and S+ to analyze and compare ROC curves. *BMC Bioinform.* **2011**, *12*, 77. [CrossRef] [PubMed]
29. Camp, R.L.; Dolled-Filhart, M.; Rimm, D.L. X-tile: A new bio-informatics tool for biomarker assessment and outcome-based cut-point optimization. *Clin. Cancer Res.* **2004**, *10*, 7252–7259. [CrossRef]
30. Chaleshi, V.; Irani, S.; Alebouyeh, M.; Mirfakhraie, R.; Aghdaei, H.A. Association of lncRNA-p53 regulatory network (lincRNA-p21, lincRNA-ROR and MALAT1) and p53 with the clinicopathological features of colorectal primary lesions and tumors. *Oncol. Lett.* **2020**, *19*, 3937–3949.
31. Melo, C.A.; Léveillé, N.; Rooijers, K.; Wijchers, P.J.; Geeven, G.; Tal, A.; Melo, S.A.; De Laat, W.; Agami, R. A p53-bound enhancer region controls a long intergenic noncoding RNA required for p53 stress response. *Oncogene* **2016**, *35*, 4399–4406. [CrossRef] [PubMed]
32. Jin, S.; Yang, X.; Li, J.; Yang, W.; Ma, H.; Zhang, Z. p53-targeted lincRNA-p21 acts as a tumor suppressor by inhibiting JAK2/STAT3 signaling pathways in head and neck squamous cell carcinoma. *Mol. Cancer* **2019**, *18*, 38. [CrossRef] [PubMed]
33. Lee, H.; Kim, C.; Ku, J.-L.; Kim, W.; Yoon, S.K.; Kuh, H.-J.; Lee, J.-H.; Nam, S.W.; Lee, E.K. A long non-coding RNA snaR contributes to 5-fluorouracil resistance in human colon cancer cells. *Mol. Cells* **2014**, *37*, 540–546. [CrossRef]
34. Zhou, Q.; Zhan, H.; Lin, F.; Liu, Y.; Yang, K.; Gao, Q.; Ding, M.; Liu, Y.; Huang, W.; Cai, Z. LincRNA-p21 suppresses glutamine catabolism and bladder cancer cell growth through inhibiting glutaminase expression. *Biosci. Rep.* **2019**, *39*, 29. [CrossRef] [PubMed]
35. Ning, Y.; Yong, F.; Haibin, Z.; Hui, S.; Nan, Z.; Guangshun, Y. LincRNA-p21 activates endoplasmic reticulum stress and inhibits hepatocellular carcinoma. *Oncotarget* **2015**, *6*, 28151–28163. [CrossRef] [PubMed]

Article

Detection of Aberrant Glycosylation of Serum Haptoglobin for Gastric Cancer Diagnosis Using a Middle-Up-Down Glycoproteome Platform

Seunghyup Jeong [1,2], Unyong Kim [3], Myung Jin Oh [1,2], Jihyeon Nam [1,2], Se Hoon Park [4], Yoon Jin Choi [5], Dong Ho Lee [6], Jaehan Kim [7] and Hyun Joo An [1,2,*]

1. Asia-Pacific Glycomics Reference Site, Chungnam National University, Daejeon 34134, Korea; shjeong0512@cnu.ac.kr (S.J.); mjoh@cnu.ac.kr (M.O.); namjihyeon97@o.cnu.ac.kr (J.N.)
2. Graduate School of Analytical Science and Technology, Chungnam National University, Daejeon 34134, Korea
3. Biocomplete Inc., Seoul 08389, Korea; unyong.kim@biocomplete.co.kr
4. Division of Hematology/Oncology, Department of Medicine, Samsung Medical Center, Sungkyunkwan University School of Medicine, Seoul 06351, Korea; hematoma@skku.edu
5. Department of Internal Medicine, Yonsei University College of Medicine, Seoul 03722, Korea; erica0007@gmail.com
6. Department of Internal Medicine for Gastroenterology, Seoul National University Bundang Hospital, Seongnam 13620, Korea; dhljohn@snubh.org
7. Department of Food and Nutrition, Chungnam National University, Daejeon 34134, Korea; jaykim@cnu.ac.kr
* Correspondence: hjan@cnu.ac.kr

Abstract: Gastric cancer is a frequently occurring cancer and is the leading cause of cancer-related deaths. Recent studies have shown that aberrant glycosylation of serum haptoglobin is closely related to gastric cancer and has enormous potential for use in diagnosis. However, there is no platform with high reliability and high reproducibility to comprehensively analyze haptoglobin glycosylation covering microheterogeneity to macroheterogeneity for clinical applications. In this study, we developed a middle-up-down glycoproteome platform for fast and accurate monitoring of haptoglobin glycosylation. This platform utilizes an online purification of LC for sample desalting, and an in silico haptoglobin glycopeptide library constructed by combining peptides and N-glycans to readily identify glycopeptides. In addition, site-specific glycosylation with glycan heterogeneity can be obtained through only a single MS analysis. Haptoglobin glycosylation in clinical samples consisting of healthy controls ($n = 47$) and gastric cancer patients ($n = 43$) was extensively investigated using three groups of tryptic glycopeptides: GP1 (including Asn184), GP2 (including Asn207 and Asn211), and GP3 (including Asn241). A total of 23 individual glycopeptides were determined as potential biomarkers ($p < 0.00001$). In addition, to improve diagnostic efficacy, we derived representative group biomarkers with high AUC values (0.929 to 0.977) through logistic regression analysis for each GP group. It has been found that glycosylation of haptoglobin is highly associated with gastric cancer, especially the glycosite Asn241. Our assay not only allows to quickly and easily obtain information on glycosylation heterogeneity of a target glycoprotein but also makes it an efficient tool for biomarker discovery and clinical diagnosis.

Keywords: gastric cancer; middle-up-down; haptoglobin; glycopeptide; biomarker; mass spectrometry

1. Introduction

Gastric cancer caused by genetic factors, dietary habits, smoking, alcohol consumption, and infection with *Helicobacter pylori* is one of the most frequently occurring cancer, with approximately 1,000,000 new cases each year and more than 750,000 deaths worldwide [1–3]. Serum protein markers, such as CEA (carcinoembryonic antigen), CA19-9 (carbohydrate antigen 19-9), CA72-4 (carbohydrate antigen 72-4), and CA125 (carbohydrate antigen 125) have been widely used in clinical practice for gastric cancer detection [4,5]. However, since these serological markers do not have sufficient sensitivity and specificity, an apparent but invasive gastroscopy is often used for gastric cancer diagnosis. Therefore, there is a need in clinic for a new analytical platform that is fast, accurate, and non-invasive with high sensitivity and high specificity [6].

Glycosylation is one of the most important post-translational modifications (PTMs) and plays a pivotal role in various biological processes, such as protein function and cell–cell interaction [7,8]. In addition, glycosylation has great potential as a biomarker for cancer and infectious diseases because it is highly sensitive to biological environments [9]. Therefore, studies based on serum and cell glycomic profiling have been conducted for biomarker discovery and cancer diagnosis [10–12]. In particular, cancer progression, including angiogenesis, cell–cell adhesion, and tumor metastasis, is known to be associated with glycosylation in various cancers [13–15]. For example, high mannose type N-glycans have been found to be specific molecules in breast cancer patients, and sialylated O-glycans, such as Tn antigen, sialyl- Le^x, and sialyl-Le^a, on the surface of tumor cells are known to be one of the important molecules for metastasis [16–20]. Recently, glyco-biomarker studies are moving towards in-depth glycosylation characterization of a target glycoprotein that can improve diagnostic specificity and sensitivity for efficient clinical applications [21–24].

Haptoglobin is one of the major serum components that accounts for 0.4–2.6% of total blood proteins, and a highly sialylated glycoprotein containing four N-glycosylation sites (Asn184, 207, 211, and 241) on β-subunit [25], whose glycosylation changes in several types of cancer, such as hepatic, prostate, ovarian, and pancreatic [26–29]. Interestingly, recent studies clearly suggest that there is an overt correlation between aberrant glycosylation of haptoglobin and gastric cancer through glycomic and glycoproteomic approaches [30–33]. Despite the tremendous potential of haptoglobin glycosylation as a cancer biomarker, there is no clinically compatible assay platform that offers high reliability and reproducibility based on extensive characterization of haptoglobin glycosylation, including the distribution of glycans present at a specific site (microheterogeneity) and the occupancy of glycans at individual sites (macroheterogeneity) [34].

Mass spectrometry (MS) is a powerful tool used in various omics, from proteomics to glycomics [35–37]. The introduction of MS in glycomics has accelerated the study of biomarkers [38–42]. Initially, biomarkers of several types of diseases, particularly immune-related diseases and cancers, have been successfully determined through overall glycan profiling [43–48]. In recent years, the target of analysis is shifting from the conventional global glycan profiling to the target glycoprotein, and various MS-based analytical tools have been developed to monitor abnormal glycosylation of a glycoprotein. It ranges from glycopeptide analysis, which provides information about site-specific glycosylation, to complete intact glycoprotein analysis, which provides intuitive information on the degree of glycosylation of the whole protein [31,32,49–51]. In particular, glycoproteomic analysis at the glycopeptide level is an effective method that can simultaneously acquire information on glycans and glycosylation sites [52–54], but it is not utilized in clinics due to the time required for the sample preparation step, difficulties of tandem MS analysis, and complex data interpretation. Therefore, there is a need for an easy and fast glycoproteomic analysis tool that can be used in the clinical field [7].

In this study, we developed a new middle-up-down glycoproteome platform that can quickly and accurately monitor the abnormal glycosylation on the target glycoprotein, haptoglobin, for the diagnosis of gastric cancer. With this approach, samples can be purified online in LC and then directly separated and detected on a diphenyl column without further purification of the glycopeptide, resulting in faster sample analysis. In addition, the in silico glycopeptide library allowed us to obtain information ranging from microheterogeneity to macroheterogeneity of the target glycoprotein without tandem MS analysis. For biomarker discovery, a middle-up-down glycoproteome platform was applied to gastric cancer patients ($n = 43$) and healthy controls ($n = 47$), and apparent differences in glycosylation were found even in a small sample set. A total of 23 individual glycopeptide biomarkers that were statistically significant were determined, which showed high sensitivity and specificity (AUC 0.783 to 0.901). Interestingly, most of the markers were complex type N-glycan decorated with sialylation, of which more than half were simultaneously fucosylated. In addition, potential biomarkers were classified into three groups according to the peptide sequence to determine the biological association between haptoglobin glycosylation heterogeneity and gastric cancer, indicating that the glycosite Asn241 is more closely related. The middle-up-down glycoproteoform approach enables easy and fast analysis of site-specific glycosylation as well as glycan heterogeneity for specific target glycoprotein, making it a powerful platform for biomarker discovery and diagnosis through large clinical samples.

2. Materials and Methods

2.1. Materials and Reagents

Commercial human serum, ammonium bicarbonate (NH_4HCO_3), and iodoacetamide (IAA) were purchased from Sigma-Aldrich (St. Louis, MO, USA). Sequencing grade modified trypsin and dithiothreitol (DTT) were purchased from Promega (Madison, WI, USA). Anti-human haptoglobin was obtained from Dako (Carpinteria, CA, USA). All other solvents used in LC–MS analysis were purchased from Sigma-Aldrich, which were analytical grade or higher.

2.2. Serum Samples from Gastric Cancer Patients and Healthy Control Subjects

The clinical information of the serum samples is summarized in Table S1. A total of 90 serum samples were used, and the population consisted of 47 healthy controls and 43 gastric cancer patients (Stage IV, adenocarcinoma type). The research design and protocol were reviewed and approved by the Institutional Review Board of the participating hospital, the Samsung Medical Center, Seoul, Republic of Korea (IRB# SMC2015-07-146-001). Cancer diagnoses and stage determinations were examined based on endoscopic ultrasound, biopsy, and gastrectomy for each patient. All participating subjects, including the healthy control, were Korean and provided informed consent for obtaining the serum samples. The samples were stored at $-80\ ^\circ C$ until further processing.

2.3. Haptoglobin Purification from Serum Samples

Serum haptoglobin was purified using an anti-haptoglobin immunoaffinity column as described in a previous study [30]. In brief, 450 μL serum from each sample subject was diluted with 4.5 mL phosphate-buffered saline (PBS: 10 mM phosphate buffer, 2.7 mM potassium chloride, 137 mM sodium chloride, pH 7.4), and then applied to the anti-haptoglobin immunoaffinity column. After a binding reaction, the unbound components were washed by 30 mL PBS, the haptoglobin was eluted with elution buffer (0.1 M glycine, 0.5 M NaCl, pH 2.8), and the eluent was fractionated into a tube containing neutralization buffer (1.0 M Tris-HCl, pH 9.0). A centrifugal filter (MWCO 10,000, Amicon Ultra, Millipore; Billerica, MA, USA) was used to remove the detergent from the eluent and the quantification of the purified haptoglobin was assayed using a Quanti-iT Assay Kit (Invitrogen; Carlsbad, CA, USA). To confirm the purity of the haptoglobin, the eluent was randomly applied to 12.5% SDS-PAGE with Coomassie Brilliant Blue staining. Each purified sample was lyophilized and kept at $-80\ ^\circ C$ until enzymatic digestion.

2.4. Enzymatic Digestion for Glycopeptide Production

The purified serum haptoglobin (20 μg) was dissolved in a buffer, which consisted of 50 mM NH_4HCO_3 and 10 mM DTT. The haptoglobin dissolved in the buffer solution was placed in a 95 $^\circ C$ water for 10 min to reduce the disulfide bond and separate the α- and β-subunits of the haptoglobin, and then alkylated with 50 mM IAA to prevent reassembly of the disulfide bond. Finally, trypsin was added to the digestion and the mixture was incubated in a 37 $^\circ C$ water bath for 16 h.

2.5. LC–MS Analysis of Haptoglobin Glycopeptide

After trypsin digestion, 6.0 µL of haptoglobin peptides (corresponding to 2 µg protein) were directly injected by an autosampler into the LC–MS system, which consisted of a 6550 iFunnel Q-TOF coupled to a 1290 Infinity II UHPLC system (Agilent Technologies, San Jose, CA, USA). First, the samples were desalted on a 2.1 × 12.5 mm narrow-bore C8 guard column and delivered to a 2.1 × 100 mm Rapid Resolution High Definition (RRHD) diphenyl column (Agilent Technologies) for separation. A rapid elution gradient for haptoglobin peptides was applied at 200 µL/min using mobile phases of (A) 0.3% formic acid in nanopure water, and (B) 0.3% formic acid in acetonitrile, ramping up from 5 to 95% over the course of 27 min. The column was flushed with 95% solvent B for 10 min and then re-equilibrated for 3 min before analyzing the next sample. The column temperature was maintained at 30 °C during the analysis. Following LC separation, the haptoglobin peptides were ionized and detected on the positive ion mode over a mass range of m/z 500 to 3200, with an acquisition rate of 2 spectra per second.

2.6. Data Processing, Glycopeptide Identification, and Statistical Analysis

All raw LC–MS data were processed by a molecular feature extraction algorithm included in the MassHunter Qualitative Analysis software (version B.07.00 SP1, Agilent Technologies). MS peaks were filtered with a signal-to-noise ratio of 5.0 and glycopeptide compounds were founded from deconvoluted masses by the theoretical accurate mass of the in silico haptoglobin glycopeptide library with a 10 ppm mass tolerance. The in silico glycopeptide library was built by the combination of theoretical tryptic peptides of haptoglobin and N-glycans obtained experimentally. An individual t-test analysis by Microsoft Excel 2016 (Microsoft, Seattle, WA, USA) was used to identify the statistical differences between gastric cancer patients and healthy controls, and all p-values were applied with a two-tailed analysis. The receiver operation characteristic (ROC) curve of each potential glycopeptide biomarker and logistic regression analysis for the combined biomarkers were performed by IBM SPSS Statistics (version 24, IBM, Armonk, NY, USA). Hierarchical Clustering Explorer (version 3.5, HCIL, University of Maryland, College Park, MD, USA) was used to confirm the reproducibility.

3. Results and Discussion

3.1. Analytical Strategy Using a Middle-Up-Down Glycoproteome Approach

The overall experimental workflow for middle-up-down glycoproteomic analysis is shown in Figure 1: (i) purification of the targeted serum haptoglobin via immunoaffinity chromatography; (ii) trypsin treatment of haptoglobin for glycopeptide generation; (iii) glycopeptide profiling via UHPLC Q-TOF MS with online purification; (iv) glycopeptide identification through in silico haptoglobin library; (v) biomarker discovery by a group of glycopeptides based on glyco-heterogeneity. The most notable points of this platform are three things. First, haptoglobin glycopeptides were injected into LC–MS without further enrichment and purification and desalted directly online through a C8-packed guard column by a quaternary pump, and then transferred to the diphenyl analytical column by valve switching [31]. This online enrichment allows the analysis of large numbers of samples with minimal preparation steps and saves time, cost, and labor, which are important considerations in clinical applications. Second, the glycan microheterogeneity and macroheterogeneity of the haptoglobin were quickly and accurately monitored through tryptic glycopeptides analyzed by only a single MS. We created an in silico haptoglobin glycopeptide library to facilitate the identification of glycopeptide by combining the theoretical tryptic peptides of haptoglobin with the experimentally obtained mass values of 41 N-glycans (Table S2). Site-specific glycopeptides, including four individual glycosylation sites via trypsin treatment, cannot be fully produced compared with a multiple enzyme reaction or non-specific protease treatment. However, this approach has the distinct advantage of significantly reducing the time, cost, and labor required for sample preparation, and further simplifying data processing by facilitating the identification of glycopeptides of Hp via an in silico library. Finally, unlike the discovery of individual molecular markers in typical biomarker studies, we reclassified individual potential markers according to their glycosylation site and determined group biomarkers to monitor changes in the glycosylation site of haptoglobin in gastric cancer. Group-by-group analysis of glycopeptides not only enables efficient detection of gastric cancer markers but also provides information on the glycosylation site that is biologically more sensitive to gastric cancer. In summary, the middle-up-down glycoproteomic platform analyzes the target glycoprotein at the glycopeptide level (middle-up) with minimal sample preparation and single MS analysis and also provides glyco-heterogeneity information (middle-down) as well as biological links between cancer and glycosylation sites.

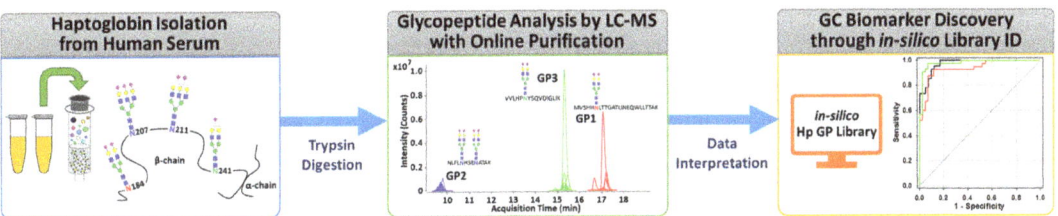

Figure 1. The overall experimental workflow of the middle-up-down glycoproteomic approach for aberrant glycosylation monitoring and gastric cancer diagnosis.

Prior to the analysis of clinical samples for biomarker discovery, the reproducibility of our platform was validated with standard haptoglobin purified from a commercial human serum. As shown in Figure S1, the Pearson correlation coefficient (R) values for the comparison of replicates were 1.000 for GP1 (16 pairs) and GP3 (19 pairs), and from 0.984 to 0.998 for GP2 (36 pairs), indicating high reproducibility and reliability.

3.2. Identification of Glycopeptides Using In Silico Haptoglobin Glycopeptide Library

In order to facilitate the interpretation of MS data and to efficiently identify glycopeptides, the in silico haptoglobin glycopeptide library was constructed by the combination of haptoglobin N-glycans obtained experimentally and the theoretical mass of tryptic peptides. The top 41 N-glycans, accounting for 99% of the total haptoglobin N-glycan quantity, were used for the library [30]. The peptide sequence of haptoglobin was referenced from the UniProt human protein database (P00738) and the theoretical mass of haptoglobin peptides was calculated using PeptideMass tool of ExPASy (https://www.expasy.org/, accessed on 9 December 2020). The amino acid sequence of the protein may be generated isoform by alternative splicing [55], but the isoform of haptoglobin does not change the sequence of β-subunit where the glycosylation site exists. Therefore, the in silico haptoglobin library was constructed using only a canonical sequence. Four potential N-glycosylation sites in the β-subunit of haptoglobin were divided into groups of three kinds of tryptic glycopeptide classified by the same peptide sequence: GP1 (Asn184, MVSHHN[184]LTTGATLINEQWLLTTAK), GP2 (Asn207 and Asn211, NLFLN[207]HSEN[211]ATAK), and GP3 (Asn241, VVLHPN[241]YSQVDIGLIK). Based on 41 N-glycan compositions, each GP1 and GP3 with one glycosylation site could have 41 possible glycoforms. Since GP2 has two glycosylation sites, Asn207 and Asn211, there are theoretically 902 glycoforms possible. However, if the total composition of N-glycans that may be present in the two glycosylation sites is the same (e.g., Asn207-$Hex_6HexNAc_5Fuc_1NeuAc_3$ + Asn211-$Hex_5HexNAc_4NeuAc_2$ and Asn207-$Hex_5HexNAc_4Fuc_1NeuAc_2$ + Asn211-$Hex_6HexNAc_5NeuAc_3$), they were recognized as duplicates and excluded from the glycopeptide library. As a result, GP2 could have a total of 416 possible glycoforms in the library. The full list of the in silico haptoglobin glycopeptide library is provided in Table S3. Representative extracted compound chromatograms (ECCs) and their mass spectra of three glycopeptide groups of haptoglobin purified from a commercial serum are shown in Figure 2. Glycopeptide groups were detected in the order of GP2, GP3, and GP1, and the elution times of each group were completely separated without overlapping. Each elution time was 10 to 11 min for GP2, about 16 min for GP3, and 17 to 18 min for GP1. The diphenyl column employed in this study separates and elutes tryptic glycopeptide according to the characteristics of the peptide, like the C18 reverse phase column commonly used in proteomics. Thus, even if the glycans were different, glycopeptides with the same peptide sequence were co-eluted at adjacent retention times. In addition, glycopeptides were somewhat separated according to the properties of the glycans even within the same peptide group under the diphenyl column [56]. On average, 16 and 23 glycopeptides were detected in the GP1 and GP3 groups, respectively. For GP2, since it contains two glycosylation sites, an average of 52 glycopeptides were identified. In all GP groups, most glycopeptides had sialylated N-glycans, of which the bi-antennary di-sialylated glycan was the most abundant (Figure 2(A-1,B-1,C-1)). Other N-glycopeptides containing undecorated or fucosylated/sialylated complex type glycans are not abundant, but are sufficiently identifiable in the magnified spectra (Figure 2(A-2,B-2,C-2)). In the deconvoluted spectrum, each peak represents a glycopeptide, and the spacing between adjacent peaks corresponds to the glycan residue difference. This clearly indicates that glycosylation within the same group is interrelated in the process of biosynthesis (Figure 2(A-3,B-3,C-3)).

Figure 3 shows a schematic diagram of site-specific glycosylation mapping of the haptoglobin purified from commercial sera. From the viewpoint of macroheterogeneity, glycosylation was the most abundant in GP3 and least detected in GP2. Bi-/tri-antennary sialylated complex types N-glycans were abundantly present in all glycosylation sites and mono-fucosylated glycans were also significantly observed, albeit in small amounts. Interestingly, the glycans of the glycopeptide are consistent with those obtained from the glycan profiling of the haptoglobin. Site-specific glycan mapping of the haptoglobin was implemented through the middle-up-down glycoproteome platform that provides information on glycan microheterogeneity and macroheterogeneity.

3.3. Gastric Cancer Biomarker Discovery via Middle-Up-Down Glycoproteome Platform

Based on the frequency and abundance of individual glycopeptides, we discovered potential biomarkers that could differentiate gastric cancer patients from healthy controls. Student's t-test was performed using 71 glycopeptides detected with a frequency of 70% or higher in all the samples tracked in the glycan heterogeneity monitoring. Subsequently, significant glycopeptides were selected according to two criteria, the p-value ($p < 0.00001$) and frequency (more than 90% in all samples). A total of 23 potential biomarkers were determined from three different glycopeptide groups (Table 1). Each glycopeptide biomarker presented an area under the curve (AUC) of 0.783 to 0.901 in the receiver operation characteristic (ROC) analysis. As a result of classifying according to the glycopeptide group, 6 markers were found in GP1, 8 markers in GP2, and 9 markers in GP3. Interestingly, the markers with the highest and lowest AUC values all belonged to the GP3 group. From the number of glycopeptide markers found and the AUC values, it can be expected that the glycosite Asn241 present in GP3 is most associated with gastric cancer. To complement and improve the sensitivity and specificity of the markers, we applied a logistic regression model to potential biomarkers and calculated a combined ROC curve [57–59]. The AUC improved significantly to 0.950 (GP1), 0.929 (GP2), and 0.977 (GP3), respectively (Figure 4). In particular, the abundance of all potential glycopeptide biomarkers was increased in gastric cancer patients, and most of the markers had differences in monosaccharide residues, clearly indicating that they correlated with each other. Changes in the glycan compositions of the haptoglobin glycopeptide suggested that in gastric cancer, the composition of one specific glycan not only changes independently but also affects the glycan synthesis itself. Compared with previous haptoglobin studies for gastric cancer biomarkers [33], the middle-up-down approach simultaneously provides the middle-up-down approach simultaneously provides information on the microheterogeneity and macroheterogeneity of glycosylation, enabling the discovery of potential biomarkers with high sensitivity and high specificity.

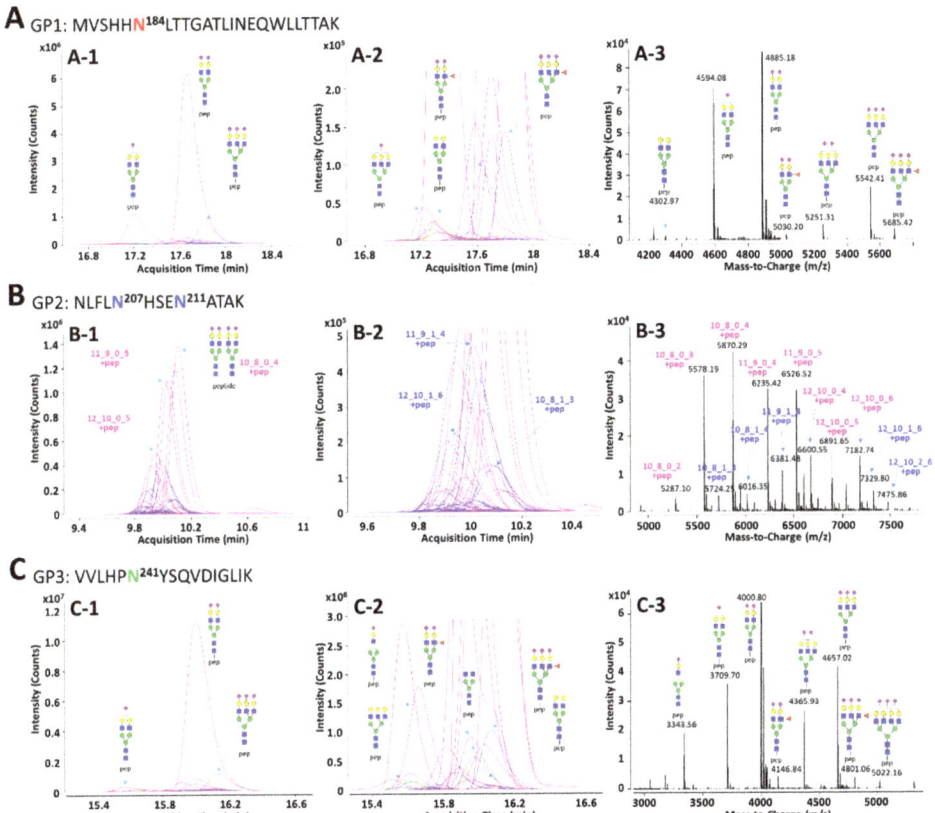

Figure 2. Representative extracted compound chromatograms (ECCs) and deconvoluted spectra of GP1 (**A**), GP2 (**B**), and GP3 (**C**) of haptoglobin purified from a commercial serum. The peptide sequence of each GP is inserted in the figure. The left column, (**A-1**), (**B-1**), and (**C-1**), represents the ECCs of Hp glycopeptides identified using the in silico haptoglobin glycopeptide library. The middle column, (**A-2**), (**B-2**), and (**C-2**), shows magnified views of low abundant glycopeptides on ECCs. The right column, (**A-3**), (**B-3**), and (**C-3**), represents deconvoluted spectra of the retention time range that glycopeptides are eluted. Glycans on GP2 were represented by composition rather than structure since there are two glycosylation sites in one peptide. For feasible interpretation and visualization of low abundant glycans, the y-axes of GP1 and GP3 are displayed at about $3\times$ magnification. Purple color—only sialylation; blue color—fucosylation and sialylation.

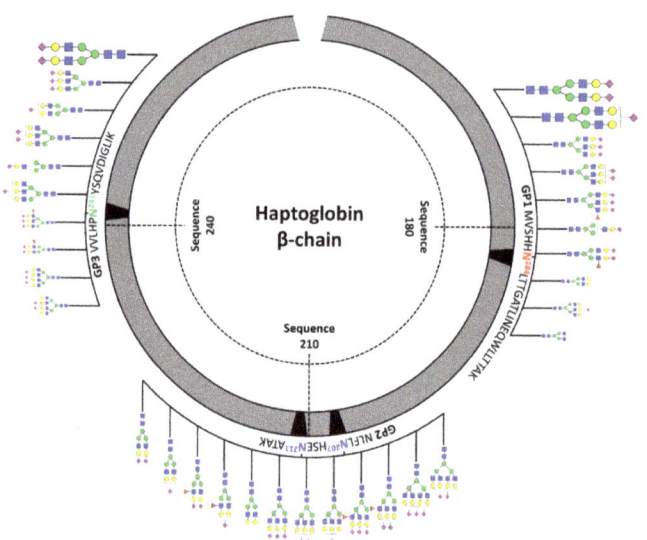

Figure 3. Site-specific glycosylation mapping of the haptoglobin standard purified from commercial human sera. For GP1 and GP3, the top 10 representative major glycoforms for each glycosylation site were indicated. Bi-antennary di-sialylated N-glycan occupies more than 50% of the total for each site. The sizes of N-glycan cartoon on GP1 and GP3 represent their relative abundance in each glycopeptide group. The biggest is more than 10%; medium is more than 1%; the smallest is less than 1%. In GP2 with two glycosylation sites, the denoted N-glycan cartoon is the composition most probable to constitute the top 10 glycopeptides of GP2 based on the abundance of haptoglobin glycan profiling.

Table 1. The list of glycopeptides found in serum haptoglobin representing a significant difference between healthy controls and gastric cancer patients. Glycopeptides were selected based on the Student's t-test ($p < 0.00001$). A complete list of glycopeptides is included in Table S4.

GP Group [a]	Mass	N-Glycan Composition [b]				p-Value	AUC
		Hex	HexNAc	Fuc	NeuAc		
GP1	4226.930	4	3	0	1	3.11×10^{-11}	0.895
	4738.120	5	4	1	1	1.17×10^{-8}	0.873
	4754.115	6	4	0	1	8.63×10^{-11}	0.864
	4883.157	5	4	0	2	3.13×10^{-11}	0.887
	5394.348	6	5	1	2	1.19×10^{-7}	0.831
	5685.443	6	5	1	3	1.12×10^{-6}	0.803
GP2	5722.234	10	8	1	3	3.28×10^{-9}	0.828
	5867.271	10	8	0	4	8.18×10^{-8}	0.818
	6013.329	10	8	1	4	2.32×10^{-8}	0.814
	6378.461	11	9	1	4	9.32×10^{-11}	0.848
	6669.557	11	9	1	5	6.78×10^{-9}	0.820
	6743.593	12	10	1	4	1.74×10^{-7}	0.805
	7034.689	12	10	1	5	1.86×10^{-9}	0.823
	7325.784	12	10	1	6	7.50×10^{-10}	0.846
GP3	3051.453	4	3	0	0	3.23×10^{-7}	0.803
	3342.549	4	3	0	1	1.19×10^{-8}	0.840
	3545.628	4	4	0	1	1.33×10^{-7}	0.827
	3707.681	5	4	0	1	3.25×10^{-6}	0.783

Table 1. Cont.

GP Group [a]	Mass	N-Glycan Composition [b]				p-Value	AUC
		Hex	HexNAc	Fuc	NeuAc		
	3869.734	6	4	0	1	9.63×10^{-9}	0.849
	3998.776	5	4	0	2	5.54×10^{-9}	0.844
	4144.834	5	4	1	2	3.51×10^{-9}	0.844
	4160.829	6	4	0	2	7.82×10^{-12}	0.901
	4801.062	6	5	1	3	1.75×10^{-6}	0.805

[a] GP group refers to a set of glycopeptides having the same tryptic peptide sequence with different glycan moieties. The peptide sequence of GP1 is MVSHHN[184]LTTGATLINEQWLLTTAK, of GP2 is NLFLN[207]HSEN[211]ATAK, and of GP3 is VVLHPN[241]YSQVDIGLIK.
[b] The monosaccharides of N-glycan composition represent Hex = Hexose; HexNAc = N-acetylhexosamine; Fuc = Fucose; NeuAc = N-acetylneuraminic acid.

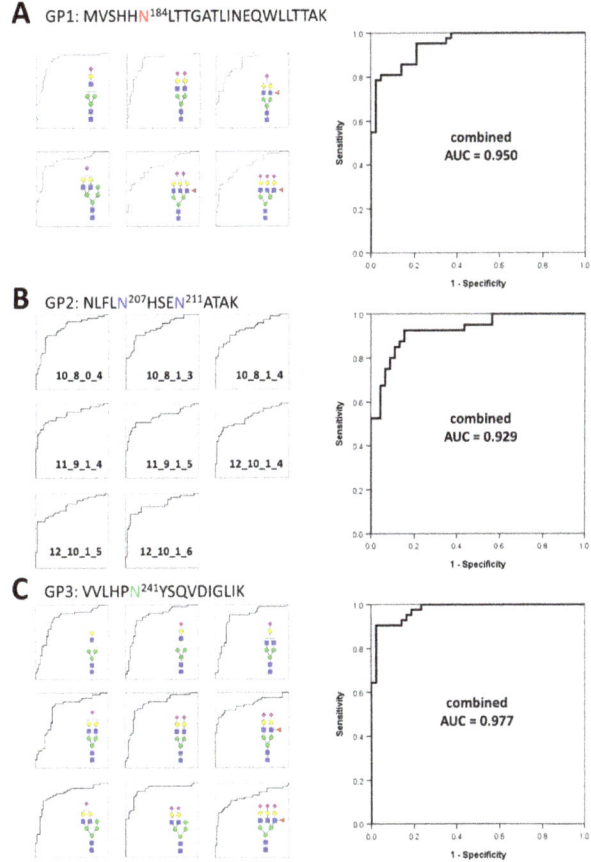

Figure 4. Individual and combined ROC curves using statistically significant glycopeptides. The small ROC curves on the left side are statistically significant individual glycopeptides in each glycopeptide group. On the right is the combined ROC curves derived using statistically significant individual glycopeptides for each glycopeptide group. The ROC curves of individual glycopeptides were calculated using their absolute abundances and combined ROC curves were derived using the logistic regression of glycopeptides. (**A**)—6 individual glycopeptides and their combined biomarker from GP1; (**B**)—8 individual glycopeptides and their combined biomarker from GP2; (**C**)—9 individual glycopeptides and their combined biomarker from GP3.

In addition, in order to verify the association between the glycan class of serum haptoglobin glycopeptides and gastric cancer, their expression levels were compared through a $\log_2$ fold change (Figure 5A). N-glycan can be classified into five biosynthetic groups: high mannose (HM) glycans; undecorated complex/hybrid (C/H) glycans; fucosylated complex/hybrid (C/H-F) glycans; sialylated complex/hybrid (C/H-S) glycans; and fucosylated-sialylated complex/hybrid (C/H-FS) glycans. The changes in the glycopeptides decorated with C/H-FS were particularly noticeable in all glycopeptide groups. The fucosylated/sialylated glycoforms increased regardless of the glycopeptide group, and the t-test also showed statistically significant values (Figure 5B).

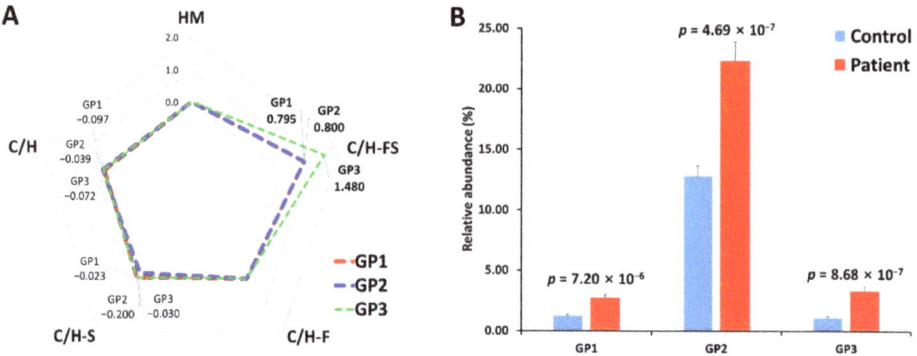

Figure 5. Comparison of N-glycosylation on haptoglobin between the healthy control and gastric cancer patient samples. (**A**) The radar chart of $\log_2$ fold changes (the ratio of cancer to control) for glycan classes of each glycopeptide group. (**B**) The relative abundance and p-value of glycoforms with both fucosylation and sialylation. HM—high mannose type; C/H—complex/hybrid type without fucose or sialic acid; C/H-S—complex/hybrid type with only sialic acid; C/H-F—complex/hybrid type with only fucose; C/H-FS—complex/hybrid type with fucose and sialic acid.

Haptoglobin primarily binds to hemoglobin, but this binding is not related to glycans because the peptide sequence of hemoglobin subunits does not contain a glycosylation site. On the other hand, haptoglobin binds to at least four receptors on leukocytes, including CD163, CD22, CCR2, and CD11b/CD18. These ligand–receptor interactions are one of the major roles of glycosylation and are reported to be affected by fucosylation and sialylation [47,48]. The fucosylated and sialylated glycans of haptoglobin were identified in a previous study as sLe^x (sialyl-Lewis x) or sLe^a (sialyl-Lewis a) epitope [30], which is the terminal structure of fucosylation and sialylation. These sLe epitopes have been reported to be associated with cancer progression [48,60]. The high expression of fucosylation/sialylation in the haptoglobin of gastric cancer patients found in our study can also be expected to have a strong correlation with cancer progression. In addition, our findings show significant consistency with previous studies of glycosylation changes in haptoglobin, suggesting that the middle-up-down glycoproteome approach is sufficient to monitor haptoglobin glycosylation.

4. Conclusions

Glycosylation is of particular interest in the field of diagnostic biomarker research because it is highly sensitive to various diseases, especially cancer. In particular, research and applications related to glycosylation are accelerating with the advancement of mass spectrometry, a high-sensitivity and high-resolution instrument. Of the various glycoform characterization methods, in-depth information of glycosylation can be obtained from site-specific glycopeptide profiling using tandem MS, but multi-step sample preparation and complex data interpretation make it difficult to expand into clinical use. We developed a middle-up-down glycoproteome platform that can rapidly and accurately obtain the microheterogeneity and macroheterogeneity of a targeted glycoprotein without tandem

MS analysis. Based on this platform, we were able to discover potential glycopeptide biomarkers with high sensitivity and high specificity that exhibit high AUC (0.783 to 0.901) at the molecular level required for actual clinical diagnosis. In addition, the biological association between specific glycosylation sites of haptoglobin and gastric cancer could be identified through glycopeptide grouping. This glycoproteome assay is a highly effective non-invasive platform that can be applied to the analysis of haptoglobin, as well as other glycoproteins with multiple glycosylation sites. Although marker validation using large-scale samples is required, this new analysis platform is expected to be widely used for biomarker discovery and clinical diagnosis based on a variety of targeted glycoprotein associated with diseases. In addition, the understanding of disease through biological association with changes in the glycosite of protein could be utilized for glycosylation-based drug development and personalized medicine.

Supplementary Materials: The following are available online at https://www.mdpi.com/article/10.3390/jpm11060575/s1, Figure S1: Pearson correlation coefficient (R) of three glycopeptides of haptoglobin composition derived from 10 commercial serum samples. (a) GP1, (b) GP2, and (c) GP3, Table S1: Clinical information of samples involved in this study, Table S2: The list of 41 N-glycans of haptoglobin used for construction of in-silico glycopeptide library, Table S3: The list of in silico haptoglobin glycopeptide library, Table S4: A complete list of glycopeptides found in serum haptoglobin.

Author Contributions: Conceptualization, S.J. and H.J.A.; methodology, S.J. and U.K., formal analysis, S.J.; data curation, S.J., M.J.O., and H.J.A., investigation, S.J. and J.N.; resources, S.H.P., Y.J.C., D.H.L.; writing—original draft, S.J.; writing—review and editing, S.J., M.J.O., and H.J.A.; visualization, S.J. and J.N., supervision, J.K. and H.J.A.; project administration, H.J.A.; funding acquisition, H.J.A. All authors have read and agreed to the published version of the manuscript.

Funding: This work was supported by a grant from the Ministry of Science and ICT (NRF-2016M3A9E1918324).

Institutional Review Board Statement: The study was conducted according to the guidelines of the Declaration of Helsinki, and approved by the Institutional Review Board of participating hospital Samsung Medical Center, Seoul, Republic of Korea (IRB# SMC2015-07-146-001).

Informed Consent Statement: Informed consent was obtained from all subjects involved in the study.

Conflicts of Interest: The authors declare that they have no competing interests.

References

1. Jung, K.W.; Won, Y.J.; Kong, H.J.; Lee, E.S. Prediction of Cancer Incidence and Mortality in Korea, 2019. *Cancer Res. Treat.* **2019**, *51*, 431–437. [CrossRef] [PubMed]
2. Torre, L.A.; Siegel, R.L.; Ward, E.M.; Jemal, A. Global Cancer Incidence and Mortality Rates and Trends—An Update. *Cancer Epidemiol. Biomark. Prev.* **2016**, *25*, 16–27. [CrossRef] [PubMed]
3. Bray, F.; Ferlay, J.; Soerjomataram, I.; Siegel, R.L.; Torre, L.A.; Jemal, A. Global cancer statistics 2018: GLOBOCAN estimates of incidence and mortality worldwide for 36 cancers in 185 countries. *CA Cancer J. Clin.* **2018**, *68*, 394–424. [CrossRef] [PubMed]
4. Matsuoka, T.; Yashiro, M. Biomarkers of gastric cancer: Current topics and future perspective. *World J. Gastroenterol.* **2018**, *24*, 2818–2832. [CrossRef] [PubMed]
5. Duraker, N.; Naci Celik, A.; Gencler, N. The prognostic significance of gastric juice CA 19-9 and CEA levels in gastric carcinoma patients. *Eur. J. Surg. Oncol.* **2002**, *28*, 844–849. [CrossRef]
6. Jeong, S.; Oh, M.J.; Kim, U.; Lee, J.; Kim, J.H.; An, H.J. Glycosylation of serum haptoglobin as a marker of gastric cancer: An overview for clinicians. *Expert Rev. Proteom.* **2020**, *17*, 109–117. [CrossRef]
7. Dube, D.H.; Bertozzi, C.R. Glycans in cancer and inflammation—Potential for therapeutics and diagnostics. *Nat. Rev. Drug Discov.* **2005**, *4*, 477–488. [CrossRef] [PubMed]
8. Fuster, M.M.; Esko, J.D. The sweet and sour of cancer: Glycans as novel therapeutic targets. *Nat. Rev. Cancer* **2005**, *5*, 526–542. [CrossRef]
9. Drake, P.M.; Cho, W.; Li, B.; Prakobphol, A.; Johansen, E.; Anderson, N.L.; Regnier, F.E.; Gibson, B.W.; Fisher, S.J. Sweetening the pot: Adding glycosylation to the biomarker discovery equation. *Clin. Chem.* **2010**, *56*, 223–236. [CrossRef]
10. Hua, S.; An, H.J. Glycoscience aids in biomarker discovery. *BMB Rep.* **2012**, *45*, 323–330. [CrossRef]
11. Powlesland, A.S.; Hitchen, P.G.; Parry, S.; Graham, S.A.; Barrio, M.M.; Elola, M.T.; Mordoh, J.; Dell, A.; Drickamer, K.; Taylor, M.E. Targeted glycoproteomic identification of cancer cell glycosylation. *Glycobiology* **2009**, *19*, 899–909. [CrossRef]

12. An, H.J.; Kronewitter, S.R.; de Leoz, M.L.; Lebrilla, C.B. Glycomics and disease markers. *Curr. Opin. Chem. Biol.* **2009**, *13*, 601–607. [CrossRef]
13. Pinho, S.S.; Reis, C.A. Glycosylation in cancer: Mechanisms and clinical implications. *Nat. Rev. Cancer* **2015**, *15*, 540–555. [CrossRef] [PubMed]
14. Anderson, N.L.; Anderson, N.G. The human plasma proteome: History, character, and diagnostic prospects. *Mol. Cell Proteom.* **2002**, *1*, 845–867. [CrossRef] [PubMed]
15. Rodrigues, J.G.; Balmana, M.; Macedo, J.A.; Pocas, J.; Fernandes, A.; de-Freitas-Junior, J.C.M.; Pinho, S.S.; Gomes, J.; Magalhaes, A.; Gomes, C.; et al. Glycosylation in cancer: Selected roles in tumour progression, immune modulation and metastasis. *Cell Immunol.* **2018**, *333*, 46–57. [CrossRef]
16. de Leoz, M.L.; Young, L.J.; An, H.J.; Kronewitter, S.R.; Kim, J.; Miyamoto, S.; Borowsky, A.D.; Chew, H.K.; Lebrilla, C.B. High-mannose glycans are elevated during breast cancer progression. *Mol. Cell Proteom.* **2011**, *10*, M110.002717. [CrossRef]
17. Josic, D.; Martinovic, T.; Pavelic, K. Glycosylation and metastases. *Electrophoresis* **2019**, *40*, 140–150. [CrossRef]
18. Oliveira-Ferrer, L.; Legler, K.; Milde-Langosch, K. Role of protein glycosylation in cancer metastasis. *Semin. Cancer Biol.* **2017**, *44*, 141–152. [CrossRef]
19. Magalhaes, A.; Duarte, H.O.; Reis, C.A. Aberrant Glycosylation in Cancer: A Novel Molecular Mechanism Controlling Metastasis. *Cancer Cell* **2017**, *31*, 733–735. [CrossRef]
20. Bull, C.; Boltje, T.J.; Wassink, M.; de Graaf, A.M.; van Delft, F.L.; den Brok, M.H.; Adema, G.J. Targeting aberrant sialylation in cancer cells using a fluorinated sialic acid analog impairs adhesion, migration, and in vivo tumor growth. *Mol. Cancer Ther.* **2013**, *12*, 1935–1946. [CrossRef] [PubMed]
21. Kim, E.H.; Misek, D.E. Glycoproteomics-based identification of cancer biomarkers. *Int. J. Proteom.* **2011**, *2011*, 601937. [CrossRef] [PubMed]
22. Zhang, Y.; Jiao, J.; Yang, P.; Lu, H. Mass spectrometry-based N-glycoproteomics for cancer biomarker discovery. *Clin. Proteom.* **2014**, *11*, 18. [CrossRef]
23. Clark, D.; Mao, L. Cancer biomarker discovery: Lectin-based strategies targeting glycoproteins. *Dis. Markers* **2012**, *33*, 1–10. [CrossRef]
24. Yuan, W.; Benicky, J.; Wei, R.; Goldman, R.; Sanda, M. Quantitative Analysis of Sex-Hormone-Binding Globulin Glycosylation in Liver Diseases by Liquid Chromatography-Mass Spectrometry Parallel Reaction Monitoring. *J. Proteome Res.* **2018**, *17*, 2755–2766. [CrossRef]
25. Kurosky, A.; Barnett, D.R.; Lee, T.H.; Touchstone, B.; Hay, R.E.; Arnott, M.S.; Bowman, B.H.; Fitch, W.M. Covalent structure of human haptoglobin: A serine protease homolog. *Proc. Natl. Acad. Sci. USA* **1980**, *77*, 3388–3392. [CrossRef]
26. Mandato, V.D.; Magnani, E.; Abrate, M.; Casali, B.; Nicoli, D.; Farnetti, E.; Formisano, D.; Pirillo, D.; Ciarlini, G.; De Iaco, P.; et al. Haptoglobin phenotype and epithelial ovarian cancer. *Anticancer Res.* **2012**, *32*, 4353–4358. [PubMed]
27. Morishita, K.; Ito, N.; Koda, S.; Maeda, M.; Nakayama, K.; Yoshida, K.; Takamatsu, S.; Yamada, M.; Eguchi, H.; Kamada, Y.; et al. Haptoglobin phenotype is a critical factor in the use of fucosylated haptoglobin for pancreatic cancer diagnosis. *Clin. Chim. Acta* **2018**, *487*, 84–89. [CrossRef] [PubMed]
28. Fujita, K.; Shimomura, M.; Uemura, M.; Nakata, W.; Sato, M.; Nagahara, A.; Nakai, Y.; Takamatsu, S.; Miyoshi, E.; Nonomura, N. Serum fucosylated haptoglobin as a novel prognostic biomarker predicting high-Gleason prostate cancer. *Prostate* **2014**, *74*, 1052–1058. [CrossRef]
29. Takahashi, S.; Sugiyama, T.; Shimomura, M.; Kamada, Y.; Fujita, K.; Nonomura, N.; Miyoshi, E.; Nakano, M. Site-specific and linkage analyses of fucosylated N-glycans on haptoglobin in sera of patients with various types of cancer: Possible implication for the differential diagnosis of cancer. *Glycoconj. J.* **2016**, *33*, 471–482. [CrossRef] [PubMed]
30. Lee, S.H.; Jeong, S.; Lee, J.; Yeo, I.S.; Oh, M.J.; Kim, U.; Kim, S.; Kim, S.H.; Park, S.Y.; Kim, J.H.; et al. Glycomic profiling of targeted serum haptoglobin for gastric cancer using nano LC/MS and LC/MS/MS. *Mol. Biosyst.* **2016**, *12*, 3611–3621. [CrossRef] [PubMed]
31. Kim, J.H.; Lee, S.H.; Choi, S.; Kim, U.; Yeo, I.S.; Kim, S.H.; Oh, M.J.; Moon, H.; Lee, J.; Jeong, S.; et al. Direct analysis of aberrant glycosylation on haptoglobin in patients with gastric cancer. *Oncotarget* **2017**, *8*, 11094–11104. [CrossRef]
32. Lee, J.; Hua, S.; Lee, S.H.; Oh, M.J.; Yun, J.; Kim, J.Y.; Kim, J.H.; Kim, J.H.; An, H.J. Designation of fingerprint glycopeptides for targeted glycoproteomic analysis of serum haptoglobin: Insights into gastric cancer biomarker discovery. *Anal. Bioanal. Chem.* **2018**, *410*, 1617–1629. [CrossRef]
33. Oh, M.J.; Lee, S.H.; Kim, U.; An, H.J. In-depth investigation of altered glycosylation in human haptoglobin associated cancer by mass spectrometry. *Mass Spectrom. Rev.* **2021**. [CrossRef]
34. Stavenhagen, K.; Hinneburg, H.; Thaysen-Andersen, M.; Hartmann, L.; Varon Silva, D.; Fuchser, J.; Kaspar, S.; Rapp, E.; Seeberger, P.H.; Kolarich, D. Quantitative mapping of glycoprotein micro-heterogeneity and macro-heterogeneity: An evaluation of mass spectrometry signal strengths using synthetic peptides and glycopeptides. *J. Mass Spectrom.* **2013**, *48*, 627–639. [CrossRef] [PubMed]
35. Di Girolamo, F.; Lante, I.; Muraca, M.; Putignani, L. The Role of Mass Spectrometry in the "Omics" Era. *Curr. Org. Chem.* **2013**, *17*, 2891–2905. [CrossRef] [PubMed]
36. Wolyniak, M.J.; Reyna, N.S.; Plymale, R.; Pope, W.H.; Westholm, D.E. Mass Spectrometry as a Tool to Enhance "-omics" Education. *J. Microbiol. Biol. Educ.* **2018**, *19*. [CrossRef] [PubMed]

37. Goldenberg, N.A.; Everett, A.D.; Graham, D.; Bernard, T.J.; Nowak-Gottl, U. Proteomic and other mass spectrometry based "omics" biomarker discovery and validation in pediatric venous thromboembolism and arterial ischemic stroke: Current state, unmet needs, and future directions. *Proteom. Clin. Appl.* **2014**, *8*, 828–836. [CrossRef] [PubMed]
38. Varghese, R.S.; Goldman, L.; An, Y.; Loffredo, C.A.; Abdel-Hamid, M.; Kyselova, Z.; Mechref, Y.; Novotny, M.; Drake, S.K.; Goldman, R.; et al. Integrated peptide and glycan biomarker discovery using MALDI-TOF mass spectrometry. In Proceedings of the 30th Annual International Conference of the IEEE Engineering in Medicine and Biology Society, Vancouver, BC, Canada, 20–25 August 2008; Volume 2008, pp. 3791–3794. [CrossRef]
39. Wuhrer, M. Glycomics using mass spectrometry. *Glycoconj. J.* **2013**, *30*, 11–22. [CrossRef] [PubMed]
40. Aizpurua-Olaizola, O.; Torano, J.S.; Falcon-Perez, J.M.; Williams, C.; Reichardt, N.; Boons, G.J. Mass spectrometry for glycan biomarker discovery. *TrAC Trend Anal. Chem.* **2018**, *100*, 7–14. [CrossRef]
41. Lebrilla, C.B.; An, H.J. The prospects of glycan biomarkers for the diagnosis of diseases. *Mol. Biosyst.* **2009**, *5*, 17–20. [CrossRef] [PubMed]
42. Zaia, J. Mass spectrometry and glycomics. *OMICS* **2010**, *14*, 401–418. [CrossRef] [PubMed]
43. Ozcan, S.; Barkauskas, D.A.; Renee Ruhaak, L.; Torres, J.; Cooke, C.L.; An, H.J.; Hua, S.; Williams, C.C.; Dimapasoc, L.M.; Han Kim, J.; et al. Serum glycan signatures of gastric cancer. *Cancer Prev. Res.* **2014**, *7*, 226–235. [CrossRef] [PubMed]
44. Hua, S.; Williams, C.C.; Dimapasoc, L.M.; Ro, G.S.; Ozcan, S.; Miyamoto, S.; Lebrilla, C.B.; An, H.J.; Leiserowitz, G.S. Isomer-specific chromatographic profiling yields highly sensitive and specific potential N-glycan biomarkers for epithelial ovarian cancer. *J. Chromatogr. A* **2013**, *1279*, 58–67. [CrossRef] [PubMed]
45. Hua, S.; Saunders, M.; Dimapasoc, L.M.; Jeong, S.H.; Kim, B.J.; Kim, S.; So, M.; Lee, K.S.; Kim, J.H.; Lam, K.S.; et al. Differentiation of cancer cell origin and molecular subtype by plasma membrane N-glycan profiling. *J. Proteome Res.* **2014**, *13*, 961–968. [CrossRef]
46. Tang, Z.; Varghese, R.S.; Bekesova, S.; Loffredo, C.A.; Hamid, M.A.; Kyselova, Z.; Mechref, Y.; Novotny, M.V.; Goldman, R.; Ressom, H.W. Identification of N-glycan serum markers associated with hepatocellular carcinoma from mass spectrometry data. *J. Proteome Res.* **2010**, *9*, 104–112. [CrossRef] [PubMed]
47. Hua, S.; An, H.J.; Ozcan, S.; Ro, G.S.; Soares, S.; DeVere-White, R.; Lebrilla, C.B. Comprehensive native glycan profiling with isomer separation and quantitation for the discovery of cancer biomarkers. *Analyst* **2011**, *136*, 3663–3671. [CrossRef] [PubMed]
48. Bones, J.; Mittermayr, S.; O'Donoghue, N.; Guttman, A.; Rudd, P.M. Ultra performance liquid chromatographic profiling of serum N-glycans for fast and efficient identification of cancer associated alterations in glycosylation. *Anal. Chem.* **2010**, *82*, 10208–10215. [CrossRef] [PubMed]
49. Schneck, N.A.; Ivleva, V.B.; Cai, C.X.; Cooper, J.W.; Lei, Q.P. Characterization of the furin cleavage motif for HIV-1 trimeric envelope glycoprotein by intact LC-MS analysis. *Analyst* **2020**, *145*, 1636–1640. [CrossRef] [PubMed]
50. Zhu, J.; Chen, Z.; Zhang, J.; An, M.; Wu, J.; Yu, Q.; Skilton, S.J.; Bern, M.; Ilker Sen, K.; Li, L.; et al. Differential Quantitative Determination of Site-Specific Intact N-Glycopeptides in Serum Haptoglobin between Hepatocellular Carcinoma and Cirrhosis Using LC-EThcD-MS/MS. *J. Proteome Res.* **2019**, *18*, 359–371. [CrossRef]
51. Baerenfaenger, M.; Meyer, B. Intact Human Alpha-Acid Glycoprotein Analyzed by ESI-qTOF-MS: Simultaneous Determination of the Glycan Composition of Multiple Glycosylation Sites. *J. Proteome Res.* **2018**, *17*, 3693–3703. [CrossRef]
52. Turiak, L.; Sugar, S.; Acs, A.; Toth, G.; Gomory, A.; Telekes, A.; Vekey, K.; Drahos, L. Site-specific N-glycosylation of HeLa cell glycoproteins. *Sci. Rep.* **2019**, *9*, 14822. [CrossRef] [PubMed]
53. Hever, H.; Darula, Z.; Medzihradszky, K.F. Characterization of Site-Specific N-Glycosylation. *Methods Mol. Biol.* **2019**, *1934*, 93–125. [CrossRef] [PubMed]
54. Cao, L.; Diedrich, J.K.; Ma, Y.; Wang, N.; Pauthner, M.; Park, S.R.; Delahunty, C.M.; McLellan, J.S.; Burton, D.R.; Yates, J.R.; et al. Global site-specific analysis of glycoprotein N-glycan processing. *Nat. Protoc.* **2018**, *13*, 1196–1212. [CrossRef]
55. Birzele, F.; Csaba, G.; Zimmer, R. Alternative splicing and protein structure evolution. *Nucleic Acids Res.* **2008**, *36*, 550–558. [CrossRef] [PubMed]
56. Rehder, D.S.; Dillon, T.M.; Pipes, G.D.; Bondarenko, P.V. Reversed-phase liquid chromatography/mass spectrometry analysis of reduced monoclonal antibodies in pharmaceutics. *J. Chromatogr. A* **2006**, *1102*, 164–175. [CrossRef] [PubMed]
57. Gomar, J.J.; Bobes-Bascaran, M.T.; Conejero-Goldberg, C.; Davies, P.; Goldberg, T.E. The Alzheimer's Disease Neuroimaging Initiative. Utility of combinations of biomarkers, cognitive markers, and risk factors to predict conversion from mild cognitive impairment to Alzheimer disease in patients in the Alzheimer's disease neuroimaging initiative. *Arch. Gen. Psychiatry* **2011**, *68*, 961–969. [CrossRef] [PubMed]
58. Laxman, B.; Morris, D.S.; Yu, J.; Siddiqui, J.; Cao, J.; Mehra, R.; Lonigro, R.J.; Tsodikov, A.; Wei, J.T.; Tomlins, S.A.; et al. A first-generation multiplex biomarker analysis of urine for the early detection of prostate cancer. *Cancer Res.* **2008**, *68*, 645–649. [CrossRef]
59. Mamtani, M.R.; Thakre, T.P.; Kalkonde, M.Y.; Amin, M.A.; Kalkonde, Y.V.; Amin, A.P.; Kulkarni, H. A simple method to combine multiple molecular biomarkers for dichotomous diagnostic classification. *BMC Bioinform.* **2006**, *7*, 442. [CrossRef] [PubMed]
60. Bones, J.; Byrne, J.C.; O'Donoghue, N.; McManus, C.; Scaife, C.; Boissin, H.; Nastase, A.; Rudd, P.M. Glycomic and glycoproteomic analysis of serum from patients with stomach cancer reveals potential markers arising from host defense response mechanisms. *J. Proteome Res.* **2011**, *10*, 1246–1265. [CrossRef] [PubMed]

Article

The Application Value of Lipoprotein Particle Numbers in the Diagnosis of HBV-Related Hepatocellular Carcinoma with BCLC Stage 0-A

Duo Zuo [†], Haohua An [†], Jianhua Li, Jiawei Xiao and Li Ren *

Department of Clinical Laboratory, Tianjin Medical University Cancer Institute and Hospital, National Clinical Research Center for Cancer, Tianjin's Clinical Research Center for Cancer, Key Laboratory of Cancer Prevention and Therapy, Tianjin 300060, China; duozuo@tmu.edu.cn (D.Z.); AHH1122@tmu.edu.cn (H.A.); lijianhua@tjmuch.com (J.L.); xiaologin233@163.com (J.X.)
* Correspondence: liren@tmu.edu.cn; Tel.: +86-022-23340123-5208
† These authors contributed equally to this work.

Abstract: Early diagnosis is essential for improving the prognosis and survival of patients with hepatocellular carcinoma (HCC). This study aims to explore the clinical value of lipoprotein subfractions in the diagnosis of hepatitis B virus (HBV)-related HCC. Lipoprotein subfractions were detected by ^{1}H-NMR spectroscopy, and the pattern-recognition method and binary logistic regression were performed to classify distinct serum profiles and construct prediction models for HCC diagnosis. Differentially expressed proteins associated with lipid metabolism were detected by LC-MS/MS, and the potential prognostic significance of the mRNA expression was evaluated by Kaplan–Meier *survival analysis*. The diagnostic panel constructed from the serum particle number of very-low-density lipoprotein (VLDL), intermediate-density lipoprotein (IDL), and low-density lipoprotein (LDL-1~LDL-6) achieved higher accuracy for the diagnosis of HBV-related HCC and HBV-related benign liver disease (LD) than that constructed from serum alpha-fetoprotein (AFP) alone in the training set (AUC: 0.850 vs. AUC: 0.831) and validation set (AUC: 0.926 vs. AUC: 0.833). Furthermore, the panel achieved good diagnostic performance in distinguishing AFP-negative HCC from AFP-negative LD (AUC: 0.773). We also found that lipoprotein lipase (LPL) transcript levels showed a significant increase in cancerous tissue and that high expression was significantly positively correlated with the poor prognosis of patients. Our research provides new insight for the development of diagnostic biomarkers for HCC, and abnormal lipid metabolism and LPL-mediated abnormal serum lipoprotein metabolism may be important factors in promoting HCC development.

Keywords: lipidomics; ^{1}H-NMR; LC-MS/MS; lipoprotein subfractions; lipoprotein lipase; cancer biomarkers; hepatocellular carcinoma

1. Introduction

Hepatocellular carcinoma (HCC) is increasingly recognized as a serious, worldwide public health concern. It is the sixth-most common malignancy and the fourth leading cause of cancer death in the world, with approximately 841,000 new cases and 782,000 deaths worldwide each year. The most common risk factors for HCC are chronic infection with hepatitis B virus (HBV) and hepatitis C virus (HCV); in the vast majority of cases, HCC occurs in individuals with cirrhosis caused by chronic infection with HBV in China [1]. China has the largest liver disease patient population in the world; according to a statistical study of Chinese HCC patients, HBV-related cirrhosis is the most important cause of HCC in China, and more than 80% of HCC patients have varying degrees of HBV infection [2].

Unfortunately, due to a lack of typical clinical manifestations, HCC is difficult to diagnose in the early stage. It has been reported that the overall median survival time of advanced-stage HCC is only 6–10 months; however, for early-stage HCC, surgical treatments such as radiofrequency ablation, selective hepatectomy, and liver transplantation can increase the 5-year survival rate to 60–80% [3]. Therefore, early diagnosis and timely surgical treatment are crucial for better patient outcomes. Pathological examination is the gold standard for diagnosing HCC; however, this method is invasive and is accompanied by risks of bleeding and needle track seeding, which are not recommended before surgery [4]. To date, alpha-fetoprotein (AFP) is the most commonly used serological test for the early diagnosis and monitoring of the development of HCC; however, owing to a lack of sensitivity and specificity, the application of AFP in the diagnosis and prognosis monitoring of HCC is limited [5].

Metabolic reprogramming is recognized as a hallmark in cancer development. Tumor cells must adjust their own metabolic states to maintain excessive proliferation rates; compared with normal cells, the metabolic activities of tumor cells are more vigorous, increasing tumor cell growth and invasion [6]. The associated changes in the metabolite network structure of tumor cells indicate that cancer biomarkers should not be assessed with regard to changes in one or several biochemical indicators but rather to changes in a set of metabolite indicators [7]. The emergence of metabolomics, a discipline that studies the small-molecule intermediates of metabolism in organisms at a certain time [8], has promoted the study of cancer metabolism. As an important branch of metabolomics, lipidomics describes spatial and temporal alterations in the content and composition of different lipid molecules and serves as a powerful tool in the development of lipid biomarkers for studying disease states [9]. Lipidomics has an extremely important position in cancer research; through the high-throughput detection and quantitative analysis of biological fluids (blood, urine, saliva, and fecal extracts), lipidomics can be used to study the mechanism of disease occurrence and development [10].

^{1}H-nuclear magnetic resonance (^{1}H-NMR) is one of the most commonly used high-throughput platforms in metabolomics research. ^{1}H-NMR spectroscopy provides an alternative method of measuring lipoprotein levels in serum, and quantitative detection by ^{1}H-NMR can determine the quality, particle number, and particle size of lipoprotein subfractions by detecting the terminal methyl protons of phospholipids, unesterified cholesterols, cholesterol esters and triglycerides [11]. As potential risk factors for HCC development, hepatitis and liver cirrhosis are often associated with serum lipid and lipoprotein aberrations, and a number of reports have illustrated that the serum levels of many kinds of lipids, lipoproteins and apolipoproteins show obvious changes in HCC patients [12,13]. Lipoprotein particle distributions have great potential for helping improve the diagnostics of metabolic disorders [14]; however, studies estimating subfractions of lipoproteins have been restricted to patients with cardiovascular disease and are rarely extended to the exploration of cancer research [15–17]. Previous serum and urine metabolomics studies have illustrated that compared with those of patients with cirrhosis or healthy controls, several small-molecule metabolites, such as glucose, glutamine, citrate, creatine, creatinine, carnitine, glycine, and acetate, show remarkable changes in HCC patients [18–21]. However, few studies have focused on lipid metabolism disturbances and serum lipoprotein subfraction changes in HBV-related HCC patients. Therefore, this study aims to develop novel diagnostic lipid biomarkers of HBV-related HCC.

In this article, we utilized ^{1}H-NMR to detect collected serum samples and performed multivariable and univariable statistical analysis to study the serum lipoprotein subfraction in patients with HBV-related HCC, patients with benign liver disease (including HBV-related hepatitis and HBV-related cirrhosis) and healthy patients. The aim of our study was to identify serum lipidome-based biomarkers as a diagnostic multivariable model for early-stage HCC. Furthermore, we obtained paired cancerous tissues and matched paracancerous tissues from HCC patients to search for differentially expressed proteins involved in lipid metabolism and explore the association with prognosis in patients with HBV-related HCC.

2. Results

2.1. Clinical Characteristics

The demographic and clinical characteristics of the study participants with HCC and LD are summarized in Table 1. In the training set, the serum levels of AFP, alanine transaminase (ALT), aspartate transaminase (AST), and total protein (TP) were significantly different between the HCC and LD groups ($p < 0.05$). In the validation set, compared with those of patients with LD, the serum AFP levels of the HCC patients were significantly different ($p < 0.05$). The clinical characteristics of the normal controls (NCs) and AFP-negative patients are summarized in Tables S2 and S3.

Table 1. Clinical characteristics of the HCC group and LD group.

Characteristics	Training Set			Validation Set		
	HCC	Liver Disease	p-Value	HCC	Liver Disease	p-Value
n	51	37		24	17	
Age (years)	58 (33.00 to 71.00)	59 (43.00 to 66.00)	0.889	60.50 (51.75 to 66.00)	49.00 (45.50 to 58.00)	0.010
Sex (male/female)	37/14	21/16	0.123	16/8	8/9	0.209
AFP (ng/mL)	23.88 (4.80 to 126.20)	2.69 (1.68 to 4.64)	<0.001	60.50 (3.42 to 481.00)	2.66 (1.98 to 3.55)	<0.001
ALT (IU/L)	36.00 (22.00 to 48.00)	25.00 (16.00 to 36.50)	0.018	24.00 (15.25 to 31.75)	21.00 (14.00 to 35.00)	0.937
AST (IU/L)	37.00 (26.00 to 59.00)	28.00 (19.00 to 36.50)	0.002	29.00 (21.50 to 38.50)	21.00 (16.00 to 29.00)	0.095
ALB (g/L)	41.60 (39.40 to 45.50)	43.50 (37.50 to 46.95)	0.422	42.55 (36.50 to 45.08)	44.90 (41.45 to 47.60)	0.058
TP (g/L)	74.50 (70.90 to 77.20)	68.40 (53.50 to 77.65)	0.016	70.45 (64.60 to 75.93)	74.40 (66.25 to 76.25)	0.404
TBIL (μmmol/L)	16.50 (13.90 to 23.60)	14.50 (11.05 to 18.63)	0.154	14.70 (11.08 to 20.98)	12.40 (9.80 to 18.25)	0.375
CRE (μmol/L)	61.00 (53.00 to 69.00)	62.00 (52.00 to 79.50)	0.244	61.50 (56.00 to 78.00)	60.00 (51.50 to 68.00)	0.255
BCLC stage						
stage 0	9	/		1	/	
stage A	42	/		23	/	
Child-Pugh class						
A	46	33		22	16	
B-C	5	4		2	1	
Tumor diameter (cm)						
≤3	25	/		10	/	
>3	26	/		14	/	

p-values: Mann–Whitney U test for continuous variables and Pearson's chi-square test for categorical variables. Continuous data are presented as medians with interquartile ranges (IQRs). Abbreviations: AFP, alpha-fetoprotein; ALT, alanine aminotransferase; AST, aspartate transaminase; ALB, albumin; TP, total protein; TBIL, total bilirubin; CRE, creatinine.

2.2. NMR Spectroscopic Multivariable Analysis

For comprehensive observation of the lipoprotein subfractions, PCA and PLS-DA were employed to explore the intrinsic differences between different groups. The score plot of PCA and PLS-DA (Figure S2) showed that lipoprotein subfractions of the three groups could be distinguished, and the model parameters of PCA ($R^2X = 0.99$, $Q^2 = 0.944$) and PLS-DA ($R^2Y = 0.447$, $Q^2Y = 0.402$, CV-ANOVA $p < 0.0001$) indicated that the constructed models have favorable robustness.

To further filter the variables, the ^{1}H-NMR serum spectra of the patients with HCC and NCs were discriminated with the OPLS-DA model, as shown in a score plot (Figure 1a), which illustrates that this model can significantly discriminate between HCC patients and NCs. The predictive ability was calculated through 7-fold cross-validation ($R^2Y = 0.843$, $Q^2Y = 0.821$, CV-ANOVA $p < 0.0001$), suggesting that the model possessed a satisfactory fit with good predictive power. The loading plot indicated a brief overview of the contribution of each lipoprotein subfraction to the OPLS-DA model (Figure 1b), and the variables responsible for significantly contributing to the separation of the two groups are indicated in the corresponding S-plot (Figure 1c) and S-line plot (Figure 1d). Using the variable importance in projection (VIP) score (>1.0) from the OPLS-DA model, a total of 17 lipoprotein subfractions were selected (Figure 1e). To further assess the robustness of the constructed OPLS-DA model and prevent it from overfitting, a 999-permutation test

(Figure 1f) was performed, and the results (intercepts: $R^2 = 0.138$, $Q^2 = -0.313$) indicated that this OPLS-DA model had high discriminability.

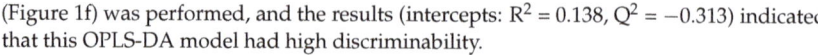

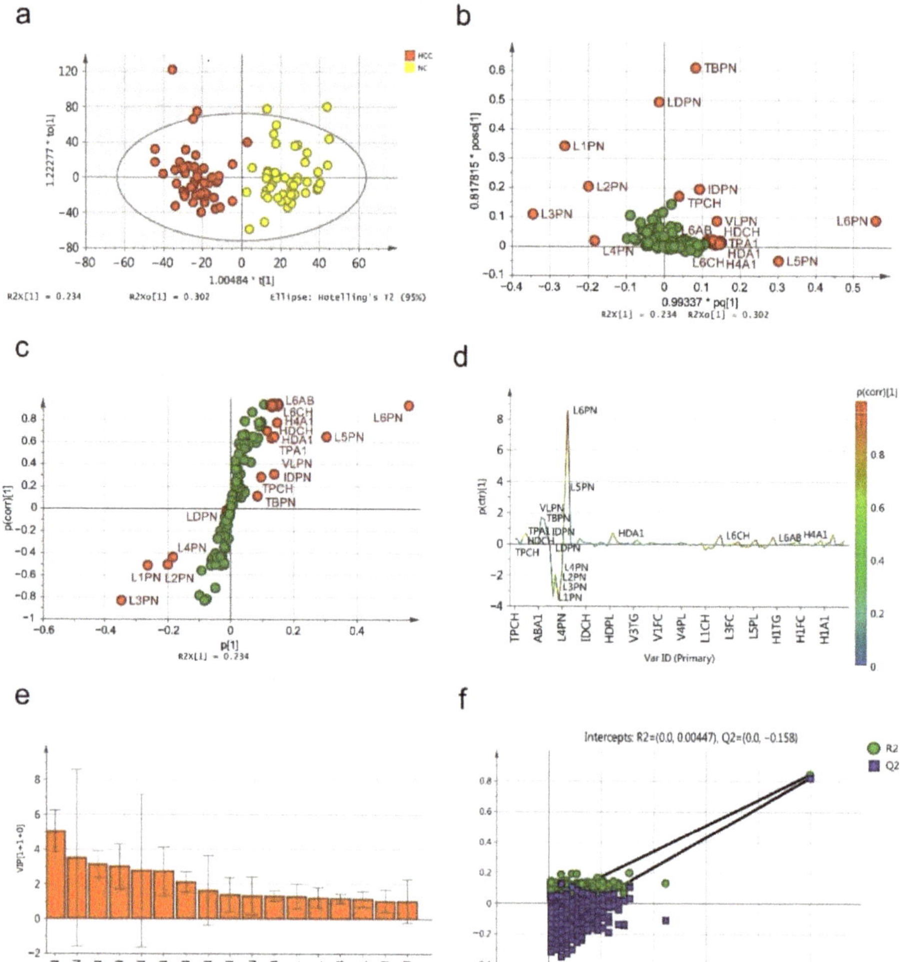

Figure 1. The serum lipidomic profile discriminates between HCC patients and normal controls (NCs). (**a**) Score plot was generated by the OPLS-DA model. The horizontal coordinate (1.00484 * t [1]) represents the score value of the main component, and the vertical coordinate (1.22277 * to [1]) represents the score value of the orthogonal component. (**b**) Loading plot was generated by the OPLS-DA model. The horizontal coordinate (0.99337 * pq [1]) represents the predicted principal component, and the vertical coordinate (0.817815 * poso [1]) represents the orthogonal principal components. The corresponding (**c**) S-plot and (**d**) S-line plot for the model displaying the discriminant variables and the associated predictive loadings. The red circles indicate selected lipoprotein subfractions with VIP scores >1.0, and other variables with no difference are referred to as green circles in (**b**,**c**). (**e**) The selected lipoprotein subfractions with VIP scores >1.0. (**f**) Permutation test (999 times) of the OPLS-DA model.

Then, we applied another OPLS-DA model to distinguish HCC patients from LD patients, and the score plot indicated that the group of HCC patients could be excellently separated from the LD patient group (Figure 2a). A loading plot (Figure 2b) illustrated the contribution of each lipoprotein subfraction in distinguishing HCC patients from

LD patients, and the S-plot (Figure 2c) and S-line plot (Figure 2d) showed the variables significantly contributing to the separation. According to the VIP score (>1.0), a total of 15 lipoprotein subfractions were selected (Figure 2e). The results of the internal validation (R^2Y = 0.530, Q^2Y = 0.343, CV-ANOVA p < 0.001) and the results of the permutation test (intercepts: R^2 = 0.004, Q^2 = −0.158) suggested that the constructed OPLS-DA model has favorable robustness and could be used in the next step of analysis (Figure 2f). The relevant lipoprotein subfractions and their statistical details are listed in Table 2.

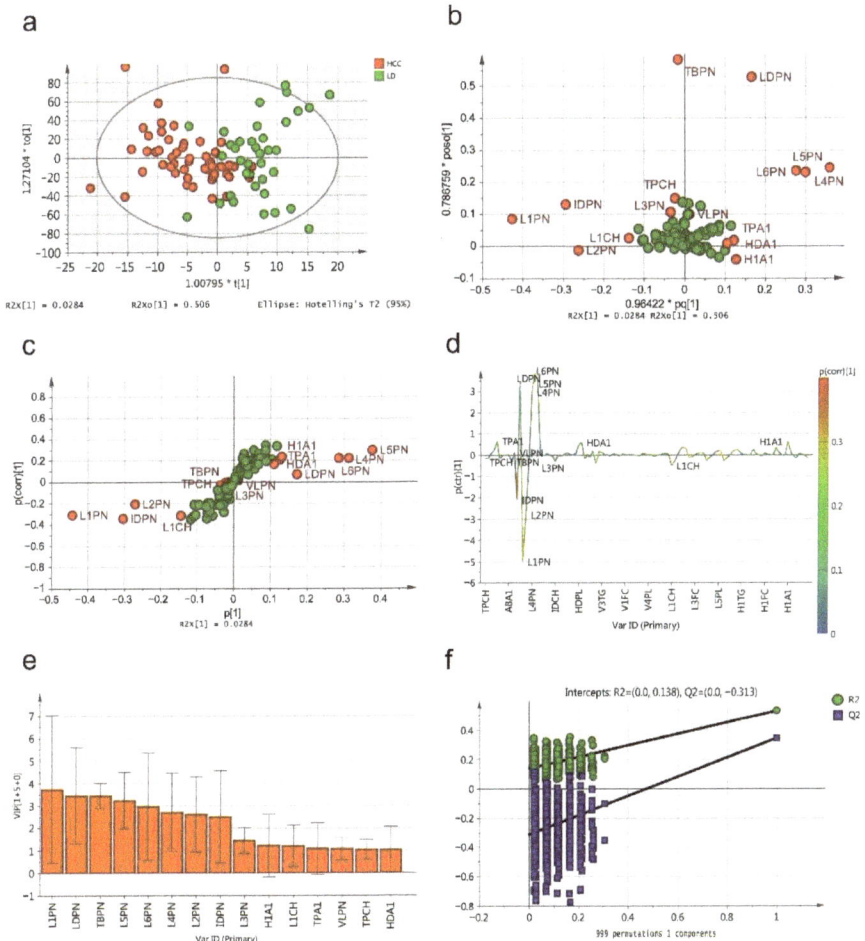

Figure 2. The serum lipidomic profile discriminates between HCC patients and liver disease (LD). (**a**) Score plot was generated by the OPLS-DA model. The horizontal coordinate (1.00795 * t [1]) represents the score value of the main component, and the vertical coordinate (1.27104 * to [1]) represents the score value of the orthogonal component. (**b**) Loading plot was generated by the OPLS-DA model. The horizontal coordinate (0.96422 * pq [1]) represents the predicted principal component, and the vertical coordinate (0.786795 * poso [1]) represents the orthogonal principal components. The corresponding (**c**) S-plot and (**d**) S-line plot for the model displaying the discriminant variables and the associated predictive loadings. The red circles indicate selected lipoprotein subfractions with VIP scores >1.0, and other variables with no difference are referred to as green circles in (**b**,**c**). (**e**) The selected lipoprotein subfractions with VIP scores >1.0. (**f**) Permutation test (999 times) of the OPLS-DA model.

Table 2. Summary of the lipoprotein subfraction statistical data from OPLS-DA analysis from HCC patients, liver disease patients, and normal controls.

Index	Description	Unit	HCC vs. LD VIP	HCC vs. LD p (corr)	HCC vs. NCs VIP	HCC vs. NCs p (corr)	p-Value
TPCH	Total Cholesterol	mg/dL	1.045	−0.026	1.030	0.165	0.596
HDCH	HDL-C	mg/dL	0.414	0.120	1.032	0.703	2.931×10^{-14}
TPA1	Apo-A1	mg/dL	1.089	0.202	1.146	0.631	3.742×10^{-10}
TBPN	Total Particle Number	nmol/L	3.451	−0.006	3.512	0.107	0.775
VLPN	VLDL Particle Number	nmol/L	1.083	0.014	1.329	0.310	0.003
IDPN	IDL Particle Number	nmol/L	2.514	−0.347	1.379	0.277	6.318×10^{-5}
LDPN	LDL Particle Number	nmol/L	3.465	0.069	2.772	−0.016	0.339
L1PN	LDL-1 Particle Number	nmol/L	3.743	−0.310	3.024	−0.512	1.399×10^{-9}
L2PN	LDL-2 Particle Number	nmol/L	2.612	−0.209	2.118	−0.511	1.712×10^{-4}
L3PN	LDL-3 Particle Number	nmol/L	1.451	−0.032	3.144	−0.838	<0.001
L4PN	LDL-4 Particle Number	nmol/L	2.708	0.221	1.645	−0.443	2.148×10^{-10}
L5PN	LDL-5 Particle Number	nmol/L	3.232	0.296	2.737	0.643	5.121×10^{-8}
L6PN	LDL-6 Particle Number	nmol/L	2.971	0.218	5.050	0.934	<0.001
HDA1	HDL Apo-A1	mg/dL	1.037	0.167	1.227	0.644	1.538×10^{-11}
L1CH	LDL-1 Cholesterol	mg/dL	1.201	−0.316	0.993	−0.563	2.456×10^{-9}
L6CH	LDL-6 Cholesterol	mg/dL	0.942	0.227	1.324	0.936	<0.001
L6AB	LDL-6 Apo-B	mg/dL	0.802	0.218	1.184	0.934	<0.001
H1A1	HDL-1 Apo-A1	mg/dL	1.229	0.230	0.146	−0.271	0.011
H4A1	HDL-4 Apo-A1	mg/dL	0.811	−0.083	1.306	0.773	<0.001

The characteristics of significantly different variables in the OPLS-DA model. p (corr) is the OPLS-DA loading scaled as a correlation coefficient. The significance of the values was assessed using the Kruskal–Wallis test. VIP, variable importance in projection; HCC, hepatocellular carcinoma; LD, liver disease; NCs, normal controls.

Next, the PCA and PLS-DA analysis of the serum AFP-negative patients showed that lipoprotein subfractions of the three groups could be distinguished (Figure S3), with parameters of PCA ($R^2X = 0.992$, $Q^2 = 0.947$) and PLS-DA ($R^2Y = 0.430$, $Q^2Y = 0.390$, CV-ANOVA $p < 0.0001$). The ^{1}H-NMR serum spectra of the patients with AFP-negative HCC and NCs were discriminated with the OPLS-DA model, as shown in a score plot (Figure 3a), and the predictive ability was calculated through 7-fold cross-validation ($R^2Y = 0.848$, $Q^2Y = 0.827$, CV-ANOVA $p < 0.0001$). The loading plot (Figure 3b) illustrated the contribution of each lipoprotein subfraction in distinguishing AFP-negative HCC patients from NCs, and the S-plot (Figure 3c) and S-line plot (Figure 3d) showed the variables significantly contributing to the separation. Using the VIP score (>1.0) from the OPLS-DA model, a total of 18 lipoprotein subfractions were selected (Figure 3e). Meanwhile, the results of the permutation test (intercepts: $R^2 = 0.00827$, $Q^2 = -0.167$, Figure 3f) suggested that the constructed OPLS-DA models have favorable robustness.

Next, we applied another OPLS-DA model to distinguish AFP-negative HCC patients from AFP-negative LD patients (Figure 4a). The loading plot (Figure 4b) illustrated the contribution of each lipoprotein subfraction in distinguishing these two groups, and the S-plot (Figure 4c) and S-line plot (Figure 4d) showed the variables significantly contributing to the separation. According to the VIP score (>1.0) and S-line plot (Figure 4e) of the OPLS-DA model, a total of 15 lipoprotein subfractions were selected. The results of the internal validation ($R^2Y = 0.142$, $Q^2Y = 0.0893$, CV-ANOVA $p = 0.057$) and the results of the permutation test (intercepts: $R^2 = 0.058$, $Q^2 = -0.0842$) suggested that the constructed OPLS-DA model has favorable robustness and could be used in the next step of analysis (Figure 4f). The relevant lipoprotein subfractions and their statistical details are listed in Table 3.

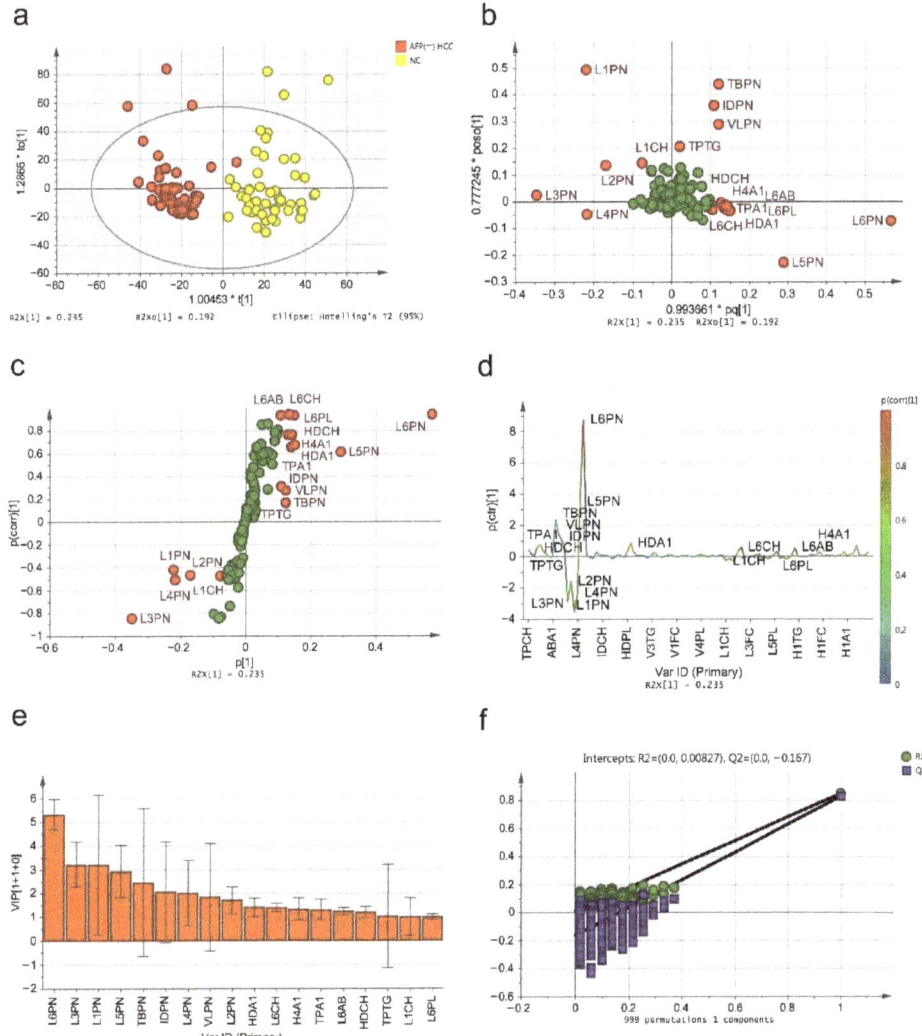

Figure 3. The serum lipidomic profile discriminates between AFP-negative HCC patients and NCs. (**a**) Score plot was generated by the OPLS-DA model. The horizontal coordinate (1.00453 * t [1]) represents the score value of the main component, and the vertical coordinate (1.2866 * to [1]) represents the score value of the orthogonal component. (**b**) Loading plot was generated by the OPLS-DA model. The horizontal coordinate (0.993661 * pq [1]) represents the predicted principal component, and the vertical coordinate (0.777245 * poso [1]) represents the orthogonal principal components. The corresponding (**c**) S-plot and (**d**) S-line plot for the model displaying the discriminant variables and the associated predictive loadings. The red circles indicate selected lipoprotein subfractions with VIP scores >1.0, and other variables with no difference are referred to as green circles in (**b**,**c**). (**e**) The selected lipoprotein subfractions with VIP scores >1.0. (**f**) Permutation test (999 times) of the OPLS-DA model.

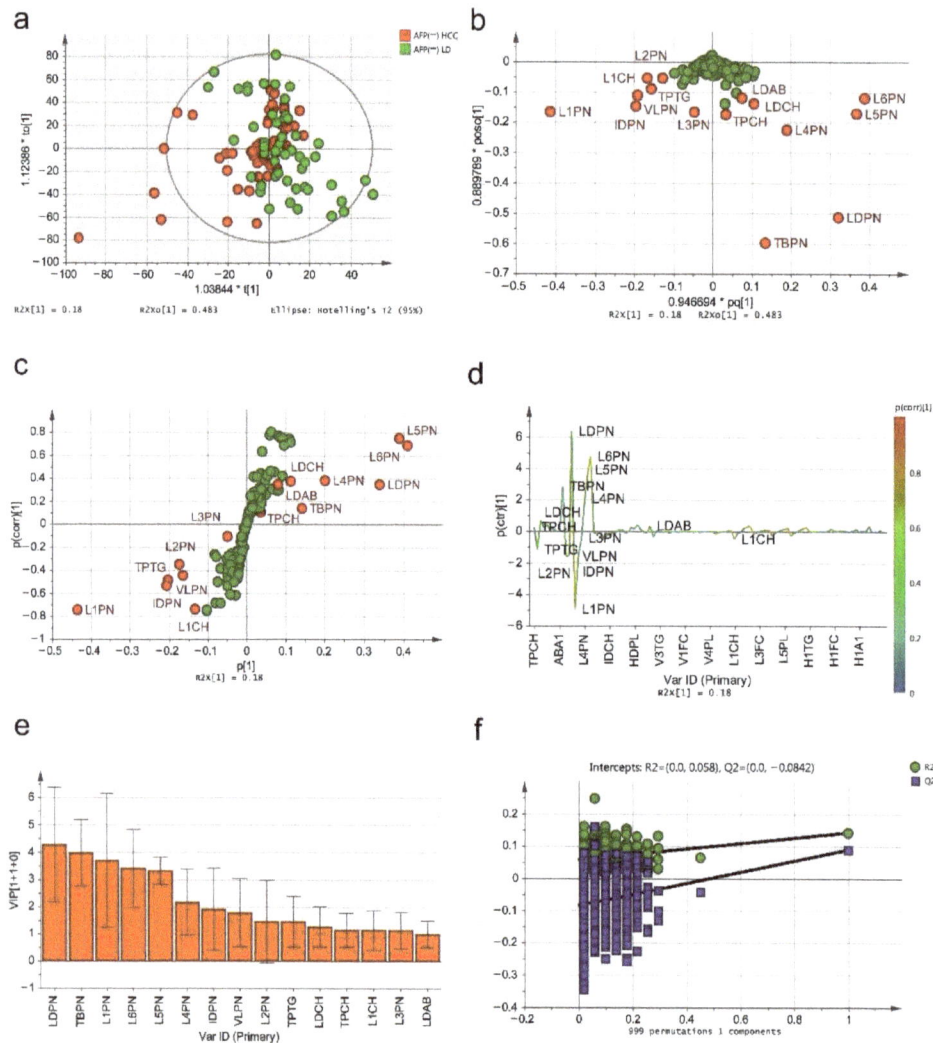

Figure 4. The serum lipidomic profile discriminates between AFP-negative HCC patients and AFP-negative LD patients. (**a**) Score plot was generated by the OPLS-DA model. The horizontal coordinate (1.03844 * t [1]) represents the score value of the main component, and the vertical coordinate (1.12386 * to [1]) represents the score value of the orthogonal component. (**b**) Loading plot was generated by the OPLS-DA model. The horizontal coordinate (0.946694 * pq [1]) represents the predicted principal component, and the vertical coordinate (0.889789 * poso [1]) represents the orthogonal principal components. The corresponding (**c**) S-plot and (**d**) S-line plot for the model displaying the discriminant variables and the associated predictive loadings. The red circles indicate selected lipoprotein subfractions with VIP scores >1.0, and other variables with no difference are referred to as green circles in (**b**,**c**). (**e**) The selected lipoprotein subfractions with VIP scores >1.0. (**f**) Permutation test (999 times) of the OPLS-DA model.

Table 3. Summary of the lipoprotein subfraction statistical data from OPLS-DA analysis from AFP (-) HCC patients, AFP (-) liver disease (LD) patients, and normal controls (NCs).

Index	Description	Unit	AFP (-) HCC vs. AFP (-) LD		AFP (-) HCC vs. NCs		p-Value
			VIP	p (corr)	VIP	p (corr)	
TPCH	Total Cholesterol	mg/dL	1.160	0.105	0.836	0.281	0.463
TPTG	Total Triglycerides	mg/dL	1.456	−0.450	1.047	0.060	0.075
LDCH	LDL-C	mg/dL	1.274	0.373	0.204	−0.013	0.214
HDCH	HDL-C	mg/dL	0.414	0.261	1.207	0.763	<0.001
TPA1	Apo-A1	mg/dL	0.690	0.294	1.307	0.655	5.091×10^{-14}
TBPN	Total Particle Number	nmol/L	3.986	0.138	2.477	0.163	0.594
VLPN	VLDL Particle Number	nmol/L	1.794	−0.490	1.850	0.271	0.005
IDPN	IDL Particle Number	nmol/L	1.919	−0.537	2.066	0.309	1.076×10^{-5}
LDPN	LDL Particle Number	nmol/L	4.282	0.350	0.463	0.035	0.084
L1PN	LDL-1 Particle Number	nmol/L	3.702	−0.747	3.206	−0.422	9.240×10^{-8}
L2PN	LDL-2 Particle Number	nmol/L	1.467	−0.354	1.715	−0.465	1.573×10^{-4}
L3PN	LDL-3 Particle Number	nmol/L	1.137	−0.110	3.224	−0.849	<0.001
L4PN	LDL-4 Particle Number	nmol/L	2.181	0.382	2.018	−0.511	1.325×10^{-11}
L5PN	LDL-5 Particle Number	nmol/L	3.338	0.746	2.940	0.610	7.372×10^{-9}
L6PN	LDL-6 Particle Number	nmol/L	3.419	0.685	5.325	0.939	<0.001
HDA1	HDL Apo-A1	mg/dL	0.703	0.298	1.418	0.675	4.663×10^{-14}
LDAB	LDL Apo-B	mg/dL	1.004	0.350	0.109	0.035	0.084
L1CH	LDL-1 Cholesterol	mg/dL	1.149	−0.738	1.013	−0.475	1.382×10^{-8}
L6CH	LDL-6 Cholesterol	mg/dL	0.942	0.717	1.398	0.930	<0.001
L6PL	LDL-6 Phospholipids	mg/dL	0.709	0.757	1.006	0.932	<0.001
L6AB	LDL-6 Apo-B	mg/dL	0.802	0.685	1.249	0.939	<0.001
H4A1	HDL-4 Apo-A1	mg/dL	0.811	0.421	1.335	0.760	<0.001

The characteristics of significantly different variables in the OPLS-DA model. p (corr) is the OPLS-DA loading scaled as a correlation coefficient. The significance of the values was assessed using the Kruskal–Wallis test. VIP, variable importance in projection; AFP (-) HCC, AFP-negative hepatocellular carcinoma; AFP (-) LD, AFP-negative liver disease; NCs, normal controls.

To further determine which lipoprotein subfractions could be used as biomarkers for HCC diagnosis, the common variables in these OPLS-DA models (VIP scores > 1.0 and p-values < 0.05) were selected for the subsequent analysis, including VLPN, IDPN, and L1-L6PN (Table 4 and Figure 5). Detailed information on all lipoprotein subfractions is listed in Tables S4 and S5. The absolute numbers of each lipoprotein subfraction are listed in Table S6.

Table 4. Changes in relative levels of lipoprotein subfractions in serum samples from HCC patients, liver disease patients and normal controls.

	HCC vs. LD		HCC vs. NCs			AFP (-) HCC vs. AFP (-) LD		AFP (-) HCC vs. NCs		
Index	VIP	Trend	VIP	Trend	p-Value	VIP	Trend	VIP	Trend	p-Value
VLPN	1.083	up	1.329	down	3.023×10^{-3}	1.794	up	1.850	down	4.733×10^{-3}
IDPN	2.514	up	1.379	down	6.318×10^{-5}	1.919	up	2.066	down	1.076×10^{-5}
L1PN	3.743	up	3.024	up	1.398×10^{-8}	3.702	up	3.206	up	9.240×10^{-8}
L2PN	2.613	up	2.118	up	1.712×10^{-4}	1.467	up	1.715	up	1.573×10^{-4}
L3PN	1.451	down	3.144	up	<0.001	1.137	down	3.224	up	<0.001
L4PN	2.708	down	1.645	up	2.148×10^{-10}	2.181	down	2.018	up	1.325×10^{-11}
L5PN	3.232	down	2.737	down	5.121×10^{-8}	3.338	down	2.940	down	7.372×10^{-9}
L6PN	2.971	down	5.050	down	<0.001	3.419	down	5.325	down	<0.001

The common variables between the two constructed OPLS-DA models were selected if their VIP scores > 1.0 and univariable p-values < 0.05. HCC, hepatocellular carcinoma; NCs, normal controls; VIP, variable importance in projection.

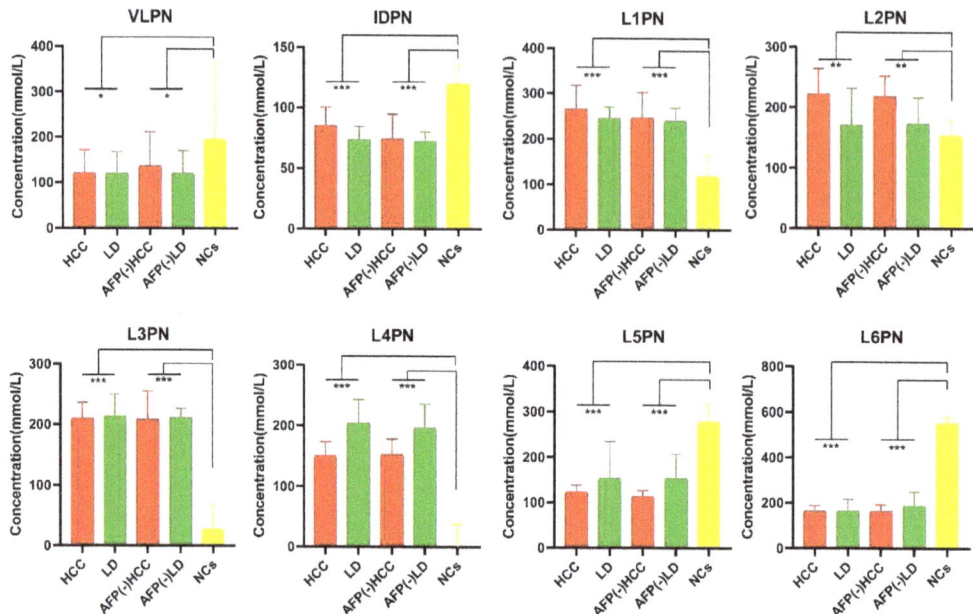

Figure 5. Comparisons between the groups of diagnostic biomarkers. Histograms indicate the median, upper, and lower quartiles of the eight lipoprotein particle numbers used to construct the diagnostic panel. The significance of the values was assessed using the Kruskal–Wallis test (* $p < 0.05$, ** $p < 0.001$, *** $p < 0.0001$). HCC, all HCC patients in the training set; LD, all liver disease patients in the training set; AFP (-) HCC, serum AFP-negative expression HCC patients; AFP (-) LD, serum AFP-negative expression liver disease patients; NCs, normal controls.

2.3. Biomarker Selection and Validation of the Diagnostic Model

To judge the diagnostic performance of selected variables, binary logistic regression analysis was employed to construct the best diagnostic model. Meanwhile, the correlation analysis showed that serum AFP levels had no significant correlation with the selected variables, indicating that lipoprotein particles and AFP are independent of each other (Figure S4). Meanwhile, we analyzed the correlation of selected variables with clinical features by the nonparametric Spearman correlation test. According to the results, L1PN, L2PN, and L3PN were strongly positively associated with age (the range of Spearman's rank correlation coefficient was from 0.3 to 0.5, p-value < 0.05) and negatively associated with male gender (the range of Spearman's rank correlation coefficient was from -0.3 to -0.5, p-value < 0.05). Furthermore, VLPN was strongly positively associated with tumor size (Spearman's coefficient $r = 0.326$, $p = 0.004$) and negatively associated with liver cirrhosis (Spearman's coefficient $r = -0.372$, $p = 0.001$). IDPN was strongly positively associated with tumor size (Spearman's coefficient $r = 0.306$, $p = 0.008$). The results are shown in Figure S5 and Table S7. According to the results of ROC curve analysis, the panel composed of VLPN, IDPN, and L1-L6PN reached excellent diagnostic performance in discriminating HCC patients from NCs with an AUC of 1.000 (95% CI: 0.964–1.000) (Figure 6a). Furthermore, the panel showed better diagnostic performance than serum AFP alone in discriminating HCC patients from LD patients, as indicated by an AUC of 0.850 (95% CI: 0.758–0.917) vs. 0.831 (95% CI: 0.736–0.902), respectively, in the training set, and combining the lipidomic biomarkers with AFP increased the AUC to 0.861 (95% CI: 0.771–0.926) (Figure 6b).

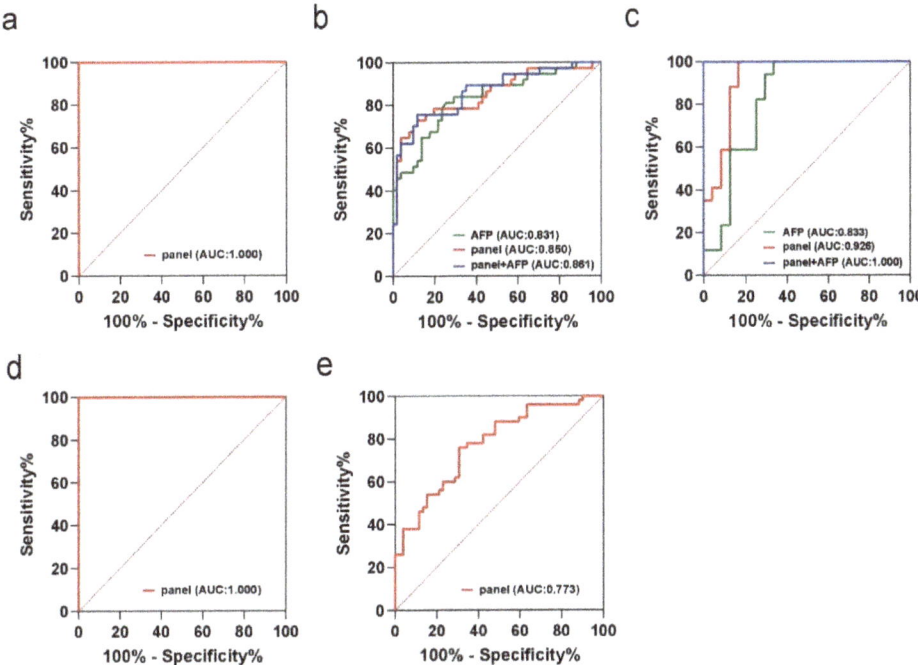

Figure 6. ROC curve analysis following binary logistic regression. The diagnostic performance of the lipidomic biomarker panel in discriminating HCC from (**a**) normal controls (NCs) in the training set. Comparison of the diagnostic performance of the lipidomic biomarker panel with that of serum alpha-fetoprotein (AFP) and their combination in discriminating HCC from liver disease (LD) in the (**b**) training set and (**c**) validation set. The ROC curve of the lipidomic biomarker panel also showed excellent discriminability in discriminating AFP-negative HCC patients from (**d**) NCs and (**e**) AFP-negative LD patients. The red-, green-, and blue-colored lines indicate lipid biomarkers, AFP, and lipid biomarkers with AFP, respectively.

The diagnostic performance of the panel was further confirmed in the external validation set. The diagnostic accuracy of this panel in the validation cohort also demonstrated a superior performance to serum AFP alone (AUC: 0.926; 95% CI: 0.800–0.984 vs. AUC: 0.833; 95% CI: 0.684–0.931), and their combination increased the AUC to 1.000 (95% CI: 0.914–1.000) (Figure 6c). Meanwhile, this panel also achieved good diagnostic accuracy in discriminating AFP-negative HCC patients from NCs and AFP-negative LD patients, with AUCs of 1.000 (0.964–1.000) (Figure 6d) and 0.773 (0.680–0.850) (Figure 6e), respectively. The ROC results are shown in Table 5. Furthermore, to investigate whether the diagnostic panel can more realistically reflect the diagnostic approach in regular patient care, we unified LD patients and NCs into the non-HCC group ($n = 104$) and performed a differential diagnosis analysis of the non-HCC group versus the HCC group ($n = 75$). According to binary logistic regression and ROC curve analysis, the panel constructed by these eight indicators achieved good diagnostic accuracy (AUC: 0.842; 95% CI: 0.780–0.892). Meanwhile, we unified AFP-negative LD patients and NCs into the AFP (−) non-HCC group ($n = 100$) and compared them with AFP-negative HCC patients ($n = 52$). The panel constructed by these eight indicators also achieved good diagnostic accuracy (AUC: 0.837; 95% CI: 0.769–0.892) in determining serum AFP-negative expression populations. The ROC results are shown in Table S8 and Figure S6.

These results indicate that the panel constructed by VLPN, IDPN, and L1-L6PN has strong potential in the diagnosis of HBV-related HCC. ^{1}H-NMR-based quantitative analysis of serum lipoprotein subfractions thus has potential in clinical applications for discovering specific novel diagnostic biomarkers of HBV-related HCC.

Table 5. Test performance characteristics for the signature panel.

Experiment Set	Group	Dataset	AUC (95% CI)	Sensitivity (%)	Specificity (%)
Training set	HCC vs. LD	AFP	0.831 (0.736 to 0.902)	74.51	81.08
		panel	0.850 (0.758 to 0.917)	88.24	72.97
		Panel + AFP	0.861 (0.771 to 0.926)	88.24	75.68
	HCC vs. NCs	panel	1.000 (0.964 to 1.000)	100.00	100.00
Validation set	HCC vs. LD	AFP	0.833 (0.684 to 0.931)	66.67	100.00
		panel	0.926 (0.800 to 0.984)	83.33	100.00
		Panel + AFP	1.000 (0.914 to 1.000)	100.00	100.00
AFP-negative	HCC vs. LD	panel	0.773 (0.680 to 0.850)	69.23	76.00
	HCC vs. NCs	panel	1.000 (0.964 to 1.000)	100.00	100.00

AUC, area under the receiver operating curve; CI, confidence interval; HCC, hepatocellular carcinoma; LD, liver disease; NCs, normal controls.

2.4. Lipoprotein Lipase (LPL) Is Upregulated in HCC and Associated with Poor Prognosis

VLDL is hydrolyzed by LPL to generate smaller denser particles and subsequently IDL in the peripheral circulation, which is converted to LDL by further hydrolysis. Our results showed that the serum VLDL and IDL levels of HCC patients decreased significantly, while the serum LDL1, LDL2, LDL3, and LDL4 levels of HCC patients increased significantly. The reason for this phenomenon may be related to the increased secretion of LPL into the peripheral blood by tumor cells in HCC patients. We analyzed the LPL transcript level data of HCC patients in the TCGA database and found that LPL mRNA expression level in cancerous tissues of HCC patients showed a significant increase compared with paracancerous tissues (Figure 7a), and the high expression of LPL showed a significant positive correlation with the poor prognosis of patients (Figure 7b), suggesting that abnormal lipoprotein metabolism due to upregulation of LPL mRNA expression in liver tissues may be related to the development of HCC.

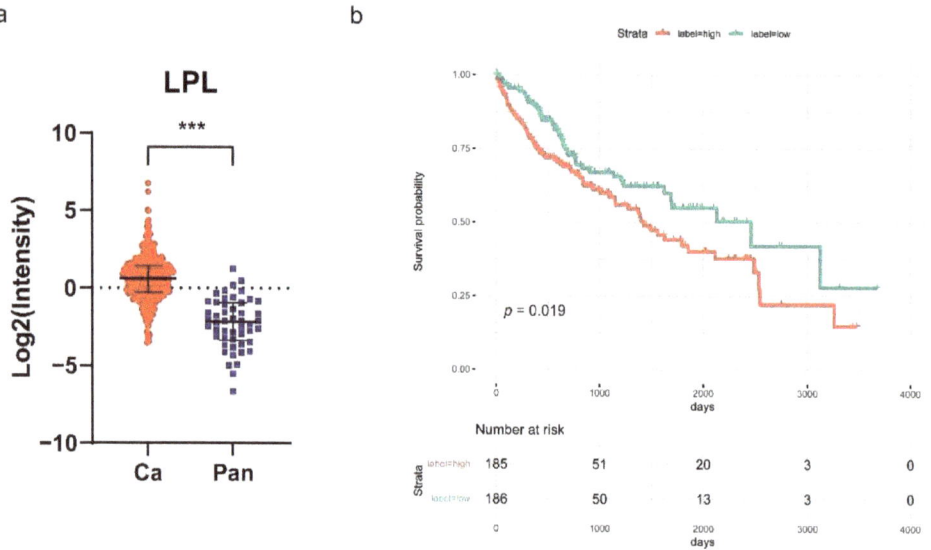

Figure 7. LPL levels are upregulated in HCC tissues and associated with poor prognosis. (a) LPL mRNA levels in HCC (red circle) and adjacent normal (blue square) tissues in the TCGA-LIHC dataset. The p-value was assessed using the Mann–Whitney U test (*** $p < 0.0001$). (b) Overall survival analysis was plotted using the TCGA database at the threshold p-value of < 0.05.

2.5. Identification of Differentially Expressed Protein and Lipid Metabolism-Related Pathways

Through LC-MS/MS platform analysis, a total of 5393 proteins were identified in eight cancerous and paracancerous tissue samples. By employing an FDR adjusted *p*-value of 0.01, a fold-change value >1.5 or <0.5 and a *p*-value < 0.05 as cutoff values, a total of 11 differentially expressed proteins (DEPs) associated with lipid metabolism were detected. Among these, four proteins were significantly upregulated in HCC tissue, including ACSL4, MBOA7, ACLY, and GPDM. In contrast, seven proteins were downregulated, including GPDA, ACOX2, ECHM, ACADS, CP2C9, H17B6, and CP39A (Figures S7 and S8). Through further pathway enrichment analysis, we found that the genes that regulate these differential proteins were highly enriched in fatty acid biosynthesis, glycerophospholipid metabolism, primary bile acid biosynthesis, arachidonic acid metabolism and steroid hormone biosynthesis (Figure S9 and Table S9). Moreover, we evaluated the potential prognostic significance of the mRNA expression of the genes encoding these proteins using data from the GEPIA database (http://gepia2.cancer-pku.cn/ accessed on 23 August 2021). Kaplan–Meier survival analysis revealed that high expression of MBOAT7 and GPD2 and low expression of ACADS were associated with a poor prognosis (Figure S10).

3. Discussion

A definite differential diagnosis between early-stage HBV-related HCC and HBV-related benign LDs, such as HBV-related hepatitis and HBV-related cirrhosis, is often difficult due to a lack of obvious clinical, serological, or radiological evidence. At present, AFP remains a widely used tumor-specific serological biomarker in the diagnosis and management of HCC. However, high AFP expression may be detected in certain pathological conditions, such as deterioration of chronic liver disease, pregnancy, and the presence of germ cell tumors or gastric cancer [5].

^{1}H-NMR spectroscopy is the most commonly used detection platform in the application of lipidomics; despite its lower sensitivity, NMR spectroscopy has several unique advantages over mass spectrometry (MS). ^{1}H-NMR is a noninvasive testing technology that has excellent cross-laboratory reproducibility and does not require elaborate sample preparation or fractionation [8,22]. Routine lipid detections (such as the tests of serum levels of total cholesterol (TC), triglycerides (TG), LDL cholesterol (LDL-C), and HDL cholesterol (HDL-C)) are conventionally used in the clinical analysis of circulating lipid metabolites. ^{1}H-NMR spectroscopy is a rapid, alternative method for quantifying lipoproteins; through the detection of amplitudes of spectral signals emitted by lipoprotein subfractions, one can obtain a direct indication of subclass particle concentration [11]. In this study, we utilized a ^{1}H-NMR high-throughput platform to detect serum lipoprotein subfractions in HBV-related HCC patients with BCLC stage 0-A, at-risk populations (HBV-related hepatitis and cirrhosis) and a healthy control population. In the selection of biomarkers and validation of the diagnosis model, OPLS-DA showed a distinct separation of HCC patients from benign LD patients and NCs. Furthermore, the OPLS-DA model achieved good accuracy for HCC patients relative to normal controls. However, in the OPLS-DA model constructed by the HCC and LD subjects, the Q^2 value failed to reach the desired cutoff level, which might be because of the fewer differences in the metabolic patterns due to the similarity of liver function status between early-stage HCC and LD patients. After multivariable and univariable statistical analyses, a total of eight lipoprotein particle numbers, including VLPN, IDPN, L1PN, L2PN, L3PN, L4PN, L5PN, and L6PN, were selected to build the diagnostic panel. Compared with the serum AFP level alone, the panel constructed from the different lipoprotein particle numbers achieved a higher accuracy in discriminating HCC in the training set and validation set than AFP alone. We also found that the panel achieved excellent diagnostic performance in discriminating AFP-negative HCC patients from AFP-negative LD patients and NCs.

The liver is the major organ of energy metabolism and plays a central role in lipoprotein metabolism by regulating the balance between β-oxidation and lipid synthesis [23]. Most serum endogenous lipids and lipoproteins are synthesized in the liver. The main function of lipoproteins is to transport lipids between cells, which are critical in maintaining energy homeostasis as well as the pathogenesis of atherosclerosis [24]. Under normal physiological conditions, the liver ensures homeostasis of lipid and lipoprotein metabolism, which depends on the structural and functional integrity of hepatocytes [25,26]. However, due to their increased demand for lipids, tumor cells show increased extracellular lipid uptake and a high de novo lipid synthesis rate, which is necessary for HCC tumorigenesis, survival, and progression [27]. De novo lipogenesis starts with the conversion of citrate to oxaloacetate and acetyl-coenzyme A (CoA), which is mediated by ATP-citrate lyase (encoded by ACLY) [28]. Acetyl-CoA is converted to malonyl-CoA via acetyl-CoA carboxylase (ACC) and then to saturated fatty acids (FAs) through the action of fatty acid synthase (encoded by FASN) [28]. HCC is typically characterized by the aberrant overexpression of enzymes in this process, such as ACLY, ACC, and FASN [28,29]. The mass spectrometry results showed that ACLY and MBOA7 were significantly elevated in HCC patients' cancerous tissues, indicating that the upregulation of de novo lipid synthesis was associated with HCC tumorigenesis [30]. In particular, we noted that the level of long-chain ACSL4 expression in HCC cancerous tissue was significantly higher than that in paracancerous tissue (fold change = 15.59), and members of the ACSL family are key enzymes involved in the initial steps of FA metabolism, converting FA to fatty acyl-CoA esters [31]. As a member of the ACSL family, ACSL4 is poorly expressed in the organs of the gastrointestinal system, such as the liver. Chen et al. found that ACSL4 is frequently upregulated in HCC tissues compared with normal samples and promotes HCC progression via c-Myc stability mediated by the ERK/FBW7/c-Myc axis [32].

Under physiological conditions, lipid components such as triglycerides and cholesterol are transported as lipoproteins in the peripheral blood. Among them, exogenous lipids are absorbed through the intestinal epithelium and synthesized as celiac particles (CMs), endogenous lipids entering the liver and synthesized as VLDL, both collectively known as triglyceride-rich lipoproteins (TRLs) [33,34]. The newly secreted TRL enters the bloodstream and needs to be marginalized along the luminal surface of capillaries and hydrolyzed by lipoprotein lipase (LPL) expressed on the surface of vascular endothelial cells from TG within the neutral core of CMs and VLDL to produce CMs residue and IDL, respectively, and release free fatty acids (FFA) for use by peripheral tissues, where IDL can be absorbed by the liver or through further TG hydrolysis to LDL [35,36]. Adipocytes, cardiomyocytes, and skeletal muscle cells are the main sites for producing LPL. Because these cells are far away from the capillary cavity and need to be transported through the subendothelial space, recent studies have shown that glycosylphosphatidylinositol anchors high-density lipoprotein binding protein 1 (GPHIBP1), which captures LPL and binds to form the LPL–GPDIBP1 complex to mediate LPL entry into the lumen through capillary endothelial cells and specifically binds to ApoCII in TRL to exert a hydrolytic effect [36,37]. Recent studies have shown that LPL expression appears upregulated in several types of tumor cells and is associated with cancer progression and poor prognosis. In our study, the serum VLDL and IDL levels of HCC patients decreased significantly, while the serum LDL1, LDL2, LDL3, and LDL4 levels of HCC patients increased significantly. We speculated that this phenomenon may be related to the increased secretion of LPL into the peripheral blood by tumor cells in HCC patients. Therefore, we analyzed and found that the LPL mRNA expression level in cancerous tissues of HCC patients showed a significant increase compared with paracancerous tissues, and the high expression of LPL showed a significant positive correlation with the poor prognosis of HCC patients from TCGA database. Cao et al. found that the mRNA and protein expression levels of LPL were upregulated in mouse and human HCC tissues and positively correlated with poor prognosis, and in vitro experiments further showed that culturing cells in the absence or silencing of LPL significantly reduced cell proliferation [38]. This is consistent with our findings. Wu et al. found that the expression of the antioncogene ZHX2 was significantly reduced in nonalcoholic fatty liver disease (NAFLD)-associated HCC and that overexpression of ZHX2 inhibited the uptake of exogenous lipids and the ability of HCC cells to proliferate by suppressing

LPL promoter activity, thereby delaying the progression of NAFLD-associated HCC [39]. Manupati et al. found that LPL transcript levels were upregulated 16-fold in CD44-positive breast cancer stem cells. LPL, as a unique downstream target of CD44 signal transduction, can activate endothelial cell-mediated angiogenesis during tumor growth. In addition, knockdown of CD44 or intratumoral injection of tetrahydrolipostatin (LPL inhibitor) can inhibit breast cancer progression and angiogenesis [40]. LPL also plays an important role in the production of IDL and LDL in the human body. This suggests that the significant differences in some indicators between the HCC group and the control group may be related to LPL expression, such as VLPN, IDPN, L1PN, L2PN, L4PN, and L5PN. We will explore the correlation in the future study.

The diagnostic panel constructed from serum lipoprotein particle numbers effectively improved the detection of patients with early-stage HCC, illustrating that ^{1}H-NMR lipoprotein subfraction testing plays an important role in the diagnosis of early-stage HCC. Lu et al. observed that the L1 and L5 subfractions of LDL and VLDL promoted breast cancer cell migration and invasion through increased Akt Ser473 phosphorylation [41]. Further angiogenic assays in vitro indicated that the L1 and L5 subfractions and VLDL enhanced the secretion of angiogenic factors and promoted angiogenic activity [41]. There are few studies on the mechanisms of lipoprotein subfractions in the tumorigenesis and development of hepatocarcinoma, and we will explore this aspect in the future.

In addition, we recognize some limitations in our research. First, the sample size of the external validation set was relatively small, and the patients were not equally distributed between the HCC and LD groups. Therefore, a sufficiently sized external validation set is required to further confirm our research conclusions. Second, since lipidomics is a branch of systems biology, circulating lipoprotein subfractions mainly reflect an overall metabolic shift in cancer patients and may not reflect the metabolic states of the tumor cells alone. Therefore, in future studies, we plan to determine the relationship between abnormal lipoprotein metabolism and HCC development at the cellular level.

4. Materials and Methods

4.1. Ethical Statement

Prior to commencing the study, ethical approval was sought from the Research Ethics Committee of Tianjin Medical University Cancer Institute and Hospital in accordance with the 1964 Helsinki Declaration ethical standards (NO. bc2020098). Written informed consent was obtained from all participants, and the study was approved by the local Ethical Board.

4.2. Patients and Sample Collection

A total of 197 serum samples were enrolled at Tianjin Medical University Cancer Institute and Hospital (Tianjin, China) from July 2018 to December 2020. All serum samples were collected from 7:00 to 8:00 in the morning after the participants had fasted for at least 6 h. The samples were collected from all the patients who were initially diagnosed without liver disease-related treatment. The Barcelona Clinic Liver Cancer (BCLC) staging system was used to assess tumor stage. In the training set, we collected 51 patients with early-stage HBV-related HCC (BCLC stage 0-A) before surgical treatment, 37 patients with HBV-related hepatitis and HBV-related cirrhosis (hereafter referred to as liver disease, LD), and 50 NCs (with normal liver biochemistry, no type of malignancy or history of other benign disease, alcohol abuse and viral hepatitis). To identify the lipoprotein profile and establish a diagnostic model of HCC, a validation set was built from independent early-stage HCC ($n = 24$) and LD ($n = 17$) patient serum samples collected in the same way as those used in the training set. In addition, we selected serum AFP-negative patients (AFP level < 20 ng/mL) in the HCC and LD groups and collected 18 HCC patients with negative serum AFP expression for the next analysis. Next, eight pairs of cancerous and paracancerous tissue samples of HBV-related HCC patients from the validation set were obtained from surgical resections at Tianjin Medical University Cancer Institute and Hospital from May 2020 to August 2020. The inclusion and exclusion criteria, sample collection and storage are shown in the Supplementary Materials and methods. The collected test tube containing the blood sample was placed in a centrifuge at 4 °C and centrifuged at 3000 rpm for 15 min. Then, 400 µL of serum was collected from the upper layer of the test tube and stored at −80 °C until required for NMR detection. Tissue samples were fixed in 10% formalin and embedded in paraffin. The paraffin-sectioned tissues were serially cut into 5 µm sections and preserved at room temperature until required for mass spectrometric measurement.

4.3. Inclusion Criteria and Exclusion Criteria

The diagnoses of HCC, hepatitis and cirrhosis were based on the American Association for the Study of Liver Diseases (AASLD) Practice Guidelines.

The inclusion criteria were as follows:

1. Primary HCC diagnosed by histological or cellular examination.
2. Single tumor (regardless of size) or the number of tumors is less than 3 and the maximum diameter is ≤ 3 cm, and no history of portal invasion or extrahepatic spread.
3. HCC, cirrhosis and hepatitis with a history of HBV infection confirmed by virological assay.
4. Age > 18 years.
5. No previous treatment for HCC.
6. Knowledge of the study and agreement to follow-up.

Participants were excluded from the study if they met any of the following conditions:

1. History of other diagnosed malignancies.
2. History of anticancer treatment for HCC.
3. History of hepatitis virus infection without HBV.
4. Factors can cause abnormal elevation of serum AFP in normal controls, including pregnancy and any type of liver disease.
5. Participants with severe illnesses, including cardiovascular disease, endocrine disease and renal impairment.
6. Participants with lactation, current smoking and drug dependence.
7. Participants were taking lipid-lowering, hyperglycemic, anti-inflammatory, antithrombotic medications, dietary supplements, or antihypertensive treatment.

4.4. Magnetic Resonance Experiments

A Bruker 600 MHz NMR spectrometer was applied to estimate the lipoprotein subfractions. The Bruker IVDr lipoprotein subclass analysis (B.I.-LISA) method was used to predict the subfractions of lipoproteins for the analysis. Bruker's Quant Ref manager within Top Spin was used to normalize the spectra to the same quantitative scale, and the spectral intensity was normalized to the proton concentration in units of millimoles per liter. First, Topspin 3.6.0 was used to calibrate the chemical shift to the methyl signal of trimethylsilyl propanoic acid (TSP), and then the alanine doublet was calibrated to 1.48 ppm; this method requires integration of the lipoprotein -CH3 and CH2- signals appearing in the 1D ^{1}H NMR spectrum with chemical shifts of 0.8 and 1.25 ppm, respectively (Figure S1). The ^{1}H-NMR platform has good intralaboratory repeatability and interlaboratory repeatability [11], and all tests were blind to the disease status of participants.

Lipoprotein subfractions were determined based on one-dimensional nuclear Overhauser effect spectroscopy (NOESY) magnetic resonance (MR) spectra using a partial least-squares regression model. Each lipoprotein class was further subdivided into subfractions according to its density: very-low-density lipoprotein (VLDL) was divided into VLDL 1–5, low-density lipoprotein (LDL) into LDL 1–6 and high-density lipoprotein (HDL) into HDL 1–4, with larger numbers indicating increasing density. Serum lipoprotein particle numbers (PNs) and serum concentrations of TG, cholesterol (CH), free cholesterol (FC), phospholipids (PL), apolipoprotein A1 (Apo-A1), apolipoprotein A2 (Apo-A2), and apolipoprotein B (Apo-B), as well as in each of the lipoprotein classes of VLDL, LDL, intermediate-density lipoprotein (IDL) and HDL, were estimated using a regression model developed by Bruker BioSpin. Finally, a dataset constructed from 112 variables was used in this study. Four-letter abbreviations were used to represent the variables; for example, the estimated VLDL-1 content of phospholipids was named V1PL, and the estimated total serum cholesterol was named TPCH. The NMR lipoproteins and subfractions are shown in Table S1.

4.5. Nanoscale Liquid Chromatography-Tandem Mass Spectrometry (Nano-LC-MS/MS) Analysis

Orbitrap Q-Exactive HF mass spectrometry (Thermo Fisher Scientific, Waltham, MA, USA) accompanied by a Thermo Scientific UltiMate 3000 UHPLC system was used to acquire lysed peptide sample data. Peptides were redissolved in loading buffer (2% ACN) with iRT standards (Biognosys, Schlieren, Switzerland) and separated using a 150-min gradient method (0–3 min, 3 to 9% buffer B; 3–127 min, 9 to 63% buffer B; 127–131 min, 63% buffer B; 131–149 min, 63 to 3% buffer B). The digested peptides were ionized at 2 kV, and mass spectrometry analysis data were collected using data-independent acquisition (DIA) mode. Full-scan MS1 acquisition was performed by an Orbitrap mass analyzer (scan range 300–1400 m/z) at a high resolution of 120,000. For MS2 acquisition, the spectra were recorded in top speed mode with a duty cycle time of 3 s. Precursor ions were selected and fragmented using higher-energy collisional dissociation (HCD) with 32% normalized collision energy. The maximum ion injection time for the MS2 scan was set to 35 ms, and the dynamic exclusion for the selected ions was 60 s. All tests were blind to the disease status of participants.

4.6. Statistical Analysis

4.6.1. Multivariable and Univariable Statistical Analysis of NMR Data

Due to the hypothesized biological mechanisms between lipid fractions and HCC development, multivariable data analysis based on the projection principle was applied for statistical analysis of the ^{1}H-NMR dataset. A pattern-recognition method that can discriminate between groups even in the presence of highly structured noise or confounding factors, unsupervised principal component analysis (PCA) and supervised partial least squares-discriminant analysis (PLS-DA), were implemented to analyze the raw data and classify the samples. Then, orthogonal partial least-squares discriminant analysis (OPLS-DA) was used to extract the correlated variables and optimize the maximum separation by using the Simca version 14.1 software package (UmetricsAB, Umea, Sweden).

The models were validated using 7-fold cross-validation to quantitatively assess their generalization ability and acquire robust statistical models. In 7-fold cross-validation, the dataset is split into seven equal-sized subsets. In each round, one subset is used for validation, and the remaining six subsets are used for training; this process is repeated seven times. The goodness-of-fit parameters and R^2 and Q^2 values calculated with 7-fold cross-validation as well as with cross-validated analysis of variance (CV-ANOVA, where $p < 0.05$ suggests the model is superior to one chosen at random) were obtained to measure the robustness and quality of the models. The associated R^2 and Q^2 parameters represent the interpretation rate of the matrix and model predictive capability; the closer the metrics are to 1, the larger the variance explained by the model and the more reliable its predictive power. Furthermore, a permutation test (999 permutations) was performed to validate the degree of overfitting based on the values of the R^2-intercept and Q^2-intercept. The reproducibility and robustness of each model were validated by the Q^2-intercept; the more negative the value of the Q^2-intercept was, the better the performance of the model.

In the OPLS-DA model, most of the variables related to the classification were concentrated in the direction of the first predicted principal component. To identify the differential lipoprotein subfractions, the VIP scores calculated by the OPLS-DA model were used to reflect the most influential contribution of each variable to the model. When VIP > 1.0, the variable was considered potentially relevant. Differences in lipoprotein subfractions between the three groups were assessed by the Kruskal–Wallis test (nonnormally distributed data) in the training sets, and $p < 0.05$ was considered statistically significant. Lipoprotein subfractions with VIP scores > 1.0 and $p < 0.05$ were selected and entered into a binary logistic regression model to design the best lipoprotein subfraction combination. To further evaluate the diagnostic performance of the potential biomarkers, receiver operating characteristic (ROC) curves were analyzed to evaluate the accuracy of this model. Each biomarker panel's diagnostic performance was evaluated by using the area under the ROC curve (AUC) and the sensitivity and specificity at the optimal cutoff point defined by the minimum distance to the top-left corner of the ROC curve graph. For the participants' clinical characteristics, the Mann–Whitney U test was used to compare continuous variables, and Pearson's chi-square test was used to compare categorical variables. Correlations were calculated by Spearman rank correlation analysis, and $p < 0.05$ was considered statistically significant. Logistic regression and statistical analysis were performed by using IBM SPSS version 26.0 (SPSS Inc., Armonk, NY, USA).

4.6.2. Quantification and Statistical Analysis of LC-MS/MS Data

The DIA data were searched using the Human-specific UniProt database (20,365 sequences), and LC-MS/MS data were analyzed by Spectronaut (v14.5.200813.47784). The library was generated using the default settings for trypsin/P digestion rules and high protein and peptide confidential levels [false discovery rate (FDR) = 0.01]. The output-quantified protein intensities were processed using Spectronaut, and a median normalization procedure was applied to normalize the data. Proteins with at least 30% appearance in all samples were chosen for the subsequent analysis, and missing values were replaced with half of the minimum value of each protein intensity. A fold-change of >1.5 or <0.5 and a p-value < 0.05 (The Mann–Whitney U test) were set as cutoff values for the differential proteins. The protein corresponding gene and OS information of 371 cancer samples from TCGA were applied to generate survival curves with the survival and survminer packages in the R package (version 3.6.0). Gene Ontology (GO) functional annotations and Kyoto Encyclopedia of Genes and Genomes (KEGG) pathway enrichment were performed using the R package (clusterProfiler, v3.16.1) and the org.Hs.eg.db (v3.11.4) annotation database. The background genes were set to all quantified genes, and the differential genes were input to generate the enrichment pathway list and figures.

5. Conclusions

In conclusion, this study aimed to objectively assess the clinical applicative value of serum lipoprotein subfraction testing in the diagnosis of HBV-related HCC patients with BCLC stage 0-A. The results clearly indicate that the lipidomic biomarker panel constructed with the particle numbers of VLDL, IDL, LDL-1, LDL-2, LDL-3, LDL-4, LDL-5, and LDL-6 could be used in the diagnosis of HCC. Meanwhile, we found that LPL transcript levels in cancerous tissues of HCC patients showed a significant increase compared with paracancerous tissues, and the high expression of LPL showed a significant positive correlation with the poor prognosis of patients by bioinformatic analysis. Moreover, LC-MS/MS analysis indicated that abnormal lipid metabolism is an important influential factor in potentially promoting HBV-related HCC development. Our study focuses on an innovative combination of alterations in the lipid profile of cancer patients and ^{1}H-NMR-based lipidomics research, which provides new insight for the development of diagnostic and prognostic biomarkers for HBV-related HCC with BCLC stage 0-A.

However, several limitations to this pilot study need to be considered. For example, this study lacks a large number of external verification samples to further verify the generalizability of the diagnostic panel. Despite the relatively limited sample size, the study certainly adds to our understanding of the use of serum lipoprotein biomarkers in the diagnosis of HCC. Further large prospective studies with external validation should be undertaken to determine whether this lipidomic panel may improve surveillance and management strategies in patients with HCC.

Supplementary Materials: The following are available online at https://www.mdpi.com/article/10.3390/jpm11111143/s1. Figure S1: Stacked view of the ^{1}H-NMR spectra of lipoprotein subfractions. HCC patient (blue), benign liver disease patient (green) and normal control (red). Figure S2: The PCA and PLS-DA score plot in the training set. HCC patient (red), benign liver disease patient (green) and normal control (yellow). Figure S3: PCA and PLS-DA score plot of serum AFP-negative patients. HCC patient (red), benign liver disease patient (green) and normal control (yellow). Figure S4: The correlation of serum AFP and selected lipoprotein particles was analyzed using Spearman rank correlation analysis. Figure S5: The correlation of selected lipoprotein particles and clinical features was analyzed using Spearman rank correlation analysis. Figure S6: ROC curve analysis following binary logistic regression in distinguishing HCC patients from non-HCC patients. The diagnostic performance of the lipidomic biomarker panel in discriminating (a) HCC from non-HCC patients and (b) AFP (-) HCC from AFP (-) non-HCC patients. Figure S7: Comparison of DEPs associated with lipid metabolism between cancerous and paracancerous tissue samples. In eight patients, 11 differential proteins showed significant differences. The Mann–Whitney U test was performed to determine whether the differences were statistically significant. N: paracancerous HCC tissues, T: cancerous HCC tissues. Figure S8: Relative expression levels of 11 differentially expressed proteins between cancerous and paracancerous HCC tissues. The data are expressed as medians with interquartile ranges and were analyzed by the Mann–Whitney U test (* $p < 0.05$, ** $p < 0.001$, *** $p < 0.0001$). N: paracancerous HCC tissues, T: cancerous HCC tissues. Figure S9: Results of KEGG pathway enrichment analysis. The enriched pathways of upregulated (red) and downregulated (blue) DEPs associated with lipid metabolism. Figure S10: The prognostic value of DEPs for HCC patients. The overall survival curves (a) and disease-free survival curves (b) for HCC patients with high (red) and low expression levels of (blue) MBOAT7, GDP2, and ACADS were plotted using the GEPIA database (http://gepia2.cancer-pku.cn/ accessed on 23 August 2021) at the threshold p-value of < 0.05. Table S1: NMR lipoproteins & subfractions. Table S2: Clinical characteristics of the HCC group and normal control group. Table S3: Clinical characteristics of the AFP-negative HCC group and AFP-negative LD group. Table S4: List of calculated parameters in the training set from multivariable and univariable statistical analysis. Table S5: List of calculated parameters of AFP-negative patients from multivariable and univariable statistical analysis. Table S6: The absolute numbers of each lipoprotein subfraction. Table S7: Correlation between selected variables and clinical features of HCC patients. Table S8: Test performance characteristics for the signature panel. Table S9: Differentially expressed proteins between cancerous tissues and paracancerous tissues.

Author Contributions: L.R., D.Z., H.A., J.L. and J.X. had full access to all the data in the study and take responsibility for the integrity of the data and the accuracy of the data analysis. All authors approved the final manuscript as submitted and agreed to be accountable for all aspects of the work. Study concept and design: L.R. and D.Z. Acquisition of clinical samples and completion of experiments: D.Z., H.A., J.L. and J.X. Acquisition, analysis, or interpretation of data: L.R., D.Z. and H.A. Drafting of the manuscript: L.R., D.Z. and H.A. Statistical analysis: D.Z. and H.A. Obtained funding: L.R. Administrative, technical, or material support: L.R. Study supervision: L.R., D.Z. and H.A. contributed equally to this work. All authors have read and agreed to the published version of the manuscript.

Funding: This work was supported by a grant from the Natural Science Foundation of Tianjin City in China (Grant No. 18JCZDJC32600).

Institutional Review Board Statement: This study was conducted according to the guideline of the Declaration of Helsinki and approved by the Research Ethics Committee of Tianjin Medical University Cancer Institute and Hospital (NO. bc2020098).

Informed Consent Statement: Written informed consent has been obtained from all subjects involved in the study.

Acknowledgments: The authors thank all patients for their important contribution and all public databases for providing data.

Conflicts of Interest: The author declares that they have no competing interest.

References

1. Zhang, B.; Zhang, B.; Zhang, Z.; Huang, Z.; Chen, Y.; Chen, M.; Bie, P.; Peng, B.; Wu, L.; Wang, Z.; et al. 42,573 cases of hepatectomy in China: A multicenter retrospective investigation. *Sci. China Life Sci.* **2018**, *61*, 660–670. [CrossRef]
2. Yang, J.D.; Hainaut, P.; Gores, G.J.; Amadou, A.; Plymoth, A.; Roberts, L.R. A global view of hepatocellular carcinoma: Trends, risk, prevention and management. *Nat. Rev. Gastroenterol. Hepatol.* **2019**, *16*, 589–604. [CrossRef]
3. Bruix, J.; Reig, M.; Sherman, M. Evidence-Based Diagnosis, Staging, and Treatment of Patients with Hepatocellular Carcinoma. *Gastroenterology* **2016**, *150*, 835–853. [CrossRef] [PubMed]
4. Silva, M.A.; Hegab, B.; Hyde, C.; Guo, B.; Buckels, J.A.C.; Mirza, D.F. Needle track seeding following biopsy of liver lesions in the diagnosis of hepatocellular cancer: A systematic review and meta-analysis. *Gut* **2008**, *57*, 1592–1596. [CrossRef] [PubMed]
5. Wong, R.J.; Ahmed, A.; Gish, R.G. Elevated Alpha-Fetoprotein. *Clin. Liver Dis.* **2015**, *19*, 309–323. [CrossRef]
6. Pavlova, N.; Thompson, C.B. The Emerging Hallmarks of Cancer Metabolism. *Cell Metab.* **2016**, *23*, 27–47. [CrossRef]
7. Johnson, C.H.; Ivanisevic, J.; Siuzdak, G. Metabolomics: Beyond biomarkers and towards mechanisms. *Nat. Rev. Mol. Cell Biol.* **2016**, *17*, 451–459. [CrossRef] [PubMed]
8. Markley, J.L.; Brüschweiler, R.; Edison, A.S.; Eghbalnia, H.R.; Powers, R.; Raftery, D.; Wishart, D.S. The future of NMR-based metabolomics. *Curr. Opin. Biotechnol.* **2016**, *43*, 34–40. [CrossRef]
9. Han, X. Lipidomics for studying metabolism. *Nat. Rev. Endocrinol.* **2016**, *12*, 668–679. [CrossRef] [PubMed]
10. Yang, K.; Han, X. Lipidomics: Techniques, Applications, and Outcomes Related to Biomedical Sciences. *Trends Biochem. Sci.* **2016**, *41*, 954–969. [CrossRef]
11. Jiménez, B.; Holmes, E.; Heude, C.; Tolson, R.F.; Harvey, N.; Lodge, S.L.; Chetwynd, A.J.; Cannet, C.; Fang, F.; Pearce, J.T.M.; et al. Quantitative Lipoprotein Subclass and Low Molecular Weight Metabolite Analysis in Human Serum and Plasma by 1H NMR Spectroscopy in a Multilaboratory Trial. *Anal. Chem.* **2018**, *90*, 11962–11971. [CrossRef]
12. Jiang, J.; Nilsson-Ehle, P.; Xu, N. Influence of liver cancer on lipid and lipoprotein metabolism. *Lipids Health Dis.* **2006**, *5*, 4. [CrossRef]
13. Privitera, G.; Spadaro, L.; Marchisello, S.; Fede, G.; Purrello, F. Abnormalities of Lipoprotein Levels in Liver Cirrhosis: Clinical Relevance. *Dig. Dis. Sci.* **2017**, *63*, 16–26. [CrossRef]
14. Deprince, A.; Haas, J.T.; Staels, B. Dysregulated lipid metabolism links NAFLD to cardiovascular disease. *Mol. Metab.* **2020**, *42*, 101092. [CrossRef]
15. Mihaleva, V.V.; van Schalkwijk, D.B.; de Graaf, A.A.; van Duynhoven, J.; van Dorsten, F.A.; Vervoort, J.; Smilde, A.; Westerhuis, J.A.; Jacobs, D.M. A Systematic Approach to Obtain Validated Partial Least Square Models for Predicting Lipoprotein Subclasses from Serum NMR Spectra. *Anal. Chem.* **2013**, *86*, 543–550. [CrossRef]
16. El Harchaoui, K.; van der Steeg, W.A.; Stroes, E.S.; Kuivenhoven, J.A.; Otvos, J.D.; Wareham, N.J.; Hutten, B.A.; Kastelein, J.J.; Khaw, K.-T.; Boekholdt, M. Value of Low-Density Lipoprotein Particle Number and Size as Predictors of Coronary Artery Disease in Apparently Healthy Men and Women: The EPIC-Norfolk Prospective Population Study. *J. Am. Coll. Cardiol.* **2007**, *49*, 547–553. [CrossRef] [PubMed]
17. Kostara, C.E.; Papathanasiou, A.; Psychogios, N.; Cung, M.T.; Elisaf, M.S.; Goudevenos, J.; Bairaktari, E.T. NMR-Based Lipidomic Analysis of Blood Lipoproteins Differentiates the Progression of Coronary Heart Disease. *J. Proteome Res.* **2014**, *13*, 2585–2598. [CrossRef] [PubMed]
18. Casadei-Gardini, A.; Del Coco, L.; Marisi, G.; Conti, F.; Rovesti, G.; Ulivi, P.; Canale, M.; Frassineti, G.L.; Foschi, F.G.; Longo, S.; et al. 1H-NMR Based Serum Metabolomics Highlights Different Specific Biomarkers between Early and Advanced Hepatocellular Carcinoma Stages. *Cancers* **2020**, *12*, 241. [CrossRef]
19. Fages, A.; Duarte-Salles, T.; Stepien, M.; Ferrari, P.; Fedirko, V.; Pontoizeau, C.; Trichopoulou, A.; Aleksandrova, K.; Tjønneland, A.; Olsen, A.; et al. Metabolomic profiles of hepatocellular carcinoma in a European prospective cohort. *BMC Med.* **2015**, *13*, 242. [CrossRef]
20. Nahon, P.; Amathieu, R.; Triba, M.N.; Bouchemal, N.; Nault, J.C.; Ziol, M.; Seror, O.; Dhonneur, G.; Trinchet, J.-C.; Beaugrand, M.; et al. Identification of Serum Proton NMR Metabolomic Fingerprints Associated with Hepatocellular Carcinoma in Patients with Alcoholic Cirrhosis. *Clin. Cancer Res.* **2012**, *18*, 6714–6722. [CrossRef] [PubMed]
21. Cox, I.J.; Aliev, A.E.; Crossey, M.M.; Dawood, M.; Al-Mahtab, M.; Akbar, S.M.; Rahman, S.; Riva, A.; Williams, R.; Taylor-Robinson, S.D. Urinary nuclear magnetic resonance spectroscopy of a Bangladeshi cohort with hepatitis-B hepatocellular carcinoma: A biomarker corroboration study. *World J. Gastroenterol.* **2016**, *22*, 4191–4200. [CrossRef] [PubMed]
22. Ranjan, R.; Sinha, N. Nuclear magnetic resonance (NMR)-based metabolomics for cancer research. *NMR Biomed.* **2018**, *32*, e3916. [CrossRef]
23. Alves-Bezerra, M.; Cohen, D.E. Triglyceride Metabolism in the Liver. *Compr. Physiol.* **2017**, *8*, 1–22. [CrossRef]
24. Ramasamy, I. Recent advances in physiological lipoprotein metabolism. *Clin. Chem. Lab. Med.* **2014**, *52*, 1695–1727. [CrossRef]
25. Perez-Matos, M.C.; Sandhu, B.; Bonder, A.; Jiang, Z.G. Lipoprotein metabolism in liver diseases. *Curr. Opin. Lipidol.* **2019**, *30*, 30–36. [CrossRef] [PubMed]
26. Lewis, G.F.; Rader, D.J. New Insights Into the Regulation of HDL Metabolism and Reverse Cholesterol Transport. *Circ. Res.* **2005**, *96*, 1221–1232. [CrossRef] [PubMed]
27. Zaidi, N.; Lupien, L.; Kuemmerle, N.B.; Kinlaw, W.B.; Swinnen, J.V.; Smans, K. Lipogenesis and lipolysis: The pathways exploited by the cancer cells to acquire fatty acids. *Prog. Lipid Res.* **2013**, *52*, 585–589. [CrossRef]

28. Satriano, L.; Lewinska, M.; Rodrigues, P.M.; Banales, J.M.; Andersen, J.B. Metabolic rearrangements in primary liver cancers: Cause and consequences. *Nat. Rev. Gastroenterol. Hepatol.* **2019**, *16*, 748–766. [CrossRef]
29. Sangineto, M.; Villani, R.; Cavallone, F.; Romano, A.; Loizzi, D.; Serviddio, G. Lipid Metabolism in Development and Progression of Hepatocellular Carcinoma. *Cancers* **2020**, *12*, 1419. [CrossRef]
30. Bianco, C.; Jamialahmadi, O.; Pelusi, S.; Baselli, G.; Dongiovanni, P.; Zanoni, I.; Santoro, L.; Maier, S.; Liguori, A.; Meroni, M.; et al. Non-invasive stratification of hepatocellular carcinoma risk in non-alcoholic fatty liver using polygenic risk scores. *J. Hepatol.* **2020**, *74*, 775–782. [CrossRef]
31. Tang, Y.; Zhou, J.; Hooi, S.C.; Jiang, Y.M.; Lu, G.-D. Fatty acid activation in carcinogenesis and cancer development: Essential roles of long-chain acyl-CoA synthetases (Review). *Oncol. Lett.* **2018**, *16*, 1390–1396. [CrossRef] [PubMed]
32. Chen, J.; Ding, C.; Chen, Y.; Hu, W.; Lu, Y.; Wu, W.; Zhang, Y.; Yang, B.; Wu, H.; Peng, C.; et al. ACSL4 promotes hepatocellular carcinoma progression via c-Myc stability mediated by ERK/FBW7/c-Myc axis. *Oncogenesis* **2020**, *9*, 42. [CrossRef]
33. Dash, S.; Xiao, C.; Morgantini, C.; Lewis, G.F. New Insights into the Regulation of Chylomicron Production. *Annu. Rev. Nutr.* **2015**, *35*, 265–294. [CrossRef]
34. Heeren, J.; Scheja, L. Metabolic-associated fatty liver disease and lipoprotein metabolism. *Mol. Metab.* **2021**, *50*, 101238. [CrossRef]
35. Kumari, A.; Kristensen, K.K.; Ploug, M.; Winther, A.L.-M. The Importance of Lipoprotein Lipase Regulation in Atherosclerosis. *Biomedicines* **2021**, *9*, 782. [CrossRef] [PubMed]
36. Young, S.G.; Fong, L.G.; Beigneux, A.P.; Allan, C.M.; He, C.; Jiang, H.; Nakajima, K.; Meiyappan, M.; Birrane, G.; Ploug, M. GPIHBP1 and Lipoprotein Lipase, Partners in Plasma Triglyceride Metabolism. *Cell Metab.* **2019**, *30*, 51–65. [CrossRef] [PubMed]
37. Kristensen, K.K.; Leth-Espensen, K.Z.; Kumari, A.; Grønnemose, A.L.; Lund-Winther, A.-M.; Young, S.G.; Ploug, M. GPIHBP1 and ANGPTL4 Utilize Protein Disorder to Orchestrate Order in Plasma Triglyceride Metabolism and Regulate Compartmentalization of LPL Activity. *Front. Cell Dev. Biol.* **2021**, *9*, 1900. [CrossRef]
38. Cao, D.; Song, X.; Che, L.; Li, X.; Pilo, M.G.; Vidili, G.; Porcu, A.; Solinas, A.; Cigliano, A.; Pes, G.M.; et al. Bothde novosynthetized and exogenous fatty acids support the growth of hepatocellular carcinoma cells. *Liver Int.* **2016**, *37*, 80–89. [CrossRef]
39. Wu, Z.; Ma, H.; Wang, L.; Song, X.; Zhang, J.; Liu, W.; Ge, Y.; Sun, Y.; Yu, X.; Wang, Z.; et al. Tumor suppressor ZHX2 inhibits NAFLD–HCC progression via blocking LPL-mediated lipid uptake. *Cell Death Differ.* **2019**, *27*, 1693–1708. [CrossRef]
40. Manupati, K.; Yeeravalli, R.; Kaushik, K.; Singh, D.; Mehra, B.; Gangane, N.; Gupta, A.; Goswami, K.; Das, A. Activation of CD44-Lipoprotein lipase axis in breast cancer stem cells promotes tumorigenesis. *Biochim. Biophys. Acta (BBA)—Mol. Basis Dis.* **2021**, *1867*, 166228. [CrossRef]
41. Lu, C.-W.; Lo, Y.-H.; Chen, C.-H.; Lin, C.-Y.; Tsai, C.-H.; Chen, P.-J.; Yang, Y.-F.; Wang, C.-H.; Tan, C.-H.; Hou, M.-F.; et al. VLDL and LDL, but not HDL, promote breast cancer cell proliferation, metastasis and angiogenesis. *Cancer Lett.* **2016**, *388*, 130–138. [CrossRef] [PubMed]

MDPI
St. Alban-Anlage 66
4052 Basel
Switzerland
Tel. +41 61 683 77 34
Fax +41 61 302 89 18
www.mdpi.com

Journal of Personalized Medicine Editorial Office
E-mail: jpm@mdpi.com
www.mdpi.com/journal/jpm

www.ingramcontent.com/pod-product-compliance
Lightning Source LLC
LaVergne TN
LVHW070052120526
838202LV00102B/2218